Energy and the New Reality 1

Energy Efficiency and the Demand for Energy Services

L. D. Danny Harvey

publishing for a sustainable future
London • Washington, DC

First published in 2010 by Earthscan

Earthscan Ltd, Dunstan House, 14a St Cross Street, London EC1N 8XA, UK

Earthscan LLC, 1616 P Street, NW, Washington, DC 20036, USA

Earthscan publishes in association with the International Institute for Environment and Development

For more information on Earthscan publications, see www.earthscan.co.uk or write to earthinfo@earthscan.co.uk

ISBN: 978-1-84407-912-4 hardback
      978-1-84971-072-5 paperback

Typeset by Domex e-Data, India
Cover design by Susanne Harris

A catalogue record for this book is available from the British Library

Library of Congress Cataloging-in-Publication Data
Harvey, Leslie Daryl Danny, 1956-
  Energy and the new reality / L. D. Danny Harvey.
     v. cm.
  Includes bibliographical references and index.
  Contents: 1. Energy efficiency and the demand for energy services.
  ISBN 978-1-84407-912-4 (hardback) -- ISBN 978-1-84971-072-5 (pbk.) 1. Energy conservation. 2. Energy consumption. 3. Climatic changes--Prevention. I. Title.

TJ163.3.H38 2010

333.79--dc22

                                                                    2009032851

At Earthscan we strive to minimize our environmental impacts and carbon footprint through reducing waste, recycling and offsetting our CO_2 emissions, including those created through publication of this book. For more details of our environmental policy, see www.earthscan.co.uk.

This book was printed in the UK
by The Cromwell Press Group.
The paper used is FSC certified.

Mixed Sources
Product group from well-managed
forests and other controlled sources
www.fsc.org Cert no. TT-COC-2082
© 1996 Forest Stewardship Council

Energy and the New Reality 1

Energy Efficiency and the Demand for Energy Services

To all those dedicated to changing the world for the better

Contents

List of Figures, Tables and Boxes

Figures

Tables

Boxes

Preface

This book and the accompanying Volume 2 (*Carbon-Free Energy Supply*) are an attempt to objectively, comprehensively and quantitatively examine what it would take to limit the atmospheric carbon dioxide (CO_2) concentration to no more than 450 parts per million by volume (ppmv). By the time this book is published, the CO_2 concentration will have risen to 390ppmv from a pre-industrial concentration of about 280ppmv. With only 60ppmv to go, stabilization at no more than 450ppmv might seem like an impossible task, yet 450ppmv is already a dangerously high concentration. However, part of the CO_2 emitted by humans is quickly absorbed by the terrestrial biosphere and the oceans – the two 'sinks' of anthropogenic CO_2 – and the rest temporarily (for hundreds to thousands of years and longer) accumulates in the atmosphere. To stabilize the atmospheric CO_2 concentration requires reducing total sources only to the point where they equal the total rate of absorption by the sinks.

In 2005, industrial activities (primarily the combustion of fossil fuels) released about 8 billion tonnes (gigatonnes, Gt) of carbon (C) (in the form of CO_2) to the atmosphere, while land use changes (primarily deforestation) in recent years released another 1–2GtC. The total annual emission was therefore about 9–10GtC/yr, while the observed annual increase in the amount of CO_2 in the atmosphere amounted to 4–5GtC/yr. In the near term (two to three decades), stabilization requires reducing total emissions by 4–5GtC/yr, as this would bring emissions in line with sinks. If net deforestation and the associated 1–2GtC/yr emission can be eliminated, fossil fuel emissions would need to be reduced by 2–4GtC/yr, and by less if net deforestation can be turned into net reforestation. On a longer timescale, the sinks themselves would weaken, which would require further emission reductions. Indeed, given a modest positive feedback between climate and the carbon cycle, fossil fuel emissions would need to go to zero, but not until near the end of this century.

Stabilization at 450ppmv is still an enormous challenge. Until 2009, fossil fuel emissions had been growing rapidly, and if the global economy resumes growing, emissions can be expected to grow again as well. The stabilization task naturally divides itself into two parts. The first is to dramatically slow (and, in some regions, reverse) the growth in energy demand. The second is to dramatically increase the rate of deployment of C-free energy sources. The difference between total demand and C-free energy supply is what must be satisfied by fossil fuels, leading to emissions of CO_2 into the atmosphere.

Volume 1 examines the prospects for reducing the growth in, or reducing, energy demand. The generation of electricity from fossil fuels and all the major energy end use sectors (buildings, transportation, industry, agriculture, municipal services) are systematically examined. In each sector, an overview of recent trends and patterns of energy use is given, along with a description of the underlying physical processes involved in using energy, followed by a thorough discussion of the potential for reducing energy use through more efficient energy-using devices and systems. Practical issues, environmental impacts and benefits, and various other co-benefits are also discussed where these are potentially important considerations. However, the solutions to reducing future energy demand are not entirely technological. Behavioural and lifestyle factors are also important, as are the future human population and gross domestic product (GDP) per person. The penultimate chapter of Volume 1 synthesizes the findings of the preceding chapters through the construction of a number of different scenarios that incorporate close to the maximum potential efficiency improvements as determined here, combined with a variety of assumptions concerning population and economic growth. The final chapter of Volume 1 outlines the strategies and policies needed to slow population growth, to slow economic growth without destabilizing the world economy, and to achieve the dramatic improvements in the efficiency with which energy is used that are shown here to be possible.

Volume 2 examines the prospects for rapid deployment of C-free energy sources, with a systematic examination of physical principles, technical potential and cost of each of the major and some not-so-major C-free energy sources. The demand scenarios from Volume 1 are combined with scenarios of deployment of C-free energy (constrained by what are thought to be the limits of feasibility) to develop scenarios of fossil fuel CO_2 emissions. These in turn are used as an input to a coupled climate–carbon cycle model to determine the resulting CO_2 concentration and global mean temperature change.

Some of the ideas presented here and in Volume 2 will be considered to be radical by some readers, and others will be considered to be politically impossible. However, recent history has shown that what is politically 'impossible' one day can become accepted practice the next. I refrain here from making any judgements concerning what is politically feasible or not. I am interested only in showing what is technically feasible and at what estimated cost, and in comparing this with what is required to achieve the 450ppmv concentration target. The onus is on politicians and other leaders to respond with the urgency dictated by the situation that the world now faces. The well-being of future generations, and indeed of much of the life on this planet, depends on their ability to do so.

These two books are a comprehensive blueprint concerning what needs to be done to solve the global warming problem (that is, which will stabilize climate at a warming that will still preserve much that is valuable and beautiful in the world). Nothing less than a complete and rapid transformation of our energy system, and indeed, of our deep-seated ways of thinking is required. However, the political and (in some cases) business response so far has been to consider incremental changes – adjustments – to what is still fundamentally a business-as-usual (BAU) trajectory. There is still little evidence of a political acceptance of the nature and the magnitude of the changes needed. Global warming changes all the old rules about energy, economic growth and the jostling for perceived comparative advantage in international negotiations. There is a new 'reality', but it is a new reality that we by and large have not yet faced up to. It is high time that we did.

L. D. Danny Harvey
Department of Geography
University of Toronto
harvey@geog.utoronto.ca
March 2010

Online Supplemental Material

The following supplemental material can be accessed by visiting www.earthscan.co.uk/resources and selecting the link for this book:

- Powerpoint presentations for each chapter, containing figures, bullet points suitable for teaching purposes, selected tables and supplemental photographs;
- Excel-based problem sets for many chapters;
- Supplemental tables;
- Excel spreadsheets used to generate the demand scenarios presented here.

Acknowledgements

I would like to thank the people listed below for kindly reviewing parts of (or, in some cases, all) of the indicated chapters, in many cases also providing additional information that I would not otherwise have obtained. As well, numerous people responded to questions about their work and in so doing contributed greatly to the final product. I accept responsibility, however, for any errors, misconceptions or important omissions that may remain.

Chapter 2   Alfred Cavallo (Department of Homeland Security, New York)
Chapter 3   Jacob Klimstra (Wärtsilä Corporation, Finland)
Chapter 4   This chapter is a factor-of-eight condensation and update of parts of *A Handbook on Low-Energy Buildings and District Energy Systems*, which was extensively reviewed prior to its publication by Earthscan in 2006.
Chapter 5   Richard Gilbert (Transportation Consultant, Toronto)
Chapter 6
    Sections 6.1–6.5, 6.7 (Metals, cement) Hendrik van Oss (US Geological Survey)
    Section 6.3 (Iron and steel) Andreas Orth (Outokumpu Technology, Germany)
    Section 6.4 (Aluminium) Ken Martchek (Alcoa Corporation)
    Section 6.5 (Copper) Damien Giurco (University of Technology Sydney)
    Section 6.6 (Stainless steel and titanium) Jeremiah Johnson (Yale University)
    Section 6.7 (Cement) Ellis Gartner (Lafarge Cement, France)
    Section 6.9 (Pulp and paper) Kenneth Möllersten (Swedish Energy Agency)
Chapter 7   Stefan Wirsenius (Chalmers University of Technology, Gothenburg, Sweden)
Chapter 10   Paul York (University of Toronto), Helmut Haberl (Institute of Social Ecology, Alpen Adria Universität, Vienna)
Chapter 11   Don Dewees (University of Toronto)

Chapter Highlights

Chapter 1 Prospective climatic change, impacts and constraints

Four independent lines of evidence (simulations with three-dimensional coupled atmosphere–ocean models, observations of temperature changes during the past century, inferences concerning temperature changes and driving factors during the geological past, and inferences concerning natural variations in atmospheric CO_2 concentration during the geological past) indicate that the long-term, global average temperature response to a sustained doubling of the atmospheric CO_2 concentration is very likely to be a warming of 1.5 Celsius degrees (°C) to 4.5°C. Under typical BAU emission scenarios, the concentration of greenhouse gases (GHGs) will rise to the equivalent of three to four times the pre-industrial CO_2 concentration by the end of this century and the climate will have warmed by 3–9°C on average. Likely impacts include an eventual sea level rise of at least 10 metres, reductions in food production in key food-producing regions, reductions in the availability of water in regions already subject to water stress, eventual extinction of the majority (up to 90 per cent) of species of life on this planet, and acidification of the oceans (with potentially catastrophic consequences for marine life). To stabilize the atmospheric CO_2 concentration at 450ppmv (which is the rough equivalent of a CO_2 doubling when the heating effect of other GHGs is taken into account) requires the near elimination of human emissions before the end of this century. Even at 450ppmv, significant ecosystems losses and negative impacts on humans cannot be avoided.

Future CO_2 emissions depend on the future human population, average gross domestic product (GDP) per person, the average energy intensity of the global economy (primary energy required per dollar of GDP) and the average carbon intensity of the global economy (kg of C emitted per gigajoule of primary energy use). The carbon intensity in turn depends on how rapidly C-free energy sources grow compared to the growth in total energy consumption. For middle population and GDP/person scenarios, stabilization of atmospheric CO_2 at a concentration of 450ppmv requires either that the rate of decrease in the global mean energy intensity increase from 1.1 per cent/year (the average over the period 1965–2005) to 3 per cent/year until 2050 with no increase in C-free power, or that annual average C-free power supply increase from 3.3TW in 2005 to 21TW in 2050 (almost 1.5 times total world primary power supply in 2005) with no increase in the rate of reduction in energy intensity, or some less stringent combination of the two.

Chapter 2 Energy basics, usage patterns and trends, and related greenhouse gas and pollutant emissions

A peaking in the global supply of oil by 2020 or sooner is a near certainty. Rates of discovery of new oil have been steadily declining over the last few decades, while the rate of decline in the supply of oil from individual utilized oilfields has occurred at progressively faster rates for oilfields peaking progressively later in time. Data for gas supply are much more uncertain than for oil, but it is likely that gas supply will peak soon after oil supply peaks. Until recently, supplies of coal were thought to be sufficient to last several hundred years. However, recent re-evaluations indicate that the supply of mineable coal is much less than previously thought, and it is possible that the global availability of coal could peak as early as 2050.

Chapter 3 Generation of electricity from fossil fuels

Technologies exist to dramatically improve the efficiency with which electricity is generated from fossil fuels. The global average efficiency in generating electricity from coal powerplants is about 34 per cent, that of new state-of-the-art plants is about 45 per cent, and both advanced pulverized coal powerplants and integrated gasification combined cycle (IGCC) powerplants are expected to achieve efficiencies of 48–52 per cent. State-of-the-art natural gas combined cycle powerplants have an efficiency of 60 per cent. With cogeneration (the co-production of electricity and useful heat), the effective efficiency of electricity generation can exceed 100 per cent. The key to high effective electrical efficiency in cogeneration is to make use of almost all of the waste heat that is produced.

Chapter 4 Energy use in buildings

Technologies already exist to reduce the energy use in new buildings by a factor of two to four compared to conventional practice for new buildings. This is true for buildings of all types and in all climate zones of the world. The keys to achieving such large reductions in energy use are: (1) to focus on a high-performance thermal envelope, (2) to maximize the use of passive solar energy for heating, ventilation and daylighting, (3) to install energy-efficient equipment and especially energy-efficient systems, (4) to ensure that all equipment and systems are properly commissioned and that building operators and occupants understand how they are to be used, and (5) to engender enlightened occupant behaviour. In order to design buildings with factors of two to four lower energy use, an integrated design process is required, in which the architects and various engineering specialists and contractors work together simultaneously in an iterative fashion before key design decisions are finalized. Attention to building form, orientation, thermal mass and glazing fraction is also critical. With regard to renovations of existing buildings, factors of two to four reductions for overall energy use, and up to a factor of ten reduction in heating energy use, have frequently been achieved. In many parts of the world, the cost of reductions in energy use of this magnitude is already justified at today's energy prices.

Chapter 5 Transportation energy use

Urban form (in particular, residential and employment density, and the intermixing of different land uses) and the kind of transportation infrastructure provided are the most important factors affecting future urban transportation energy use. Today there is almost a factor of 10 difference in per capita transportation energy use between major cities with the lowest and largest transportation energy use per capita.

Existing or foreseeable technologies could reduce the fuel requirements of gasoline automobiles and light trucks (sport utility vehicles (SUVs), vans, pickup trucks) by 50–60 per cent with no reduction in vehicle size or acceleration. With a modest reduction in vehicle size and acceleration (to that of the 1980s), a factor of three reduction in fuel consumption could be achieved. Due to the inherent high efficiency of electric drivetrains compared to gasoline or diesel drivetrains, plug-in hybrid electric vehicles (PHEVs) would reduce the onsite energy requirements per kilometre driven by about a factor of three to four when compared to otherwise comparable gasoline vehicles. The economic viability of PHEVs depends on significant reductions in the cost of batteries and verification that they will maintain adequate long-term performance, but the prospects look good. Use of hydrogen in fuel cells could reduce onsite energy requirements by up to a factor of two compared to advanced gasoline–electric hybrid vehicles (depending on the performance of the latter) and by a factor of four compared to current gasoline vehicles, but significant problems remain concerning the cost of fuel cells and onboard storage of hydrogen. The global supply of platinum (Pt) would probably be a significant constraint on the development of a global fleet of hydrogen fuel cell automobiles.

The foreseeable feasible reductions in the energy intensity (energy use per passenger kilometre or tonne kilometre) of other modes of transportation are as follows: transport of freight by trucks, 50 per cent; transport of freight by ship, 45 per cent; diesel freight trains with conversion to trains using hydrogen in fuel cells, 60 per cent; air travel, 25–30 per cent; urban buses, 25–50 per cent (through use of diesel–electric hybrids); interurban buses, 50 per cent.

Chapter 6 Industrial energy use

Compared to the current world average energy intensity, improved technology could reduce the primary energy requirements per unit of output by almost a factor of three for iron and steel, by almost a factor of two for aluminium and cement, by 25 per cent for zinc, and by 20 per cent for stainless steel. Technical improvements in the production of refined copper should roughly balance the tendency for increasing energy requirements as poorer grades of copper ore are exploited. However, much larger reductions in primary energy requirements for metals are possible through recycling combined with projected technical advances: a reduction in primary energy requirements by a factor of 7 for aluminium, a factor of 4.5 for regular steel, a factor of 2.5 for zinc, and a factor of 2 for copper and stainless steel (these savings pertain to uncontaminated materials and assume that the recycled fraction reaches 90 per cent for steel, aluminium and copper, and 80 per cent for zinc and stainless steel). Yet larger reductions in primary energy requirements would occur in combination with ongoing improvements in the efficiency with which electricity is generated. Increasing the recycled fraction of new glass from 25 per cent to 60 per cent reduces on-site energy requirements by about 10–15 per cent compared to production of glass from virgin materials. If the world population and material stock stabilizes by the end of this century, then the production of metals would be used almost exclusively for replacement of existing materials and so could be largely based on recycling, with attendant energy savings. The pulp and paper industry can become energy self-sufficient or a net energy exporter through the efficient utilization of all biomass residues. The potential energy savings in the plastics industry is unclear, but is probably at least 25 per cent through improved processes and at least 50 per cent through recycling of plastics. Potential energy reductions in the chemical industries appear to be very large but cannot be specifically identified at present. Better integration of process heat flows through pinch analysis and better organization of motor systems can save large amounts of energy in a wide variety of different industries.

Chapter 7 Agricultural and food system energy use

Energy in the food system is used for the production of food, for transportation, processing, packaging, refrigerated storage and cooking. Energy use for the production of food consists of direct on-farm energy use and the energy used to produce fertilizers, pesticides and machinery used in farm operations. Fertilizers and pesticides are energy-intensive products. Fertilizer energy use can be reduced through substitution of organic fertilizers for chemical fertilizers, more efficient use of chemical fertilizers (30–50 per cent savings potential in the case of nitrogen (N) fertilizer), and more efficient production of chemical fertilizers (40 per cent savings potential in the case of N fertilizer). Pesticides are particularly energy intensive, but many jurisdictions have targets of reducing pesticide use in agriculture by 50 per cent or so through integrated pest management techniques. Organic farming systems reduce energy use per unit of farm output by 15 per cent to 70 per cent, but can also reduce yields by up to 20 per cent. However, rebreeding of crop varieties to maximize growth under organic farming systems could result in no yield reduction compared to current varieties with conventional methods. The biggest potential for energy savings in the food system is with a shift toward diets with lower meat content. Low-meat diets (and especially vegetarian and vegan diets) reduce direct and indirect fossil fuel energy inputs, and free up land that can be used to produce bioenergy crops.

Chapter 8 Municipal services

Energy is used in the supply of municipal water through pumping and water treatment. Per household, this energy use is comparable to that of major individual household appliances. It can be reduced through measures to use water more efficiently (up to 50 per cent savings potential), through the reduction of leakage in water distribution systems (up to 30 per cent of input water is lost), and through optimization of distribution system pressures and flow rates (10–20 per cent savings potential). The biggest opportunity to reduce net energy requirements at sewage treatment plants is through the recovery and use of methane from anaerobic digestion of sludge. Installation of systems (toilet, plumbing, storage tanks) to separately collect minimally diluted urine in new housing developments would facilitate energy-efficient recycling of nutrients from human wastes, something that will eventually be necessary. With regard to solid wastes, recycling of metals, plastics, paper and paper products is preferred to other management options. Dedicated anaerobic digestion with recovery and use of methane is the preferred option for organic wastes. Incineration with energy recovery is not particularly efficient but is preferred to landfilling for wastes that cannot be recycled further.

The energy requirements of new recreation facilities such as indoor skating arenas, swimming pools and gymnasia can generally be reduced by at least 50 per cent compared to current typical practice for new facilities.

Chapter 9 Community-integrated energy systems

Community-integrated energy systems involve district heating and/or district cooling systems consisting of underground networks of pipes to distribute hot or cold water, respectively, from centralized sources. Waste heat from electricity production can be provided to district heating networks as part of cogeneration, or heat can be captured from low-temperature heat sources (such as sewage water) and upgraded to a higher temperature with heat pumps. Cogeneration at the community scale can displace more centrally generated electricity and yield greater overall energy savings than cogeneration at the building scale (25–45 per cent instead of 10–25 savings, depending on the efficiency of central electricity generation). Centralized production of chilled water with electric chillers can save 30 per cent compared to cooling with separate chillers in each building. Due to economies of scale and reduced backup requirements, the investment cost of district cooling systems can be no greater than having separate chillers in each building. Use of cogeneration greatly reduces the impact of higher natural gas prices on the cost of electricity. Trigeneration (the production of cold water using steam from power generation to drive an absorption chiller) does not save energy compared to operating a powerplant to maximize electricity production and using the extra electricity so produced in highly efficient electric chillers.

Chapter 10 Energy demand scenarios

Scenarios of energy use as fuels and as electricity to the year 2100 are constructed for ten different world regions, taking into account differences in per capita income, floor area and travel today, and are then summed to give a scenario of global demand for fuels and electricity. A low population scenario (global population peaking at 7.6 billion around 2035) combined with modest growth in GDP per capita is considered along with a high population scenario (global population reaching 10.3 billion by 2100) and high growth in GDP per capita. Slow (by 2050) and fast (by 2020) implementation of stricter standards for new and renovated buildings are considered along with the assumption that all existing buildings are either replaced or undergo a major renovation between 2005 and 2050 and that all buildings existing in 2050 undergo a major renovation between 2050 and 2095. Relatively slow and fast rates of improvement in automobile and industrial efficiencies to the potentials identified here are considered, but replacement of existing fossil fuel powerplants with the current state-of-the-art is not assumed to be completed before 2050.

For the low population and GDP scenario, global fuel demand peaks at 20 per cent and 35 per cent above the 2005 demand for fast and slow implementation of energy efficiency measures, respectively, before dropping to about half the 2005 demand by 2100, while global electricity demand rises to and stabilizes at about 70 per cent above the 2005 demand. For the high population and GDP scenario, global fuel demand peaks at 30 per cent and 60 per cent above the 2005 demand for fast and slow rates of efficiency improvement, respectively, then returns to about the 2005 level, while global electricity demand rises to 2.6 times the 2005 demand. When additional structural shifts in the economy (50 per cent of baseline value-added for industry and freight transport shifted to commercial services by 2100), the resulting annual average compounded rate of decrease in the primary energy intensity of the global economy is 2.7 per cent a year between 2005 and 2050 and 1.8 per cent a year between 2005 and 2100.

Chapter 11 Policies to reduce the demand for energy

This chapter outlines strategies for non-coercively promoting lower fertility rates, and gives examples of recent rapid reductions in fertility rates in several countries. Strategies for promoting slower economic growth while simultaneously improving human well-being are discussed. Foremost among these are the channelling of increasing labour productivity into shorter working weeks and greater investment in public transportation systems, affordable housing and other public services. Policies for promoting much greater energy efficiency in all sectors of the economy and for promoting diets low in meat and with less embodied energy are outlined. Important areas where additional research and development are needed are identified. Strategies for reducing or reversing the rebound effect (the tendency for the energy cost savings due to more efficient use of energy in specific applications to result in greater energy use in other areas) are also outlined. Some final reflections on the present overarching policy goal of promoting economic growth and of the urgent need to figure out how to build stable economies that don't depend on continuous growth are offered.

List of Abbreviations

°C	degrees Celsius
μm	micron
AC	alternating current
ACH	air changes per hour
Adt	air-dried tonne
AFC	alkaline fuel cell
AIC	advanced insulation component
AIP	advanced insulation panel
AOGCM	atmosphere–ocean general circulation model
APU	auxiliary power unit
ASHP	air-source heat pump
ASHRAE	American Society of Heating, Refrigeration, and Air Conditioning Engineers
ASK	available seat-kilometre
BaP	benzo-a-pyrene
BAU	business-as-usual
BEV	battery-electric vehicle
BFG	blast furnace gas
BIGCC	biomass integrated gasification combined cycle
BLGCC	black liquor gasification combined cycle
BOF	basic oxygen furnace
BWS	balanced wind stack
C	carbon
$CaCO_3$	calcium carbonate
CAFE	corporate automobile fuel economy
CAV	constant air volume
CBIP	Commercial Building Incentive Program
CC	chilled ceiling
CCFL	cold cathode fluorescent lamp
CCS	carbon capture and storage
CEE	Consortium for Energy Efficiency
CFC	chlorofluorocarbon
CFL	compact fluorescent lamp
cgs	centimetres, grams, seconds
CH_4	methane
CHP	combined heat and power
CI	compression ignition
CMH	ceramic metal halide
CO	carbon monoxide
CO_2	carbon dioxide
COG	coke oven gas
COP	coefficient of performance
CRF	cost recovery factor
CRT	cathode ray tube
CSA	calcium sulphoaluminate

CV	conventional vehicle
CVT	continuously variable transmission
dB	decibel
DC	direct current
DCV	demand-controlled ventilation
DECC	diesel engine combined cycle
DHW	domestic hot water
DME	dimethyl ether
DMFC	direct methanol fuel cell
DOAS	dedicated outdoor air supply
DSF	double-skin façade
DSM	demand-side management
DV	displacement ventilation
DVR	digital video recorder
EAF	electric arc furnace
EAHP	exhaust-air heat pump
ECF	elemental chlorine free
ED	electrodialysis
EER	energy efficiency ratio
EF	energy factor
EI	energy intensity
EIFS	external insulation and finishing system
EJ	exajoules
EMC	energetically modified cement
EOR	enhanced oil recovery
EPBD	Energy Performance of Buildings Directive
EPRI	Electric Power Research Institute
EROEI	energy return over energy invested
ESCO	energy service company
EU	European Union
EURIMA	European Mineral Wool Manufacturers Association
FACE	Free Air Concentration Enhancement
FBC	fluidized bed combustor
FCV	fuel cell vehicle
FGR	flue-gas recirculation
FO	forward osmosis
FOM	Farmers Own Market
FSU	former Soviet Union
GBS	granulated blast furnace slag
GDP	gross domestic product
Gg	gigagram
GHG	greenhouse gas
GIS	Greenland ice sheet
GJ	gigajoule
gm	gram
GRP	gross regional product
GSHP	ground-source heat pump
Gt	gigatonne
GTCC	gas turbine combined cycle

GW	gigawatt
GWP	global warming potential
H_2	hydrogen
HAT	humid air turbine
HC	hydrocarbon
HCFC	hydrochlorofluorocarbon
HDD	heating degree day
HERS	Home Energy Rating System
HEV	hybrid electric vehicle
HFC	hydrofluorocarbon
Hg	mercury
HHV	higher heating value
HIR	halogen infrared-reflecting
HRSG	heat recovery steam generator
HRV	heat recovery ventilator
HSPF	heating season performance factor
HTS	high-temperature superconducting
Hz	Hertz
ICE	internal combustion engine
IDP	integrated design process
IEA	International Energy Agency
IGCC	integrated gasification combined cycle
IPCC	Intergovernmental Panel on Climate Change
IPM	integrated pest management
IRR	internal rate of return
IT	information technology
J	joule
K	kelvin
kHz	kilohertz
kg	kilogram
kph	kilometres per hour
kt	kilotonne
kW	kilowatt
kWh	kilowatt hour
LCD	liquid crystal display
LDV	light-duty vehicle
LED	light-emitting diode
LEED	Leadership in Energy and Environmental Design
LFG	landfill gas
LHV	lower heating value
LiBr	lithium bromide
lm	lumen
LNG	liquefied natural gas
m	metre
mb/d	million barrels per day
MBT	mechanical biological treatment
MCFC	molten carbonate fuel cell
MEB	multi-effect boiling
MEF	Modified Energy Factor

mg	milligram
MIT	Massachusetts Institute of Technology
MJ	megajoule
mks	metres, kilograms, seconds
mpg	miles per gallon
MSF	multi-stage flash
MSW	municipal solid waste
MTOW	maximum take-off weight
MV	mixing ventilation
N	nitrogen
NA-SI	naturally aspirated spark ignition
NGCC	natural gas combined cycle
NGSC	natural gas simple cycle
NH	northern hemisphere
NiMH	nickel-metal hydride
NIR	near-infrared
NO_x	nitrogen oxide
N_2O	nitrous oxide
NPV	net present value
NREL	National Renewable Energy Laboratory
NSPS	New Source Performance Standards
O_3	ozone
O&M	operation and maintenance
OECD	Organisation for Economic Co-operation and Development
OEW	operating empty weight
OLED	organic light emitting diode
OPEC	Organization of Petroleum Exporting Countries
P	phosphorus
PAFC	phosphoric acid fuel cell
PC	personal computer
PCB	polychlorinated biphenyl
PCM	phase-change material
PEM	proton exchange membrane
PEMFC	proton exchange membrane fuel cell
PET	polyethylene terephthalate
PFC	perfluorinated hydrocarbons
PHEV	plug-in hybrid electric vehicle
pkm	passenger-kilometre
PM	particulate matter
POU	point-of-use
ppbv	parts per billion by volume
PPE	polyphenylene ether
ppmv	parts per million by volume
Pt	platinum
PUF	polyurethane foam
PV	photovoltaic
PVC	polyvinyl chloride
RDF	refuse-derived fuel
RH	relative humidity

RL	recirculation loop
rms	root mean square
RO	reverse osmosis
RPK	revenue passenger-km
rpm	rotations per minute
S	sulphur
SAR	specific air range
SCM	supplementary cementitious materials
SEER	seasonal energy efficiency ratio
SHC	Solar Heating and Cooling
SHGC	solar heat gain coefficient
SI	spark ignition
SiO_2	silica
SLR	sea level rise
SO_2	sulphur dioxide
SOFC	solid oxide fuel cell
SPFC	solid polymer fuel cell
SRES	Special Report on Emission Scenarios
SSF	single-skin façade
SSP	single superphosphate
STAF	Stocks and Flows
STB	set-top box
SUV	sport utility vehicle
SX-EW	solvent-extraction electro-winning
TCF	totally chlorine free
TGV	Train à Grande Vitesse
TI	transparent insulation
TSFC	thrust specific fuel consumption
TSP	triple superphosphate
TVSP	TV service provider
TW	terawatt
UHDG	Uppsala Hydrocarbon Depletion Group
UNFCCC	United Nations Framework Convention on Climate Change
UPS	uninterruptible power supply
USGS	United States Geological Survey
V	volt
V2G	vehicle-to-grid
VAV	variable air volume
VCR	variable compression ratio
VIP	vacuum insulation panel
VOC	volatile organic compound
VSD	variable speed drive
WAIS	West Antarctic ice sheet
WF	Water Factor
WTW	well-to-wheels
W/m^2	watts per square metre
XPS	extruded polystyrene
yr	year

1

Prospective Climatic Change, Impacts and Constraints

The earth's average temperature is governed by the balance between the absorption of shortwave radiation from the sun and the emission of longwave radiation (also referred to as *infrared radiation*) from the earth. A surplus in the absorption of shortwave radiation over the emission of longwave radiation will cause the climate to become warmer, while an excess of emitted longwave radiation will cause the climate to cool. Longwave radiation is emitted from the earth's surface and clouds, and to a lesser extent, from the atmosphere itself. The ease with which the earth can emit longwave radiation to space depends on how much of the surface emission can pass through the atmosphere. Certain naturally occurring gases in the atmosphere – water vapour foremost, followed by carbon dioxide (CO_2), ozone (O_3) and methane (CH_4) – absorb some of the longwave radiation emitted from the surface, thereby retaining energy within the atmosphere and making the climate warmer than it would be otherwise. This warming effect is popularly referred to as the 'greenhouse effect', and the gases contributing to it are called greenhouse gases (GHGs).

The naturally occurring greenhouse effect makes the earth's climate 33°C warmer than it would be otherwise. However, during the past 200 years, humans have caused large increases in the concentrations of a number of important GHGs: the CO_2 concentration has increased by over 36 per cent (from 280 parts per million by volume, or ppmv, to 387ppmv by 2010), the methane concentration has increased by a factor of 2.5 (from 0.7ppmv to 1.75ppmv), the nitrous oxide (N_2O) concentration has increased by about 15 per cent

(from 275 ppbv to 315ppbv) and the lower-atmosphere (or tropospheric) ozone concentration has increased by a factor of two to five (depending on location) in the northern hemisphere. As well, several classes of entirely artificial GHGs have been created and added to the atmosphere – the chlorofluorocarbons (CFCs) and their replacements: the hydrochlorofluorocarbons (HCFCs) and hydrofluorocarbons (HFCs).

These increases have strengthened the natural greenhouse effect by trapping infrared radiation, thereby upsetting the prior balance between absorbed solar radiation and emitted infrared radiation. A perturbation in the radiative balance is referred to as *radiative forcing*. It is what 'forces' or drives subsequent changes in climate. For a doubling in the concentration of CO_2 (something that could occur by the middle of this century under business-as-usual (BAU) scenarios), the radiative forcing is just under 4 watts per square metre (W/m^2). This is equivalent to the heat output of two incandescent Christmas-tree lights over every square metre of the earth's surface. The radiative forcing due to the collective effect of all the increases in all of the GHGs that have occurred so far is around $3.0W/m^2$. About half of this forcing is due to CO_2 alone, with all the other gases adding up to the other half.

Figure 1.1 shows the CO_2 and CH_4 increase during the past 200 years in the context of natural variations during the past 400,000 years (as determined from the composition of air bubbles trapped in snow in Antarctica as the snow was transformed into ice). The recent increases in the concentrations of these two GHGs have been rapid (compared to natural

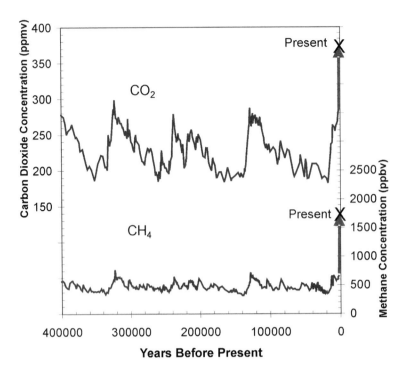

Note: The Vostok ice core is discussed in Petit et al (1999). Data referring to the last 400,000 years are represented by the thin lines, while data referring to the last 200 years are represented by the thick lines.
Source: Data in electronic form were obtained from the US National Oceanographic and Atmospheric Administration (NOAA) website (www.ngdc.noaa.gov/paleo)

Figure 1.1 *Variation in atmospheric CO_2 and CH_4 concentration during the last 400,000 years as deduced from the Vostok ice core and during the past 200 years*

fluctuations) and far outside the bounds of natural variability during the past 400,000 years.

CO_2 is emitted to the atmosphere during the combustion of fossil fuels (natural gas, oil and coal), to a small extent from additional chemical releases during the production of cement, and as a result of deforestation (commonly referred to as land use changes). Fossil fuel and cement emissions (henceforth referred to as industrial emissions) grew from 6.4GtC/yr in 1995 to 8.0GtC/yr in 2005, while land use changes caused a further emission of 1.6±1.1GtC/year during the 1990s. The rate of industrial emission could grow several-fold during the coming century, while long-term land use emissions are constrained by the availability of remaining forests and are unlikely to rise much above the current rate before declining. Thus, in the longterm, industrial emissions are of much greater concern than land use emissions as far as the buildup of atmospheric

CO_2 is concerned. Furthermore, projected future increases in CO_2 concentration will dominate the overall further increase in the greenhouse effect under BAU scenarios. Since the natural processes that remove CO_2 from the atmosphere are quite slow (requiring thousands of years to remove most of the emitted CO_2), the present and future increase in concentration is irreversible for all practical purposes.[1] For these reasons, industrial CO_2 emissions, which are overwhelmingly from the combustion of fossil fuels, will be the single largest factor in future human-induced warming of the climate.

1.1 Past changes of climate

Figure 1.2 shows the variation in global average surface temperature over the period 1856–2009. This curve is a composite of sea surface temperature variations in the

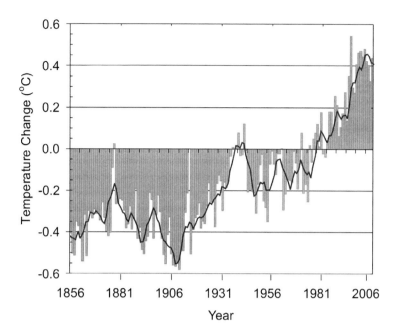

Figure 1.2 *Variation in global average surface temperature during the period 1856–2009*

ocean portion of the earth and surface air temperature variations over the land portion of the earth (a detailed discussion of the methods used in constructing this curve can be found in Harvey, 2000, Chapter 5). There is no remaining doubt that the earth's climate has warmed, by 0.6–0.8°C on average, during the past 140 years or so. However, one cannot judge whether the warming seen in Figure 1.2 is unusual or not, without a longer record of global-scale temperature variation.

Fortunately, such records are available through temperature variations that can be reconstructed from the chemical composition of annual layers in polar ice caps, through the annual variations in tree ring width (largely at middle-latitude sites), and from the chemical composition of annual layers in coral reefs (from low latitudes). Figure 1.3 shows several such reconstructions of the variation in northern hemisphere (NH) or global average temperature from AD 1000 to AD 1980. Also shown in Figure 1.3 are the directly measured variations in global average temperature from Figure 1.2. All scientifically credible reconstructions agree that the late 20th century is the warmest period during the past 1000 years. There can be little doubt that this warming is largely due to the concurrent buildup of GHG

concentrations during this time period. Other factors, such as solar variability, have been thoroughly and repeatedly assessed and consistently found to be only a small contributor to climatic variations over the past century (see Forster et al, 2007, Section 2.7.1, and Hegerl et al, 2007, Section 9.4.1).

1.2 Prospective future changes of climate

Given a scenario of CO_2 emissions, the buildup of atmospheric CO_2 concentration can be computed with reasonable accuracy using a computer *carbon cycle model.* Carbon cycle models compute the absorption of a portion of the emitted CO_2 by the terrestrial biosphere (primarily by forests) and by the oceans. At present, about half of the annually emitted CO_2 is absorbed by these two sinks, while the balance accumulates in the atmosphere. Given a scenario for the buildup of CO_2 and other GHGs, future climatic changes can be computed using a *climate model.* However, changes in climate will alter the natural flows of CO_2 into and out of the atmosphere, leading to further changes in the concentration of atmospheric

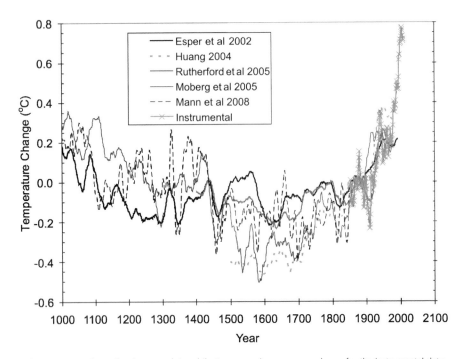

Note: 25-year running means are shown for the proxy data, while 5-year running means are shown for the instrumental data.

Source: The various proxy data were obtained from the US National Oceanographic and Atmospheric Administration (NOAA) website (www.ncdc.noaa.gov/paleo) and adjusted to be relative to the 1871–1900 mean

Figure 1.3 *Variation in NH or global average surface temperature based largely on ice core, tree ring and coral reef (proxy) data, as reconstructed by various sources, and the directly observed (instrumental) temperature variation of Figure 1.2*

CO_2 beyond those due to direct human emissions. Calculation of these changes requires a *coupled climate–carbon cycle model*. In the following subsections, some of the key processes relevant to determining future changes in climate are outlined.

1.2.1 Climate sensitivity and climate–carbon cycle feedbacks

The single most important parameter in determining the climatic response to a given GHG buildup is the *climate sensitivity* – the long-term global average warming in response to a doubling of the atmospheric CO_2 concentration.[2] For over 30 years, the consensus has been that the climate sensitivity is highly likely to fall between 1.5–4.5°C. The climate sensitivity is the net effect of a variety of positive and negative feedbacks that tend to amplify or diminish the initial warming that is triggered by the initial radiative forcing. The most

important of these feedbacks involves water vapour – as the climate warms, the atmosphere holds more water vapour, and as water vapour itself is a GHG, this leads to further warming. This constitutes a positive feedback, as much as doubling the warming that would occur in its absence.[3] Another feedback involves the retreat in the extent of ice and snow as the climate warms. Ice and snow have a high reflectivity to solar radiation, while ice- and snow-free surfaces absorb more solar radiation, so the retreat of ice and snow leads to more solar heating of the surface, which amplifies the warming at high latitudes. Changes in the properties, amount, location and timing of cloud cover exert a variety of positive and negative feedbacks. The uncertainty concerning the net effect of competing cloud feedbacks is the single largest contributor to the factor-of-three uncertainty in the likely climate sensitivity.

The estimated 1.5–4.5°C range for the climate sensitivity is based on a number of *independent*, *observationally based* or *observationally constrained* estimation methods. These methods include

- Simulations of climate feedback processes with three-dimensional coupled atmosphere–ocean general circulation models (AOGCMs). Although the feedback predicted by AOGCMs for long-term climatic changes cannot be observed and verified, AOGCMs have successfully simulated a number of observed feedbacks operating at shorter timescales in response to seasonal and inter-annual variations in temperature, volcanic forcing and decadal trends, thereby providing at least partial validation of some of the very processes that are responsible for long-term feedbacks (see Randall et al, 2007, Section 8.6.3.1.1).
- Constraints based on the observed warming during the past century and estimates of GHG and non-GHG radiative forcings. The observed global mean warming over the past century (about 0.6–0.8°C) can be thought of as the product of three factors:

$$(\text{net radiative forcing, W/m}^2) \times (\text{climate sensitivity, °C/(W/m}^2))$$
$$\times (\text{ocean lag factor}) = 0.6\text{--}0.8°C \qquad (1.1)$$

The ocean lag factor is about 0.7–0.8 for the current rate of increase in net radiative forcing. The net radiative forcing is the net effect of increasing GHG concentrations, increasing aerosol concentrations in the atmosphere, and small effects of variation in solar irradiance and volcanic activity. Aerosols are small particles suspended in the atmosphere, and humans have emitted a number of compounds that produce aerosols that have a net cooling effect on climate, thereby offsetting a portion of the heating effect due to the buildup of GHGs. Were it not for this aerosol cooling effect, the warming during the last 100 years would have been larger. The largest uncertainty in the net forcing concerns the extent to which aerosols have offset some of the radiative forcing by GHGs. The larger the aerosol offset, the smaller the net forcing, and so the greater the permitted climate sensitivity while still producing the same observed warming, as can be seen from Equation (1.1). The magnitude of the permitted aerosol cooling is constrained by the fact that most

of the aerosol cooling effect is in the northern hemisphere. If too large a climate sensitivity is offset by too large an aerosol cooling, this will suppress the warming in the NH more than observed (even if the correct global average warming is simulated). On this basis, Harvey and Kaufmann (2002) conclude that aerosols have offset at most about half of the GHG heating so far, and that the climate sensitivity is likely to be 2–3°C, with a possible extreme range of 1–5°C. This is consistent with most other estimates based on a variety of different observations (Hegerl et al, 2007, Section 9.6.2.1).[4]

- Comparisons of estimates of the radiative forcing and global mean temperature changes at various times in the geological past. During the peak of the last ice age (around 20,000 years ago), the radiative forcing was a cooling tendency of 4.6 to 7.2W/m² while the inferred global mean decrease in temperature was 3–5°C. This implies a climate sensitivity for a CO_2 doubling of 1.2–4.3°C (Hegerl et al, 2007, Section 9.6.3.2).
- Constraints based on past variations in atmospheric CO_2. The atmospheric CO_2 concentration varied in the geological past over periods of million of years in response to imbalances between volcanic outgasing and changing rates of weathering due to variations in plate tectonic activity and associated mountain uplift. The magnitude of the CO_2 variation depended in part on a negative feedback between CO_2 concentration and removal rates by chemical weathering, whereby higher atmospheric CO_2 leads to a warmer climate that in turn accelerates chemical weathering, leading to more rapid removal of CO_2 from the atmosphere. This feedback in turn depends on how large the climate sensitivity is. Simulations with models of the global carbon cycle appropriate for multi-million year timescales indicate that if climate sensitivity is too small (less than about 1.5°C), impossibly high CO_2 peaks are obtained at certain times, while if it is too large (greater than about 3°C), unreasonably low values are obtained at certain times (Royer et al, 2007).

In summary, there is a wide range of independent evidence that indicates that the climate sensitivity is very likely to fall between 1.5 and 4.5°C. This is the global mean warming that can be expected for each successive doubling in the atmospheric CO_2 concentration. To a good approximation, the climate response to increases in CO_2 and other GHGs depends only on the net heat trapping of all the gases, and not on the specific gases contributing to the heating. Thus, it is possible to speak of the *equivalent CO_2 concentration* – the concentration of CO_2 alone that would have the same heat-trapping effect (or radiative forcing) as that from the mixture of gases present in reality. Under most BAU scenarios, GHG concentrations rise beyond the equivalent of four times the pre-industrial atmospheric CO_2 concentration by 2100 (see Prentice et al, 2001, Figure 3.12). Even under stringent constraints on the emissions of all GHGs, concentrations will peak at close to a CO_2-doubling equivalent.

Given a global mean warming for each CO_2 doubling of 1.5–4.5°C, a quadrupling in the effective CO_2 concentration would lead to an eventual warming of 3.0–9.0°C. However, long before the climate warmed this amount, the warming would induce further, natural fluxes of CO_2 into atmosphere (from, for example, the thawing and decomposition of organic-rich permafrost in the Arctic, or dieback of mid-latitude and/or tropical forests), leading to yet further increases in atmospheric CO_2 concentration and further warming. Current coupled climate–carbon cycle models indicate that feedbacks between climate and the carbon cycle augment the increase in atmospheric CO_2 concentration by 5–20 per cent compared to the increase that would otherwise occur, although these models omit several poorly understood processes that could lead to an even larger increase (Denman et al, 2007, Section 7.3.5).

1.2.2 Scenarios of future global mean temperature change

Given an estimate of past GHG+aerosol radiative forcing and a scenario of future forcing, past and future globally average temperature change can be computed using simple (one- or two-dimensional) climate models (Harvey et al, 1997). The climate sensitivity can be prescribed in such models to have any reasonable value one wishes, with a compensating adjustment of the aerosols forcing so as to give approximately the correct global average warming up to the present. The effect of variations in volcanic activity and solar luminosity can also be incorporated when fitting the simulated climate to the observed variation.

Aerosol particles remain in the atmosphere only a few days before being washed out with rain, so they need to be continuously replenished. The primary aerosol – sulphate – is associated with acid rain, but sulphur emissions are being reduced in much of the world in order to reduce the damage due to acid rain. As emissions are reduced, the aerosol cooling effect will weaken, leading to a temporary acceleration in global warming. Aerosol emissions would also fall as a byproduct of reductions in the use of fossil fuels. The magnitude of the acceleration depends on how large the aerosol cooling effect is at present, something that is highly uncertain. However, the heating effect of a number of short-lived GHGs (such as methane, various halocarbons and tropospheric ozone) would also quickly diminish as fossil fuel use decreases. These competing effects can all be easily incorporated in simple models.

Figure 1.4 shows past and projected future global average warming for a typical BAU scenario of increasing GHG emissions but falling aerosol emissions for climate sensitivities of 2°C and 4°C, using a simple climate–carbon cycle model with a very weak positive climate–carbon cycle feedback. Fossil fuel CO_2 emissions grow from 7.7GtC/yr in 2005 to 20.8GtC/yr by 2100 for the BAU scenario considered here. The globally averaged warming by the end of this century ranges from 3.5°C to as high as 6.5°C. Even the optimistic case (produced with a rather low climate sensitivity of 2°C) produces a warming by the end of this century comparable to the difference between the last ice age and the ensuing interglacial climate – a globally averaged difference of 4–6°C. However, the majority of the warming that occurred at the end of the last ice age occurred over a period of 7000 years, whereas comparable GHG-induced warming could occur over the next 100 years – about 70 times faster. Also shown in Figure 1.4 are results where fossil fuel emissions are reduced to zero by 2100, which is likely to be sufficient to stabilize the atmospheric CO_2 concentration at 450ppmv in the face of positive climate–carbon cycle feedbacks (discussed below in subsection 1.4.2). In combination with stringent reductions in emissions of other GHGs, this corresponds closely to the equivalent of a CO_2 doubling, so the global mean warming by 2100 approaches 2°C and 4°C for the 2°C and 4°C climate sensitivities.

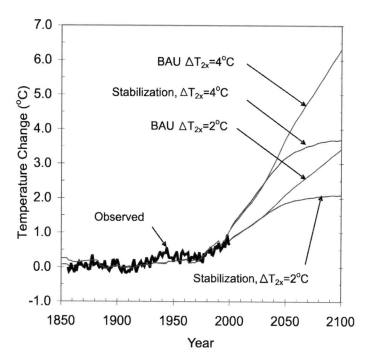

Note: The assumed aerosol cooling effect in 1990 is 30 per cent of the GHG heating for a 2°C sensitivity and 50 per cent of the GHG heating for a 4°C sensitivity. In this way we obtain a reasonable fit to the observed warming for both assumed climate sensitivities. The simulations were extended to 2100 for a scenario in which fossil fuel CO_2 emissions increase by a factor of four while total aerosol cooling decreases by a factor of two (labelled 'BAU'), and for simulations in which the CO_2 concentration is stabilized at 450ppmv.

Figure 1.4 *Comparison of the observed temperature record of Figure 1.2 (up to 2002) and global average temperature changes (up to 2100) as simulated using a simple climate model using prescribed climate sensitivities for a CO_2 doubling (ΔT_{2x}) of 2°C and 4°C*

The unprecedented and abrupt nature of prospective climatic change during the next century is dramatically illustrated in Figure 1.5, which combines the data shown in Figures 1.2, 1.3 and 1.4.

1.3 Drivers of energy use and fossil fuel CO_2 emissions

Future CO_2 emissions, expressed as a mass of carbon per year, can be written as the product of the following factors:

population × GDP per year per capita × primary energy per unit GDP × carbon emission per unit of primary energy used

or, in terms of units,

total emission per year =
P × ($/yr)/P × MJ/$ × kgC/MJ   (1.2)

where MJ/$ (primary energy use per unit of GDP) is referred to as the *energy intensity* of the economy and kgC/MJ (carbon emission per unit of energy) is referred to as the *carbon intensity* of the energy system.[5] This decomposition of CO_2 emission is referred to as the *Kaya identity*, in honour of the Japanese scientist who first proposed it.

The above equation is a straightforward mathematical identity but it hides a series of complex interactions. For example, population and GDP per capita are not really independent. Faster growth in the economy can lead to a faster decrease in birth rates and hence a smaller future population, so the product of the first two terms might not change as much as one might think. That is, we could have a large population of poor people, or a smaller population of relatively wealthy people. Faster growth of the economy may also lead to faster rates of innovation and improvement in energy

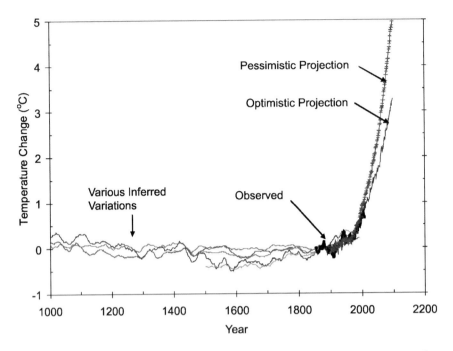

Figure 1.5 *Proxy temperature variations of Figure 1.3, directly observed temperature variations of Figure 1.2, and BAU model simulations of Figure 1.4*

efficiency, leading to a faster decline in energy intensity. Energy intensity reflects not only changes in the efficiency with which energy is used, but also progressive changes in the structure of the economy and in purchasing behaviour and lifestyles. As a country develops, less of its resources are spent building infrastructure (pipelines, sewers, power systems, roads) and buildings, and relatively more resources are used to produce or purchase computers or CDs. The average energy use per dollar spent therefore decreases. As a population ages, the mix of different types of consumption will change as well, so energy intensity will depend in part on the age structure of the population. The fourth factor, carbon intensity, reflects the energy sources used – the proportions of fossil fuel and C-free energy and, within the fossil fuel portion, the proportions of coal, oil and natural gas. Among the fossil fuels, natural gas has the lowest CO_2 emission per unit of primary energy, followed by oil and then coal (see Chapter 2, subsection 2.7.1). Thus, a shift from coal to oil or natural gas, along with an increase in the proportion of non-fossil energy sources, leads to a decrease in the average carbon intensity.

In 2005, the world's population reached 6.47 billion (US Census Bureau, 2008), world GDP was US$55.7 trillion on a purchasing-power-parity basis (IMF, 2008), primary energy demand was 483EJ (exajoules; this and other energy and power units are defined in Appendix A) (Chapter 2, Table 2.3), and global fossil fuel emission was 7.7GtC (Marland et al, 2008). Thus, global average GDP per person was $8609, energy intensity was 8.68MJ/$ (0.00868GJ/$), and carbon intensity was 15.9 kilograms (kg) of carbon per gigajoule (GJ). Figure 1.6 shows the trends in world population, world GDP, world primary energy use and total fossil fuel CO_2 emissions from 1965–2005. Also given are trends in GDP/P (GDP per capita), energy intensity and carbon intensity. Decadal average rates of change in each of the four terms of the Kaya identity are given in Table 1.1, as well as the average rate of change. Population has been growing at a decreasing rate, while GDP/P grew at a decreasing rate for the first three decades (2.9 per cent/year, 1.6 per cent/year and 1.2 per cent/year during the periods 1965–1975, 1975–1985 and 1985–1995, respectively), but accelerated to 2.2 per cent/year during the period 1995–2005 due to sudden rapid growth in China. Over the period 1965–2005,

GDP/P grew at an average rate of 2.0 per cent/year. Energy intensity fell at a steadily increasing rate, with an average rate of decrease of 1.07 per cent/year over the period 1965–2005 and 1.34 per cent/year during 1995–2005, while carbon intensity fell by only 0.35 per cent/year from 1965–1995 and by an imperceptible 0.01 per cent/year from 1995–2005.

If world population peaks at 9 billion by 2100 (the median projection of Lutz et al, 2001) and GDP per capita were to grow at 2.0 per cent/year, the world

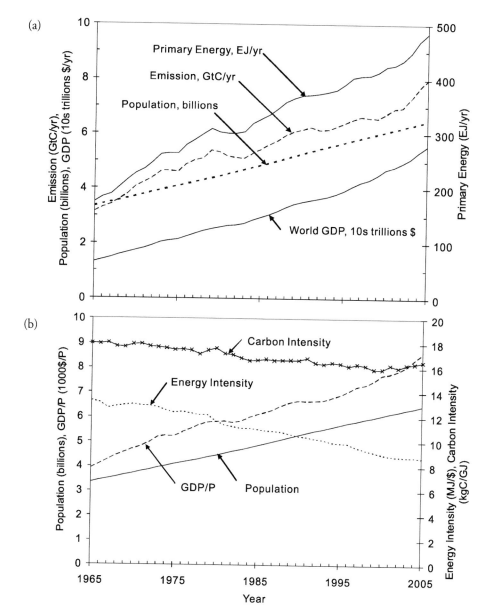

Source: World population data from www.census.gov/ipc/www/worldpop.html; world GDP data from WWI (2008); world primary energy use data from BP (2007); world fossil fuel CO$_2$ emission data from Marland et al (2008)

Figure 1.6 *(a) Variation in world population, aggregate GDP, primary energy use and fossil fuel CO$_2$ emissions from 1965–2005; (b) Variation in world average GDP/capita, energy intensity and carbon intensity from 1965–2005*

Table 1.1 *Decadal trends (per cent/yr) in the four terms of the Kaya identity, based on the data presented in Figure 1.6*

Decade	Population	GDP/P	Energy intensity	Carbon intensity
1965–1975	2.00	2.87	–0.73	–0.32
1975–1985	1.71	1.57	–1.16	–0.51
1985–1995	1.59	1.24	–1.05	–0.23
1995–2005	1.28	2.22	–1.34	–0.01
Overall	1.65	1.97	–1.07	–0.27

economy would be 9.2 times larger in 2100 than in 2005, and there would be a corresponding increase in CO_2 emissions if the energy system did not change. Continuation of the 1.07 per cent/year average decrease in energy intensity of the global economy to 2100 would reduce the energy use and CO_2 emissions by almost a factor of three ($0.9893^{95}=0.360$), giving an increase in emissions by a factor of $9.2 \times 0.360 = 3.3$. However, under typical BAU assumptions, the share of coal in the global energy mix would increase compared to at present, due to the depletion of oil and natural gas. This would tend to drive emissions higher, due to the much greater CO_2 emission per unit of energy for coal compared to oil or natural gas. Failure to maintain 1.1 per cent/year improvement in global average energy intensity would also produce higher emissions. Altogether then, fossil fuel CO_2 emissions by 2100 could be a factor of four or five higher than in 2005, assuming no limitations in the supply of coal.

1.4 The United Nations Framework Convention on Climate Change

The broad principles that should be adhered to in responding to the threat of global warming are laid out in the United Nations Framework Convention on Climate Change (UNFCCC), a document signed and ratified by 186 of the world's countries. The UNFCCC declares its 'ultimate objective' to be:

stabilization of greenhouse gas concentrations in the atmosphere at a level that would prevent dangerous anthropogenic interference with the climate system. Such a level should be achieved within a time frame sufficient to allow ecosystems to adapt naturally to climate change,

to ensure that food production is not threatened and to enable economic development to proceed in a sustainable manner. (UN, 1992)

By speaking of adaptation to climatic change, it is implied that the ultimate climatic change (related to the chosen GHG stabilization levels) is small enough and hence benign enough that adaptation is possible in the first place. The three subsidiary conditions (allowing ecosystems to adapt, maintaining food production and enabling sustainable economic development) are restrictions on the *rate* at which non-dangerous GHG concentrations are reached. They are related to that fact that climatic change that is not harmful (that is, sufficiently limited that adaptation is possible), were it to occur slowly, could be highly disruptive (harmful) if it were to occur too fast. These conditions thus set a constraint on rates of allowable GHG emissions, while the overall goal of capping GHG concentrations at non-dangerous levels largely represents a constraint on cumulative CO_2 emissions.

Inasmuch as the UNFCCC has been accepted and ratified by almost all nations in the world, a number of value judgements have already been implicitly accepted and do not need further debate. For example, the value judgement has already been made that ecosystems are worthy of protection (by ensuring that rates and magnitudes of climatic change are such as to allow their natural adaptation), irrespective of their economic value to humans.

In the following subsections, the various threats to ecosystems, food production and socio-economic systems posed by various degrees of warming are outlined, followed by a conclusion concerning the lowest CO_2-equivalent concentration that would constitute dangerous interference in the climate system.

1.4.1 Threats to ecosystems, food production and human security

A CO_2 concentration of 450ppmv roughly corresponds to a doubling of the pre-industrial atmospheric CO_2 concentration of 280ppmv when the heating effect of other GHGs is taken into account, assuming *stringent* limitations in emissions of other GHGs.[6] As discussed above, the climate sensitivity – and thus the eventual warming for a mere 450ppmv CO_2 concentration – is very likely to fall between 1.5 and 4.5°C, so we shall take the extreme limits to be 1°C and 5°C.

There is a continuously increasing risk of serious impacts as the projected warming increases from 1–5°C. In particular:

- widespread extinction of coral reef ecosystems with 1–2°C warming above preindustrial levels;
- possible collapse of the Greenland ice sheet with as little as 1–2°C sustained global average warming, and almost certainly with 3–4°C sustained warming;
- likely extinction of 15–30 per cent of species of life on earth due to a mere 2°C warming happening by 2050, with greater losses with greater warming;
- reduced agricultural productivity in some major food-producing regions once local warming exceeds 1–3°C (depending on the crop and location);
- increasingly severe water shortages in semi-arid regions, including eventual loss of glacier meltwater that supplies summer water to 25 per cent of the population of China;
- acidification of the oceans as CO_2 is absorbed by the oceans, with severe and still poorly understood impacts.

I will briefly elaborate on each of these impacts below. For a comprehensive discussion, the report of the Second Working Group to the Fourth Assessment Report of the Intergovernmental Panel on Climate Change (IPCC) can be consulted, particularly Chapter 19 (Schneider et al, 2007).

Coral reef ecosystems

Coral reefs are a great storehouse of biodiversity, housing about 100,000 species that have been described

and an estimated total of 0.5–2.0 million species (Reaka-Kudla, 1996). They provide direct sustenance to an estimated 100 million people, with additional benefits to local societies through tourism revenues (Hoegh-Guldberg, 2005). They are particularly sensitive to temperature increases, will be adversely affected by decreasing carbonate supersaturation, and may not be able to keep up with rising sea level.

Coral organisms live in a symbiotic relationship with a photosynthetic organism, and grow in ocean waters ranging from 26–34°C in temperature. When water temperatures increase by only 1°C beyond the normal peak seasonal temperature in a given region, the symbiont is expelled, causing the coral to lose colour. This process is referred to as *coral bleaching*. Prolonged expulsion of the symbiont leads to the eventual death of the coral. Seasonal temperature peaks in tropical regions generally occur during El Niño years. The most severe El Niño to date occurred in 1998, when 16 per cent of the world's coral reefs experienced severe bleaching, in some cases killing corals that were 1000 years old (Goldberg and Wilkinson, 2004). The decadal global mean temperature was only 0.6°C warmer than that of the late 1800s/early 1900s at this time, with 1998 being about 0.9°C warmer than the late 1800s/early 1900s mean. As long-term average temperatures increase, future El Niño peaks will be progressively warmer, leading to more severe and more frequent beaching events. In the Indian Ocean, more than 90 per cent of shallow corals were killed by the 1998 El Niño, and an event of this magnitude is expected one year in five by the mid-2020s (Sheppard, 2003). Donner et al (2005) have assessed the impact of climatic change as projected by two different AOGCMs for the frequency and severity of coral bleaching. They find that severe bleaching occurs at the majority of the world's coral reefs in the 2030s and becomes a biannual event by the 2050s, by which time global mean temperature has increased by about 2°C above that of the late 1800s in the AOGCMs under consideration.

Adaptation could delay the regular occurrence of severe bleaching. The large range of temperatures at which coral reefs grow today implies that there is some genetically based diversity in the temperature tolerance of the symbionts. Warmer temperatures would favour organisms with a greater tolerance of heat, leading to a shift in the composition of reef symbiont populations toward tolerance of warmer temperatures over successive generations. The critical issues concern: (1) the rate at which such shifts can occur, (2) the temperature limits of adaptation, and (3) the occurrence of adverse tradeoffs in other areas (such as productivity) in exchange for greater temperature tolerance. Although there may be a substantial genetic ability for adaptation to warmer temperature, natural rates of adaptation are likely to be quite slow and will be further inhibited by other stresses on coral reefs, including reduced carbonate supersaturation in surface waters (Hughes et al, 2003; Hoegh-Guldberg, 2005). More temperature-tolerant symbionts are observed to have slower and less vigorous growth because of the energetic cost of enhanced protective machinery (Donner et al, 2005). This in turn would limit the reef's ability to keep up with rising sea level or to deal with other stresses (such as nutrient loading, sedimentation and disease). With adaptation, it is conceivable that significant and widespread harm could be delayed until the global mean temperature increase exceeds 2°C. Thus, it seems reasonable to assume that the likely threshold for significant harm to coral reefs occurs somewhere between global mean temperature changes of 1–2°C.

Sea level rise

With regard to sea level rise (SLR), the critical processes include the melting of the Greenland ice sheet (GIS) and the destabilization and collapse of the West Antarctic ice sheet (WAIS). There are two critical issues here: (1) the temperature thresholds at which the irreversible melting of the GIS and the collapse of the WAIS could be provoked, and (2) the rate of collapse (and associated SLR). Either event would raise sea level by about 5–6 metres (m).

In the case of the GIS, the critical processes are increased surface melting and increased flow of ice in concentrated ice streams to the edge of the ice, next to the ocean, where the ice would break off into the water and melt. The process of breaking into the water is called *calving*. Ice flow and subsequent calving would be accelerated by the lubricating effect at the base of the ice of meltwater that percolates down from surface melting, and would produce much more rapid rates of SLR than if surface melting processes alone were at work (Alley et al, 2005; Hansen, 2005). As the elevation of the interior ice sheet is drawn down by surface melting and outflow, the surface temperature would increase, accelerating the melting. At some point – after (hopefully many) hundreds of years – a

point of no return would be reached, such that the ice would continue to melt even if atmospheric GHG concentrations returned to their pre-industrial levels, because the lowered ice sheet surface would be warmer than today or during pre-industrial times.

The WAIS presently terminates in ocean water, with the base of the ice frozen to underwater sills (high points in the ocean bed), which serve to buttress the ice shelf. The ice shelf is presently flanked by floating ice shelves that are vulnerable to disintegration. Indeed, a progressive disintegration of ice shelves floating along the Antarctic Peninsula has occurred during the past 20 years (see Kaiser, 2002), and this has been accompanied by up to a factor of five acceleration in the rate of flow of small land-based glaciers on the Antarctic Peninsula within two years of the loss of floating ice shelves (Scambos et al, 2004). Melting of floating ice shelves themselves does not contribute to sea level rise, but could permit warming of the ocean water next to the WAIS, eventually thawing the ice where it is frozen to underwater sills and permitting a rapid flow (over a period of centuries) of much of the WAIS into the ocean. As with the GIS, it is possible that warming this century could, if sustained, push the WAIS into an irreversible disintegration.

Evidence for this is found in geological data, which indicate that sea level was likely to have been 4–6m higher during part of the previous interglacial period (130,000 to 127,000 years ago) but global mean temperature was only about 1°C warmer than during the mid-1900s (Stirling et al, 1998; Cuffey and Marshall, 2000; McCulloch and Esat, 2000; Overpeck et al, 2006). Simulations with a state-of-the-art coupled ice sheet atmosphere–ocean climate model of conditions during the last interglacial indicate that summer temperatures along the coast of Greenland were about 3°C warmer than at present and were sufficient to provoke partial but not complete melting of the GIS (Otto-Bliesner et al, 2006). The simulated partial melting of the GIS contributes 2.2–3.4m SLR; however, the Greenland ice cap will probably be more susceptible to future melting than during the last interglacial, due to the reduction in snow albedo today from anthropogenic soot. Some contribution to the last interglacial sea level peak from the WAIS or from the margins of the East Antarctic ice sheet is implied, probably induced by the initial sea level rise from partial melting of the Greenland ice cap. During the middle Pliocene (3 million years ago), global mean temperatures are estimated to have been 3°C warmer than today (Crowley, 1996; Dowsett et al, 1996) and sea level 25±10m higher than today (Barrett et al, 1992; Dowsett et al, 1994; Dwyer et al, 1995). This suggests that the threshold for a mere 5–6m sea level rise is less than 3°C global mean warming. Using a semi-empirical approach based on observations during the past century, Rahmstorf (2007) projects a sea level rise of 0.5–1.4m by 2100, while Pfeffer et al (2008) estimate the maximum possible sea level rise by 2100 to be 2.0m and a likely range of 0.8–2.0m.

Based on the above, it is concluded here that an increase in global mean temperature of somewhere between 1°C and 3°C relative to pre-industrial times will probably destabilize either the Greenland ice cap or the West Antarctic ice cap or both.

Threats to forests and biodiversity

Leemans and Eickhout (2004) performed a global-scale assessment of the impact of 1°C, 2°C and 3°C warming on terrestrial ecosystems, using a series of computer simulation models. Their simulations indicate that a 1°C global mean warming will cause changes in the type of ecosystem over about 10 per cent of the earth's land surface, increasing to about 22 per cent for a 3°C global mean warming. However, even at 1°C warming over a period of a century, only 36 per cent of impacted forests can shift in step with the climatic change (that is, 'adapt'), due to their long generation times and relatively slow rate of dispersal. At 3°C warming by 2100, only 17 per cent of impacted forests can keep up. Thus, even a 1°C warming over a period of 100 years (0.1K/decade) can be regarded as dangerous climatic change in that many terrestrial ecosystems are not likely to be able to adapt naturally to the changing climate.

A global mean warming of 3–4°C over the course of the next century could have highly disruptive effects on tropical forest ecosystems (Arnell et al, 2002). Of particular concern are simulations by the HadCM3 atmosphere–ocean model at the Hadley Centre in the UK, in which almost the entire Amazon rainforest is replaced by desert by the 2080s due to a drastic reduction in rainfall in the Amazon basin in association with a global mean warming of about 3.3°C (White et al, 1999; Cox et al, 2004). Non-climatic human

disturbances of tropical forests are increasing and disturbed forests are particularly vulnerable to drought as logging removes deep-rooted trees, edges around cleared areas are subject to drying, and the pastures in logged areas are burned regularly and thus serve as a fire hazard to the remaining forest (Laurance and Williamson, 2001). This further increases the vulnerability of tropical forests to climatic change.

A group of 19 ecologists from five continents (Thomas et al, 2004) has recently assessed the impact of projected future climatic change on the rate of extinction of terrestrial animal species.[7] Species loss was estimated based on empirical relationships between habitat area and the number of species, using three different approaches that give similar answers. Temperature changes for 2050 as obtained by the HadCM2 model for a variety of different emission scenarios were used as input, with two limiting assumptions concerning plant and animal species: (1) uninhibited dispersal, and (2) no dispersal. The case of no dispersal represents the impact of habitat fragmentation and human barriers; reality will fall somewhere between the full-dispersal and no-dispersal cases. For scenarios in which the global mean warming reaches 2.1–2.3°C by 2050, between a sixth and a third of land animal species will be committed to extinction – that is, although not yet extinct, will have passed the point of no return in terms of habitat loss and its implications for survival. Similar losses averaged over 25 biodiversity hotspots and in association with a CO_2 doubling were independently obtained by Malcolm et al (2006). There is growing evidence that some species have already been pushed to or close to extinction by recent warming (Blaustein and Dobson, 2006; Pounds et al, 2006). Not included in either assessment are possible synergistic interactions between climatic change and other factors (such as further reductions in habitat area due to land use changes or the impact of new invasive species on ecosystems already stressed by climatic change). Inasmuch as a loss of a sixth to a third of terrestrial land species is unacceptable, the slightly more than 2°C warming associated with this outcome can be regarded as too much. Thus, we again come to the conclusion that an allowable warming for ecosystems is well below 2°C, probably around 1°C or less (depending on how much species loss is accepted).

Threats to food production, water supplies and sustainable socio-economic systems

It is increasingly recognized that a global mean warming poses a great risk to sustainable development in many ways, by undermining agricultural and forest productivity, reducing water supplies and water quality, and endangering productive coastal marine ecosystems (Beg et al, 2002; Swart et al, 2003; Munasinghe and Swart, 2005). Estimates of the number of people at risk from water shortages by Parry et al (2001), as a function of global average warming, show an abrupt increase between 1°C and 2°C global mean warming, with a more gradual increase in other risks (flooding, malaria, hunger). Barnett et al (2005) review potential impacts of global warming on water supplies in regions dependent on melting of winter snow cover and on glacier meltwater. Currently, more than one sixth of the world's population is dependent on such water sources, including one quarter of China's population. Serious water shortages are likely to arise in such regions within a few decades. Warming and precipitation changes due to anthropogenic climatic change may already be claiming 150,000 lives annually, with the prospect of a significantly greater toll in the future, especially in regions with many climate-sensitive diseases such as Africa (Patz et al, 2005).

A warmer climate will affect food crops in a number of different, and sometimes competing, ways. Where growing conditions are colder than optimal, warming will tend to have beneficial effects on food production, whereas where temperatures are already optimal or warmer than optimal, further warmth will reduce production. Food production depends critically on soil moisture, which depends on the amount, timing and nature of precipitation on the one hand, and evaporative losses on the other hand. Warmer temperatures will tend to increase evaporative losses, thereby increasing the frequency of drought conditions. Adaptation (such as shifting the time when crops are planted, changing the kinds of crops planted and conserving soil moisture) can do much to mitigate the adverse effects of warmer and dryer conditions. Nevertheless, assessments indicate aggregate losses of global food production in the order of 20 per cent caused by a global mean warming of 2–5°C.

A higher atmospheric CO_2 concentration also has direct, beneficial effects on crops by increasing the rate of photosynthesis and reducing the rate of water loss through stomata. Nevertheless, assessments of the impact of a doubled CO_2 climate (2–5°C global mean warming) frequently show decreases in agricultural yields in specific regions of 10–30 per cent and more, even after allowing for the beneficial physiological effects of higher CO_2 and allowing for adaptation (see Gitay et al, 2001, Table 5.4, and Parry et al, 2004). Furthermore, the latest generation of experiments on the direct effects of higher CO_2 on crops (so-called Free Air Concentration Enhancement or FACE experiments) indicate that earlier experiments – on which currently available assessments are based – have overestimated the beneficial effects of higher CO_2 by about a factor of two (Schimel, 2006). Thus, rather than a global mean yield increase of 20 per cent due to the beneficial physiological effects of a CO_2 doubling, the global mean yield increase is only 10 per cent. When combined with the roughly 20 per cent loss due to the effects of climatic change alone, this implies a net loss of food production in the global aggregate. Thus, the threshold for significant harm is likely to lie between 2–3°C global mean warming.

Direct effects of higher CO_2 on ocean chemistry

Carbon in the oceans occurs as dissolved organic matter and as dissolved inorganic carbon. The latter occurs in three forms that can be transformed from one to the other and back: as aqueous CO_2, HCO_3^- (the bicarbonate ion) and CO_3^{2-} (the carbonate ion). The buildup of atmospheric CO_2 concentration is accompanied by absorption of CO_2 from the atmosphere by the surface waters of the oceans. This is accompanied by the following chemical reactions:

$$CO_2 + H_2O \rightarrow H_2CO_3 \qquad (1.3)$$

$$H_2CO_3 \rightarrow H^+ + HCO_3^- \qquad (1.4)$$

$$CO_3^{2-} + H^+ \rightarrow HCO_3^- \qquad (1.5)$$

If the above reactions were to proceed fully, the net result would be the reaction:

$$CO_2 + H_2O + CO_3^{2-} \rightarrow 2HCO_3^- \qquad (1.6)$$

Thus, the absorption of anthropogenic CO_2 from the atmosphere by the oceans decreases the concentration of carbonate in ocean water. The reactions do not proceed fully to the right, so a somewhat smaller depletion of carbonate is accompanied by an increase in dissolved CO_2 concentration and in ocean acidity – through reaction (1.4) not being fully offset by reaction (1.5).

Surface waters of the oceans are presently supersaturated with respect to calcium carbonate ($CaCO_3$). Calcium carbonate occurs in two mineral forms – aragonite and calcite – and is used as the structural material of coral reefs and of many other organisms (such as various phytoplankton, zooplankton, echinoderms, crustaceans and molluscs) throughout the marine food chain. The absorption of CO_2 by the oceans, by reducing the carbonate concentration, will make it more difficult for calcareous marine organisms to build their skeletal material. Figure 1.7 shows the variation with latitude in the concentration of CO_3^{2-} in surface waters of the oceans for pre-industrial conditions, present conditions and in association with various future atmospheric CO_2

concentrations, as simulated by Orr et al (2005). Also shown is the variation in surface-water pH. For a doubling of the pre-industrial CO_2 concentration (to 560ppmv), the carbonate concentration decreases from about 660 per cent to about 460 per cent of the saturation value with respect to calcite in tropical regions, but is driven to the point of undersaturation with respect to aragonite in southern polar regions. This would be accompanied by a decrease in surface-water pH by about 0.2, from about 8.1–8.2 to 7.9–8.0. The large decreases in the supersaturation with respect to calcite associated with a mere doubling of atmospheric CO_2 concentration and the associated increase in ocean acidity are likely to have profoundly negative impacts on marine ecology and productivity, and imply that a 'safe' atmospheric CO_2 concentration is well below a doubling of the pre-industrial concentration.

The higher concentration of aqueous CO_2 associated with decreasing carbonate concentration is likely to adversely affect marine macro-fauna

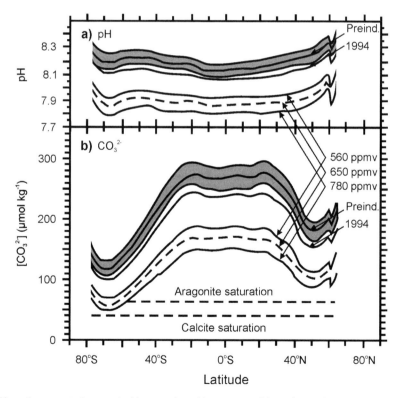

Note: Also shown in (b) are the concentrations required for saturation with respect to calcite and aragonite.
Source: Orr et al (2005)

Figure 1.7 *Variation with latitude in (a) the pH, and (b) the concentration of CO_3^{2-} in surface waters of the oceans, as simulated by Orr et al (2005), for pre-industrial conditions, present conditions, and for future atmospheric CO_2 concentrations of 560ppmv, 650ppmv and 780ppmv*

(Pörtner et al, 2005). A mere 200ppmv increase in CO_2 has been observed in controlled experiments to reduce the rate of growth of gastropods and sea urchins by up to 40 per cent (Shirayama and Thornton, 2005). Essentially nothing is known about the long-term effects on fish of sub-lethal concentrations of CO_2, but fish are likely to be sensitive to higher aqueous CO_2 concentrations because the difference between the CO_2 partial pressure in their body fluids and that of the ambient medium is an order of magnitude smaller than that of terrestrial animals (Ishimatsu et al, 2005).

Synthesis

It follows from the above that, *if* the climate sensitivity is only 1°C, then GHG concentration increases that are collectively equivalent to a CO_2 doubling might be permissible, although there are still important concerns about impacts through changes in marine chemistry and impacts on coral reefs. If, by contrast, the climate sensitivity is close to 5°C, then GHG concentrations must be kept substantially below a CO_2 doubling equivalent. Since, at this point in time, we do not know what the true climate sensitivity is, and the repercussions of changes in marine chemistry are uncertain, a CO_2 doubling equivalent (450ppmv) must be regarded as dangerous. It is not necessary to show that 450ppmv CO_2 leads to *certain* impacts that violate the UNFCCC principles of protecting ecosystems and food security; it is simply necessary to show that such a concentration represents dangerous interference in the climate system, that is, that there are non-negligible risks to important ecosystems and to food security. This is clearly the case.

Since the goal of the UNFCCC is to *avoid* dangerous anthropogenic interference in the climate system, and since 450ppmv CO_2 is dangerous (given current uncertainties and heating effects from other gases), it follows that current policy should be directed at stabilization in the range of 350–400ppmv or even less. By the end of 2005, the CO_2 concentration had reached 380ppmv.

1.4.2 Implications of climate–carbon cycle feedbacks

A change in climate due to an initial increase in atmospheric GHGs will lead to changes in the natural flows of CO_2 and methane between the atmosphere and the terrestrial biosphere, soils and oceans. If, as a result of climatic change, there is a net flux of CO_2 and CH_4 to the atmosphere, this will augment the original increase in these gases, thereby serving as a positive feedback. Unfortunately, there are many ways in which interaction between climate and the carbon cycle will likely serve as a positive feedback.

A larger reduction in anthropogenic emissions of CO_2 and other GHGs is required for stabilization at a given atmospheric concentration in the presence of positive climate–carbon cycle feedbacks than in their absence. If the additional CO_2 fluxes to the atmosphere from natural fluxes exceed human emissions, the atmospheric CO_2 concentration would continue to increase even if emissions were to be reduced to zero. In the HadCM2 model, a positive climate–carbon cycle feedback results in an atmospheric CO_2 concentration in 2100 of 1000ppmv for an emission scenario that otherwise would have led to 750ppmv by 2100 (Cox et al, 2000).[8] This model predicts a larger positive feedback involving CO_2 from the terrestrial biosphere than any other model, and so probably overestimates this particular feedback. Alternatively, the model does not include high-latitude methane in frozen soils, and so cannot, for example, include the recently observed release of methane from thawing yedoma soils in Siberia (Walker, 2007). As discussed later (Chapter 2, subsection 2.6.5), methane is about 26 times stronger than CO_2 as a GHG on a molecule-per-molecule basis, so release of even relatively small amounts of methane from thawing soils would constitute an important positive feedback.

To clarify, potential climate–carbon cycle feedbacks do not directly influence the impacts for a given set of GHG concentration limits and so would not directly affect the choice of GHG concentrations that constitute dangerous anthropogenic interference in the climate system. However, if strong positive feedbacks occur and negative anthropogenic emissions are required in order to maintain CO_2 and/or methane at the desired concentration limits, and if the required negative emissions are impossible to achieve, one would have to set lower permitted GHG concentrations in order to prevent the climate system from spiralling out of control. In this sense, the permitted GHG concentrations depend on potential climate–carbon cycle feedbacks (as well as on potential impacts).

1.5 Stabilization of atmospheric CO$_2$ concentration

To stabilize the atmospheric CO$_2$ concentration requires reducing emissions from their current levels. The lower the stabilization concentration in the case of CO$_2$, the sooner and the faster that emissions need to be decreased, while the greater the positive climate–carbon cycle feedback, the faster that emissions need to be reduced for a given stabilization level. Given a set of pathways to hypothetical concentrations at which CO$_2$ is stabilized, the permitted fossil fuel CO$_2$ emissions can be determined by running a carbon cycle model in *inverse* mode. Figure 1.8 shows possible pathways for stabilizing at CO$_2$ concentrations ranging from 350ppmv (i.e. less than the present concentration) up to 750ppmv, beginning with the observed concentration of 370ppmv at the end of 2000.

The top panel of Figure 1.9 shows the fossil fuel carbon emissions that are permitted for the stabilization scenarios of Figure 1.8, as computed using the coupled climate–carbon cycle model of Harvey (2001). These emissions were computed assuming that net deforestation ends by 2015, that cement-related emissions of 0.32GtC/yr initially increase by 50 per cent, then decrease to half of present emissions by

2100, and that there is a weak positive climate–carbon cycle feedback.[9] Also shown is the total emission for a BAU scenario constructed using Equation (1.2) with the following inputs: the median population scenario of Lutz et al (2001), in which global population peaks at 8.97 billion in 2070, then decreases to 8.41 billion by 2100; a 1.6 per cent/year growth in GDP/person; a 1.0 per cent/year decrease in global energy intensity; and a 0.5 per cent/year increase in the supply of carbon-free power. The breakdown of the BAU emission between natural gas, oil and coal is also given. Given a limit on fossil fuel emissions, it makes sense to use as much natural gas as possible, since natural gas has the lowest emissions per unit of energy (see Chapter 2, subsection 2.6.1). Any remaining permitted emissions can then be allocated to oil, and if any further emissions are permitted after using all the oil that one might want to or can use, they can be allocated to coal. Thus, the BAU natural gas emission is shown as the lowest layer in Figure 1.9, followed by the BAU oil emission, and lastly, the coal emission. For stabilization at 450ppmv, the permitted emission drops to the combined natural gas+oil BAU emission by 2040, which means that no coal emissions are permitted if the amount of fossil fuel energy derived within the emission limit is to be maximized. By 2085, no further

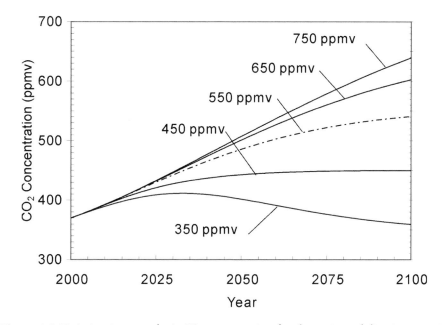

Figure 1.8 *Variation in atmospheric CO$_2$ concentration for alternative stabilization scenarios*

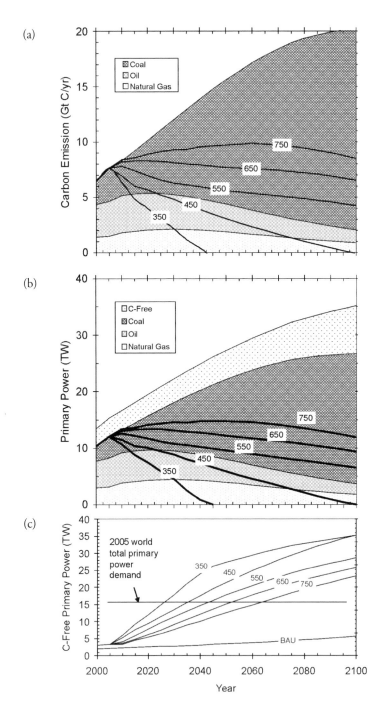

Figure 1.9 *(a) BAU fossil fuel CO$_2$ emission and emissions permitted when the atmospheric CO$_2$ concentration follows the stabilization pathways shown in Figure 1.8. Also shown is the breakdown of BAU emissions between natural gas, oil and coal. (b) Primary power in the BAU scenario and primary power permitted from fossil fuels for the stabilization scenarios, assuming that any permitted emissions are assigned to the least carbon-intensive fossil fuel first so as to maximize the power obtained from fossil fuels. (c) Carbon-free power supply needed for the various stabilization scenarios. This is the difference between the BAU power demand shown in (b) and the permitted fossil fuel power supply*

oil emissions would be permitted. Even for stabilization at 750ppmv, fossil fuel emissions have to be greatly constrained by the end of this century compared to the BAU scenario.

The middle panel of Figure 1.9 shows the primary power that is provided by natural gas, oil, coal and carbon-free energy sources in the BAU scenario, as well as the permitted primary power from fossil fuels for the stabilization scenarios.[10] Assuming the same total demand for energy (or power) for the BAU and each stabilization scenario, the difference between the total power demand and the permitted fossil fuel primary power gives the required carbon-free power supply. Carbon-free power could come from a variety of renewable energy sources, from nuclear energy, or from fossil fuel energy with burial of CO_2 in permanent underground reservoirs or in the deep ocean. The required carbon-free power supply is shown as the lower panel in Figure 1.9. For stabilization of CO_2 at 450ppmv, *the amount of carbon-free power required by 2050 (21.4 terawatts (TW)) is 40 per cent greater than the total global primary power supply in 2005 of 15.3TW.* The current carbon-free primary power supply is 3.3TW, so the required carbon-free power supply in 2050 is over six times that at present (about half of which is nuclear or hydro power, the balance largely biomass).

The above calculations assume that the energy intensity of the global economy improves by 1 per cent per year (comparable to the average improvement of 1.1 per cent per year from 1965–2005). If a greater rate of energy intensity improvement occurs, the growth in energy demand will be smaller, so, for a given CO_2 emission, less carbon-free energy is needed. These tradeoffs are illustrated in Figure 1.10 for the case of stabilizing at 450ppmv. To stabilize at 450ppmv requires *either* 21TW of carbon-free power by 2050 and continuous 1 per cent/year energy intensity improvement, *or* increasing the average rate of improvement of energy intensity to 3 per cent/year (from 1.5 per cent during the 1995–2005 period) and expanding carbon-free power supply by about 60 per cent (from 3.3TW to 5.2TW). Figure 1.11 is an alternative representation of the same information shown in Figure 1.10; shown is the growth in carbon-free power required over the period 2000–2100 if the energy intensity improves by 3.0 per cent per year, and the *additional* carbon-free power required for each 0.5 per cent per year slower rate of improvement of energy intensity. The magnitude of the energy intensity reductions implied by different annual rates of decrease is shown in Figure 1.12. Compounded reductions in energy intensity of 1 per cent/year, 2 per cent/year, or 3 per cent/year correspond to energy intensity reductions by a factor of 1.6, 2.6 and 4.1 by 2050, respectively, and by factors of 2.7, 7.0 and 18.2 by 2100, respectively.

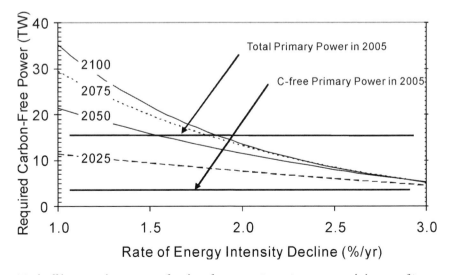

Figure 1.10 *Tradeoff between the amount of carbon-free power in various years and the rate of improvement in the energy intensity of the global economy (MJ/$) required in order to stabilize the CO_2 concentration at 450ppmv*

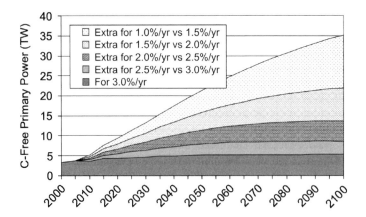

Figure 1.11 *Carbon-free power required during 2000–2100 for stabilization of the CO_2 concentration at 450ppmv if the energy intensity of the global economy decreases by 3.0 per cent per year, and the additional carbon-free power required for each 0.5 per cent per year slower rate of decrease in energy intensity*

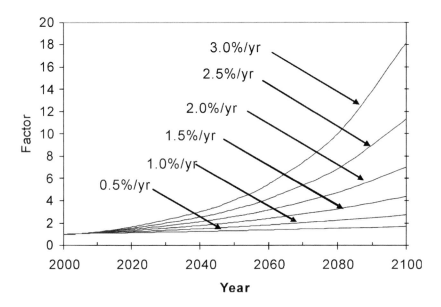

Figure 1.12 *Factor by which energy intensity will have decreased during the period 2005–2100 for various annual rates of decrease in energy intensity, beginning in 2005*

Given that much of the developing world still has to build its infrastructure, which tends to increase rather than reduce energy intensity, some may doubt whether the world can maintain a 1 per cent per year reduction in average energy intensity over a period of 100 years, much less double or triple it and sustain this rate of reduction for a century. Neither is it clear that renewable power sources, which tend to be diffuse, intermittent and often in the wrong place and at the wrong time, can be ramped up to 21TW of power by 2050. However, much of the pessimism concerning the prospects for very large (factors of eight to ten) reductions in energy intensity arises from focusing on individual end-use technologies (i.e. cars or heating and cooling equipment in buildings) rather than the entire societal system for providing a given set of energy services (i.e. urban form and transportation systems, or all the factors that determine how much heating and cooling buildings need in the first place). Thus, it might indeed be possible to achieve some combination of accelerated energy intensity reduction and accelerated deployment of carbon-free power adequate to stabilize the atmospheric CO_2 concentration at 450ppmv, while still providing a good standard of living for all people.

We should recognize from the outset that GDP is a poor indicator of material welfare (much less total welfare), since it sums up the services rendered by the stock of manmade material goods, additions to the stock of material goods, and the throughput of energy and material needed to maintain the stock of material goods, the latter being a cost rather than a benefit (see Daly, 1996, Chapters 6 and 7). It includes some activities that are undesired consequences of the production of useful goods and services (such as health care costs related to air pollution), which should be subtracted rather than added when summing up total GDP. Not included are losses in services rendered by natural ecosystems as a result of damage to ecosystems, or depletion of natural capital. That is, GDP is a mixture of the partial costs and benefits of economic activities, and reflects changes in manmade but not natural inventories. It should also be recognized that exponential growth of per capita GDP forever is impossible, and it is even questionable whether 1.6 per cent/year growth in global average per capita GDP can be sustained for the next 100 years. The single-minded pursuit of increasing GDP per capita, rather than pursuing that which GDP is supposed to measure (i.e. human well-being) leads, if it can be achieved in practice, to extraordinary upward pressure on CO_2 emissions, as seen above. As discussed in the closing chapter, the single-minded pursuit of increasing GDP per capita is also associated with decreasing human happiness once some basic level of services is provided.

At the same time as GDP increases, the exponential decrease in energy intensity per unit of GDP must at some point begin to weaken as the technical possibilities for efficiency improvements at both the micro and macro scales are gradually exhausted. Indeed, increasing efficiency in the use of energy and resources will tend to reduce GDP for the simple reason that the throughput of energy and materials decreases as efficiency increases, and GDP is in part a measure of this throughput.

To sum up, atmospheric CO_2 concentration had risen from a pre-industrial concentration of 280ppmv to 380ppmv by 2005, and is projected to rise far above a doubling level of 560ppmv under BAU assumptions. When the heating effect of other GHGs is accounted for, a CO_2 concentration of only 450ppmv (or less) is the climatic equivalent of a doubling of CO_2 concentration, which in turn will very likely eventually warm the climate by 1.5–4.5°C. Increasingly grave consequences can be expected as the actual warming increases from 1.5 to 4.5°C. Thus, even 450ppmv CO_2 is likely to be too high a concentration, so it will probably be necessary to draw down atmospheric CO_2 to as low as 350ppmv after peaking at 450ppmv or higher. This in turn requires creating artificial carbon sinks, either by building up soil carbon or through deep burial of carbon captured from the use of biomass energy in place of fossil fuels. Sustainable use of biomass itself would be C-neutral, as the CO_2 released from combustion of biomass would be balanced by the prior absorption of CO_2 from the atmosphere by photosynthesis when the biomass was grown. If, instead of being released to the atmosphere, this CO_2 is captured and buried, negative emissions would be created. However, fossil fuel use with subsequent capture and burial of the released CO_2 – either in geological formations or in the deep ocean – is one of the carbon-free options that could be used to limit peak atmospheric CO_2 to 450ppmv. This option is referred to as *carbon sequestration*. Since the amount of carbon that can be sequestered in geological formations is limited, and ocean sequestration presents its own set of environmental problems; heavy reliance on carbon sequestration in order to achieve 450ppmv may preclude the additional sequestration required to restore atmospheric CO_2 to a lower, more acceptable concentration.

1.6 Purpose and organization of this book and of Volume 2

The purpose of this book and the companion book, *Energy and the New Reality 2: Carbon-Free Energy Supply*, is to critically assess the prospects for a long-term acceleration in the rate of reduction in energy intensity of the global economy, on the one hand, and for large-scale deployment of carbon-free power, on the other hand. That is, this book and Volume 2 assess how far the world can move along each of the axes in

Figure 1.10. In so doing, the two books address the following central questions:

- Is there a technically and economically feasible combination of energy intensity reduction and deployment of carbon-free power that is sufficient to stabilize the atmospheric CO_2 concentration at 450ppmv, given the baseline population and economic growth assumptions?
- If the answer to the above question is no, or if we assume that the full economic and technical potential is not achieved to various degrees, then to what extent must the product of population and GDP per capita be constrained in order to stay within emission limits needed for stabilization at 450ppmv?

Implicit in these questions, which assume a CO_2 stabilization target of 450ppmv, is the assumption that carbon sequestration can and will be used to achieve the additional emission reductions (or negative emissions) needed to stabilize atmospheric CO_2 concentration as low as 350ppmv, should this be deemed necessary.

This rest of this book is organized as follows. Chapter 2 introduces relevant energy concepts and presents data on energy use and energy intensities in the world today. Chapters 3 to 9 assess options to reduce the energy intensity of the global economy, beginning with the generation of electricity, then proceeding to residential, commercial/institutional, transportation, industrial and agricultural energy end use, and finishing with a consideration of municipal energy services (waste management in particular) and district energy systems. Chapter 10 develops estimates of the maximum rate at which energy intensity for fuels and electricity can be reduced in developed and developing groups of countries. The final chapter considers – in very broad terms – the sets of policies needed to reduce population growth through non-coercive means, to shift from a single-minded focus on growth in GDP per person to a focus on human well-being, and to achieve rapid reductions in energy intensity. Throughout, the book strives to give a 'nuts and bolts' understanding of how energy-using equipment or energy production methods work, without getting consumed with technical details. Volume 2 assesses the major options for carbon-free power: solar, wind, biomass, geothermal, hydroelectric, oceanic and nuclear energy, along with sequestration of

CO_2 from fossil fuel energy sources. Volume 2 then constructs scenarios of future energy supply and demand that achieve the goal of limiting atmospheric CO_2 to a concentration of no more than 450ppmv.

Notes

1   As discussed in Volume 2 (Chapter 4, section 4.8, and Chapter 9, subsections 9.3.6 to 9.3.8), there is a limited capacity to create negative emissions by burying CO_2 produced from the gasification of sustainably grown biomass, used as an energy source in place of fossil fuels. This provides a limited possibility of drawing down atmospheric CO_2 if we overshoot safe levels, as we almost certainly will.

2   More strictly, climate sensitivity is the warming per W/m^2 of radiative forcing, and we will sometimes use it in that way here.

3   Water vapour is also released from the combustion of fossil fuels, but the direct effect of such emissions on the atmospheric water vapour concentration is utterly negligible. Rather, the important water vapour changes are driven by the warming itself. For this reason, the increase in water vapour concentration is not regarded as a radiative forcing, but rather, as a climatic feedback.

4   Indeed, many individual observationally based estimates of climate sensitivity indicate a non-negligible possibility of much more than a 5°C warming for a CO_2 doubling. However, when these estimates are considered collectively, 4.5°C emerges as a very likely upper limit to the climate sensitivity (Annan and Hargreaves, 2006).

5   As discussed more fully in Chapter 2 (section 2.1), primary energy refers to energy as it is found in nature, prior to conversion to other energy forms (such as electricity or gasoline) by humans.

6   A CO_2 doubling gives a radiative heating of about $3.75W/m^2$ and a concentration of 450ppmv gives a heating of $2.57W/m^2$, which means that additional heating of $1.18W/m^2$ will give the equivalent of a CO_2 doubling. The current radiative heating by non-CO_2 GHGs is around $1.4W/m^2$, so their collective heating effect would need to be reduced by about 15 per cent in order for a CO_2 concentration of 450ppmv not to correspond to more than a doubling.

7   See also the exchange of viewpoints in *Nature*, vol 427, pp145–148.

8   The large feedback is due in part to a collapse of the Amazon rainforest and the ensuing release of CO_2.

9   A stronger climate–carbon cycle feedback would require more rapid emission reductions than shown in Figure 1.9.

10 Primary power is given by annual primary energy use in joules divided by the number of seconds in a year.

2

Energy Basics, Usage Patterns and Trends, and Related Greenhouse Gas and Pollutant Emissions

This chapter outlines important background information concerning the occurrence of energy, its transformation from one form to another, the relative importance of different sources of energy in the world today, and linkages between energy use and GHG and air pollutant emissions. Definitions of commonly used terms for energy and power, basic physical units, derived units and the relationships between them are given in Appendix A.

2.1 Energy and efficiency concepts

2.1.1 Primary, secondary and tertiary energy

Energy as it occurs in nature is referred to as *primary energy*. Examples of primary energy are coal, oil, natural gas and uranium as they occur in the ground. To be useful to humans, however, these forms of energy need to be extracted and transformed into *secondary energy*. Examples of secondary energy are electricity and refined petroleum products, and processed natural gas ready for use by the customer. However, even these energy forms are not really of interest to us. What we really want are end-use energy services – things like warmth, motion, mechanical power or process heat for industrial manufacturing. These are referred to as *tertiary energy*. The relationships between primary, secondary and tertiary energy are summarized in Figure 2.1.

2.1.2 First and second laws of thermodynamics

The first law of thermodynamics states that energy is neither created nor destroyed. Thus, when transforming energy from primary to secondary, and from secondary to tertiary, no energy is lost in the processes. However, the second law of thermodynamics states that there will be an irreversible degradation of energy into unusable forms. In particular, some concentrated or high-quality energy is unavoidably dissipated as low-grade heat, which is random molecular motion. Thus, although no energy disappears when we convert from primary to secondary energy or from secondary to tertiary, there is an unavoidable loss in useful energy. These losses are indicated in Figure 2.1.

The efficiency of an energy conversion process is the ratio of the useful output energy to the input energy. For example, the efficiency of a powerplant is the ratio of electrical energy produced to the energy content of the input fuel, while the efficiency of a furnace is the ratio of heat delivered to the building to the energy content of the input fuel. The efficiency of an electric motor is the ratio of mechanical energy created by the motor to the electrical energy input. The fuel energy not converted to electricity or useful heat, or the electrical energy not converted to mechanical energy, is lost as waste heat.

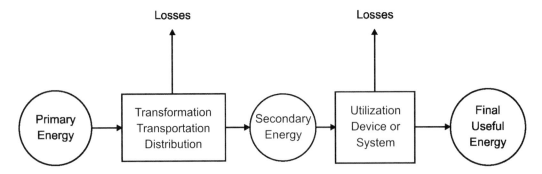

Figure 2.1 *The transformation from primary to secondary to tertiary energy, and examples of each form of energy*

2.1.3 Linkage to reduction in energy intensity

In Chapter 1, we introduced the concept of energy intensity as a key determinant of fossil fuel CO_2 emissions. We defined it as the ratio of *primary* energy use to GDP (joule/$). We can now see that there are three general ways in which we can reduce our primary energy requirements:

*   increasing the efficiency of conversion from primary to secondary energy;
*   increasing the efficiency of conversion from secondary to tertiary energy;
*   reducing tertiary energy demand.

Building houses with better insulation and better windows, so that less heat is lost and therefore less needs to be replenished by the furnace, is an example of a measure to reduce tertiary energy demand. Using light-coloured roofs along with better insulation to reduce cooling requirements in summer is another example. So is turning off unused lights. Building more compact cities and a better mix of land uses to reduce the need to travel reduces tertiary transportation energy demand, as does shifting people from cars to public transportation and bicycles. Building more efficient cars, alternatively, improves the secondary to tertiary energy-conversion efficiency, while reducing the energy losses in the refining of crude petroleum into transportation fuels improves the primary to secondary energy-conversion efficiency.

The savings in each of these three links from primary energy to human satisfaction are multiplicative. Thus, a doubling in the efficiency in the generation of electricity, a doubling in the end-use efficiency, and a halving of tertiary energy demand reduces the primary energy use by a factor of eight. As we shall see in later chapters, potential reductions of this magnitude in at least two of these three factors are the norm rather than the exception. This means that, in many energy-using sectors, energy intensity reductions by factors of four to eight are feasible. Further reductions in energy intensity, at the scale of an entire economy, could also occur through a decrease in the share of GDP arising from the production of energy-intensive products such as steel and cement, and an increase in the importance of service industries. This occurs naturally as countries develop and finish building most of their infrastructure.

2.1.4 Heating values of fuels

The energy content of a fuel is the thermal energy released when the fuel is burned, and is referred to as the *heating value* of the fuel. Heating values can be given assuming that any water vapour produced is either condensed (adding the released latent heat to the available thermal energy) or not condensed. The resulting heating values are referred to as the *higher heating value* (HHV) and *lower heating value* (LHV), respectively, or as the *gross heating value* and *net heating value*. LHVs and HHVs for a variety of fuels are given in Appendix B. Inasmuch as efficiencies for heating or electricity generation are given by the energy output divided by the energy content of the fuel, the efficiencies will be smaller if based on the HHV of the fuel than if based on the LHV. The difference is negligible for coal, but the HHV of natural gas is about 10 per cent greater than the LHV. In electricity generation, exhaust conditions are such that the water vapour cannot be condensed, so basing efficiencies on HHVs would make natural gas powerplants appear to less efficient relative to coal plants than if efficiencies were based on LHVs. However, when fuels are used for space or water heating, it is possible to condense water vapour in the exhaust by cooling the exhaust with cold water or return air, thereby preheating the water or air and capturing some of the latent heat that is released during condensation. Thus, for furnaces and boilers it is appropriate to give efficiencies in terms of the HHV.

In cogeneration applications, where fuels are first used to generate electricity, and waste heat in the exhaust (and perhaps in the cooling water) is used for heating purposes, it is possible to capture the latent heat of condensation, so it would be appropriate to use HHVs in computing electricity, thermal and overall efficiencies.

Data on primary energy consumption can be computed based on the mass of fuel used times the HHV or the LHV.[1] Consistency would require providing energy-consumption statistics in terms of the same heating values used for computing efficiencies, but since LHV efficiencies are usually given for powerplants and HHV efficiencies for heating applications, there will be an inconsistency somewhere. My preference is to consistently use HHVs. Although this makes efficiencies of power generation using natural gas appear to be less advantageous compared to conventional coal powerplants, the CO_2 emission factor (kgC/GJ) will also be smaller if computed using the HHV. However, wherever published efficiencies or other data are given in terms of the LHV (as they normally are for powerplant efficiencies), LHVs will be given in this book but flagged as such.

Finally, data on fuel costs ($/GJ) can also be given as the cost of a physical unit (such as $/m³ for natural gas) divided by either the HHV or LHV (as GJ/m³). The difference is only 10 per cent for natural gas and much less for other fuels, whereas the uncertainty in future energy costs is very large, so the distinction between HHV and LHV is neglected here as far as energy costs are concerned.

2.2 Trends in energy use and cost

2.2.1 Primary energy use and cost by region and by fuel

Figure 2.2 shows the growth in the use of non-biomass primary energy in major world regions and by fuel. Over the period 1965–2006, non-biomass primary energy use grew by amounts ranging from as low as 67 per cent in the former Soviet Union (a compound growth rate of 1.28 per cent/year) to as high as 887 per cent in the Middle East (a compound growth rate of 5.62 per cent/year). World coal use roughly doubled (an increase of 109 per cent) over this time period, while use of oil increased to 2.7 times the 1965 level by 2006 and natural gas use increased to 4.3 times the 1965 level by 2006. Rates of growth during the ten-year period from 1996 to 2006 were smaller than for the entire period for oil and gas (1.6 per cent/year versus 2.4 per cent/year for oil, 2.4 per cent/year versus 3.6 per cent/year for natural gas) but larger for coal (2.7 per cent/year versus 1.8 per cent/year). The increase in the global mean rate of growth in the use of coal is due to steady but consistently high growth in the use of coal in Asia (4.4–5.0 per cent/year) that has had an increasing effect on the global mean growth rate due to Asia's increasing share of total coal consumption.

Figure 2.3 shows the variation in the market price of crude oil since 1861, both in terms of $-of-the-day and in terms of 2008$, while Figure 2.4 shows the variation in the worldwide price of oil and in regional prices of natural gas and coal in 2008$ per GJ. Oil is currently (mid-2009) the most expensive fossil fuel energy, at around $10/GJ ($70/barrel), while natural gas has recently been in the $4–12/GJ range and coal costs about $2/GJ.

2.2.2 Electricity generation by region and by energy source

Figure 2.5 shows the growth in the generation of electricity since 1990 in major world regions and by energy source. Also shown are the average compound growth rates for each region and energy source. Over the period 1990–2006, global electricity use has grown by an average compounded rate of 3.0 per cent/year, compared to 2.6 per cent/year for global primary energy use over the same period. Growth rates range from –1.1 per cent/year in the former Soviet Union, to 6.1 per cent/year in the Asia-Pacific region and 6.6 per cent/year in the Middle East.

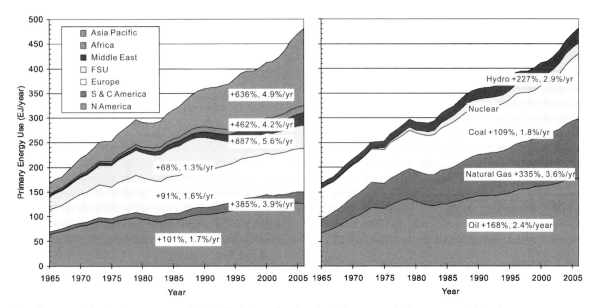

Note: Biomass, wind and solar energy are excluded, while hydro and nuclear electrical energy production are converted to primary energy equivalents assuming a conversion efficiency of 0.38. FSU = former Soviet Union.
Source: Data from *BP Statistical Review 2007*, www.bp.com

Figure 2.2 *Growth in the use of primary energy from 1965–2006 in major world regions (left) and by fuel (right)*

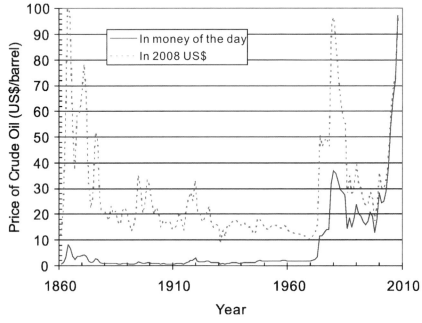

Source: Data from *BP Statistical Review 2009*, www.bp.com

Figure 2.3 *Variation in the price of crude oil from 1861–2008 in $-of-the-day and in 2008$*

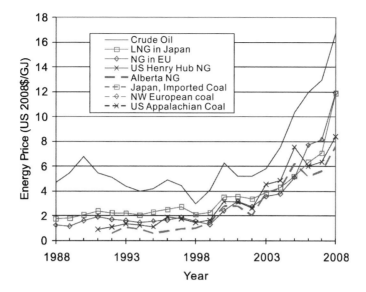

Note: Natural gas and coal prices in 2008$ are computed from $-of-the-day data in *BP Statistical Review 2009* using the implicit inflation factors taken from the oil price data.
Source: Data from *BP Statistical Review 2009*, www.bp.com

Figure 2.4 *Variation in the world price of oil and regional prices of natural gas and coal from 1988–2008 in 2008$*

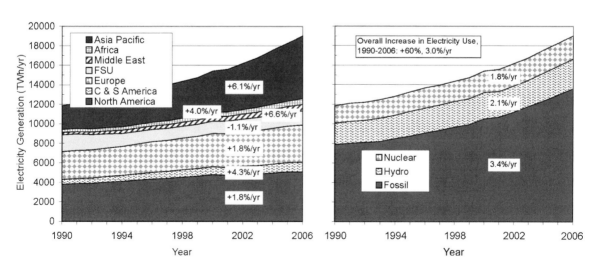

Source: Data from *BP Statistical Review 2007*, www.bp.com

Figure 2.5 *Variation in the annual generation of electricity in major world regions (left) and by energy source (right), plus average annual compound growth rate*

2.3 Overview of present-day energy supply and demand

Tables 2.1, 2.2 and 2.3 present data on primary energy supply and use in 2005 for Organisation for Economic Co-operation and Development (OECD) countries, non-OECD countries and for the world as a whole, respectively (see Appendix C for the list of OECD countries). World primary energy use in 2005 amounted to 483EJ, equivalent to an average primary power supply of 15.3TW. Of this, 173EJ were used to generate 64.6EJ of electricity (with losses of 108.1EJ), giving an average conversion efficiency of 37 per cent, and 17.9EJ were used to generate 13.0EJ of district heat (with losses of 4.9EJ), giving a conversion efficiency of 73 per cent.[2] Of the remaining 292.7EJ of primary energy, a further 29.3EJ were lost during the transformation into 263.4EJ of secondary energy other than electricity and district heat, giving an average conversion efficiency of 90 per cent. The total amount of secondary energy delivered to end-users in 2005 was about 324EJ. The average conversion efficiency from primary energy to delivered secondary energy was thus 67 per cent (324/483).

Of the 483EJ of world primary energy use in 2005, about 232EJ (48 per cent) were used in OECD countries and 251EJ (52 per cent) in non-OECD countries. Of the 334EJ of secondary energy delivered globally, about 168EJ (about 50 per cent) were used in OECD countries and about 166EJ (about 50 per cent) were used in non-OECD countries. Thus, the OECD countries, with 18 per cent of the world's population,[3] account for about 50 per cent of both primary and secondary energy use worldwide.

2.3.1 Primary energy use and sources in OECD and non-OECD countries

Figure 2.6 shows the relative contributions of different energy sources to total primary energy supply in 2005 (based on data in Tables 2.1–2.3). For the world as a whole, the largest sources of primary energy are oil (33 per cent), coal (25 per cent) and natural gas (21 per cent), followed by biomass (10 per cent), hydro (6 per cent) and nuclear (5 per cent). The shares of coal and biomass are significantly smaller in OECD than in non-OECD countries (20.2 per cent versus 29.4 per cent for coal, 3.5 per cent versus 15.9 per cent for biomass) while the oil and nuclear shares are noticeably larger (40 per cent versus 27 per cent for oil, 8.7 per cent

versus 1.8 per cent for nuclear).[4] Figure 2.7 shows how coal, oil and natural gas energy use are allocated among the major economic sectors. In OECD countries, about 80 per cent of all coal is used to generate electricity and most of the rest is used by industry; 56 per cent of all oil is used for transportation and 16 per cent is used for non-energy purposes (as a feedstock (raw material) for production of plastics and other petroleum-derived materials), with much smaller amounts used to heat buildings, for industry and to generate electricity; and natural gas is used (in decreasing order of importance) to heat buildings and provide hot water, to generate electricity and by industry. In non-OECD countries, coal use is less concentrated in electricity generation because of a large direct use by industry, and less than half of oil consumption is for transportation.

2.3.2 Energy use by aggregate sectors in OECD and non-OECD countries

Figures 2.8 and 2.9 show the ways in which primary energy and electricity were used in 2005 in OECD and non-OECD countries. In both cases, generation of electricity represents the single largest direct use of primary energy. The direct use of primary energy in buildings is relatively small – only about 12 per cent of total primary energy use in OECD countries and 18 per cent in non-OECD countries. However, 53 per cent of all electricity is used in buildings in OECD countries and 27 per cent in non-OECD countries, with industry accounting for most of the rest (see Figure 2.9). The primary energy used to generate electricity (shown later in Figure 2.12) can be allocated among different sectors in proportion to their electricity use and added to the direct use of fuels (divided by the average efficiency in producing the fuels from primary energy) to give the breakdown of primary energy use among all the sectors except electricity generation. This is shown in Figure 2.10. It is seen that buildings account for just over one third of primary energy use in both OECD and non-OECD countries when indirect consumption (through the use of electricity) is included.

Agricultural use of primary energy appears to be quite small in Figure 2.10 – about 1.2 per cent of total primary energy use in OECD countries and 3.5 per cent in non-OECD countries, or 2.7 per cent

Table 2.1 *Primary and secondary energy use in OECD countries 2005, calculated from fuel HHVs*

	Primary energy (EJ)								Secondary energy (EJ)		
	Coal	Oil	Gas	Nuclear	Hydro	Biomass & waste	Other renew	Total	Electricity	Heat	Total
Primary energy supply	46.80	93.53	50.80	20.06	10.87	8.15	1.67	231.88			
Primary energy for electric powerplants	34.59	4.14	11.03	20.06	10.87	1.17	1.34	83.19			
Primary energy for combined heat and power (CHP) plants	3.66	0.73	4.75	0.00	0.00	1.22	0.04	10.40			
Electricity portion	2.28	0.37	2.71	0.00	0.00	0.69	0.03	6.08			
Heating portion	1.38	0.36	2.05	0.00	0.00	0.53	0.01	4.32			
Primary energy for heating plants	0.17	0.07	0.23	0.00	0.00	0.14	0.01	0.61			
Total primary energy for electricity	36.87	4.51	13.74	20.06	10.87	1.86	1.37	89.27			
Total primary energy for heat grids	1.55	0.43	2.27	0.00	0.00	0.67	0.02	4.93			
Remaining primary energy	8.39	88.60	34.79	0.00	0.00	5.62	0.28	137.68			
Conversion to electricity or grid heat									37.71	3.64	41.35
Subtractions from primary or secondary energy											
Gas works[a]	0.10	0.10	-0.13	0.00	0.00	0.00	0.00	0.07	0.00	0.00	0.07
Liquefaction[b]	0.00	-0.02	0.05	0.00	0.00	0.00	0.00	0.03	0.00	0.00	0.03
Other transformations[c]	0.00	0.04	0.00	0.00	0.00	0.01	0.00	0.04	0.58	0.08	0.70
Petroleum refining	0.00	0.00	0.03	0.00	0.00	0.00	0.00	0.03	0.40	0.05	0.49
Distribution and distribution losses	0.04	0.00	0.98	0.00	0.00	0.00	0.01	1.03	2.43	0.24	3.70
Own use	0.48	5.39	3.71	0.00	0.00	0.00	0.00	9.58	2.07	0.06	11.37
Delivered secondary energy	7.77	83.81	30.16	0.00	0.00	5.61	0.27	127.62	32.23	3.20	163.05
Uses of delivered secondary energy											
Other industry	6.97	6.09	10.69	0.00	0.00	2.73	0.02	26.49	11.15	0.97	38.61
Transportation	0.00	52.58	0.06	0.00	0.00	0.48	0.00	53.13	0.41	0.00	53.54
Agriculture and fishing	0.06	2.18	0.22	0.00	0.00	0.07	0.01	2.53	0.29	0.01	2.83
Commercial and public buildings	0.18	3.30	5.99	0.00	0.00	0.16	0.03	9.66	9.49	0.30	19.45
Residential buildings	0.48	4.99	11.47	0.00	0.00	2.17	0.20	19.30	10.31	0.65	30.26
Non-specified	0.00	0.09	0.22	0.00	0.00	0.00	0.01	0.33	0.57	1.27	2.17
Non-energy use[d]	0.07	14.60	1.51	0.00	0.00	0.00	0.00	16.18	0.00	0.00	16.18

Note: [a] Gas from coal and oil operations is counted toward the natural gas supply, with a production efficiency of 0.13/(0.10+0.10). [b] Refers to liquefied petroleum products of natural gas processing, which add to the oil energy supply with a production efficiency of 0.02/0.05. [c] Consists of energy use in heat plants, for fuel mining and extraction, and 'other energy sector'. [d] Includes feedstock to chemical fertilizers and plastics, so should be counted as industrial energy use. Other renew = predominately geothermal, wind and solar. OECD = Organisation for Economic Co-operation and Development.

Source: IEA (2007a) for primary energy columns, IEA (2007b) for electricity and heat

Table 2.2 *Primary and secondary energy use in non-OECD countries in 2005, calculated from fuel HHVs*

	Primary energy (EJ)								Secondary energy (EJ)		
	Coal	Oil	Gas	Nuclear	Hydro	Biomass & waste	Other renew	Total	Electricity	Heat	Total
Primary energy supply	73.94	66.71	48.18	4.46	17.45	39.86	0.80	251.41			
Primary energy for electric powerplants	36.11	6.48	10.96	4.46	17.45	0.40	0.76	76.62			
Primary energy for CHP plants	3.63	0.53	7.41	0.00	0.00	0.08	0.00	11.64			
Electricity portion	2.26	0.27	4.22	0.00	0.00	0.04	0.00	6.79			
Heating portion	1.37	0.26	3.19	0.00	0.00	0.03	0.00	4.85			
Primary energy for heating plants	3.66	0.55	3.75	0.00	0.00	0.16	0.00	8.11			
Total primary energy for electricity	38.37	6.74	15.18	4.46	17.45	0.44	0.76	83.41			
Total primary energy for heat grids	5.03	0.81	6.94	0.00	0.00	0.19	0.00	12.96			
Remaining primary energy	30.54	59.16	26.07	0.00	0.00	39.23	0.04	155.03			
Conversion to electricity or grid heat									28.14	9.71	37.85
Subtractions from primary or secondary energy											
Gas works[a]	0.46	0.06	−0.32	0.00	0.00	0.00	0.00	0.20	0.00	0.00	0.20
Liquefaction[b]	0.75	−0.33	0.23	0.00	0.00	0.00	0.00	0.64	0.00	0.00	0.64
Other transformations[c]	0.02	0.01	0.09	0.00	0.00	2.15	0.00	2.26	0.89	0.51	3.66
Petroleum refining	0.00	2.31	0.00	0.00	0.00	0.00	0.00	2.31	0.34	0.57	3.22
Distribution and distribution losses	0.07	0.39	2.78	0.00	0.00	0.00	0.00	3.24	3.31	0.42	6.97
Own use	2.08	4.12	4.40	0.00	0.00	0.01	0.00	10.61	1.74	0.13	12.48
Delivered secondary energy	27.18	52.60	18.89	0.00	0.00	37.06	0.04	135.77	21.85	8.09	165.72
Uses of delivered secondary energy											
Other industry	21.89	7.64	7.42	0.00	0.00	4.79	0.00	41.73	11.11	3.64	56.48
Transportation	0.18	26.62	0.30	0.00	0.00	0.30	0.00	27.40	0.52	0.00	27.91
Agriculture and fishing	0.57	2.67	0.06	0.00	0.00	0.21	0.00	3.50	1.09	0.15	4.74
Commercial and public buildings	0.54	1.58	0.79	0.00	0.00	0.44	0.00	3.34	3.04	0.87	7.25
Residential buildings	2.44	4.98	5.43	0.00	0.00	30.28	0.04	43.16	5.00	3.26	51.41
Non-specified	0.46	1.00	0.89	0.00	0.00	1.04	0.00	3.40	1.09	0.19	4.68
Non-energy use[d]	1.11	8.12	4.01	0.00	0.00	0.02	0.00	13.25	0.00	0.00	13.25

Note: [a] Gas from coal and oil operations is counted toward the natural gas supply, with a production efficiency of 0.32/(0.46+0.06). [b] Refers to liquefied petroleum products of coal and natural gas processing, which add to the oil energy supply with a production efficiency of 0.33/(0.75+0.23). [c] Consists of energy use in heat plants, for fuel mining and extraction, and 'other energy sector'. [d] Includes feedstock to chemical fertilizers and plastics, so should be counted as industrial energy use. Other renew = predominately geothermal, wind and solar.

Source: IEA (2007c) for primary energy columns, IEA (2007d) for electricity and heat

Table 2.3 *World primary and secondary energy use in 2005, given by the sum of Table 2.1 and Table 2.2*

	Primary energy (EJ)								Secondary energy (EJ)		
	Coal	Oil	Gas	Nuclear	Hydro	Biomass & waste	Other renew	Total	Electricity	Heat	Total
Primary energy supply	120.74	160.24	98.98	24.53	28.32	48.00	2.47	483.29			
Primary energy for electricity	75.24	11.25	28.91	24.53	28.32	2.30	2.13	172.68			
Primary energy for heat grids	6.57	1.23	9.21	0.00	0.00	0.86	0.02	17.89			
Remaining primary energy	38.93	147.76	60.86	0.00	0.00	44.85	0.32	292.72			
Conversion to electricity or grid heat									65.85	13.35	79.20
Subtractions from primary or secondary energy											
Gas works	0.56	0.16	-0.45	0.00	0.00	0.00	0.00	0.27	0.00	0.00	0.27
Liquefaction	0.75	-0.36	0.28	0.00	0.00	0.00	0.00	0.67	0.00	0.00	0.67
Other transformations	0.02	0.04	0.09	0.00	0.00	2.16	0.00	2.30	1.47	0.59	4.36
Petroleum refining	0.00	1.95	0.03	0.00	0.00	0.00	0.00	1.98	0.75	0.62	3.71
Distribution and distribution losses	0.11	0.39	3.75	0.00	0.00	0.00	0.01	4.26	5.75	0.66	10.67
Own use	2.56	9.15	8.11	0.00	0.00	0.02	0.00	19.84	3.81	0.19	24.19
Delivered secondary energy	34.95	136.41	49.05	0.00	0.00	42.67	0.32	263.39	54.08	11.30	328.76
Uses of delivered secondary energy											
Other industry	28.86	13.72	18.10	0.00	0.00	7.51	0.02	68.22	22.26	4.61	95.09
Transportation	0.18	79.20	0.36	0.00	0.00	0.78	0.00	80.52	0.93	0.00	81.45
Agriculture and fishing	0.63	4.85	0.27	0.00	0.00	0.28	0.01	6.03	1.38	0.15	7.57
Commercial and public buildings	0.71	4.87	6.77	0.00	0.00	0.60	0.04	12.99	12.53	1.17	26.69
Residential buildings	2.92	9.96	16.90	0.00	0.00	32.45	0.24	62.46	15.31	3.90	81.68
Non-specified	0.47	1.09	1.12	0.00	0.00	1.04	0.01	3.73	1.66	1.46	6.85
Non-energy use	1.18	22.71	5.52	0.00	0.00	0.02	0.00	29.43	0.00	0.00	29.43

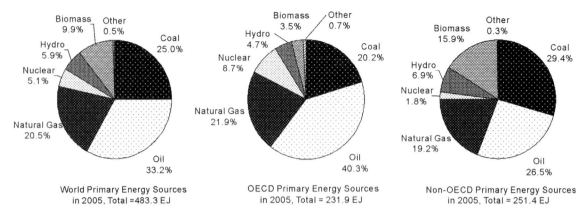

Source: Derived from data in IEA (2007a, 2007b, 2007c, 2007d)

Figure 2.6 *Relative importance of different primary energy sources for the world, OECD countries and non-OECD countries in 2005*

worldwide (13.0EJ out of 483EJ). However, 5.4EJ out of the total primary energy allocated to industry in 2005 was for the manufacture of fertilizers and pesticides, which increases the agricultural share to 3.8 per cent (18.4EJ).

In OECD countries, 5 per cent of the gross electricity production is consumed by electricity powerplants themselves on average (6–8 per cent by coal plants but only 2–3 per cent by natural gas plants) and 7 per cent is lost in distribution. Distribution losses are much larger (11 per cent) in non-OECD countries. Industry uses a much larger share of total electricity supply in non-OECD than in OECD countries (40 per cent versus 31 per cent). Within the industrial sector, the major consumers of energy are the production of iron and steel, the chemical and petrochemical industry, non-ferrous metals, non-metallic minerals, pulp and paper, and food processing (see Chapter 6 for more information).

2.3.3 Electrical powerplant capacity and electricity production

Table 2.4 presents data on present installed electricity generating capacity in gigawatts (GW) and on annual electricity generation (TWh/yr) for all countries with at least 50GW capacity and for major world regions for 2005. Also given in Table 2.4 are populations in 2005 and the resulting annual per capita energy use. The per

capita energy use in different countries is compared in Figure 2.11, where it can be seen that the three most electricity-intensive countries are Canada (19,500kWh/yr/person), the US (14,400kWh/yr/person) and Australia (12,500kWh/yr/person), with substantially smaller electricity use in Europe (5000–9000kWh/yr/person) and Japan (8600kWh/yr/yr).[5] World average electricity consumption is about 2850kWh/yr/person (7.8kWh/day/person).

The *capacity factor* of a powerplant is the average output as a fraction of the peak output, and Table 2.5 gives the world average capacity factors for different kinds of powerplants. Most types of powerplants run on average at 50 per cent or less of their peak capacity, the exception being nuclear powerplants, where the world average capacity factor in 2005 was 83 per cent. Total world electrical capacity and electricity generation in 2005 were about 4267GW and 18,450TWh, respectively. The average power output was thus (18,450 TWh/yr)/(8760 hr/yr) = 2.107TW, so the world average capacity factor averaged over all electricity sources was in 2005 was 100 per cent × (2107/4267) = 49.4 per cent.

An overview of the data presented in Table 2.4 is given in Figures 2.12–2.14. Thermal generation (largely coal, with some natural gas and oil) accounts for 67 per cent of installed capacity (Figure 2.12) and generation (Figure 2.13), with coal alone accounting for 40 per cent of all electricity generated. Nuclear

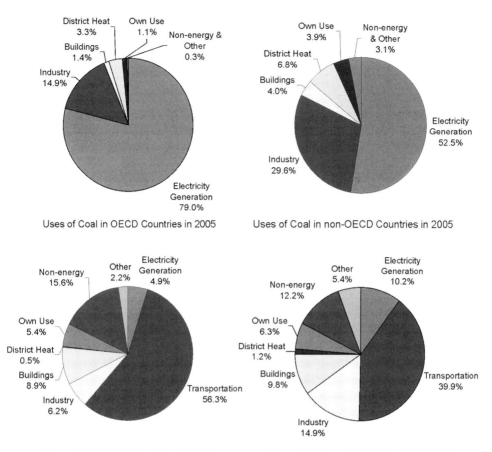

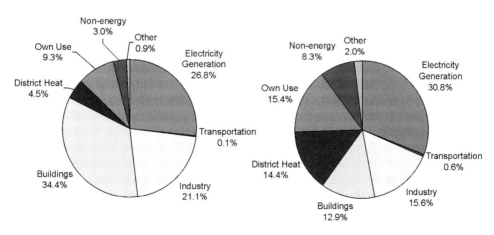

Source: Derived from data in IEA (2007a, 2007b, 2007c, 2007d)

Figure 2.7 *Absolute provision of primary energy in OECD and non-OECD countries in 2005*

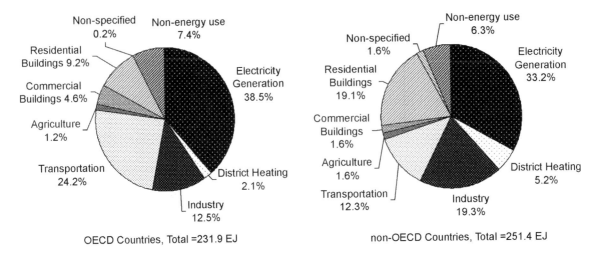

Source: Derived from data in IEA (2007a, 2007b, 2007c, 2007d)

Figure 2.8 *Direct uses of primary energy in OECD and non-OECD countries in 2005*

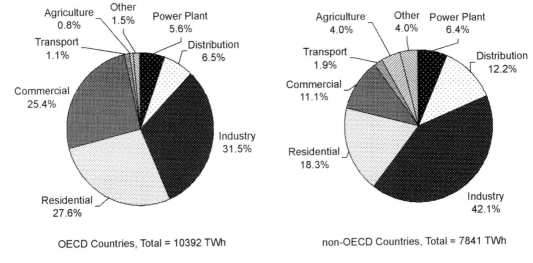

Source: Derived from data in IEA (2007a, 2007b, 2007c, 2007d)

Figure 2.9 *Electricity use by sector for OECD and non-OECD countries in 2005*

accounts for only 9.2 per cent of worldwide capacity but 15.2 per cent of worldwide electricity generation. This is because nuclear powerplants, being expensive to build, cheap to operate and incapable of rapidly powering up or down, are run at full capacity as much as possible. This gives nuclear powerplants the relatively large capacity factor noted above (0.83), so they make a larger contribution to electricity generation than to

capacity. The average hydro capacity factor (0.39), by contrast, is less than the average for all electricity, so its relative contribution to total electricity generation is less than its relative contribution to total capacity. Europe, North America and Asia each account for about 30 per cent of the worldwide installed generating capacity, as shown in Figure 2.14 (this is also true for electricity generated).

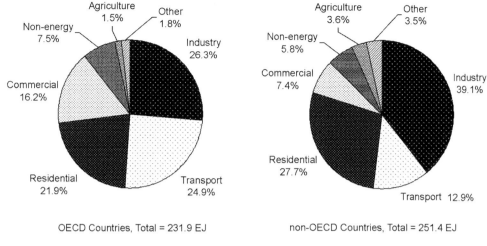

Source: Derived from data in IEA (2007a, 2007b, 2007c, 2007d)

Figure 2.10 *Primary energy use by sector for OECD and non-OECD countries in 2005*

Based on data presented in IEA (2007a, 2007b, 2007c, 2007d), the average efficiency in generating electricity in OECD and non-OECD countries can be calculated. Results are given in Table 2.6, based on gross powerplant output (i.e. not accounting for the 6–8 per cent of output used to run coal powerplants or the 2–3 per cent used to run natural gas powerplants) and fuel HHV. Coal-fired powerplants have an average gross efficiency of 39 per cent in OECD countries but only 32 per cent in non-OECD countries, while natural gas powerplants have average gross efficiencies of 51 per cent and 39 per cent, respectively. Efficiencies using biomass and waste are smaller (27–37 per cent) as a result of lower turbine inlet temperatures than in fossil fuel powerplants and the limitation imposed by the Carnot Cycle (see Chapter 3, subsection 3.2.1).

Also given in Table 2.6 is the system efficiency, based on total primary energy inputs compared to the amount of electricity delivered to end-users (which takes into account electricity use by powerplants themselves and losses during transmission and distribution). The average system efficiency is 37 per cent in OECD countries but only 21 per cent in non-OECD countries. Part of the low transmission efficiency in non-OECD countries does not represent a real loss, but rather electricity that is stolen from the grid.

2.3.4 Efficiency in conversion from primary fuel energy to secondary fuel energy

Secondary energy occurs as electricity and heat from district heating grids, but also as coal, natural gas and refined petroleum products delivered to the point of use. The data in Tables 2.1 and 2.2 can be used to compute the average efficiencies in OECD and non-OECD countries in converting from primary to secondary energy as fuels delivered to the point of use. Results are given in the lower part of Table 2.6. Very little coal is used or lost by the coal industry itself, so the overall conversion efficiency is rather large – 97.1 per cent in the global average. A world average energy loss of about 8.4 per cent is incurred in transporting oil by pipeline and in processing of crude oil into refined energy products, plus additional small amounts due to transport by supertanker and local distribution of refinery products by truck (that are not accounted for in the calculation in Table 2.6). Substantial amounts of energy are expended in transmitting and processing natural gas, such that the average conversion efficiency from raw natural gas to the delivered product is about 87.7 per cent (more information is found in Chapter 3, subsection 3.2.10).

Table 2.4 *Electricity generation data for 2005*

Region		Electricity generation capacity (gigawatt)						Electricity generation (TWh/yr)						Population (millions)	kWh/yr/person
		Thermal	Hydro	Nuclear	Geo-thermal	Wind	Total	Thermal	Hydro	Nuclear	Geo-thermal	Wind	Total		
World		2894.0	866.8	381.8	65.1	59.08	4266.8	12,413.3	2995.8	2767.9	158.0	119.0	18,454.0	6477.0	2849
Africa		84.3	23.1	1.8	0.4	0.26	109.9	453.9	90.7	11.3	1.8	0.0	558.2	906.0	616
Asia		1082.6	276.4	82.7	6.6	6.98	1455.3	5209.9	824.5	567.1	24.6	0.0	6640.2	3921.0	1693
	China	383.5	117.9	6.6	0.0	1.26	509.3	2047.0	397.0	53.1	0.0	2.6	2500.0	1315.8	1900
	India	108.1	32.3	3.4	0.0	4.43	148.2	580.2	99.9	17.3	0.0	7.8	705.2	1103.4	639
	Japan	177.3	47.3	49.6	3.2	1.08	278.5	706.4	86.4	304.8	4.8	2.2	1104.6	128.1	8623
	South Korea	45.4	3.9	17.2	0.1	0.10	66.7	237.3	5.2	146.8	0.1	0.0	389.6	47.8	8147
Europe		666.4	244.8	174.0	42.3	40.90	1168.4	2685.7	735.6	1259.2	80.6	0.0	4843.5	730.0	6635
	France	27.4	25.3	63.4	1.0	0.76	117.9	65.9	56.4	451.5	1.5	1.5	576.8	60.5	9534
	Germany	76.4	8.3	20.4	19.9	18.42	143.4	402.0	26.7	163.1	28.5	33.9	654.2	82.7	7911
	Italy	65.5	22.6	0.0	2.3	1.72	92.1	25.3	42.9	0.0	7.7	0.0	79.4	58.1	1366
	Russia	164.5	45.8	22.7	0.0	0.00	233.0	628.6	174.6	149.4	0.4	0.0	953.0	143.2	6655
	Spain	47.8	18.2	7.6	8.4	10.03	92.0	192.2	23.0	57.5	21.3	21.1	315.1	43.1	7317
	UK	34.9	4.7	12.8	0.0	1.35	53.8	84.8	12.5	88.8	0.0	3.8	189.9	59.7	3182
	Ukraine	64.1	4.3	11.9	1.6	0.08	82.0	308.1	7.9	81.6	2.9	0.0	400.7	46.5	8620
North America		950.6	184.1	120.3	14.4	9.96	1279.4	3601.8	702.3	913.6	46.6	0.0	5284.4	515.0	10,261
	Canada	35.5	72.0	13.3	0.7	0.68	122.2	171.0	363.6	92.0	1.5	1.4	629.5	32.3	19,507
	Mexico	38.2	10.6	1.4	1.0	0.00	51.2	189.6	27.7	10.8	7.3	0.0	235.4	107.0	2199
	USA	852.2	96.9	105.6	12.3	9.15	1076.2	3150.0	290.4	810.7	35.3	18.4	4304.8	298.2	14,436
South America		65.0	123.3	3.0	0.0	0.09	191.4	204.4	601.3	16.7	0.1	0.0	822.6	373.0	2205
	Brazil	20.3	70.9	2.0	0.0	0.03	93.2	55.6	337.5	9.9	0.0	0.1	403.1	186.4	2162
Oceania		45.1	15.1	0.0	0.9	0.89	62.0	257.5	41.4	0.0	4.4	0.0	305.1	33.0	9245
	Australia	40.5	9.3	0.0	0.3	0.71	50.8	234.3	15.9	0.0	0.9	1.4	252.5	20.2	12,526

Source: UN (2008) for electricity data except for wind capacity data; an April 2006 press release of the Global Wind Energy Council (www.gwec.net) for wind capacity data; Population Research Bureau (www.prb.org) for populations in 2005. Electricity generation data for wind are estimates based on installed capacity at the end of 2005 combined with capacity factors given by BTM Consult (2001) for individual countries (where available), or using the world average capacity factor of 0.23 (for other countries).

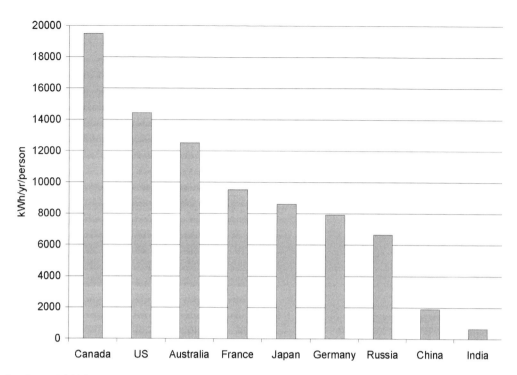

Source: Data from UN (2008)

Figure 2.11 *Annual per capita electricity use in various countries in 2005*

Table 2.5 *Global average capacity factors in 2005
for different kinds of electricity powerplants*

Electricity source	Capacity factor
Thermal	0.49
Hydro	0.39
Nuclear	0.83
Geothermal	0.28
Wind	0.23
Overall	0.49

Source: Derived from data given in Table 2.4

2.3.5 Markup factors from secondary energy at the point of use to primary energy

Given the efficiency in converting from primary energy to secondary energy and delivering it to the point of use, a *markup factor* can be computed – that is, the factor by which a given secondary energy use needs to be multiplied in order to determine how much primary energy was required in order to provide the given secondary energy. Thus, given an efficiency in producing transportation fuels from crude oil of 91.6 per cent (from Table 2.6), the markup factor would be 1.0/0.916 = 1.09. However, this markup factor does not include the energy expended in exploring for oil, in drilling wells, and in extracting the oil from the ground. Cleveland (2005) calculates that, in the US, these energy inputs amount to about 9 per cent of the energy

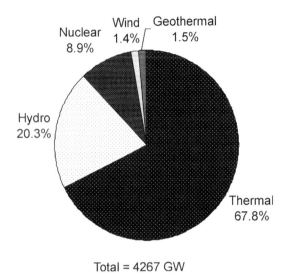

Total = 4267 GW

Note: See Table 2.4 for more information.
Source: Data from UN (2008)

Figure 2.12 *Distribution of electricity generating capacity by source in 2005, all countries*

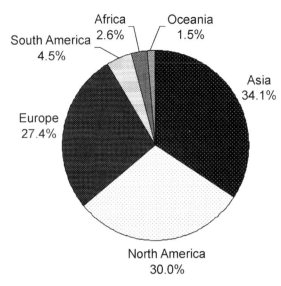

Total = 4267 GW

Note: See Table 2.4 for more information'
Source: Data from UN (2008)

Figure 2.14 *Geographical distribution of electricity generating capacity in 2005*

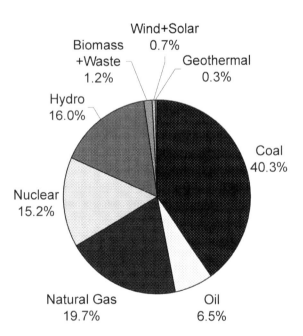

Total = 18223 TWh

Note: See Table 2.4 for more information.
Source: Data from IEA (2007a,c).

Figure 2.13 *Distribution of electricity generated by source in 2005, all countries*

content of the crude oil and natural gas together on a primary-energy basis (that is, the energy return over energy invested (EROEI) is about 11 for unprocessed oil + natural gas in the US). Thus, for refined petroleum products in the US, the markup factor would be 1.09 × 1.09 = 1.19 (assuming that the world average refinery efficiency of 91.6 per cent applies to the US).

With regard to the production of fuels from the Canadian tar (bitumen) sands in Alberta, two techniques are possible for the initial extraction of the bitumen: surface mining (applicable to about 20 per cent of the deposits) and in situ heating with steam followed by pumping (applicable to the remaining 80 per cent). The external energy requirement for the extraction, upgrading and refining of tar sands to a 50:50 mixture of gasoline and diesel fuel is estimated to be equal to about 21 per cent the energy content of the gasoline and diesel with surface mining and 42 per cent using the in situ route (see Chapter 6, Table 6.32), giving markup factors of 1.21 and 1.42, respectively (any exploratory drilling in this case would be negligible). These markup factors assume that refining of the crude oil produced from the

Table 2.6 *Average HHV-based efficiencies of electricity generation and of producing fuel secondary energy, derived from data in Tables 2.1 and 2.2[a][b]*

Energy source or fuel	Efficiency (%)		
	OECD countries	non-OECD countries	Illustrative calculation (OECD case, using data from Table 2.1)
		Gross generation of electricity	
Coal	38.5	31.9	(Gross electrical output)/(Fuel energy)
Oil	43.0	35.4	(Gross electrical output)/(Fuel energy)
Natural gas	51.3	38.9	(Gross electrical output)/(Fuel energy)
Biomass + waste	37.0	27.4	(Gross electrical output)/(Fuel energy)
Fossil average	42.1	34.1	(Gross electrical output)/(Fuel energy)
Overall average	41.8	32.7	37.35/89.27
		Net generation and distribution of electricity	
Average	36.8	20.6	(32.22+0.21+0.40)/89.27
		Production of district heat	
Average	67.4	75.0	(3.20+0.05+0.05)/4.93
		Processing and delivery of fuels	
Coal	98.9	96.0	(36.87+1.55+0.10+7.77)/46.80
Oil products	91.7	91.6	(4.51+0.43+0.10+0.04+83.81–2.0[c])/
			$(93.53+0.02/\eta_{liq}+0.03/\eta_{ng}+0.40/\eta_{el}+0.05/\eta_{th})$
Natural gas	90.7	71.6	(13.74+2.27+30.16+0.05+0.03)/
			$(50.80+0.13/\eta_{go})$
Fossil fuel weighted mean	77.3	77.3[d]	(23.19+2.86+7.77+83.81+30.16)/
			(46.80+93.53+50.80)

Note: η_{liq}, η_{ng}, η_{go}, η_{el}, and η_{th} are efficiencies in producing liquefied petroleum products, secondary natural gas, gas from coal and oil, electricity, and district heat, respectively. [a] Efficiencies are based on gross powerplant output by central and CHP plants, divided by total primary energy use. Total primary energy use is computed as that used by central powerplants plus that used by CHP plants times the ratio of electricity output to electricity + useful heat output. [b] An alternative approach is to use total fuel input to CHP plants but to take into account the loss of electrical generating capacity when useful heat is produced alongside electricity in a cogeneration facility, as discussed in Chapter 3 (subsection 3.3.2) and utilized by Graus et al (2007). This results in lower effective electrical efficiencies from cogeneration. [c] Imports of petroleum products. The import and export values given in IEA (2007a, 2007c) are not consistent between OECD and non-OECD groups, and in the case of non-OECD countries, result in a conversion efficiency >100 per cent. In the absence of imports or exports, the inferred efficiency in producing oil from primary energy is 93.4 per cent for OECD countries and 88.7 per cent for non-OECD countries. A small transfer of 2EJ/yr, assumed here, makes the efficiencies in the two regions essentially the same. [d] This is not a typographical error.

bitumen occurs in Edmonton, whereas plans are underway to build pipelines for the processing of some of the crude in Texas. Construction and operation of the pipeline would increase these markup factors.

For coal and natural gas, the world average processing and delivery efficiencies given above result in markup factors of 1.03 and 1.14, respectively. As with the conversion from crude oil to refined products, these markup factors do not include the energy expended in exploring for new resources or in mining (in the case of coal) or drilling (in the case of natural gas) (mining energy use is included as part of 'Other industry' in Tables 2.1–2.3). For coal, this additional markup factor is likely to be very small (about 1.02), but for natural gas it could be larger (about 1.05). For average electricity in OECD and non-OECD countries the markup factors,

based on the system efficiencies given in Table 2.6, would be 2.72 and 4.85, respectively.

Clearly, the appropriate markup factor will vary strongly from region to region, and will also vary over time. In the case of oil, the efficiency of both refinery and tar sands operations has improved over time and can be expected to continue to improve, all else being equal. However, requirements for lower sulphur content in gasoline and diesel fuel tend to increase refinery energy requirements. The EROEI in finding and extracting unprocessed crude oil has been falling over time (from about 100 in 1930 to about 11 at present according to Cleveland, 2005), and can be expected to continue to fall (thereby increasing the markup factor) as more difficult oil reserves are exploited. Conversely, large improvements in the efficiency in generating electricity by new fossil fuel powerplants have occurred recently and can be expected to continue (reaching 50 per cent or better, as discussed in Chapter 3), reducing the large markup factor for electricity. Large improvements in transmission efficiency and reductions in electricity theft can be expected in non-OECD countries.

In light of the above, the following standardized markup factors will be used here and in Volume 2 in computing the primary-energy equivalent of various forms of secondary energy at the point of use: for electricity, 2.5; for petroleum products, 1.2; for natural gas, 1.25; and for coal, 1.05. Multiplication of the amount of secondary energy delivered to the end-user by these markup factors gives a rough estimate of the amount of primary energy required to find, extract, process and deliver the energy. In principle, the markup factor for the production of electricity from fossil fuels, and its subsequent transmission to the end-user, should be multiplied by the markup factor in providing fuels to the electric powerplant. As most fossil fuel electricity is

produced from coal, and the markup factor for coal is small, this additional markup factor will be neglected here.

2.4 Energy resources and energy reserves

An important distinction is between energy *resources* and energy *reserves*. For non-renewable forms of energy, an energy resource is the amount that is physically present (in the ground). The energy reserve is that portion of the energy resource that is worth extracting, given the current technology and prices. As technology improves, it becomes easier (and hence less expensive) to extract an energy resource, so the reserve grows in size. Similarly, if the market price of an energy source increases, it become profitable to extract more difficult and more expensive resources, so again, the reserve increases in size.

Figure 2.15 shows the estimated increase in the cost of the next unit of natural gas in the US, as the cumulative extraction increases.[6] The left curve shows

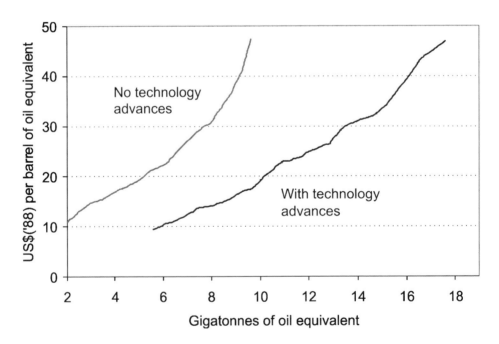

Note: Estimates are given assuming constant technology, and assuming technological progress (based on past trends) associated with learning as the cumulative extraction increases.
Source: Rogner (1997)

Figure 2.15 *Speculative projected increase in the cost of the next unit of natural gas extracted in the US as the cumulative extraction increases*

the relationship between cumulative extraction and cost for present technology, while the curve on the right shows the expected relationship given reasonable improvements in technology. Technological progress is expected to almost double the amount of natural gas that is available up to a given cost threshold, but technological progress occurs only through the cumulative extraction of the resource. The real impact of technological progress on the amount of oil and gas that can ultimately be recovered is highly uncertain.

Of course, it will be economically worthwhile to continue extracting energy as the cost of extracting increases only if the cost of extracting (including taxes) is less than the market price of the energy, which depends on the interaction between supply and demand. If the rate of extraction decreases due to increasing cost, this reduces supply, which drives up the price. Physical limits in the rate at which the resource can be extracted (discussed in the next section) as the amount remaining decreases will also limit the supply and drive up the price. As the price increases, demand drops, as consumers seek less expensive ways to meet their tertiary energy demand (either by switching to alternative fuels, by improving efficiency, or by reducing tertiary energy demand). As this happens, the rate of extraction of the remaining energy resource decreases, and could even drop to zero. In this sense, we will never run out of non-renewable fossil fuel energy. However, we will certainly run out of inexpensive fossil fuel energy.

In the case of renewable energy, there is not a fixed stock of energy, but rather, a continuous flow of energy (power). Thus, resources and reserves of renewable energy apply to the rate at which energy can be supplied (power), rather than to the amounts of energy available. Thus, in Chapter 1, we spoke of the need for carbon-free *power* reaching up to 21TW by 2050 (to stabilize CO_2 at 450ppmv).

2.5 Peaking in the rates of extraction of fossil fuels

The rate of extraction of a finite resource grows rapidly at the beginning, since it is relatively plentiful and therefore relatively easy to find. As the resource is depleted, it is harder, and takes longer, to find further deposits that can be exploited. Thus, at some point the rate of extraction peaks, then gradually declines. This behaviour was predicted by M. King Hubbert in the 1950s, and has been borne out by the temporal variation in extraction of oil in many regions. For example, oil extraction in the US (excluding Alaska) peaked in the early 1970s and has been on a steady decline ever since. Intensified exploration activity cannot reverse this inexorable trend.

One way to mathematically represent the behaviour described above, and used by Hubbert, is through the logistic growth function. This function and its properties are described in Box 2.1. The logistic growth function is such that the peak rate of extraction occurs when exactly half of the ultimately recoverable resource has been consumed, with the decline in production after the peak being the mirror image of the rise in production up to the peak. The key questions with regard to the exploitation of energy resources are, what is the ultimate potential extraction? To what extent can advances in extraction technology increase the ultimately recoverable fraction? And to what extent will technological advances lead to an asymmetric extraction curve, with a delayed peak at the expense of a potentially faster decline after the peak? These questions are addressed below with regard to oil, natural gas and coal resources.

Box 2.1 The logistic growth function

The logistic growth function is derived as follows. Let C be the cumulative extraction and consumption of a resource. Then the rate of extraction, E, is equal to the rate of growth of C with time, dC/dt. Assume that C is initially able to grow exponentially. Exponential growth in C is represented by the differential equation:

$$\frac{dC}{dt} = aC \tag{2.1}$$

which has the solution $C(t) = C_o e^{at}$, where C_o is the initial cumulative consumption at time $t = 0$ and a is the exponential growth constant and is equal to the fraction by which C grows in one time period. C in this case grows without limit. For the logistic growth curve, the exponential growth equation is modified by assuming that a negative feedback on E occurs as C becomes large. The growth equation in this case is:

$$E = \frac{dC}{dt} = aC - bC^2 \qquad (2.2)$$

where a is the initial fractional growth rate and b causes the growth rate to decline as C gets larger. The ultimate cumulative consumption (C_U) occurs when there is no further growth in cumulative consumption, that is, when $dC/dt = 0$, which occurs when $C = a/b$. Thus, $b = a/C_U$. The solution to Equation (2.2) is:

$$C(t) = \frac{a}{b + ((a - bC_0)/C_0)e^{-a(t - t_0)}} \qquad (2.3)$$

where C_0 is the cumulative consumption at some starting time t_0. Equation (2.3) can be rewritten as:

$$C(t) = \frac{C_U}{1 + ((C_U - C_0)/C_0)e^{-a(t - t_0)}} \qquad (2.4)$$

Thus, given an estimate of the ultimately recoverable resource (C_U) and historical data on the initial growth in the rate of extraction over time (in order to estimate a), the variation in $C(t)$ over time can be estimated. An easy way to estimate these parameters from historical extraction data is to plot E/C versus C. From Equation (2.2) we get:

$$\frac{E}{C} = a - \left(\frac{a}{C_U}\right)C \qquad (2.5)$$

so when $C(t)$ follows a logistic curve, E/C versus C will plot as a straight line, with E/C decreasing linearly with increasing C. The point on the C axis where E/C extrapolates to zero gives an estimate of C_U (this can be seen from Equation (2.5) and also intuitively: when $E = 0$, there is no further extraction). Such an estimate of C_U should be compared with geologically based estimates.

The properties of the logistic growth function are illustrated in Figure 2.16 for a general case. Shown in the upper panel is the cumulative extraction C and the rate of extraction E vs time, while E/C vs C is shown in the lower panel.

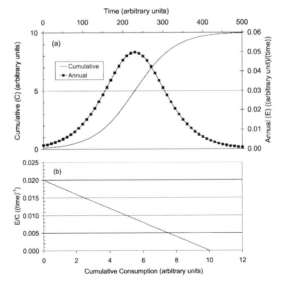

Figure 2.16 *(a) Variation in cumulative extraction* (C) *and in the rate of extraction* (E) *of a non-renewable resource as modelled by the logistic growth function with* a = 0.02 *and* C_U = 10, *and (b) plot of* E/C *versus* C

2.5.1 Near-term peaking in oil supply

There is a general consensus that the original physical oil resource consisted of about 6 trillion barrels, of which about 1.13 trillion had been consumed by 2006. There is disagreement over how much of the original endowment can ultimately be extracted; conventional opinion is that only about a third, or two trillion barrels can be extracted, in which case we have consumed about half of the total recoverable resource and (based on the logistic growth curve), we should be close to the peak in the rate of extraction. Others argue that advances in technology should permit recovery of half of the original endowment, or 3 trillion barrels. Application of the logistic growth function to historical global oil supply data by Nel and Cooper (2009) yields an ultimate extraction of 2.6 trillion barrels and a peak in oil supply in 2014.[7]

Figure 2.17 shows the forecast of global oil demand by the International Energy Agency (IEA) in its *World Energy Outlook 2008*. Global oil demand is expected to grow from about 85 million barrels per day in 2007 to 106 million barrels per day by 2030. Oil supply from existing wells in existing reserves is expected to fall sharply over this timeframe, but is expected to be slightly more than offset by further development of existing reserves. Improved techniques for extracting oil (called 'enhanced oil recovery' or EOR), unconventional oil (such as the Canadian and Venezuelan tar sands and US oil shales) and new discoveries are expected to supply the 25 per cent increase in global demand forecast by 2030. EOR techniques include injection of water or CO_2 into the field so as to maintain the oil pressure as the oil is extracted, or injection of steam or a chemical surfactant into the field so as to reduce the viscosity of the oil. Typically, only about 35 per cent of the oil in an oilfield can be extracted in the absence of EOR. Also shown in Figure 2.17 are the forecasts for 2010 and 2020 made in the 2000 *Outlook* and the forecasts for 2020 and 2030 made in the 2004 *Outlook*. As can be seen from Figure 2.17, the forecasts of future energy demand have been falling over time, largely due to rising price expectations.[8]

There are a number of reasons for doubting whether these new oil supplies can be brought on line as forecast. First, IEA forecasts (and those of other

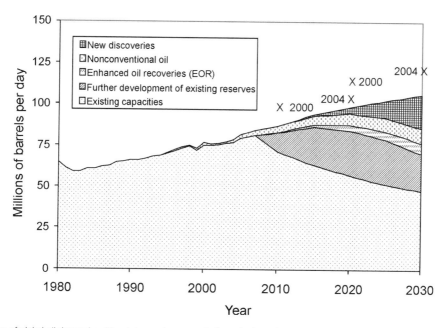

Source: Projection of global oil demand and breakdown of sources of oil supply through to 2030 as forecast in the International Energy Agency's *World Energy Outlook 2008*, plus projections for 2010 and 2020 in the 2000 *Outlook* and for 2020 and 2030 in the 2004 *Outlook*

Figure 2.17 *Projection of global oil demand and breakdown of sources of oil supply through to 2030*

government agencies) rely on official government statements of the size of the remaining reserves. The official reserve estimates of the member states of the Organization of Petroleum Exporting Countries (OPEC), however, seem to be dictated more by political considerations than by objective assessments of how much oil is in the ground. Table 2.7 shows the history of official estimates of remaining reserves in several OPEC nations. All of the countries increased their reserve estimates by 50–200 per cent at about the same time (1985–1989), shortly after the system of setting extraction quotas was changed so as to be based on the size of a country's estimated reserves. Part of the increase could be due to underestimation of the reserves prior to implementation of the new quota system, but because outside analysts do not have access to the data used to estimate national reserves, it is not possible to independently assess the size of the reserves.

Second, increased rates of drilling of conventional oilfields are unlikely to significantly slow the rate of decline in oil supply once the peak occurs (much less delay the peak). This is borne out by experience in the lower 48 states of the US, where drilling rates increased several-fold between 1973 and 1985, but had only a minor effect on oil supply. This is shown in Figure 2.18, which also nicely illustrates the adherence of US oil supply to the logistic function.

Third, an analysis by Gowdy and Juliá (2007) of two large oilfields where EOR techniques have been applied indicates that EOR results in a short-term increase in the rate at which oil can be extracted (slowing the initial decline after the peak) that is compensated by a faster decline later, with negligible effect on the total amount of oil that can be extracted. This effect, which is similar to that of increased rates of drilling, is illustrated in Figure 2.19 for the Forties oilfield in the North Sea. Extraction rates from oilfields tend to rise rapidly to a peak that is dictated by the pipeline capacity, are maintained at that rate for a few years, and then begin to decline in inverse proportion

Table 2.7 *Official size of oil reserves (billions of barrels) in various OPEC countries*

Year	Country					
	Abu Dubai	Iran	Iraq	Kuwait	Saudi Arabia	Venezuela
1980	28	58	31	65	163	18
1981	29	58	30	66	165	18
1982	31	57	30	65	164	20
1983	31	55	41	64	162	22
1984	30	51	43	64	166	25
1985	31	49	45	**90**	169	26
1986	30	48	44	90	167	26
1987	31	49	47	92	167	25
1988	**92**	**93**	**100**	92	170	**56**
1989	92	93	100	92	**258**	58
1990	92	93	100	92	259	59
1991	92	93	100	95	259	59
1992	92	93	100	94	259	63
1993	92	93	100	94	259	63
1994	92	93	100	94	259	65
1995	92	93	100	94	259	65
1996	92	93	112	94	259	65
1997	92	93	113	94	259	72
1998	92	90	113	94	259	73
1999	92	90	113	94	261	73
2000	92	90	113	94	261	77
2001	92	90	113	94	261	78
2002	92	90	113	94	259	78
2003	92	126	115	97	259	78

Note: Estimates for years in which reserve sizes were abruptly revised upward are indicated in bold.
Source: Colin Campbell, Oil Depletion Analysis Centre, Edinburgh, 25 April 2005

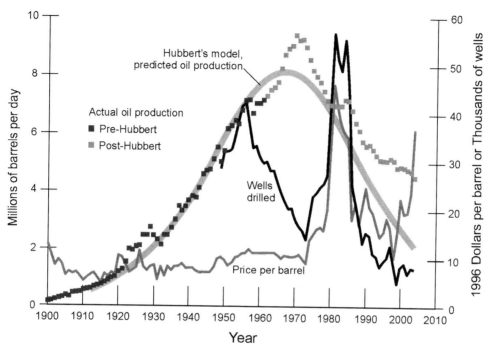

Source: Kaufman (2006)

Figure 2.18 *Hubbert's prediction of the variation in rate of oil extraction from the lower 48 states of the US, actual extraction rates, the price of crude oil, and the number of wells drilled per year*

to the cumulative extraction (as in the logistic growth curve). This linear decline is illustrated by the straight line drawn through the data points in Figure 2.19. EOR in the Forties oilfield began in 1987 but caused only a temporary deviation from the long-term declining trend. With regard to the largest oil-extracting country, Saudi Arabia, oil output is generated by a few large fields that were first developed 40–50 years ago and have been subjected to EOR techniques (in particular, injection of water into the field, which at some point will make further extraction of oil difficult if not impossible due to an increasing proportion of water in the extracted oil). However, data on individual oilfields in Saudi Arabia (and in most other countries) are not publicly available, so it is not possible to assess the prospects for continued extraction at current (much less accelerated) rates.

Fourth, extrapolation of the past trend of a decreasing rate of discovery of new oilfields suggests that there is little remaining oil to be discovered. This is shown in Figure 2.20, alongside the growth in annual oil

consumption. In cases where the amount of oil in a field was revised upward after the initial discovery, the increased amount of oil is assigned to the year in which the field was discovered. This is contrary to the usual practice, but gives a better picture of the variation over time in the rate of discovery of oil. Figure 2.20 suggests that there are about 100 billion barrels of oil yet to be discovered. The cumulative supply of oil during the period 2008–2030 from undiscovered oil is about 50 billion barrels according to Figure 2.17, but the rate of extraction from a pool of only 100 billion barrels of as-yet undiscovered oil would begin to decline by about 2030, rather than continuing to grow (as in the IEA forecast).

An analysis of the decline in the rate of oil extraction from individual major oilfields after they reach their peak by Höök et al (2009) indicates that the later an oilfield reaches its peak, the faster the rate of decline after the peak. Thus, land-based oilfields that peaked in the 1960s declined at a rate of 5 per cent/year after their peak, whereas those that peaked in the 2000s have been declining at an average rate of almost

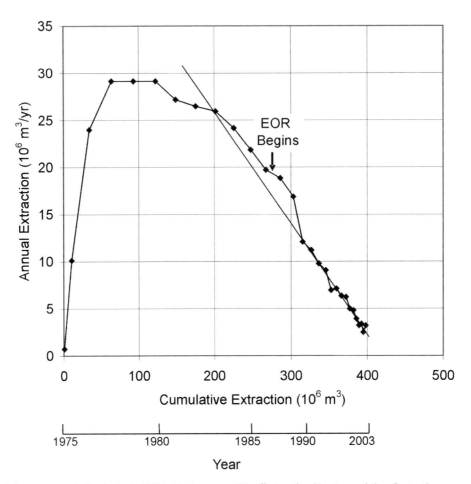

Note: EOR techniques were applied beginning in 1987 but had no perceptible effect on the ultimate cumulative oil extraction
Source: Gowdy and Juliá (2007)

Figure 2.19 *Historical variation in the rate of extraction of oil from the Forties oilfield in the North Sea, plotted against cumulative extraction*

11 per cent/year. Over time, increasingly effective methods of enhanced oil recovery have been applied, so the faster decline rate of the fields that were exploited and have peaked more recently is further evidence that the effect of EOR in extending the period of growth is compensated by faster post-peak declines. The combination of rapidly falling rates of discovery of new oil and ever faster rates of decline once existing fields reach their peak guarantees that the peak in total world oil supply cannot be far in the future.

An alternative assessment of possible future oil supply prepared by the Uppsala Hydrocarbon Depletion Group (UHDG, 2006) is shown in Figure 2.21.

According to this assessment, oil supply will peak around 2015. Other analysts conclude that oil supply is peaking about now, while others see the peak delayed until 2025 or later. However, Bentley et al (2007) present a strong case that peaking of conventional oil supplies is imminent (sooner than 2020) and that there are serious and shocking flaws in most economically derived estimates of future oil supply and in official forecasts by most government agencies, including the IEA. Hirsch (2008) presents alternative scenarios of how the decline in oil supply might unfold (multi-year peak followed by 2–5 per cent/year decline, gradual transition to 2–5 per cent/year decline, or sharp peak

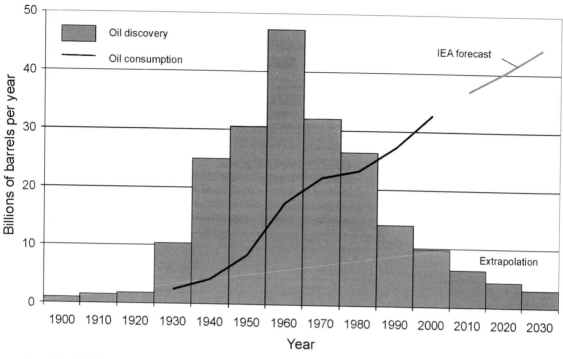

Source: Campbell and Siobhan (2009)

Figure 2.20 *Historical variation in the amount of new oil discovered per year and in the consumption of oil per year*

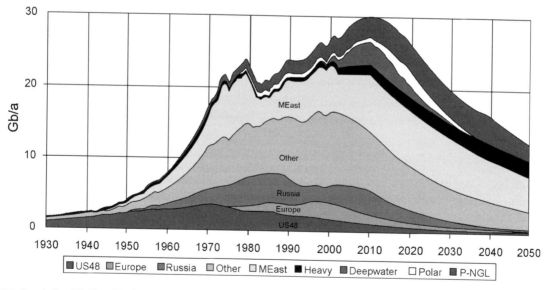

Source: Campbell and Siobhan (2009)

Figure 2.21 *As assessment of future oil supply based on geological considerations rather than on abstract and hypothetical economic considerations*

followed by 2–5 per cent/year or faster decline) and how exporting countries might react.

Meng and Bentley (2008) provide rebuttals to the common argument that there is not an imminent peak in oil supply. Among these is the argument that unconventional sources of oil, such as the Canadian tar sands and various oil shales, can make up for the declining supply of conventional oil. Even in the IEA forecast, unconventional oil is not a large factor. There are several reasons for this. About 80 per cent of the tar sands are amenable to extraction only through injection of steam underground so as to liquefy the tars prior to pumping them to the surface. This requires large amounts of both water and natural gas. Nuclear energy has been considered as an alternative energy source for making steam. The energy in the fuels produced from the tar sands through initial in situ processing is about 2.5 times the external energy inputs (see Chapter 6, Table 6.32), which is a very poor EROEI. Current tar sands production of oil is about 1.4 million barrels/day, and even a crash programme would not see production rise above 5 million barrels/day (mb/d) in 2030 (Söderbergh et al, 2007).[9] As for oil shale, oil can be produced either by heating the shale in situ electrically for about two years, resulting in an EROEI of 2.4–15.8 assuming the electricity to be generated at 45 per cent efficiency

(Brandt, 2008), or by mining and crushing the shale followed by heating in a retort, resulting in an EROEI of only 2.9–6.9 (Brandt, 2009). Large amounts of water are used in both cases, and half of all known oil shales worldwide occur in the semi-arid US states of Colorado, Utah and Wyoming.

Figure 2.22 compares the IEA demand projection (Figure 2.17) with the UHDG supply projection. The only things that can be stated with certainty are that (1) the actual demand and supply curves will coincide, and (2) the two curves will be brought together through an increase in price. Increasing price will force demand to be lower while also increasing supply. The more responsive consumers are to increasing price, or the more that supply can be increased in response to increasing price, the less that the price needs to rise in order to bring the two curves together.

2.5.2 Medium-term peaking in natural gas supply

Figure 2.23 shows the distribution of the sum of known natural gas reserves, estimated additions to known reserves and estimated undiscovered natural gas. Reserve additions and undiscovered natural gas are based on the 50th probability percentile as estimated by the US

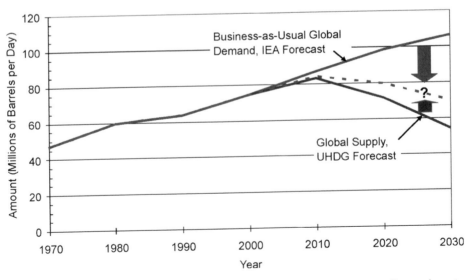

Figure 2.22 *Comparison of the IEA demand projection of Figure 2.17 with the Upsalla supply projection of Figure 2.21, and of the role of price increases in bringing the two curves together*

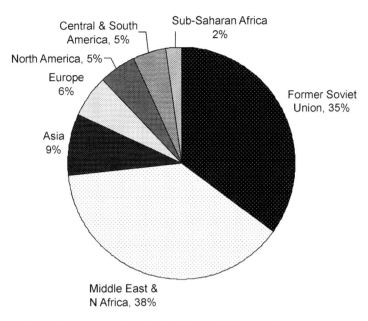

Note: Reserve additions and undiscovered natural gas are based on the 50th probability percentile.
Source: Data from USGS (2000)

Figure 2.23 *Distribution of the sum of known natural gas reserves, estimated additions to known reserves, and estimated undiscovered natural gas*

Geological Survey (USGS, 2000). The total remaining amounts of natural gas estimated with 95 per cent, 50 per cent and 5 per cent probability are 8031EJ, 13,739EJ and 21,729EJ, respectively. By comparison, global natural gas demand in 2005 was about 100EJ (see Table 2.3), which implies sufficient natural gas to last for 80 years, 137 years and 217 years for the three supply estimates, at current rates of use. However, demand is growing, and the relevant date is not when the supply will be exhausted, but rather, when the rate of supply will reach a peak and then begin to decline.

Based on application of the logistic growth function to historical natural gas extraction data, Nel and Cooper (2009) estimate the ultimate natural gas extraction to be 324 trillion m³ (about 11,600EJ). Given that cumulative extraction to 2006 was about 84 trillion m³, this corresponds to a remaining recoverable resource of about 8500EJ, which is close to the lowest of the three USGS estimates given above. The corresponding peak in supply occurs around 2025.

Figure 2.24 shows a recent projection of the supply of natural gas by Colin Campbell (personal communication, July 2009). Gas is more difficult to assess than oil because large amounts are flared or re-injected into the ground in some countries. As well, the depletion profile is different (gas being a gas and not a liquid), normally having a long plateau followed by a cliff rather than a peak and gentle decline. For this reason, the projection terminates at 2030, by which time the total gas supply (conventional + non-conventional) is expected to have peaked.

At some point, the price of both natural gas and oil may increase enough that synthetic gaseous and liquid fuels produced from coal will become economically viable. This would be accompanied by a large increase in CO_2 emissions. Although the cost of these synthetic fuels is high at present, some analysts believe that once large-scale production begins, costs could fall significantly through learning-by-doing. It is generally assumed that the exploitable coal resource is so large that both conventional uses of coal (primarily for power generation) and as a source of gaseous and liquid fuels to replace oil and natural gas could be supported for a century or more. However, recent analyses suggest

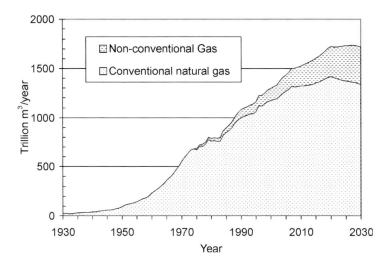

Source: Colin Campbell, personal communication, July 2009

Figure 2.24 *Projected future supply of conventional and conventional + non-conventional natural gas*

that the usable coal resource may be much smaller than commonly assumed. These analyses are discussed in the next section.

2.5.3 Doubts concerning the size of the recoverable coal resource

The distinction between resource and reserve is not clear for coal. Table 2.8 gives recent estimates of the size of what the authors call the global coal reserve, but these numbers need to be multiplied by the recoverable fraction to give amounts that can be extracted, with the recoverable fraction depending in part on the expense that mining companies are prepared to incur. Thus, the numbers in Table 2.8 are more akin to resource estimates, although they certainly do not represent the total amount of coal in the ground. Data are given for different grades of coal: anthracite, bituminous and subbituminous coal (hard, high-quality coals), and lignite (a soft, low-quality coal, also called brown coal). Also given are carbon contents. The total estimated remaining coal reserve is 909 billion tonnes, corresponding to a potential CO_2 emission of about 600GtC. The world consumes far more hard coal (about 4.2 billion tonnes in 2005) than soft coal (about 900 million tonnes in 2005). Six countries (USA, Russia, China, India, South Africa, Australia) account for 84 per cent of hard coal reserves, and four countries (USA, Russia, China, Australia) account for 78 per cent of soft coal reserves. Most of the world's hard coal is extracted through

deep mining, while most of the world's lignite is extracted through open-pit mining. The fraction of coal that can be extracted ranges from 50–60 per cent for inexpensive underground mining techniques, to 75 per cent for expensive underground techniques, and 90 per cent or more for open-pit mines (Kavalov and Peteves, 2007). The lifespans given in the final column of Table 2.8 were derived assuming a recoverable fraction of 0.65 for hard coals and 0.9 for lignite.

There are a number of reasons for suspecting that the available coal resource is much smaller than the reserve estimates given in Table 2.8. First, the basis for resource and reserve estimates is highly uncertain, but the estimates of both have been declining over time with improved data. Figure 2.25 shows assessments of the available world coal resource made from 1976 to 2005. The size of the coal resource as estimated in 2005 (4500 billion tonnes) is about half of the size that was estimated in 1976. Estimates of the size of the coal reserves in various countries have also usually gone down in time, sometimes quite drastically, and by amounts that are much larger than the cumulative extraction between the dates of successive reserve estimates. For example, the estimated reserve of hard coal in China was downgraded from 156Gt to 62.2Gt in 1992, but has remained constant since then. Reserve estimates were downgraded from 23,000Mt to 183Mt in 2004 for Germany, from 20,300Mt to 14,000Mt in 2004 for Poland, and from 2000Mt to 1000Mt in

Table 2.8 *Information on different grades of coal*

Type of coal	% carbon	Amount (Gt)	Rate of extraction (Gt/yr)	Lifespan of reserve (years)
Anthracite	80–94	479		
Bituminous	70–85		4.2	179
Subbituminous	60–80	272		
Lignite	20–40	158	0.9	158
Total or weighted average	approx. 66	909	5.1	178

Note: Lifespans are for the current rate of extraction and take into account recovery fractions less than 100 per cent, as explained in the main text.

Source: EWG (2007) for amounts and extraction rates, Hiete et al (2001) for carbon contents

1998 and then to 220 Mt in 2004 for the UK (EWG, 2007). These changes and the persistent downward trend in resource estimates, combined with the underlying uncertainty, suggest resource and reserve estimates are still likely to be too high.

Like oil and natural gas, the rate of extraction of coal in a given region reaches a peak, then begins an inexorable decline that usually nicely fits the logistic function. As an example, Figure 2.26 is a plot of the annual extraction (E) vs cumulative extraction (C) for British coal. The point on the cumulative extraction axis where E/C extrapolates to zero is an estimate of the ultimate extraction. The estimated ultimate extraction based on all the data up to the present is about 28Gt, and it can be readily seen that there is little remaining British coal to be extracted. Figure 2.27 shows the extrapolated total extraction that would have been obtained by applying the

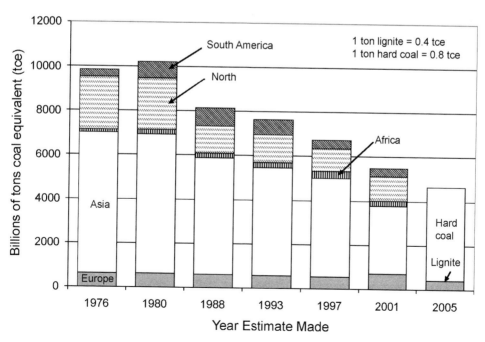

Source: EWG (2007)

Figure 2.25 *Estimates of the global coal resource made from 1976–2005*

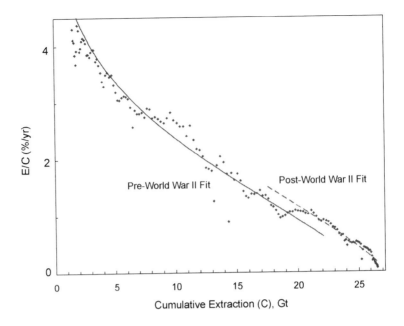

Source: Rutledge (2007)

Figure 2.26 *Variation in the ratio of rate of extraction of British coal* (E) *divided by cumulative extraction* (C) *with the cumulative extraction*

logistic curve analysis at various times in the past, based on the extraction data up to that point in time. Also shown in Figure 2.27 as diamonds are various past estimates of the ultimately recoverable British coal. All except the most recent estimates grossly overestimated what it is now evident the ultimate extraction will be. A similar story is seen for Pennsylvania anthracite in Figure 2.28.

Rutledge (2007) constructed similar curves for other coal supply regions and used these to estimate the ultimate extraction. Results are given in Table 2.9, along with cumulative extraction up to the present and conventional estimates of the amount of remaining recoverable coal. The ultimate remaining world extraction based on the extrapolation of historical extraction data using the logistic function is 414Gt, compared to 909Gt of estimated remaining reserves according to official estimates and 4000–5000Gt ultimate extraction assumed in many CO_2 emission scenarios. However, the US National Academy of Sciences, in its June 2007 report on coal (US NAS, 2007), states that:

Present estimates of coal reserves are based upon methods that have not been reviewed or revised since their inception in 1974, and much of the input data were compiled in the early 1970s. Recent programs to assess reserves in limited areas using updated methods indicate that only a small fraction of previously estimated reserves are actually minable reserves. (p3)

It is possible that, with the abrupt introduction of radically different technologies or mining techniques, the straight line plot of annual over cumulative extraction versus cumulative extraction (seen in Figures 2.16 and 2.26) will be broken into two segment, with the second having a smaller slope than the first. This happened with Appalachian coal, where the ratio of annual over cumulative extraction decreased much more slowly after the 1950s than before due to the adoption of the controversial technique of removing entire mountain tops, thereby allowing exploitation of the remaining coal seams (see Höök and Aleklett, 2009, Figure 8). This extends the ultimately recoverable coal resource.

The normal assumption is that the ultimately extractable fossil fuel resource is greater than the current reserve estimate, because the resources that are currently not part of the reserved base are added to the reserve as the price of the resource resources and technology improves (so that more of the resource can

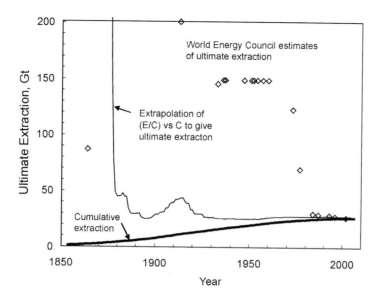

Source: Rutledge (2007)

Figure 2.27 *Variation in the cumulative extraction of British coal with time and in the estimated ultimate extraction based on extrapolation of the* (E/C) *versus* C *line, along with estimates made at various times of the ultimate extraction*

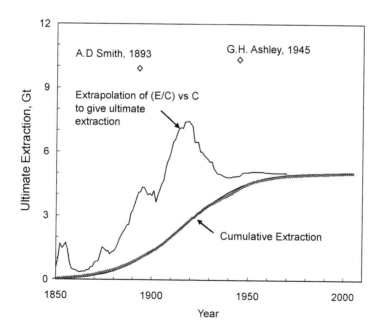

Source: Rutledge (2007)

Figure 2.28 *Variation in the cumulative extraction of Pennsylvania anthracite coal with time and in the estimated ultimate extraction based on extrapolation of the* (E/C) *versus* C *line, along with estimates made at various times of the ultimate extraction*

be economically extracted). However, if the extrapolations given in Table 2.9 and the conclusion quoted above by the National Academy of Sciences are correct, the ultimately extractable coal resource will be considerably *less* than current reserve estimates because much of the coal classified as part of the reserve is in fact not mineable at any affordable price.

Given a cumulative coal consumption of 285Gt and remaining extraction of 414Gt derived from logistic analysis, the ultimately extractable coal amounts to about 700Gt. This implies that about 38 per cent of the ultimately recoverable coal has already been used. Figure 2.29 compares the already used and remaining coal (based on the extrapolation of historical data) in different world regions.

The 700Gt estimate of the global recoverable resource is derived by extrapolating historical extraction data for individual regions using the logistic function, then summing (aggregating) over all regions. An alternative approach, used by Nel and Cooper (2009), is to aggregate the production data globally first, then apply logistic curve analysis. Given uncertainty in the fitting of the linear trend to the historical data, they

estimate an ultimately recoverable coal resource of 1100–1700Gt, with the peak in the annual global coal supply occurring between 2050 and 2070. If the ultimately recoverable resource is only 700Gt, the peak in world coal supply will occur well before 2050. At the regional level, Tao and Li (2007) predict a peak in Chinese coal supply around 2030, assuming that the ultimately recoverable resource is 223Gt, as given by the Chinese Ministry of Land and Natural Resources (this estimate is almost twice the recoverable resource of 114 Gt given for China in Table 2.9). In the US, Croft and Patzek (2009) estimate that extraction of bituminous coal from existing mines is 80 per cent complete and can be carried out at the current rate for only another 20 years, while extraction of subbituminous coal from existing mines can be carried out at the current rate for only another 40–45 years. However, as noted by Croft and Patzek (2009), expansion of existing mines or development of new mines would be politically difficult and costly (both economically and environmentally).

Whatever the amount of remaining extractable coal, there are a number of constraints on expansion in the use of coal. Overcoming these constraints should

Table 2.9 *Comparison of ultimate coal extraction in various regions based on extrapolation of historical data, and according to recent assessments, along with the rate of extraction in 2005 and the lifetime of the remaining coal if extracted at the 2005 rate*

Region	Amount of coal used so far (Gt)	Estimate of remaining coal based on extrapolation of historical data (Gt)	Conventional estimate of remaining reserve at end of 2005 (Gt)	Extraction in 2005 (Gt)	Years remaining based on extraction in 2005 and historically based and conventional estimates of remaining coal	
					Historical	Conventional
Eastern US	43	37	96	1.027	100	203
Western US	13	66	12	0.0653	67	107
Canada	2.9	4.4	7.0	0.0857	187	
Central & South America	1.3	16		0.432		
China	44	70	189	2.205	32	86
South Asia	15.5	68		0.432	157	
Australia and New Zealand	9.5	50	77	0.384	130	201
FSU	37	31	226	0.468	66	483
Europe	78	21	55	0.610	34	90
Africa	8	16	30	0.250	64	120
World	**252**	**414**	**903**	**5.886**	**70**	**153**

Source: Rutledge (2007) for estimates based on historical extrapolations and BP (2007) for conventional estimates

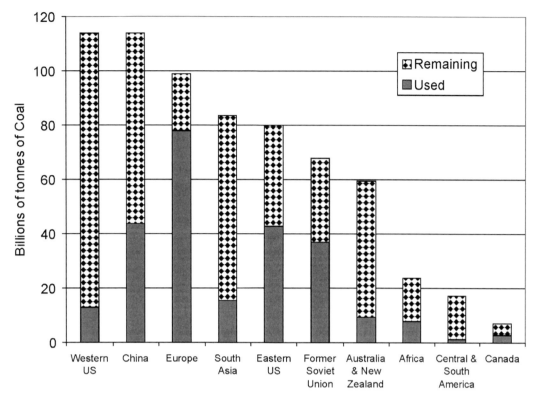

Source: Rutledge (2007)

Figure 2.29 *Amount of coal already used in various world regions, and the estimated amount of ultimately recoverable remaining coal based on extrapolation of historical data in each region in the manner shown in Figures 2.27 and 2.28*

act to increase the price of coal in the medium and long term. The constraints, summarized from Kavalov and Peteves (2007), are as follows:

- USA – productivity of US coal mines (in tonnes/worker/year) has been steadily decreasing as less accessible coal is tapped, which will tend to increase cost.
- Russia – most of the coal reserves in Russia are in the central and eastern parts of the country, far from the main demand centres, so expensive (compared to ship) long-distance transport by train is required.
- China – most of China's coal deposits are in the north and northeast of the country, while the major demand centres are in the south and southeast, so expensive long-distance transport by train is required. For this reason the southern region imports some of its coal from Indonesia, South Africa and Australia. Chinese coal has a high

sulphur and ash content, which reduces its heating value and poses pollution problems (if sulphur and ash are emitted) or disposal problems (if sulphur and ash are captured).[10] Chinese coal mines have the worst safety record in the world (6000 deaths per year by official estimates, several times more according to some unofficial estimates), a record that can be improved only through greater investment in safety procedures and hence greater cost. In addition, worker productivity (tonnes coal mined/person/year) is among the lowest in the world, so increasing productivity through the application of modern technology will act to reduce the cost of coal in China.

- India – most coal in India is of low quality (30–50 per cent ash and 4–7 per cent water), and most of it is in the northeast part of the country, while the major demand centres are in the west and southwest. Transport of unwashed coal is particularly expensive, and is hampered by the existence of three different railway gauges in India. Improving rail transport so as to handle greater coal demand would require a huge investment, which is not likely to be feasible during the foreseeable future. For these reasons, some coal is imported to western and south-western India.

- South Africa – most coal mines in South Africa are near the end of their economic life, and development of new mines is expected to be much more costly and the coal extracted of lower quality.
- Australia – Australia is the single largest exporter of coal, accounting for about 20 per cent of world trade, with more than 80 per cent of its annual extraction exported. Australia is likely to maintain this position, but its exports are small compared to Chinese coal demand.

The aforementioned US National Academy of Sciences report states that:

> almost certainly, the coals mined in the future will be lower quality because current mining practices result in higher quality coal being mined first… The consequences of relying on poorer quality coal for the future include (1) higher mining costs (e.g., the need for increased tonnage to generate an equivalent amount of energy, greater abrasion of mining equipment), (2) transportation challenges… (3) beneficiation challenges (e.g., the need to reduce ash to acceptable levels, creation of more waste), (4) pollution control challenges (e.g., capturing higher concentrations of particulates, sulphur, and trace elements, and dealing with increased waste disposal), and (5) environmental and health challenges. (p47)

Höök and Aleklett (2009) discuss how environmental constraints have limited the amount of coal in the US that is mineable. Limitations include rules against mining in or near water streams, limitations on mountaintop mine dumping, limitations on sulphur emissions, and limitations in order to protect agricultural and ranching land uses. Much of the remaining coal in the Appalachian region and in Illinois has a high sulphur content, while the coal in the western US has a low sulphur content but also a one third lower heating value than other US coals, so more needs to be mined for a given energy value.

In summary, the usable coal resource is probably nowhere near as large as thought two or three decades

ago. Official estimates of the ultimately extractable coal resource have fallen from about 10,000Gt to about 4500Gt, while official estimates of the recoverable reserve (based on present technology and prices) sum to about 900Gt globally. However, much of the coal currently classified as a reserve may not in fact be mineable. An alternative estimate based on the extrapolation of regional extraction data and subsequent aggregation indicates that the ultimately extractable resource is about 700Gt, implying that about one third of the extractable coal resource has already been used (about 285Gt have been mined to date) and that peak supply will occur well before 2050. Another estimate, based on the analysis of globally aggregated data, indicates an ultimately extractable resource of about 1100–1700Gt, with the peak in the annual global coal supply occurring between 2050 and 2070. In any case, a significant increase in the rate of use of coal is subject to a number of constraints, such that the cost of coal delivered to the customer is likely to increase significantly over the next few decades in many if not most parts of the world.

2.6 Emissions of GHGs and aerosols, radiative forcing and present global warming commitment

This book is motivated by the need to limit the peak CO_2 concentration to the lowest possible level through the rapid reduction and eventual elimination of fossil fuel CO_2 emissions. However, there is not a direct 1:1 relationship between fossil fuel energy use and CO_2 emissions. Furthermore, methane and nitrous oxides are also emitted in association with the use of fossil fuels, while tropospheric ozone is produced from emissions of hydrocarbons and nitrogen oxides produced by the burning of fossil fuels. Methane, nitrous oxide and ozone are GHGs whose collective impact is at least half that of CO_2 alone.

2.6.1 CO_2 emission factors, breakdown by fuel and country, and historical trends

Table 2.10 lists the emission of CO_2 per unit of primary energy for coal, oil and natural gas.[11] The chemical

Table 2.10 *CO_2 emission factors for combustion of various fossil fuels*

Fuel	CO_2 emission, kgC/GJ	
	LHV basis	**HHV basis**
Anthracite coal	25.0–27.3	24.1–26.7
Bituminous coal	25.0–26.1	23.7–26.7
Subbituminous coal	25.0–26.2	24.1–25.3
Lignite	25.0–26.2	24.1–27.6
Crude oil	20.0–20.9	16.9–19.9
Natural gas[a]	15.0–15.7	13.5–14.0
Methane	15.0	13.5
Ethane	16.8	15.4
Propane	17.6	16.2
Butane	18.1	16.7

Note: [a] Natural gas is 90–98 per cent methane, with most of the balance being ethane but also containing minor amounts of propane and butane.

Source: Hiete et al (2001)

Table 2.11 *Global anthropogenic CO_2 emissions and perturbation to the global carbon cycle*

Anthropogenic emissions (GtC/yr) in 2005	
Coal	3.10
Oil	3.03
Natural gas use	1.48
Flaring of natural gas	0.06
Cement chemical reaction	0.32
Total	7.99
Perturbations to the global carbon cycle (GtC/yr) averaged over the period 2000–2005	
Industrial CO_2 emissions	7.2±0.3
Land use changes	1.6±1.1[a]
Observed atmospheric CO_2 increase	4.1±0.01
Estimated oceanic sink	2.2±0.5
Terrestrial biosphere sink (residual)	2.5±2.0

Note: [a] This is the most up-to-date estimate, and pertains to the 1990s.
Source: Data on emissions from Marland et al (2008); data on carbon cycle from Denman et al (2007)

composition of coal, oil and to some extent natural gas varies regionally, so there is some variability in the emission factors. The emission factors for coal and oil are about 90 per cent and 30 per cent greater, respectively, than that of natural gas. Thus, in order to maximize that amount of fossil fuel primary energy that can be used for a given emission limit, the least carbon-emitting fuel (natural gas) should be used first, with higher carbon-emitting fuels (oil, then coal) used to the extent that additional emissions are permitted. This sequence was assumed in computing the permitted fossil fuel energy supply that is shown in Figure 1.9.

CO_2 is also produced and emitted through a number of chemical transformations, in particular, the production of cement from limestone, the reduction of iron, aluminium and copper ores, and the manufacture of glass from sand (see Chapter 6, Table 6.2). However, except for emissions from the production of cement and glass, these CO_2 emissions are already accounted for in the use of fossil fuels to drive the chemical reactions.

Table 2.11 gives the breakdown of global CO_2 emissions from the use of fossil fuels and from the chemical reaction to produce cement (the emission from the reaction to produce glass is very small). For comparison, an estimate of emissions related to changes in land use (primarily deforestation) is also given along with the estimated overall budget related to anthropogenic CO_2. Energy and industrial CO_2 emissions averaged 7.2±0.3GtC/yr over the period

2000–2005, compared to an estimated land use emission of 1.6±1.1GtC/yr during the 1990s. Approximately half of the total emission accumulated in the atmosphere, with the other half absorbed by the oceans and terrestrial biosphere (through enhanced photosynthesis) in roughly equal but uncertain proportions.

Figure 2.30 shows the variation from 1850 to 2005 in global CO_2 emissions from combustion of coal, oil and natural gas, and from production of cement. Oil accounted for the largest share of total emissions in 2005 (38.8 per cent), following closely by coal (38.0 per cent), but the use of coal is rising rapidly and may soon overtake oil. Figure 2.31 gives CO_2 emissions in 2005 for the ten largest CO_2-emitting countries in the world, while Figure 2.32 gives per capita emissions from selected countries.[12] Topping the list are the US, Australia and Canada (per capita emissions of 5.3tC/yr, 4.7tC/yr and 4.5tC/yr, respectively). Emissions from Western European countries tend to be 2–3tC/person/yr, while China and India come in at 1.16 and 0.35tC/person/yr, respectively.

2.6.2 Methane emissions

Methane is released to the atmosphere through the mining of coal (it leaks from the coal seams), through the extraction and distribution of natural gas (which is largely methane), and through the extraction of oil

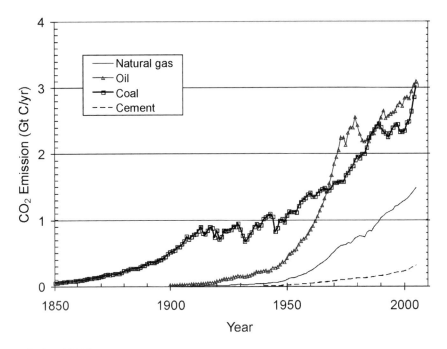

Source: Data from Marland et al (2008)

Figure 2.30 *Variation in CO_2 emissions between 1850 and 2005 from the combustion of coal, oil and natural gas, and released chemically during the production of cement*

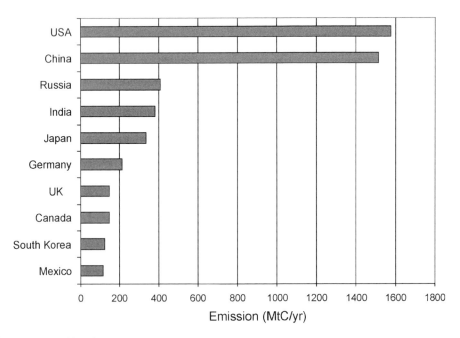

Source: Data from Marland et al (2008)

Figure 2.31 *Industrial CO_2 emissions in 2005 from the ten largest CO_2-emitting countries*

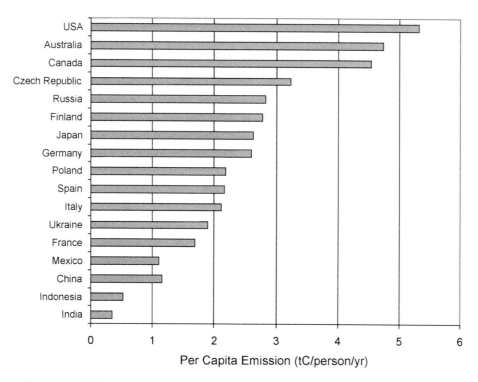

Source: Data from Marland et al (2008)

Figure 2.32 *Per capita industrial CO$_2$ emissions in 2005 from selected countries*

(which often occurs in association with natural gas). The emission from coal seams is much greater for deep seams than for shallow seams. The emission through the natural gas system depends on how well maintained the distribution system is. Table 2.12 gives emission factors per unit of primary energy. Given an efficiency in generating electricity from coal of about 33 per cent, the emission factor per unit of electrical energy will be three times greater. Although there is considerable uncertainty in the appropriate emission factor in any given region (due to differences in the nature of the coal seams or in the rate of leakage from the natural gas system), it is almost always the case that switching from electricity generated from coal to onsite use of natural gas will entail a reduction in methane emissions, particularly if this switch is accompanied by an increase in electricity generation efficiency.

Table 2.13 gives an estimate of the overall budget of atmospheric methane. The total sink strength is constrained to be 581Tg(teragram)(CH$_4$)/yr, and the observed atmospheric increase is 0.2ppbv/yr (about 0.6Tg/yr), so the total source (natural + anthropogenic)

is constrained to be 582Tg(CH$_4$)/yr. There is a large uncertainty in the estimated emissions by individual natural and human sources. Methane emissions from coal mining, the oil and gas industry, landfills and waste, ruminant animals, rice agriculture and biomass

Table 2.12 *Methane emission factors for various fossil fuels*

Fuel	CH$_4$ emission factor	
	kgC/GJ	kgCH$_4$/GJ
Coal mining		
– underground	0.35–0.37	0.47–0.49
– surface	0.08–0.09	0.11–0.12
Oil	≤ 0.025	≤ 0.032
Natural gas	0.14–0.15[a]	0.19–0.20

Note: [a] This is the emission factor for each 1 per cent leakage of natural gas, from the point of extraction to the point of use. The leakage rate is reported to be much less than 1 per cent in modern, well-maintained systems, but was as high as 10 per cent in the Soviet Union prior to its collapse.

Source: Harvey (2000)

Table 2.13 *Global anthropogenic emissions of methane and the global methane budget*

Source	Emission or sink (Tg(CH_4)/yr)
Anthropogenic and natural emissions during the 1990s	
Coal mining	32–46
Oil and gas industry	52–68
Landfills and waste	35–69
Ruminant animals	76–92
Rice agriculture	31–83
Biomass burning	40–88
Anthropogenic total	264–381
Natural emissions	168–220
Constrained total emissions	582
Sinks	
Reaction with hydroxyl (OH) in the troposphere	511
Biological oxidation in drier soils	30
Loss to the stratosphere	40
Constrained total sinks	581

Source: Denman et al (2007)

burning are all important contributors to total anthropogenic emissions, with ruminant animals likely to be the single largest source (with twice the emissions as from coal mining).

2.6.3 Other GHGs related to energy use

N_2O is another GHG emitted in association with the use of energy or energy products. Table 2.14 gives an estimate of anthropogenic emissions of N_2O during the 1990s. The single largest source is estimated to be from

Table 2.14 *Anthropogenic and total natural emissions of N_2O*

Source	Emission (Tg/yr)
Fossil fuel combustion and industrial processes	0.7 (0.2–1.8)
N fertilizer	2.8 (1.7–4.8)
Biomass and biofuel combustion	0.7 (0.2–1.0)
Human excreta	0.2 (0.1–0.3)
Rivers, estuaries and coastal zones	1.7 (0.5–2.9)
Atmospheric deposition	0.6 (0.3–0.9)
Anthropogenic total	6.7
Natural	11.0
Total	17.7 (8.5–27.7)

Note: Given are best-estimate values and the uncertainty range.
Source: Denman et al (2007)

the application of N fertilizer in agriculture, an energy-intensive product that may see increased use as part of biomass energy systems. Combustion of fossil fuels and biomass for energy are also important sources of N_2O.

Halocarbons are compounds containing carbon and one or more halogen gases – chlorine, fluorine and bromine. Those containing chlorine and fluorine are CFCs, those containing hydrogen, chlorine and fluorine are HCFCs, and those containing hydrogen and fluorine are HFCs. All three groups are GHGs, while those containing chlorine (the CFCs and HCFCs) have an ozone-depleting effect as well. CFCs were originally the refrigerant of choice in air conditioners and other cooling equipment, but their use has been phased out in order to protect the stratospheric ozone layer. They have been temporarily replaced by the HCFCs, which in turn will be replaced by the HFCs or non-halocarbon refrigerants. The halocarbons have also been used as blowing agents in various kinds of foam insulation for building exteriors, refrigerators and hot-water tanks (see Chapter 4, subsection 4.10.6).

2.6.4 Emissions of aerosols or precursors to aerosols

Humans are affecting the earth's climate not only through the emission of GHGs, but also through the emission of a variety of aerosols or of compounds that react to form aerosols. Aerosols are very small solid or liquid particles in the atmosphere that scatter and, in some cases, absorb sunlight. The principle aerosols produced by humans and their effect on climate are: sulphate aerosols (cooling), nitrate aerosols (cooling), organic carbon aerosols (cooling), sooty aerosols (warming). These aerosols are emitted or produced from compounds that are emitted in association with the use of fossil fuels and biomass for energy. The net effect of all aerosol emissions is probably a cooling effect, *offsetting up to half of the heating effect from GHGs currently in the atmosphere.* The dominant aerosol is the sulphate aerosol, produced largely from the release of sulphur oxides produced during the combustion of coal, the smelting of sulphur-containing ores, and the refining of petroleum. The sulphate aerosol is the major contributor to acid deposition and is also an air pollutant with direct effects on humans, so sulphur (S) emissions have been sharply reduced in recent decades in North America and Europe, but have increased sharply in Asia. The regional trends from 1850–2000

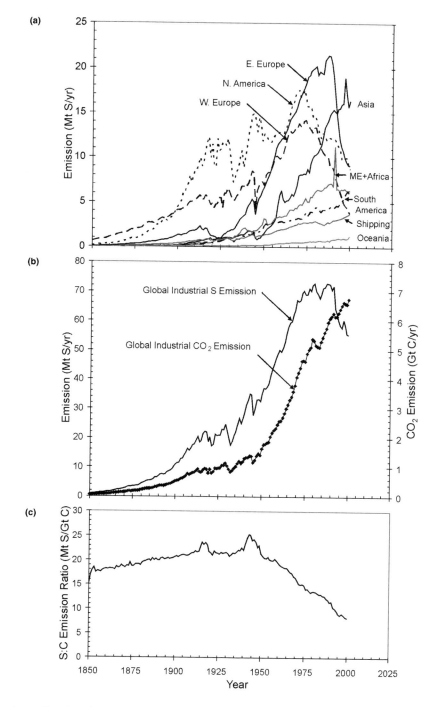

Source: S emission data are from Stern (2006) and were kindly provided in electronic form for use here, while CO$_2$ emission data are from Marland et al (2008) and are available at cdiac.ornl.gov

Figure 2.33 *Variation in (a) regional S emissions from 1850–2000, (b) global industrial S and CO$_2$ emissions, and (c) the ratio of S:C emissions from 1850–2000*

are shown in Figure 2.33a, while Figure 2.33b compares the global emission trends of S and C (as CO_2) and Figure 2.33c gives the trend in the ratio of S:C emissions. The S:C emission ratio has fallen by a factor of three since 1950 and can be expected to continue to fall. This in turn will contribute to accelerated warming (particularly in currently polluted regions) in the near term, and may be a factor in the rapid global mean warming during the past three decades.

2.6.5 Global warming potential

The impact on climate of the emission of a given mass of a GHG depends on the effectiveness of the gas in trapping heat, on a molecule-by-molecule basis, and on the average lifespan of molecules of that gas in the atmosphere. If we consider the sudden emission of a pulse of gas into the atmosphere, then for all GHGs except CO_2, a constant fraction of the gas remaining after a given time is removed during a given time period. Thus, the fraction of the initial pulse remaining in the atmosphere exponentially approaches zero at a fixed percentage rate of decrease. The removal of a pulse of

CO_2, however, involves multiple processes spanning a wide range of timescales: air–sea gaseous exchange at a timescale of one year, exchange with the terrestrial biosphere on a timescale of years to decades, mixing into the deep ocean on a timescale of decades to centuries, reaction with marine carbonate sediments on a timescale of thousands of years, and accelerated rock weathering on land removing CO_2 on a timescale of 100,000 years.

Figure 2.34 illustrates the decrease over time in the amount of a pulse of methane, N_2O and CO_2 remaining in the atmosphere. Methane has an average lifespan in the atmosphere of 12 years and N_2O about 114 years, but about 20 per cent of the emitted CO_2 remains in the atmosphere essentially 'forever' from a human perspective – that is, it still remains 500 years after emission. As the amount of gas in the atmosphere decreases after a pulse emission, the heat trapping decreases, but the amount of gas in the atmosphere and the heat trapping decrease at different rates for different gases, so the ratio of heat trapping by two different emitted gases will change over time. The integral (or summation) of this heat trapping over some arbitrary time horizon can be computed and compared with that for CO_2; the ratio of the two forms

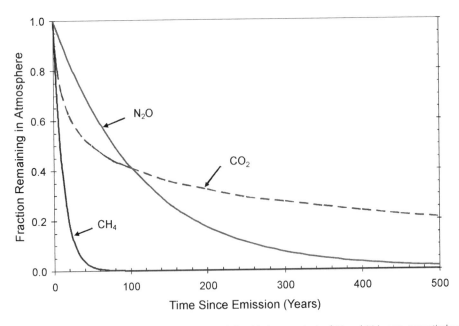

Note: For methane and N_2O, the amount remaining decreases exponentially with time constants of 12 and 114 years, respectively, while the amount of CO_2 remaining in the atmosphere was computed using the carbon cycle model of Harvey (2001). Potential climate–carbon cycle feedbacks are not included for this calculation.

Figure 2.34 *Decrease over time in the amount of methane, N_2O and CO_2 remaining in the atmosphere following the sudden emission of a pulse of each gas into the atmosphere at time zero*

Table 2.15 *GWP of various GHGs or groups of GHGs on a mass basis on 20-year and 100-year time horizons*

Gas	20-year GWP	100-year GWP
CO_2	1	1
CH_4	72	25
N_2O	289	298
CFCs	5000–11,000	7000–14,000
HCFCs	429–5500	77–2310
HFCs	437–12,000	124–14,800

Source: Forster et al (2007)

an index called the *global warming potential* (GWP). This is a rough but adequate measure of the relative contribution of equal emissions (in terms of mass) of different gases to global warming (see Harvey, 1993, for a critique of the GWP index).

Table 2.15 gives GWPs for different gases and groups of gases when the heat trapping is integrated over 20-year and 100-year time horizons. GWP for methane is 72 over a 20-year period, meaning that avoiding the emission of 1kg of methane is considered to be 72 times as effective in limiting global warming over a 20-year time period as avoiding the emission of 1kg of CO_2.[13] However, the amount of methane remaining in the atmosphere after it has been emitted decreases more rapidly than the amount of CO_2, so over a 100-year time horizon, limiting methane emissions is considered to be only 25 times as effective in limiting CO_2 emissions. These GWPs include indirect effects of methane through increased tropospheric ozone and stratospheric water vapour. The GWP of N_2O over a 100-year time horizon is estimated to be 298. Methane and N_2O are important considerations in the climatic effect of several different agricultural activities (as discussed in Chapter 7). Note that the CFCs (which are being phased out) have 100-year GWPs of 7000–14,000, but some of their HCFC and HFC replacements also have very large GWPs.

2.6.6 Radiative forcing, equivalent CO_2 concentration and present warming commitment

The amount of a given GHG in the atmosphere at a given time is a reflection of emission up to that point in time and the rate of removal of that gas. The observed concentration combined with the heat-trapping ability of the gas (as determined from laboratory measurements, and accounting for overlap

between the added GHG and pre-existing gases in the atmosphere) gives the net trapping of infrared radiation, which is the *radiative forcing* of the GHG.[14] As previously noted, humans have also altered the radiative balance of the climate system through the addition aerosols to the atmosphere. As well, changes in land use (primarily deforestation) have increased the albedo (reflectivity) of the earth's land surface, which has had a cooling effect on climate, while deposition of sooty aerosols on snow in polar regions has reduced the albedo in polar regions, which has had a warming effect that would have been amplified through enhanced seasonal melting of snow cover.

Table 2.16 gives the radiative forcings in 2005 relative to 1750 due to the increase in GHG and aerosol concentrations and to changes in surface albedo as given in the Fourth Assessment Report of the IPCC. The best estimate of the total GHG radiative forcing is $3.01W/m^2$ (with an uncertainty range of $2.49W/m^2$ to $3.73W/m^2$). By comparison, the radiative forcing for a CO_2 doubling is $3.75W/m^2$. Thus, current GHG concentrations are equivalent to an increase in CO_2 concentration of about 60–100 per cent.[15] At present, a large fraction of the heating effect due to increasing GHGs could (in the global mean) be offset by the combined effect of increases in aerosols and changes in the earth's surface albedo. The best estimate of the combined effect of these changes is a radiative forcing of $-1.3W/m^2$ (offsetting 43 per cent of the GHG heating), with an uncertainty range of $-3.1W/m^2$ to $-0.2W/m^2$. Thus, the best estimate of the net effect of all human perturbations to the radiative balance of the climate system is a global mean radiative forcing of $1.71W/m^2$, which corresponds to a 37 per cent increase in CO_2 concentration. However, aerosols have a disproportionately large effect in suppressing precipitation along with cooling the climate, so they do not represent a simple offset of GHG radiative forcing.[16]

The lifespan of aerosol particles in the atmosphere is in the order of a few days; thus, maintenance of the existing concentrations requires a continuous emission source, and if emissions of aerosols or aerosol precursors were to cease, the aerosol loading in the atmosphere would almost instantly revert to natural conditions. As noted, aerosols are largely produced in association with the use of fossil fuels, so if fossil fuel use were to be eliminated, so would the aerosol offset. This in turn would cause a temporary acceleration in the rate of warming.

Table 2.16 *Global mean radiative forcing in 2005 relative to 1750*

Source of forcing	Radiative forcing (W/m²)	
	Best estimate and uncertainty	Resulting range
GHGs		
CO_2	1.66±0.17	
CH_4	0.48±0.05	
N_2O	0.16±0.02	
Halocarbons	0.34±0.03	
Stratospheric O_3 loss	−0.05±0.10	
Tropospheric O_3	0.35 [−0.1, 0.3]	
Stratospheric H_2O from, CH_4	0.07±0.05	
Total GHG forcing	**3.01 [−0.52, 0.72]**	**2.49 to 3.73**
Direct aerosol effects		
Sulphate	−0.4±0.20	
Fossil fuel organic carbon	−0.05±0.05	
Fossil fuel black carbon (soot)	+0.20±0.15	
Biomass burning	+0.03±0.12	
Nitrate	−0.10±0.10	
Mineral dust	−0.10±0.20	
Total	**−0.5±0.40**	**−0.9 to −0.1**
Indirect aerosol effect and changes in surface albedo		
Indirect aerosol effects	−0.70 [−1.1, 0.4]	−1.8 to −0.3
Surface albedo due to land use changes	−0.20±0.20	
Surface albedo due to deposition of black carbon on snow	+0.10±0.10	
Total non-GHG forcing	**−1.30 [−1.8, 1.1]**	**−3.1 to −0.2**

Source: Forster et al (2007)

Tropospheric ozone is a short-lived gas whose buildup in the atmosphere also requires a continuous emission of various precursors, all of which are associated with the combustion of fossil fuels or biomass (some of which is for energy purposes, some as part of deforestation activities). Elimination of fossil fuels and cleaner combustion of biomass for energy would largely eliminate the radiative forcing due to tropospheric ozone. Finally, methane has a relatively short lifespan in the atmosphere (about 12 years at present) and about half of anthropogenic methane emissions are related to energy use and biomass burning. Reduction in these emissions would be accompanied by a parallel decrease in excess methane in the atmosphere, with two thirds of the decrease occurring within 12 years of the emission reduction.

With these considerations in mind, I present in Table 2.17 the radiative forcing commitment that would remain for the next few decades if the use of fossil fuels and deforestation were to instantly cease. This commitment involves the current radiative forcing due to CO_2 and N_2O, half of that due to CH_4, about a third of that due to the halocarbons (a few of which have

lifespans of 100 years or more), and that due to changes in surface albedo due to land use changes and mineral dust loading (which cannot be quickly nor largely reversed). The resulting best estimate of the radiative forcing commitment is 1.98W/m² (corresponding to a 44 per cent increase in CO_2 concentration), with an uncertainty range of 1.34W/m² to 2.62W/m² (corresponding to a CO_2 increase of 28–62 per cent). Given a likely range for the climate sensitivity (the long-term global mean warming associated with a CO_2 doubling) of 1.5–4.5°C (see Chapter 1, subsection 1.2.1), and given that the climate response is expected to vary close to linearly with the radiative forcing for the range of forcings considered here, this radiative forcing commitment corresponds to a warming commitment (relative to pre-industrial) of 0.6°C (i.e. about equal to that observed so far) to 3.1°C. The real commitment of course is larger because the CO_2 concentration will increase further before it can be stabilized, due to the long time required to change the world's energy system. As discussed in Chapter 1 (section 1.4), there is a high risk of serious impacts beginning with a global mean warming of 1–2°C.

Table 2.17 *Present medium-term (many decades) radiative forcing commitment*

Source of forcing	Radiative forcing (W/m²)	
	Best estimate and uncertainty	Resulting range
CO_2	1.66±0.17	
CH_4	0.24±0.03	
N_2O	0.16±0.02	
Halocarbons	0.12±0.02	
Total GHG forcing	**2.28±0.24**	**2.04 to 2.52**
Surface albedo due to land use changes	20.20±0.20	
Mineral dust	20.10±0.20	
Overall total	**1.98±0.64**	**1.34 to 2.62**
Equivalent CO_2 increase	44%	28 to 62%

2.7 Non-climatic environmental costs of fossil fuel use

Many of the efficiency measures that reduce primary energy requirements are economically attractive in their own right, in that the fuel cost savings at present market prices will pay for the extra upfront cost of more efficient practice (should the more efficient practice be more costly in the first place). In other instances, efficiency measures – and carbon-free or less carbon-intensive energy sources – are more expensive based on the market prices of the energy sources. However, fossil fuel energy sources impose significant and real costs on society that are not reflected in their market price. These costs include the environmental damage and adverse health impacts caused by pollution from fossil fuel combustion; disability costs and deaths of coal miners; significant landscape destruction associated with surface coal mining and tar sands developments (see online Figures S2.1 to S2.4); significant fragmentation of wildlife habitats associated with exploring and drilling for natural gas (see online Figures S2.5 and S2.6); and military expenditures to maintain secure oil supplies. Although underground (in situ) liquefaction of tar sand deposits entails less direct landscape destruction than surface strip mining, in situ liquefaction indirectly disturbs a larger area due to the much greater natural gas requirements and the landscape disturbance associated with providing the natural gas (Jordaan et al, 2009). Unless these costs are included in the price of fossil fuels, fossil fuels will sometimes be chosen over renewable energy sources even when the latter are less costly, consumers will

underinvest in energy-efficiency measures, and manufacturers will have less incentive to develop more efficient equipment.

Societal costs that are not included in the market price of energy are referred to as *externalities*. Estimation of the magnitude of fossil fuel externalities is fraught with enormous uncertainty and is, to some extent, subjective. This is because the major externalities (excluding climatic change) are health impacts on humans, including premature death. Various techniques have been used to try to quantify these impacts in monetary terms. Mortality-related impacts are often quantified based on the declared willingness of individuals to pay extra money in order to reduce the risk of death by a given amount. Wealthy individuals tend to be more willing than poor people to pay in order to reduce the risk of death, so this approach has the perverse effect of valuing human life more highly in developed countries than in developing countries. Health care costs and the cost of reduced agricultural and forest productivity can be more easily quantified.

2.7.1 Pollutant emission rates using fossil fuels

The social costs of fossil fuel use depend on the rates of emission of various pollutants times the social cost per unit of emission for each pollutant. Emission rates depend on the end-use application, the technology used, and the effectiveness of emission controls. Figures 2.35 and 2.36 show the historical variation in emission standards for gasoline and diesel-fuel passenger cars, respectively, in the US, Europe and Japan for the three regulated pollutants:

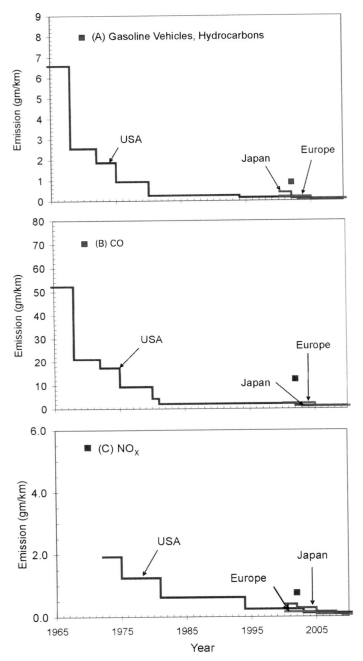

Source: www.dieselnet.com/standards for Europe. Solid squares are the estimated average emissions from all light-duty vehicles on the road in the US in 1970 and 2002, as given in *US National Emissions Inventory,* available from www.epa.gov/ttn/chief/trends

Figure 2.35 *Historical variation of gasoline automobile emission standards for HCs, CO and NO$_x$ in the US, Europe and Japan*

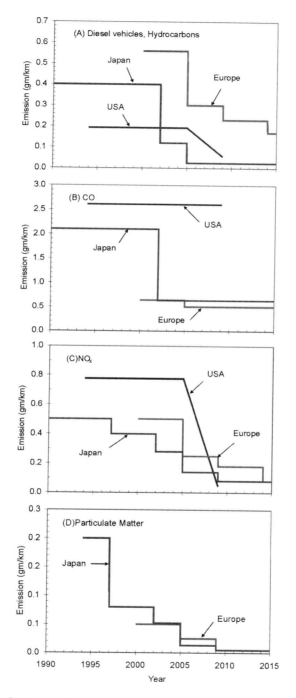

Source: www.dieselnet.com/standards for Europe and Japan

Figure 2.36 *Historical variation of diesel automobile emission standards for HCs, CO, NO$_x$ and particulate matter in the US, Europe and Japan*

hydrocarbons (HCs) (usually based on the amount of particles of 10μm diameter or less, referred to as PM_{10}), carbon monoxide (CO) and nitrogen oxides (NO_x). Also shown as solid squares, for gasoline vehicles, are the estimated average emissions from all light-duty vehicles on the road in the US in 1970 and 2002. For gasoline vehicles there is little difference in permitted emissions now between the three regions, although the US permits greater emissions for diesel vehicles than Europe or Japan (but diesel vehicles make up a very small fraction of the passenger vehicle market in the US, unlike in Europe and Japan). Compared to unregulated emissions (prevalent in the 1960s), emissions of HCs, CO and NO_x from gasoline vehicles are factors of 85, 50 and 20 lower now in the US.

Until recently, absolute emissions from diesel vehicles were substantially higher than for gasoline vehicles (this is still true in the US). However, in some cases absolute emissions are lower for diesel, or will soon be lower. This is shown in Figure 2.37, which gives the current ratio of diesel to gasoline emissions in the US, Europe and Japan, and the ratio of best diesel over best gasoline emissions. Diesel emissions have been or will be brought down to gasoline emissions by equipping the diesel vehicles with advanced filters to largely remove HCs and with two-way catalytic converters to reduce CO and NO_x emissions. Gasoline vehicles rely on three-way catalytic converters. Catalytic converters in turn require platinum to operate, with the required amount

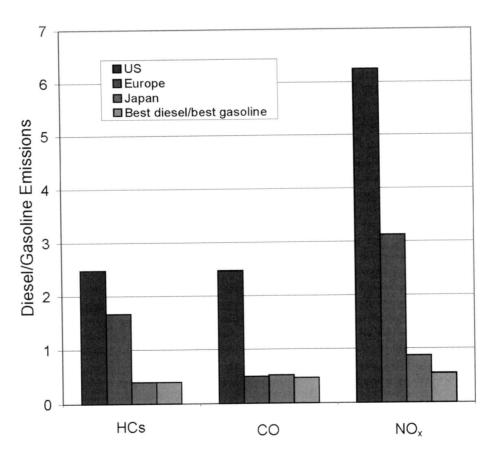

Source: Derived from data in Figures 2.35 and 2.36

Figure 2.37 *Ratio of permitted diesel-automobile emissions to permitted gasoline-automobile emissions in 2005 in the US, Europe and Japan, plus the ratio of the best planned emission standard in any of the three regions for diesel to the best planned emission standard for gasoline*

greater for diesel vehicles (Yang, 2009). As discussed in Chapter 5 (subsection 5.4.8), Pt is a scarce resource found in rare geological settings, and as discussed by Yang (2009), it is a legitimate investment vehicle and so could be subject to large, speculatively driven fluctuations in price in the future.

The positive trends shown in Figures 2.35–2.36 need to be tempered by the fact that (i) emissions under real-world conditions can be several times the emissions obtained under test conditions for new vehicles (and this gap was expected by Williams et al (2000) to increase); and (ii) emissions of ultrafine particulate matter (PM) (less than 2.5μm diameter or less, referred to as $PM_{2.5}$) and of trace toxic chemicals (such as benzene, formaldehyde and toluene) are poorly addressed by current emission control programmes.

Attainment of the very strict emission limits being adopted in the US, Europe and Japan requires a very low S content in automobile fuels (no more than about 10ppm) so that the pollution-control technologies work properly. As shown in Table 2.18, the permitted S content in gasoline ranges from 10ppm in Europe and Japan to over 250ppm in many parts of the world. The EU-27 (defined in Appendix C), US and other jurisdictions are also moving toward 'ultra-low' emission standards for diesel fuel (10–50ppm), but in many parts of the word the permitted S level in diesel fuel is 500ppm or more. Thus, widespread attainment of the emission limits set for the US, Europe and Japan requires significant improvements in the quality of

Table 2.18 *Maximum permitted S content in gasoline in different countries of the world as of March 2009*

Permitted S content	Applicable countries
10ppm	EU-27, Japan, South Korea
30ppm	Canada, Chile, US
50–80ppm	Morocco, New Zealand, Turkey
150ppm	Australia, Ukraine
250–600ppm	Argentina, China, Egypt, India, Indonesia, Libya, Mexico, Papua New Guinea, Philippines, South Africa, Tanzania, Venezuela
1000–2500ppm	Most other countries

Note: The permitted S limit given for at the International Fuel Quality Centre website for Canada and the US is incorrectly stated to be 50–80ppm. There may be errors for other countries too.
Source: International Fuel Quality Centre (www.ifqc.org/NM_Top5.aspx)

transportation fuels as well as improvements in vehicle technology. However, reducing the S content of fuels increases the energy required for refining, while the more stringent pollution controls on diesel trucks (at least) increase fuel consumption by 4 per cent according to Duleep (2007) and by 7–10 per cent according to Tokimatsu (2007).

Figure 2.38 gives the emissions of sulphur dioxide (SO_2), NO_x and PM permitted for new coal powerplants in the US under successively more stringent regulations, as well as for advanced powerplants and the average of existing powerplants in the US. Permitted emissions for new powerplants have decreased by factors of 50, 7 and 20 for SO_2, NO_x and PM, respectively, compared to unregulated emissions of the 1960s. However, emissions averaged over all operating powerplants have fallen by factors of only 3.6, 1.2 and 5.5, respectively, while total electricity generation from coal has increased by about 80 per cent since 1970. As with automobile emissions, there is a wide range of emissions of toxic heavy metals and other toxic compounds that are not regulated. As with tighter automobile and truck emission standards, the reduced emissions from coal powerplants comes with a price in terms of reduced efficiency – increasing fuel use by about 1.5 per cent for the most stringent controls according to Graus and Worrell (2007).

Table 2.19 presents data on emissions of SO_2, NO_x, PM, CO and mercury (Hg) associated with coal powerplants and various systems for heating and cogeneration (simultaneous heating and generation of electricity). According to these data, advanced systems for the use of coal to generate electricity achieve emissions comparable to or in some cases better than using natural gas.

2.7.2 Estimated cost of air pollution and of other externalities

The estimated external costs of gasoline and diesel automobiles in Paris (population density 7500/km²), and for rural population densities between Paris and Lyon (population density 400/km²), are given in Table 2.20. Costs in other cities would be larger or smaller, depending on the local pollutant emission factors and their population density (population densities in other groups of cities are given in Table 5.5 of Chapter 5). The externality for gasoline-powered automobiles is over 50 cents/litre, while the externality for diesel vehicles is over $5/litre! These costs are for vehicle emissions pertaining

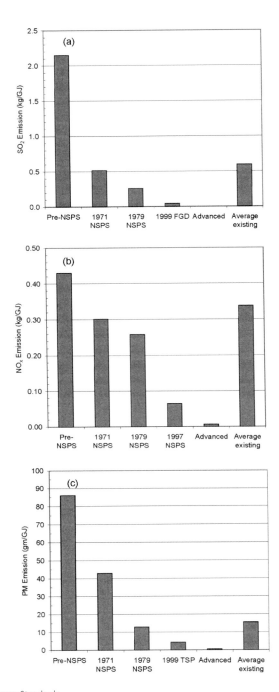

Note: NSPS = New Source Performance Standards
Source: Data from Rubin et al (2001) for regulated emissions, Table 2.19 for advanced plants, and Table 2.22 for the existing average

Figure 2.38 *Pollutant emissions permitted from new coal-fired powerplants in the US under successively more stringent regulations, for advanced plants, and for the average of existing plants*

Table 2.19 *Pollutant emission factors for various technologies using fossil fuels for power generation or heating*

	Emissions as grams (gm)/kWh					Emissions as gm/GJ of fuel				
	SO_2	NO_x	PM	Hg	CO	SO_2	NO_x	PM	Hg	CO
	Electricity generation									
Pulverized coal powerplant										
Uncontrolled	22.716	2.735	35.2336	3.5E-05		2208.54	265.93	3425.49	0.00343	
Controlled	0.556	0.051	0.1344	2.6E-06	0.1[a]	54.09	4.96	13.07	0.00025	9.7
Controlled + CCS	0.004	0.065	0.0834	3.4E-06		0.36	6.31	8.11	0.00033	
IGCC powerplant										
Baseline	0.556	0.088	0.0046	4.1E-05		54.09	8.56	0.45	0.00397	
Controlled	0.603	0.051	0.0046	3E-06		58.59	4.96	0.45	0.00029	
Controlled + CCS	0.695	0.065	0.0056	2.4E-06		67.61	6.31	0.54	0.00023	
Natural gas turbine										
Uncontrolled	0.0001	0.994	0.0190	7.9E-07	0.236	376S[b]	138	2.64	0.00011	32.8
Water-steam injection		0.374			0.027[c]		51.9			12
Lean premix		0.285					39.6			6
Advanced		0.068					1.4			
Microturbines										
30kW		0.23	< 0.08		0.63					
100kW		0.33	< 0.08		0.20					
Reciprocating engines										
100kW		20.9	1.0		16.8					
1MW		1.4	1.4		2.8					
5MW		0.7	0.7		3.4					
	Space heating									
Large boilers (>30MW$_{th}$)										
Uncontrolled	0.0001	0.577	0.023	7.9E-07	0.255	376S[d]	80.1	3.2	0.00011S[d]	35.4
Low-NO$_x$ burner		0.425					59			35.4
FGR		0.304					42.2			35.4
Small boilers (<30MW$_{th}$)										
Uncontrolled	0.0001	0.304	0.0230	7.9E-07	0.255	376S[d]	42.2	3.2	0.00011S[d]	35.4
Low-NO$_x$ burner		0.153					21.2			35.4
Low-NO$_x$ burner + FGR		0.097					13.5			35.4
Residential furnaces (<90kW$_{th}$)	0.0001	0.285	0.0230	7.9E-07	0.122	376S[d]	39.6	3.2	0.00011S[d]	16.9

Note: CSS = carbon capture and storage; FGR = flue-gas recirculation. Coal emissions are given per kWh in the original source and were converted to gm/GJ assuming a coal powerplant efficiency of 35 per cent, while emissions from natural gas turbines are given per GJ of fuel in the original source and were converted to gm/kWh here assuming an efficiency of 50 per cent (which would apply if the gas turbine is combined with a steam turbine in a combined cycle powerplant). [a] Spath et al (1999). [b] S is the percentage of sulphur, typically about 0.005 per cent, giving an SO_2 emission factor of less than 2gm/GJ. [c] Base case of Spath and Mann (2000). [d] Assumed to be the same as for uncontrolled natural gas turbines.
Source: For pulverized coal and IGCC powerplants, Bergerson and Lave (2007), based on the Integrated Environmental Control Model developed at Carnegie Mellon University (see www.iecm-online.com); for natural gas turbines, boilers and furnaces, US EPA (1995) except where indicated otherwise; for microturbines and reciprocating engines, Goldstein et al (2003)

to the mid-1990s. Between 1993 and 2002, emission standards for new vehicles in the US dropped by factors of two, three and five for CO, HCs and NO_x, respectively (see Figure 2.35), although, as noted above, the gap between regulated and actual emissions is expected to grow, such that the emissions and external costs will fall by much less.

The non-climatic external costs of electricity generation using different fossil fuel technologies have been estimated for the European Commission's ExternE programme, originally only for Europe (Rabl and Spadaro, 2000) and later for parts of China (Hirschberg et al, 2004; Zhang et al, 2007). Table 2.21 compares the estimated cost per tonne of various pollutants in the two regions. The damage costs are largely due to human health impacts, and so pertain to the population density found in the regions under consideration. Estimated external costs in China are three to seven times less than in Europe, quite the opposite of what one would expect if the impacts were valued equally. Table 2.22 gives the emission rates for different kinds of powerplants; these multiplied by the costs given in Table 2.21 give the costs

Table 2.20 *Median estimate of the costs of pollution from automobiles*

	External costs ($ per litre of fuel)		
	NO$_x$	PM	Total
Gasoline			
– Urban	0.128	0.408	0.58
– Rural	0.224	0.029	0.30
Diesel			
– Urban	0.172	5.000	5.20
– Rural	0.212	0.360	0.64

Note: The gasoline vehicle is assumed to be equipped with a catalytic converter. The urban cost estimate is for a population density of 7500/km² (as found in Paris), while the rural cost estimate is for an average population density of 400/km² (as found between Paris and Lyon). Total costs include the impact of CO, volatile organic compounds, SO_2 and benzo-a-pyrene (BaP), as well as NO$_x$ and PM.
Source: Williams et al (2000)

Table 2.21 *Estimated cost per tonne of various pollutants when emitted from electric powerplants in Europe and in Shandong province of China*

Pollutant	External cost ($/t)	
	Europe	Shangdong, China
SO_2	10,200	3680
NO$_x$	16,000	2438
PM$_{10}$	15,400	2625

Source: Rabl and Spadaro (2000) for Europe (assuming 1€ = 1$) and Hirschberg et al (2004) for Shandong province

per kWh of electricity generated, which are also given in Table 2.22 for Europe. For a typical coal-fired powerplant, the median estimate of the external cost is 12.0 cents/kWh – about *three to four times* the wholesale market price of coal-fired electricity (i.e. the true cost is about four to five times the market price). For new coal plants with the best technology available today, the external cost is 2.1 cents/kWh. For yet-to-be developed coal IGCC powerplants, the external cost in Europe is 0.2 cents/kWh, while for natural gas combined cycle powerplants (NGCC), the cost is 0.15 cents/kWh.

Yushi et al (2008) have attempted to quantify in economic terms the cost of all externalities associated with the use of coal in China. These externalities include air pollution, mining deaths (about 6000[17] in 2005, or five deaths per Mt of mined coal), disabilities due to mining accidents (about 18 per Mt of mined coal), deaths due to pneumocomiosis (2500 per year), additional cases of pneumocomiosis (5000 per year), damage to water systems and ecological degradation. The authors estimate that the annual costs are equivalent to 7 per cent of China's annual GDP, although as noted above, any quantification of these costs is partly subjective.

2.7.3 Implications for the cost of reducing CO_2 emissions

The use of fossil fuels entails costs paid by society that are not included in the price of fossil fuels. Reduction in the use of fossil fuels in order to reduce GHG emissions will therefore produce cost savings that can be credited against the cost of energy efficiency measures or renewable energy sources. However, as the pollutant

Table 2.22 *Median estimate of the external costs in Europe (averaged over the entire domain extending from Sicily to Scotland and from Portugal to Poland) of electricity generation from fossil fuels*

Type of powerplant	Efficiency (%)	Emission rate (gm/kWh)			External costs (cents/kWh)			
		SO_2	NO$_x$	PM$_{10}$	SO_2	NO$_x$	PM$_{10}$	Total
Average US coal plant	33	6.1	3.47	0.16	6.2	5.6	0.3	12.0
New coal plant	35.5	0.46	0.87	0.15	0.5	1.4	0.2	2.1
Coal IGCC	43.8	0.075	0.082	0.0025	0.08	0.13	0.00	0.21
NGCC	54.1	0.000	0.092	0.0000	0.00	0.15	0.00	0.15

Note: It is assumed that 1€ = 1 US$. Costs of CO, VOCs, As, Cd, Cr, Ni and dioxins have also been estimated by Rabl and Spadaro (2000) but, due to their low emission factors, do not affect the total costs given here. The costs given below pertain to an average population density of 80 people/km². Costs are calculated using emission factors from Williams et al (2000) and the median cost estimates of Rabl and Spadaro (2000): €10.2/kg for SO_2, €16.0/kg for NO$_x$, and €15.4/kg for PM$_{10}$
Source: Powerplant efficiencies and emission factors are from Williams et al (2000); median cost estimates are from Rabl and Spadaro (2000)

emissions of fossil fuels decrease, the external costs of fossil fuels and the associated cost savings from reduced fossil fuel use will be smaller. However, the reduced emissions (especially with regard to coal for electricity) are produced through advanced technologies that are often also more efficient, or are produced through the use of natural gas in place of oil or coal. If the advanced technologies or the use of natural gas is more costly, the avoided societal costs through reduced pollutant emissions can be credited against the cost of the advanced technologies or natural gas, thereby reducing the cost assigned to the reduction in CO_2 emissions. A number of studies indicate that the annual cost savings due to reduced pollution as CO_2 emissions are reduced is a large fraction of or can even exceed the annualized cost (defined in Appendix D) of the investments needed to achieve the CO_2 emission reduction (Aunan et al, 2004; van Vuuren et al, 2006).

2.8 A glimpse of technological changes over the past 100 years

This book and the accompanying Volume 2 discuss what is possible in terms of increasing the efficiency with which energy is used and in the large-scale deployment of entirely new energy systems over the next 100 years. Since the potential changes may seem radical to some, it is only fitting to look at some of the remarkable changes that have occurred in our use of energy over the *preceding* 100 years. In this respect, the natural complement to this book is Vaclav Smil's *Energy in World History* (Smil, 1994a), which covers not only the preceding 100 years but energy use from the earliest agricultural societies to the late 20th century.

Figure 2.39 illustrates energy efficiency-related trends in four representative areas: the efficiency of electricity powerplants in England from 1900–1990,

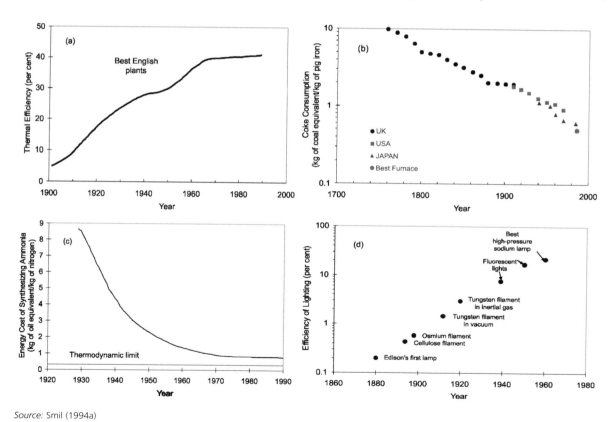

Source: Smil (1994a)

Figure 2.39 *(a) Variation in thermal efficiency of the best English electric powerplants from 1900 to 1990; (b) variation in the consumption of coal in the production of pig iron from 1760 to 1990; (c) variation in the energy cost of synthesizing ammonia fertilizer from 1930 to 1990; (d) variation between 1880 and 1970 in the efficiency of producing light from electricity*

the consumption of coal in the production of pig iron from 1760–1990, the energy intensity of ammonia fertilizer from 1930–1990, and the efficiency of lighting from 1880–1970. Dramatic reductions in the amount of energy needed for a given product have occurred over the last one or two centuries, and it may appear that we are close to the practical limits of what can be achieved. However, even where efficiency improvements have reached an apparent plateau (e.g. electricity generation, Figure 2.39a) or are approaching inherent thermodynamic limits (e.g. nitrogen production, Figure 2.39c), there is often scope for further dramatic reductions in the energy needed for a given energy service. This is possible by using energy-intensive products (such as fertilizer) more efficiently (so that less needs to be produced), by reconfiguring entire energy-using systems (such as transportation systems in cities, or ventilation systems in buildings), or by developing entirely new processes (as in manufacturing, lighting or the generation of electricity) subject to different thermodynamic constraints. Thus, as interesting as past trends and improvements may be, they often provide little guidance as to the scope for further energy savings. This is because the most important opportunities are not through small, incremental improvements in existing systems and methods, but through radical and revolutionary change.

Notes

1   Adding to the potential confusion, energy use data in the annual *Energy Statistics* reports and Table 9 (Fuel use for electricity and heat production) of the annual *Electricity Information* reports of the IEA are in terms of HHV, while the data in the annual *Energy Balances* published by the same agency are in terms of LHV.

2   In preparing Tables 2.1–2.3, the primary energy equivalent of nuclear, hydro, geothermal, solar and wind electricity was assumed to be equal to the amount of fossil fuel primary energy that would be required to produce the same amount of electricity, using the average primary-to-secondary conversion efficiencies for OECD and non-OECD countries, as computed from the data sources used to prepare Tables 2.1 and 2.2. These average efficiencies are 0.421 and 0.341, respectively. Thus, a unit of non-fossil electricity is taken to be equivalent to 2.38 units of primary energy in OECD countries and 2.94 units in non-OECD countries. This differs from IEA (2007a, 2007b, 2007c, 2007d), where nuclear and hydro electricity are converted to primary energy equivalents using conversion factors of 3.0 and 1.0, respectively.

3   1.172 billion out of 6.477 billion in 2005.

4   All of the energy shares given for OECD and non-OECD countries are averages for these two groups; energy shares for individual countries within each group could differ significantly from the group average.

5   The high Canadian electricity use is due in part to the location of energy-intensive export industries in Canada using inexpensive hydropower.

6   It is customary to refer to the extraction of oil and gas as 'production', a completely inappropriate term as it is nature that 'produced' the oil and gas over periods of millions of years, while all humans are doing is extracting and using it.

7   This projection assumes continued 2 per cent/year growth in global demand, but with the economic slowdown that began in 2008, the exhaustion of the remaining resource and the corresponding time of peak supply will be delayed. Geopolitical instability and the extent of investment in new oil supplies are other factors that can alter the timing of the peak.

8   In its 2000 *Outlook*, the IEA was projecting oil prices of $16.5/barrel in 2010 and $22.5/barrel in 2020 (in 1990$), and in its 2004 *Outlook* it was forecasting prices of $22/barrel in 2010, $26/barrel in 2020 and $29/barrel in 2030 (in 2000$), but it also considered a 'high' price scenario whereby the average price from 2004–2030 is $35/barrel. In its 2008 *Outlook*, the IEA projects prices of $100–122/barrel between 2010–2030 (in 2007$).

9   In this case, the term 'production of oil' is appropriate, because oil is not initially in the ground but is produced through an energy-intensive process.

10 One quarter of Chinese coal has more than 1 per cent sulphur and half has more than 20 per cent ash according to Zhao and Gallagher (2007).

11 All CO_2 emission factors given here are given in terms of the mass of C emitted rather than the mass of the entire CO_2 molecule. This is consistent with the convention in the carbon-cycle scientific community, which is to track the mass of C as it moves from one carbon-cycle reservoir to another. This leads to simple and transparent accounting of C fluxes.

12 Excluded from Figure 2.30 are countries with small populations and large exports of oil or natural gas, as the associated emissions are more properly attributed to the consuming nations.

13 In applying these GWPs, masses of CO_2 and CH_4 emitted rather than the mass of C alone should be used.

14 More generally, the radiative forcing is the change in the global mean net radiation at the boundary between the troposphere and stratosphere, which can be due to a change in the emission of infrared radiation to space or in the absorption of solar radiation by the troposphere or the earth's surface.

15   This is derived based on the fact that the radiative forcing due to increasing CO_2 varies with the natural logarithm of CO_2 concentration, and that the global mean temperature response to different combinations of GHG increases is closely proportional to the total radiative forcing for forcings of the magnitude experienced so far.

16   With present GHG heating as small as $2.5W/m^2$ and aerosol and land albedo cooling as large as $3.1W/m^2$, it is possible that the aerosol radiative forcing entirely offsets the GHG increases in terms of global mean radiative forcing. However, this outcome can be ruled out because (1) in that case there would have been no significant warming during the 20th century, and (2) the aerosol cooling effects are concentrated over northern hemisphere land areas, while the GHG heating effects are relatively uniform globally, so there would have been much larger regional variations in the warming than observed (and even some regions of cooling) if the aerosol forcing were large enough to entirely offset GHG heating in the global mean (Harvey and Kaufmann, 2002).

17   This is the official tally of annual coal mining deaths in China. Some observers believe that the true death toll is much higher.

3

Generation of Electricity from Fossil Fuels

Fossil fuels such as coal, oil and natural gas (primary energy) are converted into electricity (a form of secondary energy) with substantial losses during the conversion process. Reducing these losses, thereby increasing the overall efficiency of electricity generation, is the first opportunity in the chain of energy use for reducing the primary energy intensity of the global economy.

An overview of global electricity production and use was presented in Chapter 2 (subsection 2.3.3). As noted there, 68 per cent of global electricity generating capacity (GW) consists of coal, oil or natural gas powerplants, and 67 per cent of the generated electricity (GW-hours or TW-hours) comes from these plants. Under almost any business-as-usual scenario, fossil fuels – coal in particular –

dominate future electricity production. Even in scenarios that stabilize atmospheric CO_2 at 450ppmv, a significant amount of electricity would still be produced from fossil fuels for a few decades. The efficient generation of electricity from fossil fuels is therefore an important consideration in reducing fossil fuel-related CO_2 emissions, and is the subject of this chapter.

Frequent reference will be made to energy, power and heat; to the efficiency of power generation; and to the cost of electricity using different technologies and fuels. The basic relationships between energy, power, heat and work are explained in Box 3.1. The efficiency in using a given fuel can be expressed in terms of either the lower or higher heating value of the fuel, as discussed in Appendix B.

Box 3.1 Power, energy, heat and efficiency

Power and energy

Energy is defined as the ability to do work and has the unit *joule*. Work done is equal to the force applied times the distance over which it is applied. Force is equal to mass times acceleration. Thus:

$$1 \text{ joule} = 1 \ (\text{kg m/s}^2)(\text{m}) = 1 \text{kg m}^2/\text{s}^2 \tag{3.1}$$

Power is the rate of doing work and has the unit *watt*:

$$1 \text{ watt} = 1 \text{ joule per second (J/s)} = 1 \text{kg m}^2/\text{s}^3 \tag{3.2}$$

To convert the power production of a powerplant to energy production, or to convert the power use of a light bulb or appliance to energy used, multiply the average power output or use times the length of time (in seconds, so that the time units match and can be cancelled). A convenient unit for electrical energy is the *kilowatt-hour* (kWh). One kWh is the energy resulting from a power of 1kW over a period of one hour. Thus:

$$1 \text{kWh} = 1000 \text{J/s times } 3600 \text{ seconds} = 3,600,000 \text{ joules} = 3.6 \text{MJ (MJ=megajoule)} \tag{3.3}$$

Heat

Energy can exist in many forms – as heat, motion and potential energy, among others. The thermal energy in a given volume of something is equal to its absolute temperature (on the kelvin scale) times its density times a constant (called the *specific heat*) that depends on the molecular properties of the particular substance under consideration. Thus:

$$\text{thermal energy (joules)} = \rho \ (\text{density, kg/m}^3) \ c_p \ (\text{specific heat, J/kg/K}) \ T \ (\text{K}) \ V \ (\text{volume, m}^3) \qquad (3.4)$$

A large mass at a low temperature could have the same thermal energy content as a smaller mass but at a larger temperature. However, it is possible to do more with thermal energy at higher temperature, so thermal energy at higher temperature is said to be of higher quality.

Efficiency

When a fossil fuel is burned, the chemical energy of the bonds in the fuel is converted to thermal energy. Some of this thermal energy is converted to electricity in the powerplant, and the rest is lost as heat that is dissipated to the environment (the surroundings). The reported energy content of a fossil fuel is the thermal energy that is produced when the fuel is burned. The efficiency of the powerplant is the fraction of this thermal energy that is converted to electricity. Thus:

$$\text{electricity energy production} = (\text{powerplant efficiency}) \text{ times (fuel energy use)} \qquad (3.5)$$

Conversely, the primary energy (fuel energy) required to produce a given amount of electricity is given by the electricity production divided by the powerplant efficiency. The higher the efficiency, the less primary energy required to produce a given amount of electricity.

When there is a chain of processes with losses at each step, compute the efficiency for each step and multiply the efficiencies together in order to get the overall efficiency. For example, if the transmission of electricity entails a 7 per cent loss, the transmission efficiency is 0.93 (93 per cent). If the powerplant efficiency is 35 per cent, the efficiency of generation + transmission is 0.35 × 0.93 = 0.326.

3.1 Some basics on electricity and electricity generation

Before discussing how electricity is generated using fossil fuels, it is useful to outline some of the basics of electricity.

3.1.1 DC electricity

When a wire is connected to a battery, the electrons flow continuously in one direction. This is called *direct current (DC)* electricity. The force that makes electrons flow is called *voltage potential* or simply the *voltage*, and has units of *volts*. The electrical current is measured in units called *amperes*. The rate of energy carried (power, watts) by an electrical current is given by the product of voltage (V) and current (I):

$$\text{power} = \text{voltage} \times \text{current} = VI \qquad (3.6)$$

There is a resistance within the wire to the movement of electrons, given in units of *ohms*; this resistance causes some of the electric energy to be dissipated as heat, so there is a loss in electric energy. The resistance is normally assumed to be constant, although it increases slightly with temperature. The voltage drop over a resistance R is proportional to the current, that is:

$$\Delta V = I \cdot R \qquad (3.7)$$

Combining Equations (3.6) and (3.7), the power loss through a resistor varies with the square of the current. Thus, in order to minimize the resistance losses, one should transmit the electrical power at the highest possible voltage, since this requires the smallest current for a given power.

3.1.2 AC electricity

Most of the electrical grids in the world use *alternating current* (AC) rather than DC electricity. In AC electricity, the instantaneous values of voltage and current vary sinusoidally, which means that the electrons oscillate back and forth in position rather than flowing continuously in one direction. The voltage and current are given by:

$$V(t) = V_a \cos(\omega t) \tag{3.8}$$

and:

$$I(t) = I_a \cos(\omega t + \varphi) \tag{3.9}$$

respectively, where V_a and I_a are the amplitudes of the variation in V and I, respectively, and φ is the phase shift between the voltage and current variation. The current can be decomposed into a part that is in phase with voltage and a part that is shifted 90 degrees from the voltage. The part of I that is in phase with V gives the *real power* (henceforth just called power) while the part that is 90 degrees out of phase with V gives the *reactive power*. Reactive power does not carry energy but subtracts from real power, so a goal in electric power systems is to have zero reactive power, that is, to have zero phase difference between voltage and current variation. This is an issue when renewable energy sources such as wind power are added to a grid. The power in an AC system at any instant when $\varphi = 0$ is given by:

$$P(t) = V(t)I(t) = V_a I_a \cos^2 \omega t \tag{3.10}$$

Inasmuch as the average of $\cos^2$ is ½, the average power is given by:

$$\overline{P} = 1/2 V_a I_a \tag{3.11}$$

and the average of the square of the voltage is given by:

$$\overline{[V(t)]^2} = (1/2)(V_a)^2 \tag{3.12}$$

Thus, the root mean square (rms) of the voltage, V_{rms}, is given by:

$$V_{rms} = V_a / \sqrt{2} \tag{3.13}$$

A similar expression applies to I_{rms}. The average power is then given by:

$$\overline{P} = V_{rms} I_{rms} \tag{3.14}$$

an expression identical in form to Equation (3.6). More generally, the average power is given by:

$$\overline{P} = V_{rms} I_{rms} \cos \varphi \tag{3.15}$$

In most of the world, the AC grid has a frequency of 50Hz (Hertz) and an rms voltage of 230V. This means that the voltage oscillates from −325V to +325V and back 50 times per second. In North America, the AC grid has a frequency of 60Hz and an rms voltage of 115V. Since the AC voltage oscillates up and down, a generator cannot be connected to the power grid unless it has exactly the same frequency and phase (timing of peaks and troughs) as the grid.

The power grid contains three separate power lines with AC electricity, with a phase offset of a third of a cycle or 120 degrees between any two lines (this is called *3-phase* AC). The sum of the voltages in the three lines always sums to exactly zero, while the difference between any two lines fluctuates as an alternating voltage. This is illustrated in Figure 3.1.

Small domestic customers receive one active phase and a zero (neutral) line to half the circuits in a fuse box, and another active phase and a zero line to the other half of the circuits. These circuits are then either 115V (North America) or 230V (elsewhere). The three phases of the grid are distributed evenly among all customers. The rms difference between any two phases of the 3-phase grid is a factor of √3 larger than the rms voltage of any one phase, namely, 200V in North America and 400V elsewhere. Heavy-duty equipment (dryers, ovens, baseboard electric heaters) run between two active lines, giving 1-phase AC power at 200V (in North America) or 400V (elsewhere). In commercial buildings, 3-phase wiring is used for heavy-duty equipment such as chillers and ventilation system fans.

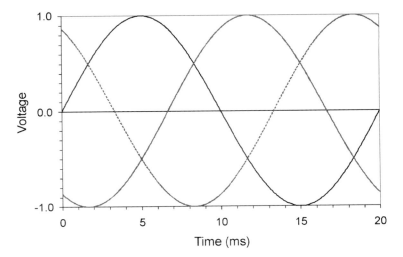

Figure 3.1 *Variation of voltage in 3-phase AC electricity*

Three-phase power is also used by industry, generally at a much higher voltage.

3.1.3 Synchronous generators

A synchronous generator has a central magnet called a *rotor* that rotates inside a set of three coils called a *stator* that are placed on a circle 120 degrees apart, as illustrated in Figure 3.2. Each of the coils is connected to its own phase of the 3-phase power grid. When more turning force (torque) is applied to the rotor, more electricity is generated, but the rotation speed does not on average deviate from the grid frequency. Hence, this kind of generator is called a *synchronous* generator. To generate 60Hz power, the rotor magnet rotates at 3600 rotations per minute (rpm) (60 cycles per second × 60 seconds per minute).

The two ends of a magnet are called poles (there is a north pole and a south pole). If the number of magnets in the rotor is doubled, so is the number of poles, and the rotor magnet will rotate at half the speed to produce the same output frequency, since in that case the coils in the stator will see a given pole of the magnet twice per rotation. Thus, with a 60Hz grid, a 4-pole generator would rotate at 1800rpm.

3.1.4 Asynchronous (induction) generators

In asynchronous generators, the rotor rotation speed differs slightly from the rotation rate of a synchronous

generator. The power production depends on the difference between the synchronous frequency and the generator frequency. However, asynchronous generators are not widely used for generating electricity except in wind turbines (where they provide distinct advantages in dealing with fluctuations in wind speed), so we will

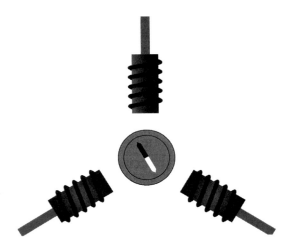

Note: The 3 electromagnets, each connected to its own phase of the 3-phase electrical grid, form the stator. Between them they create a single N-pole and a single S-pole, which rotate with the oscillating electrical current. In the centre is the rotor, with a rotating magnet having a single N-pole and a single S-pole that is aligned with the rotating stator poles.
Source: Danish Wind Industry Association, www.windpower.org, *The Guided Tour.*

Figure 3.2 *A 2-pole synchronous generator*

defer further discussion of these generators until our discussion of wind energy in Volume 2 (Chapter 3).

3.1.5 Transformers

As noted above, the resistance loss in power transmission lines is smaller at higher voltage (because the current required for a given power flow is smaller if the voltage is larger). Thus, power must be stepped up to a higher voltage for long-distance transmission, then stepped down prior to use, in order to minimize the losses during transmission.

The principle behind a transformer is as follows. A wire carrying AC is wrapped as a coil around a bar of magnetically permeable material (such as iron). An alternating magnetic field is generated. This induces a current in a second coil of wire wrapped around the same iron bar. The ratio of voltages in the two coils is equal to the ratio of windings for the two wires. The efficiency of the transformer (or conversely, the energy loss in stepping from one voltage to the next) depends on the details of the transformer design – in particular, how the core is built (the most efficient transformers have cores made of many thin laminations of non-crystalline iron). The energy lost in stepping from one voltage to the next is dissipated as heat. The energy loss in transforming electricity from one voltage to another offsets some of the savings in transmitting the electricity at higher voltage. These tradeoffs are explored in Volume 2, Chapter 3 in the context of long-distance transmission of wind-generated electricity.

3.1.6 Baseload, fluctuating load and spinning reserve

Electricity demand (the 'load' placed on the system) fluctuates over time periods of seconds, hours, days and entire seasons as consumers turn lights and equipment on and off, and as heating and cooling energy use varies in response to changing weather conditions. Electricity utilities thus have the task of rapidly changing their power output to match continually changing power demand. To the extent that the power output does not exactly match the varying demand, there will be fluctuations in the frequency and voltage, but the frequency fluctuations are usually less then ±0.5 per cent in developed countries. The minimum electricity demand is referred to as the *baseload*. The baseload demand is commonly met through steady power production from those powerplants in the utility's portfolio that were most expensive to build but are inexpensive to operate, and are operated as close to their full output as possible in order to recoup the investment cost more quickly. Baseload plants would be nuclear and hydro powerplants if available, or otherwise coal powerplants with steady output. This choice minimizes overall cost, although nuclear powerplants must be operated as baseload plants for technical reasons – it simply is not possible to rapidly vary their power output (it takes about two weeks to go from a cold start to full power, and a comparable time to shut down).[1]

Some relatively slow part of the variation in power demand can be anticipated in advance and will be met by powerplants (such as coal or natural gas powerplants) whose output can be gradually altered. Simple cycle natural gas powerplants (described below) are a common choice to meet fluctuating loads, as they are inexpensive to build but expensive to operate (due to high fuel costs). Very rapid fluctuations in demand, however, have to be met with *spinning reserve* – fossil fuel units that are running at part (and perhaps very low) load so that they are able to rapidly increase their output when needed. Fossil fuel powerplant efficiencies are smaller the smaller the load, so the greater the fraction of spinning reserve, the lower the overall efficiency of electricity generation. In jurisdictions where hydropower is available, some portion of the hydro capacity can be used to meet rapid fluctuations in demand rather than meeting baseload demand. This reduces the requirement for fossil fuel spinning reserve.

3.2 Electricity from fossil fuels

In 2005, 40 per cent of the world's electricity was generated from coal, 20 per cent from natural gas and another 6.5 per cent from oil, for a total fossil fuel contribution of 66.5 per cent (Figure 2.13). Table 3.1 compares the efficiencies in 2002 of electricity generation from coal, oil and natural gas in a variety of countries. These efficiencies are based on the LHV of the fuels and, in the case of natural gas, would be about 10 per cent lower if based on the HHVs. They are also based on gross electricity production, that is, excluding electricity consumption by the powerplant itself. In the case of coal plants, this uses 6–8 per cent of the output of the plant, but only 2–3 per cent in the case of natural gas powerplants. The efficiency in generating electricity from coal is particularly low in India because Indian coal contains 30–35 per cent ash.

Table 3.1 *Average gross efficiencies (that is, neglecting electricity consumption by the powerplant itself) for the generation of electricity in central fossil fuel powerplants in 2002 in various countries, based on the fuel LHV*

Country	Fuel		
	Coal	Oil	Natural gas
Nordic countries	42	42	46
UK and Ireland	38	40	51
France	36	37	–
Germany	37	37	40
USA	36	36	42
Japan	42	45	44
China	33	34	48
India	27	30	52
Weighted average	34.4	38.3	43.8

Source: Graus and Voogt (2005)

3.2.1 Pulverized coal powerplants

The *pulverized coal* powerplant is the coal-based technology overwhelmingly used today. The layout of a pulverized coal powerplant is shown in Figure 3.3. Coal is pulverized to a consistency of talcum powder (at least 70 per cent passes through a 200-mesh sieve) and pneumatically injected through the burners into a boiler. Combustion takes place almost entirely while the coal is suspended in the boiler volume. The boiler heats water under pressure to produce pressurized steam (steam is vaporized water, although in common terminology, the term refers to the mist that condenses above boiling water as the hot vapour cools). The steam is carried to a turbine, where it expands and cools, causing the turbine to rotate. There is a shaft along the axis of rotation of the turbine that is attached to the rotor in a generator, which is where electricity is produced (as explained above). The low-pressure and low-temperature steam exiting from the steam turbine is carried to a condenser, which is a coil through which cooling water circulates. When the steam comes into contact with the coil, it condenses to liquid water at very low pressure. This water is pressurized to the boiler pressure and pumped back to the boiler, where the cycle begins again (pressurizing the water requires relatively little energy). The cooling water that circulates through the condenser cooling coil flows to a cooling tower, where heat is released to the atmosphere.

The heat produced from burning the fuel ends up in one of two forms: as electricity and as heat dissipated to the surrounding environment. Clearly, the lower the temperature of the exhaust gases from the powerplant or of the condenser, the less heat that is being

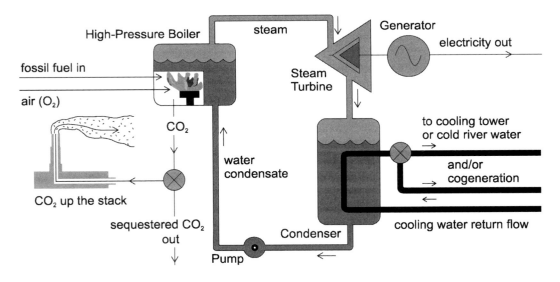

Source: Hoffert et al (2002)

Figure 3.3 *Layout of a coal-fired powerplant, showing the boiler, turbine, condenser and generator*

discharged to the environment through the exhaust or the cooling tower, and the higher the potential efficiency (ratio of electricity production to thermal energy produced by combustion). The maximum possible efficiency of a generic thermal cycle is given by the *Carnot Cycle*, for which the efficiency is:

$$\eta = \frac{T_{in} - T_{out}}{T_{in}} \qquad (3.16)$$

where T_{in} is the turbine inlet temperature and T_{out} is the turbine outlet temperature.

The maximum turbine inlet temperature that is allowed is limited by the properties of the steels used to make the boiler and by the rapid increase in steam pressure with increasing temperature. Higher pressure in turn requires thicker (or stronger) boiler walls. Higher temperature tends to reduce the lifetime of the turbine components; alloys of steels that can accommodate higher temperatures are more costly. Thus, the design temperature – and hence efficiency – represents a tradeoff between the added cost of stronger steels or thicker walls and the reduced cost of less fuel consumption for a given electricity production. However, increasing the thickness of the boiler walls has practical limits. The stress on the boiler walls varies with pressure × (boiler radius/wall thickness), and the steam pressure increases exponentially with temperature, so the radius to wall-thickness ratio must fall at an ever-increasing rate to compensate for increasing pressure as the boiler temperature increases. This poses a fundamental limitation on the maximum potential efficiency of steam turbines.

Coal-fired steam turbines in operation today overwhelmingly operate at 290 times sea level atmospheric pressure (290 atmospheres) and 580°C (McMullan et al, 2001). These are *subcritical* powerplants and have efficiencies ranging from less than 30 per cent (in smaller turbines, common in developing countries) to 33–35 per cent typical of many developed countries. More recent coal powerplants are *supercritical*, with a steam temperature greater than 600°C and an efficiency of 35–40 per cent (Vuorinen, 2007).[2] The next step will be operation at 375 atmospheres and 720°C using nickel-based superalloys that are yet to be developed (McMullan et al, 2001). IEA (2004a) projects coal powerplant efficiencies of 48–50 per cent by 2010 and 50–55 per cent by 2020.

Efficiencies at part load are smaller than the efficiencies given above, which pertain to full load. For example, a powerplant with a full-load efficiency of 38 per cent would have a part-load efficiency as low as 28 per cent. As well, about 6–8 per cent of the electricity generated by coal plants is needed to run the plant itself, so the efficiencies (given above) need to be multiplied by 0.92–0.94 to give the net powerplant efficiency.

Steam turbines in developed countries typically come in 250–650MW sizes, which tend to be more efficient than smaller turbines. A single coal-fired powerplant will typically consist of several steam turbines, and thus might produce up to 2–3GW of electric power. The potential reduction in CO_2 emissions by upgrading existing powerplants to the most efficient powerplants envisaged at present is particularly large in China because China has many small (<50MW) coal-fired powerplants that consume up to 60 per cent more coal per kWh than large, state-of-the-art plants (Martinot, 2001). Over half of its plants have a capacity of less than 200MW, with only 12 per cent at 300MW or larger capacity (compared to 80–90 per cent in industrialized countries). The central Chinese government decided to retire all plants of less than 50MW capacity by 2002 but has had some difficulty in enforcing this directive (the majority of the small plants are owned by municipal governments who want to keep them running).

Recent pulverized-coal powerplants with state-of-the art pollution controls have generally cost in the range of $1200–1400/kW (IEA, 2005), although large increases in the costs of steel and other materials (such as occurred during 2007–2008) can lead to substantially higher powerplant costs. These powerplants are a mature technology, so the tendency for future cost decreases due to further improvements will be small.

3.2.2 Integrated gasification/combined cycle using coal

A promising alternative to the pulverized coal powerplant is the integrated gasification/combined cycle (IGCC) powerplant. Coal is gasified to a mixture of CO_2, CO and hydrogen (H_2) by heating it to high temperature (around 1000°C) at high pressure (up to 25 atmospheres) in the presence of steam or pure oxygen.[3] The CO can be converted to CO_2 and H_2 by reaction with steam, or can combusted as a fuel along with the H_2 already present. The H_2 and CO (if any remains) are directly combusted in a gas turbine (which typically operates at 1250°C). The hot exhaust gases from the gas turbine are used to produce steam for use in a steam turbine. As in natural gas/combined cycle systems (described below), the gas turbine blades can be cooled with air or with steam from the steam turbine. Oxygen-blown IGCC efficiencies (based on LHV) are expected to range from 42–48 per cent, depending on the specific technology used (McMullan et al, 2001). These efficiencies are, at best, only slightly better than the best expected pulverized-coal efficiency.

Gasification of the coal allows separation of the sulphur prior to combustion, eliminating the need for flue gas desulphurization (the standard treatment for removing sulphur from the exhaust gases of pulverized coal powerplants). However, sulphur is recovered from an IGCC plant as marketable elemental sulphur rather than as wet scrubber sludge or as a dry spent sulphur sorbent, as in conventional coal powerplants equipped with flue gas desulphurization. This reduces the volume of solid waste by a factor of two to three (Williams et al, 2000). Emissions of most other pollutants are also very low, and this used to be a significant advantage of IGCC over pulverized coal powerplants. However, advanced pulverized coal powerplants can achieve comparable emissions to IGCC powerplants, as indicated in Chapter 2 (Table 2.19).

The first IGCC powerplant was the 94MW Coolwater demonstration project in southern California, which operated between 1984 and 1989, and several large commercial-scale demonstrations plants have been built in a number of countries since then. The primary disadvantage of IGCC compared to pulverized coal powerplants is cost and cost uncertainty. At present, the US government provides a 20 per cent tax credit and other incentives to encourage the construction of IGCC powerplants instead of conventional powerplants, which are still overwhelmingly chosen. As a result, several IGCC projects are in the planning stage, including a 680MW plant in New York State for $1.5 billion ($2206/kW), scheduled to go into operation in 2013. As well, plans are underway for conversion of existing coal powerplants in Connecticut, Delaware and New York to IGCC. A number of analysts expect IGCC costs to decrease to the $1150–1400/kW range by 2020 (assuming constant material costs), while pulverized coal plants with flue gas desulphurization are expected to

decrease in cost by a smaller amount, from $1200–1400/kW at present to the $1100–1200/kW range. An IGCC demonstration project involving co-production of electricity and methanol was completed in China in 2006, and several more are in the planning stages (Zhao and Gallagher, 2007; Liu et al, 2008). Costs for projects in China based on bids in the early 2000s were expected to be around $1000/kW, the lower cost being due in part of cheaper construction costs in China, but costs turned out to be much higher than expected, which has delayed further projects. The efficiency and cost estimates given here for IGCC account for the energy required to separate oxygen from air for the gasifier operating with pure oxygen (12 per cent of gross electrical output) and the associated capital cost.

IGCC plant design is very sensitive to characteristics of the fuel such as heating value and sulphur, ash and moisture content. This makes the plant rather inflexible with regard to the fuel, at least not without greater capital cost. Costs themselves are rather uncertain because so few IGCC plants have been built, whereas costs for conventional pulverized coal powerplants are well established. To develop enough data for firm construction budgeting and financing requires an upfront engineering assessment that costs in the range of $4–6 million. Many published studies of the cost of IGCC include technologies that are not yet commercially available (such as hydrogen turbines and membrane-based oxygen systems), and many assume a higher level of integration than might be advisable in early plants (Anonymous, 2006).

The gasifier in an IGCC plant makes use of waste heat from the combustion phase, and so cannot be readily turned on and off (48 hours are required to go from a cold start to full capacity). Thus, IGCC plants cannot follow variations in the load to the same extent as pulverized coal powerplants. As noted above, pollutant emissions from advanced pulverized coal plants have dropped to levels comparable to those of IGCC powerplants, while the efficiency has risen to comparable levels. The primary advantage of IGCC is thought to be its ability to capture CO_2 with a much smaller energy penalty and potentially at lower cost than from pulverized coal powerplants (see Volume 2, Chapter 9, subsection 9.3.5). However, as discussed in Volume 2, combustion of pulverized coal in pure oxygen could push the energy penalty for CO_2 capture to as low as 15 per cent – comparable to the lowest penalty with IGCC plants.

In summary, there are two main options at present for the use of coal to generate electricity at substantially greater efficiency than at present – advanced conventional pulverized coal powerplants and IGCC powerplants. Both promise gross efficiencies of around 50 per cent, comparable and vastly reduced pollutant emissions compared to conventional powerplants with existing pollution controls, comparable costs (although IGCC is currently much more expensive and future costs for IGCC are quite uncertain), and comparable (and relatively small) energy penalties if CO_2 is captured from the flue gases. Which one is a better choice from an environmental and economic point of view will depend on the characteristics of the available fuel, operational constraints (whether the plant will be largely used to satisfy baseload or the fluctuating portion of the total electricity demand), and local market conditions.

3.2.3 Natural gas turbines and combined cycle powerplants

A gas turbine is an internal combustion engine that operates with rotary rather than reciprocating (up and down) motion (as in an automobile engine). A gas turbine consists of three parts: a *compressor* – this draws air in and compresses it to many (typically 30) times the outside air pressure, then directs the air to a *combustor*. Fuel and air are mixed in the combustor, ignited and burned. Hot gases from the combustor are diluted with additional air from the compressor, and sent to the *power turbine*. The hot gases expand in the turbine, which turns a shaft. About 50 per cent of the shaft power is needed to drive the compressor, and the rest is available to drive the electric generator. Since hot combustion gases rather than steam are directly used in a gas turbine, the limitation on the turbine inlet pressure for steam due to the exponential increase of the steam pressure with increasing temperature, does not apply to gas turbines. This allows gas turbines to operate at higher temperatures – around 1250°C. This would be expected to give much higher efficiencies for gas turbines than for steam turbines, based on the Carnot Cycle limitation (Equation 3.16), except that half or more of the shaft power is needed to run the compressor, rather than producing electricity.

Several different things can be done with the heat from the exhaust gases of a gas turbine. The easiest is to discard the heat. This gives a *simple cycle* gas turbine, with a typical efficiency of 30–40 per cent but as low as 16 per cent (in very small units) to as high as 43 per cent. This is illustrated in Figure 3.4, which gives the power output and LHV efficiency of most of the commercially available gas turbines. The power output and to some extent the

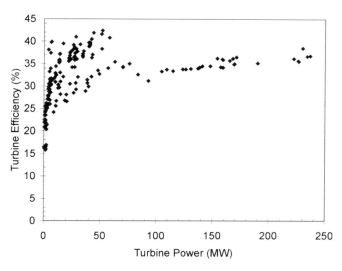

Source: Based on data from Poullikkas (2005)

Figure 3.4 *Efficiencies (LHV basis) of simple cycle gas turbines with different capacities*

efficiency of the simple cycle gas turbine can be increased by injecting water or steam into the combustion chamber. This is done in several commercially available turbines, and the power and efficiency with and without steam or water injection are compared in Figure 3.5. Four of the turbines achieve efficiencies between 40 per cent and 45 per cent. This is a substantial improvement over many existing older gas turbines, which have efficiencies in the 27–30 per cent range.

The remaining options for gas turbines all require the use of a *heat recovery steam generator* (HRSG) to produce steam by using the exhaust to heat water. The steam can either be used for industrial processes or as input into a steam district heating system (giving *simple cycle cogeneration*), or can be fed into a steam turbine for production of additional electricity (giving *combined cycle* power generation, so-called because gas and steam turbines are combined). Another option is to take some of the steam from the steam turbine in a combined cycle system to use for heating (this gives *combined cycle cogeneration*). The energy flows for these different options are illustrated in Figure 3.6. The fraction of fuel input that is converted to electricity or useful heat is typically 75–85 per cent in simple cycle cogeneration but can approach 92 per cent if, in addition to making steam, the remaining waste heat

exiting from the HRSG unit is used to make hot water that can be used somewhere. Thus, overall efficiencies as high as 92 per cent are possible. The energy flow in this case is illustrated in Figure 3.7.

Natural gas turbines are available at scales ranging from less than 1MW to hundreds of megawatts, with the lowest cost (per kW) at the largest scale. The natural gas combined cycle powerplant (referred to as NGCC or GTCC, the latter standing for gas turbine combined cycle) provides the best electrical efficiencies, at about 50 per cent with air-cooled gas turbines and 55–60 per cent with steam-cooled gas turbines, which permit turbine inlet temperatures of up to 1500°C (Williams et al, 2000). However, combined cycle powerplants, with their greater efficiency, are economically feasible only at 25–50MW sizes and larger. These powerplants are more complicated than simple cycle turbines, which also makes them difficult to deploy at small scales. Thus, there is interest in developing efficient simple cycle turbines at small scales, and a number of options that could increase the efficiency of simple cycle turbines to 43–57 per cent are discussed by Poullikkas (2005).

NGCC powerplants are the least expensive fossil fuel powerplants, with capital costs in the early 2000s in OECD countries, as reported by IEA (2005), ranging from \$380/kW (for a 791MW unit with

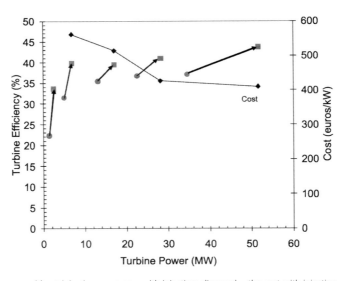

Note: Circles = efficiency and power without injection; squares = with injection; diamond = the cost with injection.
Source: Based on data from Poullikkas (2005)

Figure 3.5 *Impact of injecting water or steam into the compressor of a simple cycle gas turbine, and the cost with injection*

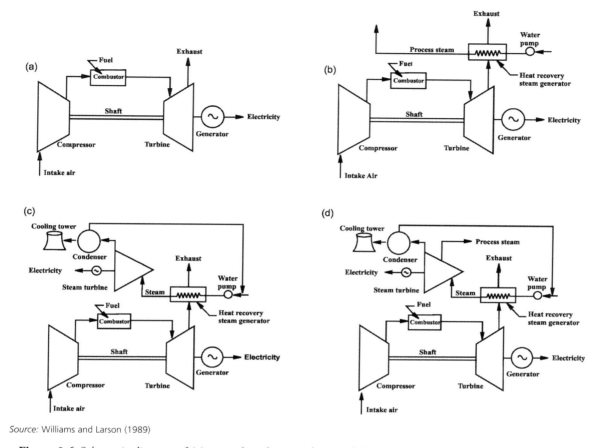

Source: Williams and Larson (1989)

Figure 3.6 *Schematic diagram of (a) a simple cycle gas turbine and electric generator, (b) simple cycle cogeneration, (c) combined cycle (gas and steam turbines) power generation, and (d) combined cycle cogeneration*

56 per cent efficiency in Italy) to about $1000/kW (for a 500MW unit with 60 per cent efficiency in The Netherlands), with most costs between $400/kW and $600/kW. Costs have been higher in less mature markets, with average costs in 2000 of $803/kW in Southeast Asia, $875/kW in India and $923/kW in Africa, compared to $573/kW in North America, according to Sathaye and Phadke (2006).

3.2.4 Diesel and natural gas reciprocating engines

A *reciprocating engine* is an engine with pistons that move back and forth. The back and forth motion is converted to rotational motion with a crankshaft. The automobile internal combustion engine is an example of a reciprocating engine. Small diesel generators that are used for backup power generation (or sometimes for dedicated or peaking power generation) are also reciprocating engines. Efficiencies range from 30 per cent for 100kW units to 37 per cent for 5MW units, but pollutant emissions are higher than from gas turbines or pulverized coal powerplants (as shown in Table 3.5). However, natural gas-powered reciprocating engines are now available with full-load mechanical efficiencies greater than 46 per cent, even when the oil and water pumps are directly driven by the engine shaft (Klimstra and Hattar, 2006). When multiplied by a possible generator efficiency of 98 per cent, this gives an electrical generation efficiency of 45 per cent.

Modern reciprocating engines can operate on either diesel fuel or natural gas. In the *dual-fuel* engine a small

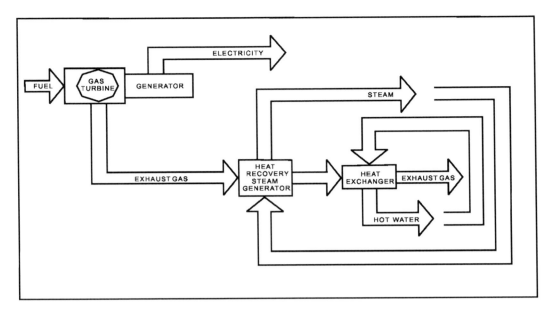

Source: Malik (1997)

Figure 3.7 *A cogeneration system with steam and hot water production*

amount of oil is used to ignite the gas/air mixture, and 1 per cent of the fuel requirement is met with diesel at full load. In the event that the gas supply fails, the engine can immediately switch to full diesel operation, even at full load (Klimstra, 2008). In the gas-diesel engine, up to 95 per cent of the fuel can be natural gas (with the balance diesel), which must be compressed to 350 bar before injection at the end of the compression stroke. Capital costs of reciprocating engines are in the $600–1200/kW range.

Diesel engines can be combined with steam turbines to produce diesel engine combined cycle (DECC) powerplants, analogous to NGCC powerplants. However, the full-load efficiency of DECC is only slightly better than that of a simple diesel engine (45 per cent versus 40 per cent at best, on a HHV basis) and is less than that of NGCC (55–60 per cent). This is because the steam turbine in DECC produces relatively little electricity (5–15 per cent of the total output) compared to the steam turbine in NGCC (30–33 per cent of the total output), due to the relatively low temperature of the exhaust from diesel engines (350–450°C) compared to gas turbines (up to 1250°C).

3.2.5 Fuel cells

A fuel cell produces electricity through an *electrochemical* process rather than through combustion (as in steam and gas turbines). In fuel cells, hydrogen gas or syngas (H_2 + CO) in the case of molten carbonate and solid oxide fuel cells, is combined with oxygen (from the atmosphere) to produce water (and CO_2 in the case of syngas), DC electricity and heat. A fuel cell element consists of an anode and a cathode separated by an electrolyte (an ion-conducting substance). A hydrogen-rich fuel is added to the anode, clean air is added to the cathode and an external current is created that runs from the anode to the cathode. This circuit is completed by the flow of ions within the electrolyte. The arrangement of the parts of a fuel cell is illustrated in Figure 3.8. The voltage potential created by a single fuel cell element is 0.7–0.9V (as explained in detail in Chapter 5, Box 5.2), so fuel cell elements must be joined in series to get higher voltages, creating a *fuel cell stack*. The power generated by a fuel cell is given by the product of voltage and current, the latter given by the product of current density (amps/m^2) across the fuel cell plates and the plate area.

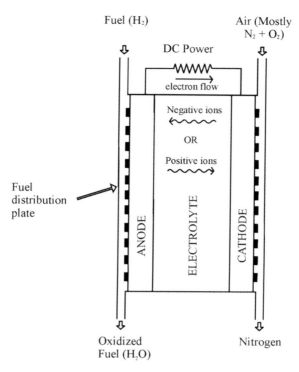

Figure 3.8 *Layout of a fuel cell*

There are six types of fuel cells in use or under development, which are listed in Table 3.2 along with the electrolyte-conducting ion and the reactions occurring at the anode and cathode. Alkaline fuel cells (AFCs) require pure hydrogen and have been used in the space programme, but the other fuel cells can be fed on pure hydrogen or a hydrogen-rich gas, the latter produced by coal gasification or by steam reforming of methane.[4] A 200kW phosphoric acid fuel cell (PAFC), produced by United Technologies Company (formerly International Fuel Cells Inc.) and powered by natural gas, has been commercially available for stationary power applications for several years already. The unit, illustrated in Figure 3.9, is 3m wide × 3m high × 5m

Table 3.2 *Chemical reactions occurring at the anode and cathode of different fuel cells*

Fuel cell type	Anodic reaction	Cathodic reaction	Electrolyte conducting ion
AFC	$H_2 + 2OH^- \rightarrow 2H_2O + 2e^-$	$\frac{1}{2}O_2 + H_2O + 2e^- \rightarrow 2OH^-$	OH^-
PEMFC	$H_2 \rightarrow 2H^+ + 2e^-$	$\frac{1}{2}O_2 + 2H^+ + 2e^- \rightarrow H_2O$	H^+
DMFC	$CH_3OH + H_2O \rightarrow CO_2 + 6H^+ + 6e^-$	$\frac{1}{2}O_2 + 2H^+ + 2e^- \rightarrow H_2O$	H^+
PAFC	$H_2 \rightarrow 2H^+ + 2e^-$	$\frac{1}{2}O_2 + 2H^+ + 2e^- \rightarrow H_2O$	H^+
MCFC	$H_2 + CO_3^{2-} \rightarrow H_2O + CO_2 + 2e^-$ $CO + CO_3^{2-} \rightarrow 2CO_2 + 2e^-$	$CO_2 + \frac{1}{2}O_2 + 2e^- \rightarrow CO_3^{2-}$	CO_3^{2-}
SOFC	$2H_2 + 2O^{2-} \rightarrow 2H_2O + 4e^-$ $2CO + 2O^{2-} \rightarrow 2CO_2 + 4e^-$	$O_2 + 4e^- \rightarrow 2O^{2-}$	O^{2-}

Note: AFC = alkaline fuel cell, PEMFC = proton exchange membrane fuel cell (also called solid polymer fuel cell, SPFC), DMFC = direct methanol fuel cell, PAFC = phosphoric acid fuel cell, MCFC = molten carbonate fuel cell, SOFC = solid oxide fuel cell.

Source: Harvey (1995)

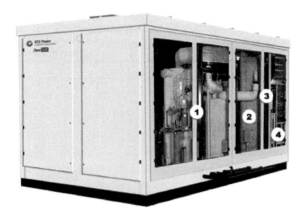

Source: www.utcfuelcells.com

Figure 3.9 *An inside view of the United Technologies Company 200kW phosphoric acid fuel cell, showing (1) the fuel processor, (2) the fuel cell stack, (3) the power conditioner and (4) electronics and controls*

long in size. The proton exchange membrane fuel cell (PEMFC) is of great interest for transportation applications, and is discussed further in Chapter 5 (subsection 5.4.8).

Fuel cell systems powered by coal or natural gas have three main components: a fuel-processing component, in which a hydrogen-rich gas is produced, the fuel cell stack and a power conditioning unit to convert from DC to AC electricity. The direct methanol fuel cell (DMFC) also uses a hydrocarbon fuel (methanol, CH_3OH) but does not need a fuel-processing unit because it directly uses methanol in the anode reaction (see Table 3.2). This is of interest because methanol is a liquid at normal temperatures and so is a potential alternative transportation fuel.

Broader issues related to the integration of hydrogen-powered fuel cells in a renewable energy-based energy system are discussed in Volume 2, Chapter 10. In this chapter, however, our primary interest is in the use of fuel cells as a means of increasing the efficiency with which electricity can be generated from fossil fuels.

Efficiency of hydrocarbon-powered fuel cells

The efficiency of a fuel cell system depends on the electrical output (and hence efficiency) of the fuel cell stack minus the electricity used to operate auxiliary devices such as pumps. The system efficiency involves tradeoffs with operating lifetime and capital costs. Waste heat from the fuel cell stack can be used for processing of hydrocarbon-based fuels, thereby increasing the system efficiency, but with increased capital cost and complexity. Increasing operating temperature leads to better use of waste heat in fuel processing and hence higher system efficiency, but results in a shorter lifetime. Reducing the current density leads to higher efficiency because resistance losses are reduced, but requires a greater cell area and hence greater cell capital costs for the same power output (and a larger building to house the larger units). Fundamentals concerning the efficiency of a fuel cell are discussed further in Box 5.2 of Chapter 5.

Fuel processing is simpler using natural gas than using coal, such that natural gas systems will have greater efficiency and lower capital cost compared to coal-based systems. Efficiencies are higher still and capital costs lower using pure hydrogen, and efficiencies are higher using pure oxygen rather than air at the cathode. Table 3.3 compares the range of efficiencies and operating temperatures for different fuel cells using coal, natural gas and pure hydrogen as input fuels. Efficiencies are highest at lower power output at peak output, as the current density and associated resistance losses are smaller at low outputs (the variation of efficiency with output for the PEMFC is illustrated Chapter 5, Box 5.2).

In the high-temperature fuels cells (SOFCs and MCFCs), internal rather than external methane reforming is possible, so a separate fuel processor is not needed, leading to a substantially higher efficiency. SOFCs have the added advantage that the electrolyte is a solid ceramic made of low-cost materials, and they are more compact than MCFCs. The major developer of SOFCs is Siemans-Westinghouse, which has adopted a tubular design. This makes sealing inherently easy but the manufacturing process is inherently expensive. Reducing costs with this design is a major challenge. At least one company (Global Thermoelectric, Canada) is developing thin-film, flat-plate SOFCs (having a 1mm-thick zirconium oxide anode, a 10μm electrolyte and a 50μm cathode). A problem with SOFCs is the sensitivity of the ceramic elements to thermal transients, resulting in cracking and limiting fluctuations in output (Klimstra, personal communication, 2008). The MCFC is the only fuel cell that can use coal as a fuel, via

Table 3.3 *Technical characteristics of fuel cells, with efficiencies based on HHV of the fuel*

Fuel cell type	Electricity generation efficiency (%)	Fuel/ oxidant	Tolerance for			Electro catalyst used	Operating temperature
			CO	CO_2	H_2S		
SPFC (PEM)	55–60	H_2/O_2	<10ppm	Yes	Yes	Pt using pure H_2, Pt and Ru if the H_2 feed contains trace CO	50–90°C
	40	NG/air				Pt and Ru	
DMFC	28	methanol/air	Yes	Yes	Yes	Pt	50–90°C
AFC	55–65	H_2/O_2	No	No	Yes	Pt, Ni/NiO$_x$	50–260°C
PAFC	50	H_2/O_2	<1%	Yes	<50ppm	Pt	180–200°C
	35–40	NG/air					
MCFC	60	H_2/O_2	Yes	Yes	<0.5ppm	Ni/LiNiO$_x$	650°C
	40–50	NG/air					
	45	coal/air					
SOFC	60	H_2/O_2	Yes	Yes	<1ppm	Ni/ Perovskites	750–1000°C, perhaps down to 600°C
	45–50	NG/air, MW scale					
	30–40	NG/air, 3–10kW scale					
	45	coal/air					

Note: NG = natural gas.
Source: Harvey (1995), Srinivasan et al (1999), Zogg et al (2006, 2007)

coal gasification. Carbon monoxide (a gasification product) is a fuel input (along with H_2), rather than a poison. No noble metal catalysts are needed due to the high operating temperature.

Of particular interest here is the hybrid fuel cell turbine power system, in which either a steam or a gas turbine uses waste heat from the fuel cell, with additional use of waste heat in a cogeneration configuration. Hybrid systems will be based on either SOFCs or MCFCs, as only these fuel cells produce waste heat at a high enough temperature (≥650°C) to be used in a steam or gas turbine. In the case of a fuel cell/gas turbine hybrid, unused fuel from the fuel cell exhaust is burned in the turbine combustor to provide additional heat for the gas turbine. The flow diagram for a SOFC/gas turbine hybrid is shown in Figure 3.10 and illustrates the cascading of energy flows so as to maximize overall efficiency. For a system modelled by Calise et al (2006), the exhaust from the catalytic burner is at 1290°C. It passes through a first turbine, which supplies electricity needed by the plant itself, then through a second turbine that generates electricity for export. Exhaust from the second turbine (at 847°C) is used to preheat the air (at heat exchanger 1) and fuel (at heat exchanger 2) for the fuel cell, then makes steam in a HRSG that is used for the internal steam reforming of methane. The exhaust is at a temperature of 224°C, which is more than adequate to supply heat to a district heating system. For a hot-water system with a distribution temperature of 70–90°C, most of the heat in the exhaust (relative to ambient conditions) could be extracted, giving an overall efficiency well in excess of 90 per cent. The electricity-generation efficiency is expected to reach 70–75 per cent (Srinivasan et al, 1999; Calise et al, 2006), which is about twice the current average in industrialized countries and more than twice the current average in developing countries. Fuel Cell Energy Inc. is testing a molten carbonate fuel cell/microturbine combination with a near-term efficiency goal of 70 per cent using natural gas in the single-MW size and a long-term goal of 80 per cent electricity generation efficiency. A 40MW powerplant is in the design phase, but one never knows how long that phase will last.

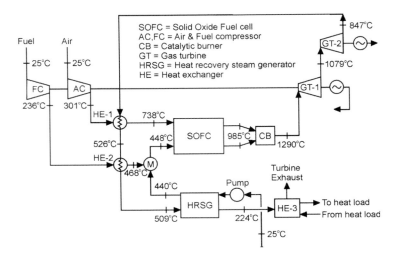

Source: Modified from Calise et al (2006)

Figure 3.10 *Flow diagram of a hybrid solid oxide fuel cell/gas turbine system powered by natural gas*

Cost of fuel cells

Fuel cells are currently much more expensive than combined cycle turbines (around \$3000–5000/kW versus \$400–600/kW for 100MW and larger combined cycle systems) but are already being used in niche applications where their special benefits (very low pollutant emissions, no noise and vibrations) make them competitive. The projected long-term costs (excluding installation costs) are \$1200–1500/kW for MCFCs, \$1000–1500/kW for SOFCs and \$1000 or less for PEMFCs (however, prices have recently risen due to higher raw material prices and labour scarcity). It is felt that fuel cells will become highly competitive in the stationary power market, particularly in the few-kW to 1MW range for cogeneration of heat and electricity for apartment buildings, office complexes and individual houses. The cost of natural gas turbine cogeneration systems is highly dependent on the size of the system, increasing sharply as one drops to scales of 10MW and less. The cost of fuel cell powerplants is largely independent of the size of the plants, so fuel cells will first be competitive with alternative cogeneration systems at small scales. They will also be most competitive in terms of efficiency at smaller scales, as combined cycle systems are not available below capacities of about 25MW and the efficiency of the gas turbine decreases with decreasing size (see Figure 3.4).

Conversely, at larger scales it will be difficult to handle and maintain fuel cells due to their low power density (Klimstra, personal communication, 2008).

At present, intensive efforts are underway to develop fuel cells (primarily PEMFCs) for the automotive market, as discussed in Chapter 5 (subsection 5.4.8). This will require bringing costs down to the \$50/kW range if they are to be viable alternatives to the internal combustion engine (which costs as little as \$10/kW). Fuel cells for stationary power production would be designed for longer life (cell component replaced every five years or after 40,000 hours operation, versus 5000 hours for automotive applications), so the associated costs would be greater. Nevertheless, it would appear that at least some kinds of fuel cells could eventually become competitive with gas combined cycle powerplants for large-scale generation of electricity (where costs are about \$450–600/kW).

Carbon dioxide disposal with fuel cells

When fuel cells are powered by fossil fuels, the fossil fuel (whether coal or natural gas) is converted to a hydrogen-rich gas (or to almost pure hydrogen), in the same way that coal IGCC powerplants begin with conversion of coal to a hydrogen-rich gas. As with IGCC, this lends itself to relatively inexpensive separation of CO_2, followed by its sequestration in subsurface geological strata (see Volume 2, Chapter 9).

3.2.6 Rates of increase in output

As noted above, spinning reserve is provided by running some generating units at part load. The value of the unit as spinning reserve depends in part on how rapidly the power output can be increased if needed. If a given unit fails and the output of spinning reserve units increases to compensate, a standby unit needs to be brought on line and synchronized with the grid frequency in order to serve as new spinning reserve. Table 3.4 gives ramp rates and times to synchronization from a cold start for various fossil-fuel powered generators and for hydroelectric power. Hydro powerplants, gas and diesel engines and gas turbines can all increase their output relatively quickly (≥20 per cent/minute), while gas combined cycle powerplants and gas- and coal-fired steam turbines are slow to increase their output (≤5 per cent/minute). The ramp rates and synchronization times, as well as efficiency losses at part load, are relevant to the ability of fossil fuel units to compensate for fluctuations in the output of solar and wind powerplants.

3.2.7 Part-load behaviour

The efficiency of steam and gas turbines, combined cycle power generation and reciprocating engines decreases with decreasing load, while that of fuel cells increases. This is illustrated in Figure 3.11 for an advanced combined cycle system, gas turbine, microturbine, dual-fuel reciprocating engine and

SOFC. If the operation of the cogeneration system follows the thermal load (rather than providing baseload power), such that power generation is often at a small fraction of peak capacity, there could be a significant energy efficiency penalty using gas or steam turbines. However, the mechanical efficiency of reciprocating engines only drops from 46 per cent at full load to 43 per cent at 70 per cent of full load, so the unit can provide 30 per cent of its output as spinning reserve with very little impact on efficiency. Combined cycle powerplants also have a small efficiency penalty at part load (almost identical to that of the reciprocating engines shown in Figure 3.11), but their output cannot be increased as fast (see subsection 3.2.6). The thermal and overall efficiency of SOFCs decrease at part load while electrical efficiency increases, but SOFCs are not yet commercially available.

3.2.8 Impact of temperature and elevation on electricity generation efficiency

The volumetric rate of flow of combustion air into a gas turbine is constant, but the density and hence rate of mass flow decreases with increasing air temperature or altitude. Thus, the power output of a gas turbine decreases. Outputs and efficiencies are normally given for an air temperature of 15°C and sea level pressure. A decrease in output in turn causes a proportional increase in the capital cost per kW of capacity. The output decreases from 110 per cent of the rated output at 0°C to 83 per cent at 40°C (Vuorinen, 2007) – a decrease of 25 per cent even though absolute temperature increases

Table 3.4 *Dynamic properties of different kinds of electric powerplants*

	Ramp rate (% of peak power per minute)	Time to frequency synchronization from cold start (minutes)
Gas-fired steam turbine	3	120
Coal-fired steam turbine	3	150
Gas turbine	20	5
Gas combined cycle	5	8
Diesel engine	100	0.5
Gas engine	30	0.5
Hydro	40	0.2

Source: Klimstra (2007)

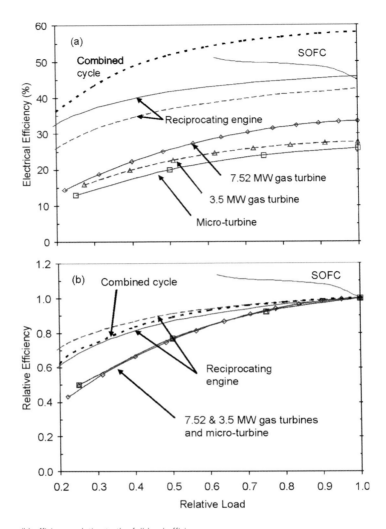

Note: (a) Absolute efficiency, (b) efficiency relative to the full-load efficiency.

Sources: Data from Kim (2004) for combined cycle and Li et al (2006) for simple cycle gas turbines, Osman and Ries (2007) for microturbines and solid oxide fuel cells, and Klimstra and Hattar (2006) for reciprocating engines

Figure 3.11 *Variation in electrical generation efficiency with load for a natural gas combined cycle turbine, simple cycle gas turbine, microturbine, advanced reciprocating engine and solid oxide fuel cell power generation*

by only 15 per cent. The impact on efficiency is much smaller – a decrease from 102 per cent of the rated efficiency at 0°C to 94 per cent at 40°C. As elevation increases from 0m to 3000m, the output of a gas turbine decreases to 70 per cent of the rated output.

In contrast to gas turbines, reciprocating engines experience almost no decrease in output or efficiency at higher temperatures or elevations. Gas turbines and reciprocating engines are of particular interest because

both can be used to offset rapid fluctuations in electricity production by other units or in electricity demand.

3.2.9 Auxiliary electricity use

All of the efficiencies given are gross efficiencies, that is, based on the electricity production by the electrical generator. However, some of the electricity produced by the generator is needed by the powerplant itself to operate pumps, fans and lights. This so-called auxiliary electricity use is about 2–3 per cent of the gross output for gas turbines, 3–5 per cent for reciprocating engines and 5–8 per cent for steam turbines (the last category also applies to nuclear powerplants) (Vuorinen, 2007).

3.2.10 Relative GHG emissions for coal- and natural gas-based electricity

Electricity production is given by the fuel energy input times the powerplant efficiency, so the amount of fuel energy E_f required to generate an amount of electrical energy E_{el} (in the same units as E_f) is given by:

$$E_f = \frac{E_{el}}{\eta} \quad (3.17)$$

where η is the efficiency of electricity generation. If one wishes to express electrical energy in kWh and fuel energy in GJ, a conversion factor of 0.0036 has to be used, as 1kWh is equal to 3.6MJ or 0.0036GJ. That is:

$$E_f(GJ) = 0.0036(GJ/kWh)\frac{E_{el}(kWh)}{\eta} \quad (3.18)$$

Thus, increasing the efficiency from 33 per cent (typical of present powerplants) to 50 per cent (in the future) would reduce the required fuel energy to 33/50 = 0.67 of present use – a savings of 33 per cent.

The CO_2 emitted (as kgC) in generating a kWh of electricity from coal or natural gas is given by the fuel use (GJ) times the emission factor (kgC/GJ). However, energy is required in finding, extracting and delivering the fuel to the powerplant, and methane emissions are associated with the extraction and delivery these fuels, both of which must be taken into account for a proper comparison. Emissions occurring prior to the delivery of the fuel to an end-user are referred to as 'upstream' emissions.

Energy uses associated with the delivery of fossil fuels are:

- For the delivery of coal by rail, about 1.3 per cent of the energy content of coal as diesel fuel per 1000km transported, according to Bergerson and Lave (2005). Given an energy content of coal of 19.4GJ/tonne, this is equivalent to 0.25MJ/tonne-km, which should be increased by about 20 per cent to account for energy losses in producing diesel from crude oil.

- For the delivery of coal by ship, 0.11–0.25MJ/tonne-km as diesel fuel, based on the entry in Table 5.30 for bulk cargo. This is equivalent to 0.6–1.3 per cent of the energy content of coal per 1000km transported.
- For the delivery of natural gas by high-capacity pipeline, about 2.5 per cent of throughput per 1000km due to the use of natural gas to power compressor stations that are located every 60–150km along gas pipelines.
- For the delivery of liquefied natural gas (LNG) by supertanker, an amount of natural gas equal to 8–13 per cent of the amount of LNG produced is used to chill the gas down to –161°C (depending on the efficiency of the equipment used for liquefaction) (Darley, 2004), followed by boiloff during transit of some LNG as heat is absorbed, and finally, use of about 1.5 per cent of the LNG energy for regasification at the receiving terminal. Over a 14-day, 12,500km journey from Qatar to the eastern coast of the US, for example, about 2.0–3.5 per cent of the LNG energy is lost through boiloff, although in modern LNG tankers the boiloff gas is used in duel-fuel engines for propulsion, with liquid fuels used during the empty return trip. Altogether, transportation of natural gas from the Middle East to North America consumes 10–18 per cent of the original energy content of the natural gas.

These upstream uses of energy can be accounted for by multiplying the energy use at the powerplant by a markup factor. For example, if the primary energy required to get the fuel to the powerplant is equal to 15 per cent of the delivered fuel energy, the fuel use at the powerplant should be multiplied by a factor of 1.15 when making comparisons with renewable energy options or in calculating the impact on total primary energy use of reduced demand for electricity due to end-use efficiency or conservation measures. As discussed in Chapter 2 (subsection 2.3.5), the markup factor for delivery needs to be multiplied by another markup factor to account for the energy used for exploration and extraction of the fuel. Worldwide, this additional markup factor is about 1.02 for coal 1.05 for natural gas.

Upstream emissions of methane also need to be accounted for if one is interested in total GHG emissions as well as primary energy use. This can be done by multiplying the fuel energy use (GJ) by the methane emission factor (kgC/GJ, from Table 2.12) times the molar global warming potential for methane (9.1), and adding this to the CO_2 emissions (kgC) to get the combined equivalent CO_2 emission (as kgC). Okamura et al (2007) present estimates of the upstream energy use and CO_2 and CH_4 emissions associated with delivery of natural gas to Japan from the Middle East in liquefied form. Total CO_2-equivalent emissions from combustion plus upstream emissions are 23 per cent greater than the emissions at the point of combustion, with about three quarters of the additional emission due to energy use and venting and flaring of methane during liquefaction.

3.3 Cogeneration and trigeneration

Cogeneration is the co-production of electricity and *useful* heat. All electricity generation produces heat as a byproduct, of course, but in many cases the heat is not used because it is not of sufficiently high *quality* (i.e. not of high enough temperature) to be useful. Cogeneration is widely used in industries that need both heat and electricity (such as the pulp and paper and petrochemical industries) and in district heating systems. There is also a growing interest in installing small cogeneration systems in individual buildings. Trigeneration is the co-production of electricity, useful heat and chilled water, either by using some of the mechanical power from a gas turbine to directly drive the compressor of a chiller, or, more commonly, by taking some of the waste heat from a turbine and using it to drive an absorption chiller rather than using it for heating purposes. In this chapter we discuss a number of issues related to cogeneration, while the merits of trigeneration are discussed in Chapter 4 (subsection 4.5.4) and Chapter 9 (subsection 9.5.2). Further information concerning cogeneration in industry and district heating systems is provided in Chapter 6 (subsection 6.9.4 and section 6.14) and Chapter 9 (section 9.2), respectively.

Cogeneration can occur in a number of different ways:

- using a simple steam turbine, as part of a coal or nuclear powerplant;
- using a simple gas turbine combined with a HRSG (illustrated in Figure 3.6b);
- as part of a combined cycle powerplant, in which exhaust heat from a gas turbine is first used to

produce steam for a steam turbine, and some of the steam from the steam turbine is used to provide useful heat instead of generating electricity (illustrated in Figure 3.6d).

Cogeneration can also be done using microturbines (defined as gas turbines with an electrical power output of less than 500kW), small reciprocating engines (such as diesel engines) and fuel cells. These are being considered at the scale of individual large buildings or even for individual houses. Residential-scale cogeneration with microturbines, reciprocating engines and fuel cells is reviewed in Onovwiona and Ugursal (2006). One of the advantages of cogeneration in individual buildings or at the scale of small district heating systems (20MW and up) is that it provides electricity where it is needed, thereby reducing transmission bottlenecks and energy losses.

Cogeneration units are normally sized to meet the average heat demand, rather than the peak demand, as otherwise the unit will be running at a small fraction of peak capacity (which increases cost) or wasting heat. A backup boiler is then used to meet peak heat demand, and can also operate in place of the cogeneration plant when the heat load is below acceptable operation levels for the cogeneration unit. Addition of even small amounts of heat storage (1–2 hours worth of output) can greatly increase the number of hours per year that the cogeneration facility can operate, and allows for more continuous operation (which increases efficiency) (Haeseldonckx et al, 2007).

Figure 3.12 gives the proportion of total electricity generation that is produced decentrally in the ten countries with the largest decentralized fraction. The decentralized production consists of cogeneration and PV (photovoltaic) power, but the latter constitutes less than 3 per cent of the decentralized portion. Cogeneration accounts for just over 50 per cent of total electricity production in Denmark and about 10 per cent of global production according to data in IEA (2007a, 2007c).

There are four important and interrelated issues with regard to cogeneration:

- the impact of withdrawing useful heat on the amount of electricity produced;
- the ratio of electrical energy to useful heat produced;
- the overall efficiency in using fuel in cogeneration;

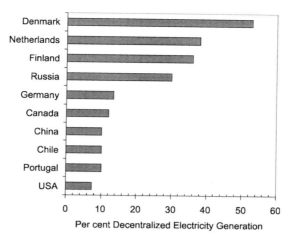

Source: Data from World Alliance for Decentralized Energy, www.localpower.org

Figure 3.12 *Fraction of total electricity supply that is generated decentrally in various countries*

- the marginal efficiency of electricity produced in cogeneration.

I elaborate on these issues below.

3.3.1 Impact of cogeneration on electricity production

The withdrawal of useful heat from the exhaust of a simple gas turbine, reciprocating engine or fuel cell for cogeneration has no impact on the production of electricity. However, for cogeneration using a steam turbine – whether a standalone turbine in a coal-fired powerplant or the steam turbine of an NGCC powerplant – an increase in the withdrawal of useful heat leads to a decrease in the production of electricity. To understand why this is so, we need to know a little more about steam turbines.

The steam turbine in a conventional coal powerplant, discussed above, can be more precisely called a *condensing steam turbine*. The steam vaporizes and expands through the turbine to well below atmospheric pressure, condenses and is recycled to the boiler. The high vacuum level allows maximal extraction of mechanical energy from the steam (to rotate the turbine), but the exhaust heat is of too low a quality (pressure and temperature) to be useful in most cases. About half the heat in the fuel ends up in the cooling water and another 15 per cent is lost through

the stack with the flue gases, resulting in an efficiency of 35 per cent in conventional systems based on condensing steam turbines. In a *condensing-extraction steam turbine* some of the steam is drawn off the turbine before it has fully expanded and is used elsewhere for heating before it is returned to the boiler. In a *back-pressure steam turbine*, all of the steam is expanded through the turbine but it exits above atmospheric pressure. Instead of passing to a condenser, the steam is used for heating elsewhere (where it condenses), then it is returned to the boiler. In either case, the steam is not fully expanded, so the work that it can do in generating electricity is less. Thus, less electricity is produced.

The loss in electricity production is greater, the greater the temperature at which the steam (or hot water) is extracted, as a greater temperature requires extraction at greater pressure. This is illustrated in Figure 3.13, which shows the loss in electrical energy output per unit of steam energy withdrawn from a stream turbine, as a function of the steam temperature. The loss in electrical output ranges from 11 per cent of the thermal energy recovered if it is recovered at a temperature of 80°C, to 35 per cent for recovery at 240°C. From a total energy point of view, one is better off extracting some heat (as long as there is a ready use for it, of course), as the loss in electricity production is smaller than the gain in useful heat. However, whether this increases the *system* efficiency depends on how

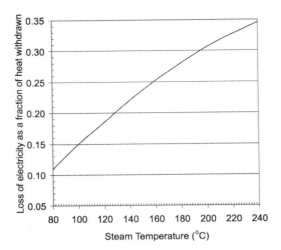

Source: Bolland and Undrum (1999)

Figure 3.13 *Loss of electricity generation when thermal energy is extracted from a steam turbine, as a fraction of the amount of energy extracted, depending on the extraction temperature*

efficiently the electricity that has been sacrificed would have been used. This is something that we will consider in Volume 2, Chapter 11, where we discuss community energy systems. It can also be seen that it will be advantageous to design whatever system is using the heat such that it can use heat at the lowest possible temperature, so that there will be the smallest possible loss in electricity production.

One might ask, why not use cogeneration with a simple cycle gas turbine, since there is no loss in electricity production in that case? The reason is because the electricity production is already very low (35 per cent of fuel energy) using a simple turbine compared to combined cycle generation (55–60 per cent of fuel energy).

3.3.2 Ratio of electricity to useful heat in cogeneration

In most cogeneration situations it will be desirable to have as high a ratio of electricity production to useful heat production as possible. There are several reasons for this:

1   electricity is a more valuable product than heat;
2   if the cogeneration system is sized to meet a given heat demand, more electricity will be produced – and hence more inefficiently generated electricity

from central powerplants can be displaced – the greater the electricity to heat production ratio; and
3   in highly insulated buildings, heat demands are very small.

The ratio of electricity to useful heat output for different cogeneration systems is summarized in Figure 3.14 for systems producing steam at ten atmospheres pressure (T = 180°C). A simple back-pressure steam turbine produces only 0.25 units of electricity for every unit of heat. This is often used for cogeneration in the pulp and paper and sugarcane industries, where electricity requirements are small (and excess electricity often cannot be exported to the electrical grid due to regulatory impediments) and part of the motivation is simply to dispose of biomass residues rather than to maximize electricity production (see Chapter 6, subsection 6.9.4, and Volume 2, Chapter 4, subsection 4.4.6). With a gas turbine using a HRSG – which can be used with natural gas or with coal and biomass fuels if they are gasified first – the electricity to useful heat ratio rises to 0.8:1.0. With fuel cells and reciprocating engines, the ratio is about 1:1. With a combined cycle system (gas turbine and back-pressure steam turbine), the ratio is 1.25:1, and in advanced systems still under development involving fuel

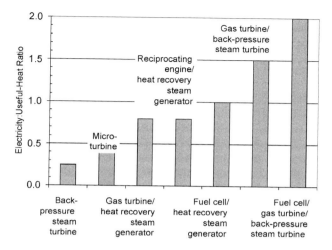

Note: Where a steam turbine is involved, the steam turbine is a back-pressure turbine without a steam condenser. The ratios are for steam produced at a pressure of ten atmospheres
Source: Based on Williams et al (2000)

Figure 3.14 *The ratio of electricity to useful heat produced in various cogeneration systems*

cells, gas turbines and back-pressure steam turbines, an electricity to useful heat ratio of 2:1 is foreseen.

3.3.3 Overall efficiency, marginal efficiency and capital cost

The electrical energy output from cogeneration divided by the total fuel energy use gives the electrical efficiency η_{el}, while the output of useful heat divided by the total fuel use gives the thermal efficiency η_{th}. The sum of the two gives the overall efficiency. The electrical efficiency can be quite low, but the relevant parameter in comparing cogeneration with standalone electricity generation is the *marginal efficiency* of electricity generation. This is the electricity produced divided by the *extra* fuel energy used compared to the generation of heat alone. The term 'marginal' is used in the sense used by economists – in reference to the benefits and costs of the last unit of production. The marginal efficiency can also be thought of as an effective efficiency. It is given by:

$$\eta_{marginal} = \frac{\eta_{el}}{1 - \eta_{th}/\eta_b} \qquad (3.19)$$

where η_b is the boiler efficiency. The more efficient the system for standalone heat production, the less fuel that

the heating system alone would use, so the greater the additional fuel use by cogeneration compared to heating alone, and the lower the marginal efficiency of electricity production.

To illustrate with a specific example, suppose that 100 units of fuel are used in a cogeneration system that produces 30 units of electricity and 35 units of useful heat (for an overall efficiency of 65 per cent, characteristic of some microturbines). To produce the same amount of heat in a condensing boiler at 92 per cent efficiency would require 38 units of fuel. Thus, an extra 62 units of fuel are used to produce 30 units of electricity, giving a marginal or effective efficiency of 30/62 = 48.4 per cent. If the efficiency of the central powerplant times the transmission efficiency is smaller than this, we save primary energy using cogeneration. Thus, computation of the marginal or effective efficiency of electricity production in cogeneration and comparison with that of the central powerplant (times transmission efficiency) automatically takes into account the energy that would otherwise be used for heating and the savings in primary energy at the central powerplant resulting from cogeneration.

Figure 3.15 shows how the marginal efficiency of electricity generation varies with the thermal efficiency for various boiler and electrical efficiencies. If $\eta_{th} = 0$, then $\eta_{marginal} = \eta_{el}$, but when η_{th} is such that the overall

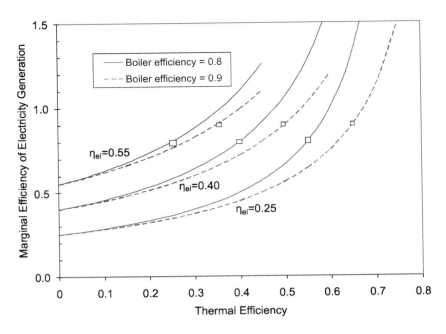

Source: Computed from Equation (3.19)

Figure 3.15 *Variation in the marginal efficiency of electricity generation in cogeneration with the cogeneration thermal efficiency, for cogeneration electrical efficiencies of 0.25, 0.40 and 0.55, each for efficiencies of 0.8 and 0.9 for the boiler that would otherwise be used to produce heat*

efficiency is equal to that of the boiler that would otherwise be used for heating, then $\eta_{marginal} = \eta_b$. Thus, if a cogeneration system with an overall efficiency of 90 per cent is used in place of a boiler with an efficiency of 90 per cent, then the marginal efficiency is 90 per cent irrespective of the direct electrical efficiency. For fixed electrical efficiency η_{el}, the marginal efficiency increases with the thermal efficiency, and can exceed 100 per cent if the overall efficiency exceeds the boiler efficiency, so maximizing η_{th} rather than η_{el} is the key to getting a large $\eta_{marginal}$.

The waste heat is produced in a cogeneration facility at a variety of temperatures. The highest temperature waste heat will be easy to use, but in order to make use of a majority of the waste heat, there must be some load that can make use of lower temperature (30–50°C) heat. This would be the case in district heating systems where heat is distributed in separate low- and high-temperature heating networks, or in some industrial applications. However, for buildings to be able to utilize 30–50°C heat for space heating applications, they require a high-performance envelope (discussed in Chapter 4, section 4.2) combined with large radiator areas (such as

underfloor radiant heating), as in this case the radiator needs to be only a few degrees Celsius warmer than the interior air temperature in order to provide adequate heat. During the summer season the only heat load would normally be for domestic hot water, which requires warmer waste heat (60–70°C), but a high utilization of waste heat (and hence, a high $\eta_{marginal}$) would be possible if low-temperature heat is used to preheat incoming cold water, and higher-temperature heat is used to bring it to the final temperature.

Table 3.5 gives the cost, electrical, thermal, overall and marginal efficiencies (all on a HHV basis; see Chapter 2, subsection 2.1.4); and pollutant emissions for cogeneration systems using natural gas or diesel fuel. From Table 3.5 it can be seen that:

- electricity generation efficiencies are lowest for microturbines (23–26 per cent) and simple cycle gas turbines (22–37 per cent); intermediate for reciprocating engines (30–37 per cent) and fuel cells (36–46 per cent); and best for combined cycle cogeneration (47–55 per cent);

Table 3.5 *Characteristics of cogeneration technologies available for use at the scale of individual large buildings (microturbines, fuel cells, reciprocating engines) and in district heating networks (simple and combined cycle turbines)*

Electric capacity	Cost ($/kW)		Efficiency (%, HHV basis)				Power to heat ratio	Emissions (gm/kWh)		
	Elect only	CHP	Electrical	Thermal	Overall	Marginal		NO$_x$	CO	Hydro-carbons
Microturbines										
30kW	2263	2636	23	44	67	44	0.52	0.23	0.63	<0.08
70kW	1708	1926	25	36	61	41	0.70	0.20	0.12	<0.08
80kW	1713	1932	24	39	63	42	0.63	0.57	0.69	<0.08
100kW	1576	1769	26	36	62	43	0.73	0.33	0.20	<0.08
Fuel cells										
200kW PAFC		5200	36	36	72	59	1.00	0.02	0.02	<0.01
5–10kW PEMFC		5500	30	39	69	52	0.79	0.05	0.03	<0.01
200kW PEMFC		3800	35	37	72	59	0.95	0.05	0.03	<0.01
250kW MCFC		5000	45	20	65	58	1.95	0.03	0.02	<0.01
2000kW MCFC		3250	46	24	70	62	1.92	0.02	0.02	<0.01
100–250kW SOFC		3620	45	25	70	62	1.79	0.02	0.02	<0.01
Diesel reciprocating engines										
100kW	1030	1350	30	48	78	63	0.61	20.9	16.8	1.0
300kW	790	1160	31	46	77	62	0.67	2.8	2.8	1.4
1MW	720	945	34	37	71	57	0.92	1.4	2.8	1.4
3MW	710	935	35	34	69	56	1.04	1.0	3.5	1.8
5MW	695	890	37	36	73	61	1.02	0.7	3.4	0.7
Gas turbines (simple cycle cogeneration)										
1MW	1403	1910	22	43	65	41	0.51	1.09	0.32	2–3
5MW	779	1024	27	40	67	48	0.68	0.50	0.27	2–3
10MW	716	928	29	40	69	51	0.73	0.50	0.23	2–3
25MW	659	800	34	36	70	56	0.95	0.41	0.18	2–3
40MW	592	702	37	35	72	60	1.07	0.36	0.18	2–3
Gas and steam turbines (combined cycle cogeneration)										
20–50MW		860	47	43	90	88	1.09	0.33	0.15	1–2
50–100MW		770	49	41	90	88	1.20	0.30	0.15	1–2
>100MW		600	55	35	90	90	1.57	0.13	0.08	1–2

Note: As discussed in the main text, electrical and overall efficiencies of the latest reciprocating engines, and overall efficiencies of the latest microturbines, are better than indicated here.
Source: Lemar (2001); Goldstein et al (2003)

- overall efficiencies for the conditions assumed in Table 3.5 are also lowest for microturbines and small gas turbines (61–67 per cent), intermediate for reciprocating engines and fuel cells and largest for combined cycle systems (90 per cent);
- as a result of the above, marginal efficiencies for electricity production are not large (around 40–45 per cent for microturbines, 52–62 per cent for fuel cells and 56–63 per cent for reciprocating engines), with the exception of combined cycle cogeneration systems (88–90 per cent);
- the power to heat ratio tends to increase with increasing size, and is largest for fuel cells and combined cycle systems;
- costs fall dramatically with increasing size;
- reciprocating engines are the least expensive building-scale option but also entail comparatively high pollutant emissions when using diesel fuel.

The marginal efficiency for electricity generation given above for building-scale microturbines is much less than that of state-of-the-art NGCC powerplants (55–60 per cent), given the electrical and thermal efficiencies in Table 3.5. As well, the marginal efficiencies for microturbines, reciprocating engines and fuel cells are all

substantially less than what can be achieved in large-scale combined cycle cogeneration (88–90 per cent). However, as noted above, much higher marginal efficiencies – in the order of 80–90 per cent – are possible in all cases if the buildings or other heat loads are designed so as to be able to use low-temperature (30–50°C) heat.

As noted earlier (subsection 3.2.3), NGCC powerplants (without cogeneration) typically cost $400–600/kW in mature markets and more in other markets. Table 3.5 gives costs for NGCC cogeneration systems of about $600–900/kW, with the lower cost pertaining to large (>100MW$_e$) systems. Figure 3.16 provides data on the expected construction cost of NGCC cogeneration facilities in various OECD countries. Costs range from $609/kW to $1568/kW, with a tendency for lower costs in the largest (around 500MW) units.

In summary, cogeneration saves energy used for electricity if the marginal efficiency of electricity generation is greater than the efficiency of supplying electricity centrally. Marginal efficiencies as high as 80–90 per cent are possible if waste heat at the temperature of condensation (30–50°C) can be utilized. By comparison, the best efficiencies for standalone powerplants are around 45 per cent using

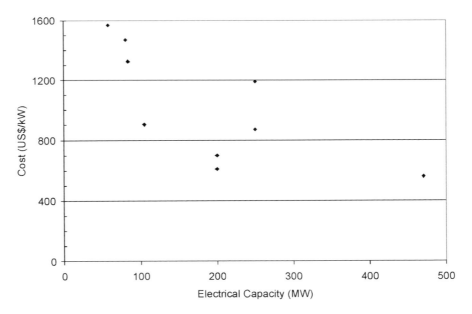

Source: Data from IEA (2005)

Figure 3.16 *Cost of recent NGCC cogeneration facilities in OECD countries versus electrical capacity*

coal and 55–60 per cent using natural gas. However, central powerplants might attain electricity-generation efficiencies of 70–75 per cent in the future through fuel cell–steam turbine hybrid powerplants powered by natural gas, in which case the efficiency benefit of cogeneration will be much smaller than today. Where cogeneration is used in district heating systems, the efficiency gain will be eroded by losses in distributing heat (Chapter 9, section 9.2), whereas there would be no distribution losses from cogeneration inside individual buildings. Alternatively, unit costs and pollutant emissions per kWh of electricity generated are smaller and the ratio of electricity to useful heat is larger at the scale of small district heating systems than at the scale of individual buildings. This in turn favours cogeneration as part of a district heating system rather than in individual buildings wherever this is feasible, with efforts to minimize distribution losses.

Two additional important points to keep in mind are: (1) cogeneration does not result in greater overall efficiency if a use cannot be found for the waste heat, and (2) however high the marginal efficiency in generating electricity with cogeneration, there is no saving in CO_2 emissions if the alternative electricity supply comes from carbon-free sources such as hydropower (unless the cogeneration plant is powered with sustainably grown biomass energy, in which case there is little or no net CO_2 emission, as discussed in Volume 2, Chapter 4, section 4.8).

3.3.4 Overall energy savings

As noted above, cogeneration reduces the amount of energy required to produce electricity compared to a centralized fossil fuel powerplant producing electricity only if the marginal or effective efficiency of electricity production is greater than that of the central powerplant. However, if there is no difference in the overall efficiency of cogeneration and the efficiency of the boiler that would otherwise be used to produce heat, then cogeneration does not save any heating energy. Thus, the overall energy savings – averaged over both heat and electricity production – depend on the ratio of electricity to heat production.

Figure 3.17 plots the savings in overall energy use with cogeneration as a function of the electricity to

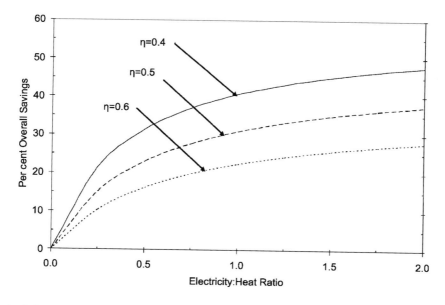

Note: Transmission loss is 5 per cent while overall cogeneration efficiency and the efficiency of the alternative for heating are both assumed to be 90 per cent.

Figure 3.17 *Savings in overall heating + electricity-generation energy use for cogeneration compared to separate production of heat and electricity for central electricity generation efficiencies of 40, 50 and 60 per cent*

heat production ratio. The overall cogeneration efficiency and the efficiency of separate heat production are both assumed to be 90 per cent. Results are given for three different efficiencies of the central alternative for producing electricity (40, 50 and 60 per cent) with 5 per cent transmission loss. When the most efficient central alternative is considered, cogeneration can reduce the combined heating and electricity-generation energy use by 25 per cent for systems that can be applied at the district heating scale (combined cycle systems with an electricity to heat production ratio of 1.5) but by only 15 per cent for systems that can be applied at the scale of individual buildings (micro-turbines with an electricity to heat production ratio of 0.4). However, there would be losses of 3–5 per cent during the distribution of heat in district heating systems, or more if pipes are not well insulated and transmission distances are large.

To summarize this and the preceding subsections: a high marginal efficiency of electricity generation requires a high overall efficiency of cogeneration and hence a high thermal efficiency, which in turn requires that a use be found for most of the waste heat. It is possible to achieve very high marginal efficiencies of electricity generation even with a low direct electricity generation efficiency if the thermal efficiency is so high that the overall efficiency is close to 100 per cent. However, to achieve large overall savings in energy use (considering both heating and electricity generation), a large electricity to heat generation ratio is required, which in turn requires a large direct electricity generation efficiency. The highest direct electricity generation efficiencies are possible at present with NGCC cogeneration, but this can only be implemented at the scale of small district heating systems and larger, rather than at the building scale.

3.4 Cost of generating electricity

When electricity is generated, it must be sold to the transmission grid (the distributor) at a price that covers the cost of the powerplant (the so-called *capital cost*) plus interest, the fuel cost, operating and maintenance costs and insurance. The price of electricity to the consumer contains additional costs to cover the cost of the power transmission lines and transformers, the loss of power during transmission, administrative costs and taxes.

A powerplant can be financed in two ways – by borrowing money (debt), on which interest must be paid, or by selling shares (equity) in the company that owns the powerplant. The rate of return that shareholders expect on their equity is much higher than the rate of interest that a credit-worthy client would have to pay to a bank or to bondholders. This is because, when individuals or institutions buy equity, they own part of the powerplant, and if the plant loses money, the shareholders lose money. The risk is greater, so a higher return is expected in compensation for the greater risk. Thus, the cost of financing a new powerplant depends critically on the *debt-to-equity ratio*. The higher the proportion financed as debt, the less the financing costs.

3.4.1 Levelized cost of electricity

A powerplant entails a large upfront expenditure before any electricity is generated, then revenues occur over the life of the powerplant (typically 40 years for fossil fuel powerplants). During this time, inflation can be expected to increase the annual fuel and operation and maintenance costs, but also to increase the price at which the electricity can be sold. A common procedure is to calculate the single price at which the electricity would have to be sold in order to just cover all the costs, including financing charges, over the lifetime of the powerplant (i.e. to just pay off the initial investment with interest or return on equity when the powerplant is retired). This is referred to as the *levelized* cost of electricity, since it is the single price of electricity that would have to be charged over the life of the powerplant, without taking into account inflation. Since inflation is ignored in all the costs, consistency requires that the interest rate that is used in calculating the financing costs be the *real* rate of interest rather than the *nominal* rate of interest. The real interest rate is the nominal interest rate (that which is quoted) minus the rate of inflation.

The levelized cost of electricity ($/kWh) from a powerplant will be given by the annual revenue requirements divided by the number of kilowatt-hours produced per year. The annual revenue requirements are given by the fixed fraction of the initial capital cost that must be recovered each year plus the fuel cost and the operation and maintenance (O&M) costs. The fixed fraction of the initial investment that must be paid back every year is referred to as the *cost recovery factor* (*CRF*) and includes interest and payments against the original loan (principle). It depends on the real interest rate per year expressed as a fraction, *i*, and the powerplant lifespan in years, *N*, as follows:

$$CRF = i / \left(1 - (1+i)^{-N}\right) \qquad (3.20)$$

Capital costs of powerplants are given as the cost per kW of peak output (per kW of capacity). If the powerplant were to run at full output all year, it would produce 8760kWh of electricity per year for each kW of capacity (8760 being the number of hours in a year). The average output as a fraction of the peak output is called the *capacity factor*. If C_{cap} is the capital cost and *CF* the capacity factor, then the component of the total electricity cost needed to pay back the capital cost is:

$$(CRF \times C_{cap}) / (8760 \times CF) \qquad (3.21)$$

If the cost of the fuel, C_{fuel}, is given in \$/GJ (1GJ = 10^9 joules) and the powerplant efficiency is η, then the fuel cost per kWh of electricity generated is:

Table 3.6 *Fixed and variable O&M costs expected for new fossil fuel and other powerplants in 2005 or 2010*

Powerplant	Fixed O&M ($/kW/yr)	Variable O&M (cents/kWh)	Reference
Coal, pulverized	30–60 (10–110)	0.20	IEA (2005)
Hard coal, pulverized	16[a]	0.2–0.74[b]	Klimstra (2007)
Lignite coal, pulverized	18[a]	0.5–2.5[b]	Klimstra (2007)
Coal, IGCC	38	0.05	Price et al (2007)
	25	0.26	Short and Denholm (2006)
	12–20[a]	0.4–1.0[b]	Klimstra (2007)
NGSC	15–35 (5–47)		IEA (2005)
	6.6	0.28	Price et al (2007)
	10.3	0.31	Short and Denholm (2006)
NGSC cogeneration, <1MW	20[a]	1.5b	Klimstra (2007)
NGSC cogeneration, <20MW	7–10[a]	0.3–0.8[b]	Klimstra (2007)
NGCC	14.4	0.30	Price et al (2007)
	11.0	0.19	Short and Denholm (2006)
	10[a]	0.35–3.0[b]	Klimstra (2007)
NGCC (300MW$_e$ scale)	12.5	0.15	Uddin and Barreto (2007)
NGCC, cogeneration (50MW$_e$)	18.0	0.17	Uddin and Barreto (2007)
Reciprocating engine, <1MW	24[a]	0.5–1.2[b]	Klimstra (2007)
Reciprocating engine, <20MW	15[a]	0.7–1.3[b]	Klimstra (2007)
Biomass steam turbine	18	0.45	Uddin and Barreto (2007)
	30[a]	0.8–2[b]	Klimstra (2007)
Biomass steam turbine, cogeneration	48	0.6	Uddin and Barreto (2007)
BIGCC	43	0.45	Uddin and Barreto (2007)
BIGCC, cogeneration	39		Uddin and Barreto (2007)
Hydro	20–60 (3–125)		IEA (2005)
Nuclear	50–75 (46–108)		IEA (2005)
Nuclear	90	0.05	Price et al (2007)
	21[a]	0.7–3.5[b]	Klimstra (2007)
Wind	15–45 (14–132)		IEA (2005)
Wind, onshore	7.5	0.38	Short and Denholm (2006)
Wind, shallow offshore	10	1.50	Short and Denholm (2006)
Wind, deep offshore	10	1.80	Short and Denholm (2006)

Note: [a] €/kW/yr; [b] € cents/kWh; where values are given inside and outside brackets, the values in brackets are extreme ranges, while values outside brackets are more typical values. BIGCC = biomass integrated gasification combined cycle; NGSC = natural gas simple cycle.

$$C_{fuel}/\eta \ (\$/GJ) \times 0.0036 \ (GJ/kWh) \qquad (3.22)$$

Annual O&M costs will have a fixed component and a component that varies with the amount of electricity generated. The total levelized cost of electricity, C_{elec} ($/kWh), is given by:

$$C_{elec} = \frac{CRF \times C_{cap} + OM_{fixed}}{8760CF} + OM_{variable} \\ + \frac{0.0036C_{fuel}}{\eta} \qquad (3.23)$$

where OM_{fixed} is the fixed annual O&M cost per kW of power capacity ($/kW/yr) and $OM_{variable}$ is the variable cost ($/kWh). For a 40-year lifespan and 7 per cent real interest rate, $CRF = 0.075$, meaning that 7.5 per cent of the investment cost has to be recouped every year in order to cover payments of interest plus principle. Table 3.6 gives the range of fixed and variable O&M costs encountered for coal, natural gas, nuclear, hydro and biomass powerplants.

The capital costs for various powerplants given earlier in this chapter are referred to as *overnight* construction costs, that is, not taking into account interest that accumulates on the initial cost expenditures between the time when the costs are incurred and when the plant is completed. The C_{cap} used in Equation (3.23) should include direct cost outlays plus interest accumulated during construction, as it is this total cost that will need to be paid back over the lifetime of the plant once it starts to generate electricity. The total cost by the time the plant is ready to start generating electricity depends on the interest rate, the time required for construction and the distribution of costs over the construction period. Table 3.7 gives the fraction of total construction costs occurring in each year during the construction of coal, natural gas and nuclear powerplants as given by IEA (2005). Construction times are typically five years for coal and nuclear powerplants and three years for natural gas powerplants, although much longer construction times have occurred for nuclear powerplants. Construction times for some other powerplants are: wind and solar, less than a year; biomass, one to two years; geothermal, two to three years; and hydroelectric, two to ten years.

Equation (3.23) is adequate if the annual O&M cost is fixed over time (excluding the effects of inflation). However, powerplants often require a midlife refurbishing, or replacement of certain subcomponents, before the lifetime of the powerplant is reached. The procedure for calculating the required levelized cost of electricity in this case is to perform a *discounted cash flow analysis*. This technique is explained in Box 3.2.

Table 3.7 *Percentage of total construction and development costs of coal, natural gas and nuclear powerplants incurred in years prior to the completion of the powerplants, based on questionnaires sent to experts in various OECD countries*

Time (Years)	Coal Range	Coal Typical	Natural gas Range	Natural gas Typical	Nuclear Range	Nuclear Typical
−9					0–1	
−8					0–2	
−7					0–2	
−6	0–0.2		0–25		0–10	
−5	0–4.7	3	0–25		8–20	10
−4	0–45	10	0–25		15–25	20
−3	25–50	30	0–57	15	17–38	30
−2	20–40	35	10–50	45	18–33	25
−1	10–35	22	5–50	40	3–23	15

Note: Costs for nuclear six to nine years prior to the completion of the powerplant are for engineering studies.
Source: IEA (2005)

Box 3.2 Discounted cash flow analysis

Discounted cash flow analysis should be used to compute the required cost of electricity when the expected operation and maintenance costs vary from year to year (such as when a major midlife refurbishing of the powerplant is required). In discounted cash flow analysis, the *net present value* (NPV) of all of the annual costs is computed as explained in Appendix D. The future costs are said to be *discounted*, and the discount rate (the fraction by which future costs are reduced for each further year into the future that they occur) is the interest rate used in computing the NPV. The amount of kWh sold each year in the future also has to be discounted, as if it were a dollar amount. This is done because we cannot discount the selling price because we do not know what it is until we calculate it, but discounting the number of kWh is equivalent to discounting the cost per kWh, since revenue equals number of kWh sold times revenue per kWh.

The constant cost at which electricity has to be sold in order to break even at the end of the lifespan of the powerplant is given by the sum of all discounted costs (including initial investment costs) divided by the sum of the discounted electricity sales over the life of the plant. An example is given below. Investment costs and O&M costs are all discounted to the time when construction begins, at year zero, so interest accrued during construction but before the plant is completed is accounted for in the final cost of electricity.

Table 3.8 *Example of a net present value analysis of a hydroelectric project with refurbishing costs at years 35 and 60 out of a 70-year lifespan, for an interest rate of 0.1*

| Year | Annual costs (£) | | | | | | | | Output (TWh) | |
	Civil works	Turbines and generators	Soft Costs	Other	O&M	Offsite	Total	NPV	Actual	NPV
1	328	119		277			724	690.31		
2	952	352		105			1409	1221.30		
3	1060	410	27	173			1670	1315.94		
4	928	266	61	171			1426	1021.52		
5	657	275	74	226	5	1	1238	806.22		
6	658	275	78	226	15	4	1256	743.58		
7	313	278	78	140	20	7	836	449.94		
8		278	49	61	29	18	435	212.84	8	3.91
9		164	17	57	34	24	296	131.66	17	7.56
10					40	30	70	28.31	17	6.87
34					40	30	70	2.87	17	0.70
35		90			40	30	160	5.97	17	0.63
36					40	30	70	2.37	17	0.58
59					40	30	70	0.27	17	0.06
60		90			40	30	160	0.55	17	0.06
61					40	30	70	0.22	17	0.05
74					40	30	70	0.06	17	0.02
75					40	30	70	0.06	17	0.01
Total	4896	2597	384	1436	2743	2034	14090	6907.75	1147	86.95

Cost of electricity (discounted cost/discounted output): £0.079/kWh

Source: Walker (1996)

3.4.2 Application to baseload and peaking powerplants

Equation (3.23) can be used to illustrate the sensitivity of electricity costs to capital costs, interest rate, capacity factor and fuel costs. From Equation (3.23), it can be seen that high capital costs and fixed operating costs will be amplified by a low capacity factor (amortizing these costs over fewer annual kilowatt-hours drives up the cost per kWh generated). At the same time, high fuel costs will be amplified by low efficiency (low efficiency means that more fuel is required to produce a kWh than if efficiency is higher). More efficient powerplants tend to be more expensive, and so will be more sensitive to higher interest rates but less sensitive to higher fuel costs than inefficient and inexpensive powerplants.

The variation in the cost of electricity with capacity factor is illustrated in Figure 3.18 for cases representing a relatively inexpensive and inefficient coal powerplant ($1000/kW, 40 per cent), a relatively expensive and efficient coal powerplant ($1500/kW, 50 per cent), an inexpensive and inefficient natural gas powerplant ($500/kW and 40 per cent), and an expensive and efficient natural gas powerplant ($2000/kW and 70 per cent, representing a solid oxide fuel cell/gas turbine hybrid). Costs are given assuming annual fixed O&M costs equal to 2 per cent of the investment cost, a variable O&M cost of 0.5 cents/kWh, 5 per cent real interest, a 40-year lifespan and fuel costs of $4/GJ for coal and $10/GJ for natural gas (both costs are almost twice current costs, reflecting my expectation of long-term scarcity for both fuels). It can be readily seen that, at low capacity factors (such as would be applicable if fossil powerplants are used merely as backup for wind and solar energy), inefficient and inexpensive powerplants are favoured over efficient but expensive powerplants, whereas the reverse is true at high capacity factors, especially when fuel costs are high. The low-capital-cost natural gas powerplant ($500/kW at 40 per cent efficiency) is not able to overcome the penalty resulting from the high assumed cost of natural gas compared to coal ($10/GJ versus $4/GJ). A carbon tax in the order of $166/tC ($45/tCO$_2$) would be required to make the expensive but efficient natural gas option competitive with coal if the price of natural gas is $10/GJ; such a tax would raise the price of coal by about $4/GJ to $8/GJ.

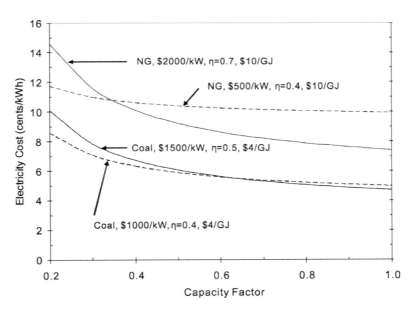

Figure 3.18 *Variation in the cost of electricity with powerplant capacity factor for various combinations of fuel cost and powerplant capital cost and efficiency, as computed from Equation (3.23) with 5 per cent interest and a 40-year lifespan*

3.4.3 Cost of electricity from cogeneration plants

The capital cost of a cogeneration powerplant is greater than that of a similar powerplant producing only electricity, but the overall efficiency is greater and there will be a revenue stream due to the sale of heat. Assume that heat is sold to a district heating grid at a selling price, P_h (cents/kWh thermal), that is dictated by the heating market. The cost of electricity generation is then given by:

$$C_{elec} = \frac{CRF \times C_{cap} + OM_{fixed}}{8760CF} - \frac{\eta_h}{\eta_e} P_h$$
$$+ OM_{variable} + \frac{0.0036 C_{fuel}}{\eta_e} \quad (3.24)$$

where η_h is the fraction of fuel energy converted to useful heat that is sold (i.e. the thermal efficiency) and η_e is the fraction of fuel energy converted to electricity (i.e. the electrical efficiency). The ratio η_h/η_e is the number of units of heat sold per unit of electricity generated. The market price of heat will vary with C_{fuel}, assuming that competitors use the same fuel for heating as is used by the cogeneration facility. The fuel contribution to the market cost of heat will be equal to $0.0036 C_{fuel}/\eta_b$, where η_b is the efficiency of the boilers used by competitors to generate heat.

Figure 3.19 illustrates the fuel contribution to the cost of heat for fuel costs of \$2–10/GJ and a boiler efficiency of 90 per cent. This will be taken as the cost at which heat from a cogeneration facility can be sold, and will be less than the market cost of heat as it does not include the cost of the heating equipment. Also shown in Figure 3.19 is the cost of electricity from a dedicated powerplant with a capital cost of \$750/kW and an efficiency of 53 per cent, as well as from a cogeneration facility with a capital cost of \$850/kW, an electricity efficiency of 47 per cent and a thermal efficiency of 43 per cent. Both gross electricity cost (i.e. before accounting for credits from the sale of heat) and net cost (accounting for such credits) are shown. Because the selling price of heat increases with the cost of fuel, the credit from heat sales serves to greatly reduce the dependence of the electricity cost on the fuel cost. With cogeneration, electricity with fuel at \$10/GJ is comparable in cost to electricity from a slightly cheaper non-cogeneration facility with fuel at \$5/GJ.

This comparison applies to onsite power generation at the 25MW scale with and without cogeneration. However, if the comparison is between small onsite cogeneration and distant, large central powerplants, then additional costs for the central powerplant need to

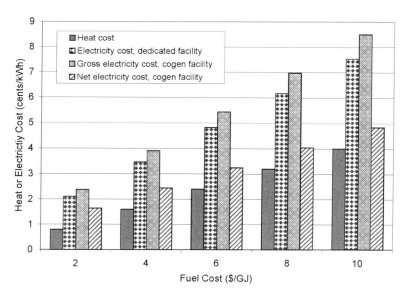

Figure 3.19 *Variation in the cost of heat (based on fuel costs only) and of electricity from dedicated electrical powerplants or from cogeneration facilities, as a function of the fuel cost and as computed from Equation (3.24)*

be considered: the greater backup requirements (typically 15 per cent instead of 5 per cent for small onsite powerplants), the greater transmission and distribution costs ($1300/kW versus $100–200/kW) and transmission losses from the central powerplant (typically 10–15 per cent during times of peak demand). If L_t is the fractional transmission loss, C_{pp} and C_T the powerplant and transmission capital costs ($/kW), respectively, and R the reserve margin as a fraction of the peak demand, then the capital cost of central power is given by:

$$C_{cap} = \frac{C_{pp}(1+R)}{1-L_t} + C_T \qquad (3.25)$$

An adjustment should also be made in the fixed O&M cost. Figure 3.20 compares the cost of electricity from central coal plants using coal at $2/GJ and central and onsite natural gas powerplants using natural gas at $10/GJ, with a 5 per cent real interest rate, a 40-year lifespan and additional assumptions as given in Table 3.9. In spite of the much higher assumed cost of natural gas compared to coal, the cost of electricity from an onsite peaking natural gas powerplant is, even without cogeneration, less than that from a distant peaking coal powerplant, and the economic advantage is even greater with cogeneration.[5] With cogeneration, natural gas would need to cost $18/GJ before the cost of electricity from onsite natural gas cogeneration exceeds that from a

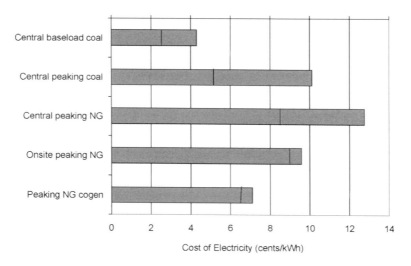

Note: The left part of each bar gives the cost using C_{pp} for C_{cap} in Equation (3.24), while the full bars give the costs using C_{cap} in Equation (3.24) as computed from Equation (3.25).

Figure 3.20 *Cost of electricity from central baseload or peaking coal powerplants and from peaking central natural gas, onsite natural gas and natural cogeneration power plants, computed for the assumptions given in Table 3.9 along with a 5 per cent interest rate and 40-year lifespan*

Table 3.9 *Assumptions used to generate the electricity costs shown in Figure 3.20*

	Fuel cost ($/GJ)	C_{pp} ($/kW)	C_T ($/kW)	R	L	C_{cap} ($/kW)	η_e	η_h	CF
Central baseload coal	2	1200	1300	0.15	0.1	2833	0.48		0.9
Central peaking coal	2	1200	1300	0.15	0.1	2833	0.43		0.3
Central peaking natural gas	10	550	1300	0.15	0.1	2002	0.60		0.3
Onsite peaking natural gas	10	750	150	0.05	0.0	938	0.53		0.3
Peaking natural gas cogeneration	10	850	150	0.05	0.0	1043	0.47	0.43	0.3

Note: R = reserve margin as a fraction of peak demand; L = fractional loss during transmission at time of peak demand.

distant coal powerplant (assuming the price at which heat could be sold increases with the cost of natural gas). Thus, highly efficient onsite power generation can absorb quite high natural gas prices and still remain competitive with distant coal-fired electricity.

3.5 Water requirements of thermal powerplants

Most steam-based powerplants make use of a flow of cool water in order to condense the steam that exits from the steam turbine, as illustrated earlier in Figure 3.3. The cool water could be either river or lake water that is withdrawn from the river or lake, passes through the condenser and returns to the river or lake. This is referred to as a *once-through* system. Conversely, cooling water could be continuously circulated between the condenser and a *cooling tower*. The cooling water is cooled in the cooling tower through partial evaporation of the cooling water or through evaporation of a separate water flow that is sprayed onto the cooling water coils (see Chapter 4, Figure 4.43). This is referred to as a *recirculating-loop* system. The amount of water withdrawn from a water source is relatively large in a once-through system, as each unit of water is used only once before being returned to the source. However, very little water is lost, so the net consumption is very small. Conversely, relatively little water is withdrawn in a recirculating-loop system, but water is continuously lost through evaporation in the cooling tower, so the water consumption is relatively large. The water that is withdrawn in a recirculating-loop system is largely to make up for evaporative losses, so the consumptive water use is close to the rate of withdrawal. Small amounts of water are also used for purposes other than cooling, such as periodically flushing the boiler with clean water to clear out impurities that accumulate. Water is also required for periodically flushing the cooling tower.

Table 3.10 compares the amount of water withdrawn and consumed per kWh of electricity generated for different kinds of powerplants and cooling systems. Gas turbines, which do not use steam and therefore have no need for a condenser and associated cooling water, have very low water requirements (0.0–0.4 litres/kWh according to Lindenberg et al, 2008). In combined cycle powerplants (whether based on natural gas or IGCC), about two thirds of the electricity comes from the gas

Table 3.10 *Water withdrawals and water consumption by steam and combined cycle powerplants*

Powerplant and cooling system	Water use (litres/kWh)	
	Withdrawal	Consumption
Steam		
Once-through	80–190	1.1
Recirculating	1–3	0.9–2.4
Dry	0.15	0.00
Combined cycle		
Once-through	30.0	0.38
Recirculating	0.87	0.68
Dry	0.15	0.00
IGCC		
Recirculating	1.40	0.76

Source: CATF (2003)

turbine and one third from the steam turbine. Thus, the average water use by the cooling tower per kWh of electricity generated is about a third that of steam turbines. However, IGCC requires additional water (as steam) for the gasification process and for suppression of coal dust during the transport of coal. All else being equal, the cooling water use by steam turbines will be smaller the more efficient the turbine, as less steam will be used (and less cooling required) per kWh of electricity generated.

Lack of water can be a serious constraint on further development of electric powerplants in arid regions. It is possible to use air rather than water for cooling purposes, but at the cost of reduced efficiency in generating electricity. Water is also required for nuclear, solar thermal and biomass powerplants, with water use per kWh of electricity generated roughly proportional to the thermal efficiency of the steam turbine. Altogether, thermoelectric powerplants account for about 52 per cent of total freshwater withdrawals in the US (see Hutson et al, 2004). Wind and PV electricity generation are advantageous over other forms of electricity generation in that they have no water requirements (except very small amounts for cleaning of PV modules).

3.6 Summary

The efficiency, capital cost and cost of electricity (neglecting transmission, backup and administrative costs) for current and advanced systems for generating electricity from fossil fuels are summarized in Table 3.11.

Table 3.11 *Summary of efficiency, capital cost, CO_2-equivalent emission relative to typical existing coal plants, and cost of electricity*[6]

	Efficiency (LHV basis) (%)		Overnight capital cost ($/kW)	Relative CO_2 emission	Cost of electricity (cents/kWh)	
	Electrical	Thermal			@ 5%	@ 10%
Pulverized coal						
Developing country average	28–30					
Developed country average	33–35			1.00		
Best current	45		1200–1400	0.76	3.8–5.4	4.8–6.4
Future	45–48		1100–1200	0.73	3.5–5.1	4.4–6.0
Integrated gasification combined cycle						
Current	38–47		1400–2200	0.80	4.7–6.4	6.1–7.8
Future	41–55		1150–1400	0.71	3.7–5.2	4.7–6.2
Natural gas combined cycle						
Best current	56–60		400–600[a]	0.42	5.1–8.8	5.7–9.4
Other natural gas						
FC, current	35–45		3000–5000	0.61	11.8–17.2	14.9–20.3
FC, projected future	40–60		1000–1500	0.49	6.5–10.8	7.4–11.7
FC/GT or FC/ST hybrid	70–80		2000–3000	0.33	7.0–9.8	8.9–11.8
Cogeneration using natural gas						
Microturbine, 100kW	23–27	40–50	1800–2600	0.55	7.6–11.5	9.3–13.2
Combined cycle, >20MW	45–55	30–40	600–900	0.27	3.4–5.4	4.6–6.0
FC, current	35–45	45–55	3000–5000	0.27	8.1–9.7	11.1–12.8
FC, projected future	40–60	40–50	1500–2000	0.27	4.6–6.2	5.9–7.6
Reciprocating engine	40–46	40	600–1200	0.32	5.9–8.2	7.7–9.9

Note: FC = fuel cell, GT = gas turbine, ST = steam turbine. Future costs are projections by proponents of the technologies in question.
[a] In mature markets, but $600–900/kW in most developing countries.

Also given are the relative CO_2 emissions. There is a factor-of-four improvement in efficiency in going from present central coal-fired powerplants to natural gas cogeneration. However, as noted in Chapter 2, natural gas supplies are already limited in parts of the world, so high natural gas prices are expected. Nevertheless, natural gas cogeneration with gas at $12/GJ is competitive with advanced coal without cogeneration when coal costs $4/GJ. When transmission costs and losses, and the difference in required backup are taken into account, natural gas cogeneration in urban centres for peaking can compete with electricity from distant coal-fired powerplants for natural gas prices of up to $18/GJ when the cost of coal is only $2/GJ. If one has peaking plants in combination with cogeneration, heat storage systems have to be added, but these are not costly.

Notes

1   In France, where 78 per cent of all electricity is supplied by nuclear reactors (see Table 2.4), the output is varied by discarding some steam (that is, wasting steam energy) when less power is needed, but the heat production from the nuclear reactor does not vary.

2   A supercritical steam generator operates above the highest pressure at which steam bubbles can form. This pressure is called the *critical pressure* and is equal to 22MPa or 220atm. As there is no generation of steam bubbles within the water (that is, boiling does not occur), the steam generator in this case is not a 'boiler'.

3   Gasifiers using regular air also exist but are not expected to be competitive on economic or environmental grounds, for reasons explained by Williams et al (2000).

4   Methane reforming involves two steps, the first being:

$$CH_4 + H_2O \rightarrow CO + 3H_2$$

followed by a two stage water shift reaction:

$$CO + H_2O \rightarrow CO_2 + H_2$$

Coal gasification can also be pushed to an output consisting essentially of CO_2 and H_2.

5   Operation of a cogeneration facility for providing peak power requires that buildings be sufficiently well

insulated that they can handle periods without heat supply, or that heat generated during times of electricity demand can be stored for use when it is needed. However, storage systems are not costly.

6   Relative CO_2 emissions were computed using the mid-range of the CO_2 emission factors from Table 2.10 and assuming upstream CO_2-equivalent emissions of 5 per cent and 25 per cent of the onsite emissions for coal and natural gas, respectively. Electricity costs have been computed from Equations (3.23) or (3.24) assuming the midpoint for the range of capital costs and efficiencies given here, 5 per cent or 10 per cent financing over 40 years, fixed annual O&M equal to 3 per cent of the capital cost, variable O&M of 0.2 cents/kWh, a heat credit (in cogeneration) equal to $1.25(\eta_h\eta_e)P_h$, and a 65 per cent capacity factor. Lower and upper costs are for coal at \$2/GJ and \$4/GJ, respectively, or for natural gas at \$6/GJ or \$12/GJ, respectively.

4

Energy Use in Buildings

The building sector is often subdivided into residential, commercial and institutional buildings. Commercial buildings include low-rise retail buildings and high-rise office towers, but could also include (depending on one's definition) multi-unit residential buildings – condominiums and apartments. The institutional sector includes government buildings, hospitals, schools and universities. Many buildings in the institutional sector share similarities with commercial buildings. Thus, the distinctions between residential, commercial and institutional buildings are somewhat blurred and arbitrary. For this reason we shall lump institutional buildings with the commercial sector, so that henceforth we shall refer to just the residential and commercial sectors.

As seen in Chapter 2, energy use in buildings accounts for 53 per cent of total electricity use and 38 per cent of total primary energy in OECD countries (see Figures 2.9 and 2.10, respectively). Energy is used in buildings for heating, cooling and ventilation, for producing hot water, for lighting, and as electricity to power appliances and/or consumer goods and/or office equipment. Figure 4.1 shows the breakdown of total onsite residential energy use in the US, the 15-nation European Union (subsequently referred to as the EU-15, with the countries listed in Appendix C) and in China. In all three regions, space and hot water heating together account for two thirds to four fifths of total residential energy use. Figure 4.2 shows the breakdown of onsite energy use in the commercial sector of the US, the EU-15 and China. Space heating accounts for a third to a half of total energy use in commercial buildings in all three regions. Lighting accounts for 15–30 per cent of total onsite energy use and 30–60 per cent of electricity

use. The commercial sector differs from the residential sector in that lighting, cooling and ventilation are large energy uses.

The most effective strategy to reduce energy use in buildings is to first focus on reducing heating and cooling loads through a high-performance envelope, then to meet as much of the reduced load as possible through passive solar heating, ventilation and cooling while optimizing the use of daylighting, and finally, to use the most efficient mechanical equipment and configurations of mechanical equipment possible to meet the remaining loads. This sequence is followed in the discussion of heating, cooling and lighting in this chapter. Full realization of the potential energy savings requires an integrated, team approach to the design of buildings. This chapter shows that, for all building types in all climate zones, application of an integrated design process can achieve reductions in energy use in new buildings of 50–75 per cent and sometimes even more compared to buildings built according to local building codes. Comprehensive retrofits of existing buildings are also addressed in this chapter, and it will be seen that 50–90 per cent reductions in energy use have been achieved for many different types in buildings in many different parts of the world.

Active uses of solar energy (building-integrated PV, solar hot-water heaters, solar absorption chillers) are discussed in Volume 2, Chapter 2 (Solar Energy). Active solar heat collectors can be combined with underground seasonal thermal energy storage, but this is best done at a community scale as part of a district heating system, so yet another layer of analysis and integration is also deferred to Chapter 9 (Community-Integrated Energy Systems).

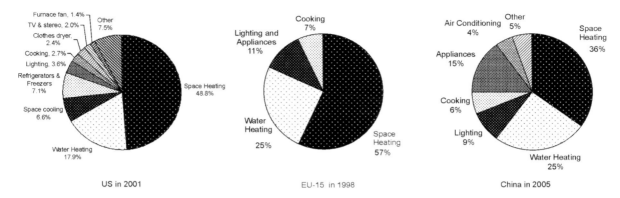

Source: Data from (a) Residential Energy Consumption Survey, www.eia.doe.gov/emeu/recs/recs2001, (b) EC (2001a), (c) Lin et al (2008)

Figure 4.1 *Breakdown of residential secondary (onsite) energy use in (a) the US in 2001,*
(b) the EU in 1998, and (c) China in 2005

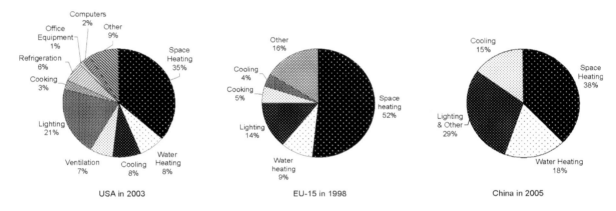

Source: Data from (a) Energy Information Administration, www.eia.doe.gov, (b) EC (2001a), and (c) Lin et al (2008)

Figure 4.2 *Breakdown of commercial sector secondary (onsite) energy use in (a) the US in 2003,*
(b) the EU in 1998, and (c) China in 2005

4.1 Physical principles

Physical principles to be discussed are: processes of heat transfer between the inside and outside of a building; processes for distributing heat, coldness and fresh air inside a building; and the human perception of temperature. These principles lay the foundation for understanding the substantial scope for reducing energy use in buildings.

4.1.1 Processes of heat transfer

Heat can be transferred across the building envelope through molecular conduction, convective mixing, the passage of air through leaks in the envelope, and through radiative exchange.

Conduction and convection

Molecular conduction involves transferring molecular kinetic energy (heat) to nearby molecules through collisions, while convection involves movement of air parcels from one place to the next either as turbulent eddies, airflow or a combination of the two. Convection can occur in the air on either side of a wall or window, in cavities inside hollow-core walls or concrete blocks, and in the air gap inside double-glazed windows. In the last of these, an overturning circulation cell will occur, with air sinking adjacent to the cold glazing and rising adjacent to the warm glazing. Immediately next to a wall, or on either side of a pane of glass, will be a thin layer where heat transfer occurs by molecular conduction only. The heat flux across a wall due to conductive and convective heat transfer, Q_c, is given by:

Q_c = temperature difference × heat transfer
coefficient = $\Delta T \times U$ (4.1a)

or:

Q_c = temperature difference/thermal
resistance = $\Delta T/R$ (4.1b)

and has units of W/m². This times surface area (m²) times time (in seconds) gives one component of the heat loss (J) that must be supplied by the heating system during that time period if a constant temperature is to be maintained. The heat transfer coefficient (U or h, W/m²/K) for a given layer is given by the *thermal conductivity* (k, W/m/K) for that material divided by the thickness of the layer (D, m), while the resistance R is given by the reciprocal of U. Thus:

$U = k/D$ (4.2a)

$R = 1/U = D/k$ (4.2b)

Thus, the heat loss through a layer will be smaller the thicker it is and the smaller its thermal conductivity.

Exchange of air between inside and outside

The heat loss due to exchange of warm indoor air with cold outdoor air depends on the indoor-to-outdoor temperature difference (the greater this difference, the more heat that needs to be added to the incoming outdoor air in order to maintain the same indoor temperature), as well as on the rate of air exchange (Q, m³/s), density (ρ, kg/m³) and *specific heat* of air (c_{pa}, J/kg/K).[1] That is:

$Q_e = \rho c_{pa} Q (T_{indoor} - T_{outdoor}) = \rho c_{pa} Q \Delta T$ (4.3)

where T_{indoor} is the temperature of indoor air (which leaves the building) and $T_{outdoor}$ is the temperature of outdoor air (which enters the building). Q_e has units of watts. Multiplication by time (in seconds) gives the heat loss (J) due to air exchange over that time period.

Exchange of air occurs through unintentional infiltration of outside air and exfiltration of inside air, as well as through deliberate air exchange through the ventilation system in order to provide fresh air.

Radiative transfer

All objects above 0K (–273.15°C) emit electromagnetic radiation. The warmer the object, the shorter the wavelengths of the radiation that it emits. The sun (with a surface temperature of 5773K) emits radiation largely between wavelengths of 0.1µm and 4.0µm (1µm = 10^{-6}m). The peak solar emission occurs at 0.55µm, and visible light – radiation to which our eyes are sensitive – occurs between wavelengths of 0.4 and 0.7µm. For objects typical of earth and atmospheric temperatures (200–300K), the emitted radiation falls in the longwave or *infrared* part of the spectrum. The maximum amount of radiation that can be emitted depends on temperature (warmer objects emit more radiation, as well as at shorter wavelengths). This maximum amount of radiation is called *blackbody emission*, B, and is given by:

$B = \sigma T^4$ (4.4)

where σ (= 5.670400 × 10^{-8} W/m²/K⁴) is the Stefan-Boltzmann constant and T is in K.

An object that emits the maximum possible amount of radiation is called a *blackbody*. The *emissivity* ε is the ratio of actual emission to maximum possible emission; thus, actual emission E is given by:

$E = \varepsilon \sigma T^4$ (4.5)

Most building materials have emissivities of 0.9–0.95, that is, they behave almost as blackbodies. However, it is possible to achieve emissivities of 0.05–0.20 with many metal surfaces if they are highly polished or galvanized. The absorption of infrared radiation by a body is given by the incident infrared radiation (emitted by the surroundings) times the *absorptivity* (the fraction absorbed). The absorptivity is always equal to the emissivity. Infrared radiation that is not absorbed by solids (such as window panes or walls) is reflected.

Thus:

• a blackbody emits the maximum amount of radiation for its temperature, and absorbs all of the incident infrared radiation from the surroundings (which is why it is called a 'black' body – real black objects absorb all the incident visible light);

- for objects with $\varepsilon<1$, the emission of radiation is only ε times the blackbody emission (so it cools less), while $(1-\varepsilon)$ times the incident infrared radiation from the surroundings is reflected and only ε times the incident radiation is absorbed.

When an object absorbs radiation (whether solar or infrared), it gains energy and warms up. Conversely, when an object emits infrared radiation it losses energy, so it cools.

Figure 4.3 shows the distribution with wavelength of solar radiation and of infrared radiation emitted by blackbodies for typical earth surface temperatures. Also given is the variation in the sensitivity of the human eye to radiation. Solar radiation is divided into ultraviolet radiation (wavelengths less than 0.1μm), visible radiation (wavelengths of 0.4–0.7μm) and near-infrared radiation (wavelengths of 0.7–4.0μm). Objects at typical earth atmosphere temperatures emit radiation almost exclusively at wavelengths greater than 4.0μm. The human eye is most sensitive to radiation at the peak of solar emission, near a wavelength of 0.55μm, corresponding to green light. The tendency for warmer objects to emit more radiation but shift to shorter wavelengths is evident from the blackbody curves.

Computation of overall window and wall U-values

An outline of how the U- and R-values of individual components of a window or wall are combined in computing the overall window or wall U-values is given in Box 4.1.

4.1.2 Heat transfer by circulating air or water

Heat can be distributed to individual rooms either by supplying air that is warmer than the room through the ventilation system, or by supplying hot water to some sort of radiator. As long as the supply air or radiator is warmer than the room, it will give off some of its heat, thereby returning to the furnace or water heater at a cooler temperature. It is then reheated and completes the circuit again. The temperature at which the air or hot water leaves the heating system is referred to as the *supply temperature*, while the temperature when it returns to be reheated is referred to as the *return temperature*. Systems based on hot water are called *hydronic* systems. The rate at which heat is given off (in W) by an air or water flow, Q_H, is given by:

$$Q_H = \rho c_p\, Q\, (T_{supply} - T_{return}) = \rho c_p\, Q\Delta T \qquad (4.6)$$

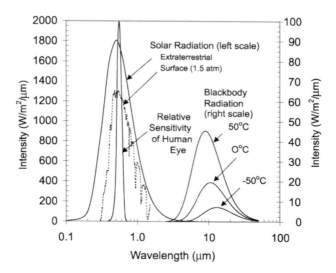

Note: The left scale is for solar radiation and the right scale is for blackbody radiation at –50 to 50°C

Figure 4.3 *Variation in the amount of solar radiation with wavelength, in the blackbody emission of radiation with wavelength, and in the sensitivity of the human eye to radiation of different wavelengths*

Box 4.1 Computation of heat flow through a wall or window

To get the total resistance of a wall to the flow of heat, the resistances of the individual layers (such as exterior cladding, insulation, interior drywall) are simply added up. However, for layers that consist of different elements in different regions (such as studs and insulation between studs), one must compute the separate U-values for the studs and the insulation (or air gap) between the studs, then compute the area-weighted average U-value (based on the thickness of the studs and the distance between stud surfaces), and finally, take the reciprocal of the area-weighted U-value to get the effective mean resistance for that layer. Once the effective mean resistance for the stud layer has been computed, the resistances for any other layers and for the air next to the inside and outside of the wall are added together to get the total resistance. The reciprocal of the total resistance gives the overall wall U-value. The calculation procedure is outlined in Figure 4.4. In this figure, h_1 and h_6 are heat transfer coefficients with units of $W/m^2/K$ (that is, having the same units as U-values).

In the case of windows, heat flow occurs by molecular conduction through the glass layers and by conduction or convection and radiative heat exchange between glass layers, between the outer glazing layer and the outside air, and between the inner glazing layer and the inside air. At the edge of the

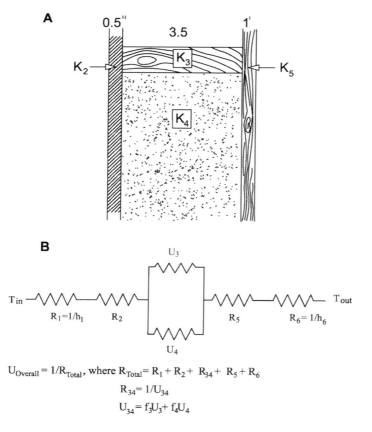

$$U_{Overall} = 1/R_{Total}, \text{ where } R_{Total} = R_1 + R_2 + R_{34} + R_5 + R_6$$
$$R_{34} = 1/U_{34}$$
$$U_{34} = f_3 U_3 + f_4 U_4$$

Source: Sherman and Jump (1997)

Figure 4.4 *Heat flow through a wood stud, insulated wall*

window, heat transfer also occurs through the frame and through the spacers that hold multiple glass layers apart. The resistance diagram for heat flow away from the edge of the window is shown in Figure 4.5. As conductive and radiative flows through air occur simultaneously over the entire window area, the corresponding heat transfer coefficients for a given layer are simply added, then the reciprocal is taken to get the resistance for that layer. The total resistances are added up, and the reciprocal is taken to get the overall centre-of-glass U-value.

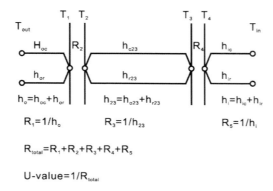

Figure 4.5 *Heat flow through a double-glazed window*

where Q is the volumetric flow rate (m³/s), c_p is the specific heat (J/kg/K), ϱ is density, and the ΔT in this case is the difference between the supply and return temperatures. For air, $c_p = c_{pa} = 1004.5$J/kg/K and $\varrho = 1.25$kg/m³ at 10°C and 1atm, while for water, $c_p = c_{pw} = 4186$J/kg/K and $\varrho = 1000$kg/m³ at 4°C. From this it can be seen that, for a *given* volumetric flow rate and temperature drop, water delivers 3333 times as much heat as air.

To keep a room at a given temperature, the rate of heat delivered by the heating system must equal the rate of heat loss. For a given flow of a given fluid (air or water), the rate of heating is proportional to the temperature drop of the fluid. A larger temperature drop will occur the warmer the air or water relative to the space being heated, the larger the radiator, or the slower the flow rate. If the building has a good thermal envelope, the rate of heat loss is small, so heat does not need to be supplied as rapidly by the heating system, and a lower supply temperature and/or a smaller flow rate is possible. However, most heating systems are rather inflexible, with a fixed supply temperature and a fixed flow rate, so part load is satisfied with intermittent on/off operation (and often overheating the building, requiring windows to be opened in winter).

Energy used to move air or water

Energy is required for fans or pumps to move air or water. The following discussion applies to both pumps and fans, moving water and air through pipes and ducts, respectively, but to keep the wording less onerous, the term 'pump' will be taken to mean pumps or fans, the term 'pipe' will be taken to mean pipes or ducts, and the term 'fluid' will be taken to mean air or liquid. When a fluid flows through a pipe, it encounters a resistance that leads to a drop in pressure (ΔP) along the flow path. The pump supplies pressure to the fluid that exactly compensates for the pressure drop during the roundtrip circuit from the pump, through the pipes and back to the pump. The rate at which energy needs to be supplied to a fluid by the pump (i.e. the power input to the fluid, P_{fluid}) is equal to the pressure drop times the rate of flow. However, for turbulent flow, the pressure drop varies with the flow velocity squared, so for a given pipe diameter, the pressure drop will vary with the flow rate squared. That is:

$$P_{fluid} = \Delta P\, Q\ \alpha\ Q^3 \qquad (4.7)$$

Thus, the fluid power (W) varies with the flow rate to the *third power* – a cubic relationship. The required electrical power is given by:

$$P_{electric} = \frac{P_{fluid}}{\eta_m \eta_p} \tag{4.8}$$

where η_m and η_p are the motor and pump efficiencies, respectively. The motor efficiency is the ratio of shaft power to electrical power, while the pump (or fan) efficiency is the ratio of power supplied to the fluid to shaft power. One way to vary the output of a pump is to vary the speed of the motor that drives the pump. This requires an electronic interface called a *variable speed drive* (VSD), which converts the input AC electricity from 50 or 60Hz to some other frequency. This entails some energy loss, so the required electric power is:

$$P_{electric} = \frac{P_{fluid}}{\eta_{VSD} \eta_m \eta_p} \tag{4.9}$$

where η_{VSD} is the VSD efficiency (ratio of electrical power out to electrical power in).

The pressure drop along a pipe depends on the pipe diameter, as well as on the flow rate. For a given flow rate, ΔP varies *inversely* with the diameter of the pipe to the *fifth power*. Thus, a very small increase in the pipe diameter will dramatically decrease the energy loss due to friction. Significant pressure drops occur where there are bends, but these can be minimized in ducts through turning vanes (metal fins) inside the duct.

Based on Equation (4.7), a 50 per cent reduction in the required flow would reduce the energy used by a pump by a factor of *eight* if ducts or pipes are not changed in sized. However, the motor and pump efficiencies also decrease with decreasing fluid flow. This occurs both for peak flow for different systems (smaller pumps and motors have smaller efficiencies) and as the flow rate decreases for a given system (part-load efficiencies are smaller than peak-load efficiencies).

Data on peak-load efficiencies for different-sized fans and motors are given in Table 4.1, where it can be seen that there is more than a factor of two difference between the smallest and largest systems in the combined efficiency, mostly due to differences in the

Table 4.1 *Typical fan and motor peak-load efficiencies*

Volumetric flow rate (m³/h)	Fan efficiency	Motor efficiency	Combined efficiency
Up to 300	0.4–0.5	0.80	0.32–0.40
300–1000	0.6–0.7	0.80	0.48–0.56
1000–5000	0.7–0.8	0.80	0.56–0.64
5000–100,000	up to 0.85	0.82	up to 0.70

Source: Hastings and Mørck (2000)

fan efficiency. Thus, the decrease in electric power required as the flow rate decreases is much less than expected based on the cubic relationship, but is still substantial. For example, a 50 per cent reduction in the peak flow rate can be expected to reduce the peak electric power by about a factor of six for constant-sized ducts.[2]

With regard to part-load operation in a given system, the change in energy use depends on how the flow is made to vary. The usual method to reduce the flow is by *throttling* – partially blocking the flow while the pump continues to pump as if it needed to provide the peak flow (this is like driving with the accelerator pedal to the floor, and using the brakes to prevent yourself from going too fast). This is obviously highly inefficient. There is some reduction in the power used by the pump when the flow is throttled, but not much. An alternative is to vary the rotation rate of the pump or fan (with a VSD) or to change the pitch of the blades, in which case close to the cubic dependence of power on flow is obtained. This is illustrated in Figure 4.6. For fans or pumps with VSDs, the required power decreases several times faster than if throttling is used to control the flow, but not as rapidly as expected based on the cubic relationship. This is because the efficiency of a fan or motor at part-load is smaller than the peak-load efficiency. For example, a high-efficiency motor might have an efficiency of 72 per cent at full load and 20 per cent at 30 per cent of full load (Metwally, 2001). A pump with a full-load efficiency of 85 per cent might have an efficiency of 60 per cent at 20 per cent of full load. The efficiency of a VSD also decreases with decreasing load, easily falling from 95 per cent at full load to 75 per cent at 30 per cent of full load (Bernier and Lemire, 1999). Since the net efficiency is the product of these three efficiencies, there is a substantial efficiency penalty at part load. Thus, it is important to

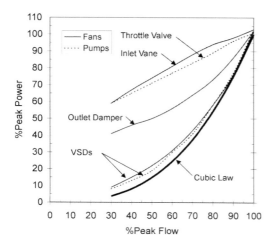

Note: The curve labelled 'cubic law' assumes that power α (flow)$^{2.75}$.
Source: Smith (1997)

Figure 4.6 *Variation of fan or pump power with flow for various ways of reducing the flow*

avoid oversizing the fans or pumps relative to the design flow (although, as noted above, the pipes can be oversized), so that they operate at as large a fraction of full load as possible. Although the VSD introduces a substantial efficiency loss at part load, the energy consumption at part load with a VSD is still less than without it because the pump is attempting to provide only the fluid power (P_{fluid}) that is needed.

In summary, the keys to reducing the energy required to circulate air or water are:

- to *oversize* pipes or ducts (because the pressure drop decreases with pipe diameter to the fifth power), something that will be easier for pipes than for ducts due to space limitations;
- to lay out the heating and cooling elements so as to minimize the length of pipe or duct and to minimize the number of turns, and to use turning vanes in the bends of ducts;
- to design the heating and cooling system to require as little water flow or airflow as possible (because the power that must be supplied to the flow varies with flow rate to the third power);
- to utilize VSDs for efficient part-load operation;
- to *avoid oversizing* the pumps, motors or fans so that the system is operating on average as close to full load as possible (because the efficiencies of

pumps, motors, fans and VSDs are all greatest at or close to full load).

Comparison of hydronic and air systems

As noted above, the heat that can be transferred for a given volumetric flow rate and temperature drop is over 3000 times greater for water than for air. However, the energy required to produce a given volumetric flow rate is also greater for water than for air. The relevant quantity in each case is the required pump or fan power divided by the rate of heat flow that is accomplished, that is, Equation (4.7) divided by Equation (4.6):

$$E = \frac{\Delta P}{\rho c_p \Delta T} \tag{4.10}$$

where ΔT is the difference between supply and return temperatures. Using the values of ρ and c_p given above, and taking ΔP to be 1400Pa in an air system and 50,000Pa for an equivalent hydronic system (as computed by Niu et al, 2002, for a given building), it can be seen that heat transfer by water requires about 100 times less energy for a given ΔT than heat transfer by air, neglecting differences in pump, fan or motor efficiencies. However, if the temperature drop ΔT in the air system is four times larger (i.e. 8K rather than 2K), the energy used to transfer a given amount of heat with water would be 25 times smaller than using air. This is still a substantial difference!

4.1.3 Sensible and latent heat

Heat in air occurs in two forms: as sensible heat and as latent heat. *Sensible heat* refers to the heat that we can sense, as a warmer temperature. *Latent heat* refers to the heat that is released when water vapour condenses. The amount of moisture that air can hold depends on its temperature; the warmer the air, the more it can hold. The *saturation vapour pressure* is the water vapour pressure when the air is holding the maximum amount of water vapour possible, and it increases rapidly with increasing temperature. The *relative humidity* (RH) of an air parcel is the ratio of the actual vapour pressure to the saturation value, times 100 to give it as a percentage. The amount of moisture in air can also be represented by its *mixing ratio, r* (kg/kg), which is the ratio of mass of water vapour to mass of dry air. Quantitative relationships involving vapour pressure, mixing ratio and sensible and latent heat are given in Box 4.2. If an air parcel is cooled, the saturation vapour pressure decreases but the actual vapour pressure remains the same, so the relative humidity increases. Further cooling causes water vapour to condense, releasing latent heat that has to be removed along with sensible heat. Condensation begins when the relative humidity reaches 100 per cent, and the

temperature at which this occurs is called the *dewpoint temperature*, T_{dp}.

If we allow liquid water in an air parcel to evaporate without adding heat from the surroundings, the temperature of the parcel will decrease (as heat energy is used to evaporate water) and the humidity of the air will increase. This will continue until the parcel becomes saturated, at which point no further evaporation or cooling will occur. The temperature at which this occurs is called the *wetbulb temperature*, T_{wb}, because it is given

---

Box 4.2 Vapour pressure, sensible heat, latent heat and enthalpy

The mixing ratio r (kg/kg, the ratio of mass of water vapour to mass of dry air) is related to the *vapour pressure*, e_a, by the relation:

$$r = 0.622 \frac{e_a}{P_a - e_a} \tag{4.11}$$

where 0.622 is the ratio of molecular weights for water vapour and air, and P_a is the total atmospheric pressure. The *relative humidity* of an air parcel is the ratio of the actual vapour pressure (or mixing ratio) to the saturation value, times 100 to give it as a percentage. That is:

$$RH = \frac{e_a}{e_s} 100\% = \frac{r}{r_s} 100\% \tag{4.12}$$

The latent heat content of an air parcel per unit mass (J/kg) of dry air is given by:

$$L = L_c r \tag{4.13}$$

where L_c is the *latent heat of condensation* – the amount of heat released when 1kg of water vapour condenses. It depends weakly on temperature, with a value of 2.501×10^6 J/kg at a temperature of 0°C.

The sensible heat content of air per unit mass (J/kg) is given by:

$$S = c_{pa} T + r c_{pwv} T \tag{4.14}$$

where T is in kelvin and c_{pa} is the specific heat of air (1004.5 J/kg/K) – the amount of heat that must be added to warm 1kg of air by 1K – and c_{pwv} is the specific heat of water vapour (1860 J/kg/K). The combination of sensible heat and latent heat per unit mass of dry air gives the *specific enthalpy H* of an air parcel, where:

$$H = c_{pa} T + r (L_c + c_{pwv} T) \tag{4.15}$$

In practice, it is more convenient to compute enthalpies relative to the enthalpy of an air parcel at a temperature of 0°C, that is, to use T in Celsius degrees rather than in kelvin.

by the temperature of a thermometer bulb with a wet cloth over it, in equilibrium with a steady airflow. The initial air temperature is referred to as the *drybulb temperature*, T_{db}, because it is the temperature that a dry thermometer measures. The wetbulb temperature is the lowest temperature to which an air parcel can be cooled by evaporating water into the air parcel, or to which water can be cooled through partial evaporation. The lower the relative humidity, the greater the difference between T_{db} and T_{wb} – that is, the greater the cooling potential through evaporation.

The relationships between temperature, mixing ratio, dewpoint and wetbulb temperatures and relative humidity are given in a specially constructed diagram called a *psychrometric chart*, which is shown in simplified form in Figure 4.7. Temperature and mixing ratio are given as the vertical and horizontal lines, respectively.

The leftmost concave upward line gives the variation of saturation mixing ratio (r_s) with temperature; as noted above, e_s and hence r_s increases rapidly with increasing temperature. Any point on this line corresponds to 100 per cent relative humidity. The temperature at which a given mixing ratio line intersects this curve gives the dewpoint temperature. The other concave upward lines correspond to various other RHs. When water evaporates into air, the mixing ratio of the air increases but the air is cooled; this is represented by the lines sloping upward to the left, which are lines of constant wetbulb temperature. Evaporation can continue only until the air has become saturated, which is where a given wetbulb line intersects the saturation mixing ratio line. The temperature at this point is the wetbulb temperature. Thus, given the mixing ratio r and temperature T of an air parcel:

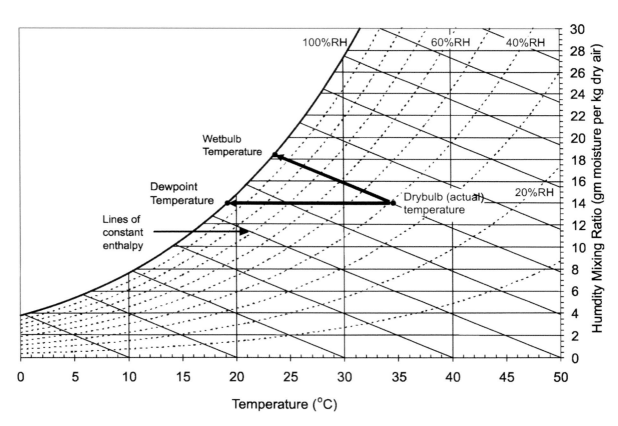

Source: Computed by author. A more detailed version is found in ASHRAE (2001, Chapter 6)

Figure 4.7 *The psychrometric chart, showing wetbulb temperature (dashed lines sloping upward to the left) and relative humidity (concave upward lines) as a function of drybulb temperature (vertical lines) and humidity ratio (horizontal lines)*

- trace the temperature line vertically until it intersects the r_s line, read off the r_s value and compute RH using Equation (4.12) of Box 4.2, or estimate the RH by interpolating between the RH lines given on the psychrometric chart;
- trace the r line to the left until it intersects the r_s line – the temperature at that point is the dewpoint temperature;
- trace a path parallel to the constant wetbulb temperature lines until it intersects the r_s line – the temperature at that point is the wetbulb temperature.

4.1.4 Conventional dehumidification

The commonly used method to remove water vapour from air is to cool the air until the desired amount of water vapour has condensed and fallen out, then to reheat the air to the desired final temperature. If the desired final temperature of air supplied to a room is 16°C with an RH of 50 per cent or 70 per cent (a common practice today), then the air must initially be cooled to 5°C or 10°C, respectively, irrespective of the initial temperature and RH (unless the initial RH is so low that no moisture needs to be removed).

To illustrate the relative energy requirements, suppose that we begin with outside air at 40°C and a mixing ratio of 30gm/kg. From Equation (4.15) of Box 4.2, the initial enthalpy is 117.4kJ/kg. To cool the air to 16°C with a mixing ratio of 7.5gm/kg (which gives an RH of 65 per cent, as can be seen from the psychrometric chart shown in Figure 4.7) requires removing 26.1kJ/kg of sensible heat and 56.3kJ/kg of latent heat (a total load of 82.4kJ/kg), as calculated using Equations (4.14) and (4.13). However, to condense out enough water vapour to produce a mixing ratio of 7.5gm/kg would require cooling the air to 6°C (see Figure 4.7), thereby removing an additional 10.2kJ/kg of sensible heat, which would then be added back. Thus, the total cooling load would be 92.6kJ/kg and the heating load 10.2kJ/kg.

4.1.5 Human comfort and the perception of temperature

The human perception of warmth or coldness depends on four environmental parameters (air temperature, radiant temperature, humidity, air movement) and two personal parameters (metabolic rate and clothing). The radiant temperature is the temperature of a blackbody

that would emit the same amount of radiation as is emitted and reflected by the surroundings (see Equation (4.4)). Putting aside humidity and air movement, the perceived temperature is very close to the average of the air temperature and the radiant temperature. Thus, if the ceiling and floor are kept at a warmer temperature in winter (or a lower temperature in summer), the air temperature does not need to be as high in winter (or as low in summer), which will reduce the heat loss (or gain) through the walls and windows, or due to exchange of outdoor air with indoor air. A lower relative humidity also reduces the perceived temperature by allowing a greater rate of evaporative cooling of the skin as a person perspires. Not only is the skin temperature reduced, but an individual feels more comfortable due to the very fact of being drier. Air movement reduces the perceived temperature as long as the air temperature is less than the skin temperature (usually a few degrees cooler than the body temperature, which is normally 37°C). This is because moving air can readily remove heat from a body.

In addition to these physical effects, there is also a psychological or adaptive component to the temperatures that are considered to be acceptable (de Dear and Brager, 1998, 2002). The warmer the outdoor air temperature, the warmer the indoor temperature that test subjects consider to be acceptable. Consideration of psychological adaptation has led to a proposal for a new standard, whereby the range of acceptable temperatures inside buildings depends on the outside temperature, as shown in Figure 4.8. This standard indicates that indoor temperatures of 28°C and 31°C are acceptable to 80 per cent of subjects on days with outdoor temperatures of 25°C and 35°C, respectively. Warmer interior temperatures are acceptable on hot days and colder interior temperatures are acceptable on cold days if an individual knows what to expect. In contrast, most air-conditioned buildings are operated to maintain a temperature in the lower part of the 23–26°C range set by the American Society of Heating, Refrigeration, and Air Conditioning Engineers (ASHRAE) in its comfort standard (Standard 55), irrespective of outdoor conditions.

The psychological adaptation to warmer temperatures is enhanced if an individual can *control* his or her environment by being able, for example, to open or close windows, or to activate or deactivate a fan. Research in Denmark indicates that a temperature of 28°C with personal control over air speed is

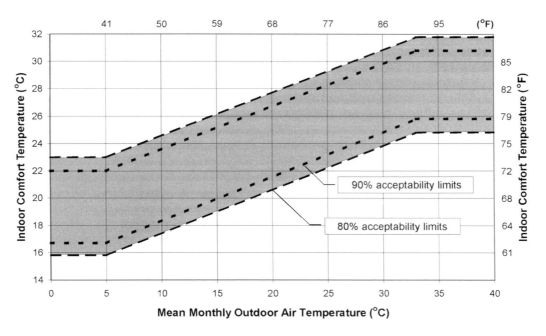

Source: Brager and de Dear (2000)

Figure 4.8 *A standard for acceptable indoor temperatures that takes into account psychological adaptation to different outdoor temperatures*

overwhelmingly preferred to a temperature of 26°C with a fixed air speed of 0.2m/s (de Dear and Brager, 2002). In Thailand, Busch (1992) found that the maximum temperature accepted by 80 per cent of survey respondents is about 28°C in air-conditioned offices and 31°C in naturally ventilated offices. The higher temperature setting for air-conditioning that is permitted under an adaptive standard would reduce the thermal shock often encountered in moving between indoor and outdoor spaces on hot days.

Increasing the thermostat from 24 to 28°C in summer will reduce annual cooling energy use by more than a factor of three for a typical office building in Zurich and by more than a factor of two in Rome (Jaboyedoff et al, 2004), and by a factor of two to three if the thermostat setting is increased from 23 to 27°C for night-time air conditioning of bedrooms in apartments in Hong Kong (Lin and Deng, 2004). The percentage savings in air conditioning energy use, however, will diminish the greater the cooling load of a building, and hence the warmer the outdoor temperature.

4.2 Thermal envelope and the role of building shape, form, orientation and size

The term *thermal envelope* refers to the shell of the building as a barrier to the loss of interior heat or to the penetration of unwanted outside heat into the building. It refers to the walls, windows, roof and basement floor of the building. The effectiveness of the thermal envelope depends on:

- insulation levels in the walls, ceiling and basement;
- the thermal properties of windows and doors;
- the rate of exchange of inside air with outside air through infiltration and exfiltration;
- the presence of shared walls with other buildings.

A better thermal envelope reduces the amount of heat that needs to be supplied by the heating system in winter, or the amount of cooling that is needed in summer.

4.2.1 Insulation

The thermal properties of insulation, doors and windows are rated using three different parameters:

- the *U-Value* (W/m²/K), equal to the rate of heat flow per unit area and per degree of inside-to-outside temperature difference;
- the *RSI-Value* (W/m²/K)⁻¹, a term used in Canada and here to denote the resistance to heat flow and equal to $1/U$ when U is expressed in metric units (SI denotes Système International);
- the *R-Value* (Btu/ft²/hr/°F)⁻¹, which in Canada and the US is a resistance to heat flow equal to $1/U$ when U is expressed in British units, but outside North America means the metric resistance. To convert from RSI-values to North American R-values, multiply by 5.678.

The U-value of a layer of insulation is equal to its thermal conductivity (having units of W/m/K) divided by the thickness of the layer (see Equation (4.2a)). The R-value, being equal to the reciprocal of the U-value,

increases in direct proportion to the thickness of the insulation. However, there are diminishing returns to adding ever more insulation because the heat flow varies with $1/R$. This is illustrated by Figure 4.9, which shows how the relative heat loss for a given temperature difference varies with the R- or RSI-value for the range of values that characterize walls and roofs. Insulation thickness is directly proportional to the insulation R-value, while heat loss is directly proportional to the U-value.

Conventional insulation

Conventional insulation will be taken here to mean one of the following:

- fibreglass insulation batts;
- mineral fibre batts;
- blown-in, loose cellulose;
- sprayed adhesive cellulose fibre;
- other wood-based fibre products;
- solid foam panels;
- blown-on foam insulation.

Table 4.2 lists the thermal conductivity and density for these and for two unconventional kinds of insulation (discussed later), along with the thickness of insulation required to achieve a U-value of 0.1W/m²/K. The most highly insulated houses in the world have wall U-values of 0.1–0.2W/m²/K (R28–R60) and roof U-values of 0.10–0.15W/m²/K (R40–R60), which is two to three times better than required in most cold-climate countries. Except when insulated with foam insulation, such houses normally require more wood, but high levels of insulation can be combined with advanced framing systems so at to minimize the amount of wood required and the amount of wood waste generated (Baczek et al, 2002). Optimized wood framing can reduce the cost of framing by 40 per cent and reduce the generation of wood waste by 15 per cent compared to conventional (non-optimized) framing.

As seen from Table 4.2, solid foam insulation has the lowest thermal conductivity of any conventional insulation and so provides the greatest insulation value for a given thickness. It is widely used in commercial buildings, as it can be directly glued to the outside of concrete or masonry walls. There are several different kinds of solid foam insulation, all of which are manufactured from chemical products produced from

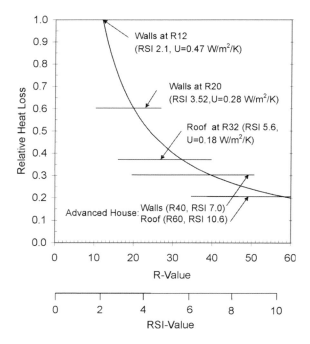

Figure 4.9 *Heat flow versus R-value for the range of R-values encountered in insulated walls and ceilings, relative to the heat flow at R12*

Table 4.2 *Thermal conductivity and density of different insulating materials, and thickness required for a U-value of 0.1W/m²/K*

Insulating material	Conductivity (W/m/K)	Density (kg/m³)	Thickness (cm) for U = 0.1W/m²/K
Fibreglass	0.042	20	42
Mineral fibre	0.037	40	37
Cellulose	0.04	32	40
Flax, hemp	0.04	30	40
Wood fibre	0.038	40	38
Kenof	0.038	40	38
XPS	0.030–0.034	32	34
Spray-on PUF	0.030–0.035	32	35
Solid PUF	0.022–0.024	32	24
Aerogel	0.012	100	12
Vacuum	0.004		4

Note: XPS = extruded polystyrene; PUF = polyurethane foam. For XPS and PUF, the lower thermal conductivity pertains to insulation made using an HFC blowing agent, while the higher thermal conductivity pertains to use of a non-halocarbon blowing agent. See subsection 4.10.6 for the significance of the choice of blowing agent.
Source: Harvey (2006, 2007); Ardente et al (2008)

oil or natural gas and require a substantial amount of energy to produce (as discussed in subsection 4.10.5). They also require some gaseous blowing or expanding agent in order to make the liquid raw materials expand into a foam as they are heated during the manufacturing process. At first, CFCs were used as the expanding agent in many foams, then these were replaced with HCFCs, which in turn are being replaced with HFCs. HFCs have no adverse affect on the stratospheric ozone layer (unlike CFCs and HCFCs) and might be promoted as 'green', but the most popular ones are powerful GHGs. As discussed in subsection 4.10.6, leakage of HFC blowing agents from the foam insulation during its lifespan can cancel the climatic benefit of needing less energy for heating and thereby emitting less CO_2. A number of foam insulation products are available that use water, pentane or CO_2 as blowing agents, the leakage of which has negligible climatic effects.

Cellulose insulation is recycled newsprint, treated with borate to provide fire and insect resistance. Insulation batts can be made of fibreglass or mineral fibres. Insulation batts and blown-in cellulose are usually applied between wood studs. As wood is a relatively poor insulator, the studs constitute thermal 'bridges'. The bridges can be eliminated and the insulation level between studs increased by adding a layer of solid foam (or wood fibre) insulation on the

outside of the stud wall, crossing over the studs. Conversely, thermal bridges can be minimized through the use of engineered wood I-beams. These are illustrated in Figure 4.10 and have the advantage that they are stronger than solid studs with the same outside dimensions, can be manufactured from wood waste or from small but fast-growing tree species, and the exact lengths needed for a given job can be ordered from the factory, thereby eliminating wood waste from the construction site. Sloppy workmanship (leaving gaps in the insulation or compressing the insulation to fit around wiring, plumbing and electrical boxes rather than cutting it to size) can seriously undermine the benefit of high levels of insulation, and this seems to be a widespread problem. An advantage of blown-in cellulose insulation is that it will fill the various irregularly shaped gaps in the wall structure, thereby providing a more continuous insulation barrier.

There are a number of insulation materials available that are made from various wood fibres or from hemp, flax or straw, with thermal conductivities comparable to that of fibreglass insulation or mineral fibre (0.04W/m/K).

Vacuum insulation panels

Vacuum insulation panels (VIPs) have received considerable attention since the 1990s as they provide the possibility of high thermal resistance at low

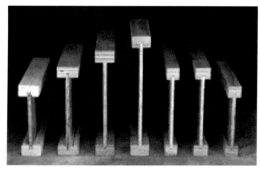

Figure 4.10 *Engineered I-beams that can be used as columns, beams and floor joists*

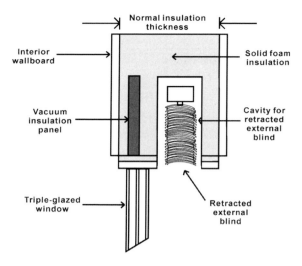

Figure 4.11 *Use of a vacuum insulation panel to maintain high thermal resistance behind the housing for retractable external window shading blinds*

thickness. The core of the panel consists of a microporous material (such as fumed silica) under a soft vacuum (10^{-4} atmospheres), which yields a thermal conductivity eight to ten times less than that of cellulose (0.004–0.005W/m/K versus 0.04W/m/K). The evacuated core is wrapped in an airtight envelope that forms a thermal bridge, the extent of which depends on the thickness and thermal conductivity of the envelope. For 1 × 1m, 3cm thick panels, Wakili et al (2004) measured an overall RSI (including thermal bridges) of 6.2 (U-value = 0.16W/m²/K; R35). Conventional insulation with the same RSI-value would be 20–25cm thick. Even if the vacuum is lost, the thermal resistance in the core of a VIP is still twice that of cellulose (Nussbaumer et al, 2006).

About 10,000m² of VIPs, covered by a protective layer, water barrier and concrete plate, had been installed by 2004 on terrace roofs in Switzerland (Simmler and Brunner, 2005). VIPs are preferred in such applications

because the floor height inside and outside can be made equal while providing stringent (RSI>5) resistance to heat loss on the outside part of the roof. Another niche application of VIPs is next to the housing for external windows shading blinds, as illustrated in Figure 4.11.

One concern with VIPs is the possibility that such panels could be damaged during the construction process. For this reason, prefabricated assemblies protecting the VIP are preferred. Prefabricated concrete wall slabs are available in the Swiss and German markets with a total wall thickness (including interior finish) of 27cm and an average U-value of 0.15W/m²/K (Binz and Steinke, 2005). An example is shown in Figure 4.12. VIPs are also available in prefabricated wood-frame roof and wall units and in VIP doors, and have been used in retrofit applications (as discussed in subsection 4.12.1).

Aerogel

Aerogel insulation panels consist of silica (SiO_2) granules. They are highly porous and thus have a density of only about 100kg/m³ (compared to 2200kg/m² for glass, which is also silica). The thermal conductivity of one product is 0.013W/m/K – about three times that of vacuum insulation but one third that of cellulose.

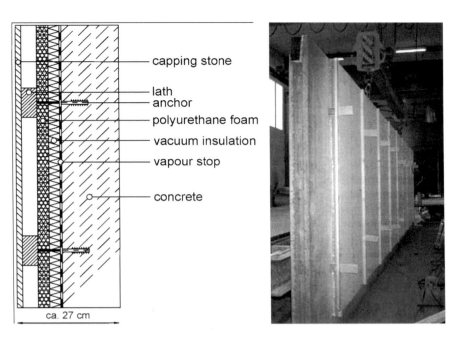

Source: Binz and Steinke (2005)

Figure 4.12 *On the left is a cross-section of a prefabricated wall unit with a VIP, having a total thickness of 27cm and a U-value of 0.15W/m²/K. On the right is a photographic example*

4.2.2 Windows

Windows are intended to permit light to enter into a building and to provide a view to the outside, but offer substantially less resistance to the loss of heat than insulated walls. Minimizing this heat loss is important in cold climates. At the same time, windows permit solar energy to enter a building. This can be an asset in winter (reducing the heating requirements) but a problem in summer (increasing the cooling requirements).

Heat flow through windows

A single-glazed window has a centre-of-glass U-value (that is, excluding the frame and the area next to the frame) of about 5W/m²/K. Thus, when the indoor temperature is 20°C and the outdoor temperature is −20°C, heat is lost through the window at a rate of 200W per m² of window area. Compared to a single-glazed window, heat loss can be reduced by:

• Adding extra layers of glass (glazing). The thermal resistance of glass is essentially zero. In a single-glazed

window, the resistance to heat flow arises from the thin motionless layer of air on either side of the glass, as motionless air is a relatively good insulator. In double- and triple-glazed windows, additional layers of motionless air are created, increasing the thermal resistance. Double and triple glazing alone will reduce the U-value to about 2.5 and 1.65W/m²/K, respectively.

• Using alternative gases between the window glazings. Gases with a heavier molecular weight than air have a lower molecular thermal conductivity than air. Argon has a thermal conductivity a third less than that of air, while krypton and xenon have thermal conductivities of just over one third and just over one fifth that of air, respectively. In a double-glazed window, argon and krypton will reduce the U-value from 2.5W/m²/K to 2.4 and 2.3W/m²/K, respectively. Argon- and krypton-filled windows are commercially available in many countries, while xenon-filled windows have been used in Germany.

• Using low-emissivity coatings. The loss of heat to the outside by emission of infrared radiation can be reduced if the emissivity of the glazing surface is reduced by coating the glazing with a low-emissivity film. Such windows are said to be 'low-e'. Emissivities as low as 0.04 can be achieved in this way (compared to 0.845 for regular uncoated glass). A double-glazed, argon-filled window with two low-e coatings can achieve a U-value of as low as $1.0W/m^2/K$, whereas triple-glazed windows with krypton fill and two low-e coatings have centre-of-glass U-values as low as $0.4W/m^2/K$ – more than 12 times less than that of single-glazed windows.

- Creating a vacuum between the window glazings. This completely eliminates conductive (and any convective) heat transfer, except at small, almost-invisible pillars that are used to prevent the glass layers from collapsing together. If combined with low-e coatings, the two processes of heat transfer through the non-frame part of the window are largely eliminated. The gap between the glazings in vacuum windows is only about 0.15mm, so vacuum windows are very thin (6–8mm). This makes it possible, in some cases, to retrofit them onto existing windows without having to remove the pre-existing glazing. Double-glazed vacuum windows are commercially available in Japan, with a U-value of about $1.2W/m^2/K$, but triple-glazed vacuum windows should achieve a centre-of-glass U-value of $0.2W/m^2/K$ (Manz et al, 2006).

- Using frames and spacers with low thermal conductivity. The heat loss through the glazed area in high-performance windows is so small that heat loss from the window frame and spacers can be a large fraction of the total heat loss. Thus, in high-performance windows, insulated frames and spacers need to be used. The worst choice is an aluminium frame (U-value of $10W/m^2/K$) and the best choice is fibreglass (U-value of $1.2–1.8W/m^2/K$, depending on the design). Fibreglass is used in many high-performance windows not only because it is a good insulator, but also because it can be made thinner in profile (due to its strength) than some alternatives, thereby providing more glazing area and allowing more solar heat to enter.

Transmission of solar radiation through windows

The *solar heat gain coefficient* (SHGC) or *g-value* (as it is known in Europe) is the fraction of solar radiation incident on a window that passes through the window, taking into account absorption of some solar radiation by the window and the transfer of some of this absorbed energy to the interior through emission of infrared radiation and through conduction. An uncoated double-glazed window has a SHGC of about 0.7. Low-e coatings absorb some solar radiation and thus reduce the SHGC.[3] Thus, the reduced heat loss through the window comes at the expense of less useful solar heat gain in winter. However, reducing the solar heat gain in summer is desirable, as it reduces the need for air conditioning. Vacuum windows can achieve the same reduction in heat loss as from using two or three low-e coatings, but without a reduction in solar heat gain. Vacuum windows also have much higher visible transmittance than non-vacuum windows with the same heat loss, so there is a greater opportunity to use daylight instead of electric lights.

Only about half of the solar radiation reaching the ground consists of visible light, the remainder consisting largely of near-infrared (NIR) radiation (see Figure 4.3). To minimize summer air conditioning requirements while allowing for daylighting, one would prefer a window with adequate visible transmittance but minimal NIR transmittance. Conversely, in cold climates we want to maximize the transmittance at all wavelengths. Neither uncoated clear glass nor regular reflective glass are suitable for minimizing the cooling load while maximizing daylighting because the visible and NIR transmittances are similar in both. However, glazing systems can be designed to have high visible transmittance and low NIR transmittance by varying the iron content of the glass, or by using a specially engineered low-e coating.

Table 4.3 compares the U-values, SHGC and visible transmittance for a number of commercially available high-performance windows. One window has a U-value of $0.4W/m^2/K$ (minimizing winter heat loss), a SHGC of 0.23 (minimizing summer solar heating) and a visible transmittance of 0.41 (better than tinted glass for daylighting).

Windows as net heat sources during cold winters

The heat loss through high-performance windows ($U<1.0W/m^2/K$) is so small that it is less than the usable solar heat gain (not all of the solar heat gain is

Table 4.3 *Properties of selected high-performance, commercially available windows*

Window product	U-value (W/m²/K)			SHGC	Transmittance	
	Air	Argon	Krypton		Visible	UV
Sunlite (www.sunlite-ig.com), U-values pertain to glazed area only						
DG, clear	2.689	2.547		0.70	0.79	0.50
DG, hard low-e	1.878	1.650		0.67	0.73	0.35
DG, soft low-e	1.666	1.407		0.38	0.70	0.14
TG, low-e	0.903	0.710	0.613	0.32	0.54	0.04
HM TC-88	1.016	0.835	0.738	0.48	0.63	<0.005
HM TC-88, low-e	0.948	0.761	0.659	0.34	0.55	<0.005
HM 44	1.209	1.028	0.937	0.28	0.38	<0.005
QG, clear	0.687	0.545	0.454	0.29	0.50	<0.005
QG, low-e	0.636	0.494	0.397	0.23	0.41	<0.005
Interpane (www.interpane.net)						
DG, IPLUS ²S		1.1		0.56	0.80	
DG, IPLUS ᶜS			1.1	0.64	0.81	
TG, IPLUS ³S		0.6		0.52	0.72	
TG, IPLUS ³ᶜS			0.5	0.52	0.72	
Steindl Glas (www.steindlglas.com/isog_wd_ws.html#)						
DG	1.7 (1.8)	1.4 (1.5)	1.0 (1.1)	0.58	0.77	
TG		0.6 (0.8)		0.44	0.66	
TG			0.7 (0.8)	0.60	0.75	
Nippon Sheet Glass (www1.nsg.co.jp/en)						
SPACIA normal		1.5 (0.2mm vacuum)		0.76	0.76	
SPACIA low-e		1.2 (0.2mm vacuum)		0.50	0.68	

Note: DG, TG and QG = double-, triple- and quadruple-glazed windows; HM=heat mirror. There are probably differences in the test conditions used for calculating U-values, so the results from different companies may not be strictly comparable.
Source: Manufacturers' websites

usable because at times heat may be absorbed while the room temperature has already reached the thermostat setting). This is true for climatic conditions in southern Canada for windows facing east, west and south, but not necessarily for windows facing north. However, in commercial buildings, the energy value of the admitted daylight will exceed the heat energy lost through high-performance windows, so moderate glazing areas on north-facing façades can be justified as long as high-performance windows are used and the lighting system is designed to take advantage of daylighting.

Equatorward-, east- and west-facing high-performance windows are not only a net heat source averaged over the course of the heating season, they can also be a net heat source on the coldest days of the year up to a latitude of about 50°N, given that the coldest days tend to occur under calm sunny conditions, created by stagnant Arctic air masses. An insulating blind (RSI 0.35) lowered at night would add to the net energy gain.

Savings in perimeter heating equipment

The heat loss through high-performance windows is so small that perimeter heating units, usually placed below windows to prevent draughts, can be eliminated. Figure 4.13 shows the window U-value below which perimeter heating is not needed, for 2m high windows, as a function of the coldest winter temperature for which the heating system is designed. Even for a design temperature of −30°C, windows are available that eliminate the need for perimeter heating. When perimeter heating is eliminated, ductwork or hot-water piping can be made shorter, as all the radiators can be located closer to the central core of the building, with associated cost savings but also savings in fan and pump size and energy use. If the default design involves floor-mounted fan-coil units, their elimination will increase the amount of usable floor space. The new courthouse in Denver (US) is an example of a building where triple-glazed windows were used to eliminate perimeter heating in public corridors (Mendler and Odell, 2000).

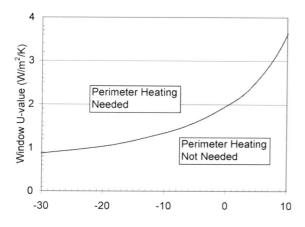

Source: Geoff McDonell, Omicron Consulting, Vancouver

Figure 4.13 *Window U-values below which perimeter heating is not needed for a 2m high window, as a function of the winter design temperature*

4.2.3 Curtain walls in commercial buildings

A curtain wall is a wall on the exterior of a building that carries no roof or floor loads. It commonly consists entirely or largely of glass and other materials supported by a metal framework, although precast concrete panels have also been used. The frame is referred to as a *mullion*, while the opaque panels between the glazed portions are referred to as the *spandrel*. Mullions can be provided with or without thermal breaks. Mullion U-values range from 16.8W/m²/K for a double-glazed curtain wall with an aluminium frame without a thermal break, to 4.3W/m²/K for a triple-glazed curtain wall using recently available insulated framing systems. The mullion can easily account for 10 per cent of the total wall area, so the large mullion U-values are a significant factor in the overall U-value.

Double-glazed and triple-glazed curtain walls are commercially available, with overall U-values (including the frame) for the 7500 Kawneer series as low as 1.8W/m²/K and 0.8W/m²/K, respectively, for large units (see www.kawneer.com). The triple-glazed units represent a 70 per cent reduction in heat loss compared to the glazing value of 2.6W/m²/K that is permitted under the ASHRAE 90.1-2004 commercial building code in moderately cold climates. These units entail thermal breaks and insulation covering the

shoulder of the mullion, thereby reducing the otherwise substantial heat loss through the mullion, so the overall U-value is dependent on the details of the construction. The spandrel portion can achieve U-values of 0.40W/m²/K. In high-performance curtain walls, the spandrel will contain additional insulation on the inside. Alternatively, vacuum insulation panels are available for the spandrel from the German company Okalux (www.Okalux.com), with U-values of 0.08–0.16W/m²/K. These U-values are a factor of three to eight smaller than permitted for the spandrel under the ASHRAE 90.1-2004 building code.

4.2.4 Air leakage

When air inside a building is heated above that of the outside air, a pressure variation is established such that the interior air pressure in the upper part of the building is slightly greater than the outside air pressure, while the interior pressure in the lower part of the building is less than the outside pressure. As a result, cold inside air is sucked into the lower part of the building through various cracks and openings in the walls and windows, and warm air exits through the upper part of the building. This temperature-induced air exchange is referred to as the *stack effect*. Up to 40 per cent of the heating requirement for houses in cold climates is to heat the outside air that continually replaces the inside air. The rate of air exchange will be greatest when it is coldest outside, which is the very time that a given rate of air exchange will cause the greatest heat loss. In hot humid climates, air leakage can be a significant source of indoor humidity.

In residential construction, careful application of a continuous impermeable barrier can reduce rates of air leakage by a factor of five to ten compared to standard practice. This is not a matter of advanced technology, but rather a matter of attention to detail and the enforcement of careful workmanship during construction. In buildings with very low air leakage, a mechanical ventilation system is required that circulates fresh outdoor air through the building and then exhausts. Up to 95 per cent of the available heat in the warm exhaust air can be transferred to the cold incoming air in winter using a heat exchanger. Air leakage is more difficult to measure and control in commercial buildings but can nevertheless be greatly reduced through simple measures that are part of a higher overall quality of construction. In office buildings in the central and northern US, these measures can reduce natural gas energy use by more than 40 per cent and electricity use by more than 25 per cent (Emmerich et al, 2007). Improving the airtightness of commercial buildings is essential to the proper operation of a number of low-energy ventilation and cooling techniques (discussed later).

Achieving low rates of air leakage requires a continuous impermeable barrier just inside the interior wall and ceiling finish of the entire house, with all breaks (such as for electrical boxes, wiring and plumbing) carefully sealed and all joints between walls and window or door joists, walls and ceilings, and walls and floors sealed. To achieve low rates of air leakage, particular attention is needed to delineate the responsibilities of the various trades involved in house construction (for example, deciding who is responsible for sealing breaks in an airtight envelope made by plumbers or electricians). Again, these are behavioural and organizational issues, not technological issues.

4.2.5 Double-skin façades

A *double-skin façade (DSF)* is a façade with an inner and outer wall separated by an air space that is not actively heated or cooled. The outer façade consists of a single- or double-glazed glass wall with fixed or adjustable openings, and the inner façade may also consist of a single-, doubled- or triple-glazed glass wall with operable windows, or may be partially opaque. Almost all examples of DSFs are found in Europe, particularly in Germany, and are amply illustrated by Herzog (1996), Oesterle et al (2001), Wigginton and Harris (2002) and Pasquay (2004). Design strategies to use DSFs to maximize passive ventilation across on office floor plan are reviewed by Gratia and de Herde (2007a). One early notable example is the Daimler-Chrysler Building on Potsdamer Platz in Berlin (the Debis Building), completed in 1996 and illustrated in Figure 4.14.

All-glass façades have become very popular among architects and their clients. These provide plenty of daylighting but the issues are then to minimize winter heat loss and summer heat gain and to avoid glare. With regard to winter heat gain, it will never be possible to match the insulative properties of a well-insulated wall with a glass façade. However, glass permits solar energy to enter the building and for high-performance glazing systems, the glazing is a net heat source during the heating season for most climates and façade orientations. This can be achieved without building a second façade

Source: Author

Figure 4.14 *The multi-storey DSF on the Daimler-Chrysler Building, Potsdamer Platz, Berlin*

over the first, so the DSF does not provide a particular advantage with regard to winter heat loss. However, the main issue with any all-glass façade is to avoid overheating, and this is where the DSF is advantageous (excess heat gain can even be a problem at times on cold but sunny winter days). Avoidance of overheating without excessively decreasing daylighting or rejecting solar heat when it is desired requires adjustable shading devices, but internal shading devices reduce the heat gain by only 50 per cent, compared to 90 per cent for external devices. However, external devices, especially on tall buildings, are subject to wind damage and cleaning them is difficult. A major benefit of the DSF is that it permits the installation of external, adjustable shading devices. DSFs are most appropriate on buildings subject to large external noise and wind loads, where external shading and natural ventilation would not otherwise be possible.

Quantitative examples of the savings in cooling energy use with DSFs, in comparison with other measures, are given in subsection 4.5.1.

4.2.6 Role of building shape, form, orientation and glazed fraction

Building shape (the relative length, width and depth), form (small-scale variations in the shape of a building) and orientation are architectural decisions that have significant impacts on heating and cooling loads (as well as on daylighting and the opportunities for passive ventilation, passive solar heating and cooling, and for active solar energy systems). For rectangular buildings, the optimal orientation is with the long axis running east to west, as this simultaneously maximizes passive solar heating in the winter and minimizes solar heating in the summer (due to the fact that the large

south-facing façade will be exposed to the low sun in the winter but can be easily shaded from the higher sun in the summer, while the area exposed to the low late-afternoon sun in the summer is minimized).

High glazing fractions, as found in much modern architecture, significantly increase the energy requirements for heating and cooling. Figure 4.15 compares the impact of glazing fraction and glazing upgrades on heating and cooling energy use for an office building in Sweden. For this particular building and climate, increasing the glazing fraction from 30 per cent to 100 per cent with base case windows increases the heating, cooling and total energy use by about 80 per cent, 160 per cent and 50 per cent, respectively. Upgrading to higher-performance windows (from a U-value of $1.85W/m^2/K$ to $1.11W/m^2/K$, and from a g-value of 0.69 to 0.22) cannot compensate for this increase in the glazing fraction; total energy use still increases by 20 per cent. This may seem surprising, given that high-performance glazing can function as a net heat source (see subsection 4.2.2). However, as the glazing fraction increases the proportion of the solar heat gain that can be used decreases (as, otherwise, overheating occurs) while the conductive heat loss increases in proportion to the glazing area. High glazing fractions do increase the opportunities

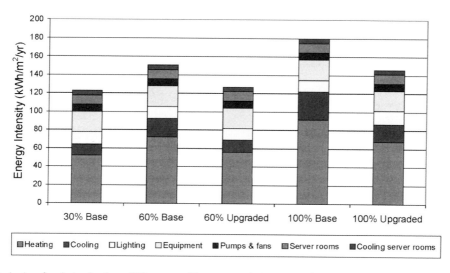

Note: Energy use is given for glazing fractions of 30 per cent, 60 per cent and 100 per cent, for base case windows (U-value = $1.85W/m^2/K$, g-value = 0.69, venetian blinds between glazings) and for upgraded windows (U-value = $1.11W/m^2/K$, g-value = 0.22, venetian blinds on the room side of the window).
Source: Data from Poirazis et al (2008)

Figure 4.15 *Impact of glazing fraction and glazing upgrades for an office building in Sweden*

for daylighting but there is little additional daylighting benefit once the glazed fraction increases beyond 30–50 per cent of the total façade area.

4.2.7 Role of house size

House size and the complexity of the shape (in particular, the surface to volume ratio) is an important factor in total energy use. In the US, the living area in new houses per family member increased by a factor of three between 1950 and 2000 (Wilson and Boehland, 2005). This is due in part to declining average family size (from an average of 3.67 to 2.62 members) and in part due to larger houses (from an average of 100m² to 217m²). As illustrated in Figure 4.16, a moderately insulated 3000ft² (around 300m²) house in Boston requires more heating + cooling energy than a poorly insulated 1500ft² house in the same location. The larger house also requires substantially more materials. According to a designer-builder quoted by Wilson and Boehland, the growth in house size is due to: (1) the loss of a sense of community and public life, so that the house becomes more of a fortress that needs to provide multiple forms of entertainment instead of basic shelter; (2) the promotion of the idea that 'bigger is better' by the building industry; and (3) the diminishing craftsmanship in house construction and design, leading to a substitution of greater size to counteract the sterility of modern homes.

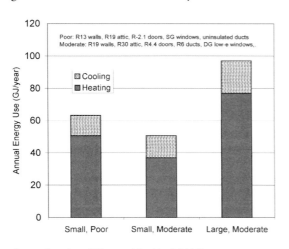

Source: Data from Wilson and Boehland (2005)

Figure 4.16 *Comparison of heating and cooling energy use for a small (approximately 150m²) poorly insulated house in Boston and for a large (approximately 300m²) moderately insulated house in Boston*

Wilson and Boehland (2005) list various strategies to make more efficient use of space, so that smaller houses provide the same services. These are:

- to eliminate the formal dining room in favour of a larger kitchen that provides both dining space and some informal living space;
- to provide built-in furnishings and storage space;
- to eliminate single-use hallways;
- to design multiple uses into rooms (such as guest room and office);
- provide for both television viewing and music functions in the living room;
- to keep the 'master bedroom suite' simple;
- to make use of attic spaces by insulating the roof;
- to design windows and doors to increase the visual connection to the outside;
- to provide visual, spatial and textual contrasts to make spaces feel larger than they really are;
- to use light colours for larger areas;
- to provide natural daylight;
- to keep some structural elements (structural beams, posts and timber joists) exposed;
- to create usable outdoor living space through careful landscaping;
- to make use of interior windows;
- to design spaces for visual flow (through a continuous moulding line and continuity of flooring, for example);
- to design for flexibility and change (so that houses are not built big just to allow for changing needs in the future).

4.2.8 Benefits of multi-unit versus single-family housing

Multi-floor, multi-family housing is significantly more energy efficient than single-family housing, especially one-floor single family housing. This is due to the sharing of walls and reduction in roof area, with concomitant reduction in heat loss. When six square one-storey single-family houses are combined into a triplex (three units side by side), wall area drops by about one third, while if six one-storey units are combined into two floors of three units, wall area drops by one third and roof area by half. When rectangular units are joined together along the long wall, the reduction in external wall area is larger still. Stacking units vertically, or designing single-family

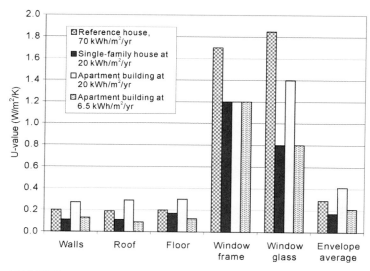

houses as two- or three-storey houses rather than as one-storey houses, will increase the opportunities for passive ventilation in the summer by increasing the stack effect and will protect the lower floors from the hot sun. By reducing the surface to volume ratio and the relative importance of the external envelope to the total cost, multi-family housing reduces the building cost per unit of floor area. Material resource requirements (wood, insulation) are also reduced, and public transportation, walking and cycling alternatives to the automobile are enhanced and land is spared because a more compact urban form can be created. Thus, multi-family (and multi-unit office and retail buildings) simultaneously reduce energy use and investment costs. Conversely, choosing shallow-plan floor designs (narrow buildings) in order to increase the opportunities for daylighting and ventilation can increase costs per unit of floor area.

The advantages of multi-unit over single-unit residential buildings in terms of costs and energy savings are illustrated in Figure 4.17, which compares the wall, roof and window U-values allowed in conventional single family houses in Sweden (with an annual heating energy use of about 70kWh per m² of floor area per year (70kWh/m²/yr) and a total onsite

energy use of about 100kWh/m²/yr), and that are required in order to achieve a heating energy use of 20kWh/m²/yr for single-family and multi-unit housing in Stockholm (latitude 59.2°N). To achieve the same energy performance as an apartment building, much lower window, wall and roof U-values are required for a single-family house. Conversely, adoption of about the same insulation levels and window performance in an apartment building as in a single-family house reduces the annual energy use to 6.5kWh/m²/yr – about three times smaller than for the energy-efficient house and more than ten times smaller than for the reference house. Another benefit of multi-unit housing is that, even when having low energy use per unit of floor area, the connection to district heating and cooling grids is more likely to be economically justifiable, which then opens up a number of possibilities for further energy savings or use of renewable energy (as discussed in Volume 2, Chapter 11).

4.3 Heating systems

In the previous section we discussed various ways to reduce the heat loss from a building, thereby reducing the heating requirement. We now consider various

Source: Data from Smeds and Wall (2007)

Figure 4.17 *Comparison of envelope U-values associated with new conventional single-family housing in Sweden (with 70kWh/m²/yr heating energy use), as required in order to achieve an annual heating energy use of 20kWh/m²/yr for single-family and apartment dwellings in Stockholm and as required in order to achieve an energy use of 6.5kWh/m²/yr in apartments in Stockholm*

ways of providing the required heat, beginning with passive solar heating, followed by mechanical heating systems.

4.3.1 Passive solar heating

Passive solar heating occurs when a building is heated by direct absorption of sunlight. This can occur in a variety of ways, three of which are described below. Technical details, real-world examples and data on energy savings are provided in Hastings (1994) and Hastings and Mørck (2000). For European climatic conditions, 20–30 per cent of seasonal heating requirements can be routinely met with passive solar heating, and sometimes much more, particularly if the building is well insulated (so that the overall heat demand is small).

Direct gain

This is the simplest passive heating system, and involves large window areas on sunlit sides of the building. The most advanced windows (whole-window U-values less than 1.0W/m²/K) produce a net seasonal heat gain in winter in southern Canada for most orientations.

Solar collectors

Passive solar collectors consist of a black absorbing surface, an air gap and a glass cover. Air inside the gap is heated and flows on its own accord (or with fan assistance) into the building. Alternatively, metal pipes can be welded to the absorbed surface and water flowing through the pipe will be heated. Solar collectors can be mounted on a roof or on walls and integrated into the structure of the roof or wall as part of the building envelope. They can be used to preheat ventilation air that is drawn into the building naturally in winter. In the summer, the heated air is vented directly to the outside while drawing ventilation through an earth coil or from open north-facing windows. Details concerning a roof-mounted collector system and the associated circulation of air and the distribution of temperatures in a school in northern Japan are given in Figure 4.18. When snow accumulates on the roof, the fans are reversed so that warm exhaust air warms the roof sufficiently for the snow to slide off, thereby making the roof ready to collect solar heat when the sun shines. Solar-collection efficiencies range from 30 per cent to 70 per cent, which is sufficient to meet 80 per cent or more of the

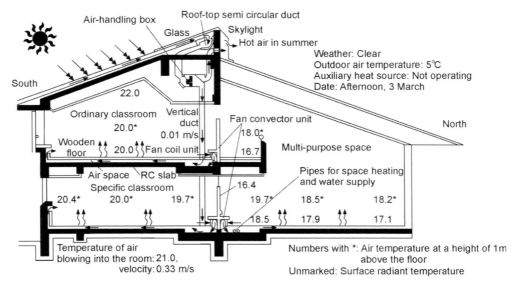

Source: Yoshikawa (1997)

Figure 4.18 *Airflow and temperature distribution in a school in northern Japan with roof-based collection of solar heat*

annual heating load for well-insulated buildings with a collector area equal to 16 per cent of the floor area in northern Europe.

Airflow windows

An airflow window consists of an outer double-glazed window, an interior single- or double-glazed window, gaps allowing airflow from the inside or outside of the building into the space between the glazings at the bottom of the window, another gap at the top to allow air to exit the space between the glazings, and an adjustable absorbing/reflecting blind within the air gap.

Two variants of the airflow window are found. In the first variant, known as the *supply air* window, air from outside is drawn between the window glazings, then either enters the interior of the building to satisfy some of the ventilation requirements (winter) or is directed back outside (summer), as shown in Figure 4.19a,b. During winter the blind would be lowered with the absorbing side turned outward. The incoming air would pick up some of the heat that would otherwise be lost through the window, as well as picking up heat absorbed between the inner and outer glazings by the blind. During the summer, the blind would be lowered with the reflective side turned outward, with outdoor air

serving to remove heat from the window assembly that would otherwise enter the building.

In the second variant, known as the *exhaust-air* window, indoor air passes through an inner air gap at the base and is either vented directly to the outside at the top (summer operation) or (during the winter) passes through hollow-core concrete floor slabs or is directed to a heat exchanger in order to transfer the collected heat to the incoming ventilation air before being vented (see Figure 4.19c,d). In any case, heat conducted to the outside through the window comes largely from the exhaust air rather than from the interior air, thereby reducing heat loss from the building. During both winter and summer, the exhaust air window would tend to draw fresh outside air into the building through other openings.

Triple-glazed airflow windows can be designed as a counterflow heat exchanger, with both fresh air and exhaust air flowing through the window but in opposite directions, as illustrated Figure 4.20.

Airflow windows were first used in Finland in the 1950s, and a window produced by the Finnish company Domlux (www.domus.fi) is illustrated in Figure 4.21. Sir Norman Foster's Commerzbank in Frankfurt, completed in 1997 and profiled in Fischer et al (1997) and Melet (1999), was the first office tower to use airflow windows.

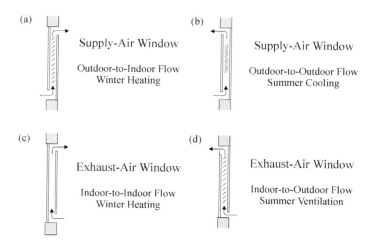

Source: Harvey (2006)

Figure 4.19 *Alternative flow configurations for airflow windows functioning as both supply-air or exhaust-air windows, and the applicable season*

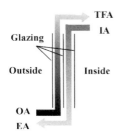

Dual Airflow Window
(Supply and Exhaust)

TFA = Tempered Fresh Air
OA=Outside Air
IA=Inside Air
EA=Exhaust Air

Source: Gosselin and Chen (2008)

Figure 4.20 *Airflow window functioning as counterflow heat exchanger*

Source: Domlux (www.domus.fi)

Figure 4.21 *An airflow window showing (1) external air inlet, (2) sound-attenuating filter, (3) gap between glazings, where the air warms, and (4) adjustable discharge vent*

4.3.2 Furnaces in residential buildings

Furnace efficiencies during steady operation at peak output range from 60 per cent (i.e. 40 per cent of the heat from burning the fuel is lost out the chimney) to 92 per cent (only 8 per cent loss). The highest-efficiency furnaces have an electronic ignition rather than a pilot light, have a fan to assist the removal of exhaust gases, directly use outside air for combustion rather than drawing air from the house, have a larger or more

effective heat exchanger to transfer more heat from the combustion gases to the air that is to be heated, and condense some of the water vapour in the combustion gas. Efficiencies are lower during part-load operation due to heat losses associated with on–off cycling, so it is important to avoid oversizing the furnace, as this will lead to more frequent on–off cycling. Alternatives are: (1) two-stage furnaces, which operate most of the time at lower capacity, with steadier operation, and resort to the full capacity only when needed, and (2) modulating furnaces, which can alter their heat output by varying the rate of fuel consumption. In modulating furnaces there is no efficiency loss at part-load operation down to about 25 per cent of full load.

4.3.3 Boilers in commercial buildings

A boiler is a pressure vessel in which heat is transferred from combustion gases to a liquid. To obtain high efficiency, water vapour in the exhaust gas must be condensed and the latent heat that is released must be used as the first step in reheating the return water. To do this requires that the return-water temperature be below the dewpoint of the exhaust gas (about 55°C for natural gas boilers with 10 per cent excess air). The colder the return water temperature, the greater the amount of water vapour that can be condensed and the higher the boiler efficiency. Colder return temperatures are possible if the water used for heating does not need to start out as hot, which is possible in better-insulated buildings with large radiators. Non-condensing boilers generally have full-load combustion efficiencies of 75–85 per cent, while condensing boilers have full-load combustion efficiencies of 88–95 per cent. Condensing boilers require corrosion-resistant materials and are generally limited to using clean-burning fuels, such as natural gas.

Non-condensing boilers generally use on–off cycling to achieve part-load operation, which is inefficient, so the overall efficiency falls with decreasing load. In condensing boilers, the flow of fuel and air are reduced so that the boiler can operate continuously with reduced heat output. Efficiency increases at part load. Because a boiler operates at part load most of the time, the decrease in non-condensing-boiler efficiency and the increase in condensing-boiler efficiency at part load amplifies the difference in seasonal-average efficiency between condensing and non-condensing boilers. The dependence of boiler efficiency on load and return temperature for a modern boiler is illustrated in Figure 4.22, which shows

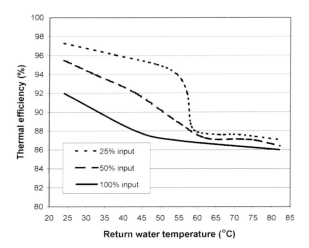

Source: Durkin and Kinney (2002)

Figure 4.22 *Efficiency of the AERCO 2.0 condensing boiler as a function of return-water temperature and output as a fraction of peak output*

boiler efficiency versus return-water temperature for the AERCO 2.0 boiler at 100 per cent, 50 per cent and 25 per cent of full load.

Condensing boilers are 25–100 per cent more expensive than non-condensing boilers in North America due to an immature market, but the extra cost can be entirely offset by the simpler and lower cost of plumbing required for connection to the heating-water loop (Rishel and Kincaid, 2007).

4.3.4 Pellet-burning boilers

Although biomass is bulky as a fuel compared to heating oil or natural gas (which can be conveniently delivered by pipe), the heating load in highly insulated buildings is so low that it can be practical to use pellet-burning boilers to provide supplemental heat on the coldest days, when passive solar and internal heat gains are not sufficient.

Fiedler (2004) reviews state-of-the-art pellet-based heating systems in Sweden, Austria and Germany. By 2001, there were 30,000, 12,000 and 10,000 pellet boilers or stoves in these three countries, respectively. Pellet boilers have a maximum heat output of 10–40kW, and some can automatically modulate from 30–100 per cent of full output, while pellet stoves are used to heat single rooms, compact apartments or entire low-energy houses, have a maximum heat output

of 10kW and can be regulated manually or automatically based on room temperature. Helical screws automatically deliver pellets to the combustion area as needed, and ash is automatically removed and compressed before being stored in an ash container. Austrian pellet boilers have efficiencies of 86–94 per cent. Examples of commercially available products can be found at www.pelletstove.com. Standardized biomass pellets are being produced in increasing quantities in Europe, largely from dry sawdust and wood shavings, but also from bark, straw and crops.

4.3.5 District heating

Connection of a building to a district heating system provides large energy savings if the heat supplied to the district heating system is waste heat that would otherwise be discarded to the environment. In some cases, some or part of the heat supplied by a district heating system is provided by dedicated centralized boilers, but these can generally be operated more efficiently than individual boilers in each building. A brief discussion of district heating systems, including energy losses in distributing heat to individual buildings and the opportunities for using heat that would otherwise be wasted, is found in Chapter 9, section 9.2.

4.3.6 Electric-resistance heating

The efficiency of electric-resistance heating, at the point of use, is 100 per cent – all of the electricity used is converted into heat inside the building. However, given that electric heating contributes to peak electric load, which is invariably met by fossil fuels at present, the low efficiency of converting primary energy into electricity must be taken into account. In the present energy system, with much of the world's electricity generated from coal at low efficiency, switching from electric heating to direct use of oil or natural gas in a high-efficiency furnace would lead to a significant reduction in primary energy use, thereby contributing to a decline in primary-energy intensity. However, in super-insulated houses with good thermal mass, peak and overall heating loads will be very small, so it will be possible to take advantage of intermittent sources of carbon-free electricity (as explained in subsection 4.4.5). Electric resistance heating would then be a reasonable choice where heat pumps are not practical.

4.3.7 Onsite cogeneration

Cogeneration – the simultaneous production of electricity and useful heat – was introduced in Chapter 3. Cogeneration has a long and extensive history of use in industrial facilities and to a lesser extent in district heating systems, but has seen very limited use at the scale of individual buildings due largely to economics. However, microturbines – defined as gas engines with an electrical power output of 30–500kW – have fallen in price to the point where they are or will soon be attractive for cogeneration in large individual buildings or groups of buildings (Brandon and Snoek, 2000).

Cogeneration improves overall energy efficiency only if the waste heat that is captured can be utilized. In microturbines, the electricity generation efficiency is only 23–26 per cent (see Table 3.5), meaning that about 75 per cent of the fuel energy is converted into heat. In fuel cells, the electricity generation efficiency is 35–45 per cent (using natural gas as fuel), meaning that a use for less waste heat needs to be found. Since new buildings can be built to require almost no winter heating even in cold climates, the scope for effective use of cogeneration in new, high-performance buildings is limited. Furthermore, heating loads are seasonal in character, meaning that a cogeneration system sized for the winter heating load will have excess heat during the summer. Hot-water heating loads, by contrast, are more uniform seasonally and therefore provide a better match for cogeneration.

One option, then, is to size a cogeneration system to match the hot water load and design a building to largely or entirely (depending on the climate) eliminate the heating load. The residual heating load could be met with a ground-source heat pump or high-efficiency boiler with low-temperature radiant heating. A second option is to size the cogeneration system to meet the combined heating + hot water load in winter, with both electricity (and heat) production scaled back in summer so as to retain high overall efficiency.

4.4 Heat pumps

Heat pumps can be used for heating, air conditioning and production of hot water. Residential air conditioners and commercial chillers operate on the same principles as heat pumps, so they can be thought of as heat pumps that operate in only one direction, to cool buildings.

4.4.1 Operating principles

The natural tendency of heat is to flow from warm to cold. A heat pump transfers heat against its natural tendency, from cold to warm, in the same way that a bicycle pump moves air against its natural tendency, from low pressure (outside the tyre) to high pressure (inside the tyre). In both cases, work must be done (requiring energy).

There are two broad types of heat pumps, based on either a vapour-compression cycle or an absorption cycle. An outline of how a vapour-compression heat pump works is found in Box 4.3. The transfer of heat from cold to warm is accomplished through a compression–expansion cycle involving a refrigerant or *working fluid*. If heat needs to be transferred from the outside to the inside of a building, the refrigerant must be cooled (through expansion) to a temperature colder than the outside, so that it can absorb heat from the outside. This absorption occurs through a heat exchanger – a coil through which the refrigerant flows as outside air is blown past it. As the refrigerant absorbs heat, it evaporates rather than increasing in temperature, so the heat exchanger is called an *evaporator*. Once inside the building, the refrigerant must be warmed (by compressing it with a *compressor*) to a temperature warmer than the medium to which the heat is transferred (either air or water), so that it will release heat to the inside. Once again, the heat transfer occurs through a heat exchanger, which maximizes the contact area between the warmed refrigerant and the air or water to be heated. As heat is released from the working fluid, the working fluid condenses rather than decreasing in temperature, so this heat exchanger is called a *condenser*. The difference between the evaporator and condenser temperature is referred to as the *temperature lift*. By reversing the direction of flow of the working fluid, the former evaporator serves as a condenser, and the former condenser serves as an evaporator, and heat is transferred in the opposite direction. Thus, a heat pump can act as a heater in winter and an air conditioner in summer.

4.4.2 Heat pump performance

The critical parameter measuring the performance of a heat pump is the *coefficient of performance (COP)*. When the heat pump is used for heating, the COP is the ratio of heat supplied to energy used.[4] In cooling mode, the COP is the ratio of heat removed from the building to energy used.[5] The performance of a heat pump depends

Box 4.3 Principles behind the operation of an electric vapour-compression heat pump

Vapour-compression heat pumps, refrigerators and air conditioners operate on the basis of two key principles: (1) a gas cools as it expands (or a liquid cools as it evaporates) but warms as it is compressed (or releases heat and thus becomes warmer if condensation occurs), and (2) heat flows from warm to cold. Although a heat pump (or refrigerator or air conditioner) appears to be violating the second principle (by transferring heat from the cold region to the warmer region), this is done by using principle (1) to create smaller scale reversals of the larger-scale temperature gradient, such that there is an overall heat transfer from the cold exterior to warm interior (or from a cool interior to a warm exterior) although, at each point in the process, the heat transfer is from warm to cold.

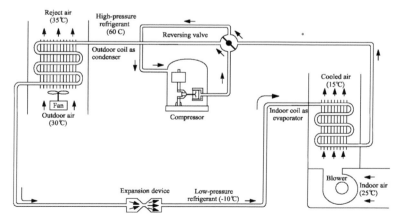

Figure 4.23 *Operation of a heat pump in (a) heating mode and (b) cooling mode*

How this is done is shown in the upper part of Figure 4.23. A compressor increases the pressure of a working fluid on the discharge side and creates low pressure on the suction side. As the working fluid (Freon in older systems, CFC replacements in new systems) is compressed, it is heated to a temperature in excess of the indoor air temperature. This allows heat to be transferred to the indoor airstream in an indoor heat exchanger, thereby cooling (and condensing) the refrigerant. The more the gas is compressed (i.e. the greater the pressure), the more it warms up. The liquid refrigerant travels through an expansion valve to a heat exchanger that is connected to the suction side of the compressor. The low pressure there induces evaporation and hence cooling of the refrigerant. The lower the pressure, the greater the cooling that occurs. The refrigerant must be cooled to below the temperature of the outdoor air in order to absorb heat from the outside air. The cool, low-pressure refrigerant, now in the gaseous state, returns to the compressor, where the cycle is repeated.

By simply reversing the direction of fluid flow, a heat pump can act either as a heating unit (transferring heat from the outside to inside) or as an air conditioner (transferring heat from the inside to outside). This is illustrated in the lower part of Figure 4.23.

Clearly, the colder the outdoor temperature, the more difficult it will be to transfer heat from the outside to the inside because the refrigerant must be cooled more in order to maintain the same temperature difference from the outside air. Thus, more work will be required for a given rate of heat transfer. Indeed, a greater temperature difference would be needed to drive a larger heat flow to match the larger heating demand when it is colder. Similarly, a warmer air-distribution temperature or a greater airflow rate, and a greater temperature difference between the condenser and airflow would be needed to mathch the larger heat demand when it is colder outside. This requires a warmer condenser temperature and hence greater compression of the refrigerant. Thus, a heat pump will work less efficiently but harder the colder the outdoor temperature while, at the same time, the heating requirement of the building will increase. To handle the heating load on the coldest days, an auxiliary electric heating coil will be used.

Most heat pumps operate on the vapour-compression cycle, as described above, with electrically driven compressors. In industrial applications there are many mechanically driven compressors (the compressor is driven by a turning shaft from a gas or steam turbine, rather from a shaft driven by an electric motor). A growing minority of heat pumps use the absorption cycle instead of a compression cycle.

on the performance of the compression–expansion cycle and on the performance of the heat exchangers.

Performance of the compression–expansion cycle

An ideal heat pump is equivalent to a heat engine running in reverse. A heat engine takes high-temperature heat and releases low-temperature heat, and in so doing does work, while a heat pump requires work to transfer heat from a low temperature to a higher temperature. An example of a heat engine is the steam turbine. The COP of an ideal heat pump in cooling mode is thus given by the reciprocal of the Carnot Cycle efficiency for a turbine (Equation 3.16). That is:

$$COP_{cooling,ideal} = \frac{T_L}{T_H - T_L} \qquad (4.16)$$

where T_L is the evaporator (or lower) temperature and T_H is the condenser (or higher) temperature. In heating mode, 1.0 is added to the above expression to account for the energy input to the heat pump, which is ultimately dissipated as heat and, along with the heat from outside, is part of the heat supplied to the building. Thus:

$$COP_{heating,ideal} = \frac{T_L}{T_H - T_L} + 1.0 \qquad (4.17)$$

No heat pump can achieve a COP greater than that of an ideal heat pump, given above. The ratio of actual COP to that of an ideal heat pump is called the *Carnot efficiency* (η_c). Thus:

$$COP_{cooling,real} = \eta_c \left(\frac{T_L}{T_H - T_L} \right) \qquad (4.18)$$

The Carnot efficiency ranges from 0.3 in conventional electric heat pumps, to 0.5 in more advanced residential units and 0.65 in large, advanced electric heat pumps.

As can be seen from the above equations, the heat pump COP will be larger the smaller the required temperature lift $(T_H - T_L)$. Figure 4.24a shows the variation of heating COP with evaporator temperature for various condenser temperatures, while Figure 4.24b shows the variation with condenser temperature for various evaporator temperatures, assuming a Carnot efficiency of 0.65. For heating with a given condenser temperature, the COP is smaller the colder the evaporator temperature. Since the evaporator must be colder than the outside in order to draw heat from the outside, this means that the heat pump COP decreases when it gets colder (and at the same time, the heating requirement for the building increases). Similarly, the greater the outside air temperature when the heat pump is used for air conditioning, the greater the condenser temperature required in order to be able to reject heat to the environment, and the smaller the cooling COP.

Performance of the heat exchangers

The evaporator and condenser are heat exchangers, with the heat flow between the heat exchanger and the heat source or sink driven by the temperature difference between the two. The relationships between the heat source, heat sink, evaporator and condenser temperatures are shown in Figure 4.25. Larger heat exchangers, or exchangers that are more effective in transferring heat, will minimize the temperature differentials ΔT_L and ΔT_H and thereby increase the COP. To illustrate, suppose that the source temperature is 0°C and the distribution temperature in a heating system is 30°C. The apparent Carnot Cycle COP in this case would be 273/30 + 1 = 10.1. If, however, the evaporator is 5K below the source temperature and the condenser 5K above the distribution temperature, in order to achieve adequate rates of heat transfer, then the real Carnot COP is 268/40 + 1 = 7.7 – one third lower. The performance of the heat exchangers is thus a critical factor in the performance of the heat pump.

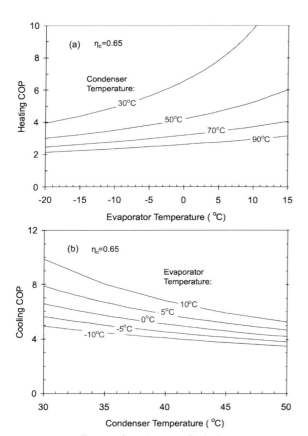

Note: A Carnot efficiency of 0.65 is assumed in both cases.
Source: Harvey (2006)

Figure 4.24 *Variation of (a) the COP of a heat pump in heating mode with evaporator temperature for various condenser temperatures, and (b) the COP of a heat pump in cooling mode with condenser temperature for various evaporator temperatures*

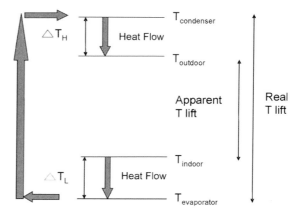

Figure 4.25 *Relationships between real and apparent temperature lifts, temperature differentials (ΔT_L and ΔT_H) and heat flow in a heat pump*

4.4.3 Air-source, ground-source and exhaust-air heat pumps

The evaporator coil of the heat pump used for heating purposes can be placed in the air outside the building, with a fan to ensure a steady flow of air past it. The outside air serves as the heat source for the heat pump. This is referred to as an *air-source heat pump* (ASHP). Alternatively, the evaporator coil can be connected

(through a heat exchanger) to a closed pipe loop that is buried underground, so that heat is absorbed from the ground rather than from the air. The ground thus serves as the heat source for the building, so this kind of heat pump is referred to as a *ground-source heat pump* (GSHP). The ground loop can be placed horizontally (if adequate space is available) or vertically, as illustrated in Figure 4.26. In vertical systems, a series of 10–15cm diameter boreholes is drilled to a depth of up to 50m. U-shaped,

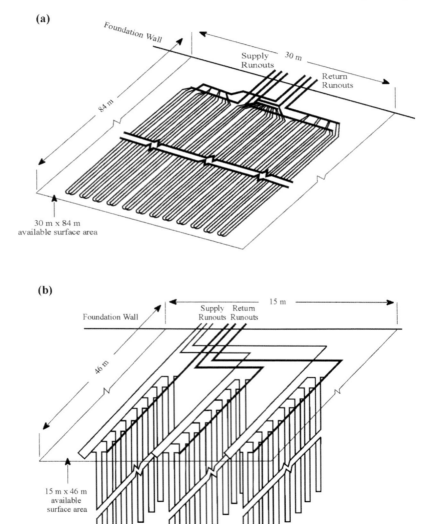

Source: Caneta Research Inc. (1995)

Figure 4.26 *Illustration of a ground-source heat pump with (a) a horizontal piping network, and (b) a vertical piping network*

high-density polyethylene pipes with a typical inside diameter of 30mm are inserted into each borehole and connected at a depth of 1–2m below the ground surface.

Because ground temperatures are warmer than the air temperature in winter and cooler than the air temperature in summer, GSHPs require less energy for heating and cooling than ASHPs. Depending on the climate, heating energy use can be reduced by up to 60 per cent and cooling energy use by 40 per cent compared to ASHPs (Johnson, 2002).

Another attractive option is an *exhaust-air heat pump* (EAHP). An EAHP is used to extract heat from the exhaust air and transfer it to the fresh air supply during winter, and to transfer heat from the hot incoming fresh air to the comparatively cool exhaust air during the summer. Today, almost all new single-family houses in Sweden are equipped with EAHPs (about 4000/year by 1997), and another 5000–10,000/year are installed in Germany, with heating COPs in the order of 3.0–4.0 (Fehrm et al, 2002). This is not much larger than for a GSHP, but is less expensive than a GSHP and better than an ASHP. However, much larger COPs have been demonstrated theoretically and in practice. Using an EAHP in combination with a ground preheating loop and a radiant floor heating system (so that the heat supply temperature need be no warmer than 35°C), Halozan and Rieberer (1997, 1999) calculate seasonal mean COPs of 6–7 for the climate of Graz, Austria. In an experimental setup in Quebec, Minea (2003) measured a seasonal mean COP of 7.5 for heating and 16.3 for cooling.

4.4.4 Effect of heating and cooling distribution temperatures

During the heating season, the COP of a heat pump will be smaller the warmer the temperature at which heat is distributed by the heating system within a building. This is because the condenser will have to achieve a temperature somewhat greater than the heat distribution temperature in order to reject heat to the distribution system, so the heat pump will have to work harder. In a forced-air heating system, which is common in houses in North America, heat is typically supplied at 50–60°C. In Europe it is common to distribute heat with hot water (a *hydronic* system), traditionally with a supply temperature of 90°C and a return temperature of 70°C. Even for a more modest distribution temperature of 70°C (which might require a condenser temperature of 80°C), the

COP would be reduced by 1.0–2.0 (depending on the temperature of the heat source) compared to a 50°C distribution temperature (which might require a condenser at 60°C). If the insulation is improved and air infiltration reduced in old north-European buildings, the required supply temperature can be reduced to as low as 45–55°C. The lowest distribution temperatures can be achieved through floor radiant heating systems; in new, thermally tight buildings, a distribution temperature of 30–35°C can be used (residential heat-distribution systems are discussed further in subsection 4.6.2). For an evaporator at –10°C (corresponding to an outside temperature of about 0°C), reducing the distribution temperature from 70°C to 30°C increases the heat pump COP by about 75 per cent. There is thus a double benefit from better thermal envelopes: the amount of heat that needs to be supplied decreases, and the efficiency in supplying the required heat with heat pumps increases, because the heat can be distributed at a lower temperature and thus with a higher heat pump COP.

In cooling mode, maximization of the heat pump COP requires that the evaporator temperature be as high as possible, which in turn requires as warm a distribution temperature for cooling as possible (this also applies to chillers, which are discussed in subsection 4.5.3). As explained in subsection 4.6.3, cooling can be provided by circulating cool air as part of the ventilation system, or by circulating cold water (hydronic system). Hydronic systems can involve a fan blowing air past a cold-water coil (a *fan-coil system*) or chilling the entire ceiling or floor (a *radiant chilling* system). In a fan-coil system, a typical cold-water supply temperature is around 5–7°C, with a return temperature of around 12–14°C. This requires an evaporator temperature of around 0–2°C. In ceiling radiant-cooling systems, water at temperatures of 18–20°C has been used, so the evaporator temperature can be much higher.

If humidity is high, dehumidification of the ventilation air will be required whether or not radiant cooling is used. In conventional systems, this is accomplished by cooling the air to a low enough temperature to condense sufficient moisture. However, active desiccant dehumidification (discussed in subsection 4.6.3) provides an alternative method of dehumidification, without having to cool the air to below the final distribution temperature. This allows a warmer evaporator. As seen from Figure 4.24, increasing the evaporator temperature from 0 to 10°C increases the COP by 1.0–3.0 (depending on the condenser temperature).

In summary, lowering the temperature at which heat is distributed or increasing the temperature at which chilled water is distributed leads to significant improvements in the COP of heat pumps. This can be achieved through radiant floor or ceiling heating or cooling, and by reducing the required heat flows to or from the radiant floor or ceiling by reducing the building heating or cooling loads. Avoiding the need to overcool ventilation air in order to remove moisture also leads to a significant improvement in COP (as well as saving on energy used to reheat the air).

4.4.5 Heat-pump + powerplant system efficiency

For a COP of 3.0 (close to the highest at present for an air-to-air heat pump), the heat supplied is three times the energy used (this is equivalent to a furnace with an efficiency of 300 per cent). However, if that energy is supplied as electricity from a coal-fired plant with an efficiency of 33 per cent (typical of many plants today) there is a factor of three energy loss in going from primary energy (coal) to secondary energy (electricity). Thus, the overall efficiency of primary energy use will be only modestly better than with a high-efficiency natural gas furnace (85–92 per cent efficiency). As the efficiency in generating electricity from fossil fuels and the COPs of electric heat pumps improve, heat pumps become more favourable from a primary energy point of view. If the heat pump is operated with a COP of 6.0 (by using the ground as a heat source and with a low heat distribution temperature) and electricity is generated at 60 per cent efficiency (as in state-of-the-art NGCC powerplants) then primary energy use is reduced by about a factor of four compared to direct use of natural gas.

Heat pumps can be particularly advantageous in regions where substantial amounts of intermittent electricity generation have been added to the grid. As long as the building is well insulated with high thermal mass, so that it retains most of its heat (during the winter) or coldness (during the summer) for several hours, the heat pump could be turned on when there is available carbon-free power on the grid and turned off when there is not. Electrical water heaters are already being used in this way in some jurisdictions – remotely turned on or off by the power utility in order to smooth out overall electricity demand. Heat pumps would become a *dispatchable* electricity end use.[6] In effect, the building thermal mass becomes a way of storing carbon-free electrical energy in the form of building heat or coldness. If electricity is supplied on the margin by fossil fuel power during the day and by carbon-free power during the night, the heat pumps could be operated largely during the night. Thus, electric heat pumps could provide substantial savings in GHG emissions even if fossil fuels are still part of the electric generation mix.

4.5 Cooling loads and cooling devices

The first step in meeting a cooling load is to reduce the load compared to what it would be under standard practice, then to consider passive techniques for meeting some or all of the load, and lastly to consider efficient mechanical systems to meet any remaining load.

4.5.1 Reducing the cooling load

Analysis of the opportunities for reducing the cooling load requires identifying the relative importance of the heat sources, which will vary with climate, building size and shape, the activities carried out in the building, and the properties of the thermal envelope. Figure 4.27 shows the breakdown of heat gains for a large commercial building in Los Angeles (hot and dry climate) and in Hong Kong (hot and humid climate). For the Los Angeles building, the three largest heat sources are lighting (28 per cent), windows (21 per cent) and fans (13 per cent), while conditioning of incoming outside air accounts for only 10 per cent of the total cooling load and heat conduction through walls accounts for only 3 per cent. For the Hong Kong building, the three largest loads are people (27 per cent), lighting (18 per cent) and conditioning of outside air (20 per cent), with windows and walls accounting for 12 per cent. Most of the cooling load in large commercial buildings comes from internal heat gains. In smaller buildings, heat gains through the envelope are relatively more important.

Options to reduce the cooling load include:

- orienting a building to minimize the wall area facing east or west;
- clustering buildings to provide some degree of self-shading (as in many traditional communities in hot climates);

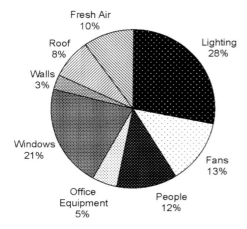

Cooling Loads in a Los Angeles Office Building

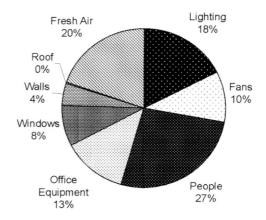

Cooling Loads for a Generic Office Building
in Hong Kong

Source: Los Angeles: Feustel and Stetiu (1995); Hong Kong: data provided by Joseph Lam (personal communication, 2003) in Lam and Li (1999)
and Lam (2000)

Figure 4.27 *Relative contribution of different heat sources to the cooling needs of larger
commercial buildings in various cities*

- using high-reflectivity building materials (Parker et al (2002) monitored six side-by-side houses in Florida that differed only in the reflectivity of the roof, and found that houses with white reflective roofs have a cooling-energy consumption about 20–25 per cent lower and peak-power demand about 30–35 per cent lower than houses with dark shingles);
- increasing insulation (Florides et al (2002) found that, for a single-storey house in Cyprus, adding 5cm of polystyrene insulation to the roof reduces the cooling load by 45 per cent (and the heating load by 67 per cent), while addition of 5cm of polystyrene insulation to the walls reduces the remaining cooling load by about 10 per cent (and the remaining heating load by 30 per cent));
- providing fixed or adjustable shading (external shading devices are more effective than internal devices, as only about 10 per cent of the sunlight absorbed by the device is transferred to the interior of a room in the former case, compared to almost 50 per cent in the latter case (Baker and Steemers, 1999));
- using windows with a low solar heat gain and avoiding excessive window area (particularly on east- and west-facing walls);

- utilizing thermal mass to minimize daytime interior temperature peaks, combined with night-time cooling;
- using vegetation to directly shade buildings and to indirectly reduce cooling loads by reducing ambient air temperature.

Thermal mass is provided by stones, bricks and concrete blocks, or concrete slab floors. It reduces the daytime peak air-conditioning requirements by absorbing heat with minimal temperature increase. In cases where air conditioning is needed only during times of peak conditions, thermal mass itself serves to eliminate a cooling load. Porta-Gándara et al (2002) simulated the cooling load for housing built with traditional adobe bricks and modern hollow concrete blocks (having minimal thermal mass) in Baja California, and found the air conditioner load of the former to be a quarter that of the latter during the hottest summer months. However, unless the heat is removed at night, the temperature of the thermal mass will build up over a period of days, and so it will become less and less effective in limiting daytime temperatures. As well, thermal mass inhibits night-time cooling. This can be beneficial if indoor temperatures would otherwise become too cold at night (as in many desert climates), but is a problem if indoor temperatures remain too warm for comfortable sleep. Thermal mass is thus most effective if combined with night-time ventilation (at times when night-time air is sufficiently cool) and external insulation to inhibit daytime penetration of outside heat into the thermal mass while leaving it exposed to cool night air. The thermal mass would be most effective if it contains a hollow core through which cool air can be circulated by night while cooling incoming ventilation air by day. This is possible with concrete slabs.

Thermal mass can also be provided through *phase-change materials* (PCMs), the most common being a paraffin wax that melts at around 25°C. The PCM can be embedded in drywall or plaster inside 50μm capsules, as illustrated in Figure 4.28 (left), or might consist of larger spheres placed inside a canister which in turn is placed in the ventilation airflow (illustrated in Figure 4.28, right). The waxes will not rise in temperature above their melting point, just as ice will not rise above 0°C as it melts. Air in contact with the plaster or spheres will rise only a few degrees above the melting point of the wax. At night the waxes refreeze if they can be cooled to below their melting point with cool night air, releasing the heat that they absorbed during the day as they melted.

The combination of switching to a high albedo surface and planting shade trees can yield dramatic energy savings. The benefits of trees arise both from direct shading and by cooling the ambient air. Rosenfeld et al

Source: Gypsum PCM: Schossig et al (2005); canister: Arkar and Medved (2007)

Figure 4.28 *Scanning electron microscope image of microencapsulated PCM in gypsum plaster (left), and PCM spheres inside an experimental canister that is placed within the ventilation airflow (right)*

(1998) calculate the impact on cooling loads in Los Angeles of increasing the roof albedo of all 5 million houses in the Los Angeles basin by 0.35 (a roof area of 1000km^2), increasing the albedo of 250km^2 of commercial roofs by 0.35, increasing the albedo of 1250km^2 of paved surfaces to 0.25 (by using whiter, limestone-based aggregates in pavement whenever roads are resurfaced), and planting 11 million additional trees. In the residential sector, they computed a total savings of 50–60 per cent, with a 24–33 per cent reduction in peak air-conditioning loads. For Toronto, Akbari and Konopacki (2004) calculated potential savings in cooling energy use of about 25 per cent for residential buildings and 15 per cent for office and retail buildings through similar measures. Growing vegetation on building walls can also provide important reductions in cooling energy use; simulations by Kikegawa et al (2006) indicate a saving of 10–30 per cent for residential buildings in Tokyo.

DSFs for shading and passive ventilation

The benefits of a DSF from a cooling point of view are:

(1) a shading device can be placed outside the inner façade, where heating of the shading device by the sun does not contribute to heat gain inside the building;
(2) as air between the two façades is heated, it will rise and can be used to induce passive ventilation of the building, as described in subsection 4.3.1 for airflow windows; and
(3) by addressing night-time security concerns, cool night air can be used for cooling through night ventilation. Manz (2004) computed the flow of solar heat through a DSF and found that solar heat gain can be reduced to 4–7 per cent of the incident solar irradiance, but the details of the design are important.

Figure 4.29 compares the heating and cooling loads for a five-storey office building in Belgium with different façades, as simulated by Gratia and de Herde (2007b). The base case is a conventional single-skin façade with moderate insulation (wall, roof and window U-values of 0.65W/m^2/K, 0.56W/m^2/K and 2.8W/m^2/K, respectively). With no change in the thermal envelope but implementation of sensible operating strategies (using window blinds and opening windows when appropriate and optimizing daytime versus night-time

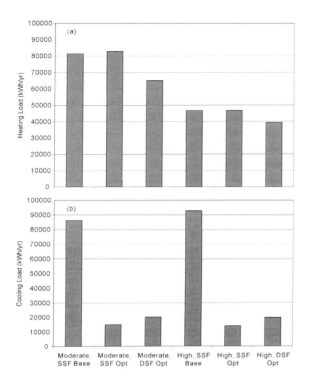

Note: SSF = single-skin façade.
Source: Data from Gratia and de Herde (2007b)

Figure 4.29 *Comparison of simulated heating and cooling loads for a five-storey office building in Belgium with a single-skin façade that does (Base) or does not (Opt) have an optimized operating strategy, and with moderate insulation or high levels of insulation, plus results for optimized with moderate and high levels of insulation*

16 per cent but increases the cooling load by 40 per cent. However, the absolute changes in heating and cooling loads are about the same, so depending on the relative efficiencies of the heating and cooling equipment and in generating electricity, this tradeoff could reduce or increase total primary energy use. In any case, the final result is a 52 per cent reduction in heating load and a 77 per cent reduction in cooling load relative to the base case. With more stringent insulation and window standards (including glazing with a low solar heat gain coefficient), even larger savings would be possible. Similar results have been obtained by Stec and van Paassen (2005), who simulated the energy use for alternative south-facing façade systems in The Netherlands and found savings in cooling loads (excluding dehumidification) of 85–88 per cent.

In summary, in the central European climate, a DSF can increase the cooling load compared to a building with a normal façade and ventilation with outdoor air for cooling. However, day and night ventilation may not be practical in buildings with normal façades, due to exterior noise, dust or the risk of intrusion. The value of DSF construction, then, is not that it results in the lowest possible energy use, but that it facilitates passive or hybrid ventilation in the first place. The DSF is unlikely, however, to be adequate in very hot climates, at least if the inner façade is largely of glass and with minimal thermal mass. That is, we must recognize limits to the appropriateness of buildings with predominantly glass façades. In hot arid climates, the traditional concepts of large thermal mass and night-time ventilation are likely to be more appropriate, along with external shading that could nevertheless be part of a DSF.

4.5.2 Passive and passive low-energy cooling

Having reduced the cooling load through the techniques described above (often by a factor of two or more), the next strategy in priority is to use passive and/or passive low-energy cooling strategies. A purely passive cooling technique requires no mechanical energy input at all. It includes such techniques as designing a building to maximize natural ventilation (particularly during the day). Other techniques involve small inputs of mechanical energy to enhance what are largely passive cooling processes. The major passive and passive low-energy cooling techniques are discussed below.

ventilation), there is almost no change in the annual heating load but the annual cooling load (in the Belgian climate) is reduced by a factor of almost six. Addition of a second skin reduces heating energy use by about 20 per cent (due to the utilization of solar heat collected between the two skins) but doubles cooling energy use. Increasing the insulation level in a SSF (giving wall, roof and window U-values of $0.37W/m^2/K$, $0.30W/m^2/K$ and $1.8W/m^2/K$, respectively) without a sensible operating strategy reduces heating energy use by 43 per cent but increases cooling energy use by 8 per cent. Optimization of the operating strategy preserves the heating energy savings but reduces the cooling load by a factor of 6.2 relative to the base case. Relative to this case, addition of a second skin reduces the heating load by

Natural ventilation

In the absence of ventilation, the interior air temperature in a building will rise considerably above the outside air temperature, such that the building is quite uncomfortable even when the outside temperature is pleasant. As the ventilation rate increases, the interior temperature will approach the outside temperature. At the same time that the real interior air temperature decreases, the *perceived* temperature will decrease further due to the greater ability of moving air to remove heat from a warmer body. Finally, with natural ventilation, the *acceptable* air temperature increases due to enhanced psychological adaptation to warmer conditions compared to buildings with mechanical ventilation.

Natural ventilation can be achieved in a variety of ways:

• Cross-ventilation and wind suction. This technique has been widely employed in traditional architecture around the world (see Al-Temeemi, 1995, for Kuwait; Lee et al, 1996, for Korea; Krishan, 1996, for India; Oktay, 2002, for Cyprus; and Tantasavasdi et al, 2001, for Thailand). Airflow over an opening at the top of a domed roof, common in the Arab world, creates a suction effect that removes hot air that accumulates inside the top of the dome (Gallo, 1998). Passive *ventilation stacks* depend on a similar effect, and are commonly used in north European residential buildings. *Balanced wind stack* (BWS) systems provide both inlet and exhaust airflow through co-axial stacks (one inside the other), with the possibility of latent and sensible heat recovery, and are being promoted by European manufacturers for use in larger buildings (Axley, 1999).
• Atria. These provide an excellent opportunity to induce natural ventilation, through proper placement of air inlets and outlets, along with shading controls over the rooftop glazing or passive measures (such as the geometry of laser-cut glazing) as to avoid overheating in summer.
• Solar chimneys. Natural ventilation can also be induced by creating a 'stack' or 'chimney' effect – that is, by creating air that is warmer and therefore less dense than its surroundings, so that it will rise up and out of a building, forcing cooler outside air to enter the lower part of the building. A tower constructed for this purpose is referred to as a *solar*

chimney. Solar chimneys are finding particularly wide application in the UK. An example is the Building Research Establishment offices in Garston (Figure 4.30).
• Cooltowers. These are the opposite of a solar chimney. Water is pumped into a honeycomb medium at the top of a tower and allowed to evaporate, thereby cooling the air at the top of the tower, which then falls through the tower and into an adjoining building under its own weight. This technique is also referred to as *passive downdraught evaporative cooling* and has apparently been used for hundreds of years. It was applied in the new Visitor Center at Zion National Park, US (Torcellini et al, 2002) and at the Torrent Pharmaceutical Research Centre in Ahmedabad, India (Ford et al, 1998), the latter is illustrated in Figures 4.31 and 4.32.
• Airflow windows. These were described in subsection 4.3.1. In the TEPCO (Tokyo Electric Power Company) R&D centre, constructed in

Source: Dennis Gilbert, London

Figure 4.30 *Use of solar chimneys in the Environmental Building, Building Research Establishment, Garston (UK)*

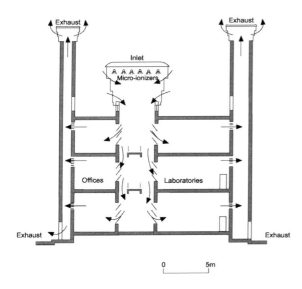

Source: Baird (2001)

Figure 4.31 *Schematic illustration of the Torrent Pharmaceutical Research Centre in Ahmedabad, India, showing the central downdraught tower and peripheral solar chimneys*

Source: Baird (2001)

Figure 4.32 *The Torrent Pharmaceutical Research Centre in Ahmedabad, India*

Tokyo in 1994, this technique reduced the inward heat transfer by a third to two thirds, compared to double-glazed windows with interior or built-in blinds, and eliminated the need for perimeter air conditioning (Yonehara, 1998). The airflow window in this example is combined with monitoring of solar irradiance and the use of computer-controlled blinds. The blind control in turn is coupled to a daylight compensation lighting-control system. Other examples of computer-controlled blinds inside airflow windows or DSFs, programmed to optimize the balance between ventilation, heat gain and daylighting, are found in several of the case studies of 'intelligent' skins presented in Wigginton and Harris (2002).

• Roof solar collectors. These consist of a solar-energy-absorbing outer layer, a gap in the order of 14cm and an insulated lower layer. The lower end of the air gap is connected to the interior air. As the outer layer of the roof and the air in the gap are heated, the air rises, drawing out interior air.

Some degree of automatic control can be achieved through the use of self-regulating *trickle ventilators*, in which a flap moves to create a larger inlet area when the pressure difference falls, thereby maintaining a near-constant airflow. An exterior metal mesh prevents intrusion of dust, rain and insects.

Night-time passive and mechanical ventilation

Where the day–night temperature variation is at least 5–7K, cool night air can be mechanically forced through hollow-core ceilings or through the occupied space to cool the building prior to entering the building the next day. Where artificial air conditioning is still needed by day, external air can be pre-cooled by passing it through the ceiling that has been ventilated at night. Effective night ventilation requires a high exposed thermal mass, an airtight envelope, minimal internal heat gains and a stack-ventilation building configuration so that minimal fan energy is required. In such buildings in southern UK, energy savings of 30–40 per cent can be achieved in this way (Kolokotroni, 2001). External insulation should be used in order to inhibit the inward penetration of daytime outside heat while leaving the thermal mass exposed to the cooling effect of night-time ventilation and free to absorb internal heat during the day.

For Beijing, da Graça et al (2002) find that thermally and wind-driven night-time ventilation eliminates the need for air conditioning of a six-unit apartment building during most of the summer (an extreme outdoor peak of 38°C produces a 31°C indoor peak), but there is a high risk of condensation during the day due to moist outdoor air (relative humidity as high as 95 per cent in summer, and usually above 70 per cent) coming into contact with the night-cooled indoor surfaces. A solution would be to close all openings during the day and dehumidify incoming air sufficient to prevent condensation.

Simulations by Springer et al (2000) indicate that night-time ventilation is sufficient to prevent peak indoor temperatures from exceeding 26°C over 43 per cent of California's geography in houses that include improved wall and ceiling insulation, high-performance windows, extended window overhangs, 'tight' construction and modestly greater thermal mass compared to standard practice in California. Thus, air conditioners – which account for almost 60 per cent of residential energy use in California – could have been eliminated altogether over almost half of the state while adhering to a very strict discomfort threshold.

Where mechanical air conditioning is used in combination with night ventilation, the energy savings from night ventilation depend strongly on the daytime temperature setpoint. For a three-storey office building in La Rochelle, France, Blondeau et al (1997) simulated energy savings due to night ventilation of 12 per cent, 25 per cent and 54 per cent for setpoints of 22°C, 24°C and 26°C, respectively. More importantly, night ventilation with a 26°C setpoint requires only 9 per cent the cooling energy of the case with a 22°C setpoint and no night ventilation for this particular building and climate.

The combination of external insulation, thermal mass and night ventilation is particularly effective in hot dry climates, as there is a large diurnal temperature variation in such climates. In contrast, low thermal mass and an open design with plenty of cross-ventilation are normally recommended in hot humid climates. However, Tenorio (2007) finds that in the humid tropical areas of Brazil, thermal mass combined with night ventilation and selective use of air conditioning can reduce cooling energy use in a two-storey house by up to 80 per cent compared to a fully air-conditioned house.

Evaporative cooling

As discussed in subsection 4.1.3, evaporation can cool water down to the wetbulb temperature (T_{wb}). The difference between T_{wb} and the ambient temperature is greater the lower the absolute humidity, so the potential cooling effect of evaporative cooling is greater in arid regions, although the amount of water available could be limiting. In *direct* evaporative cooling systems, water evaporates directly into the airstream to be cooled. If the resulting increase of air humidity is a problem, the water can evaporate into a secondary airstream that is used to cool the supply air through a heat exchanger, without adding moisture to the supply air. This is referred to as *indirect* evaporative cooling. Figure 4.33 illustrates the flow of the primary and secondary airstreams in an indirect evaporative cooling module. Figure 4.34 illustrates the variation of T and r for the two airstreams. By cooling the primary airstream (through indirect evaporative cooling) and then applying direct evaporative cooling, it is possible to achieve final temperatures that are below the original wetbulb temperature. However, the airstreams would not in practice be able to move the entire distances indicated by the direct and indirect steps in Figure 4.34. The fraction of the potential change in temperature at

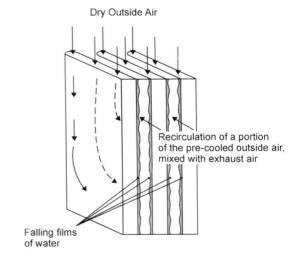

Source: Harvey (2006)

Figure 4.33 *Diagram showing folded heat-exchanger plates and the dry and wet airflows in an indirect evaporative cooling module*

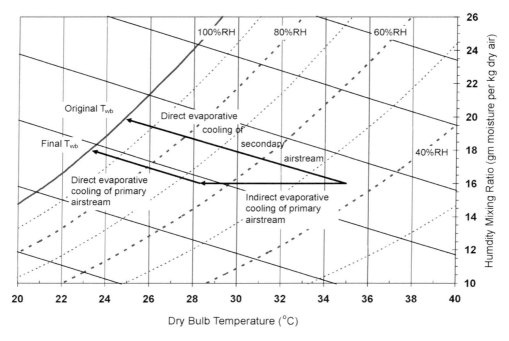

Figure 4.34 *Variation of temperature and water vapour mixing ratio during direct and indirect evaporative cooling*

each step that does occur is referred to as the *effectiveness*.

Table 4.4 compares extreme ambient conditions for selected cities in hot parts of the world: the ambient temperature that is exceeded only 1 per cent of the time, and the average T_{wb} during these times. Also given are the supply-air temperature and humidity that can be produced with direct and indirect evaporative cooling, assuming a cooling effectiveness of 85 per cent for direct cooling and 65 per cent for indirect cooling. With direct cooling, supply-air temperatures are about 18–25°C but RHs are rather high (65–90 per cent). Indirect cooling produces higher supply-air temperatures but much lower RHs (20–65 per cent), such that the temperature–humidity combination would be acceptable to most people. If indirect cooling is used as a first stage followed by direct evaporative cooling, it is possible to achieve a final temperature slightly *lower* than the initial T_{wb} even with the effectiveness <100 per cent as assumed here. On this basis, evaporative cooling can provide comfortable conditions most of the time in most parts of the world.

A number of residential direct evaporative coolers are on the market in the US. Two products from one company are illustrated in Figure 4.35. Energy is required to operate the fans, which draw outside air through the evaporative cooler and directly into the space to be cooled, or into ductwork that distributes the cooled air. Simulations for a house in a variety of California climate zones indicate savings in annual cooling energy use of 92–95 per cent, while savings are somewhat less (89–91 per cent) for a modular school classroom (DEG, 2004). In humid climates the energy savings would be much less. However, in humid climates a better approach is to enhance the evaporative cooling capacity using desiccants, as explained in subsection 4.5.4.

Underground earth-pipe cooling

Outside air can be drawn through a buried coil, cooled by the ground and used for ventilation purposes. Simulations by Lee and Strand (2008) indicate that earth pipes can reduce the June–August cooling load by 70 per cent in Illinois and by 65 per cent in Spokane (Washington).

Table 4.4 *Drybulb temperature (T_{db}, °C) exceeded 1 per cent of the time, average wetbulb temperature (T_{wb}, °C) corresponding to these conditions, and the corresponding RH*

Location	Ambient conditions			Direct cooling		Indirect cooling		Indirect + direct cooling				Diurnal ΔT (K)
	T_{db}	T_{wb}	RH	T_f	RH	T_f	RH	T_f	RH	T_f	η (per cent)	
North America												
Miami	32.2	25.1	56	26.2	91	27.6	73	24.5	95	26.6	78	6.3
New Orleans	33.1	25.7	55	26.8	91	28.3	73	25.0	95	27.2	79	8.6
New York	31.5	22.8	47	24.1	89	25.8	66	21.9	94	23.7	89	8.1
Latin America												
Mexico City	27.9	13.7	16	15.8	79	18.7	29	11.2	85	11.7	113	13.8
Caracas	32.7	28.5	73	29.1	95	30.0	85	28.2	97	30.9	42	7.0
Lima	28.8	23.2	62	24.0	93	25.2	77	22.6	96	24.8	71	6.4
Recife	32.7	25.6	56	26.7	91	28.1	74	25.0	95	27.1	78	6.3
São Paulo	30.9	20.3	37	21.9	86	24.0	56	19.0	91	20.4	98	8.3
Europe												
Athens	33.0	20.1	29	22.0	84	24.6	48	18.5	90	19.7	103	9.4
Rome	29.8	23.2	57	24.2	92	25.5	73	22.5	95	24.6	79	9.9
Budapest	30.2	19.9	38	21.4	86	23.5	56	18.6	92	20.0	98	12.2
Africa												
Casablanca	32.9	21.4	35	23.1	86	25.4	55	20.1	91	21.5	99	11.0
Cairo	36.2	20.5	23	22.9	81	26.0	41	18.5	87	19.5	106	13.3
Harare	29.1	16.3	25	18.2	82	20.8	41	14.3	88	15.2	108	11.7
Asia												
Istanbul	29.1	20.8	47	22.0	89	23.7	65	19.8	93	21.5	91	8.5
Riyadh	43.1	17.8	4	21.6	69	26.7	11	14.1	77	14.1	114	14.0
New Delhi	40.5	22.4	20	25.1	79	28.7	38	20.3	87	21.3	106	12.0
Madras	37.0	25.2	38	27.0	86	29.3	59	24.1	92	25.8	94	8.1
Seoul	30.1	24.0	60	24.9	92	26.1	76	23.4	96	25.6	74	8.0
Tokyo	31.2	25.1	61	26.0	92	27.2	77	24.5	96	26.8	72	6.2
Hong Kong	32.8	26.1	59	27.1	92	28.4	75	25.5	95	27.8	74	4.5
Ho Chi Minh City	34.2	25.2	48	26.6	89	28.4	67	24.4	94	26.3	87	8.2
Singapore	32.3	25.9	60	26.9	92	28.1	76	25.3	96	27.6	73	6.3
Manila	34.1	26.5	55	27.6	91	29.2	73	25.9	95	28.1	79	8.8
Australia												
Alice Springs	38.9	17.7	9	20.9	73	25.1	20	14.6	81	14.7	114	13.7
Cairns	32.1	25.1	57	26.1	91	27.5	74	24.5	95	26.6	78	7.3
Brisbane	30.0	22.4	52	23.5	90	25.1	69	21.6	94	23.5	85	7.6

Note: Given are the temperature (T_f) and RH that would be produced by a direct evaporative cooler with an effectiveness of 0.85, by an indirect evaporative cooler with an effectiveness of 0.65, and by a combined indirect + direct cooler with effectiveness of 0.65 and 0.85 for the indirect and direct stages, respectively. Temperature and humidity results for the last of these cases are given with no constraint on RH. The final temperature and overall effectiveness are also given with the RH constrained not to exceed 80 per cent. The final column gives the average diurnal temperature variation during the three warmest months of the year.
Source: Harvey (2006), where the calculation algorithm is given along with results for a larger set of cities

The performance of such a system can be characterized in terms of peak indoor temperatures compared to peak outdoor temperatures, and by the ratio of the rate of heat removal by the air exchange to the power used by the fans – analogous to the COP of a heat pump or air conditioner. In an experimental house in India, such a system limited the peak indoor temperature to 30–32°C with peak outdoor air temperatures of 42°C, and had a COP of 3.35 (Sawhney et al, 1999). Outdoor RH ranges from about 10 to 30 per cent, resulting in indoor RHs of 25–60 per cent. In another experimental study from India, the

Source: Product brochures at www.adobeair.com

Figure 4.35 *Evaporative coolers from Adobe Air: Arctic Circle rooftop, ducted (top); Alpine window-mounted (bottom)*

indoor temperature fluctuated between 24°C and 31°C as the outdoor temperature fluctuated between 31°C and 48°C (Thanu et al, 2001). The measured COP during the summer was 7.9, but much smaller during the monsoon, when the ambient temperature was closer to the ground temperature. In an experimental building in Italy, the measured ground-loop COP is 5.2 if night ventilation is not also used, and 4.6 if night ventilation is used (Solaini et al, 1998). The measured COP of a ground loop for a building in Germany is 35–50 (Eicker et al, 2006). Argiriou et al (2004) built and tested an earth pipe that was coupled to a PV array on a building in Greece that directly powers a 370W DC motor, thereby avoiding the need for DC to AC conversion normally associated with PV power. The fan speed increases as the incident solar radiation increases, matching the need for increased cooling. The measured average COP (based on DC power output) was 12.1.

Underground earth-pipe loops have been used in many buildings in Japan (Ray-Jones, 2000). Many of these buildings also include rooftop solar collectors and/or atria designed to function as a solar chimney so as to induce natural ventilation through the building. The combination of a solar chimney and earth-pipe cooling for a school in Norway is illustrated in Figure 4.36. Hestnes et al (1997) and Reinhart et al (2000) provide other examples from Europe.

Use of cool groundwater

In the London area, many buildings use cool (14–16°C) groundwater to cool ventilation air in displacement ventilation systems, or to produce chilled water for use in chilled-beam cooling systems (Ampofo et al, 2006).

Hybrid natural and mechanical ventilation

Hybrid ventilation systems are systems with primary reliance on wind and buoyancy to provide adequate ventilation, with assistance by fans only as needed. Delsante and Vik (2002) provide technical information from 22 buildings with hybrid systems in ten countries. Commonly used design elements in these buildings include:

- underground ducts, culverts or plenums to precondition the supply air (in six buildings);
- operable windows and/or ventilation grilles, in many cases automatically controlled;
- automatically controlled shading devices;
- solar chimneys or atria, often with backup fans;
- temperature and/or CO_2 sensors.

All the buildings rely on good thermal design and thermal mass with intensive night-time ventilation in order to limit daytime temperatures. Many of the buildings have noise attenuators. Some retrofits of purely mechanical systems have been done.

A hybrid mechanical passive ventilation and cooling system does not necessarily require more energy than a purely passive cooling system, at least for UK conditions (Wright, 1999). This is because, in a hybrid system, greater use of natural light is possible because solar heat gain does not need to be limited as much. The year-round savings in lighting energy use can offset the annual energy use for mechanical ventilation and cooling. As well, greater cooling is achieved if night

 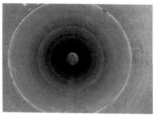 

Source: Schild and Blom (2002)

Figure 4.36 *Jaer school in Nesodden, Norway, showing the solar chimney (top), air intake (bottom left), underground culvert for pre-cooling (summer) or preheating (winter) the air (bottom middle), and fans where the underground culvert enters the basement (bottom right), which are used to assist the flow only when needed (which has rarely been the case)*

ventilation is delayed until the early morning hours, and then enhanced with fans (Eicker et al, 2006).

In a simulation study of an office floor in Belgium with double-glazed, low-e, argon-filled windows with internal shading, Saelens et al (2008) found that adding ventilation heat exchangers, night ventilation and intensive daytime ventilation when conditions warrant it reduces heating energy use by 44 per cent and cooling energy use by 63 per cent, while shifting the shading from inside to outside reduces the remaining cooling energy use by 50 per cent (giving a total saving of 82 per cent).

In the hot continental climate of Illinois, Short and Lomas (2007) calculate that a hybrid ventilation system would reduce heating, fan and cooling energy use by about 12 per cent, 70 per cent and 78 per cent, respectively, for a proposed college library and faculty building. These savings are based on a comparison of

energy use in computer simulations for the proposed design and for an entirely mechanically ventilated building.

The Panasonic Multimedia Center in Tokyo, illustrated in Figure 4.37, is a particularly interesting example of a building with a hybrid ventilation system, as it is located in an area of heavy traffic congestion (so that noise might have been thought of as a problem) and has a large internal cooling load (Ray-Jones, 2000). An atrium creates a solar chimney, through which exhaust air exits the building. Air enters through vents beneath each window, then passes beneath a raised floor, entering the rooms uniformly through floor vents and exiting through slits in the ceiling, where it passes to the atrium. Fans in the floor assist the airflow when needed, and supplemental mechanical cooling is also available when needed. Ventilation rates are increased at night to cool down the concrete-slab floors.

Source: Nikken Sekkei, Japan

Figure 4.37 *The Panasonic Multimedia Center in Tokyo, in which the central atrium serves to induce a displacement ventilation airflow through the offices*

4.5.3 Air conditioners and electric chillers

An air conditioner can be thought of as a small heat pump that operates only in cooling mode. It is used in residential and small commercial buildings. Like a heat pump, the performance of an air conditioner can be characterized by its COP, which is the ratio of heat removed from a building to the energy used.[7] The rate of heat removal, like the rate of electricity use, can be expressed as kW. To avoid possible confusion between the two, cooling capacities will be designated here with a subscript 'c'. Thus, a $1kW_c$ air conditioner is one that can remove 1000J of heat per second, not one that requires 1kW of electrical power. Air conditioners, like heat pumps, operate by alternately compressing and expanding a refrigerant or working fluid. The working fluid evaporates at a low pressure in an evaporator, where it becomes colder than the surrounding air (thereby drawing heat from the air), then condenses under pressure in a condenser, where it becomes warmer than the surrounding air (thereby releasing heat to the air).

Table 4.5 compares the estimated average COP of air conditioners sold in different parts of the world at present. These are nominal COPs, pertaining to specific test conditions, and (as seen below) are likely to be much higher than COPs under real operating

Table 4.5 *Average COP of residential air conditioners sold in various world regions at present*

Region	COP
Pacific-OECD	3.85
North America	3.37
Western and Eastern Europe, Former Soviet Union	2.80
Latin America	2.64
Centrally planned Asia	2.60
South Asia	2.55
Other regions	2.40

Source: McNeil et al (2008a) based on a compilation of various sources

conditions. In Europe and North America, new residential air conditioners generally have nominal COPs of 2.5–3.5. Minimum air conditioner COPs in Japan are comparable to or better than the best COPs in Europe and North America, with a maximum COP of 6.4 (on a small, $2.8kW_c$ unit). Murakami et al (2006) expect the average COP of air conditioners in Japan to increase to about 6 by 2020. Bringing the entire world to this level would more than double the average COP of air conditioners being sold at present. As the COP of air conditioners has been increasing over time, the average COP of the existing stock would be lower than the average of new air conditioners sold at present, so the eventual improvement when there has been complete replacement of the existing stock of air conditioners with air conditioners meeting the expected future Japanese standard (if applied to the whole world) would be larger still.

Average operating COPs are likely to be lower than the average nominal COP of the existing stock. About two thirds of air conditioners in California have the wrong amount of refrigerant and/or too low an airflow speed past the cooling coil (Downey and Proctor, 2002). This causes the average air conditioner energy use to be about 15 per cent greater than if these parameters were correct. The efficiency of most air conditioners falls dramatically during part-load operation, due to the fact that part-load operation is almost always achieved through repeated on–off cycling. For example, at 20 per cent of full load, efficiency falls to about 80 per cent of the full-load efficiency (Henderson et al, 2000). This loss can be minimized through measures (such as thermal mass and external shading) that reduce peak cooling loads relative to average loads, so that smaller air conditioners (which would on average operate at a larger fraction of their peak capacity) can be used. Some air conditioners use power even when they are turned off – 40–120W in central systems according to Henderson et al (2000)! Thus, through a combination of better maintenance of air conditioning, better sizing, reduction in standby losses and worldwide implementation of projected future Japanese standards, the operating COP of residential air conditioners can be at least doubled and perhaps tripled compared to the current stock average.

In large office buildings, cooling is achieved by circulating cold water (in addition to cool ventilation air). Air conditioners that produce cold water rather than cold air are referred to as 'chillers'. There are three major kinds of commercial chillers, based on the kind of compressor used. Ranging from smallest to largest, these are *reciprocating* chillers, *rotary* (*screw* or *scroll*) chillers and *centrifugal* chillers.

Reciprocating chillers can have air- or water-cooled condensers, while centrifugal chillers usually have water-cooled condensers. The water that cools the condenser in turn is cooled through partial evaporation in a *cooling tower* (as discussed later). Cooling capacities, full-load COPs and refrigerants used are given in Table 4.6. Within each category, the COP is larger for the largest units. Large centrifugal chillers have a cooling capacity of up to $35MW_c$ and a full-load COP of up to 7.5. This is two to three times the rated COP of the residential wall- or window-mounted air conditioners available in most parts of the world. If a few large chillers that serve the entire building are used in place of many inefficient window- or wall-mounted air conditioners, a 50–60 per cent energy savings is possible after accounting for energy used by the cooling tower. This of course entails constructing a chilled-water piping system and allocating space for the central chilling facility, and it requires metering and billing of individual apartments or offices so as to discourage waste. A further saving arises from the fact that heat from chillers is ejected at the top of the roof (through the cooling tower) rather

Table 4.6 *Types and characteristics of commercial chillers*

Chiller type		Capacity range		Full–load COP	Refrigerants used
		MW$_c$	Million Btu/hr		
		Vapour-compression chillers			
Reciprocating		Up to 1.5	up to 5.1	3.8–4.6	R134a, R717, R407c
Rotary	Screw				
	Scroll	0.3–7	1–24	4.1–5.6	R134a, R717, R407c
	Rolling piston				
	Rotating vane				
Centrifugal		0.3–35	1–120	5.0–7.5	R134a, R717, R123
		Absorption chillers			
Single-stage		0.3–6	1–21	0.7	H$_2$O
Double-stage		0.3–8	1–27	1.2	H$_2$O
Direct-fired		0.3–5.2	1–18	1.7	H$_2$O

Note: R123 = HCFC-123, H134a = HFC-134a, R717 = ammonia and R407c is a mixture whose composition is a mixture of refrigerants.
Source: IEA (1999) for vapour-compression chillers, Dharmadhikari (1997) for absorption chillers

than at the sides of the building, where it degrades the performance (by up to 20 per cent) of air conditioners on higher floors (Chow et al, 2002). As discussed in Volume 2, Chapter 11, a further saving arises if chillers and heat ejection are concentrated further in a centralized district cooling system, as in this case the thermal plumes rise high into the atmosphere without contributing to the urban heat island.

Commercial chillers, like air conditioners and heat pumps, are usually equipped with fixed-speed compressors. Part-load operation of such chillers can be achieved through the use of inlet vanes. The inlet vanes regulate the rate of flow of the refrigerant so as to maintain a constant temperature of the chiller water leaving the evaporator, regardless of the cooling load. A VSD can reduce the compressor speed (and hence the rate of refrigerant flow) down to zero, but a practical lower limit is 40 per cent of full speed due to the need to circulate lubricating oil. Lenarduzzi and Yap (1998) retrofitted a variable speed drive onto a $615kW_c$ centrifugal chiller in an office building in Toronto and monitored its energy use during one cooling season. They found a saving in energy use of 41 per cent at this particular site.

4.5.4 Heat-driven chillers

In the electric chillers described above, mechanical power from an electric motor is used to drive a compressor, needed to circulate the refrigerant between a condenser and an evaporator. These chillers are referred to as *vapour-compression* chillers. However, heat can be directly used to drive an evaporation–condensation cycle in a variety of ways. This provides a way of using waste heat, which would otherwise be discarded, and of using solar thermal energy. The most popular of the heat-driven chillers is the *absorption chiller*. Heat can also be used to produce cool air (rather than chilled water) using *solid* or *liquid desiccants*.

Absorption chillers

The principles behind the operation of an absorption chiller are explained in Box 4.4. Commercial absorption chillers, in which the heat that drives the chiller is produced by burning natural gas, have existed since around 1945. Steam can also be used to drive absorption chillers.

The performance of an absorption chiller is characterized by its COP that, in this case, is the ratio of cooling energy provided to *heat* energy input. A typical COP is about 0.7, compared to COPs of 5.0–7.5 for electric chillers. The heat output from the condenser and absorber can be used as the input to drive a second absorption chiller. This produces a *double-effect* absorption chiller, although the COP is not quite doubled. The process can be continued to produce a triple-effect chiller, a quadruple-effect chiller and so on, but with diminishing returns. As well, progressively higher input temperatures are required to operate double- and triple-effect chillers, which diminishes the possibility of using waste or solar heat. Figure 4.40 shows the variation of COP of single-, double- and triple-effect absorption chillers as a function of the temperature of the heat source. These three configurations require minimum source temperatures of about 60°C, 95°C and 140°C, respectively, and have maximum COPs of about 0.7, 1.25 and 1.75, respectively.

The energy input to an electric chiller is ultimately converted to heat that is removed (in large systems) by a cooling tower (described below) along with the heat extracted from the building. From the definition of COP as the ratio of heat removed to energy input (whether as heat or as electrical energy that is ultimately dissipated as heat), it follows that the total amount of heat that must be removed by the cooling tower per unit of heat removed from the building, CT_{load}, is given by:

$$CT_{load} = 1 + \frac{1}{COP} \qquad (4.19)$$

Some electricity is used by absorption chillers to operate the pumps in the chiller and to operate fans and pumps associated with the cooling tower. Because the COP of an absorption chiller is so much less than that of an electric chiller, the total load on the cooling tower is much greater, so a much larger cooling tower is needed, with the result that the total electricity used by an absorption cooling system (to operate just pumps and fans) is typically 20–25 per cent that of the compressor in an electric vapour-compression chiller.[8]

Primary energy requirements of absorption chillers

Absorption chillers can be driven with heat supplied by burning natural gas within the unit, or can be supplied

Box 4.4 Principles behind the operation of an absorption chiller

To explain how an absorption chiller works, it is easiest to first explain how one can use a camp fire and lithium bromide (LiBr) to make cold beer while on a canoe trip on a lake in one of Canada's vast wilderness parks.

Take a mixture of LiBr and water in a bottle. Connect this bottle to another bottle with a tube having a valve that can be closed. Check that the valve is open, and heat the LiBr/water solution near the campfire. This drives off water vapour, which enters the other bottle and condenses, releasing heat. A concentrated LiBr/water solution is left behind. Remove the first bottle from the campfire, close the valve and let both bottles cool. When the bottles have cooled, open the valve. Water will evaporate from the second bottle as water vapour is sucked into the concentrated LiBr/water solution. As water is reabsorbed, heat is released in the solution, but the bottle where evaporation occurs becomes very cold – sufficiently cold to cool your beer if placed in contact with it (of course, a much easier procedure is to put the beer bottles in the lake and wait one hour!).

The operation of an absorption chiller is illustrated schematically in Figure 4.38. The absorption chiller is seen to involve an outer cycle between a condenser and an evaporator, as in the vapour-compression chiller, and an inner cycle between a generator and an absorber in place of a compressor.

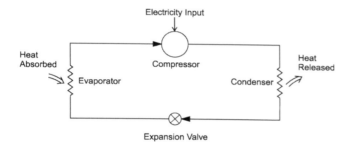

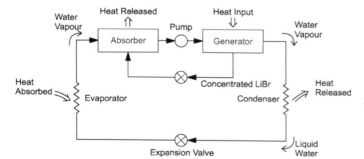

Source: Based on Reay and MacMichael (1988)

Figure 4.38 *Schematic comparison of an electric vapour-compression chiller (top) and an absorption chiller (bottom)*

Figure 4.39 shows the detailed layout of an absorption chiller. In this system, water is the *refrigerant* and LiBr is the *absorbent*. The LiBr/water solution is heated to 80–95°C in a *generator*. Water vapour is driven off, flows to a *condenser* and is cooled to about 40°C by water coming (indirectly) from a cooling tower. This is sufficient to cause the water vapour to condense. The combination of generator and condenser serves as a water distiller, since it separates pure water from the LiBr water solution, leaving concentrated LiBr behind. The condensed water is sprayed into the upper part of a cylindrical chamber having a pressure of about 1/100atm. This is called the evaporator, as the low pressure causes rapid evaporation, cooling the remaining water to as low as 4°C before it lands on the chiller-water coil. Water from the building that needs to be cooled passes through this coil. The hot concentrated LiBr solution that is left behind in the generator is sprayed into the lower part of the cylindrical chamber. Most of this solution forms a thin film on the coil of an *absorber*. The absorber coil is kept cool through the circulation of cooling water from a cooling tower, which reduces the vapour pressure of the LiBr solution. This in turn allows ready absorption of the water vapour produced in the evaporator and serves to maintain the low vapour pressure in the upper, evaporator portion of the cylinder. As water and concentrated LiBr are mixed, heat is released, but this heat is carried away by the cooling water. The cooled and diluted solution is then pumped back to the generator, where the cycle begins again. In order to increase the surface area of the diluted solution, so that water can be driven off more readily, the diluted solution is sprayed into the generator. A heat exchanger is used to preheat the absorbed liquid before it enters the generator and to pre-cool the concentrated solution before it enters the absorber.

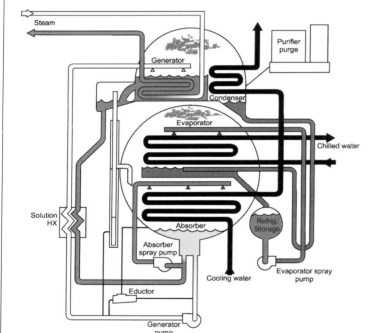

Source: Wulfinghoff (1999), reprinted by permission from the *Energy Efficiency Manual* by Donald R. Wulfinghoff

Figure 4.39 *Fluid flows in an absorption chiller*

with heat produced through the generation of electricity or supplied by solar thermal collectors. The co-production of electricity, heat and chilled water using some of the heat output in absorption or other heat-driven chillers is referred to as *trigeneration.*

Suppose that the removal of 1.0 unit of heat from a building requires x units of electricity and y units of heat using an absorption chiller (where $y = 1.0/COP_{abs}$, COP_{abs} being the COP of the absorption chiller). If η_e and η_{th} are the electrical and thermal efficiencies, respectively, in cogeneration (i.e. the fraction of input fuel converted to electricity and to useful heat, respectively), then the

production of y units of useful heat is accompanied by generation of $z = (\eta_e/\eta_{th})y$ units of electricity using z/η_e ($= y/\eta_{th}$) units of fuel. The excess electricity, $z-x$, displaces electricity produced in a central powerplant with efficiency η_{pp} and transmitted to the point of use (i.e. near the cogeneration facility) with efficiency η_{tr} (i.e. with a fractional loss equal to $(1-\eta_{tr})$). Thus, $(z-x)/(\eta_{pp}\,\eta_{tr})$ units of fuel are saved at the central powerplant. If w is the amount of heat that needs to be added back to the air after overcooling it (for humidity control by condensing out water vapour, as discussed in subsection 4.1.4), and if this heat is provided by a boiler with efficiency η_b, then

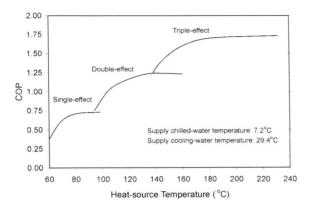

Source: Lee and Sherif (2001)

Figure 4.40 *Variation of COP with heat-source temperature for single-, double- and triple-stage absorption chillers, assuming 29.4°C cooling water and production of chilled water at 7.2°C*

the net primary energy use with absorption chilling combined with cogeneration is given by:

$$PE_{net} = \frac{w}{\eta_b} + \frac{y}{\eta_{th}} - \frac{z-x}{\eta_{pp}\eta_{tr}} \qquad (4.20)$$

Equation (4.20) applies to cogeneration using microturbines, reciprocating engines and fuel cells, all of which are applicable at relatively small scale (<500kW). In these cases, the withdrawal of useful heat does not reduce the electricity output.

An alternative is larger-scale (>20MW) combined cycle cogeneration (see Chapter 3, section 3.3). In this case, steam would be extracted from the steam turbine, reducing the electricity output. For steam withdrawn at 95°C, the sacrificed electrical energy is about 15 per cent of the heat energy withdrawn. The directly used and sacrificed electricity could be made up through additional generation in the combined cycle powerplant itself or in a central powerplant (whose efficiency in either case is represented by η_{pp}). Thus, the use of primary energy associated with the absorption chiller in this case is:

$$PE_{abs-cc} = \frac{x + 0.15y}{\eta_{pp}\eta_{tr}} + \frac{w}{\eta_b} \qquad (4.21)$$

The relative primary energy use by electric and absorption chillers depends on the COPs of the electric

and absorption chillers, as well as on all the efficiencies appearing in Equations (4.20) and (4.21) and the relative importance of overcooling and reheating (for humidity control) relative to the direct cooling requirement.

Table 4.7 gives the cooling and reheating loads and the computation of electric chiller COPs for extreme temperate and tropical conditions. In the temperate case, the reheating load is 22 per cent of the total cooling load (which includes overcooling) and the electric chiller COP is 4.9 for the assumptions given in Table 4.7, while for the tropical case the reheating load is 11 per cent of the cooling load and the COP is 3.7. Table 4.8 shows the computation of the relative primary energy requirements for electric chillers using electricity from a central powerplant and for absorption chillers directly fuelled with natural gas or powered with heat from microturbine, solid oxide fuel cell, or combined cycle cogeneration. Results are given for the relative cooling and reheating loads and electric chiller COPs given above, for an absorption chiller COP of 1.2, and for various microturbine and central powerplant efficiencies. Use of directly fired absorption chillers substantially increases the primary energy requirement compared to electric chillers for all cases. This is not surprising, as the product of the electric chiller COP and powerplant efficiency exceeds the absorption chiller COP. Use of waste heat from a microturbine decreases the primary energy requirement if the central electric powerplant efficiency is 35 per cent, but increases the primary energy requirement if the efficiency is 55 per cent.

Alternatively, use of waste heat from combined cycle cogeneration reduces the primary energy requirement by roughly 50 per cent for both central powerplant efficiencies, while use of waste heat from fuel cells to power absorption chillers in place of electric chillers using electricity from a central powerplant with low efficiency (35 per cent) results in a reduction in total energy use (savings >100 per cent).

Desiccant dehumidification and associated cooling systems

A passive desiccant wheel consists of a rotating drum that contains a solid desiccant and rotates from the incoming airflow (picking up moisture) to the outgoing airflow (releasing moisture) and back. In an active desiccant wheel, the outgoing airstream is heated before it passes

Table 4.7 *Determination of air conditioning COP and sensible and latent cooling loads for extreme temperate climate and tropical climate weather conditions*

	T_{db} (°C)	r (gm/kg)	Changes in heat content (kJ/kg)		
			Sensible heat	Latent heat	Total enthalpy
Temperate case					
Initial T_{wb} = 24°C, $T_{condenser}$ = 34°C, $T_{evaporator}$ = −4°C, η_{Carnot} = 0.65, COP = 4.6					
Starting values	30	16			
Final values and required cooling	16	7.5	−14.7	−21.3	−36.0
Overcooling step	6	7.5	−10.2		−10.2
Reheating step	16	7.5	+10.2		+10.2
Tropical case					
Initial T_{wb} = 33°C, $T_{condenser}$ = 43°C, $T_{evaporator}$ = −4°C, η_{Carnot} = 0.65, COP = 3.7					
Starting values	40	30			
Final values and required cooling	16	7.5	−26.1	−56.3	−82.4
Overcooling step	6	7.5	−10.2		−10.2
Reheating step	16	7.5	+10.2		+10.2

Note: T_{db} = drybulb temperature, T_{wb} = wetbulb temperature, r = water vapour mixing ratio. Shown are the changes in sensible and latent heat per kg of air required to go from the starting temperature and mixing ratio values to the desired final values, the additional sensible heat that must be removed to overcool the air sufficient to condense out the excess moisture, and the sensible heat that then needs to be added back. The air conditioner condenser is assumed to be cooled with water from a cooling tower, whose temperature cannot be any less than the wetbulb temperature. A temperature difference of 10K is assumed to occur between the condenser and cooling water, and between the evaporator and minimum air temperature needed to condense out the required moisture. COPs are computed from Equation (4.18) using the given condenser and evaporator temperatures and Carnot efficiency.

through the desiccant wheel so that it is able to drive moisture off of the wheel. The dry (and warm) wheel is then able to absorb moisture from the incoming air when it rotates into the incoming airstream. Gas-fired desiccant dehumidification is widely used in supermarkets, where very low humidity is needed in order to prevent frosting of refrigerated displays.

An active desiccant wheel can be combined with evaporative cooling and a wheel that exchanges sensible heat in order to achieve simultaneous cooling and dehumidification, in an arrangement illustrated in Figure 4.41. Outdoor air at state 'A' passes through the warm desiccant wheel, which dries but heats the air to state 'B'. This air is then partly cooled with the cool exhaust air using some sort of sensible heat exchanger. As the supply air is now quite dry, it is possible to achieve further cooling with direct evaporative cooling, but proceeding only part way to the wetbulb temperature so as to avoid excessive humidity, finishing at 'C'. Meanwhile, the exhaust air is cooled through evaporation from point 'D' to point 'E' (high humidity does not matter in this case, as the air is now leaving the building), then passes through the sensible heat exchanger, warming it to point 'F' while cooling the supply air. By first cooling the exhaust air evaporatively,

it is more effective in cooling the supply air. Additional heat is added to the exhaust air (bringing it to point 'G') before passing through the rotating desiccant wheel, heating the desiccant wheel and driving off the moisture that it had extracted from the supply air and finishing at point 'H'. The sensible heat wheel thus serves a dual purpose – it cools the supply air after it has been dried and it preheats the exhaust air, so that less heat needs to be added to it prior to regenerating the desiccant.

The COP of desiccant cooling systems (ratio of decrease in air enthalpy to heat energy added), taking into account free evaporative cooling, is in the order of 0.8–1.0. Interestingly, the COP increases with lower driving temperatures, although the drying power decreases. This is shown in Figure 4.42. The COP is also larger the hotter and more humid the outside air (although a warmer driving temperature is needed in order to achieve the same final result). *Most importantly, the combination of desiccant wheels with evaporative cooling extends the applicability of evaporative cooling into the hot humid regions of the world, where it otherwise cannot be used.* For a typical office building in the hot and humid climate of Hong Kong, Niu et al (2002) calculated a saving in total energy use (electricity + heat) of 30 per cent using a desiccant system instead of

Table 4.8 *Inputs used in calculating the relative primary energy used by electric and absorption chillers to remove 1.0 unit of heat for the temperate and tropical conditions given in Table 4.7*

	Electricity use	Heat input		Efficiencies		Primary energy (PE)				% savings in PE
		Reheat	Abs chiller	η_e	η_{th}	Heating	Cogen	Powerplant	Total	
Temperate case (η_{pp} = 0.35, η_{tr} = 0.95)										
Electric chiller (COP = 4.6)	0.307	0.28				0.31		0.92	1.24	
Absorption chiller										
– Directly fired	0.056	0.28	1.07			1.38		0.17	1.55	−25
– Using heat from MT	0.056	0.28	1.07	0.25	0.40		3.38	−2.37	1.01	19
– Using heat from MT	0.056	0.28	1.07	0.25	0.60		2.25	−1.53	0.73	41
– Using heat from SOFC	0.056	0.28	1.07	0.45	0.40		3.38	−4.41	−1.03	183
– Using heat from GTCC	0.056	0.00	1.07					0.62	0.62	50
Temperate case (η_{pp} = 0.55, η_{tr} = 0.95)										
Electric chiller (COP = 4.6)	0.307	0.28				0.14		0.59	0.93	
Absorption chiller										
– Directly fired	0.056	0.28	1.07			1.38		0.11	1.49	−65
– Using heat from MT	0.056	0.28	1.07	0.25	0.40		3.38	−1.51	1.87	−107
– Using heat from MT	0.056	0.28	1.07	0.25	0.60		2.25	−0.97	1.28	−42
– Using heat from SOFC	0.056	0.28	1.07	0.45	0.40		3.38	−2.80	0.58	36
– Using heat from GTCC	0.056	0.00	1.07					0.39	0.39	56
Tropical case (η_{pp} = 0.35, η_{tr} = 0.95)										
Electric chiller (COP = 3.7)	0.332	0.12				0.14		1.00	1.14	
Absorption chiller										
– Directly fired	0.060	0.12	0.94			1.07		0.18	1.26	−10
– Using heat from MT	0.060	0.12	0.94	0.25	0.40		2.65	−1.81	0.84	26
– Using heat from MT	0.060	0.12	0.94	0.20	0.60		1.77	−1.15	0.62	45
– Using heat from SOFC	0.060	0.12	0.94	0.45	0.40		2.65	−3.40	−0.75	166
– Using heat from GTCC	0.060	0.00	0.94					0.60	0.60	47
Tropical case (η_{pp} = 0.55, η_{tr} = 0.95)										
Electric chiller (COP = 3.7)	0.332	0.28				0.14		0.64	0.77	
Absorption chiller										
– Directly fired	0.060	0.12	0.94			1.07		0.12	1.19	−54
– Using heat from MT	0.060	0.12	0.94	0.25	0.40		2.65	−1.15	1.50	−94
– Using heat from MT	0.060	0.12	0.94	0.20	0.60		1.77	−0.73	1.04	−34
– Using heat from SOFC	0.060	0.12	0.94	0.45	0.40		2.65	−2.17	0.48	37
– Using heat from GTCC	0.060	0.00	0.94					0.38	0.38	50

Note: The electric chiller COPs are also taken from Table 4.7, while an absorption chiller COP of 1.2 is assumed. See text and below for additional assumptions, as well as for details of the calculation. MT = microturbine cogeneration. η_e and η_{th} are the electrical and thermal efficiencies when cogeneration is used to supply heat and electricity; η_{pp} is the efficiency of a central or onsite powerplant used instead of cogeneration to supply the required efficiency, or it is the efficiency of central electricity that is displaced by electricity produced by cogeneration in excess of the needs of the absorption chiller; and η_{tr} is the efficiency in transmitting electricity from the central powerplant. Auxiliary electricity uses by electric and absorption chillers are assumed to be 10 per cent and 20 per cent, respectively, of the electricity used by the compressors in electric chillers. Reheat with electric chillers and directly fired absorption chillers is supplied from a boiler with an efficiency of 90 per cent. With MT and SOFC cogeneration, the amount of electricity that would be generated in supplying the required heat input is computed from the ratio of the electrical and thermal efficiencies, then the electrical needs of the absorption chiller are subtracted, and the remaining electricity is assumed to displace centrally generated electricity. The displaced electricity appears as a negative entry in the column 'Powerplant'. The fuel used in producing heat and electricity through cogeneration is given in the column 'Cogen'. For the GTCC cogeneration case (the last absorption chiller option given), heat for reheat is assumed to be available without penalty, but heat used to drive an absorption chiller is assumed to be taken at a temperature such that the electrical energy output is reduced by 15 per cent of the heat energy withdrawn. Entries are blank where not applicable.

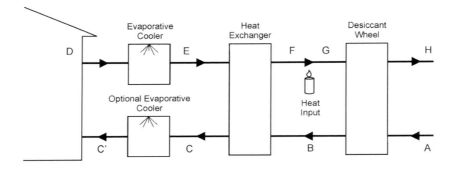

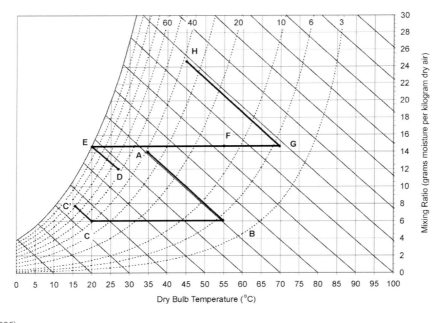

Source: Harvey (2006)

Figure 4.41 *Trajectory of process and secondary airflows in a system consisting of a solid desiccant wheel, sensible heat wheel, evaporative cooler and heat source to regenerate the desiccant*

a conventional system (with overcooling and reheating for dehumidification), and a saving of 50 per cent if the thermal energy (which can be largely provided with solar heat) is neglected.

Liquid desiccants are a second option for providing dehumidification and cooling without having to overcool and reheat the air. A liquid desiccant is a liquid solution that has a strong affinity for water vapour. Typical liquid desiccants use triethylene glycol or salts such as calcium chloride or lithium chloride

dissolved in water. The more concentrated the desiccant solution, the lower the vapour pressure of the solution and the more readily it will absorb water vapour from the air. The design of different liquid desiccant systems is reviewed by Mei and Dai (2008). Liquid desiccant systems have been used in hospitals (Florides et al, 2002), but unlike solid desiccants, they are not yet commercially available in large quantity. Like solid desiccants, they can be used to extend evaporative cooling techniques into hot humid regions

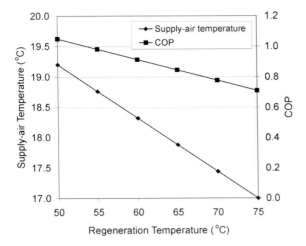

Source: IEA (1999)

Figure 4.42 *Variation of COP and supply-air temperature with regeneration temperature for a solid desiccant chiller with incoming ambient air at 32°C and 40 per cent RH*

where they could not otherwise be used, and they can be powered with solar heat. COPs of about 1.4 in humid climates and 2.2 in dry climates should be possible with a cooling thermostat setting of 24.4°C (Lowenstein and Novosol, 1995). Liquid desiccant systems have received considerable attention in China, where they have been found to be effective in purifying the air (apparently destroying the SARS virus).

Whether using a solid or liquid desiccant, a significant advantage of desiccant systems is that dehumidification is accomplished without saturation, thereby eliminating the potential for the growth of mould and bacteria and associated health hazards.

Primary energy requirements using desiccants

Desiccants can be driven with hot air heated by burning some fuel, or heated using waste heat from microturbine cogeneration. Desiccant dehumidification produces cool and dry air, and so can be applied only at individual buildings, whereas absorption chillers produce chilled water that could be produced centrally and distributed through a district cooling system. This permits use of waste heat from combined cycle cogeneration, which is the most attractive option for cogeneration (due to the high electricity to heat output ratio), but combined

cycle cogeneration is too large (>20MW) to be applied at the scale of individual buildings. Thus, trigeneration with desiccants would of necessity involve small-scale systems, either simple cycle gas turbine power generation (microturbines) or reciprocating engines.

Table 4.9 shows the calculation of the net primary energy use for cooling with conventional chillers and reheating and for various options involving onsite desiccants. Because water vapour is directly removed by the desiccant, the air does not need to be cooled to as low a temperature. This reduces the total cooling load (and especially reduces the load on the chiller), increases the chiller COP (because the evaporator does not need to be as cold) and avoids the need for reheating. For the conditions assumed in the calculation of chiller COPs in Table 4.7, but assuming an evaporator temperature of 6°C rather than –4°C, the chiller COP increases from 4.6 to 6.5 in the temperate case and from 3.7 to 4.9 in the tropical case. The primary energy required to provide 1.0 unit of cooling using electric chillers (at the original COPs) and using solid desiccants combined with electric chillers, for various assumptions concerning efficiencies, is given in Table 4.9. The cooling loads handled by the desiccant and electric chiller are given assuming that 40 per cent of the sensible heating load given in Table 4.7 in the line 'Required cooling' is handled by the desiccant system (through evaporative cooling) and the rest by the chiller. Every unit of cooling that is shifted from an electric chiller to a desiccant chiller entails a drop from the chiller COP (4–6) to the desiccant COP (0.8–1.0), but the losses associated with generating and transmitting electricity are avoided. The net result is that driving desiccants with independently produced heat (using a boiler) either slightly decreases or increases the total primary energy requirement, while use of waste heat from microturbines reduces the primary energy requirement in the temperate climate case if the reference powerplant efficiency is small (35 per cent) or if the microturbine thermal efficiency is large (60 per cent instead of 40 per cent) and the desiccant COP is large (1.0 instead of 0.8). Larger than usual heat recovery from the microturbine might be possible because desiccants can operate using relatively low-temperature heat (55–60°C, rather than the 95°C needed for double-effect absorption chillers). A higher desiccant COP is possible if air does not need to be cooled below

Table 4.9 *Inputs used in calculating the relative primary energy use by electric and desiccant chillers*

Case	Loads			COPs		Efficiencies			Primary energy	% savings
	Chiller	Desiccant	Reheat	Chiller	Desiccant	η_e	η_{th}	η_{mh}		
Temperate case ($\eta_{pp} = 0.35$, $\eta_{tr} = 0.95$)										
Conventional	1.00		0.22	4.60					0.93	
Desiccant										
– Direct heat	0.19	0.59		6.48	0.8				0.91	2
– Direct heat	0.19	0.59		6.48	1.0				0.75	20
– Heat from MT	0.19	0.59		6.48	1.0	0.25	0.40	1.40	0.51	45
– Heat from MT	0.19	0.59		6.48	1.0	0.25	0.60	2.10	0.36	61
Temperate case ($\eta_{pp} = 0.55$, $\eta_{tr} = 0.95$)										
Conventional	1.00		0.22	4.60					0.68	
Desiccant										
– Direct heat	0.19	0.59		6.48	0.8				0.88	−29
– Direct heat	0.19	0.59		6.48	1.0				0.72	−5
– Heat from MT	0.19	0.59		6.48	1.0	0.25	0.40	0.73	0.86	−26
– Heat from MT	0.19	0.59		6.48	1.0	0.25	0.40	1.10	0.59	14
Tropical case ($\eta_{pp} = 0.35$, $\eta_{tr} = 0.95$)										
Conventional	1.00		0.11	3.72					0.93	
Desiccant										
– Direct heat	0.16	0.73		4.90	0.8				1.13	−21
– Direct heat	0.16	0.73		4.90	1.0				0.93	0
– Heat from MT	0.16	0.73		4.90	1.0	0.25	0.40	1.40	0.63	32
– Heat from MT	0.16	0.73		4.90	1.0	0.25	0.60	2.10	0.46	51
Tropical case ($\eta_{pp} = 0.55$, $\eta_{tr} = 0.95$)										
Conventional	1.00		0.11	3.72					0.64	
Desiccant										
– Direct heat	0.16	0.73		4.90	0.8				1.08	−70
– Direct heat	0.16	0.73		4.90	1.0				0.88	−39
– Heat from MT	0.16	0.73		4.90	1.0	0.25	0.40	1.40	1.06	−67
– Heat from MT	0.16	0.73		4.90	1.0	0.25	0.60	2.10	0.73	−15

Note: Electricity and heat energy required to remove 1.0 unit of heat for the temperate and tropical conditions given in Table 4.7 are given here. Electricity for the conventional case is assumed to be generated with efficiency η_{pp} at a central powerplant and transmitted with efficiency η_{tr}. Heat for reheating overcooled air or for regenerating the desiccant is assumed to be produced with a boiler at 90 per cent efficiency.

19°C (see Figure 4.42), which is possible with chilled ceiling cooling and displacement ventilation. The greatest potential of desiccant dehumidification is if the heat can be provided by solar thermal collectors; in this case, primary energy use is reduced by about 90 per cent for the conditions assumed in Table 4.9. Further analysis of solar-powered desiccant (and absorption) cooling is found in Volume 2 (Chapter 2, section 2.5).

4.5.5 Cooling towers

Cooling towers cool water through evaporation, and this cooled water is used to remove heat from the condenser of a chiller. In the simplest kind of cooling tower (a *direct-contact* cooling tower), the water used to cool the condenser is sprayed into a rising air current. About 5 per cent of the water evaporates, cooling the remaining liquid water as it falls through the airstream.

The water can be cooled to within 2–3K of the ambient wetbulb temperature (defined in subsection 4.1.3). In an *indirect-contact* or *closed-circuit* cooling tower, illustrated in Figure 4.43, the cooling water circulates through a coil within the airstream, and a separate water supply is sprayed into an airstream and onto the coil, evaporating and cooling the cooling-water coil in the process. Where cooling towers are not used (perhaps due to water limitations), a closed water loop may still flow from the condenser through an air-cooled heat exchanger. The first cost and fan energy consumption of air-cooled heat exchangers are large, and the water can typically be cooled only to within about 10K of the ambient air temperature. Thus, a higher condenser temperature is needed in air-cooled systems, which reduces the chiller efficiency due to the greater temperature lift required (see subsection 4.4.2). Cooling towers and air-cooled heat exchangers are normally placed on the roof of the building.

The cooling system in a large commercial building thus consists of three fluid circulation loops, as shown schematically in Figure 4.44. A refrigerant circulates between the chiller condenser and evaporator, a chilled water loop picks up heat from the building and ejects heat to the evaporator, and a cooling water loop picks up heat from the condenser and ejects it to the atmosphere, usually with a cooling tower.

In addition to consuming water, cooling towers require energy to operate the fan (which induces the airflow) and the pump (which circulates the cooling water). Energy used by the cooling tower can be minimized by: (1) interlocking the operation of pumps and fans with the operation of the compressors, so that the former shut down when the latter are not operating

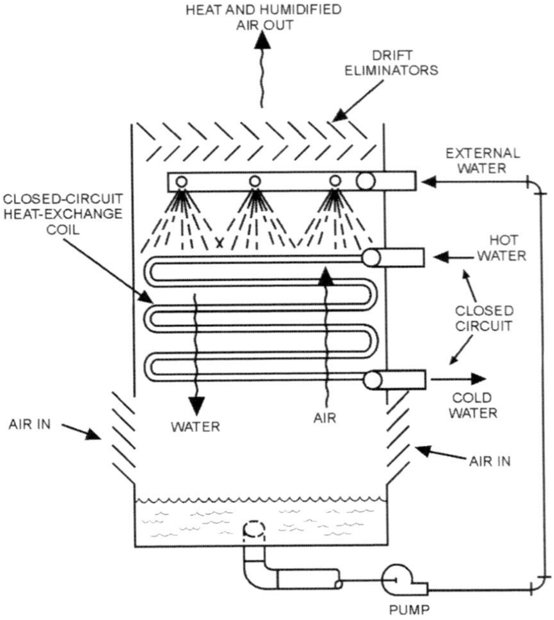

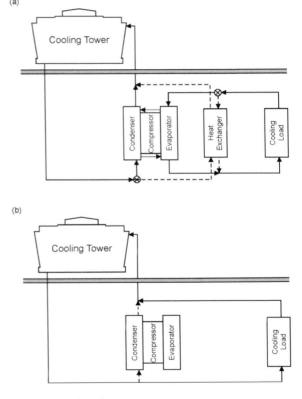

Source: Harvey (2006)

Figure 4.44 *(a) Layout of a cooling tower/chiller/chilled water loop system during operation of the chiller (solid lines), with free-cooling through bypass of the compressor (dotted lines), or with free-cooling through heat exchange between the cooling water and chilling water loops using a heat exchanger (dashed lines); (b) free-cooling through direct interconnection of the cooling water loop and chilled water loop*

Source: ASHRAE (2000, Chapter 36)

Figure 4.43 *Illustration of an indirect-contact or closed-circuit cooling tower*

(in a typical situation, this is not done); and (2) through efficient part-load operation. The combined chiller-cooling tower energy use can be reduced if the system is configured to allow maximum use of 'free-cooling' when outside temperatures are cold enough, as described next.

Free-cooling

If the cooling-tower water becomes sufficiently cold as the outside air temperature and cooling load decrease, it can be used to cool the building chilling water through a heat exchanger, or can flow directly through the building chilling-water circuit. These options are indicated by the dashed lines in Figure 4.44a and the solid lines in Figure 4.44b, respectively. In either case, the energy-intensive compressor is bypassed and turned off. Pumps are still required to circulate the cooling water and chilled water. This procedure, referred to in the industry as 'free-cooling', is really a form of evaporative cooling (previously discussed in subsection 4.5.2).

The most that the cooling water can be cooled through evaporation is to the atmospheric wetbulb temperature, which depends on the ambient air temperature and humidity. In practice, the best that one can achieve is to bring the cooling water to within 1.5K of the wetbulb temperature, and if the building chilling water is connected to the cooling-tower water through a heat exchanger, there will be another step of at least 1.5K. However, the warmer that the chilling water is allowed to be while still providing an adequate cooling effect, the more often it will be possible to chill it enough evaporatively using the cooling tower – that is, the greater the opportunity for free-cooling. As discussed in subsection 4.6.3, radiant chilled-ceiling cooling systems operate with a chilled water temperature of 16–20°C, rather than the 5–8°C (with dehumidification) or 11–14°C (without dehumidification) temperatures that are used in conventional systems today. Assuming a difference between the wetbulb temperature and chilling water temperature of 3K (which requires use of enhanced heat exchangers), and assuming the chilling water to be supplied at 18°C, a cooling tower could directly meet cooling requirements 97 per cent of the time in Dublin and 67 per cent of the time in Milan according to Costelloe and Finn (2003). If chilling water at 20°C is adequate, then evaporative cooling in a cooling tower is sufficient 99 per cent of the time in Dublin and 78 per cent of the time in Milan.

4.5.6 Correct estimation of cooling loads to avoid oversizing of equipment

In commercial buildings, internal heat gains represent a large fraction of the total cooling load (see Figure 4.27). These are often estimated using standardized guidelines that are far in excess of actual cooling loads. This results in cooling equipment that is greatly oversized, requiring it to operate at a smaller fraction of its peak capacity than would otherwise be the case, with a significant efficiency penalty. Based on a detailed analysis of six buildings in Hong Kong, W. L. Lee et al (2001) find that more realistic sizing of the cooling equipment alone would have reduced annual cooling energy use by 6–22 per cent and reduced the size of the chilling equipment by 32–46 per cent (thus significantly reducing the capital cost).

Oversized air conditioners are also a problem in residential central air conditioning systems. According to field studies reviewed by Neal (1998), central air conditioning systems in the US are oversized by an average of 25–100 per cent. Oversizing is used as a method to minimize customer complaints, that is, to cover up other problems (Mast et al, 2000). This directly leads to lower average efficiency by requiring frequent on–off cycling. Paradoxically, it creates comfort problems because the run times of the system are so short that the blower does not have time to fully mix the air, causing the occupant to lower the thermostat setting, thereby further increasing energy use. The dehumidification capacity is also reduced because water that condenses on the evaporator coil during the on cycle will partially re-evaporate during the off cycle. The net result is that an air conditioning system with a rated seasonally averaged COP of 3.5 may have an average operating COP of only 1.9 or less, implying almost a doubling in energy use!

4.5.7 Efficiency tradeoffs with off-peak air conditioning

Diurnal storage of chilled water is widely used in building cooling systems, with chilled water produced by running chillers during the night when electricity rates are lower, stored in large tanks and used for at least part of the air conditioning load during the day. The required storage volume is greatly reduced if an ice-water slurry is created in the storage tank, as ice has a very large latent heat of fusion (a lot of heat needs to be extracted from water to freeze it, and a lot of heat is absorbed when ice melts). A lower evaporator temperature (below 0°C) is required to make ice, which leads to a lower chiller COP because of the larger temperature lift between the evaporator and condenser temperature (see Equation 4.18), and therefore leads to greater energy consumption. However, a number of other factors (such as lower ambient temperature at night) tend to reduce energy consumption. The net effect on resulting energy use depends on what would have been done without ice thermal storage.

In many cooling systems, chilled water is cooled to 12–14°C and ventilation air is cooled down to 5–10°C or so for dehumidification, then reheated to about 16°C. However, there is a growing tendency to use chilled water at 5–7°C, as this permits a smaller rate of water flow (which saves on pumping energy) and smaller pumps, pipes and valves (which saves on capital costs). In this case, the additional cooling required to make ice (and the associated COP penalty) is small enough that the net change in onsite energy use is very small, generally ranging from a few per cent savings to a few per cent increase.

These differences in onsite energy use can be swamped by differences between night and day in the efficiency of generating and distributing electricity. During the day, steam powerplants must deliberately operate at part load, so as to be able to instantly meet sudden increases in electricity demand when needed. The plant efficiency is much less than at full load (fuel consumption per kWh at 30 per cent of full load is 50 per cent larger than at full load). Less part-load operation is required at night. Electricity distribution losses in California are 5 per cent smaller at night than during daytime peaks in summer, due both to smaller loads and to lower air temperatures at night. A detailed analysis by the California Energy Commission (CEC, 1996) indicates that, during summer, the energy required to produce 1kWh of electricity at night is 20–43 per cent less than during the daytime peak, depending on the region and method of calculation. Savings due to shifting from daytime to night-time in winter are about two thirds as large. To the extent that the California results are applicable to other jurisdictions, shifting a given electricity demand to night-time can result in significant primary-energy savings.

However, if systems using ice thermal storage are compared against advanced alternative systems, ice thermal storage is seen to greatly increase the overall electricity use. A chilled-ceiling cooling system (discussed later) normally operates with chilled water at 16–20°C (rather than at 5–7°C). Dehumidification can be performed without initial deep cooling using desiccants, and, if combined with a chilled ceiling and displacement ventilation, the ventilation air need be cooled down to only 20–25°C (rather than down to 5–10°C). In this case, the COP penalty required to make ice is substantially larger because the additional cooling would be substantially larger, so the total energy use is likely to increase substantially. Furthermore, systems with ice storage are designed to operate with very cold chilled water temperatures, which will prevent using the cooling tower to directly produce chilled water at times when it would be favourable to do so (see subsection 4.5.5). For many climates, this would be a significant penalty.

In order to achieve the savings (per kWh generated) at fossil fuel powerplants by shifting electricity consumption from daytime to night-time, while avoiding or minimizing the onsite energy penalty (extra kWh used) of ice thermal storage, it is necessary to store coldness at a temperature above 0°C. To do so compactly requires using some alternative to ice that changes from solid to liquid and back at a warmer temperature and which has a reasonably large latent heat of fusion. Possibilities include *eutectic salts* (a combination of inorganic salts, water and other elements that form a mixture with freezing points in the 8–10°C range) and *paraffin waxes*, some of which have freezing points in the 6–8°C range. He and Setterwall (2002) discuss one of several commercially available products, Rubitherm RT5, which has a freezing point of 7°C and a latent heat of fusion almost half that of water.

4.6 Heating, ventilation and air conditioning systems

The term HVAC refers to the combined *H*eating, *V*entilation and *A*ir *C*onditioning system in a building. The term is more commonly used in reference to commercial buildings than to residential buildings, as these three functions are closely integrated in the former. However, energy-efficient houses are almost airtight (see subsection 4.2.4), so mechanical ventilation has to be provided, often in combination with the heating or cooling system. The preceding sections of this chapter have extensively discussed furnaces, boilers, heat pumps and air conditioners – the devices that generate heat or coldness. We now discuss the missing parts – the distribution of heat or coldness throughout a building, the concurrent supply of ventilation air, and the integration of heating and cooling devices with the distribution of heat and cold, and with ventilation.

4.6.1 HVAC energy-efficiency principles

In the crudest HVAC systems, heating or cooling is provided by circulating enough air at a sufficiently warm or cold temperature to maintain the desired room temperature. The volume of air circulated in this case is normally much greater than what is needed for ventilation purposes (to remove contaminants and provide fresh air). The energy required to transport a given quantity of heat or coldness by circulating water is 25 to 100 times less than the energy required by circulating air (see subsection 4.1.2). Thus, the first principle of energy-efficient HVAC design is to separate the ventilation and heating or cooling functions by using chilled or hot water for temperature control, and to circulate only the volume of air needed for ventilation purposes.

A second principle, but one that is not always applicable, is to separate cooling from dehumidification. In most commercial buildings with air conditioning, dehumidification is accomplished by overcooling the air so as to condense sufficient water vapour, then reheating the air so that it can be supplied at a comfortable temperature. Instead, dehumidification can be done using solid or liquid desiccants, but to obtain savings in primary energy, the solid or liquid desiccants should be regenerated either with waste heat from a cogeneration system with high overall efficiency, or with heat from solar thermal collectors (see subsection 4.5.4).

A third principle is to integrate mechanical HVAC systems with passive heating, cooling and ventilation, so that passive systems can be used whenever conditions permit. This requires designing the building so that there is good passive heat gain during the winter but not during the summer, and so that exterior air will naturally flow through the building in a controlled manner in response to wind and temperature differences set up by differential solar heating (see subsection 4.5.2).

A fourth principle is to distribute heat at the coolest possible temperature and to distribute coldness at the warmest possible temperature. The required heating temperatures are minimized and the required cooling temperatures maximized if heating and cooling loads are kept small and if large radiator surfaces (such as radiant floors or ceilings) are used.

A fifth principle is to allow the temperature maintained by the HVAC system to vary seasonally with outdoor conditions, rather than remaining invariant. As previously noted, a large body of evidence indicates that the typical temperature setting of 22–24°C in air conditioned buildings is much lower than necessary. Temperatures in the range of 28–30°C are acceptable on hot days, particularly if individually controlled fans are available to create air speeds of about 0.5m/s and if natural ventilation through operable windows is allowed. Choosing temperatures in the upper part of this range can dramatically reduce cooling energy use. Computer simulations by Jaboyedoff et al (2004) indicate that increasing the thermostat from 24 to 28°C will reduce annual cooling energy use by more than a factor of three for a typical office building in Zurich and by more than a factor of two in Rome, while simulations by Lin and Deng (2004) indicate a factor of two to three reduction if the thermostat is increased from 23 to 27°C for night-time air conditioning of bedrooms in apartments in Hong Kong.

4.6.2 HVAC systems in residential buildings

Most houses heated by natural gas or oil in North America use a *forced-air* system – heat from a furnace is transferred to inside air via a heat exchanger, and the heated air is blown through ductwork. Two efficiencies are relevant here: the efficiency of the furnace in providing heat, and the efficiency of the air handler in moving air through the ductwork. The furnace efficiency is the ratio of heat delivered to the ductwork to fuel energy used by the furnace, while the air handler efficiency is the ratio of kinetic energy added to the airstream to electrical energy used by the air handler (motor + fan). In many houses, the furnace air handler represents the single largest use of electricity. The efficiency of an air handler is equal to the efficiency of the motor times the efficiency of the fan, and measured overall efficiencies are typically only 10–15 per cent (Walker and Mingee, 2003). Through a combination of more efficient motors and fans, and measures that reduce how much work the air handler needs to do (such as proper sizing of the heating system, smoother and shorter ducts, and better matching of the fan to the ductwork), the energy required to distribute warm air through a house can be reduced by a factor of 10–15 (see Harvey, 2006, Section 7.3.2).

Hot water systems, which are the norm in Europe, eliminate the need for ductwork except for ventilation. The energy used for pumping hot water is less than the fan energy used in a forced-air system. The airflow needed for ventilation alone is far less than the airflow in a typical forced-air heating system, which provides both ventilation and heat, so a significant net saving in the energy used to distribute heat is possible.

Hot water systems also indirectly reduce the amount of heat that has to be supplied to the house. This is because substantial amounts of heat are lost from the ductwork in forced-air systems, both through leakage of air at duct joints and by conduction of heat through the duct walls. In parts of the US (and perhaps in other countries) where houses are built without basements, much of the ductwork passes through unconditioned spaces (attics, crawl spaces), and in some cases the furnace is located in an attached garage. Forced-air heat distribution systems in such houses often lose 20–40 per cent of the heat prior to delivery to the intended spaces. Compensating for this heat loss by increasing the initial airflow requires a disproportionately large increase in fan power, due to the cubic relationship between airflow and power (see Equation (4.7)). As well, operation of a forced-air system (even when the ductwork is entirely within the conditioned spaces) induces increased infiltration of outside air due to pressure imbalances, which means that the furnace has to work harder. Studies in Tennessee and Florida found 80 per cent and 200 per cent increases, respectively, in the infiltration of outside air when the air handler is operating (Sherman and Jump, 1997).

Radiant floor heating

Hot water systems supply heat silently and potentially more uniformly than in a forced-air system. However, conventional systems require a high supply temperature (70–90°C), which leads to inefficiencies and poses a burning hazard. A better choice is a floor radiant-heating system, in which hot-water pipes are embedded in the floor and heat a room by heating the floor. Floor radiant heating has a number of advantages from an efficiency point of view. First, it requires the lowest distribution temperature of any system, which is particularly

advantageous when a heat pump is used to provide heat but also when using condensing furnaces. If plastic pipes are embedded in a 7cm floor slab at a density of 10m of pipe per square metre of floor, the required supply temperature is 30–35°C, while if plastic mats with integrated water ducts and a sturdy floor cover are used, a supply temperature of 27–32°C is sufficient (Halozan, 1997). Second, radiant heating is more effective than convective heating (air movement) in warming objects in a room, so that air temperature can be 1K lower with the same perceived temperature. Third, excess heat on outside walls and windows (arising from perimeter air vents or radiators) is avoided, which will reduce conductive heat loss by 10–30 per cent (however, if a floor slab directly on the ground is heated, then 30cm of insulation below the slab would be recommended under Danish climate conditions, based on calculations by Weitzmann et al, 2005). Fourth, it avoids warm air rising to the ceiling. Fifth, it provides thermal energy storage in the floor itself, so pumps do not need to run all of the time. A floor radiant system can also be used for cooling; in this case, a warmer distribution temperature (15°C) can be used then needed in a forced-air cooling system (10°C). Radiant heating was used by the Romans 2000 years ago and has been the traditional system of heating in Korea for hundreds of years (de Carli and Olesen, 2002). A less expensive alternative to radiant floors is the use of large (1m high by 2–3m long) radiant panels mounted on the lower part of walls.

Heat-recovery ventilators

In houses that are almost completely airtight, fresh air needs to be provided by a mechanical ventilation system when the windows are closed. The fresh air that needs to be provided in this way will be substantially less than the uncontrolled ventilation in leaky, poorly built houses. Furthermore, a portion of the heat in the outgoing exhaust air can be used to partially heat the cold incoming air, using an *air-to-air heat exchanger* (to form a *heat-recovery ventilator*, HRV). The heat exchangers come in two basic types: *flat plate* and *thermal wheel*. Flat-plate heat exchangers can be of the *counterflow* or *cross-flow* type, illustrated schematically in Figure 4.45. A commercial product is shown in Figure 4.46. Up to 95 per cent of the heat that would normally be lost in the outgoing air can be used to heat the incoming air. A thermal wheel consists of a wheel

(a)

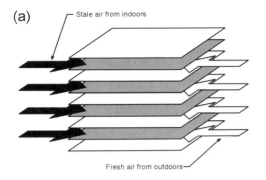

(b)

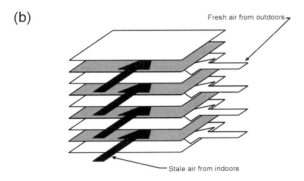

Source: Bower (1995)

Figure 4.45 *Schematic illustration of the two airstreams in (a) a counterflow heat exchanger, and (b) a cross-flow heat exchanger*

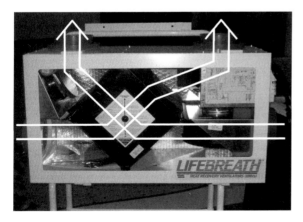

Source: Author

Figure 4.46 *A cross-flow heat exchanger*

that rotates through the incoming and outgoing airstreams. It can transfer both heat and moisture. Thus, in winter, incoming air is humidified as well as warmed, and this can eliminate the need for a humidifier. During the summer, the relative coldness and dryness of the outgoing air can be used to partially cool and dry the incoming air. The low rate of air exchange in airtight, mechanically ventilated houses, combined with partial recovery of heat or coldness from the exhaust air, serves to substantially reduce the heating and cooling energy requirements.

Ceiling fans

Ceiling fans are common in hot climate regions of the US, where up to five fans per house can be found. Moving air creates a cooling effect (by promoting evaporation and transfer of sensible heat away from the skin), thereby allowing a higher temperature to feel comfortable. James et al (1996) have examined the tradeoff between fan energy use and reduced air conditioning energy use in houses in Florida. A fan set to high speed (0.75m/s airflow) allows the air conditioner thermostat to be set 1.8K higher with the same degree of comfort, as long as the person is within the moving airstream created by the fan. This in turn will reduce the annual air conditioning energy use in Florida by about 20 per cent, after accounting for the energy used by the fan and the waste heat produced by the fan (which needs to be removed by the air conditioner). However, if the thermostat is not adjusted, air conditioning energy use increases due to the waste heat generated by the fan. Furthermore, a fan provides no benefit when the room is not occupied. Some fans on the market are equipped with occupancy sensors that turn the fan off some time after the room is vacated, and at least one model has a photo-optical sensor to prevent the fan from being turned off due to lack of motion in darkness (i.e. while the occupants of a bedroom are sleeping) (Parker et al, 1999).

There are two significant technical measures that can be taken to reduce the energy use by fans. First, most ceiling fans contain lights, and in some units, there are no separate controls for the lights! Second, ceiling fans have an efficiency of only 4 per cent (at low speed) to 12 per cent (at high speed) in converting electrical energy into moving air. Better motors and more aerodynamic fan blades (Figure 4.47) can double

Figure 4.47 *An aerodynamic ceiling fan referred to as the 'Gossamer Wind', developed at the Florida Solar Energy Center and available commercially in North America*

the efficiency at low speed and triple it at high speed (Camilleri, 2001).

4.6.3 HVAC systems in commercial buildings

A variety of methods have been used to ventilate and heat or cool commercial buildings. In an all-air system, air of sufficient coldness and in sufficient volume to remove all of the heat that needs to be removed is circulated through the building. This can be done using either a constant flow rate, or a variable flow rate – referred to as constant air volume (CAV) and variable air volume (VAV) systems, respectively. The air that is circulated through the supply ducts may be taken entirely from the outside and exhausted to the outside by the return ducts (this is mandatory in hospitals), or a portion of the return air may be mixed with fresh outside air and recirculated through the building (typically, 20 per cent will be exhausted, 80 per cent will be recirculated and blended with 20 per cent fresh air). The incoming air needs to be cooled and dehumidified in the summer, or heated and sometimes humidified in the winter. In an all-air cooling system, the amount of air that needs to be circulated in order to achieve adequate cooling is so large that fresh-air requirements are easily met; ventilation in this case is a byproduct of the air conditioning.

Existing large commercial buildings typically require simultaneous heating and air conditioning

365 days per year. Air conditioning is required on even the coldest winter days because the core of the building tends to overheat due to all of the waste heat generated by inefficient lighting and office equipment, while the periphery requires heating in winter due to heat losses through windows and walls. In some buildings, the proportion of outside air can be increased so as to avoid the need for air conditioning in the winter, but in many buildings this cannot be done. In summer, heat is needed either to reheat air after it has been overcooled for dehumidification purposes, or in buildings where ventilation air is pre-cooled to a fixed temperature and then reheated as needed based on the actual cooling load.

Below, we briefly discuss ways of making all-air cooling systems more efficient, then discuss a variety of alternative systems that are more efficient than even the most efficient all-air systems.

Conventional all-air cooling systems

In many buildings, air in a ventilation system is pre-cooled to a fixed temperature and then distributed to individual rooms. In some cases, the volumetric airflow rate is also fixed. The amount of cooling provided to a room is varied in one of two ways:

- by reheating the air (sometimes electrically!) by the amount required to maintain the desired workspace temperature. This is called *terminal reheat* and used to be quite common in North America. If combined with over-ventilation or a minimum ventilation rate that provides more cooling than needed, substantial amounts of energy can be wasted. For example, Sellers and Williams (2000) report the case of a building where the boiler was firing at 50 per cent of its wintertime design-maximum on the warmest day of the summer in order to provide the reheating demanded by the system;
- by throttling the airflow (partly closing the ducts with vanes in order to reduce the airflow). Most commercial buildings in North American cities still use either terminal reheat or throttling to control the cooling of individual rooms. This is because the motors in the fans of older buildings could run at only one speed (dictated by the frequency of the AC electricity, which is fixed).

Variable air volume, all-air cooling

An alternative to terminal reheat or throttling is to vary the speed of the fan and hence the rate of airflow using a VSD. For buildings in Washington, Chicago and Charleston (North Carolina), Franconi (1998) calculated that switching from a CAV to VAV system using a VSD leads to a 25 per cent reduction in cooling energy use, a 50–75 per cent reduction in heating energy use (due in part to less terminal reheat) and a 65 per cent reduction in fan energy use, giving an overall savings of 53–63 per cent.

Chilled-ceiling and chilled-beam cooling

An alternative to all-air cooling is to restrict the airflow to that required for ventilation and to use fan coil units (in which the air in a given room is blown past a cold pipe with fins) for additional cooling. This has become common practice. A more recent approach, pioneered in Europe in the 1980s, is to rely on a chilled ceiling to provide the additional cooling through radiant heat exchange. This can occur by circulating chilled water through radiant panels that cover a large fraction of the ceiling area, by chilling the concrete ceiling slab through cold-water pipes inside the slab, or by chilling ceiling beams. Chilled ceiling panels are illustrated in Figure 4.48. In a radiant chilled ceiling (CC), water at a temperature of 16–20°C is circulated through ceiling panels. Cooling occurs in two ways: through exchange of radiation (the cold panel emits less infrared radiation than it absorbs from the warmer room interior) and

through convection (air next to the panel is cooled and sinks). Radiant heat exchange is the more important of these two processes. The radiant asymmetry associated with a CC creates a sensation of 'freshness' that is similar to being outdoors under an open sky (Hodder et al, 1998), permitting the air temperature to be 2K warmer with the same comfort level.

The energy advantage of CC relative to an all-air cooling system arises:

- in part from the use of water rather than air to distribute coldness (an advantage that also applies to conventional hydronic fan-coil systems);
- in part from the fact that cooling can be accomplished using water at 16–20°C rather than at 14–16°C or even 5–7°C (so that the chiller can operate more efficiently);
- in part from the fact that the ventilation air does not need to be cooled as much either (it can be supplied at a temperature 5K cooler than the room air rather than 8K cooler, and the room air itself can be 2K warmer).

However, if a CC ceiling is combined with an otherwise conventional system, the second and third energy savings can be lost. In particular:

- if ice thermal storage is being used, then the COP benefit of a warmer chilled-water temperature and the greater opportunity for directly cooling the chilling water using the cooling tower (see subsection 4.5.5) are lost because initial cooling down to freezing occurs whether a CC or all-air system is used;
- if dehumidification is accomplished by over-cooling the ventilation air and then reheating it, more energy will be needed for reheating because the ventilation air is supplied at a warmer temperature in a CC system.

Stetiu and Feustel (1999) carried out simulations of buildings with all-air and combined air/CC cooling systems for a variety of US climates, assuming the same rate of intake of outside air for the two cases. They found that radiant cooling reduces energy use by an amount ranging from 6 per cent in Seattle to 42 per cent in Phoenix. The savings are smaller in hot humid or cool humid climates than in hot dry climates because relatively more of the total air conditioning energy is used for dehumidification, which is not

Source: www.advancedbuildings.org

Figure 4.48 *Chilled-ceiling cooling panels*

affected by the choice of all-air versus air/CC chilling. Stetiu and Feustel (1999) assumed the same chiller COP in both systems, dictated by the evaporator temperature needed for dehumidification. However, if dehumidification were to be done without overcooling by using desiccants (as discussed in subsection 4.5.4), then a further energy savings with CC would occur.

Chilled beams usually contain a fan that actively blows room air past the cold pipes within the beam. They provide effective cooling in rooms with large cooling loads, such as internet data centres or some laboratories. In a lab in the US with chilled beams, total HVAC energy use was reduced by 57 per cent, due mostly to avoided terminal reheat and a warmer temperature of the cold water used for cooling (Rumsey and Weale, 2007). The first-cost was reduced due to equipment downsizing, reduced size of ductwork and reduced floor-to-floor height.

Displacement ventilation

In a conventional CAV or VAV ventilation system, the ventilation air typically enters a room through an air outlet in one part of the ceiling and returns through a vent elsewhere in the ceiling. Even when the air outlet is placed near the floor, the system still relies on *turbulent mixing* of fresh air with stale air in the lower half of the room in order to provide fresh air to the occupants. An alternative is *displacement ventilation* (DV), in which fresh air is introduced from many holes in the floor at a temperature of 16–18°C, is heated by heat sources within the room, and continuously rises and *displaces* the pre-existing air. An example of the ventilation outlet used with DV is shown in Figure 4.49. Indoor air quality can be maintained with 40–60 per cent less airflow than in a conventional ventilation system, which shall henceforth be referred to as a *mixing-ventilation* (MV) system. DV is best combined with CC cooling. The required airflow is reduced both by using airflow only for ventilation (rather than also for cooling), and by increasing the ventilation effectiveness of a given airflow. DV, like CC cooling, was first applied in northern Europe. By 1989, it had captured 50 per cent of the Scandinavian market for new industrial buildings and 25 per cent for new office buildings (Zhivov and Rymkevich, 1998). It is still only rarely used in North America.

Source: Author

Figure 4.49 *Diffuser for a displacement ventilation system*

DV ventilation saves energy in the following ways:

- by requiring less airflow due to its greater effectiveness in removing stale air, thereby capturing the savings through the cubic law (Equation (4.7));
- by allowing ventilation with a warmer airflow, so that outdoor air will be cool enough more often and so can be used;
- because the total amount of heat that needs to be removed by the chillers is reduced if the ventilation air is vented directly to the outside after passing through the building only once.

With regard to the last point, the normal practice in a ventilation system is, as previously noted, to recirculate most of the air, exhausting only a small portion to the outside after each circuit through the building and blending in an equal amount of fresh outdoor air. The blended air is cooled as necessary at the start of each circuit, meaning that most of the heat picked up from the previous circuit has to be removed by the chiller. In a system where most of the cooling is done with chilled water, the airflow is reduced to that needed only for fresh-air purposes, meaning that it will be completely vented to the outside and replaced with 100 per cent fresh air after one circuit (this is referred to as *dedicated outdoor air supply* or DOAS). Because the air in a DV system rises from the occupants to the ceiling, and from there directly to the outside, heat picked up at ceiling level does not need to be removed by the chillers. Calculations by Loudermilk (1999) indicate that, for an office in Chicago, about a third of the total heat gain (including 50 per cent of the heat gain from electric lighting) can be directly rejected to the outside in this way.

The overall impact on energy use of a DV/CC system compared to a MV/CC or a VAV MV system depends on a number of competing factors, and if the overall system is not fully optimized, there can be little net saving. If overcooling and subsequent reheating for dehumidification are avoided, then cooling + ventilation energy use can be reduced by 30–60 per cent (Bourassa et al, 2002; Howe et al, 2003).

Demand-controlled ventilation

Having decoupled the ventilation and heating or cooling functions of a HVAC system using some hydronic cooling method (preferably CC), one is free to vary the ventilation rate based on actual and changing ventilation requirements, rather than using a fixed ventilation rate or varying it according to some inflexible schedule. Even in all-air cooling systems, it can be advantageous to vary the ventilation rate or the ratio of outdoor to recirculated air based on the ventilation requirement. This is referred to as *demand-controlled ventilation* (DCV). California's 2001 building code (the Title 24 standards) requires that the ventilation rate respond to changing occupancy in high-density buildings, rather than run at a fixed rate based on maximum occupancy. Depending on the kind of building and occupancy schedule, DCV can save 20–30 per cent of the combined ventilation, heating and cooling energy use in commercial buildings (Brandemuehl and Braun, 1999). The total volume of outdoor air circulated through a building during most working hours can be reduced by 30–50 per cent (Schell et al, 1998). For primary schools in Norway, Mysen et al (2005) calculate a 50 per cent saving in heating + ventilation energy use compared to a CAV system that would run only six hours a day, and a 75 per cent saving compared to a CAV system with 24-hour operation.

Sensible heat exchangers

As in residential ventilation systems, some energy savings can be achieved through sensible heat exchangers, which transfer heat from the outgoing

airflow to the incoming airflow. Depending on the kind of heat exchanger used, 55–85 per cent of the heat in the outgoing air can be recovered. The presence of a heat exchanger increases the fan energy needed to circulate the air, so provision should be made to bypass the heat exchanger when the temperature difference between incoming and outgoing air is small. This, however, will significantly reduce the fan energy use only if the fan has a variable speed drive. Again, careful integration of the different HVAC system components is needed in order to capture the full energy-savings potential.

Control systems and monitoring

Commissioning is the process of systematically checking that all of the components of a HVAC system are present and function properly. It also involves adjusting the system and its controls to achieve the best possible performance. Commissioning costs about 1–3 per cent of the HVAC construction cost, but in the US, only 5 per cent of new buildings are commissioned (Roth et al, 2002). Consequently, the control systems never operate as intended in many buildings. Even if a building is commissioned after construction, it is important to continue adequate monitoring of the HVAC system. In a study of 60 buildings carried out by Lawrence Berkeley National Laboratory, half had problems with their control systems. These problems can, for example, cause fans to operate all night that are supposed to shut down at night, cause boilers to start up on summer days when they are not supposed to, or prevent the use of cool outside air for air conditioning when it is supposed to be used (Barwig et al, 2002; Price and Hart, 2002). According to a survey of more than 450 control-related problems in buildings reported in Barwig et al (2002), the two largest causes of problems are software programming errors (32 per cent) and human operator errors (29 per cent). Hardware errors account for a modestly smaller fraction (26 per cent) of the problems.

In a programme involving over 80 buildings mentioned by Piette et al (2001), improved controls reduced total-building energy cost by over 20 per cent and heating + cooling energy cost by over 30 per cent. At issue is not only the proper functioning of the control sensors and actuators, but also the *design* of the control system and the operating strategy. Wulfinghoff (1999) and Levermore (2000) provide many examples to illustrate the principles underlying effective control strategies for heating and cooling plants so as to maximize overall efficiency.

Summary

A highly energy-efficient HVAC system in a commercial building will include the following:

- chilled-ceiling cooling, as this allows one to set ventilation airflow based on ventilation requirements, permits the highest possible distribution temperature for coldness, and increases the opportunity to directly use the cooling tower for air conditioning purposes;
- demand-controlled displacement ventilation, as this permits the minimum ventilation airflow at the warmest possible temperature while still providing adequate cooling, and permits once-through air circulation (or dedicated outdoor air supply) that in turn directly removes rather than recirculates heat from the ceiling areas, and gives much better indoor air quality;
- dehumidification using solar-regenerated desiccants, which directly satisfies the latent heat portion of the cooling load using solar energy and avoids the need for overchilling and reheating for humidity control;
- heat recovery on ventilation exhaust airflow;
- allowance for passive ventilation and direct use of outdoor air for cooling when conditions permit;
- correct sizing and commissioning of all components of the system.

4.7 Domestic hot water

The term *domestic hot water* (DHW) refers to hot water used for consumptive purposes, such as washing and showers, rather than for heating. The use of non-renewable energy to make hot water can be reduced by: (1) reducing the amount of hot water used; (2) heating it as much as possible with solar energy using hot-water panels; and (3) heating it more efficiently. Use of solar energy is discussed in Volume 2, Chapter 2. Figure 4.50 shows the breakdown of energy used for hot water by a typical family of four in the US. Showering and

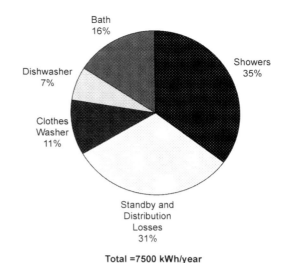

Source: Holton and Rittelmann (2002)

Figure 4.50 *Annual energy consumption for domestic hot water by a typical family of four in the US*

standby energy loss together account for about two thirds of the total energy used for hot water.

4.7.1 Reducing hot water use

Options for reducing the amount of hot water include using:

- low-flow showerheads;
- more efficient washing machines (which use less water as well as less electricity), or using cold-water washing;
- more efficient dishwashers (which use less water, as well as less electricity), or more efficient use of dishwashers (i.e. operation at full load only), or washing efficiently by hand instead (see subsection 4.9.1).

Table 4.10 compares rates of water use for standard and water-efficient shower heads and faucets. A reduction in hot water use for showering and washing by at least a factor of two is possible, if showering and washing habits do not otherwise change. However, the savings in DHW energy use will be diluted if hot water is stored in a hot-water tank because a large fraction of

Table 4.10 *Comparison of water use for standard shower heads and faucets and for water-efficient varieties*

Standard shower head	10–20 litres/minute (2.64–5.28 gallons/minute)
Low-flow shower head	5–10 litres/minute (1.32–2.64 gallons/minute)
Standard sink faucet	10–20 litres/minute (2.64–5.28 gallons/minute)
Aerated bathroom sink faucet	2–4 litres/minute (0.53–1.06 gallons/minute)
Aerated kitchen sink faucet	6–8 litres/minute (1.59–2.11 gallons/minute)

Source: City of Toronto water-efficiency program

the total energy consumption in this case is simply to offset standby losses, which are not altered if total consumption decreases.

4.7.2 Water heating

Water can be heated using electric resistance heaters, natural gas or oil heaters, hybrid solar/natural gas systems, heat pumps or by connection to a district heating system.

Electric and natural gas heaters and storage tanks

Electric resistance is 100 per cent efficient at the point of use but entails losses in generating electricity (by a factor of three if the electricity comes from a typical coal-powered plant), with further losses (typically of 5–10 per cent) in transmitting electricity to the customer. For natural gas heaters in the US, the efficiency ranges from 76–85 per cent, and for oil heaters it ranges from 75–83 per cent (Lekov et al, 2000). However, if hot water is stored in a tank between periods when it is being used (the normal situation in North America), heat will be conducted through the walls of the water tank and lost; this 'standby' heat loss has to be made up. The fractional standby loss depends on the tank surface area to volume ratio, on the temperature difference between the interior of the tank and the surroundings, and on the resistance of the insulation filling the cavity between the tank and jacket. As shown in Figure 4.50, standby energy loss is the second largest consumer of energy for domestic hot water in the US. Thus, water heaters are rated with an 'energy factor' (EF) that takes into account standby losses as well as heating efficiency. For a 150-litre tank using natural gas, the minimum EF

required under the US 2004 standard is only 0.59 (it is only 0.51 using oil) (US DOE, 2004). The best EFs for commercial and residential water heaters are 0.86 and 0.72, respectively.

Residential water heaters draw their combustion air from the house. This reduces the air pressure inside the house, causing cold outside air to be sucked into the house in winter and making the furnace work harder (the same thing happens with clothes dryers, as discussed later). This additional energy use is not accounted for in the EF rating of water heaters. It can be avoided if air is drawn directly from the outside (it should be preheated with the warm exhaust air). Direct use of outside air is rare in water heaters serving household hot-water needs only, but is standard in water heaters serving both space and hot-water needs, discussed next.

Integrated space and DHW heating

Several kinds of integrated space and DHW heating systems are available. They involve a single burner and a single hot-water tank. The stored hot water can be used partly for DHW purposes and partly for circulation through a radiant heating system, or to heat the air in a forced-air heating system by circulating hot water through a coil that is placed in the airflow next to the air handler. The heating load is large enough that it is justified going to the expense of a condensing heater with direct use of outside air for combustion. The Italian company Baxi produces a wall-mounted condensing system that can fit into a closet, with a full-load efficiency of 92 per cent and no need for storage of hot water. At any given time, the heat output is used either for space heating or for DHW, with DHW having priority over space heating. The temperature of water used for space heating and for DHW can be independently controlled.

Point-of-use tankless heaters

In Europe it is common to heat water at the point of use, only as needed. This avoids standby energy losses altogether. Because instant response is needed, a small gas burner is used. However, these systems often use a pilot light, which can account for half of the total water heating energy use. Small, tankless and pilotless point-of-use (POU) water heaters are available with efficiencies in the low 80s (Lutz et al, 2002).

Boilers in commercial buildings

Boilers can be used to heat water for multi-family buildings, athletic facilities or schools. If the boiler is also used for space heating in winter then, due to the low part-load efficiency of non-condensing boilers (see subsection 4.3.3), the efficiency for heating water in the summer will be quite low. Averaged over 30 multi-family buildings in New York City, Goldner (2000) measured a summer efficiency of only 42 per cent. The solution is to have a separate, small condensing boiler for summer DHW loads and for times of low total heating load in winter.

4.7.3 Recovery of heat from warm wastewater

To the extent that the heat in water from showers and sinks can be captured and used to preheat incoming water, the energy needs for water heating can be reduced. A variety of schemes can be used to capture heat from warm wastewater. In the *gravity falling-film* method, surface tension and gravity cause falling films of water to spread and cling to the inner wall of a vertical drainpipe. This reduces the fall velocity and enables a high rate of heat transfer. The cold incoming water passes through a coil that is tightly wrapped around the vertical drainpipe and is warmed as it flows (see Figure 4.51). This system recovers 45–65 per cent of the available heat in the wastewater, depending on water-use patterns, warming the incoming water up to about 20°C (Vasile, 1997).

4.7.4 Reducing energy use in recirculation-loop systems

Schools, hotels and multi-unit residential buildings in North American (if not elsewhere) commonly use a hot

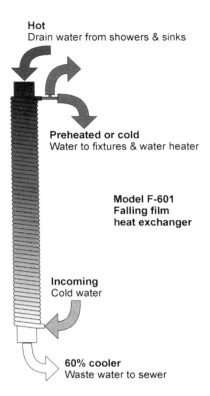

Note: The exchanger must be installed almost perfectly vertical so that the water will drain uniformly down the sides of the pipe.
Source: Vasile (1997)

Figure 4.51 *Gravity falling-film heat exchanger for recovery of heat from wastewater*

water *recirculation-loop* (RL) system in which water is heated and stored in a central tank, continuously circulated through a closed loop to all the points of use, and consumed as needed. This keeps the hot-water pipes warm, so that hot water is instantly available when the faucet is opened. Apart from convenience, this avoids wasting hot water by running the faucet until the pipes have warmed sufficiently to deliver hot water to the faucet. This hot water is lost, whereas the hot water in a recirculation loop returns to the boiler when not being used (albeit at a lower temperature). Since the purpose of recirculation is to keep the pipes warm, the required flow can be reduced by insulating the pipes well. Since pumping power varies with the flow rate to the third power (see subsection 4.1.2), dramatic reductions in pump energy use are possible along with reduced heat loss. However, even with well-insulated piping (Goldner, 1999; Lutz et al, 2002), piping heat losses can constitute 40 per cent to more than 50 per cent of the total hot-water load. Thus, wherever possible, POU water heaters should be used.

Hiller et al (2002) monitored the energy use in a new (1997 opening) school in Tennessee using an RL system, and again after it was converted to a POU system. An impressive 91 per cent savings in total (pump + water heater) energy use was achieved. The POU water heaters were operated continuously, but it is estimated that they could have been shut down at night, on weekends and during school holidays, with a further energy savings of 40 per cent (bringing the total savings to 94.3 per cent).

4.8 Lighting

Lighting energy use constitutes 25–50 per cent of the total electricity use in commercial buildings, as illustrated in Figure 4.2, and so is an important area for energy conservation. As it turns out, large (75–90 per cent) reductions in lighting energy use can be achieved in new buildings and frequently in retrofits of existing buildings. In residential buildings the role of lighting is much less, but large efficiency gains are easier to achieve.

4.8.1 Types of lighting

The main lighting technologies used in buildings and the key principle by which electricity is converted to light for each technology are summarized below (further technical details and common applications are found in Atkinson et al, 1997):

- Incandescent lamp. An electric resistance is used to heat a tungsten filament to a temperature of 2100–2800°C, such that about 10 per cent of the electromagnetic radiation that is emits falls within the visible part of the solar spectrum. A typical lifespan is 750–1000 hours.
- Halogen lamp. This is similar to an incandescent lamp, except that a small amount of a halogen is mixed with the normal inert-gas fill of an incandescent bulb. A quartz rather than a glass envelope is used, permitting a higher temperature. This increases the efficiency, as more of the emitted radiation is at visible wavelengths with a higher temperature.
- Halogen infrared-reflecting (HIR) lamp. This is a halogen lamp with a coating on the inner surface of

the capsule that reflects infrared radiation back to the filament, so that less electricity is needed to heat it.
- Fluorescent lamp. This consists of a tube filled with argon and liquid mercury, with an electrode (cathode or anode) at each end. An electric arc travels between the electrodes, vaporizing the mercury (the discharge source) and causing it to emit ultraviolet radiation. This in turn excites electrons in the *phosphors* coating the inner tube, which then emit light of various wavelengths (depending on the chemical composition of the phosphors) as they drop down in energy level.
- Compact fluorescent lamp (CFL). This is a fluorescent lamp small enough and designed such that it will screw into a regular incandescent light socket. It can be used for residential lighting and for pot lighting. CFLs have a lifespan of about 10,000 hours.
- Ceramic metal halide (CMH) lamp. This is an advance over a class of lights called high-intensity discharge lights, that are used for outdoor lighting and sports facilities, but are more compact, thereby permitting its use in place of HIR lamps in down lighting, track lighting and retail displays. CMH lamps are available in the 39–400W range, with a 39W lamp (44W with ballast) replacing a 100W HIR lamp and lasting three to four times longer (Sachs et al, 2004).
- Light-emitting diode (LED). An LED is like a PV cell running in reverse. It consists of a semi-conductor through which an electric current passes, creating positive and negative charge carriers that recombine to form light (a diode is a device that can conduct electricity in only one direction). The commonly used semi-conductors are indium gallium phosphide (InGaP), which emits amber and red light, and indium gallium nitride (InGaN), which emits near-UV, blue and green light. White light can be created in two ways: (1) by using some of the blue light from an LED to excite one or more phosphors (as in a fluorescent lamp) to produce one or more colours that are then combined with some of the remaining blue light to create white light (with a more natural light produced if more phosphors are used); and (2) by combining two or more LEDs producing different colours in the correct proportions so as to produce white light. LEDs are currently used in a variety of consumer

products (such as backlighting for colour displays in electronic goods such as cell phones), in exterior and interior automotive lighting, and in traffic lights. They have a lifespan of up to 100,000 hours.

Fluorescent lamps contain an extra component, called a *ballast*, that converts the 50 or 60Hz AC input electricity into AC electricity at a frequency of 20–40kHz (kilohertz) for use by the lamp. The ballast limits the current flow, which would otherwise be too large due to the small resistance once the discharge arc is established. Some of the electrical energy input is lost in the ballast.

4.8.2 Measuring the efficiency of lights

Light is electromagnetic radiation – a form of energy – with wavelengths between 0.4 and 0.73µm. However, the efficiency of a lamp in producing light cannot be specified as the ratio of watts of light output to watts of electrical energy input. This is because not all watts of light are equal; our eyes are more sensitive to some wavelengths of radiation than to others. The relevant quantity is the *efficacy* of the lamp, which is defined as (lumens out)/(watts in). A lumen (lm) is a measure of light weighted by the sensitivity of our eyes for various wavelengths.

The efficacy of sunlight depends on the relative amounts of solar radiation reaching the ground at different wavelengths, and so depends on cloudiness, the solar altitude and atmospheric conditions. It has an average value of 105–130lm/W (Muneer et al, 2000; Gugliermetti and Bisegna, 2003).[9]

The efficacy and operating lifespan of the major lighting technologies are given in Table 4.11 The key comparisons evident from Table 4.11 are as follows:

- CFLs are about four times as efficacious as incandescent lamps, two to three times as efficacious as HIR lamps and three times as efficacious as halogen lamps (all of which they can replace in almost all applications);
- the 80-series T8 and T5 fluorescent tubes are about 60 per cent more efficacious than T12 tubes (used in old lighting) and 25 per cent more efficacious than standard (70-series) T8 tubes;
- the CMH lamp is about twice as efficacious as the HIR lamp (which it can replace).

A four times greater efficacy implies a quarter of the electricity use for the same light output.

LEDs have the potential to be substantially more efficient than any of the above. Commercially available products currently have an efficacy of about 30lm/W, compared to 10–17lm/W for incandescent lamps, 50–70lm/W for CFLs, 105lm/W for the T5 fluorescent tubes and up to 140lm/W for high-pressure sodium lamps. However, laboratory research products have achieved an efficacy of up to 152lm/W at low current (Den Baars, 2008), and it is thought that efficacies of

Table 4.11 *Efficacy and operating lifetime of different lighting technologies*

Lamp technology	Efficacy (lm/W)	Lifetime (hours)	Reference
Incandescent	10–17	750–2000	US DOE (2002)
Halogen	12–22	2000–4000	US DOE (2002)
HIR	20–25	2000–3000	US DOE (2002)
Compact fluorescent	50–70	10,000	US DOE (2002)
T12 Fluorescent	62	20,000	McCowan et al (2002)
T8 Fluorescent, 70-series	80	20,000	McCowan et al (2002)
T8 Fluorescent, 80-series	102	20,000	McCowan et al (2002)
T8 Fluorescent, 90-series	62	20,000	McCowan et al (2002)
T5 Fluorescent	105	16,000	McCowan et al (2002)
Electrodeless induction	50	60,000	Atkinson et al (1997)
HID, mercury vapour	25–60	16,000–24,000	US DOE (2002)
Metal halide	70–115	5000–20,000	US DOE (2002)
High-pressure sodium	50–140	16,000–24,000	US DOE (2002)
CMH	56–62	9000–12,000	Sachs et al (2004)
LED, current commercial	30	up to 100,000	Den Baars (2008)
LED, best laboratory	152	up to 100,000	Den Baars (2008)
LED, projected commercial	150–200	up to 100,000	Den Baars (2008)

Table 4.12 *Lamp, fixture and lighting-system efficacy*

Luminaire	Luminaire efficacy (lm/W)	Fixture efficiency (%)	Usable lumens per watt
Halogen	17	45	8
CFL	45	33	15
HPS150-W copra head streetlight	90	50	46
Metal halide HID	70	54	38
T8 Fluorescent	80	77	62
LED	71	90	64

Source: Den Baars (2008)

150–200lm/W will be eventually achieved in commercial products. This would reduce electricity requirements by up to a factor of two compared to the best fluorescent tubes, a factor of four compared to CFLs, and a factor of 20 compared to the least-efficacious incandescent lamps (which are still widely used).

However, lamp efficacy is only part of the story with regard to lighting energy use. The light output from a lamp passes through some sort of fixture and substantial losses occur. Table 4.12 gives the typical fixture efficiency alongside lamp efficacy for various lamp–fixture combinations. Because LEDs produce light that can be directed into only the directions where it is needed, fixture efficiencies are much larger (90 per cent) than for other types of lamp (33–77 per cent). Thus, LEDs will be able to achieve energy savings compared to CFLs before they achieve parity on a lamp-efficacy basis.

LED lighting is currently several times more expensive than fluorescent lighting on a lifecycle basis (taking into account both purchase and electricity costs), but may eventually become less expensive than fluorescent lighting as the technology is improved and manufacturing is scaled up (OIDA, 2002).

4.8.3 Daylighting in buildings

As discussed in subsection 4.2.2, advances in window technology make it possible to increase window area and hence daylighting opportunities without increasing heat loss in winter (after accounting for solar heat gain) and with minimal impact on cooling loads in summer.

Techniques

Some of the techniques available for maximizing the use of daylight are:

- appropriate building form (long and narrow so that light can penetrate to all of the occupied space from both sides);
- appropriate window size, shape and position;
- use of atria (design attributes include shape and orientation to the sun, the transmittance of the atrium roof, and the reflectivities of the atrium walls and floor);
- skylights and stepped roof structure (examples are illustrated schematically in Figure 4.52);
- light shelves (horizontal or inclined reflective surfaces that are placed near the upper part of a window, either on the outside or inside and used to reflect light onto the ceiling and deep into the room, illustrated in Figures 4.53 and 4.54);
- light pipes (active light pipes with sun-tracking mirrors that reflect and possibly concentrate light onto a reflective hollow tube, as illustrated in Figures 4.55 and 4.56; or passive light pipes consisting of a clear dome to collect light on the outside, the light pipe itself and a diffuser that spreads the light at the other end, as illustrated in Figure 4.57);
- laser-cut panels (a thin panel in which a number of closely spaced laser cuts have been made, with solid sections around the edge to support the cut sections, as illustrated in Figure 4.58, deflecting a

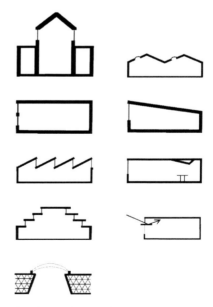

Source: Hastings (1994, Chapter 21)

Figure 4.52 *Examples of roof configurations that increase the opportunity for daylighting*

Note: Top left is semi-transparent double light shelves made out of reflective glass. Top right is an interior light shelf – note light reflected onto the ceiling, while providing some shading below. Middle is light shelves at Sacramental Municipal Utility District (SMUD) headquarters, Sacramento, California. Bottom is light shelves in the Michael Capuano Early Childhood Center in Boston.
Source: Top and middle, IEA (2000); bottom, author

Figure 4.53 *Daylighting through interior light shelves*

large portion of the incoming light onto the ceiling while providing a view to the outside).

The key to avoiding glare is to divide the window into two sections, a lower section for viewing that may be shaded, and an upper section that reflects daylight onto the ceiling and from there to the working area.

The use of daylight must be coupled with light sensors, occupancy sensors (in private spaces) and dimmable artificial lighting, so that the artificial light output can be automatically adjusted up or down as the availability of sunlight changes, and automatically turned off when private rooms or offices are not occupied. Daylighting systems can provide substantial benefits even under overcast conditions, as documented in IEA (2000), particularly anidolic (non-imaging) daylighting systems (Baker and Steemers, 2002; Scartezzini and Courret, 2002).

Savings in lighting-energy use

Detailed measurements and/or simulations demonstrate annual savings of 30–80 per cent from daylighting of perimeter offices in commercial buildings (Rubinstein et al, 1998; Jennings et al, 2000; Bodart and De Herde, 2002; Reinhart, 2002; Atif and Galasiu, 2003; Li and Lam, 2003). The economic benefit of daylighting is enhanced by the fact that it reduces electricity demand most strongly when the sun is strongest, which is when the daily peak in electricity demand tends to occur.

Impact on combined lighting and cooling-energy use

Daylighting can also reduce cooling loads. This is because the luminous efficacy of natural light averages around 105–130lm/W, compared to about 80lm/W for the best fluorescent lighting systems (105lm/W from the lamp/ballast system, times a luminaire light-transmission efficiency of 0.77) (see Tables 4.11 and 4.12), so replacing artificial light with just the amount of natural light needed reduces internal heating. The efficacy of daylight passing through a window can be increased if non-visible solar radiation is excluded through window films that are reflective at NIR wavelengths. If only visible light (wavelengths of 0.4–0.7μm) is admitted, the efficacy is about 260lm/W, and if the admitted wavelength interval is restricted slightly further (to 0.45–0.65μm), the efficacy is about 360lm/W. The higher the efficacy, the less heat produced for a given amount of light.

Systems involving automatically adjusting venetian blinds, light sensors and dimmable lights that optimize the balance between daylight admission and avoidance

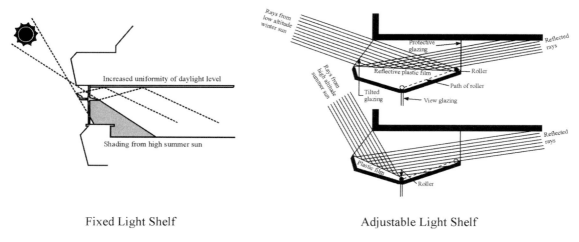

Fixed Light Shelf   Adjustable Light Shelf

Source: Hastings (1994, Chapters 21 and 22)

Figure 4.54 *Examples of a relatively simple (left) and advanced (right) exterior light-shelf system for daylighting*

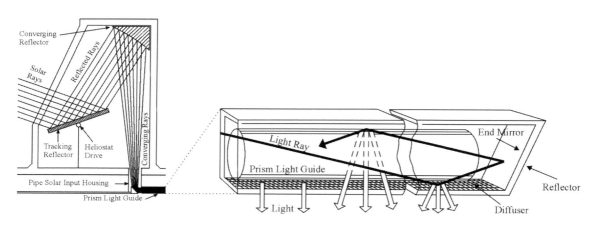

Source: Hastings (1994, Chapter 22)

Figure 4.55 *Example of a sun-tracking (active) light-pipe system used in a building in Toronto*

of solar heating have been built and tested in real office buildings, with a high degree of user satisfaction (Vine et al, 1998). Lee et al (1998) measured savings in lighting and combined lighting + cooling energy of 22–86 per cent and 23–33 per cent, respectively, compared to the case without daylighting controls and a static horizontal blind in a full-scale testbed of an office building in California. Ullah and Lefebvre (2000)

report measured savings in cooling + ventilation energy use of 13–32 per cent using automatic blinds in a building in Singapore, depending on the orientation of the external wall. Implementation of automatic blind controls reduces the peak cooling load (by 18–32 per cent for the building tested by Lee et al, 1998), thereby allowing downsizing of the cooling equipment in new buildings, with associated capital-cost savings. In Hong

Source: Anonymous (2004), from International Association of Lighting Designers

Figure 4.56 *A light pipe in a 14-storey shaft with light fed into the light pipe from a rooftop sun-tracking heliostat and refracted laterally by prismatic glass*

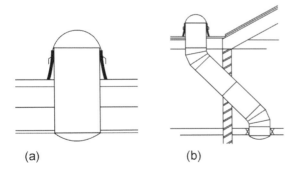

(a)   (b)

Source: Zhang and Muneer (2002)

Figure 4.57 *Illustration of a passive light pipe*

Kong offices, Li et al (2005) estimate that optimized daylighting will reduce peak cooling loads by 5 per cent and total peak electrical loads by about 9 per cent.

4.8.4 Residential lighting

Most residential lighting uses incandescent bulbs. The vast majority of these can be replaced with CFLs, with

a factor of four to five reduction in electricity use for the same light output. Other options to avoid incandescent lamps involve T8 or T5 fluorescent tubes hidden in recessed areas, and used for indirect lighting – bouncing light off of ceilings or walls rather than providing direct illumination. This provides a soft, even illumination without glare and with a minimal number of downlight interruptions in the ceilings, as illustrated in Figure 4.59. The connected lighting power density in a US house is typically 22–28W/m^2, compared to 9.8W/m^2 in the house illustrated in Figure 4.59 (Banwell et al, 2004).

4.8.5 Advanced lighting systems in commercial buildings

Lighting in a building is a system, involving individual devices and controls but also their *arrangement*. However, the dominant paradigm in building lighting is to provide uniform illumination using a rectangular grid, irrespective of the spatial variation in lighting needs. As discussed by Wulfinghoff (1999), every profession involved in the design of buildings has its

Note: Top is view through a laser-cut panel (left) and 20mm wide laser-cut panels installed venetian style between two glass panes (right). Middle is laser-cut panels in awning windows deflecting sunlight onto the ceiling of a classroom in Brisbane, Australia. Bottom is laser-cut panels in an inverted skylight, reflecting light onto the ceiling and creating a more uniform distribution of light than would occur otherwise.
Source: IEA (2000)

Figure 4.58 *Daylighting through laser-cut panels*

own set of design procedures, which are formalized and promoted by the profession's leading organization. According to Wulfinghoff,

> the official lighting design doctrine has grown in a direction that is fundamentally in conflict with energy efficiency. It wastes energy to an extent that ranges from

Source: Banwell et al (2004)

Figure 4.59 *Energy-efficient residential lighting, with small linear fluorescent tubes and CFLs*

moderate to extreme, depending on application... the basic tool of contemporary lighting design is a rectangular grid, and lighting layout is usually guided by non-lighting constraints, such as the spacing of ceiling tiles.

Wulfinghoff (1999) provides many photographs of lighting systems in contemporary buildings that can be charitably described as thoughtless, along with examples of carefully thought out and efficient systems.

In retrofits of fluorescent lighting systems, 30–50 per cent electricity savings can be routinely achieved. With considerable effort, 70–75 per cent savings in retrofits are possible. In new construction, 75 per cent and larger savings compared to current standards can be readily achieved.

These remarkable energy savings could be increased yet further through advances in the efficiency and performance of the individual components of the lighting system (Rubinstein and Johnson, 1998). Possibilities include:

- improved phosphor efficiency in fluorescent lamps, potentially doubling their efficacy;
- electrodeless fluorescent and HID lamps, permitting greater flexibility in choice of materials that in turn could be selected for greater efficacy;
- more efficient LEDs, giving efficacies of 150–200lm/W;
- improved dimmable electronic ballasts, reducing the loss of efficacy when fluorescent lights are dimmed in order to match artificial light output with daylight.

4.9 Appliances, consumer electronics and office equipment

Residential appliances and office equipment are important consumers of energy in their own right and, for office equipment in particular, can be an important source of internal heat gain that needs to be removed by the cooling system from perimeter offices during summer and year round from internal offices (see Figure 4.27). Figure 4.60 compares the per capita residential electricity consumption in various world regions in 2005 and annual growth rates from 1995 to 2005, while Figure 4.61 gives the breakdown of residential electricity use in

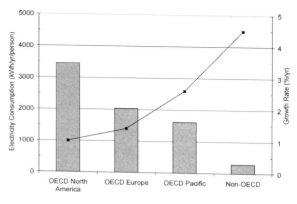

Source: Data from IEA (2009)

Figure 4.60 *Average per capita residential electricity use (bars) in different country groups in 2005, and average rate of growth in per capita residential electricity use (line) from 1995–2005*

the 27-member EU (defined in Appendix C), in the US and in India (excluding a small amount of electrical space and water heating in the EU and US). Per capita residential electricity use ranges from an average of 300kWh/yr/person in non-OECD countries to almost 3500kWh/yr/person in North America. In the US, refrigeration, air conditioning and lighting account for about 50 per cent of non-heating residential electricity use, but clothes washers, TVs and related equipment (VCR/DVD players, cable boxes, satellite dishes), ovens and furnace fans are important too. In the EU-27, refrigeration and lighting account for about 25 per cent of non-heating residential electricity use, with consumer electronics (TVs and related equipment, computers) accounting for another 20 per cent. In India, electricity use by fans takes up the largest share of the rather small residential electricity use.

4.9.1 Residential appliances

Figures 4.62 to 4.64 give the range of energy use, under standard test conditions, for the refrigerator/freezer units, freezers, ovens, ranges, cooktops (hobs), dishwashers, washing machines and clothes dryers that were commercially available in North America in 2007. Within a given size category, there is often a 30–40 per cent difference in the energy used by the least and most

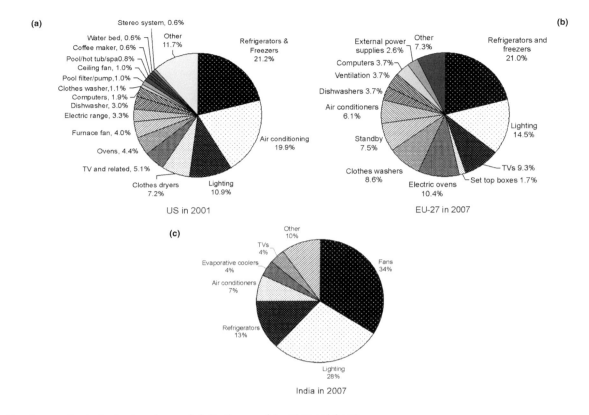

Note: Space and water heating have been excluded in the case of the EU-27 and the US.
Source: Data from (a) Residential Energy Consumption Survey (www.eia.doe.gov/emeu/recs/recs2001), (b) Atanasiu and Bertoldi (2009), (c) IEA (2009)

Figure 4.61 *Breakdown of residential electricity use in (a) the US in 2001, (b) the EU-27 in 2007 and (c) India in 2007*

efficient refrigerator/freezer and freezer units, and a factor of two difference between the least and most energy-efficient dishwashers and clothes washers. Thus, substantial energy savings would occur by bringing all existing equipment up to the energy performance of the most efficient available equipment. Technical advances and/or changes in the kind of equipment used could provide even larger energy savings, as discussed next.

Refrigerator/freezer units

Figure 4.65 shows the variation in average energy use by new refrigerators sold in the US and in the average adjusted volume from 1947 to 2000.[10] During this time, the average volume of refrigerators sold in the US increased from 8 to 20ft³. Energy use by new refrigerators peaked at 1800kWh/yr in 1975, shortly

before the first standards came into effect. The newest standard, which came into effect in 2001, saw the energy consumption drop to 479kWh/yr for a refrigerator/freezer unit with an adjusted volume of 20.7ft³ (582 litres, the current average size in the US). Figure 4.66 compares the average energy use of the refrigerator stock in various OECD countries at various times (the decline in stock average energy use lags behind the decline in the energy use of new refrigerators). Although significant reductions in the energy use of new refrigerators in the US and Canada have occurred, the average measured energy use by refrigerators in both countries still exceeds that of most other countries by 50–100 per cent and will exceed that of most other countries even after full turnover of the existing stock. This may be partly due to the use of larger refrigerators in the US and Canada, but also

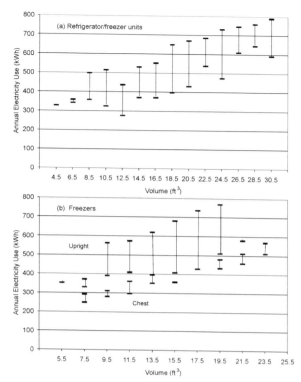

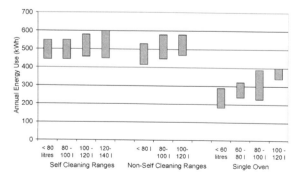

Note: The first refrigerator size category is from 4.5–6.4ft³, the second from 6.5–8.4ft³ and so on (1ft³ = 28.3 litres), with a similar convention for freezer size categories.
Source: Data from NRCan (2007)

Figure 4.62 *Range of annual energy use (under standard test conditions) of (a) refrigerators and (b) freezers in different size categories sold in Canada in 2007*

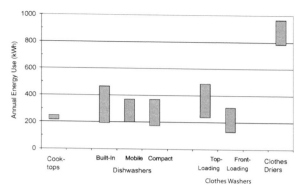

Source: Data from NRCan (2007)

Figure 4.64 *Range of annual energy use (under standard test conditions) of cooktops (hobs), dishwashers, clothes washers and clothes dryers sold in Canada in 2007*

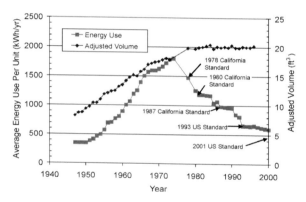

Source: Rosenfeld (1999)

Figure 4.65 *Variation in the annual energy use by new refrigerators sold in the US and in the adjusted volume, 1947–2001*

Source: Data from NRCan (2007)

Figure 4.63 *Range of annual energy use (under standard test conditions) of ranges (oven and stove) and of single ovens in different size categories sold in Canada in 2007*

partly due to design features (such as heaters inside the fridge to soften butter) that increase energy use and partly due to differences in measurement procedures.[11]

The energy used by a refrigerator/freezer unit is highly influenced by the thermal resistance of the casing and by the COP of the compressor. It is instructive to briefly compare these parameters with those typical of building systems. Table 4.13 compares the thermal resistance of the door of a conventional fridge with that of various advanced insulation panels (AIPs) and advanced insulation components (AICs) for fridges that are under development. A conventional fridge shell has

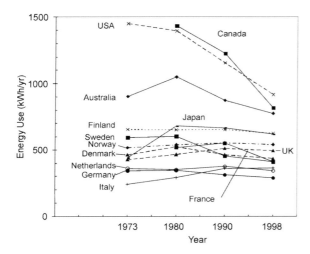

Source: Data from IEA (2004a)

Figure 4.66 *Average energy use of the entire refrigerator stock in different OECD countries in 1973, 1980, 1990 and 1998*

Table 4.13 *Average effective thermal resistances of a refrigerator door using different insulation technologies*

Refrigerator door configuration	RSI-value (R-value)
CFC blown foam with conventional steel outer shell	1.59 (9.03)
Gas-filled AIP/foam composite with conventional steel outer shell	1.71 (9.71)
Evacuated AIP/foam composite with conventional steel outer shell	1.96 (11.14)
Gas-filled AIP/foam composite with polymer outer shell	1.96 (11.15)
Evacuated AIP/foam composite with polymer outer shell	2.31 (13.09)
Gas-filler polymer-barrier AIC	2.38 (13.50)
Evacuated-powder polymer-barrier AIC	3.31 (18.80)

Note: AIP = advanced insulation panel; AIC = advanced insulation component
Source: Griffith and Arasteh (1995)

a thermal resistance of RSI 1.6 (R9), compared to RSI 3.3 (R19) for the best AICs (and RSI 5-10 for the walls of high-performance houses in cold climates). A conventional fridge/freezer uses a rotary fixed-speed compressor with a COP of 1.1–1.6 (compared to about 2.5–6.0 for residential air conditioners and up to 7.5 for large commercial chillers). The freezer is directly cooled to about −18°C and the fridge indirectly cooled (to 4°C)

through exchange of air with the freezer. This means that when cooling of the fridge only is needed, the compressor must work harder than necessary, as it is working against the much greater temperature difference between the freezer compartment and room air rather than between the fridge compartment and room air. Use of separate compressors for the fridge and freezer would directly reduce energy use by 30 per cent, with additional savings due to the need for less defrosting (Gan et al, 2000).

Apart from technical factors, lifestyle factors are also important in refrigerator energy use. An analysis for the US by Deumling (2004) indicates that, between 1957 and 2001, the number of refrigerators per household increased by 65 per cent (due largely to a second fridge in the basement, often used only to keep beer cold). During this same period the real volume (not adjusted volume) of the US refrigerator stock increased by 105 per cent. The increase in energy use from these two factors is thus about 230 per cent (multiplying energy use per household in 1957 by a factor of 3.3).

Freezers

Freezers come in chest (top-opening) and upright configurations, with chest freezers using substantially less energy than upright freezers (see Figure 4.62b). Energy use varies roughly with freezer surface area and thus roughly with volume$^{2/3}$. Ertel (2009) provides an enlightening comparison of a 1994 household chest freezer and a model to be released in 2010 (both with a volume of 195 litres or 6.9ft^3). The 1994 freezer has an annual energy use of 401kWh and requires 30 hours for a given temperature rise in the event of a prolonged power failure. The 2010 model will use 113kWh/hr and requires 110 hours for the same temperature rise. By comparison, models closest to this size sold in North America in 2007 consume 250–300kWh/yr. The 72 per cent reduction in energy use from the 1994 to 2010 models is a result of better insulation (25 per cent of total savings), a better compressor (33 per cent) and system optimization (42 per cent). The insulation improvements involve doubling the thickness (from 63mm to 120mm), using an insulation with a lower thermal conductivity, and applying the insulation more uniformly. The compressor improvements involve improved air circulation around the compressor without the use of a fan, use of a variable speed motor, and better lubricants and mechanics (which yield a 3 decibel (dB)(A) noise reduction as a side benefit). The improved system optimization includes better placement of the condenser, optimization of the evaporator, use of electronic controls instead of a mechanical thermostat and improved lid sealing. A version that can be used with DC electricity from a PV array (thereby avoiding DC to AC conversion losses) and that can serve as a fridge or a freezer will be available.

Clothes washers

Clothes washers rotate or agitate the clothes so as to induce movement of soapy water through the clothes. The agitation can occur around either a vertical axis or a horizontal axis. Clothes washers can open at the top or at the side. Top-loading clothes washers are almost always vertical-axis machines, with either an impellor (in Asia) or an agitator (in the rest of the world) (an agitator has a post in the centre of the basket that holds the clothes, while an impellor does not). In Europe, top-loading horizontal-axis machines are available with a small door in the drum that can be opened. Front-loading machines (which account for about 90 per cent of the market in Europe) are always horizontal-axis machines.

Horizontal-axis machines require less water than vertical axis machines. In vertical-axis machines, the clothes are fully immersed in water and agitated, while in horizontal-axis machines, the drum only partially fills and the clothes tumble in and out of the water as the drum turns. Test procedures in the US for measuring

energy use include the energy required to heat water by 50K, with the amount of hot water used based on a prescribed weighting of the hot water requirements for various wash and rinse temperature combinations. For example, for machines with five different settings, the prescribed weightings for purposes of computing the energy use are: hot/warm, 15 per cent; hot/cold, 12 per cent; warm/warm, 30 per cent; warm/cold, 25 per cent; cold/cold, 15 per cent (GPO, 2008). Because water is assumed to be heated and because horizontal-axis machines use less water, they are rated at lower energy use than vertical axis machines, although there is some overlap between the two groups, as can be seen in Figure 4.64. However, with cold-water washing and rinsing there would be little difference in energy use between the two, whereas with hot-water washing and warm-water rinsing, absolute energy use and the difference in energy use would be much greater than indicated by energy ratings.

Horizontal-axis machines also save energy by requiring less laundry detergent, which adds one quarter to one half to the direct energy savings as given by standard test procedures. Less wear on clothes is another benefit. Horizontal-axis machines cost substantially more than vertical-axis machines but are generally the lower-cost option over their lifetime.

Table 4.14 compares the maximum total energy and water use permitted under various standards. The total energy use is broken down here into motor and heating components assuming that 33 per cent of the water is heated by 50K for all machines except for Chinese impellor-type machines (where zero water heating is

Table 4.14 *Maximum permitted electricity and water use (per kg of clothing per wash) by clothes washers under various national standards and estimated breakdown of the total energy use into that for heating water and that for operating the motor*

Standard	Electricity (kWh/kg)	Water (litres/kg)	Energy to heat water (kWh/kg)	Motor energy (kWh/kg)
US pre-2000[a]	1.016	47	0.895	0.121
US 2007[a]	0.244	11	0.206	0.038
EU worst category	0.390	16[b]	0.307	0.083
EU best category	0.190	6[b]	0.115	0.075
Chinese impellor, 2003	0.032	36	0.000	0.032
Chinese drum, 2003	0.350	20	0.320	0.030

Note: [a]The US standard gives permitted electricity and water use per cycle, which entails a 3.2kg load according to the definition of a standard-sized clothes washer in GPO (2008). Water use is an educated guess. [b] Water use is not directly regulated, so a reasonable value is used here.

appropriate because the energy rating is based on cold-water washing) and Chinese drum machines (where a hot-water fraction of 27.5 per cent gives a more reasonable breakdown). The worst European machines use about two thirds the energy permitted under the US standard, while the best European machines use about half of the energy of the worst European machines. The motor energy use permitted by Chinese impellor machines is much lower than for any European efficiency category, but this might be at the expense of a greater moisture content of the clothes leaving the machine due to a shorter spin cycle.

Figure 4.67 compares the annual motor and heating energy use under the different standards. With cold-water washing, the only energy use is the motor energy use. The given heating energy use assumes use of hot and cold water in the proportion (about 33 per cent) implied in the energy ratings. Heating even this modest portion of water requires an amount of energy many times that used by the motor. This figure underestimates the potential impact of differences in washing behaviour, as the hot-water energy use shown in this figure (and in Table 4.14) is for a mixture of hot- and cold-water washing rather than for 100 per cent hot-water washing. Thus, discouraging hot-water washing (where it is practised) is at least as important as strengthening energy performance standards.

Lin and Iyer (2007) discuss some national differences in the use of hot or cold water for clothes washing. Traditional clothes washing in China is with cold water, but front-loading washing machines made in China obligatorily use hot water for the wash cycle. Not only does this increase energy use, it may create the cultural impression that cleanliness requires the use of hot water. Washing machines in North America and Europe have a temperature setting, while most Japanese washers have no temperature selector; instead, the machines use only cold water.

A valid comparison of the energy use of different clothes washers should take into account not only the energy directly used by the clothes washer (E_{CW}) and the energy used to heat any hot water used by the machine (E_{water}), but also the energy required to remove whatever moisture remains after the spin cycle (E_{dry}). These three terms are included in the US *Modified Energy Factor* (MEF), which is given by:

$$MEF = \frac{C}{E_{CW} + E_{water} + E_{dry}} \qquad (4.22)$$

where *C* is the capacity of the machine (GPO, 2008). Another performance indicator is the *Water Factor* (WF), which is the amount of water used for a given load. Table 4.15 compares the MEF and WF for various standards in the US and for the Tier-3 standard proposed by the Consortium for Energy Efficiency (CEE). Also given is 1/MEF, which is comparable to the energy use values given in

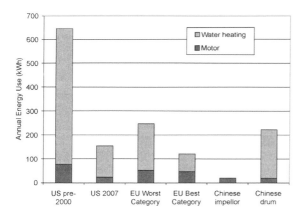

Note: Assuming 200 3.2kg loads per year.
Source: Computed from data in Table 4.14

Figure 4.67 *Annual energy use for clothes washers with and without heating about a third of the water used by 50K*

Table 4.15 *Comparison of MEF (kg of clothes washed per kWh) and WF (litres/kg) for clothes washers under the pre-2000 and 2007 US legal standards, for the Energy Star standard*

Standard	MEF	1/MEF	WF
Pre-2000 standard	1.73	0.58	—
2007 standard	2.34	0.43	—
Energy Star standard	3.20	0.31	16.7
CEE proposal	4.09	0.24	9.4

Note: Original MEFs and WFs have been converted to metric units using 1ft³ clothes = 1.86 kg, as given in GPO (2008).
Source: Consortium for Energy Efficiency (www.cee1.org)

Table 4.14 except that drying energy is included. The MEF standard proposed by the CEE amounts to almost a halving of the energy use compared to the US 2007 standard, yet CEE (2009) lists 79 models that meet their Tier-3 standard. However, as indicated above, a many-times larger energy saving occurs in going from a standard clothes washer with hot-water washing to the most efficient clothes washer with cold-water washing.

Horizontal-axis machines are capable of faster spins (1600rpm instead of 750rpm), such that the clothes are drier after washing, thereby reducing dryer energy use. Presutto (2009) has measured the extra energy required for faster spin cycles, the resulting reduction in moisture content of the clothes after the spin cycle, and the resulting savings in dryer energy. For the particular washer and dryer models and clothes that he tested, increasing the spin speed from 800 to 1200 rpm reduces the residual moisture content from 72 to 49 per cent and requires an extra 0.010kWh/kg of clothes in spin energy, but saves 0.23kWh/kg in dryer energy. Thus, the benefit to cost ratio is about 23:1. For vented dryers, there will be additional savings during the heating season due to reduced operation of the dryer. This is because venting of the dryer induces inflow of cold air into the residence which then needs to be heated. Of course, if clothes dryers are not used (as is the case for most people in many European countries),[12] there will be no savings in dryer energy resulting from faster spin speeds, but there will be savings in household heating energy when the clothes are air dried inside during the heating season that is almost as large as the savings in dryer energy use.[13]

Clothes dryers

There are three kinds of clothes dryers: (1) vented dryers, in which air is heated, passes through the clothes to pick up moisture and is vented to the outside; (2) condensing dryers; and (3) heat pump dryers. In condensing dryers, heated air that has passed through the clothes then passes through a heat exchanger, where it is cooled sufficiently to condense much of the moisture that it contains, then is reheated and repeats the circuit through the clothes rather than being vented to the outside. The heat exchanger in turn is either cooled with room air (as is the case for all units sold in Europe) or is cooled with tap water (for some of the units sold in North America). The condensed water is either stored in a tray that can be removed at the same time as the clothes, or can enter a drainpipe. In the heat pump dryer the condenser of the internal loop is cooled with a heat pump. Almost all dryers sold in North America are vented dryers, whereas both vented and condensing dryers are sold in Europe. Heat pump dryers only recently entered the market in Europe but have rapidly grown in popularity in some countries. For example, heat pump dryer sales grew from 1.7 per cent of the market in Switzerland in 2004 to 15.6 per cent in 2008 (Nipkow, 2009).

There is relatively little difference in the energy use by the least and most efficient dryers available in North America (see Figure 4.64). Table 4.16 compares the energy used to produce all of the materials in various clothes dryers and during the manufacturing process, as well as the operating energy over a 13-year lifespan, for vented, condensing and heat pump dryers in Europe.

Table 4.16 *Comparison of embodied energy and lifetime (over 13 years) operating energy of different clothes dryers and comparison of annualized purchase cost (neglecting interest) and operating costs*

	Lifetime energy (MJ)			Annualized costs (€)		
	Embodied	Operating	Total	Purchase	Operating	Total
Vented	263	3604	3867	54	50	104
Condensing	269	2835	3104	61	38	99
Heat pump	310	1622	1932	77	22	99
New heat pump	311	1322	1633	80	18	98
Savings		63%	58%			

Note: The given savings are for the new heat pump model compared to the vented model.
Source: Gensch (2009)

The lifetime energy use by the most efficient heat pump dryer is about half that of the typical vented dryer. This energy difference is amplified by the fact that the vented dryer induces infiltration of outside air into the building, and during the heating season this adds to the net heating energy use, whereas the condensing and heat pump dryers add heat to the building. During the cooling season, the heat gain from air-cooled condensing and heat pump dryers adds to the air conditioning load, but this is probably largely offset by the fact that inflow of extra outside air (which would have to be cooled and dehumidified) is not induced.

The lowest possible energy use for drying clothes is zero – if done on a clothes line outside, or inside during the summer. In the winter, indoor drying takes an amount of heat from the interior air equal to the latent heat of evaporation times the quantity of water evaporated. This heat has to be replaced with the heating system (but the air is humidified, which is often a benefit in winter), so the energy use for drying in this case is the theoretical minimum divided by the efficiency of the heating system.

Dishwashers

Energy is used by dishwashers either directly or indirectly for heating water, for drying, to operate motors and pumps, and as standby energy use. European dishwashers typically use cold inlet water that is heated by the dishwasher itself only for part of the wash/rinse cycles. This would allow a lower temperature setting on the hot water storage tanks that are common in North America, as well as reducing the energy used for heating. Table 4.17 compares the energy use permitted under various US standards and the average in Europe in 1996 and 2003. The 2007 US standard, applicable to settings of eight places or larger with six serving pieces, is an energy requirement of no more than 1.61kWh/cycle, with the Energy Star standard requiring only slightly better performance, at 1.54kWh/cycle. The EU has various energy performance categories, with the most efficient European dishwasher rating requiring ≤1kWh to wash a 12-place setting and the least efficient models requiring more than 2kWh. The sales-weighted average energy use of European dishwashers declined from 1.69kWh/cycle in 1996 to 1.20kWh/cycle in 2003 (Bertoldi and Atanasiu, 2007). Table 4.17 gives the annual electricity use assuming 215 cycles per year (as in the US standards) and an

Table 4.17 *Energy use per dishwasher cycle under various standards and annual energy use assuming 215 washing cycles per year plus an additional 8kWh/yr for standby energy use*

	kWh/cycle	kWh/yr
US 1994 standard	2.174	475
US 2007 standard	1.613	355
Energy Star 2007 standard	1.538	339
CEE, Tier-2 standard	1.471	325
EU-15 average in 1996	1.692	372
EU-15 average in 2003	1.197	265

Source: Consortium for Energy Efficiency (www.cee1.org) and Bertoldi and Atanasiu (2007)

annual standby energy limit of 8kWh/yr (as in the US 2007 standard). About half of the per-cycle energy use is for heating water.

Even the most efficient dishwashers are energy-intensive compared to water-efficient washing by hand.[14] It is questionable whether water at 60°C is required to clean dishes, as it is the soap and not the water (at least at this temperature) that disinfects. Hot water might be required in a dishwasher in order to remove food residues adhering to dishes, but mechanical scrubbing by hand when needed is surely more efficient than an indiscriminate hot-water jet. The 1kWh of energy used to wash a 12-place setting in the most efficient European dishwashers is sufficient (at 90 per cent efficiency) to heat 45 litres of water from 12 to 30°C – far in excess of what is needed with efficient hand washing. Air drying (whether in a drying rack or through the air-dry option on a dishwasher) saves dishwasher energy but takes heat from the surroundings, which must be replaced by the heating system during the winter (conversely, the heat supplied to dry dishes in a dishwasher is a building heat gain that subtracts from the required heating requirement in winter but adds to the cooling requirement (if any) during the summer).

Ovens

Oven efficiency is rated at around 12–15 per cent (Marbeck Resource Consultants, 1992), so it would appear that there is scope for large improvements in the efficiency of ovens. Factors influencing the efficiency of an oven for cooking food are: the amount of insulation (standard ovens have about 5cm of fibreglass insulation, self-cleaning ovens about 10cm); the presence of forced-air convection (which improves heat transfer to the food, giving about 20 per cent energy savings); improved door seals to reduce leakage (10 per cent savings potential); and oven interior walls that reflect heat into the interior of the oven (40 per cent savings potential).

About one quarter of the households in Europe own gas ovens rather than electric ovens, with the two highest gas oven ownership rates estimated at about 85 per cent in Spain and 68 per cent in Ireland (Kasanen, 2000, Table 5.1). For these ovens, 40–60 per cent energy loss occurs simply through venting the combustion products, but energy losses associated with the generation of electricity are avoided. Table 4.18 compares estimated average onsite energy use by electric and gas ovens in Europe, along with the primary energy required for electric ovens assuming electricity to be produced and delivered at 40 per cent overall efficiency. Onsite energy use is about 25 per cent greater for gas ovens compared to electric ovens, but primary energy use is about 50 per cent smaller. Depending on the source of electricity in the future and the availability of biogas in place of natural gas, electric or gas ovens could be preferable. Also shown in Table 4.17 is the energy use by the best available ovens in Europe. Shifting from the current average to best available would reduce energy use in Europe by about 42 per cent for electric ovens and by 25 per cent for gas ovens.

Consumer habits also have a significant influence on oven energy use. A study by the US National Bureau of Standards (Fechter and Porter, 1979) found a difference of 50 per cent in the energy use between non-professional cooks preparing the same recipe using the same oven. Kasanen (2000) reports annual average usage rates for ovens ranging from 45 times per year in

Greece to 224 times per year in France, which will also greatly influence overall energy use and which may be amenable to adjustment over time. Oven size is another factor in oven energy use, with observed energy use increasing almost in proportion to oven volume (see Figure 4.63).[15] European ovens average about 50–55 litres interior volume, which is the smallest size category found in North America.

Cooktops

Gas cooktops (also called stoves or hobs) have either a continuously burning pilot flame or an electronic spark ignition system. Spark ignition uses a negligible amount of electricity, but the pilot flame typically uses as much energy as is used for cooking (McMahon et al, 1997). Elimination of the pilot flame therefore reduces gas cooktop energy use by about 50 per cent.

4.9.2 Consumer electronics

As with household appliances, there is a wide range in the energy used by electronic goods (computer systems, home entertainment and communication equipment), most of which consume electricity even when turned off. For many common household items, the standby power could, until recently, be in excess of 10W, whereas standby power draws of much less than 1W are possible in most instances. Data summarized by Bertoldi et al (2002) indicate that, in many countries, total standby power consumption per household is 500–800kWh/yr, which is greater than that of a full-size refrigerator-freezer unit meeting the 2001 US standard (479kWh/yr). Reducing standby power draw is largely a matter of improved design, with close to zero long-term cost. Some progress has been made in this direction. For example, most home entertainment equipment (TVs, DVD players, DVD recorders and AV receivers) introduced in 2008 and 2009 has a standby power draw of less than 1W (Tippenhauer, 2009). The exception is set-top boxes, where the standby power draw of the newest models is about 7W. However, there is still a wide variation in standby power draw by products still on the market, as discussed below for specific products.

Televisions

The annual household energy use by televisions depends on the number of TVs in the household, the power draw

Table 4.18 *Energy use (kWh) under standard tests conditions for European ovens*

	Electric ovens		Natural gas ovens
	On site	At source	
Current average	1.20	3.00	1.52
Best on market	0.70	1.75	1.14
Savings	42%		25%

Note: Energy use at source is the energy required to generate electricity, assuming a generation efficiency of 40 per cent.
Source: Kasanen (2000, Table 5.2)

when the TV is turned on and when it is turned off, and the number of hours per year that the TV is turned on. The power draw when turned on depends primarily on the size of the TV screen and secondarily on the type of display technology. The power draw when turned off depends on the design of the TV, with over a factor of 100 difference between TVs of comparable size. Each of these factors is briefly discussed below.

There are three main types of technology used to create an image in televisions: the *cathode ray tube* (CRT), the *liquid crystal display* (LCD) and the *plasma* TV. CRT technology is the oldest display technology in TVs; it consists of a vacuum tube through which electrons are fired onto a phosphor-coated screen, causing the phosphors to emit light. The plasma TV consists of a panel with hundreds of thousands of tiny cells filled with a mixture of noble gases (neon and xenon) and sandwiched between electrodes. A voltage across the electrodes causes the gases to become ionized, forming a plasma that emits photons. These in turn produce light when they strike the phosphors in each cell. In an LCD display, an array of *cold cathode fluorescent lamps* (CCFLs) is placed behind liquid crystals. In response to a voltage, the liquid crystals change in such a way as to alter the amount of light that passes through them. LCD displays thus use largely the same amount of energy irrespective of the brightness of the image produced, whereas CRT and plasma TVs use modestly less energy as brightness drops below 50 per cent of maximum brightness (DCE, 2007). LCD and plasma TVs are flat TVs and so lend themselves to larger screen sizes. Figure 4.68a plots average power draw when turned on versus screen area for CRT, LCD and plasma TVs. There is a roughly linear increase in power draw with screen area, with no major difference between CRT and LCD televisions, but plasma TVs tend to draw more power for their size than the other technologies. Figure 4.68b gives the standby power draw for various TVs in Australia. The standby power ranges from 0.1W to over 20W and is largely uncorrelated with screen size.

Global electricity use by TVs is estimated by IEA (2009) to have grown from 135TWh/yr in 1990 to 300TWh/yr in 2009 and, in the absence of government efficiency policies, is projected to increase to 800TWh/yr by 2030. A number of factors are contributing to the rapid growth in electricity use by TVs:

• an increase in the average number of TVs per household in all parts of the world;

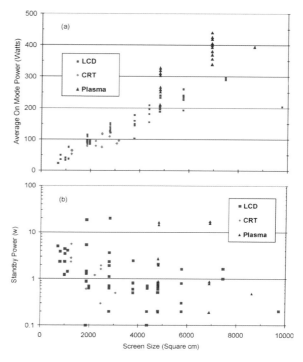

Source: DCE (2007)

Figure 4.68 *Power draw of various CRT, LCD and plasma TVs when (a) turned on and (b) turned off*

• an increase in the average size of TVs purchased;
• a shift to more energy-intensive TV technologies;
• an increase in the average number of hours per day that the TV is turned on.

Figure 4.69 shows the trend from 1990 to 2006 in the average number of TVs per household in various countries or regions. The number of TVs per household ranges from 0.3 in rural India to 1.3 in urban China, 1.7 in the EU-27 to 2.5 in the US and Japan.

Prior to the introduction of LCD and plasma TVs, the size of TVs was limited to a diagonal dimension of about 22–24 inches due to the bulk of the vacuum tube. With the development of flat-screen technologies, this limitation has been lifted. As well, the development of high-definition digital broadcasting permits high-quality images on larger screens. As a result, average TV sizes have increased. DCE (2007) indicates average TV diagonal dimensions in the global market of 56cm (22 inches), 64cm (27 inches) and 109cm (43 inches)

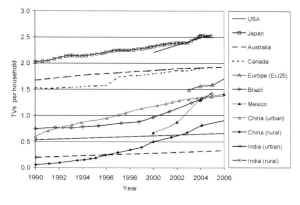

Source: IEA (2009)

Figure 4.69 *Trend from 1990 to 2006 in the number of televisions per household in selected countries*

for CRT, LCD and plasma TVs, respectively. Assuming a height to width ratio of 0.75, these diagonals correspond to screen areas of 1500cm^2, 2250cm^2 and 5700cm^2, respectively. From Figure 4.68a, the approximate average power draws are 70W for CRT, 110W for LCD and 300W for plasma TVs.

TV energy use is an area where lifestyle factors make a large difference in annual energy use. According to McNeil et al (2008a), average daily TV viewing times per person are 150 minutes in Korea, 240 minutes in Europe and Japan, and 260 minutes in the US. However, household viewing times can be longer where there are multiple TVs per household. Figure 4.70 compares the estimated household viewing times in 18 OECD countries. Estimated viewing times are between two and four hours per day in most countries, but is an astounding

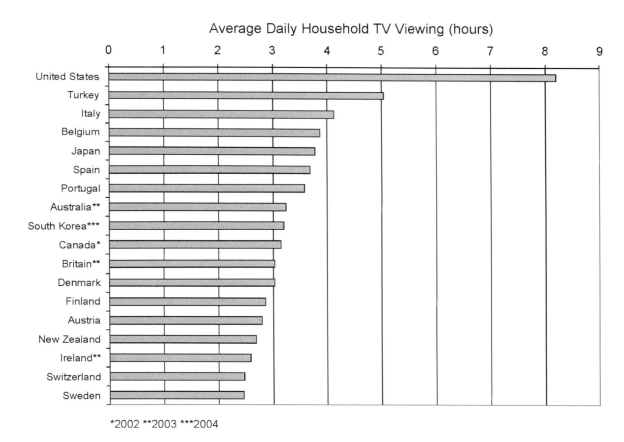

Source: OECD (2007)

Figure 4.70 *Estimated average daily household TV viewing times in 2005 (except where indicated otherwise) in various countries*

8 hours and 11 minutes per day in the US. This high viewing time is a result of there being an average of about 2.5 TVs per household in the US, with each household member often watching his/her preferred programme at the same time in different rooms (Crosbie, 2008). Indeed, the data presented in Figure 4.70 might not be viewing times, but the length of time that a TV is turned on. As discussed by Crosbie (2008), manufacturers of flat-screen TVs have been promoting the TV as an electronic picture display when not being used as a TV, such that it is never turned off. There is also a growing tendency to use TVs as a digital radio. Although drawing less power than when used as a TV, the power draw when used as a radio is still 100–180W in freeview picture mode and 60–100W with a black screen (depending on the technology and size of the TV), the latter being about ten times that of a typical digital radio (IEA, 2009). At four hours per day per TV, the power requirements given above for CRT, LCD and plasma TVs translate into energy uses of 102kWh/yr, 160kWh/yr and 438kWh/yr, respectively. If the eight hours daily TV on-time in the US were to consist of four hours of a plasma TV and four hours of LCD TVs, the energy use would be in the order of 600kWh/yr.

Options for reducing the energy use for a TV of a given size and technology are to:

- implement dimmable backlighting in LCD TVs;
- shift to more efficacious lamps, particularly in plasma TVs;
- reduce standby power draw;
- develop more efficient technologies.

LCDs apply a backlight of fixed intensity irrespective of the brightness of the image, with crystals blocking light as needed. However, the brightness of an image rarely exceeds 50 per cent of the maximum brightness. Dimmable hot cathode fluorescent lamps would reduce energy use and improve picture quality (DCE, 2007). A further energy saving would be possible by replacing CCFLs with dimmable LEDs. Although LEDs are no more efficacious than CCFLs at present, they can be more effectively dimmed. DCE (2007) estimates a potential energy saving of at least 50 per cent. This saving would grow as LEDs become more efficacious (see the discussion of LEDs in subsection 4.8.2).

Plasma panels have a luminous efficacy of only 2lm/W (far worse than even an incandescent lamp). Prototype screens have an efficacy of up to 5.7lm/W, which would cut electricity consumption approximately

Table 4.19 *Current (2005) typical standby power draw (W) and proposed standards for TVs and related equipment*

	Current	Proposed
TVs	2.8	0.8
Cable digital adaptors	16.8	9.5
Satellite digital adaptors	16.8	11.0
Terrestrial digital adaptors	9.5	1.0
Video recorders	9.4	1.0
Digital TV recorders	6.2	0.5
DVD players	4.0	0.3
VCRs	4.0	1.0
Total (TV + satellite adaptor + video recorder + DVD player)	33.0	13.1

Source: DCE (2007)

in half (IEA, 2009). Reduction in power draw during standby operation is another area of potential energy savings. Typical standby power draws at present and under proposed standards for various TV-related equipment are given in Table 4.19. Alternative technologies also promise significant savings compared to current LCD TVs, although not necessarily compared to future, more efficient LCD TVs. Foremost among these are organic light emitting diode (OLED) displays, with a potential saving of 40 per cent compared to current LCDs (IEA, 2009).

Finally, lifestyle factors should not be ignored, although these are certainly harder to change. With the growth in the size and number of TVs per household, energy use by TVs is growing rapidly and now exceeds that of many appliances for which energy use is regulated. Stringent limits on allowable power draw are clearly justified, even if this eliminates large TVs from the market for now. However, efforts could be made (by public health authorities and various advocacy groups) to promote alternative forms of entertainment, such as reading, gardening, learning to play a musical instrument, and involvement in individual or team sports or other forms of physical exercise. Improvements in the public realm (parks, streetscapes and recreation centres) would probably help to reduce dependence on TVs for entertainment.

Set-top boxes

Set-top boxes (STBs) are devices used to convert an incoming TV broadcast signal into one that can be seen on a screen. There are different STBs for aerial, cable, satellite and internet TV signals and whether the output is designed for analogue or digital display. Many new ones have built-in recording devices, such as a hard drive or a DVD recorder and are referred to as digital video recorders (DVRs). Energy consumption for basic STBs ranges from 27kWh/yr for aerial STBs to about 90kWh/yr for satellite STBs. However, the energy consumption grows rapidly as additional features are added: an extra 18kWh/yr for video processing, an extra 35kWh/yr for high definition and an extra 60kWh/yr for DVR, among others (the total can approach the average per capita total residential electricity consumption of 300kWh/yr in non-OECD countries!).

Figure 4.71 shows the range of power draws for aerial, cable, and satellite STBs in on and standby modes. There is very little difference in the power draw between these two modes, but up to a factor of four difference in the power draw among different units of a given type in on mode and up to a factor of ten difference in the power draw when on standby (but, according to IEA (2009), there is very little difference in the functionality of the different units). TV service providers (TVSPs) prefer to keep the STBs in a relatively high power mode so as to be able to send materials (such as security and software updates and promotional material) to the STB at any time.

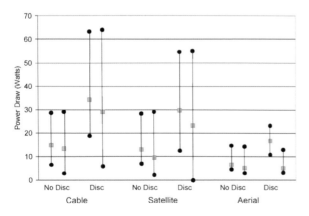

Source: Data from IEA (2009)

Figure 4.71 *Extremes in the power draw (circles) and average power draw (squares) for different STBs, classified according to the signal received and the presence or absence of built-in recording devices*

The options to reduce energy use by STBs are to improve the performance of STBs in on and standby modes to that of the best STBs on the market in each category, and to enable STBs to utilize a true low-power mode, drawing no more power than is necessary at any given time.

STBs are supplied by TVSPs as part of the monthly TV service. TVSPs have an incentive to minimize first-cost but have no incentive to reduce the electricity use of their STBs, as they do not pay the running costs. Customers normally cannot choose the STB that they receive, and where they do, they lack the information needed to make an informed choice (IEA, 2009). In the absence of regulations requiring reduced energy use, these characteristics of the STB represent significant barriers to improved efficiency, although one British TVSP is offering and promoting low-energy STBs as part of its service package.

Computers

Home computers now compete with TVs as a source of electronic entertainment, as well as providing a vehicle for online shopping (the transportation energy use implications of which are discussed in Chapter 5, subsection 5.7.7). Table 4.20 gives data on the energy use by computers and related equipment based on direct measurements made in a sample of Australian households in 2005. The average power draw of personal computers (PCs) in the on mode was 82W, while that of CRT monitors was 62W and that of LCD monitors was 30W. A shift from CRT to LCD monitors reduces the total power draw (from about 140W to 110W), and a shift from personal computers with LCD monitors to laptop computers entails a further large power reduction (from of about 110W to 34W).

The energy consumption by desktop computers (PCs) can be reduced through more efficient central processing units, more efficient internal power supplies, and implementation (by the PC users) of power management options.

Manufacturers of laptop computers have put considerable effort into reducing the energy consumption by laptops so as to extend their operating time when dependent on the battery, and these energy savings can be transferred to PCs (IEA, 2009). The best computer chips in laptops use a fifth of the energy of the Pentium 4 chip while delivering the same performance. Laptop computers come with external

Table 4.20 *Average power draw (W) by computers and related equipment as measured in a sample of Australian households in 2005*

Component	Mode of operation		
	On	Standby	Off
Individual components			
PC	82.2	35.5	3.5
CRT monitor	61.7	8.2	1.9
LCD monitor	29.3	2.6	1.0
Laptop computer	34.1	16.5	0
Speakers	6.0	6.0	2.2
Modem	5.9	4.4	2.4
Totals			
Using PC + CRT	156	54	10.0
Using PC + LCD	123	49	9.1
Using laptop	46	27	4.6

Source: Harrington et al (2006)

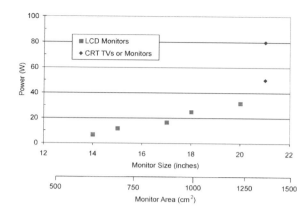

Note: The 21-inch CRT computer monitor has the same power draw as given in Figure 4.68 for CRT TVs of the same size.
Source: Roth et al (2002) and Keith Jones (personal communication, June 2009)

Figure 4.72 *Power draw of various LCD computer monitors and that expected for 21-inch CRT computer monitors*

power transformers with an efficiency of 70–80 per cent, while PCs come with internal transformers with an efficiency of 60–70 per cent. Another large difference is in the power management of laptops, which is designed to provide only the energy required for the functions being used at any given time. PCs, by contrast, tend to be shipped with minimal power management options implemented.[16]

Figure 4.72 compares the power draw of CRT and LCD computer monitors of various sizes. LCD monitors use modestly less energy then CRT monitors or LCD TVs of comparable size. LCD monitors differ from LCD TVs in a number of ways that affect energy use (Keith Jones, personal communication, June 2009). First, most LCD monitors have a fixed power consumption, irrespective of the image being shown, whereas the most recent LCD TVs have variable power consumption. Second, the backlight in an LCD monitor is usually a single CFL reflected from a parabolic mirror and designed to illuminate the entire surface of the display. This results in a variation in light level in the vertical that is not apparent in monitors but would be in TVs, so TVs use multiple CFLs. Thus, all else being equal, monitors are expected to draw less power than equivalent-sized TVs.

LED backlit LCD monitors (and TVs) are now just appearing, and these will have power consumption that varies as needed with the image being displayed.

4.9.3 Office equipment

Figure 4.73 gives the estimated breakdown of energy use by office equipment in the US. Personal computers, workstations and monitors account for about 40 per cent

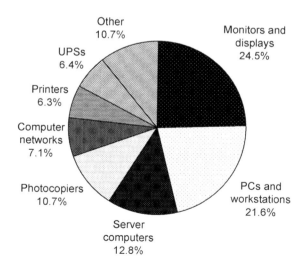

Source: Roth et al (2002)

Figure 4.73 *Breakdown of energy use by office equipment in US commercial buildings in 2000*

of the energy used by office equipment. Servers, photocopiers and printers account for another 26 per cent. In cooling-dominated climates, this energy use contributes to the cooling load. Table 4.21 compares the energy use by the most energy-intensive office equipment for different modes of operation. Options for reducing energy use by computers and monitors were discussed in subsection 4.9.2. Given that many employees do not turn off their computers and monitors when leaving, implementation of power management options with automatic hibernate and off modes for computers and monitors, respectively, is especially critical. IEA (2009) indicates that this alone can reduce the energy use of a typical PC + monitor from 500kWh/yr to 200kWh/yr.

4.9.4 Commercial refrigeration

Rosenquist et al (2006) indicate savings potentials of about 40–50 per cent for supermarket refrigerators and freezers, 50 per cent for refrigerated vending machines and 60 per cent for beverage merchandiser units. With regard to walk-in refrigerators and freezers, which are

used in restaurants, hospitals, convenience stores and supermarkets where boxes of perishable foods are stored, a standard proposed for California would reduce energy use by more than 40 per cent with simple payback times (defined in Appendix D) of 1.5 years for freezers and 3.4 years for refrigerators (Nadel et al, 2006).

4.9.5 Information technology centres

Mitchell-Jackson et al (2002) estimated that the aggregate power demand by information technology (IT) centres in the US in 2000 was less than 500MW and that electricity consumption was about 0.12 per cent of national electricity demand. Earlier estimates of the energy and power requirements of IT centres were erroneously high, for a variety of reasons that are explained by Mitchell-Jackson et al (2003). Nevertheless, Koomey (2008) estimates that worldwide electricity use by data centres doubled between 2000 and 2005, reaching 1 per cent of world electricity demand by 2005.

Table 4.22 gives the estimated breakdown of annual electricity demand by US data centres in 2000 and by two data centres in Singapore. All of the electricity produced by the equipment in data centres is converted to heat that must be removed by the air conditioning system. The ratio of equipment electricity use to the electricity used by the central chiller plants and auxiliaries gives the average COP of the air conditioning system. For the US, this is 2.0/1.6 = 1.25,

Table 4.21 *Power consumption (W) by office equipment*

Item	Mode of operation			
	Active	Standby	Suspend	Off
– Desktop	55		25	1.5
– Laptop	15		3	2
CRT monitors (14–21 inch)	61–135	26–43	9–19	3.0–4.7
LCD monitors (14–21 inch)	7–36	1.9–10.4	0.7–3.6	0.3–1.8
Workstations	110–158			
Servers	125–2520			
Printers				
– Laser, desktop	130–215	75–100	10–35	
– Laser, office	320–550	160–275	70–125	
Photocopiers				
– <12cpm	778	56	2.2	1.1
– 12–30cpm	1044	179	42	0.5
– 31–69cpm	1354	396	68	0.6
– 70+cpm	2963	673	300	2.3
Scanner	150	15		0
Fax machine	30–175	15–35		0

Note: cpm = copies per minute.
Source: Roth et al (2002)

Table 4.22 *Energy use in IT centres*

Energy end use	Energy use
US total (TWh/yr)	
Computer room, including lights and auxiliary equipment	2.6
Central cooling plant	0.6
Fans, computer room air conditioners, ventilation	1.0

Singapore examples (kWh/m²/yr)	Data Centre 1	Data Centre 2
IT equipment	1483	534
HVAC system	1243	1052
UPS loss	151	381
Lighting	82	54
Total	2960	2022

Source: US, Mitchell-Jackson et al (2002); Singapore cases, Sun and Lee (2006)

while for the two Singapore data centres the COPs are 1.38 and 0.92. These are not large COPs, as the most efficient commercial chilling systems have COPs in the order of 4.0 (including energy use by fans and pumps).

Because of the tendency to overestimate energy requirements when building IT centres, the uninterruptible power supplies (UPSs) tend to be vastly oversized. As a result, the UPSs operate at a small fraction of their capacity, with a significant decrease in efficiency. In the case of the two Singapore data centres, the UPSs were oversized by factors of 5.6 and 10.0, with average efficiencies of 83 per cent and 46 per cent, respectively. These large losses pertain to the total electricity use by the data centre (computers and HVAC system), and add to the heat that needs to be removed by the chillers. Even without oversizing, the normal practice would be to have two UPSs, each capable of handling the full power load. Thus, each unit would operate at less than 50 per cent load, which is where the large efficiency drop begins to occur. An alternative would be to have three units, each capable of handling two thirds of the full power load, thereby providing 50 per cent rather than 100 per cent backup (on the grounds that two units are unlikely to fail simultaneously) and allowing each unit to operate at closer to full load. This alone can give an efficiency gain of 5 per cent (Tschudi and Fok, 2007).

Other measures to reduce energy use in IT centres include totally enclosing the 'cold' aisles in the data centres (which alternate with 'hot' aisles, into which heat from IT equipment is ejected) and a variety of conventional techniques to greatly improve the efficiency of HVAC systems (as discussed by Tschudi and Fok, 2007, and Horvath and Shehabi, 2008). A reduction of 50 per cent in the HVAC and lighting component of IT energy use is probably achievable on average, with an additional 10–20 per cent savings in total energy use through better sizing of UPS units.

4.10 Embodied energy and operating energy

The *embodied energy* of a building is the energy used to manufacture and transport the materials used in the building, as well as the energy used during the construction process itself. The *operating energy* is the energy for heating, cooling, ventilation, lighting and appliances or office machines. The embodied energy depends on the energy intensity of the industries involved in producing building materials (which is discussed in Chapter 6 within the more general context of industrial energy use) and on transportation energy (which is discussed in Chapter 5). There is tremendous scope for reducing industrial and transportation energy intensities (i.e. energy used per unit of output or per tonne-km of transport), and as these energy intensities improve, the embodied energy in new buildings will decrease. Nevertheless, it is appropriate here to briefly identify areas where alternative designs can reduce the embodied energy of buildings, and to assess potential tradeoffs between increased embodied energy and reduced operating energy.

4.10.1 Relative energy intensity of different building materials

Table 4.23 provides estimates of the present energy intensity of different building materials, which have been classified as: (1) very high energy intensity (aluminium, plastics, copper, stainless steel), (2) high energy intensity (steel, glass, cement, plasterboard), (3) medium energy intensity (clay bricks and tiles, concrete, timber), and (4) low energy intensity (sand, aggregate, fly ash). Although the energy intensity of concrete is relatively low, enough of it is used in many buildings that it can readily account for 40 per cent or more of the total embodied energy in a building. In this respect, the foundation system used in an advanced house in

Table 4.23 *Present-day embodied energy (GJ/tonne) for building materials made from virgin materials*

Material	Embodied energy	Material	Embodied energy
Very high energy		*Medium energy*	
Aluminium	170–230	Lime	0.04–0.9
Plastics	50–160	Clay bricks and tiles	2–7
Copper	60–160	Gypsum plaster	1–4
Stainless steel	70–100	Concrete	
High energy		In situ	0.8–1.5
Steel	20–40	Blocks	0.8–3.5
Lead, zinc	25+	Precast	1.5–8
Glass	4–9	Sand-lime bricks	0.8–1.2
Cement	4–6	Timber	0.1–5
Plasterboard	8–10	*Low energy*	
		Sand, aggregate	<0.5
		Fly ash, volcanic ash	<0.5

Source: Chapter 6 and Thomas (1999)

Waterloo, Canada, is of interest (Hestnes et al, 1997). The below-grade walls are precast concrete panels that are flat on the outside but waffle-shaped on the inside. The waffle cavities are filled with cellulose insulation, and an additional 50mm of rigid insulation is applied to the outside, while wood-stud walls with additional insulation are applied on the inside. This system uses only half the concrete of conventional poured basements in Canada. In the case of the Environmental Technology Center at Sonoma State University (California), the amount of Portland cement used in construction was reduced by 50 per cent by partially substituting fly ash and rice hull husk (Beeler, 1998).

4.10.2 Comparison of embodied energy and operating energy

There are many published analyses of the embodied energy of buildings, and many published analyses of the operating energy, but very few in which embodied and operating energy are compared for the same building. Results for three such studies are summarized in Table 4.24. For office buildings in Vancouver and Toronto, the embodied energy is equivalent to only a few years of operating energy. Thus, over a 50-year timespan, reducing the operating energy is far more important than reducing the

embodied energy. For well-insulated detached houses in Sweden, the embodied energy is equivalent to six to seven years of operating energy. For an advanced house in Gothenburg (Sweden), the embodied energy is equivalent to 43 years of the very-low operating energy. However, about a third of the embodied energy could be recovered through recycling of material when the house is renovated. From this it follows that:

- it will normally be worthwhile to incur modest increases in embodied energy if this is required in order to reduce the operating energy (for example, increasing thermal mass by using more concrete in order to be able to use passive cooling techniques);
- for existing buildings with high operating energy, the payback – in energy terms – of demolishing the building and rebuilding it from scratch as an ultra-low-energy building is only a few years;
- buildings that already have very low operating energy should not be demolished unless absolutely necessary.

Of particular interest is the difference in embodied energy between buildings that are designed to require vastly different operating energies but are otherwise the same. This analysis has been carried out for a series of progressively less energy-intensive buildings in

Table 4.24 *Comparison of building embodied and operating energies*

Case	Initial embodied energy (GJ/m²)	Renovation embodied energy over 50 years (GJ/m²)	Operating energy (GJ/m²/yr)	Years of operating energy required to equal initial embodied energy	Reference
Conventional offices in Vancouver	4.5–5.1	6.6	1.05	4–5	Cole and Kernon (1996)
Conventional offices in Toronto	4.5–5.1	6.6	1.76	2.5–3	Cole and Kernon (1996)
Conventional offices in Japan	7–13	1.2–2.6[a]	1.2–1.6	6–8	Suzuki and Oka (1998)
Office buildings in Germany					
– Typical	10	16.0	0.9	11	Richard Rogers
– Good practice	5	7.0	0.45	11	Architects (1996)
– Potsdamer Platz, Berlin	7[b]	15.3	0.26	8[c]	
Well-insulated house in Sweden	2.9–3.7		0.46–0.51	6–7	Adalberth (1997)
Advanced house in Gothenburg	7.03		0.164	43	Thormark (2002)
Conventional house in Michigan	3.97–6.62		1.27	3.1–5.2	Keoleian et al (2001)
Well-insulated house in Michigan	4.13–7.32		0.41	10–18	Keoleian et al (2001)

Note: [a]Renovation energy computed over a 40-year cycle. [b]Higher than for good practice due to choice of materials. [c]Payback time if a typical building is demolished and replaced with a building having good-practice embodied energy and Potsdamer Platz operating energy.

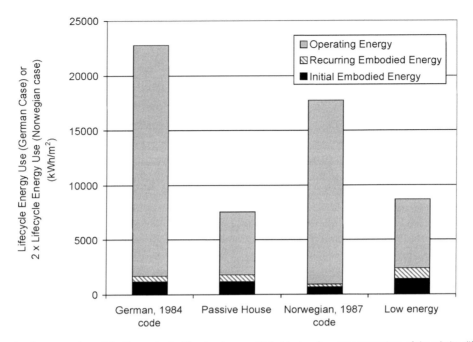

Note: Energy use for the two versions of the Norwegian building has been multiplied by two for easier comparison of the relative difference.
Source: Feist (1996) and Winther and Hestnes (1999)

Figure 4.74 *Comparison of lifecycle embodied and operating energies for standard and advanced versions of a building in Germany and for standard and advanced versions of a building in Norway*

Germany by Feist (1996) and in Norway by Winther and Hestnes (1999). The operating and embodied energies for the buildings with the highest and lowest energy demands are compared for the two cases in Figure 4.74. In both cases, designing a building to use several times less operating energy entails only a small increase in the embodied energy.

It is nevertheless important to try to minimize the embodied energy. This involves a number of tradeoffs and choices, some of which are discussed below.

4.10.3 Modern versus traditional building materials

Modern building materials – concrete, concrete blocks, glazed bricks, steel – tend to contain substantially more embodied energy than traditional materials. Tiwari et al (1996) and Tiwari (2001) have examined the impact on cost and CO_2 emissions of partial substitution of some modern building materials with traditional and locally available materials for housing in India. The latter

include compressed-mud blocks, used in place of bricks; lime, surkhi and fly ash instead of cement and sand in mortar; lime and surkhi in place of cement and sand for plastering; and filler slabs with Mangalore tiles in place of concrete-slab roofs and floors. The net result is a 60 per cent reduction in CO_2 emissions associated with the construction of housing, and a 45 per cent reduction in costs. A number of materials can be impregnated into mud bricks in order to increase their durability and resistance to water (Ren and Kagi, 1995).

Petersen and Solberg (2002) find that the energy consumed in producing steel beams is two to three times that of laminated wood beams. Furthermore, the energy released through combustion of wood beams at the end of their life is greater than the non-solar energy input required to produce the beams. Thus, the substitution of wood beams for steel beams is clearly beneficial from an energy point of view. The substitution of wood frame floors and walls for concrete floors and walls reduces the overall embodied energy of a variety of different building types in

New Zealand by about a third (Buchanan and Levine, 1999). However, the absolute difference in energy per unit of floor area, divided by an assumed building life of 50 years (which would be too small if the building is well built) amounts to about 10–30MJ/m²/yr. This is small compared to the amount of primary energy used for cooling in warm climates (which can easily exceed 100MJ/m²/yr in conventional buildings). Thus, if the thermal mass of concrete is used in combination with night ventilation (through hollow-core floor slabs) and external insulation, the reduction in cooling energy use could more than offset the additional embodied energy associated with concrete. Conversely, thermal mass could be provided through phase-change materials (see subsection 4.5.1) instead of through concrete, although perhaps not as effectively. A full accounting of the energy impact of wood versus concrete construction depends on a large number of assumptions concerning how the wood and concrete are produced and delivered to the site and concerning what is done with wood waste at the end of the building's life, as discussed by Börjesson and Gustavsson (2000).

Another factor that should be taken into account in comparing wood and concrete construction is the greater ability of concrete, due to its high density, to absorb sound. This makes it a better barrier between separate units in multi-unit housing. Lower sound transmission improves the marketability of multi-unit housing compared to detached housing, which in turn yields large savings in operating energy (as discussed in subsection 4.2.8).

4.10.4 Embodied energy of advanced windows compared to the energy saved

Table 4.25 compares the extra primary energy used in adding various energy-saving features to a 1.1m² window with the primary energy saved by that feature over a 20-year period for a mid-European climate (3000 HDDs). The addition of an extra pane of glass to the 1.1m² window increases the energy required to make the window by 242MJ but saves about 15,800MJ over a 20-year period if it is the second pane of glass and 5300MJ if it is the third pane of glass. Addition of a single low-e coating requires 8.4MJ to extract and process the materials used in the coating but negligible energy for the window manufacturing process. The 20-year energy savings is 4844MJ for the first low-e coating added to a double-glazed window, and 394MJ for the third coating added to a triple-glazed window. The energy savings when inert gas fills are used is substantially greater if the fill is added to a window that already has low-e coatings. For double-glazed windows

Table 4.25 *Incremental primary energy required to produce 1.1m² window (glazed area + frame) with the designated energy-saving feature, and annual and 20-year primary-energy savings*

Energy-saving measure	Energy required (MJ)	Decrease in U-value (W/m²/K)	Energy savings (MJ)	
			Annual	20-year
Change from SG to DG	242[a]	2.492	789.4	15,788
Change from DG to TG	242[a]	0.841	266.4	5327
Add 1st hard low-e coating to DG	8.4	0.764	242.2	4844
Add 2nd soft low-e coating to DG	8.4	0.043	13.6	272
Add 3rd hard low-e coating to TG	8.4	0.062	19.7	394
Argon instead of air in DG, clear	0.012	0.114	36.0	720
– 1S, 1H	0.012	0.194	61.4	1227
Krypton instead of air in DG, clear	508	0.194	61.6	1231
– 1S, 1H	508	0.346	109.6	2192
Krypton instead of air in TG, 2S, 1H	508	0.136	43.0	860
Xenon instead of air in DG, clear	4500	0.245	77.8	1555
– 1S, 1H	4500	0.451	142.9	2859
Xenon instead of air in TG, 2S, 1H	4500	0.183	58.0	1160

Note: DG = double-glazed, TG = triple-glazed. Energy savings are based on differences in glazing-averaged U-values (centre and edge regions), assuming 3000 heating degree days (HDDs) and a heating efficiency of 0.9.
Source: Harvey (2006)

with one soft and one hard low-e coating, use of argon in place of air requires an extra 12kJ but saves 1227MJ over a period of 20 years, a payback factor of 84,000. Use of krypton instead of air requires an extra 508MJ but saves only 2200MJ over 20 years, while use of xenon instead of air requires 4.5GJ but saves only 2.86GJ over 20 years. Thus, except in very extreme climates, it would appear that the use of xenon does not lead to a net energy saving, and that for krypton, about five years of energy savings are required before the initial energy input is paid back in a climate with 3000 HDDs.

However, this analysis does not account for the ability to eliminate perimeter heating radiators using high-performance windows (see Figure 4.13). As radiators are made of steel or aluminium, both of which have a large embodied energy, elimination of perimeter radiators gives a large reduction in embodied energy that can be credited against the embodied energy of the window. For example, use of krypton in a 2m high window would entail an extra 706MJ of embodied energy per metre of window width, while the embodied energy in a finned perimeter radiator is about 1000MJ per metre (based on the author's estimate of material use for radiators). This justifies the use of krypton fill if it makes the difference between having and not having perimeter radiators.

4.10.5 Embodied energy of insulation compared to the heating energy saved

Table 4.26 gives the embodied primary energy for different insulation materials for a 1m² panel with an RSI-value of 1. Spray-on foam insulation is the most energy intensive, both per unit mass and per unit RSI, while cellulose and mineral fibres are the least energy intensive. The embodied energy of cellulose includes the heating value of the feedstock (used newsprint), as the feedstock could otherwise be used as a source of energy through combustion. This is consistent with the energy intensities given for foams, which include process energy and the heating value of the petroleum products that are used as feedstocks. The energy intensity for flax (from Ardente et al, 2008) is surprisingly large – about twice the heating value of biomass.[17]

The length of time required to pay back the energy used to make insulation, through savings in heating energy, can be computed based on the savings in heating energy and the embodied energy in the full thickness of insulation, or can be based on the additional savings in heating energy and the additional embodied energy when an extra increment of insulation is added. In analogy to economic

Table 4.26 *Embodied primary energy for a 1m² insulation panel with an RSI value of 1(W/m²/K)⁻¹*

Type of insulation	Embodied energy (MJ/kg)	Density (kg/m³)	Conductivity (W/m/K)	Embodied energy (MJ/m²/RSI)
Spray-on polyurethane	137[a]	32	0.035	153
Solid polyurethane	137[a]	32	0.024	105
Polystyrene	127[b]	32	0.034	138
Fibreglass	58[c]	20	0.042	49
Flax[d]	40	30	0.042	50
Kenof[d]	39	40	0.038	59
Fibreboard	31[e]	40	0.038	47
Cellulose	20[f]	30	0.04	24
Mineral fibre	17[g]	40	0.037	25

Note: [a] This value is inferred from Petersdorff et al (2002), and is comparable to the value of 122–133MJ/kg inferred from Ardente et al (2008). 75/MJ/kg is given for polyurethane in Figure 6.35 but does not pertain to any specific finished product. [b] This value is inferred from Petersdorff et al (2002). 70/MJ/kg is given for polystyrene in Figure 6.35 but does not pertain to any specific finished product. [c] 58MJ/kg is taken from Table 6.1 and is very close to value of 56MJ/kg inferred from Ardente et al (2008). [d] All data in this line are given in or inferred from Ardente et al (2008). Computed from the embodied energy per unit mass, density and thermal conductivity. [e] From the value of 11.5MJ/kg given in Petersdorff et al (2002) + 19MJ/kg to account for the heating value of the biomass feedstock, but their density is 190–240kg/m³. [f] From the value of 0.9MJ/kg given in Petersdorff et al (2002) + 19MJ/kg to account for the heating value of the biomass feedstock. A value of 20.3MJ/kg is inferred from Ardente et al (2008). [g] Values of 17MJ/kg and 18MJ/kg are inferred from Ardente et al (2008) and Petersdorff et al (2002), respectively.

cost–benefit analysis, the latter will be called the *marginal payback time*. Because the absolute reduction in heat loss with successive increments of insulation decreases as more insulation is added (as seen from Figure 4.9), the marginal payback time is longer for a given increment of insulation the greater the pre-existing insulation level. At any level of insulation, the marginal payback time is longer than the overall payback time. The overall payback time is a useful

indicator of the value of a given amount of insulation, while the marginal payback time is useful in deciding when (on a lifecycle energy basis) to stop adding more insulation: the amount of insulation should not be increased beyond the point where the marginal payback time equals the expected lifespan of the insulation.

Figure 4.75 gives overall and marginal payback times based on the primary energy needed to

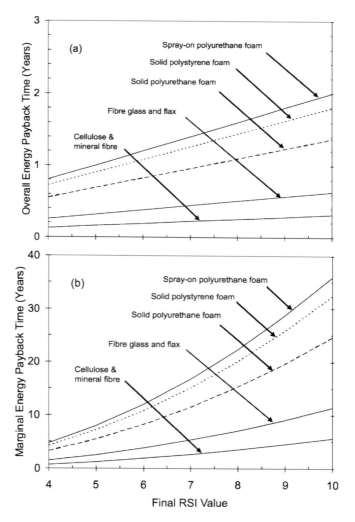

Note: Energy payback times are for a climate with 4000 HDDs and a heating system efficiency of 0.9.
Source: Computed here based on the embodied energies given in Table 4.26

Figure 4.75 *Variation in (a) the overall energy payback time as the total RSI is increased from 0.5 to values as large as 10.0, and (b) the marginal energy payback time when the RSI value is increased by 1.0 to the indicated final value*

manufacture (but not transport or install) various kinds of insulation for a climate with 4000 HDDs and a heating-system efficiency (η) of 0.9. The savings in heating energy is computed as:

$$Saving(J) = HDD \times 24 \times 3600 \times \Delta U / \eta \quad (4.23)$$

where ΔU is the difference in U-values with and without insulation (or with and without a given increment of insulation). Figure 4.75a gives overall payback times relative to an uninsulated brick or masonry wall with an RSI value of 0.5. Even for the most energy-intensive insulation, the overall payback time is less than two years for an RSI of up to 10. Figure 4.75b gives marginal payback times, based on the last RSI increment of 1.0 added. When the RSI is increased from 9 to 10, the payback time for this increment is as large as 36 years, but inasmuch as this is less than the expected lifetime of the insulation, increasing the RSI to as high as 10 using the most energy-intensive foam insulation is justified on a lifecycle energy basis with a 4000 HDD climate. It is, of course, preferable to use insulation with lower embodied energy wherever this is feasible.

4.10.6 Halocarbon GHG emissions associated with foam insulation

Halocarbon gases are used as expanding agents in the production of most solid foam insulation, and in the application of most spray-on foam insulation (subsection 4.2.1). Eventually, the only permitted halocarbon blowing agents will be the HFCs. Some of the gas is released during the manufacture of the insulation, but the majority is embedded in the foam pores and much of this can gradually leak over time. Thus, the global warming impact of any halocarbon gases used as an expanding agent needs to be considered along with emissions of CO_2 during the manufacture of the insulation and avoided through reduced use of energy for heating.

With regard to solid polyurethane foams, some currently use HCFC-141b as the expanding agent, while others use HFC-365mfc or n-pentane. Figure 4.76 gives the number of years required for the savings in heating energy CO_2 emission to completely offset the global warming impact of the CO_2 emissions associated with the manufacture of the insulation and with leakage of the expanding agent. Results are given for foam insulation sufficient to give a total RSI ranging from 4 to 10, beginning with an initial RSI-value of either 0.5 (see Figure 4.76a) as well as based on RSI increments of 1.0 up to a final RSI-value of 10 (see Figure 4.76b). The latter are marginal payback times, analogous to the marginal payback times given in Figure 4.75b for embodied energy.

As seen from Figure 4.76, payback times assuming an initial RSI-value of 0.5 and a final RSI-value of 6.5 are four and five years using HCFC-141b and HFC-365mfc, respectively. Marginal payback times are in excess of 40 years for an insulation increase from RSI 5.5 to 6.5, and in excess of 100 years from an increase from RSI 9 to 10. For polyurethane blown with n-pentane, the marginal payback at an RSI-value of 10 is only 25 years, and the average payback (based on the total thickness of foam insulation) is 1.4 years if the starting RSI-value is 0.5. Based on these results, it is seen that the net climatic benefit of halocarbon-blown foam insulation can be quite small, and that when used to build the total RSI-value up to the levels of insulation used in low-energy houses (RSI 6.5 in walls, RSI 10.0 in roofs), halocarbon-blown foam insulation can be *counterproductive* from a climatic point of view (depending on the lifespan of the insulation).

Fortunately, solid-foam (and spray-on foam) insulation produced with non-halocarbon expanding agents is available, either using a CO_2/H_2O mixture or n-pentane. Table 4.27 compares the thermal conductivity of solid and spray-on polyurethane insulation produced using different expanding agents. The thermal conductivity of polyurethane insulation made with non-halocarbon agents is 10–20 per cent greater than when halocarbons are used, meaning that either the insulation will need to be 10–20 per cent thicker for a given RSI-value or, if the thickness of the insulation is fixed (due to space constraints), the RSI will be slightly smaller. In retrofit applications, spray-on foam insulation is convenient because it is possible to insulate hard-to-reach, irregularly shaped cavities. Since the starting RSI-value in this case is normally quite small, the difference in the percentage reduction in heat loss using non-halocarbon instead of halocarbon-blown insulation, compared to the

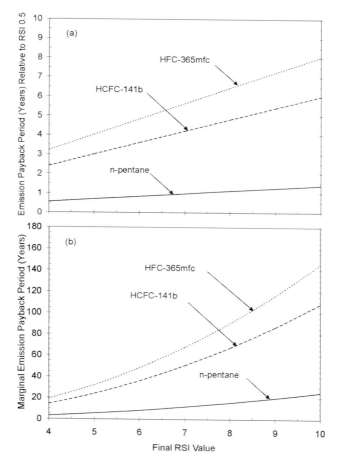

Note: Emission payback times are for polyurethane insulation using either HCFC-141b, HFC-365mfc or n-pentane as blowing agents, for a climate with 4000 HDDs and a heating system efficiency of 0.9.
Source: Computed by the author

Figure 4.76 *Variation in (a) the overall payback time for equivalent CO_2 emissions as the total RSI is increased from 0.5 to values as large as 10.0, and (b) the marginal emission payback time when the RSI value is increase by 1.0 to the indicated final value*

uninsulated case, is negligible.[18] Thus, there is significant climatic gain (through avoided halocarbon emissions) and very little loss in heating-energy savings using foam insulation with non-halocarbon blowing agents. A more detailed analysis relevant to the use of foam insulation and the choice of expanding agent is found in Harvey (2007). As previously noted, spray-on cellulose insulation is an even better choice because of its low embodied energy compared to either halocarbon or non-halocarbon blown foam insulation. It can also be easily used in retrofit applications (as well as in new buildings).

Table 4.27 *Long-term thermal conductivity of spray-on and solid polyurethane foam insulation blown with different expanding agents*

	Blowing agent	Conductivity (mW/m/K)
Sprays	HCFC-141b	29
	HFC-365mfc	30
	Water/CO_2	35
Solid	HCFC-141b	21
	HFC-365mfc	22
	HC-n-pentane	24

Source: Krähling and Krömer (2000)

4.11 Demonstrated energy savings in advanced new buildings and the role of the design process

The preceding discussion has shown how dramatic reductions in energy use in buildings are possible through improved building envelopes; more efficient heating and cooling devices; through maximizing the use of passive solar heating, cooling and ventilation; through daylighting and advanced artificial lighting systems; through clever ways of combining the components of residential and commercial HVAC systems; through better use and supply of hot water; and through more efficient household appliances, consumer goods and office equipment.

Here, the net impact on energy use in new buildings that incorporate a large number of these features is presented. The result (a saving of 75 per cent or better compared to common practice) is simply a matter of observation. More interesting and important to those who would like to see common practice move much closer to what we already know how to do is the fact that, in order to achieve these dramatic savings, major changes in the usual *process* by which buildings are designed are required – a subject also discussed here.

Our interest here is in the *gross* energy requirements of buildings, that is, the energy use prior to consideration of energy production by active solar systems (PV or solar thermal collectors). The latter gives the *net* energy consumption of the building. There is a growing interest in zero-energy buildings, in which sufficient solar PV panels are included to generate more electrical energy than used by the building, with the surplus (or rather, the savings in primary energy at the powerplants that would otherwise be used to produce electricity) offsetting fuel energy used by the building. However, it is not necessary to attain zero-energy buildings in order to achieve sustainable buildings. Rather, what is necessary is that the gross energy demand of buildings be reduced to the point that the remaining energy requirements can be reliably met through renewably based energy sources, whether these be grid energy, onsite use of biomass, or onsite generation of electricity or hot water from the sun.

Normally, renewable grid electricity (from wind now, and later from concentrating solar thermal energy) will be less expensive than onsite generation of electricity with

PV panels, even after accounting for the cost of transmission from remote but favourable sites to the edge of urban regions (see Volume 2, Chapters 2 and 3). However, onsite generation of electricity through PV avoids transmission bottlenecks within urban regions, and will become increasingly competitive with conventional electricity sources during times of peak demand in regions where the peak is determined by air conditioning loads. Minimization of costs while eliminating CO_2 emissions requires pushing measures to reduce the gross energy requirements to the point where additional measures would be more expensive than the cost of C-free energy, whether grid-based or generated on site. Where this point lies depends on the skill of the building design team in reducing energy use at low cost and on the local costs of C-free energy. Evidence presented here indicates that, considering current prices for conventional energy sources, the crossover point occurs at energy savings of 50–90 per cent (depending on the region and local energy costs). Conventional energy will be more expensive in the future, and C-free energy – despite significant projected cost reductions – will be more expensive than the current costs of conventional energy in most parts of the world, so the economically justified reduction in the gross energy requirements of buildings will be larger still in most parts of the world.

4.11.1 Advanced residential buildings

Hamada et al (2003) summarize the characteristics and energy savings for 66 advanced houses in 17 countries. The majority of the houses surveyed have external-wall U-values of 0.1–0.2W/m²/K (R28 to R57), and one third of the houses have a window U-value ≤1.0W/m²/K. For the 28 houses where the savings in heating energy use is reported, the savings compared to the same house built according to conventional standards ranges from 23–98 per cent, with eight houses achieving savings of 75 per cent or better.

Parker et al (1998) shows how a handful of very simple measures (attic radiant barriers, wider and shorter return-air ducts, use of the most efficient air conditioners with variable speed drives, use of solar hot-water heaters, efficient refrigerators, lighting and pool pumps) can reduce total energy use by 40–45 per cent in single-family houses in Florida, compared to conventional practices. These savings are achieved while still retaining black asphalt shingle roofs that produce roof surface temperatures of up to 82°C! Further measures, such as use of a white tile roof instead of asphalt shingles, can increase the savings during the warmest month to over 50 per cent (Parker et al, 2000).[19] Holton and Rittelmann (2002), Gamble et al (2004) and Rudd et al (2004) have shown how a series of modest insulation and window improvements can lead to energy savings of 30–75 per cent in a wide variety of US climates. In all three studies, alterations in building form to facilitate passive solar heating, use of thermal mass combined with night ventilation to meet cooling requirements (where applicable), or use of features such as earth-pipe cooling, evaporative coolers or exhaust-air heat pumps are not considered. Thus, the full potential is considerably greater. Demirbilek et al (2000) find, through computer simulation, that a variety of simple and modest measures can reduce heating energy requirements by 60 per cent compared to conventional designs for two-storey single-family houses in Ankara, Turkey.

Huovila et al (2007) carried out simulations of energy use for residential (and commercial) buildings in a variety of countries and climates. Simulations were performed for conditions typical of current buildings, and for buildings with modest improvements in the building envelope and in heating and cooling systems, but with no change in the underlying systems. Assumptions are summarized in Table 4.28, while residential results for four cities – New York, New Delhi, Beijing and Madrid – are presented in Figure 4.77. Heating savings range from 85 to 100 per cent and cooling energy savings range from 50 to 60 per cent. Total energy savings range from 40 to 60 per cent, but are limited by the assumption of only a 50 per cent saving in lighting energy use and 8–10 per cent savings in other energy uses.

The Passive House Standard

The gold standard for housing in cold climates is the German *Passive House Standard* – a residential building with an annual heat demand (or heating load) of no more than $15kWh/m^2/yr$ irrespective of the climate, and a total energy consumption of no more than $42kWh/m^2/yr$. In operational terms, this means that the required heat can be supplied solely by circulating the amount of air needed for ventilation only, although sometimes radiant heating is added in bathrooms for added comfort. Several hundred houses that meet this standard have been built in Europe. The heat demand is the difference between the total heat loss and the usable passive solar and internal heat gains. The heat demand divided by the efficiency of the heating system gives the heating energy requirement, so for condensing furnaces or boilers with typical efficiencies of 92 per cent or better and a distribution system designed to have negligible heat loss, the heating energy use under the Passive House Standard will be $16–17kWh/m^2/yr$. By comparison, the average heating energy use is about $60–100kWh/m^2/yr$ for new residential buildings in Switzerland and Germany, about $220kWh/m^2/yr$ for existing buildings in Germany, and about $250–400kWh/m^2/yr$ for existing buildings in central

Table 4.28 *Characteristics of the reference and low-energy buildings whose simulated energy use is shown in Figures 4.77 and 4.80*

Attribute	Reference case	Low-energy case
Residential buildings		
Average U-value (W/m²/K)	0.841	0.532
Window U-value (W/m²/K)	2.78	1.48
Window shading	None	40 deg
Heating efficiency	80%	90%
Cooling COP	2.6	3.8
Office buildings		
Average U-value (W/m²/K)	1.382	0.548
Window U-value (W/m²/K)	2.78	1.48
Window shading	None	40 deg
Heating efficiency	80%	90%
Cooling COP	2.6	3.8

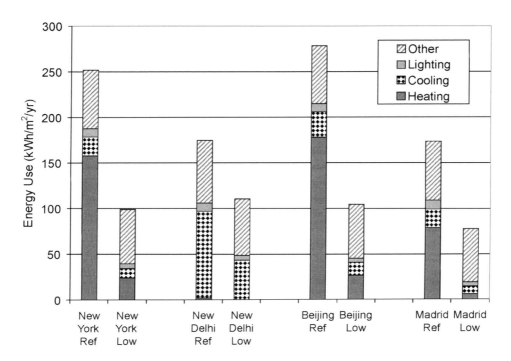

Source: Huovila et al (2007)

Figure 4.77 *Simulated energy use in reference residential buildings and buildings with modest improvements to the thermal envelope and in heating and cooling equipment (as detailed in Table 4.28) in various cities*

and eastern Europe. Thus, Passive Houses represent a reduction in heating energy use by a factor of four to five compared to new buildings, and by a factor of 15 to 25 compared to the average of existing buildings.

Achieving the Passive House Standard in central Europe requires wall, roof and floor U-values of 0.1–0.2W/m²/K (R28 to R57), a whole-window U-value ≤1.0W/m²/K, high airtightness, mechanical ventilation with recovery of 85 per cent or more of the heat in outgoing air, and meticulous attention to the avoidance of thermal bridges. Even in southern Finland (latitude 60°N), it is possible to meet the Passive House Standard in a detached house. To do so requires façade U-values of 0.07–0.1W/m²/K (R60 to R80) and whole-window U-values of approximately 0.7W/m²/K. The result (in one case) is an annual gross heat demand of 40kWh/m²/yr, half of which is met by usable solar heat gain (19kWh/m²/yr), with the balance supplied by the heating system (13kWh/m²/yr) and by internal heat gains (8kWh/m²/yr) (Pedersen and Peuhkuri, 2009).

Schnieders et al (2009) assessed the insulation levels and window performance needed to meet the Passive House Standard in south-western Europe (Portugal, Spain, southern France, Italy). Poorly insulated existing buildings in these regions have an annual heat demand of 100–200kWh/m²/yr and an annual cooling load of <30kWh/m²/yr, so even here, insulation requirements are dictated by heating rather than cooling loads. With insulation levels that meet the Passive House Standard for heat demand, comfortable summer conditions can be maintained in this part of Europe through a combination of daytime ventilation with heat recovery, night ventilation with cool air that bypasses the heat exchanger, exterior shading, and cooling and dehumidification of the supply air as needed. The resulting annual cooling load is <2kWh/m²/yr in Seville and Palermo and about 3kWh/m²/yr in Naples, while heating demand is 10–15kWh/m²/yr.

The Passive House Standard is promoted through the Passive House Institute in Darmstadt, Germany (www.passiv.de) and through regional Passive House institutes in various countries, including the UK (www.passivehouse.org.uk) and the US (www.passive house.us). A computer software package called the *Passive House Planning Package* is available for the computation of energy use, an English guidebook is available (Klingenberg et al, 2008), and training programmes in Passive House design and construction techniques are offered in many countries. Further technical details, measured performance, design issues and occupant response to Passive Houses in various countries can be found in Krapmeier and Drössler (2001), Feist et al (2005), Schnieders and Hermelink (2006) and Hastings and Wall (2007a, 2007b), while full technical reports are available at www.cepheus.de.

Cost of low-energy residential buildings

Schnieders and Hermelink (2006) report that the additional cost averaged over 13 Passive House projects in Germany, Sweden, Austria and Switzerland is 8 per cent of the cost of a standard house, but that when amortized over 25 years at 4 per cent interest and divided

by the saved energy, the cost of saved energy averages 6.2 eurocents/kWh (the range is 1.1–11 eurocents/kWh). This is somewhat more than the present cost of natural gas to residential consumers in most European countries, which ranges from 2–8 eurocents/kWh (IEA, 2004). Audenaert et al (2008) estimate extra costs of 4 per cent for low-energy houses and 16 per cent for Passive Houses in Belgium (having energy savings of 35 per cent and 72 per cent relative to current standard houses in Belgium). Figure 4.78 shows the progressive decline over time in the cost of the additional investment required to meet the Passive House Standard in central Europe. Through learning, costs have fallen to the point where the incremental cost can be justified based on 2005 energy prices and interest rates.

Behr (2009) makes the case that the energy use in apartments built to the Passive House Standard is so small that it is not worthwhile having individual metering and billing: annualized metering and billing costs of about €100/unit/yr for heating and hot water exceed the difference in annual energy costs associated with different behaviours. Rather, she proposes a billing system based on the total energy use for the building

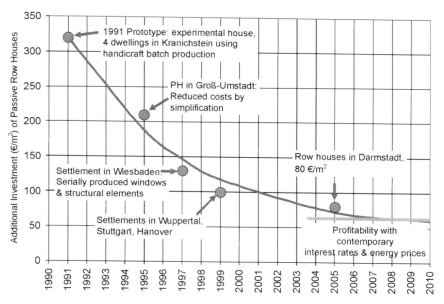

Source: Feist (2007)

Figure 4.78 *Learning curve showing the progressive decrease in the incremental cost of meeting the Passive House Standard for the central unit of terraced houses*

allocated among individual units based on either floor area or number of occupants (which will still provide an economic incentive, however diluted, for individuals to minimize their energy use). She also suggests that, by reducing the cost of achieving the Passive Standard in this way, the standard will be more readily adopted.

In the US, the National Renewable Energy Laboratory (NREL) carried out a study of the energy savings potential, along with costs, for a two-storey house (180m² floor area) facing west in a variety of US cities (Anderson et al, 2006). The additional construction costs for extra insulation were based on national average cost data, while costs from local distributors were used for better windows and more efficient furnaces and air conditioners. The financial impact of energy-saving measures entails extra mortgage costs (at 7 per cent interest over 30 years) and reduced energy costs. When the least-cost combinations of energy-efficiency measures are chosen for progressively more stringent energy-savings targets, total costs are minimized at an energy saving of about 40 per cent, but costs at 45–50 per cent energy savings are no greater than for the code-compliant houses.

Lollini et al (2006) assessed the heating energy savings and costs of 13 different insulation levels and window performance in apartment towers and single-family residential buildings in Italy. The maximum insulation considered (wall U-values of 0.08–0.35W/m²/K, depending on the climate zone) combined with the best windows considered (U-values of 2.4–5.0W/m²/K) in apartment towers saves 30–40 per cent of the heating energy compared to typical existing buildings with a 6–18 year payback, assuming natural gas at the then-current retail price in Italy (€0.6/m³ or €16/GJ), while similar insulation and window measures in the single-family house save 50–55 per cent with an 8–24 year payback. Potential cost savings due to downsizing of the heating system, shorter ductwork or hot-water lines (due to relocation of the radiators in new buildings away from the building perimeter) and reduced cooling loads were not considered.

In summary, the available published studies and demonstration projects indicate that savings in building energy use of 50–90 per cent are possible, in both hot and cold climates, through simple technical measures, and that savings of at least 50 per cent and often 75 per cent compared to current standards can be justified on a lifecycle cost basis at current energy prices.

4.11.2 Advanced commercial buildings and the pivotal role of the design process

A commercial building is a complex system, with the energy use and performance of any one part of the system affecting the energy use of the building as a whole through a complex cascade of interactions. However, the typical design process for commercial buildings is a linear, sequential process that precludes the analysis and design of the buildings as an integrated system. The architect makes a number of design decisions with little or no consideration of their energy implications, and then passes on the design to the engineers, who are supposed to make the building habitable through mechanical systems. The design of mechanical systems is also largely a linear process with, in some cases, system components specified without yet having all of the information needed in order to design an efficient system (given the constraints imposed by the architect) (Lewis, 2004).

Integrated design process

The key to achieving deep reductions in building energy use is to analyse the building as an entire system, rather than focusing on incremental improvements to individual energy-using devices. Through an integrated design approach, far deeper savings in energy use can be achieved than through an incremental approach, and at little to no additional upfront cost – or even with a saving in the initial cost compared to a standard design. As an example of the difference between an integrated and incremental approach, consider the energy used by ventilation systems. With great effort and at considerable upfront expense, it may be possible to improve the efficiency of the motors and fans by 5–10 per cent. This in turn would reduce energy use by 5–10 per cent if there is no change in the airflow. However, through the use of displacement ventilation, the required airflow can be readily reduced by a factor of two, which translates into a reduction in ventilation energy use by a *factor of six* (see subsection 4.1.2).

Figure 4.79 provides a schematic overview of the conventional design process and of a highly integrated design process (IDP). The IDP requires the setting of ambitious energy efficiency goals at the very beginning of the project, and requires an early brainstorming session involving all the members of the design team in order to

(a)   **Level 1:**

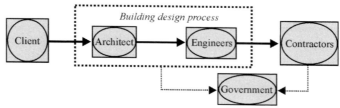

(b)   **Level 2:**

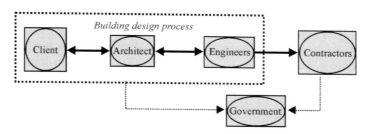

(c)   **Level 3:**

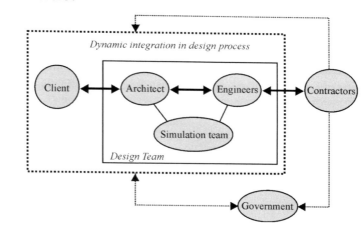

Source: Hien et al (2005)

Figure 4.79 *Conceptualization of the design process (a) as commonly practised when the client will not occupy the building, (b) as commonly practised when the client will occupy the building, and (c) for the highly integrated approach needed to produce a high-performance building*

develop a number of alternative concepts for achieving the energy target. Unlike the conventional process, the IDP includes a simulation team that specializes in the use of building simulation software and serves as a liaison between the architects and engineers. The IDP also entails two-way interaction between the design team and the contractors. Once the design is complete, the design team must be available during construction to explain details that are not clear, since no matter how thorough the plans and specifications, some details that affect the energy use by the building will be overlooked.[20]

The steps in the most basic IDP are:

- to consider *building orientation, form and thermal mass*;
- to specify a *high-performance building envelope*;
- to maximize *passive* heating, cooling, ventilation and daylighting;
- to install efficient *systems* to meet remaining loads;
- to ensure that individual energy-using *devices* are as efficient as possible and properly sized;
- to ensure the systems and devices are *properly commissioned*.

By focusing on building form and a high-performance envelope, heating and cooling loads are minimized, daylighting opportunities are maximized, and mechanical systems can be greatly downsized. This generates cost savings that can offset the additional cost of a high-performance envelope and the additional cost of installing premium (high-efficiency) equipment throughout the building. These steps alone can usually achieve energy savings in the order of 35–50 per cent for a new commercial building, compared to standard practice, while utilization of more advanced or less conventional approaches has often achieved savings in the order of 50–80 per cent.

In recognition of the importance of the design process, a number of programmes have been created that provide for the extra cost of designing and monitoring (but not constructing) highly energy-efficient buildings. These include the Canadian Commercial Building Incentive Program (CBIP), the California Savings by Design programme and the German Solar Optimized Building – SolarBau programme. CBIP was established in 1997 with a minimum required saving of only 25 per cent in order to encourage broad participation, but a sample of 43 buildings out of 300 projects completed by 2001 had an average predicted energy saving of 37 per cent (Larsson, 2001). The Savings by Design programme requires a minimum energy saving of 30 per cent compared to the California Title 24 standards, although savings as large as 60 per cent have been achieved (www.savingsbydesign.com). The SolarBau programme has as its target a *primary* energy consumption not exceeding 100kWh/m²/yr (excluding office equipment),

compared to 300–600kWh/m²/yr for typical German and Swiss office buildings (Reinhart et al, 2000).

Worldwide examples of exemplary commercial buildings

Table 4.29 gives documented examples of new commercial buildings in North America, Europe and Asia that achieved a minimum of a 50 per cent reduction in overall energy use compared to current conventional practice.[21] The largest projected saving (84 per cent) is for a Canadian building, the Center for Interactive Research on Sustainability, Vancouver (the actual saving is likely to be somewhat less than projected because buildings are usually never constructed exactly according to the design intent). Other indications of the potential energy savings based on analysis of real buildings are given below.

In the US, the NREL extracted the key energy-related parameters from a sample of 5375 buildings in the 1999 Commercial Buildings Energy Consumption Survey, and then used energy models to simulate their energy performance (Torcellini and Crawley, 2006). The results of this exercise are as follows:

- average total energy use as built is 266kWh/m²/yr;
- average energy use if complying with the ASHRAE 90.1-2004 standard is 157kWh/m²/yr, a saving of 41 per cent;
- average energy use would be 92kWh/m²/yr with improved electrical lighting, daylight, overhangs for shading and elongation of the buildings along an east–west axis (applicable only to new buildings) (a saving of 65 per cent).

With implementation of technological improvements expected to be available in the future, the gross energy use is so small that PV panels can generate more energy than the buildings consume, so that the buildings would serve as a net source of energy.

In the UK, energy consumption guidelines indicate that energy use for office buildings is about 300–330kWh/m²/yr for standard mechanically ventilated buildings, 173–186kWh/m²/yr with good practice (a saving of about 40–45 per cent) and 127–145kWh/m²/yr for naturally ventilated buildings with good practice (a saving of 55–60 per cent) (Walker et al, 2007).

Table 4.29 *Summary of exemplary (in terms of energy use) new commercial buildings where baseline and reference energy use have been published*

Building and location	Energy use	Energy savings (%)	Reference for comparison of energy use	Key features	Reference
Canadian examples					
Green on the Grand (offices), Kitchener, Ontario	81.2kWh/m²/yr (design total) Natural gas: 43.1kWh/m²/yr Electricity: 38kWh/m²/yr	50.4	ASHRAE 90.1–1989	Double-stud manufactured wood-frame wall; fibreglass-frame, triple-glazed, double-low-e, argon-filled, insulating-spacer windows; reduced lighting power densities; radiant heating and cooling panels, DOAS with enthalpy recovery ventilator, natural gas-fired absorption chiller; outdoor pond replaces conventional cooling tower	C-2000 Internal Programme Report[a] Four other C-2000 examples are found in Harvey (2006)
Father Michael McGivney Secondary School	148kWh/m²/yr	58	352kWh/m²/yr	GSHP, heat pipe type heat recovery unit	Genest and Minea (2006)
MEC Retail Store, Montreal	147.3kWh/m²/yr (design) 133kWh/m²/yr (actual 2004)	68	MNECB (466kWh/m²/yr)	High-performance envelope, daylighting, GSHP, DOAS, radiant slab heating and cooling, earth-coupled outdoor air tempering	Genest and Minea (2006)
Center for Interactive Research on Sustainability, Vancouver (proposed design)	56kWh/m²/yr without building-integrated photovoltaic (BiPV) and solar thermal (47kWh/m²/yr with solar)	84	Typical existing building (353kWh/m²/yr)	High-performance envelope, adjustable atrium shading, hybrid ventilation, daylighting, VSDs, DCV, 90% heat recovery effectiveness	Hepting and Ehret (2005)
US examples					
NREL offices and labs, Golden, Colorado		45 and 63 (two buildings)	ASHRAE 90.1		Murphy (2002)
Environmental Center, Oberlin College, Ohio	87kWh/m²/yr[b] 60kWh/m²/yr with recommended changes	48 and 64	ASHRAE 90.1-2001 (169kWh/m²/yr)	High-performance envelope, GSHP, daylighting	Pless et al (2006)
Federal Courthouse, Denver		50	ASHRAE 90.1-1989	Triple glazing, modest insulation, sunshading, daylighting, T5 lamps, VAV displacement ventilation, direct and indirect evaporative cooling, VSD on all air handlers and pumps, BiPV	Mendler and Odell (2000)

Building	Energy use	Saving (%)	Baseline	Features	Reference
Home improvement store, Silverthorne, Colorado	124kWh/m²/yr	54	ASHRAE 90.1-2001 (296kWh/m²/yr)	Higher-performance envelope, hydronic radiant floor heating, reducing lighting load and daylighting, solar thermal collectors	Torcellini et al (2004a); Deru et al (2006)
SC Johnson Wax Headquarters, Racine (WI)	<218kWh/m²/yr total	54 and 69	Ave new buildings Existing SJC buildings	Daylighting with automatic controls, fixed and adjustable shading, demand-controlled desktop personal air supply	Mendler and Odell (2000)
Academic building, University of Wisconsin, Green Bay		60	Wisconsin energy code	Wall U-value 0.16W/m²/K, roof U-value 0.11W/m²/K, skylights with suspended reflectors and motorized blackout panels BiPV	Mendler and Odell (2000)
Center for Health and Healing at the Oregon Health and Science University, River Campus		60	ASHRAE 90.1-1999	Hybrid ventilation, solar preheating of ventilation air, heat recovery, radiant heating/cooling, demand-controlled displacement ventilation, PV modules as exterior shading, commissioning	Interface Engineering (2005)
Zion National Park Visitor Center	85kWh/m²/yr	62	Code-compliant building at 222kWh/m²/yr	Modestly better insulation and windows, high thermal mass, daylighting with controls, downdraught evaporative cooling	Long et al (2006)
Cambria Office Building, Ebensburg, Pennsylvania	124kWh/m²/yr	64	Reference buildings at 322kWh/m²/yr	High-performance envelope, underfloor air distribution, heat recovery ventilators, GSHP, daylight and motion sensors	Torcellini et al (2004b)
Federal Reserve Bank, Minneapolis	<134kWh/m²/yr total 9.1W/m² connected lighting load, 7.0W/m² average lighting load	74	ASHRAE 90.1	Window U-value 0.74W/m²/K wall U-value 0.2W/m²/K, conventional VAV HVAC	Mendler and Odell (2000)
Iowa Association of Municipalities office	107kWh/m²/yr simulated, 88–91kWh/m³/yr measured	65 and 75	Iowa building code	19% window to wall ratio, high-performance envelope, daylighting, GSHP, enthalpy wheel for heat recovery	McDougall et al (2006)
Judson College Library, Illinois		69 for fans, 78 for cooling	Mechanically ventilated building	Design study to illustrate effectiveness of hybrid ventilation in reducing cooling and fan energy use in a continental climate	Short and Lomas (2007)

Table 4.29 *Summary of exemplary (in terms of energy use) new commercial buildings where baseline and reference energy use have been published (Cont'd)*

Building and location	Energy use	Energy savings (%)	Reference for comparison of energy use	Key features	Reference
Science Museum of Minnesota	64kWh/m²/yr gross, <0kWh/m²/yr net using PV arrays	78	290kWh/m²/yr for code-compliant building	Passive solar design, daylighting, GSHP for heating and cooling (with respective COPs of 3.1 and 3.7)	Steinbock et al (2007)
Environmental Technology Centre, Sonoma State University, California		80	California Title 24		Beeler (1998)
European examples					
Brundtland Centre, Denmark	50kWh/m²/yr	70	Typical comparable building (170 kWh/m²/yr)		Prasad and Snow (2005)
Center for Sustainable Building, Kassel, Germany	16.5kWh/m²/yr heating 32–42kWh/m²/yr total energy use	73 76–82	1995 German Building Code Typical office building	Wall U-value 0.11W/m²/K, window U-value 0.8W/m²/K, radiant slab heating and cooling, ground heat exchanger (COP 23), hybrid ventilation, daylighting	Schmidt (2002) and Schmidt (personal communication, 2006)
Debis Building, Potsdamer Platz, Berlin	75kWh/m²/yr total	80		DSF and passive ventilation	Grut (2003)
Ionica Building, UK	64kWh/m²/yr total	46	Good-practice air-conditioned building	Hybrid ventilation	Hybvent website (http://hybvent.civil.auc.dk)
SolarBau programme, 10 buildings in Germany	25–140kWh/m²/yr primary energy excluding office equipment	50–90	Typical office buildings, 300–600kWh/m²/yr primary energy	Mechanical night ventilation with exposed thermal mass or hydronic cooling integrated with groundwater, external shading, reduced glazing area, minimal internal heat gains, efficient lighting	Wagner et al (2004)
Schools, kindergartens, nursing homes, gymnasia	Meets the Passive House Standard (≤15kWh/m²/yr)	75–80	Current regulations for new buildings	Insulation, windows, airtightness, mechanical ventilation with heat recovery	See main text
Energy Base, office and classroom complex, Vienna	26kWh/m²/yr total onsite energy use	≥75	Current regulations for new buildings	High-performance envelope, passive ventilation, daylight, BiPV functioning as partial shading devices	Schneider (2009)

Building	Onsite energy use		Comparison	Strategies	Source
Solar Office, Doxford International Business Park, UK	85kWh/m²/yr	80	Typical new air-conditioned buildings in the UK (400kWh/m²/yr)	Passive ventilation and cooling; BiPV functioning as partial shading devices	Prasad and Snow (2005)
Elizabeth Fry Building and Zuckerman Building, University of East Anglia	30–37kWh/m²/yr heating; 93–100kWh/m²/yr total			High-performance envelope, concrete hollow-core ceiling slab with night-time ventilation, high airtightness	Cohen et al (2007); Turner and Tovey (2006)
Asian and Australian examples					
Kier Building, South Korea	68kWh/m²/yr electricity, 18kWh/m²/yr heat			DSF, ground-coupled heat exchanger, solar thermal and PV	Prasad and Snow (2005)
Liberty Tower, Meiji University, Japan		48	Japanese building code	Hybrid ventilation	Hybvent website (http://hybvent.civil.auc.dk)
Tokyo Earth Port	380kWh/m²/yr primary energy	45	Typical office building in Japan	Hybrid ventilation	Baird (2001)
Torrent Pharmaceutical Research Centre in Ahmedabad, India		64 for electricity	Conventional modern building	Evaporative cooling and hybrid ventilation (passive downdraught cooling)	Ford et al (1998)
Teri Retreat, Gurgaon, India	96kW peak electrical load versus 280kW for comparable building	66 in peak	Conventional building	Orientation, insulation, shading, ground cooling, daylighting, passive ventilation	WBCSD (2008)
Demonstration office in Beijing	65kWh/m²/yr electricity 78kWh/m²/yr total	60	Similarly equipped office in Beijing with central air conditioning	Optimized building form and orientation, improved windows and chillers, reduced window area, simple daylighting scheme	P. Xu et al (2007)
Ministry of Energy, Water & Communications Building, Putrajaya, Malaysia	100kWh/m²/yr total onsite energy use based on computer simulation	64	Conventional design (275kWh/m²/yr)	Daylighting, insulation in walls and roof, energy-efficient equipment, energy management, room temperature 24°C instead of 23°, tight building	Roy et al (2005)
Shanghai Eco-Building, National Construction Department	48kWh/m²/yr heating + cooling onsite energy use based on computer simulations	69	Conventional design (155kWh/m²/yr)	Window shading devices, advanced glazing, highly insulated envelope, natural ventilation	Zhen et al (2005)
Council House 2, Melbourne, Australia	35kWh/m²/yr total onsite energy use	82 electricity, 87 gas	Previous council house	High thermal mass, daylighting, night ventilation, PV and solar thermal panels	WBCSD (2008)

Note: [a] Available from Stephen Pope, Natural Resources Canada; [b] Gross energy use, excluding contribution from PV.

Voss et al (2007) present data on the measured energy use in 21 passively cooled commercial and educational buildings in Germany. The passive cooling techniques involve earth-to-air heat exchangers (nine cases), slab cooling directly connected to the ground via pipes in boreholes or connected to the groundwater (nine cases) and some form of night ventilation (16 cases), along with a limited window to wall ratio (0.27–0.43) and external sun shading. The buildings also have a high degree of insulation and many have triple-glazed windows. Nine of the buildings have total onsite energy use of 25–55kWh/m²/yr and ten had 55–110kWh/m²/yr energy use, compared to 175kWh/m²/yr for conventional designs, so the savings is up to a factor of seven. Three buildings have a heating energy use less than 20kWh/m²/yr and eight have a heating energy use of 20–40kWh/m²/yr, compared to a typical heating energy use of 125kWh/m²/yr.

The Passive House Standard (a heating requirement of no more than 15kWm/m²/yr, originally developed for residential buildings) has been achieved by many different kinds of commercial, institutional and educational buildings in Europe. Examples include a nursing home (Nordhoff, 2009), over 20 gymnasia (Kah et al, 2009), schools (Bretzke, 2008, 2009; Baumgärtner, 2009), a kindergarten (Jordan, 2009), a daycare centre (Bär, 2009) and a savings bank (Endhardt, 2009). Passive House envelope standards were applied to an indoor swimming pool, resulting in 60–70 per cent lower total energy use compared to a pool meeting the current German standards (Schulz, 2009).

There are few documented examples from measurements in real buildings of the energy savings that can be obtained in commercial buildings in tropical climates or from developing countries, so we augment the hot climate examples in Table 4.29 with results from computer simulations. Figure 4.80 shows the simulated energy use for reference commercial buildings and with modest improvements in the thermal envelope and in the heating and cooling systems in New York, New Delhi, Beijing and Madrid. Heating savings range from 85–100 per cent, cooling

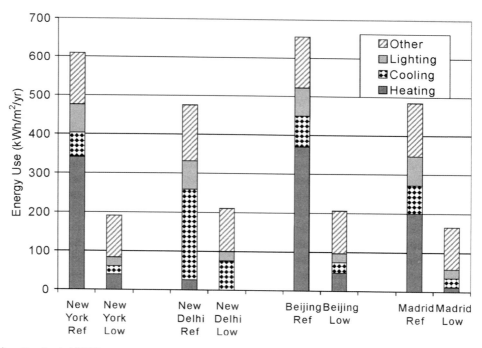

Source: Data from Huovila et al (2007)

Figure 4.80 *Simulated energy use in reference commercial buildings and buildings with modest improvements to the thermal envelope and in heating and cooling equipment (as detailed in Table 4.28) in various cities*

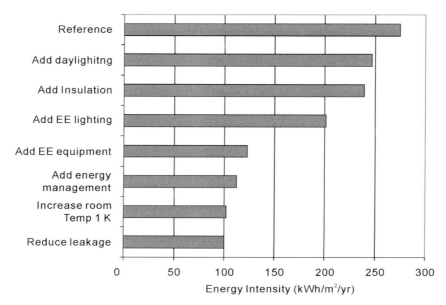

Note: EE = energy-efficient
Source: Simplified from Roy et al (2005)

Figure 4.81 *Simulated energy intensity for an office building in Malaysia as built according to conventional practice (Reference case) and for various cumulative upgrades*

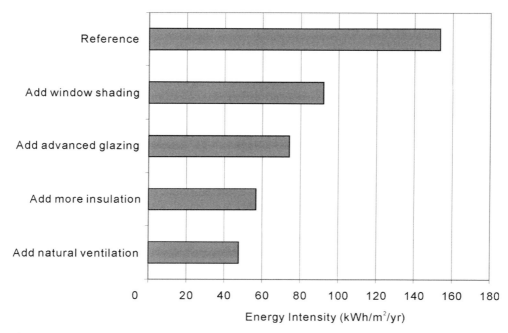

Source: Data from Zhen et al (2005)

Figure 4.82 *Simulated energy intensity for an office building in Beijing as built according to conventional practice (Reference case) and for various cumulative upgrades*

energy savings range from 50–60 per cent and the lighting saving is 75 per cent. Total energy savings are 40–60 per cent, but are limited by the assumption of only an 8–10 per cent saving in other energy uses. Figures 4.81 and 4.82 show the energy intensity for an office buildings in Malaysia and Beijing as simulated for a reference case and for cases with successively more energy efficiency features. The Malaysian building with all the energy features considered consumes 36 per cent the energy of the reference building (a saving of 64 per cent). For the Beijing building, window shading plus high-performance glazing alone reduce the total energy use by 50 per cent. When augmented insulation and natural ventilation are added, the total savings is 69 per cent.

Cost of high-performance commercial buildings

The essential first step in creating a building with very low energy use is to prescribe a high-performance envelope – a high degree of insulation, windows with low U-values in cold climates and low solar heat gain in hot climates, external shading and low air leakage through attention to the details in the design and assembly of the various envelope elements. This must be accompanied by the choice of a building form and the use of operable windows so as to facilitate passive ventilation and cooling during at least part of the year, and to maximize the opportunities for daylighting. In this way, mechanical systems (ventilation motors and fans, pumps for hot and cold water, chillers and boilers) and the associated electrical subsystems can all be vastly

downsized, leading to savings that can sometimes entirely offset the additional cost of the high-performance envelope. The high-performance envelope also permits the use of alternative, low-energy systems such as radiant floor heating, chilled ceiling cooling and displacement ventilation, which would not be viable with higher heating and cooling loads. As previously noted (see subsection 4.2.2), high-performance windows eliminate the need for perimeter radiators or heating vents and the associated plumbing or ductwork.

As an example of the cost savings with advanced, energy-efficient designs, Table 4.30 gives a breakdown of capital costs for commercial buildings in Vancouver, Canada, having conventional windows (double-glazed, air-filled, low-e with $U = 2.7\text{W/m}^2/\text{K}$ and SHGC = 0.48) and a conventional heating/cooling system, and for buildings with moderately high-performance windows (triple-glazed, low-e, argon-filled with $U = 1.4\text{W/m}^2/\text{K}$ and SHGC = 0.24) and radiant-slab heating and cooling. The high-performance building is 9 per cent less expensive to build than a comparable conventional building, while using about half the energy.

As another example, Mumma (2001) compared the cost of a conventional (VAV) HVAC system and a system with chilled ceiling panels and use of 100 per cent outside air for ventilation with sensible and latent heat exchangers, and found the alternative system to cost about the same or slightly less than the conventional system. The alternative system can readily reduce energy use by 30 per cent.

Of two recently built kindergartens in Baden bei Wien, Austria, one designed to the Passive House

Table 4.30 *Comparison of component costs for a building with a conventional VAV mechanical system and conventional (double-glazed, low-e) windows with those for a building with radiant slab heating and cooling and high-performance (triple-glazed, low-e, argon-filled) windows, assuming a 50 per cent glazing area to wall area ratio*

Building component	Conventional building	High-performance building
Glazing	$140/m²	$190/m²
Mechanical system	$220/m²	$140/m²
Electrical system	$160/m²	$150/m²
Tenant finishings	$100/m²	$70/m²
Floor-to-floor height	4.0m	3.5m
Total	**$620/m²**	**$550/m²**
Energy use	180kWh/m²/yr	100kWh/m²/yr

Note: Costs are in 2001 Canadian dollars for the Vancouver market in 2001, are given per m² of floor area and are based on fully costed and built examples over a three-year period.
Source: Geoff McDonell (Omicron Consulting, Vancouver), personal communication, December 2004, and McDonell (2003)

Standard and the other designed to a standard of 65kWh/m²/yr heating demand, the Passive House kindergarten cost less to build while meeting the same structural soundness standards (Jordan, 2009). Bretzke (2009) notes that, in other cases, the extra cost of buildings meeting the Passive House Standard is smaller than the difference in cost between design variants for the same building.

A recently completed science building at Concordia University, Montreal, with offices, classrooms and 250 fume hoods, achieved a 45 per cent reduction in energy use relative to the current practice through the following measures (Lemire and Charneux, 2005):

- motion detectors to shut off lights after an adjustable delay when the space is unoccupied, with a signal sent to the building automation system to reduce ventilation rates;
- a reduction of lab ventilation rates from 10ACH (air changes per hour) to 6ACH when unoccupied during the day and to 3ACH when unoccupied during the night;
- a reduction of non-lab ventilation rates from 6ACH to 3ACH when unoccupied during the day and to zero when unoccupied during the night;
- combining fume hood exhaust ducts into a manifold;
- heat recovery between exhaust and supply air using a run-around glycol loop;
- use of a low-temperature (30–40°C) water-heating loop to recover waste heat from various pieces of equipment;
- use of variable speed drives on all pumps and fans;
- modestly better electric lighting.

These measures increased the total cost of the building by 2.3 per cent ($1,356,000 out of $59,500,000) while yielding an annual energy cost saving of $854,000, for a simple payback time of 19 months.

Another indication that larger energy savings can cost less than smaller energy savings is provided by a survey of the incremental cost and energy savings for 32 buildings in the US by Kats et al (2003). These buildings meet various levels of the LEED (Leadership in Energy and Environmental Design) standard. Summary results are given in Table 4.31. The energy savings are broken into reductions in gross energy demand and reductions in net energy demand including onsite generation (by, for example, PV modules), which tends to be expensive. The cost premium is the total cost premium required to meet the various LEED standards and so includes the cost of non-energy features as well. Nevertheless, average costs are less than 2 per cent of the cost of the reference building and are smaller on average for buildings with 50 per cent savings in net energy use than for buildings with 30 per cent savings.

Measured performance information on ten buildings in the German SolarBau programme where at least one year of data were available by 2003 is given in Wagner et al (2004). Five of the ten buildings achieved the 100kWh/m²/yr primary energy target (compared to 300–600kWh/m²/yr for conventional designs), but no building used more than 140kWh/m²/yr of primary energy. Additional costs are reported to be comparable to the difference in cost between alternative standards for interior finishings.

The final example presented of the beneficial economics of energy-efficient buildings is one of the first buildings to be built on the new Oregon Health and Science University, River Campus, and completed in 2006. This 16-storey building is expected to achieve an energy saving of 60 per cent relative to ASHRAE

Table 4.31 *Energy savings relative to ASHRAE 90.1-1999 and cost premium for buildings meeting various levels of the LEED standard in the US*

LEED level	Sample size	% energy savings, based on		Cost premium (%)
		gross energy use	net energy use	
Certified	8	18	28	0.66
Silver	18	30	30	2.11
Gold	6	37	48	1.82

Source: Kats et al (2003)

Table 4.32 *Economics of the new Oregon Health and Science University building*

Item	
Total project cost	$145.4 million
Energy efficiency features	$975,000
PV system	$500,000
Solar thermal system	$386,000
Commissioning	$150,000
Total	**$2,011,000**
Savings in mechanical systems	$3,500,000
Value of saved space	$2,000,000
Net cost	**−$3,489,000**
Estimated annual operating cost savings	$600,000

Source: Interface Engineering (2005)

90.1-1999 through such measures as hybrid ventilation using the stack effect in stairwells, solar preheating of office ventilation air, heat recovery with laboratory ventilation, radiant heating and cooling, demand-controlled displacement ventilation, PV modules as exterior shading, accurate equipment sizing and commissioning. Incremental costs or upfront savings are given in Table 4.32. Cost savings due to downsizing of the mechanical systems permitted by the efficiency measures exceeded the cost of the efficiency measures. A further credit arises from the space saved due to more efficient and downsized mechanical systems. The net result is a construction cost saving of about $3.5 million out of an original budget of $145.4 million and operating cost savings of $600,000 per year.

4.11.3 Adoption of the Passive House Standard by municipal governments

A number of cities in Europe have adopted the Passive House Standard for some or all categories of municipal buildings. The City of Frankfurt passed a resolution in 2007 requiring that all new municipal buildings (including schools, daycare centres and nursing homes) meet the Passive House Standard. A similar resolution was passed by the City of Wels, Austria, in 2008 (Pany, 2009). The Passive House Standard is mandatory for social housing in Vorarlberg, Austria (Bähr and Sambale, 2009). Close to the Passive House Standard must be met by municipal buildings (including public housing) in Freiburg (Hoppe, 2009), while the City of Hanover resolved in 2005 to build of all new daycare centres to the Passive House Standard (Bär, 2009).

4.12 Energy savings through retrofits of existing buildings

The preceding discussion demonstrates the possibility of achieving energy savings of 50–75 per cent in new commercial buildings and of 75–90 per cent in new residential buildings, relative to current practice in most countries. However, there is a large stock of existing, inefficient buildings, most of which will still be here in 2025 and even 2050. Our long-term ability to reduce energy use depends critically on the extent to which energy use in these buildings can be reduced when they are renovated. The equipment inside a building, such as the furnace or boiler, water heater, appliances, air conditioner (where present) and lighting is completely replaced over time periods ranging from every few years to every 20–30 years. The building shell – walls, roof, windows and doors – last much longer. There are two opportunities to reduce heating and cooling energy use by improving the building envelope: (1) at any time prior to a major renovation, based on simple measures that pay for themselves through reduced energy costs; and (2) when renovations are going to be made, including replacing windows and roofs.

The term 'retrofit' refers to the extensive upgrading of the building envelope or systems some time after the building has been built (beyond the routine replacement of worn out mechanical equipment), while the term 'renovation' refers to the renewal of building components in response to deterioration over time, and may or may not be accompanied by an improvement in energy performance levels. Retrofit costs will be lowest if the retrofit is done as part of a routine renovation. Some of the energy-saving measures that might be taken during renovation might be done because of other benefits that they provide (such as providing more uniform temperatures and addressing moisture problems). Thus, some of the energy-saving measures might not be justifiable based on energy-cost savings alone, but can still be worthwhile. The time when the building envelope is upgraded is a good time to replace the heating system, as this provides an opportunity for downsizing the system, or for switching to a more efficient heating system. Conversely, if the heating or cooling system needs to be replaced, that could be a good time to undertake needed renovations if accompanied by an improved thermal envelope, as the cost savings on downsized heating and cooling equipment can offset a portion of the cost of the renovation.

4.12.1 Techniques for adding or upgrading insulation

Where cavities exit in a wall, cellulose insulation can be blown into the cavity. Where there is no wall cavity, or where additional insulation is required, there are three options: to apply external insulation; to remove the interior plaster, add interior insulation and rebuild the finished wall; or to apply internal insulation directly over the interior wall. The advantage of external insulation is that it does not subtract from the available interior space and it can be applied without disrupting the occupants of the building, but it can be awkward and expensive to apply (especially several floors above ground level). Alternatively, external insulation and finishing systems (EIFSs) provide an excellent opportunity to improve the appearance of unsightly buildings at the same time that the insulation and airtightness are improved. This is because of the wide range of external finishes that can be applied, ranging from stone-like to a finish resembling aged plaster.

Aerogel insulation can be directly applied to the interior wall surface (so entailing minimal disturbance), and it entails only a small reduction in interior space as it has one third the thermal conductivity of regular insulation (0.0012W/m/K instead of about 0.004W/m/K). Aerogel panels can be bonded to various kinds of facings to give different interior finishes (see www.spacetherm.com). Vacuum insulation panels can also be used strategically, where space is limited.

4.12.2 Conventional retrofits of residential buildings

Cost-effective measures that can be undertaken without a major renovation of residential buildings include: sealing points of air leakage around baseboards, electrical outlets and fixtures, plumbing, the clothes dryer vent and door and window joists; weather stripping of windows and doors; adding insulation in attics or wall cavities; and sealing of heating and cooling ducts. A Canadian study found that the cost-effective energy savings potential ranges from 25–30 per cent for houses built before the 1940s, to about 12 per cent for houses built in the 1990s (Parker et al, 2000). Rosenfeld (1999) refers to an 'AeroSeal' technique that he estimates is already saving $3 billion/year in energy costs

in the US. Without proper sealing, homes in the US lose, on average, about a quarter of the heating and cooling energy through duct leaks in unconditioned spaces – attics, crawl spaces and basements. Studies summarized by Francisco et al (1998) indicate that air-sealing retrofits alone can save an average of 15–20 per cent of annual heating and air conditioning energy use in US houses. Additional energy savings would arise by insulating ductwork in unconditioned spaces.

In a carefully documented retrofit of four representative houses in the York region of the UK, installation of new window and door wood frames, sealing of suspended timber ground floors, and repair of defects in plaster reduced the rate of air leakage by a factor of 2.5–3.0 (Bell and Lowe, 2000). This, combined with improved insulation, doors and windows, reduced the heating energy required by an average of 35 per cent. Bell and Lowe (2000) believe that a reduction of 50 per cent could be achieved at modest cost using well-proven (early 1980s) technologies, and a further 30–40 per cent reduction through additional measures.

Amstalden et al (2007) assess the conditions under which retrofit measures with various energy savings become profitable in buildings falling in various age categories in Switzerland. For old (pre-1948) buildings, measures that reduce energy use from 700kWh/m²/yr to about 320kWh/m²/yr are profitable (net present value >0) if the cost of heating fuel averages 80 cents/litre (1 Swiss franc/litre) or more (this is based on 40 years of energy cost savings discounted at 3.5 per cent/year). For buildings built between 1948 and 1975, a saving of about 50 per cent is profitable if fuel costs 56 cents/litre or more. For buildings constructed after 1975, no retrofit measure is profitable at current energy prices.

Verbeeck and Hens (2005) calculate that 55–60 per cent of heating energy could be typically saved through upgrades of residential buildings in Belgium, with an economic payback of six to nine years (given a heating energy cost of €17/GJ, equivalent to heating oil at 61 eurocents/litre). Tommerup and Svendsen (2006) find that insulation of a typical multi-storey residential building constructed in Denmark in the 1960s and installation of mechanical ventilation with heat recovery can reduce heating energy use by 90 per cent, while the same measures in a single-family house from the 1960s saves 75 per cent, in both cases with economic paybacks of about 30 years at current energy prices. However, the renovation in these cases provides other benefits (improved air quality and comfort due to more uniform interior temperatures) besides saving energy.

4.12.3 Renovations of residential buildings to the Passive House Standard or thereabouts

Table 4.33 lists examples of buildings in Europe that have met or almost met the Passive House Standard after major renovations. The first example is the renovation of some 1950s multi-unit residential buildings in Ludwigshafen, Germany, by the German company BASF (Badische Anilin und Soda Fabrik), which owns the buildings and makes some of the components used in EIFSs (see www.3lh.de). The EIFS in combination with other measures achieved a factor of eight measured reduction in the heating energy use. The external insulation consisted of 20cm of pentane-blown polystyrene on walls and 30cm on the roof (16cm between rafters and 14cm on top). The other measures included replacement of the existing windows with triple-glazed, krypton-filled windows having a U-value of 0.8W/m²/K, reduction in air leakage by a

factor of seven, installation of controlled ventilation with heat recovery (85 per cent effectiveness), and use of spray-on plaster with microencapsulated phase-change material (paraffin wax in 50μm bubbles, described above in subsection 4.5.1) in combination with night ventilation in order to limit daytime temperature peaks during the summer. Figure 4.83 presents photographs of one building before and after the retrofit, while Figure 4.84 shows installation of the external insulation and spray-on plaster.

The final example in Table 4.33 is from the SOLANOVA ('Solar-supported, integrated eco-efficient renovation of large residential buildings and heat-supply-systems') project of the European Commission, which started in January 2003. The project goal is to provide best-practice examples of the renovation of large, residential buildings in Eastern Europe which, at present, are being renovated with only minimal improvements in energy intensity. In 2005, one seven-storey panel building in the Hungarian town of Dunaújváros was renovated as part of this project. Heating energy consumption decreased from 220kWh/m²/yr before the retrofit to a measured consumption of 30kWh/m²/yr over a two-year period

Table 4.33 *Documented examples of deep savings in heating energy use through renovations of buildings*

Building and location	Year built	Year renovated	Energy intensity (kWh/m²/yr)			Reference
			Before	After	Metric	
Apartment buildings in Ludwigshafen, Germany	1950s	2001	250	30 (m)	System	'3-litre house' at www.basf.com
Villa in Purkersdorf, Vienna, Austria	Late 19th century	2008	–	20 (m)	System	Reinberg (2009)
Two apartment buildings on Tevesstrasse, Frankfurt	1950s	2005	290	17 (c)13.6 (m)ᵃ	Load	Kaufmann et al (2009), Peper and Grove-Smith (2009)
18-unit apartment block, Brogården, Sweden	1970	2009	115	30 (c)	System	Janson and Wall (2009)
24-unit apartment block, Zirndorf, Germany	1974	2009	116	35 (c)	Load	Bodem (2009)ᵇ
Apartment block, Ludwigshafen, Germany	1965	2006	141	18 (m)	Load	Peper (2009), www.passiv.de
50-unit apartment, Linz Austria	1958	2006	179	13.3 (c)	Load	www.hausderzukunft.at/results.html/id3951
Single-family house, Pettenbach, Austria		2005	280	14.6 (c)	Load	www.hausderzukunft.at/results.html/id3955ᶜ
Seven-storey apartment block, Dunaújváros, Hungary		2005	220	20–40	System	www.solanova.eu

Note: ᵃ Adjusted to an indoor temperature of 20°C; ᵇ €435/m², €300/m² without reconstruction of balconies; ᶜ 16 per cent greater cost than conventional renovation. Heating energy intensities before and after renovation are given, where c = calculated and m = measured. In some cases the given energy intensity is the heating demand (load) and in other cases it is the energy used by the heating system. The choice for a given case is indicated in the column labelled 'metric'.

Source: Wolfgang Greifenhagen, BASF

Figure 4.83 *View of a multi-unit residential building owned by BASF (a) before and (b) after a retrofit that reduced heating energy used by a factor of eight*

Source: Wolfgang Greifenhagen, BASF

Figure 4.84 *Installation of external insulation (upper) and of spray-on plaster with microencapsulated phase-change material (lower) in the building illustrated in Figure 4.83*

(a reduction of 86 per cent). Overheating in the summer was one of the worst characteristics of the original building, so triple-glazed windows with internal venetian blinds were used on the south- and west-facing walls, giving a g-value of 0.1 while retaining a reasonable U-value of $1.1W/m^2/K$ (in the absence of the venetian blinds, the g-value would have been 0.55). Mechanical ventilation was provided to each individual flat with a real heat recovery of 82 per cent. The investment cost was €240/m² plus taxes. The time to pay back the initial investment, based on energy-cost savings only and at current energy prices, is 17 years. However, a grungy and uncomfortable building was turned into an attractive and comfortable building at the same time, as seen from the photographs in Figure 4.85. In Eastern Europe,

Source: Andreas Hermelink, Centre for Environmental Systems Research, Kassel, Germany

Figure 4.85 *Photographs of an apartment block in the Hungarian town of Dunaújváros before (upper) and after (lower) a retrofit to close to the Passive House Standard*

about 100 million people live in large panel buildings, but results demonstrated in Dunaújváros can be transferred to the large stock of Western European panel buildings as well.

A more radical renovation for old buildings involves replacement of the existing upper floor and roof with prefabricated units that are lowered into place by crane. Figure 4.86 shows a roof unit being lowered into place on a 100-year old residential building in Zurich, along with photographs of the building before and after renovation. The roof units contain a total of 15.5m² of solar thermal collectors, which supply heat

Note: The renovation entailed removal of the original roof and its replacement with prefabricated roof elements with built-in solar thermal collectors, one of which is shown being lower by crane in the lower panel.

Source: Zimmermann (2004)

Figure 4.86 *Photographs of a 100-year old residential building on Magnusstrasse in Zurich before and after a renovation (upper panel) that reduced total energy use by a factor of six (from 233kWh/m²/yr to 37kWh/m²/yr)*

Source: Binz and Steinke (2005)

Figure 4.87 *Use of vacuum insulation panels in prefabricated dormer units used in a building renovation in Zurich, showing (a) fabrication, (b) mounting and (c) final result*

to a 2600-litre hot-water tank for space heating, supplemented by individual wood-burning stoves when the outside temperature drops below −2°C (Viridén et al, 2003). In this case, heating energy use as measured over a two-year period was reduced by 88 per cent (from 165kWh/m²/yr to 19kWh/m²/yr) and total energy use by 84 per cent (from 233kWh/m²/yr to 37kWh/m²/yr). Single-glazed windows (10 per cent of the façade area) were replaced with windows with a U-value of 0.75W/m²/K, 70 per cent of the façade was insulated to a U-value of 0.15W/m²/K, and 30 per cent of the façade could not be touched. Figure 4.87 illustrates the use of prefabricated dormer units containing VIPs in the renovation of old buildings in Zurich (VIPs are described above in subsection 4.2.1).

Bastian (2009) assessed the economics of upgrading the thermal performance of an uninsulated 1950s apartment block in Germany with a heating energy demand of 380kWh/m²/yr (including losses from heat generation and distribution in the building). Upgrading to the current German standard would reduce the heating energy demand to 85kWh/m²/yr. However, he estimates that the economically optimal upgrade

(assuming a real interest rate of 3.27 per cent/year, heating energy cost of €0.0659/kWh, and a time horizon of 20 years) would reduce the heating energy demand to 26kWh/m²/yr (a 93 per cent reduction) (he also outlines procedures for adding external insulation to walls first while making provisions to replace the windows later). The renovation of large apartment buildings to the current national building code typically costs €370–450/m² of total floor area, while renovation to the Passive House Standard or very close to it costs €450–600/m². By comparison, a new apartment costs about €1600/m² in western Austria (Ploss, 2008).

We close with a brief discussion of a retrofit programme that has been proposed for the large stock of high-rise apartments built during the 1950s and 1960s in Toronto. Table 4.34 gives the current electricity and natural gas use for a typical building and after a series of upgrades according to computer simulations. Also given is the computed payback time and internal rate of return (based on the estimated costs of the various measures and the associated energy savings) assuming an interest rate of 2.5 per cent/year and a rate of increase in energy costs of 6.0 per cent/

Table 4.34 *Current and projected energy use (kWh/m²/yr) after various upgrades of a typical pre-1970 high-rise apartment building in Toronto*

Measure	Natural gas		Electricity	Primary energy	Cost ($/m²)	Payback (years)	IRR (%/yr)
	Heating	DHW					
Current building	203	36	71	443			
Roof insulation[a]	184	36	70	420	13	11.4	11.3
Cladding upgrade[b]	167	36	69	398	44	18.1	3.4
Window upgrade[c]	122	36	64	336	73	13.5	9.2
Balcony enclosure[d]	122	36	68	345	121	21	4.3
All of the above	47	36	64	252	199	18.6	5.6
Boiler upgrade[e]	118	36	70	347	23	5.5	23
HRV[f]	136	36	68	362	17	7.8	25.8
Water conservation[g]	203	25	70	430	5	3.4	35.1
Parkade lighting[h]	203	36	70	440	0	4.4	28
All of the above	9.4	25	59	185	257	16.9	6.7
Above with 50% less tenant electricity[i]	24.1	25	29	128			

Note: [a] Addition of RSI 3.5 (R20) to roof with an initial RSI of 0.5, giving a final U-value of 0.25W/m²/K. [b] Addition of RSI 2.8 (R16) to walls with an initial RSI of 0.5, giving a final U-value of 0.30W/m²/K. [c] Replace single-glazed windows (U = 5.2W/m²/K) with standard double-glazed windows (whole window U = 2.28W/m²/K). [d] Enclose balconies with glazing having same U-value as replacement window. [e] Replace existing boilers (50–60 per cent efficiency) with condensing multi-stage boilers with 93 per cent efficiency. [f] Install central heat recovery ventilator (≥70 per cent effectiveness) with ductwork to individual suites. [g] 30 per cent savings from water-efficient fixtures and appliances. [h] Occupancy sensors and controls placed on two thirds of light fixtures in the underground parking (as is now common in new buildings in Toronto), rather than leaving all the lights on all of the time. [i] Hypothetical 50% reduction in electricity use, where half of the electricity saved is assumed to have contributed an internal heat gain during the heating season. IRR = internal rate of return. Costs are investment costs in Canadian dollars per m² of rentable floor area, where (as of June 2009), Cdn$1 = US$0.9 = €0.63.
Source: Kesik and Saleff (2009)

year (starting from $12.39/GJ for natural gas and 9.3 cents/kWh for electricity). The full package of retrofit measures (second last line) results in a 95 per cent saving in heating energy use (down to $9.4 kWh/m^2/yr$) and a 33 per cent saving in hot water energy use, but only an 18 per cent reduction in electricity use because plug loads by the apartment tenants are unchanged. If we assume a reduction of 50 per cent in electricity loads directly controlled by the tenants (based on the potentials discussed earlier), the overall reduction in primary energy use is about 75 per cent. The estimated simple payback time for the entire retrofit package is 16.9 years, with an internal rate of return (defined in Appendix D) of 6.7 per cent.

4.12.4 Conventional retrofits of institutional and commercial buildings

There are numerous published studies showing that energy savings of 50–75 per cent can be routinely achieved in commercial buildings through aggressive implementation of integrated sets of measures. These savings can often be justified in terms of the energy-cost savings alone, although in other cases full justification requires consideration of a variety of less tangible benefits. In the early 1990s, the Pacific Gas and Electric Company (PG&E) in California sponsored a $10 million demonstration of advanced retrofits. The project was entitled the 'Advanced Customer Technology Test' (ACT²). In six of seven retrofit projects, an energy saving of 50 per cent was obtained with a modest payback time; in the seventh project, a 45 per cent energy saving was achieved. For Rosenfeld (1999), the most interesting result was not that an alert, motivated team could achieve savings of 50 per cent with conventional technology, but that it was very hard to *find* a team competent enough to achieve these results!

Other, recent examples that are documented in the published literature include:

- a projected saving of 30 per cent of total energy use in 80 office buildings in Toronto through lighting upgrades alone (Larsson, 2001);
- a realized saving of 40 per cent in heating + cooling + ventilation energy use in a Texas office building through conversion of the ventilation system from one with constant airflow to one with variable airflow (Liu and Claridge, 1999);
- a realized saving of 40 per cent of heating energy use through the retrofit of an 1865 two-storey office building in Athens, where low energy use was achieved through some passive technologies that required the cooperation of the occupants (Balaras, 2001);
- a projected saving of more than 50 per cent of heating and cooling energy for restaurants in cities throughout the US by simply optimizing the ventilation system (Fisher et al, 1999);
- a projected 51 per cent saving in cooling + ventilation energy use in an institutional building complex in Singapore through simple upgrades to the existing system (Sekhar and Phua, 2003);
- a realized saving of 74 per cent in cooling energy use in a one-storey commercial building in Florida through duct sealing, chiller upgrade and fan controls (Withers and Cummings, 1998);
- a realized saving of 50–70 per cent in heating energy use through retrofits of schools in Europe and Australia (described in the March 1997 issue of the *CADDET Energy Efficiency Newsletter*, published by the International Energy Agency);
- realized fan, cooling and heating energy savings of 59 per cent, 63 per cent and 90 per cent, respectively, in buildings at a university in Texas, roughly half due to standard retrofit and half due to adjustment of the control-system settings (which were typical for North America) to optimal settings (Claridge et al, 2001);
- average realized savings of 68 per cent in natural gas use after conversion of ten US schools from non-condensing boilers producing low-pressure steam to condensing boilers producing low-temperature hot water, and an average saving of 49 per cent after conversion of ten other US schools from high- to low-temperature hot water and from non-condensing to condensing boilers (Durkin, 2006);
- projected savings of 36–77 per cent through retrofits of a variety of office types in a variety of European climates (Hestnes and Kofoed, 1997, 2002; Dascalaki and Santamouris, 2002);
- a projected saving of 48 per cent from a typical 1980s office building in Turkey through simple upgrades to mechanical systems and replacing

existing windows with low-e windows having shading devices, with an overall economic payback of about six years (Çakmanus, 2007);

- projected savings of 30–60 per cent in cooling loads in an existing Los Angeles office building simply by operating the existing HVAC system in a manner so as to make maximum use of night cooling opportunities (Armstrong et al, 2006).

Leaky ducts can be a major source of energy waste in commercial buildings. This source of energy waste can be significantly reduced with the aeroseal technique, described above for single-unit houses, as it has been successfully applied to the much larger ducts (and cracks) found in large commercial buildings (Modera et al, 2002).

A significant potential area for reduced energy use in existing buildings is through replacement of existing curtain walls, or upgrades of existing insulation and windows. Given the current frenzy constructing nearly all-glass buildings but not even using high-performance glazing, replacing existing glazing systems and curtain walls will be an essential future activity if deep reductions in heating and cooling energy use are to be achieved. Recently, the curtain walls were replaced on the 24-storey 1952 Unilever building (Lever House) in Manhattan (see www.som.com/content.cfm/lever_house_curtain_wall_replacement), so there seem to be no major technical problems in undertaking complete curtain wall replacements on high-rise office buildings.

4.12.5 Solar retrofits of residential, institutional and commercial buildings

Solar retrofitting involves reconfiguring a building so that it can make direct use of solar energy for heating, cooling and ventilation. The now-completed Task 20 of the International Energy Agency's *Solar Heating and Cooling (SHC)* implementing agreement was devoted to solar retrofitting techniques.

Solar renovation measures that have been used are:

- installation of roof- or façade-integrated solar air collectors;
- installation of roof-mounted or integrated solar DHW heating;
- installation of transpired solar air collectors;

- advanced glazing of balconies, integrated with the ventilation system of apartments;
- installation of external transparent insulation, with or without external shading;
- a transparent insulation wall in place of glazing for daylighting in an industrial facility (also eliminating problems of glare);
- construction of a second, glass façade over the original façade.

Table 4.35 provides information on a number of solar renovation projects completed in Europe under IEA SHC Task 20. Further information can be found in the indicated references or in Voss (2000). The glazed balconies include a solar collector mounted on the balcony wall below the balcony windows, so that air can be preheated before entering the balcony at the base of the window. Mechanical ventilation then draws the air into the apartments. This air is centrally vented to the outside so that heat can be recovered and used to heat water with a heat pump (this is an alternative to transferring heat from outgoing air to incoming ventilation with a heat exchanger, which is not possible nor necessary in this case). Energy savings in the projects were as high as 70 per cent, while the cost of solar heat (if all of the expenditure is treated as an energy-saving measure) is in the range of 20–40 eurocents/kWh. However, this is an artificially high cost, as the renovations would be eventually needed even if there were no energy savings.

Construction of a second façade can be used to save buildings that would otherwise need to be eventually demolished or undergo future expensive repairs due to water damage arising from poor workmanship at the time they were built. An example is the Telus Building in Vancouver, Canada, shown in Figure 4.88 before and after the construction of a second, glass façade over the original façade.[22] Transparent insulation (TI) can also be applied to old walls with substantial thermal mass but lacking interior insulation, with an air gap between the insulation and original wall that would be used in the same ways as with a glass façade. In walls without significant thermal mass, an opaque, perforated solar wall can be installed on the non-glazed portion of the façade.

Another solar renovation technique is to fully enclose balconies in glazing, reducing heat loss from the building as well as serving to preheat ventilation air during the day. An example of such a renovation in

Table 4.35 *Summary information on solar renovation projects performed as part of IEA SHC Task 20*

Location	Building description	Dates		Measures implemented	Space Heating Energy use (kWh/m²/yr)		Investment (€/m² floor area)	Cost of saved energy (€/kWh)	Reference
		Built	Renovated		Before	After			
Jambes, Belgium	Eight-storey apartment	1976	1990s	Glazed balconies	64	47			Boonstra and Thijssen (1997)
Perwez, Belgium	Single-family terraced houses	1800s	1990s	Added two level greenhouse on southeast façade	Almost 40% savings				Boonstra and Thijssen (1997)
Aalborg, Denmark	Eight apartments in four-storey building	1900	1996	Preheat ventilation air in ventilated solar walls, roof-integrated DHW solar collectors, demand-controlled ventilation, low-e glazing	230	70	2780[a]		Boonstra and Thijssen (1997)
Freiburg, Germany	Multi-family (eight units)	1950s	1989	Standard insulation of all façades, roof and cellar; low-e blinds; 120m² TI on southeast and southwest façades with adjustable shading	225	43			Haller et al (1997)
Freiburg, Germany	Multi-family Villa Tannheim	1912	1995	Insulation, new windows, 53m² TI on west façade for space heating and DHW, no shading	225	75[b]	633	0.22	Haller et al (1997); Voss (2000)
Salzgitter, Germany	Industrial hall	1940	1995–1997	7500m² TI glazed façade	300	225			Haller et al (1997)
Zaandam, Netherlands	384 apartments in 14-storeys	1968	1997	Solar DHW, glazed balconies, TI walls	145	80	59		Boonstra and Thijssen (1997)

Gardstensbergen, Gothenburg, Sweden	11-storey residential	1975	1990s	Insulation, new windows, roof-integrated DHW solar preheating, glazed balconies, TI	270	160	60		Boonstra and Thijssen (1997)
Rannebergen, Gothenburg, Sweden	188-unit apartment building	1975	1990s	Roof-mounted solar air collector	40% savings				Boonstra and Thijssen (1997)
Hedingen, Switzerland	11-unit multi-family	1971	1994	Standard insulation of all façades, roof and cellar. New low-e windows, 63m² TI on south façade with adjustable shading	245	140	86	0.36	Boonstra and Thijssen (1997); Haller et al (1997)
Niederurnen, Switzerland	12 apartments in four storeys	1971	1996	Insulation, new windows, TI on southwest façade with external blinds	175	105	79	0.43	Boonstra and Thijssen (1997)
Wollerau, Switzerland	Multi-family	1965	1996	Insulation, new windows, TI with fixed horizontal lamellae as sunshades in summer	70kWh/yr/m² of TI		€610/m² of TI		Boonstra and Thijssen (1997)

Note: [a] High cost due to intensive demonstration of advanced technologies. [b] A 75 per cent saving in space heating, along with a 50 per cent saving in DHW energy use. TI = transparent insulation.

Source: Terri Meyer-Boake, School of Architecture, University of Waterloo (Canada)

Figure 4.88 *View of the Telus Building in Vancouver, Canada, (a) before and (b) after construction of a second, glass façade over the first façade*

Zurich, that achieved a factor of four reduction in total energy use, is shown in Figure 4.89.

A final example of solar retrofits is the construction of transpired solar air collectors over window-free portions of the façade of multi-storey buildings. This consists of an exterior metal cladding that performs like a solar panel, but is perforated with millions of small holes. Outside air enters through the holes and is preheated before being fed into the ventilation system. A temperature rise of 20–35K can be achieved on a sunny winter day. A wide range of colours is possible while maintaining an absorptivity of 0.86 or more. Not only is solar energy captured, but heat loss through the wall is reduced (the two effects can be of comparable importance) and the system can be used to de-stratify the indoor air (IEA, 1996). Figure 4.90 illustrates the

Source: Zimmermann (2004)

Figure 4.89 *View of a multi-unit residential building in Zurich (a) before and (b) after a retrofit that reduced total energy used by a factor of four*

Source: Solarwall (www.solarwall.com)

Figure 4.90 *Illustration of a transpired solar collector ('Solarwall') on an apartment building in Windsor, Canada*

installation and final appearance of a transpired solar wall on an apartment building in Windsor, Canada.

4.12.6 Assessment of cost-effective energy retrofit potential

Studies for the European Mineral Wool Manufacturers Association (EURIMA) by the Dutch consulting firm Ecofys indicate that the heating energy consumption in old buildings in Western Europe (EU-15) can be reduced by more than 50 per cent, and by 60–80 per cent in new countries of the EU-27 (Petersdorff et al, 2005a, 2005b), in both cases with no additional cost over a 30-year lifetime. Further analyses by Boermans and Petersdorff (2007) show that the insulation measures consistent with achieving an 85 per cent reduction in heating energy use are similar to the set of

measures that minimizes total costs over a 30-year period. These results depend on the assumed future cost of energy, so these are summarized for the three studies in Table 4.36. The average energy prices are derived from energy prices at the time of the analyses combined with a continuous 1.5 per cent/year increase over the next 30 years. As seen from Table 4.36, the average assumed prices are very conservative; actual future prices could be much greater.

Key summary results of the analysis for the EU-15 and of eight new members of the EU are given in Tables 4.37 and 4.38, respectively. As members of the EU, all of these countries are bound by the European Energy Performance of Buildings Directive (EPBD), which requires each member country to set requirements for the upgrade of the energy performance of buildings with a useful floor area exceeding 1000m^2 when the

Table 4.36 *Average prices of energy (eurocents/kWh, including taxes) assumed for the analyses that are summarized in Tables 4.37 and 4.38*

Type of energy	EU-15, Table 4.36	New EU-8, Table 4.37	Boermans and Petersdorff (2007)	
			Central EU-15	New EU-8
	2002–2032	2002–2032	2006–2036	2006–2036
Natural gas	5.2	3.2	7.7	5.41
Heating oil	4.7	4.6	7.1	7.06
Coal		2.2		
Electricity	11.3	18.8	11.8	9.46
District heating	6.4	4.5	8.1	7.33
Wood	4.2	2.0	4.8	4.35
Mean	5.72	5.02	7.80	6.04
	Alternative measures of the assumed cost of natural gas and heating oil			
Gas as €/GJ	14.3	8.9	21.5	15.0
Oil as €/litre	0.59	0.58	0.88	0.88
Oil as US$/gallon	3.12	3.05	4.68	4.68

Source: Boermans and Petersdorff (2007)

buildings undergo a significant renovation (defined as exceeding 25 per cent of the building value or involving more than 25 per cent of the building envelope). The tables give estimated typical present energy intensities for single- and multi-unit housing as well as the energy performance that would be achieved under the requirements that experts in each country expect to be implemented under the EPBD. However, even at current energy prices, a greater improvement is economically justified in some regions. The energy use after the economically justified upgrade is also shown in Tables 4.37 and 4.38 (as the 'Optimal' case) assuming financing at 6 per cent/year interest over a 30-year period for insulation and windows and over a 20-year period for mechanical equipment, and assuming energy costs given in Table 4.36. The economically justifiable upgrade generally gives a 60–80 per cent reduction in heating energy use. If the energy-related measures are done as part of routine renovation that needs to occur anyway, the added cost has average simple payback times (that is, excluding interest) of three to four years in Southern Europe, seven to eight years in Central (moderate climate) Europe, 6–12 years in the new EU countries and 10–22 years in Northern Europe (where pre-existing insulation levels are the highest). If the energy upgrade is done independently of routine renovation, the payback times are approximately doubled.

Boermans and Petersdorff (2007) estimated the improvements in wall, window, floor and roof U-values

Table 4.37 *Estimated typical heating energy use at present in countries of the EU-15, after renovations expected after implementation of the European Energy Performance of Buildings Directive (BAU) and (for the moderate climate zone) after implementation of renovations that would be economically optimal at current energy prices*

	Energy intensity (kWh/m²/yr)	
	Single-family house	Multi-unit housing
	Cold-climate zone	
Before	165	119
BAU	68	57
Savings	59%	52%
Simple payback (years)	10	22
	Moderate-climate zone	
Before	298	179
BAU	63	43
Optimal	41	29
Savings (BAU)	79%	76%
Savings (optimal)	86%	84%
Simple payback (years)	7	8
	Warm-climate zone	
Before	239	174
BAU	38	24
Savings	84%	86%
Simple payback (years)	4	5

Note: Energy upgrades are assumed in both cases to occur as part of routine renovations.
Source: Petersdorff et al (2005a) and Carsten Petersdorff (personal communication, 2009)

Table 4.38 *Estimated typical heating energy use at present in Eastern European countries after renovations expected after implementation of the European Energy Performance of Buildings Directive (BAU) and after implementation of renovations that would be economically optimal at current energy prices*

	Energy intensity (kWh/m²/yr)	
	Terraced house	Multi-unit housing
Baltic countries		
Before	267	135
BAU	83	64
Optimal	72	54
Savings (%)	73	60
Simple payback (years)	12	11
Poland		
Before	334	163
BAU	94	66
Optimal	67	50
Savings (%)	80	69
Simple payback (years)	9	7
Czech and Slovak Republics, Hungary, Slovenia		
Before	370	176
BAU	79	55
Savings (%)	79	69
Simple payback (years)	7	6

Note: Energy upgrades are assumed in both cases to occur as part of routine renovations.

Source: Petersdorff et al (2005b)

needed in single-family and multi-unit housing in order to reduce heating and cooling energy use by 85 per cent during retrofits (which has been proposed as a possible post-Kyoto target). They also calculated the economically optimal package of U-values, taking into account the effect of lower U-values in reducing both heating and cooling energy requirements. Whether it is economically justified or not to add any insulation at all during retrofits depends on the fixed cost of insulation, but when additional insulation is justified, the optimal insulation level is the same for retrofitted buildings as for new buildings. Consideration of cooling benefits increases the optimal insulation level for walls and roofs but reduces it for floors. They find that the package of measures that achieves an 85 per cent reduction in heating + cooling energy use is generally less stringent than the economically optimal package of measures, assuming the energy costs given in Table 4.36. This implies that the economically justified package of

measures achieves greater than 85 per cent energy savings for typical buildings. However, there will certainly be circumstances where the economically optimal upgrade will produce a substantially smaller saving.

Table 4.39 summarizes a recent IEA-sponsored study by Waide et al (2006) of the potential savings in heating energy use in high-rise residential buildings in countries throughout Europe. Economically optimal upgrades were determined based on local energy costs and local material and labour costs. Countries have been grouped according to climate and socio-economic conditions, and results for a representative building in each group are shown in Table 4.39. Costs are based on the costs that would be incurred beyond those of a normal renovation (so scaffolding costs are not included in the cost of external insulation, and only the additional cost of high-performance windows and boilers compared to standard equipment is considered). The estimated cost-effective savings level is between

Table 4.39 *Characteristics of economically optimal upgrades of high-rise residential buildings in Europe*

Region	Heating savings (%)	Annualized investment cost (€/m²)	Local cost of energy (€cents/kWh)	Cost of saving energy (€cents/kWh)	Simple payback (years)
Portugal, Spain, France, Italy	71	2.61	12.4	2.0	2.7
Ireland, UK, Belgium, Netherlands	81	3.94	8.8	1.9	4.0
Austria, Germany, Denmark, Sweden, Finland	76	4.07	7.1	2.4	5.7
Cyprus, Malta	72	2.08	4.8	1.6	6.1
Czech and Slovak Republics, Hungary, Slovenia	75	2.10	3.0	1.5	8.6
Poland, Baltic states	70	1.95	2.3	1.4	10.5
Romania, Bulgaria	77	1.77	2.0	1.3	10.7
Turkey	78	1.85	1.9	1.9	16.2

Source: Waide et al (2006), with more detailed information at www.euroace.org/highrise

about 70–80 per cent in all regions, with simple payback times ranging from about 3–16 years. The energy costs assumed for this analysis (given in Table 4.36) are lower than assumed in the Ecofys work summarized above.

Finally, Table 4.40 summarizes the envelope U-values required to achieve 80 per cent and 90 per cent

reductions in heating CO_2 emissions for existing buildings in various climate zones in Europe, as determined by Boermans and Petersdorff (2007), and compares these with the values found by Petersdorff (2005a, 2005b) to be cost effective over a 30-year timeframe and with the values used in buildings constructed to the Passive House Standard in Helsinki

Table 4.40 *Envelope U-values needed to achieve 80 or 90 per cent reductions in heating CO_2 emissions in residential buildings in Europe, or found to be cost effective over a 30-year timeframe, and as required by the Passive House standard in Helsinki and Frankfurt*

Climate zone or city	HDD (Kday/yr)	U-value (W/m²/K)				Heating energy (kWh/m²/yr)
		Floor	Wall	Roof	Window	
Values giving an 80% reduction in CO_2 emission						
Southern	1800	1.00	0.65	0.50	1.60	38
Moderate	3000	0.45	0.35	0.25	1.20	65
Eastern	3400	0.40	0.30	0.20	1.20	67
Northern	4500	0.30	0.28	0.20	1.10	40
Values giving a 90% reduction in CO_2 emission						
Southern	1800	0.60	0.40	0.30	1.20	19
Moderate	3000	0.20	0.13	0.10	1.00	33
Eastern	3400	0.25	0.20	0.18	1.00	30
Northern	4500	0.20	0.20	0.12	1.00	19
Values determined to be cost effective over a 30-year timeframe						
Warm	931	0.48	0.48	0.43	2.71	
Moderate	3039	0.41	0.20	0.23	1.68	
Eastern	3747	0.46	0.35	0.23	1.70	55
Cold	4210	0.17	0.17	0.13	1.33	
Values required by the Passive House Standard						
Frankfurt		0.25	0.15	0.11	0.87	13.6
Helsinki	4898	0.08	0.10	0.07	0.67	13.8

Note: Given is the number of HDDs for the climate of each region or city.
Source: Boermans and Petersdorff (2007) for 80 and 90 per cent reduction in CO_2, Petersdorff (2005a, 2005b) for cost-effective values, Pedersen and Peuhkuri (2009) for the Passive House standard in Helsinki, and Kaufmann et al (2009) for the Passive House standard in Frankfurt

(Pedersen and Peuhkuri, 2009) and renovated to the Passive Standard in Frankfurt (Kaufmann et al, 2009). Comparison of the requirements for 80 per cent and 90 per cent emission reductions indicates that substantially greater effort is required to achieve a 90 per cent saving, but the requirements for the 90 per cent case are generally less stringent than the estimated cost-effective U-values. The Passive House values for Helsinki are substantially more stringent than the cost-effective U-values, but are only modestly more stringent in Frankfurt.

4.13 Summary

The potential exists to reduce the energy use of new buildings by factors of four to ten compared to current practice for new buildings in every part of the world. Comprehensive retrofits can usually reduce total energy use by a factor of two to four in existing commercial buildings and have reduced heating energy use by up to a factor of 20 in existing residential buildings compared to current energy use. The keys to achieving these dramatic reductions in new buildings are:

- to focus on a high-performance envelope (appropriately high levels of insulation, high-performance windows, a high degree of airtightness except for controlled openings) so as to reduce heating and cooling loads, with further reductions in heating and cooling loads where possible through optimal building shape and orientation, the use of low-reflectivity surfaces where appropriate, and the use of thermal mass coupled, during hot seasons, with night ventilation;
- to maximize the use of passive solar energy for daylighting, heating, cooling and ventilation, which in turn requires the optimization of glazing area and glazing properties, thermal mass and airflow paths within a building and between the inside and outside of the building;
- to choose the most efficient mechanical *systems* as well as the most efficient energy-using *devices* (such as boilers, chillers, pumps and lighting), appropriately sized and fully commissioned to ensure that they operate as intended;
- to facilitate energy-conscious *human behaviour*.

In cold climates, use of high levels of insulation, high-performance windows and high airtightness coupled with mechanical ventilation and heat recovery (during the heating season) can reduce heating requirements in new buildings by a factor of four or more compared to current standards. The best windows have a heat loss about one tenth that of single-glazed windows (or a fifth that of uncoated double-glazed windows) while admitting as little as a quarter of the incident solar energy into the building as heat but admitting 40 per cent of the visible radiation for daylighting purposes.

In hot dry climates, having large daily temperature variations and intense solar radiation fluxes, moderate levels of external insulation are justified and, combined with good thermal mass and night ventilation, can largely eliminate the need for air conditioning. In regions with hot but not overly humid summers and cold winters, an alternative strategy is to combine thermal mass, external insulation and pre-cooling of ventilation through underground pipes. In most parts of the world, evaporative cooling can create comfortable conditions most of the time, although water can be limiting in the driest regions. Evaporative cooling alone is not effective in hot humid climates, but can be extended to these regions if combined with desiccant dehumidification that, in turn, can be driven with solar thermal energy.

Large amounts of energy are used in modern buildings to circulate fresh air through the building. However, buildings can be designed to achieve most of the required ventilation most of the time using passive driving forces – wind and pressure differences created by differential heating of different parts of the building by the sun. The circulation of outside air through a building often provides adequate cooling, without the need for air conditioners or chillers, but many buildings are hermetically sealed and require mechanical cooling even when it is cool outside. Thus, passive ventilation with outside air can dramatically reduce both ventilation and air conditioning energy use.

Efficient mechanical systems in buildings will use displacement ventilation instead of ventilation through turbulent mixing, and will use radiant ceiling (or floor) heating and cooling instead of smaller wall-mounted radiators or fan-coil units. These systems require 50 per cent or less airflow, and permit warmer temperatures for distributing cooling and cooler temperatures for distributing heat, thereby increasing the efficiency of heating or cooling equipment, extending the range of low-energy cooling techniques (such as the use of cooling towers), and permitting district heating systems to operate at lower temperatures (which increases the overall efficiency of cogeneration at the district heating powerplant).

Appropriate sizing of mechanical equipment not only reduces capital costs but improves efficiency (by 10–20 per cent). Proper commissioning of equipment also improves efficiency by about 20 per cent.

The key to achieving deep (50 per cent or more) savings in energy use is to engage in an IDP, in which all members of the design team brainstorm together at the beginning of the design process, supported by building computer simulation and other specialists. Numerous case studies in many different countries of the world demonstrate that it is possible to create buildings which use only one half to one quarter the energy of conventional buildings, often at less cost than conventional buildings. Where energy-efficient buildings do cost more, the time required to pay back the additional construction cost through reductions in annual energy costs is usually no more than a few years.

Notes

1   The specific heat of any substance, c_p, is the amount of heat required to warm 1kg of the substance by 1K. The subscript p denotes the fact that the specific heat used here pertains to addition of heat at constant pressure, and is included for consistency with more specialized textbooks.

2   If ducts are reduced in cross-sectional area in proportion to the decrease in flow rate, then ΔP in Equation (4.7) is fixed, so the motor energy use decreases in proportion to the decrease in flow rate.

3   The absorption of solar radiation does warm the glazing layer, which increases the emission of infrared radiation by the inner glazing surface to the inside and reduces the conductive and convective heat transfer from the room to the warmed glazing surface. Thus, some of the absorbed solar radiation is indirectly transferred to the room, but the net effect of adding a low-e coating is still to reduce the transfer of solar heat to the inside.

4   An alternative measure of heat pump performance in heating mode is the *heating season performance factor (HSPF)*. This is the ratio of total heat output of a heat pump in Btu divided by electricity used in W-hr. This mixture of metric and non-metric units complicates the computation of system-scale energy use. Since 1Btu = 1055J and 1W-hr = 1J/s × 3600s = 3600J, multiply by 1055/3600 = 0.2931 to get J/J (COP).

5   As with heat pumps in heating mode, mixed Btu-metric measures of performance are, unfortunately, often used: the *seasonal energy efficiency ratio* (SEER), which is the

ratio of total heat removed (Btu) from a building to the energy input (W-hr) over the entire cooling season; and the *energy efficiency ratio (EER)*, which is the instantaneous, steady-state ratio of the rate of heat removal to the rate of energy use, in Btu/hr per Watt. As with HSPF, multiply by 0.2931 to convert to COP values.

6   The term 'dispatchable' originally referred to electricity generation sources that can be turned on or off at will ('dispatched') by the power utility.

7   Like heat pumps, it is common in North America to characterize the performance of an air conditioner using the SEER and EER values.

8   Absorption chillers, like electric vapour-compression chillers, can operate without a cooling tower, but at substantially lower efficiency.

9   With a peak solar radiation intensity of $1000W/m^2$, this implies that peak outdoor illumination can exceed $100,000lm/m^2$ (100,000lux). For reading and writing, a light intensity of 300–500 lux is required.

10   The adjusted volume is equal to the fridge volume plus the freezer volume times an adjustment factor of 1.63 to account for the greater work required to cool a given freezer volume.

11   For example, IEA (2009) reports that an apparent decrease in average electricity consumption by refrigerators sold in Japan from 650kWh/yr in 1998 to 290kWh/yr in 2004 was due to a measurement procedure that did not accurately reflect real operating conditions. After developing a new procedure, the revised numbers are about 770kWh/yr in 1998 and 600kWh/yr in 2004.

12   Ownership rates for clothes dryers in European countries range from a low of 8 per cent in Italy to a high of 70 per cent in the UK according to Presutto (2009).

13   The reduction in moisture content from 72 per cent to 49 per cent corresponds to a reduction in water content of 0.23kg/kg clothes, which would require 0.16kWh to evaporate. If the heating system efficiency is 80 per cent, the saving in heating energy is 0.20kWh.

14   Efficient hand washing, as traditionally practised in water-limited Mexico, consists of the following: scrapping off any food fragments, wetting a batch of dishes with running water, filling a small container with soapy water and using this soapy water to wash the dishes with the faucet turned off, then rinsing the dishes with running water. A water-efficient faucet should be used.

15   If energy use varies with surface area, it will vary with volume to the two-thirds power.

16   The management options are: computer to go to standby or hibernate mode and monitor to go into standby mode or to be shut off after user-specified

durations of inactivity. Selecting a black screen rather than a screen saver in monitor standby mode saves more energy. These options can be accessed by right-clicking in the desktop area.

17 As discussed in Volume 2, Chapter 4, subsection 4.6.3, the energy required to produce wood chips (harvesting, transportation, chipping) from commercial forestry residues is only a few per cent of the heating value of the chips, and one might expect a similar ratio for insulation made from wood fibres.

18 This is because the difference in heat loss is proportional to $(1/R_{before} - 1/R_{after})$, but if $1/R_{after}$ is already very small compared $1/R_{before}$, then relatively large changes in R_{after} have only a very small effect on the difference between $1/R_{before}$ and $1/R_{after}$.

19 An added benefit of tile roofs is that they are more durable than asphalt shingle roofs. The use of asphalt shingle roofs is largely a North American phenomenon.

20 Hayter et al (1998) provide examples of errors during construction that caused the energy savings in one building to be only 63 per cent instead of 70 per cent (had the design intent been followed).

21 This table will periodically be updated in the Online Supplementary material. Readers are invited to provide the author (at harvey@geog.utoronto.ca) with further documented examples of buildings achieving a minimum 50 per cent energy savings.

22 In this case, the second façade was constructed as part of measures to upgrade the resistance of the building to earthquakes.

5

Transportation Energy Use

5.1 Introduction

This chapter discusses passenger and freight transportation energy use, on land and sea and in the air. A number of different terms related to energy use and efficiency in transportation are defined in Box 5.1, while Appendices B and E contain information of the properties of various transportation fuels and on the conversion between metric and non-metric measures of transportation fuel efficiency, respectively. The general options for reducing the energy use in urban and interurban passenger transportation are: (1) to reduce the energy intensity of all modes of travel, both by reducing vehicle energy use per kilometre travelled and by increasing average passenger loadings; (2) to shift from energy-intensive to less energy-intensive modes of travel; and (3) to shift to relatively less energy-intensive choices of personal automobile (i.e. away from light trucks[1] and toward small- or medium-size cars). Energy use within cities can be further reduced by reducing the average distances travelled per person per year by creating more compact cities, while energy use for interurban travel can be further reduced by increasing the operational efficiency of air, bus and rail fleets (i.e. for aircraft, less time in the air waiting for permission to land).

Box 5.1 Terms related to transportation energy use and efficiency

The following terms will be used here in the following ways:

- *Energy intensity*. The energy required to transport one passenger one kilometre (MJ/passenger-km) or to transport a tonne of goods one kilometre (MJ/tonne-km).
- *Economic transportation energy intensity*. The amount of energy used for transportation in relation to income (ideally, on a purchasing-power-parity basis) (MJ/$).
- *Fuel economy*. The amount of fuel used to travel a given distance (typically litres per 100km) or how far one can drive with a given amount of fuel (km/litre or miles per gallon, mpg).
- *Energy efficiency*. Ratio of energy supplied as fuel to energy delivered to vehicle loads (a large, heavy vehicle can be energy efficient in terms of energy use in relation to vehicle weight, but also energy intensive).
- *Operational efficiency*. The ratio of the minimum distance required to be travelled by rail, bus or aircraft fleets compared to the actual distance travelled in transporting a given volume of goods or people between a specified set of locations (fleets with high operational efficiency will have minimal excess travel).

Technically speaking, if litres/100km is called fuel economy, then km/litre (or mpg) is an inverse fuel economy (or vice versa). An alternative would be to call km/litre (or mpg) the *fuel effectiveness*. Note that the percentage decrease in energy intensity or in fuel economy (defined as litres/100km) will be smaller than the corresponding percentage increase in fuel effectiveness. For example if the distance that can be travelled with a given amount of fuel increases by 50 per cent, the amount of fuel required to drive a given distance decreases by 33 per cent. In general, if x is the fractional improvement in fuel effectiveness, $1 - 1/(1+x)$ will be the fractional reduction in fuel intensity.

In popular terminology, the term *fuel efficient* refers to vehicles that use relatively little fuel to travel a given distance; vehicles with low fuel consumption are regarded as 'efficient', but part of that efficiency may be due to the fact that they are smaller and/or lighter. This is efficiency in the sense of using less of all the resources needed for transportation, including the amount of materials that go into the vehicle and the resulting loads that the vehicle powertrain must provide. This is a convenient concept, so the term 'fuel efficient' will be used here in this sense.

The general options to reduce energy use in transporting freight are:

(1) to reduce the movement of goods;
(2) to shift from truck to rail transport of freight;
(3) to increase loading factors (the average number of passengers or tonnes of freight compared to the maximum possible load for that vehicle); and
(4) to effect increases in the fuel efficiency of trucks, trains, aircraft and ships.

The amount and distance that goods are transported could be reduced by including the full environmental damage caused by transportation energy use in the price of fuel and by eliminating all transport subsidies. Reduced transport of goods implies greater reliance on locally produced goods and avoiding the two-way exchange of essentially identical products.

5.1.1 Trends and patterns in the movement of people and goods

As a prelude to examining options to reduce transportation-related energy use, it is worthwhile briefly examining trends and geographic patterns related to transportation energy use. Figure 5.1 gives the relative contribution of different forms of energy used for transportation worldwide. These are gasoline, diesel, jet fuel, heavy fuel oil (used largely by ships) and others (natural gas, coal, ethanol and electricity). As indicated in Table 2.3, 97 per cent of worldwide transportation energy comes from oil (but just half of world oil supply is used for transportation). Figure 5.2 gives the breakdown of energy use in 2005 between movement of freight and movement of people in OECD and non-OECD countries as projected by WBCSD (2004), and the contribution of different transport modes within the freight and passenger sectors.[2] Overall transportation energy use is about twice as large in OECD countries as in non-OECD countries. In OECD countries as a whole,

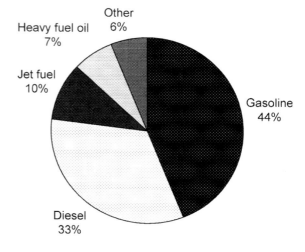

Source: Gilbert and Perl (2007)

Figure 5.1 *Proportion of different energy sources used for world transportation*

moving people accounts for about three quarters of transportation energy use and moving freight about one quarter, while in non-OECD countries as a whole the ratio is about 50:50. Cars and light trucks accounted 54 per cent of total transportation energy use in OECD countries in 2005 but only 28 per cent in non-OECD countries, while medium and heavy trucks accounted for 21 per cent of the total in OECD countries and 33 per cent of the total in non-OECD countries. Air transport is a significant share of total transportation energy use in both regions, as is bus transport in non-OECD countries.

Figure 5.3 shows the growth in the worldwide movement of people and freight from 1990–2003 (including both domestic and international movements). During this time, movement of people (in terms of person-km travelled per year) increased by 38 per cent (2.5 per cent/year) while the movement of freight (in terms of tonne-km per year) increased by 61 per cent (3.7 per cent/year). The overwhelming portion (73–75 per

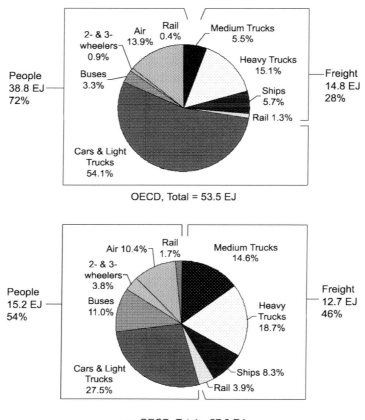

Figure 5.2 *Breakdown of transportation energy use in OECD and non-OECD countries in 2005*

Source: Data from WBCSD (2004)

cent since 1990) of freight movement occurs by water, while the overwhelming portion of motorized people movement occurs by road (87.8 per cent in 2003), followed by air (at 10.7 per cent). However, passenger air travel has been growing substantially faster than the overall movement of people, with an average growth rate of 4.4 per cent/year over the period 1990–2005 and no noticeable long-term effect of the September 2001 dip in air travel, as shown in Figure 5.4. Most of this growth would have occurred even in the absence of discount air carriers, as shown by the estimate of the growth of air travel in the absence of such carriers given in Figure 5.4.

Figure 5.5 shows the growth in the number of passenger and commercial road vehicles in the world from 1980–2002. By 2002, there were almost 700 million cars and 122 million commercial vehicles worldwide. Worldwide production of cars and light trucks reached a near-term peak of 70.9 million units in 2007, dropping to

68.1 million in 2008 and to a projected production of about 60 million in 2009. Figure 5.6 shows trends in the rates of car or light truck ownership in the US, Western and Eastern Europe and China from 1950–2004. By 2004, there were 760 cars per 1000 people in the US, 470 in Western Europe, 280 in Eastern Europe, but only 13 cars per 1000 people in China.

Figure 5.7 shows the breakdown of travel in the US in 2001 by all modes. Commuting to and from work accounted for only 18 per cent of total passenger-km travelled, with comparable contributions from travel for shopping, leisure and personal business. Travel for tourism and other long-distance travel accounted for almost 40 per cent of total distance travelled, most of that by air. The urban share of car travel in the US peaked at 64 per cent in 1994 and has slowly declined since then (Heavenrich, 2005), with the balance from long-distance travel. In low-income countries, the

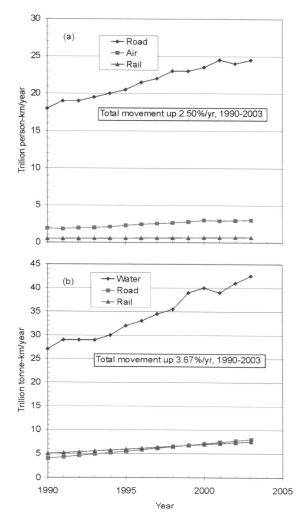

Source: Gilbert and Perl (2007)

Figure 5.3 *Variation in world (a) passenger-km movement of people, and (b) tonne-km movement of freight over the period 1990–2003*

proportion of car travel and of total travel that is within cities is surely much larger.

Corbett (2004) estimates that marine military vessels use about 14 per cent of the energy of all marine vessels, which in turn accounted for 5.7 per cent of total transportation energy use in OECD countries (see Figure 5.2). Henderson and Wickrama (1999) indicate that military aircraft accounted for about 18 per cent of all fuel use by aircraft in 1992, which in turn accounted for about 14 per cent of total transportation energy use

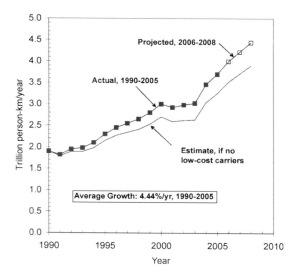

Note: Also given is an estimate of the transport that would have occurred without low-cost carriers.
Source: Gilbert and Perl (2007)

Figure 5.4 *Historical variation (1990–2005) in world passenger-km transport by aircraft, and projection to 2008*

in OECD countries in 2005. Applying these factors to the 2005 OECD transportation energy use of 53.5EJ (see Table 2.1) gives a military transportation energy use of about 1.76EJ or 6.3 per cent of total transportation energy use.

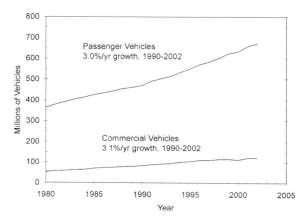

Source: Data from United Nations Common Database, unstats.un.org/unsd/cdb/cdb_help/cdb_quick_start.asp

Figure 5.5 *Growth in the number of passenger and commercial vehicles worldwide*

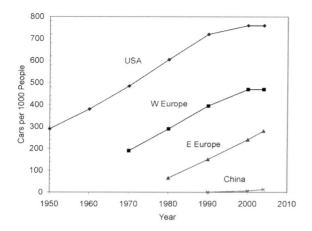

Figure 5.6 *Historical variation in the number of cars per 1000 people in the US, Western Europe, Eastern Europe and China*

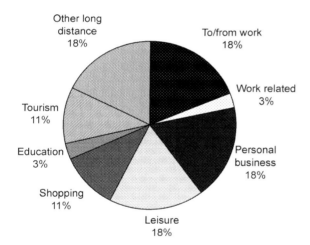

Figure 5.7 *Breakdown of travel (in terms of passenger-km/year) in US cities in 2001*

5.1.2 Energy use by different modes of transporting people

Table 5.1 compares the average primary-energy intensity and energy intensity at the point of use for different modes of transporting people in urban centres, including walking and bicycling. For car, diesel rail and

air travel, it is assumed that the energy used for refining the fuel from crude oil and delivering it to the vehicle is 20 per cent of the delivered energy, which is an approximate figure; this gives a markup factor or ratio of primary to secondary energy use of 1.2 for these modes. For electric rail, the electricity is assumed to be supplied with an efficiency of 40 per cent; this gives a markup factor of 2.5. Even if the electricity comes from renewable energy, such electricity could have instead been used to displace existing fossil fuel electricity if there are fossil fuel sources on the grid. Thus, in terms of the opportunity cost, it is appropriate to apply a markup factor of 2.5 to electric rail transport.

The energy intensities per person depend on the energy use per km travelled by the vehicle, the number of passengers carried, and the markup factors given above. Except for gas-guzzling and fuel-efficient cars, the passenger loadings given in Table 5.1 are the averages of a global sample of cities given by Newman and Kenworthy (1999). Table 5.1 indicates that the primary energy use per person-km of a single-occupant gas-guzzling car (20 litres/100km in city driving) is three to five times that of a diesel bus with typical passenger loadings and seven times that of subways. Also, fuel-efficient cars (8 litres/100km) with four persons use about two thirds the primary energy per person-km as heavy electric rail with typical passenger loadings.

Also shown in Table 5.1 is an estimate of the human metabolic energy used for walking and cycling, based on the rate of energy expenditure during these activities minus the rate of energy use while at rest. This metabolic energy is equal to the energy content of the extra food that would need to be eaten if there is to be no weight loss as a result of walking or cycling. To determine the primary energy entailed in walking or cycling, the food energy required needs to be multiplied by the energy input to the food system (excluding solar energy used for photosynthesis) per unit of food energy supplied. These energy inputs include fuel use on farms, the energy required to make fertilizers and pesticides, the energy required to transport food from the farm to where it is sold, and the energy for food processing, transport from stores to homes, preservation (in refrigerated displays where sold and in household refrigerators) and cooking. This can easily be a factor of five to ten in the current food system (see Chapter 7, subsection 7.9.3), and can

Table 5.1 *Energy intensity of different methods of transportation in cities*

Mode	MJ/ vehicle-km	Number of passengers[a]	MJ/ Passenger-km	Ratio of primary/ secondary energy	MJ of primary energy per passenger-km	Source
Car						
Gas guzzler (20 litres/ 100km, 11.8mpg)	6.46	1.0	6.46	1.2	7.75	Table 5.8
Global sample (14 litres/ 100 km, 13.1mpg)	4.52	1.5	3.01	1.2	3.62	Newman and Kenworthy (1999)
Fuel efficient (8 litres/ 100km, 29.4mpg)	2.58	4.0	0.65	1.2	0.78	Table 5.8
Bus						
Diesel, min	15	13.8	1.09	1.2	1.30	Chandler and Walkowicz (2006), Seattle
Diesel, max	30	13.8	2.17	1.2	2.61	Barnitt (2008), New York
Trolley	7.7	13.8	0.56	2.5[b]	1.39	Table 5.2
Rail						
Light	23.5	29.7[c]	0.79	2.5	1.98	Newman and Kenworthy (1999)
Electric heavy	13.2	31.0	0.43	2.5	1.06	Newman and Kenworthy (1999)
Diesel heavy	40.3	28.0	1.44	1.2	1.73	Newman and Kenworthy (1999)
Human-powered						
Walking			0.13[d]	5.75[e]	0.75	Coley (2002)
Cycling			0.093[f]	5.75	0.53	Coley (2002)

Note: [a] Includes the driver of cars; [b] Applicable to electricity generated from fossil fuels at 40 per cent efficiency; [c] Rail mode occupancies are given on the basis of average loading per rail car, not per train; [d] This is the human metabolic energy needed by a 65kg person for walking, above and beyond the energy used while at rest; [e] This is the ratio of energy required to produce and deliver the food in a typical British diet, to the energy content of the food, and does not include the energy required for preservation (in refrigerators) and cooking of food. As agricultural energy intensities decrease and if more locally produced food is consumed, this ratio will decrease. Conversely, for a diet high in meat and processed food that is transported great distances, this ratio could be 50:1 or higher (see Chapter 7, Table 7.21); [f] This is the human metabolic energy needed by a 65kg person for cycling, above and beyond the energy used while at rest.

nullify the advantage of walking or cycling compared to the more efficient modes of public transportation. Indeed, advanced diesel cars may eventually achieve on energy intensity of only 1MJ/km, so if eight units of non-photosynthetic energy are required per unit of food energy, walking (or cycling) would not save energy compared to a single-occupancy vehicle. However, if agricultural energy intensities decrease in parallel with improvement in vehicle efficiency, and if walking- and cycling-oriented lifestyles are accompanied by a conscious choice to consume food products that are less energy intensive (including locally grown foods to the extent possible), walking and cycling would still entail a substantial savings in energy use. As well, if walking and cycling lead to a lower long-term body mass, then metabolic rates of energy use at rest (typically 70–100J/s) will decrease (by about 1.5 per cent per kg of reduced mass), with savings in the required food

consumption. Finally, and as discussed below, there are substantial additional energy inputs required to build cars and the supporting infrastructure (such as roads) but not for bicycles or for walking. Thus, walking and bicycling emerge as by far the most energy-efficient modes of transport.

The energy intensities per passenger-km for public transportation are highly dependent on the number of passengers in the vehicle. Table 5.2 gives the point-of-use (POU) energy intensity for specific rail transit systems for typical and maximum passenger loadings. Fully packed electric light-transit vehicles have a primary energy requirement (assuming a markup factor of 2.5) of 0.03–0.06MJ/passenger-km, which is 12–25 times less than that of a fuel-efficient car with four persons.

Table 5.3 compares energy intensities for intercity travel. The same markup factor (2.5) is assumed for electricity as in Table 5.1, and because of this large

Table 5.2 *Secondary energy use per vehicle, passenger loadings and energy intensity for specific subway, trolley and commuter rail systems in specific world cities*

	Energy use per vehicle (MJ/km)	Passengers per vehicle			Energy intensity (MJ/passenger-km)	
		Typical	Seated	Crushed	Typical	Crushed
Light rail						
Siemens Combino, Basel	5.51	65	67	180	0.085	0.031
Siemens Combino, Potsdam	6.62	65	67	180	0.102	0.037
Siemens SD160, Calgary	11.6	145	60	200	0.080	0.058
Heavy rail						
GO Transit (diesel), Toronto[a]	289	1000	1620	3600	0.289	0.080
London Underground (electric)	10.2	19	41	152	0.537	0.067
Sky Train, Vancouver (electric)	8.69	30	40	110	0.290	0.079
Buses						
Trolley bus, Vancouver, 2005	7.7	30	34	77	0.257	0.100
Diesel bus, Vancouver	24.3	25	34	90	0.972	0.270
Diesel bus, Santa Barbara	15	20[b]	55	90[b]	0.750	0.167

Note: [a]Energy use and passenger numbers are per train; [b]Author's guess.
Source: James Strickland, http://strickland.ca/efficiency.html

Table 5.3 *Energy intensities of different modes of intercity travel, based on highway energy use for representative present-day gas-guzzling, typical and fuel-efficient cars (from Table 5.8) and for present-day intercity buses (from CEC, 2000) and trains and air (from Table 5.27 and Figure 5.40d, respectively)*

Transport mode	Secondary energy intensity (MJ/km for cars, MJ/seat-km for buses, rail and air)	Assumed ratio of primary to secondary energy	Passenger loading[a]	Primary energy intensity (MJ/passenger-km)
Car				
Gas guzzler (12 litres/100km, 19.6mpg)	3.88	1.2	4	1.16
Typical (9 litres/100km, 26.2mpg)	2.91	1.2	4	0.87
Fuel efficient (6 litres/100km, 39.2mpg)	1.94	1.2	4	0.58
Intercity bus	0.23[b]	1.2	Full	0.28
Diesel trains	0.15–0.42	1.2	Full	0.18–0.50
Low-speed electric trains (≤210km/hr)	0.102–0.162	2.5	Full	0.26–0.41
High-speed electric trains (≥270km/hr)	0.086–0.163	2.5	Full	0.22–0.41
Air	0.5–1.25[c]	1.2	Full	0.60–1.50

Note: [a] Maximum loadings are assumed for illustrative purposes and to permit easy conversion by the reader to alternative assumed loadings; [b] For 55-seat commuter buses in Santa Barbara, California (at speeds >65mph 90 per cent of the time), having a fuel economy of 6mpg; [c] The smaller energy intensity pertains to a distance travelled of 2000km, the larger to a distance of 200km.

markup factor, intercity transport in advanced cars with four people could use less primary energy per person than electric rail (depending on the efficiency of electric rail). Intercity travel in average cars with four persons is also more efficient than travel by air.

A full accounting of the energy required for transportation should include the embodied energy in the manufacture and repair of transportation vehicles

and infrastructure (such as roads, rail lines and airports). Lenzen (1999) carried out such a study for Australia in the 1990s, and his key results are summarized in Table 5.4. Shown are the energy consumption at the POU, the upstream energy associated with finding, extracting, processing and delivering fuels or electricity, and the embodied energy in the transportation vehicle and infrastructure. The

Table 5.4 *Comparison of direct (onsite) energy use by various modes of travel with the indirect energy uses (upstream fuel energy and the energy used in manufacturing and maintaining the vehicle and its supporting infrastructure) as computed by Lenzen (1999) for Australia in the 1990s using input–output tables*

Mode	Energy use (MJ/passenger-km)				Indirect/direct ratio		Total (MJ/pkm)
	Fuel energy use		Embodied energy		Excluding public infrastructure	Including public infrastructure	
	Direct	Upstream	Vehicle	Infra-structure			
Urban travel							
Urban light rail	0.37	1.0	0.60	0.06	4.3	4.5	2.0
Urban bus	1.7	0.42	0.51	0.19	0.55	0.67	2.8
Ferry	3.5	0.81	1.2	0.00	0.56	0.56	5.5
Bicycle[a]	0.076[b]	0.22	0.42	0.12	8.4	9.93	0.83
Urban heavy rail	0.6	1.3	0.5	0.44	3.1	3.9	2.8
Interurban travel							
Gasoline car	2.4	0.57	0.81	0.60	0.58	0.83	4.4
Diesel car	2.7	0.65	0.81	0.60	0.54	0.76	4.8
Interurban bus	0.86	0.18	0.23	0.09	0.47	0.58	1.4
Interurban rail	0.94	0.29	0.45	0.33	0.80	1.1	2.0
Private aviation	5.3	1.3	12.5	0.10	2.6	2.6	19.2
Charter aviation	6.9	1.7	8.9	0.04	1.5	1.5	17.6
Regional air[c]	3.5	0.84	5.3	0.04	1.8	1.8	9.7
Domestic air[d]	2.5	0.59	2.6	0.01	1.3	1.3	5.7
International air[e]	1.8	0.42	0.88	0.00	0.74	0.74	3.1

Note: Upstream energy is the energy used in finding, extracting, processing and delivering fuel. It is equal to the total energy input to the fuel cycle minus the energy content of the delivered fuel. For bicycling, it is the total energy needed to supply the amount of food energy needed to balance the bicyclists' energy consumption (for typical Australian diet and assuming no wastage) minus the energy value of the food consumed. [a] Results pertain to bicycles used for transportation purposes, and in particular, 200 trips per year of 7km distance one way. [b] Based on extra metabolic energy consumption at a rate of 190W while cycling at 9km/hr. [c] Average distance of 390km. [d] Average distance of 1100km. [e] Average distance of 7500km.

total indirect energy inputs (upstream fuel and embodied energy) are about 80 per cent of the direct energy use for cars, a little less (75 per cent) for international air travel, and significantly greater (1.3–1.8 times the direct energy use) for domestic air travel. For the US, Bin and Dowlatabadi (2005) estimate total indirect energy use for transportation as a whole to be about the same as the direct energy use, while Chester and Horvath (2009) estimate total transportation energy use to be 1.3, 1.6 and 2.6 times the direct energy use for air, road and rail modes of travel, respectively.

5.2 Role of urban form in passenger transportation energy use

Compact urban design with intermixed land uses (housing, retail, offices, light industry, recreational and education facilities) contributes to reducing transportation energy use by simultaneously minimizing the need to travel (by reducing the distances that need to be travelled) and by increasing the viability of public transportation, bicycling and walking alternatives to the private automobile. Increased economic viability for public transportation then makes it possible to provide public transportation of sufficient quality (speed, reliability, comfort) that it can compete with private modes of transport. When public transportation is used in place of the private automobile, the *total number* of trips decreases (rather than merely being shifted from one mode to another), as travellers combine formerly separate trips (that is, overall trip efficiency increases). This is known as a *multiplier effect* and, along with reduced travel distances in compact cities, also contributes to reduced distances travelled. Thus, a smaller fraction of a smaller total transportation distance is met by private motorized modes of transport in more compact cities. These and related issues are comprehensively discussed in a number of books, including Cervero (1998) and Newman and Kenworthy (1999).

5.2.1 Transportation energy use and related indicators in world cities

Table 5.5 compares US, Australian, Canadian, European and Asian cities in terms of a number of indicators

Table 5.5 *Transportation energy use in US, Australian, Canadian, European and Asian cities*

	US cities	Australian cities	Canadian cities (or Toronto)	European cities	Wealthy Asian cities	Poor Asian cities	Source of data (table or figure)
Transportation energy intensity indicators							
Per capita annual transport energy (GJ)	64.35	39.46	39.17 (33.61)	17.22	7.27	5.83	T3.2
Per capita GRP (US$, 1990)	26,822	19,761	(22,572)	31,721	21,331	2642	T3.15
Transport energy intensity (MJ/$)	2.40	2.00	(1.49)	0.54	0.34	2.20	derived
Direct contributors to transportation energy intensity							
Annual per capita gasoline use (GJ)	58.54	33.45	24.45	19.12	11.0		T3.3
Auto fuel efficiency MJ/km	5.03	5.11	4.85	3.79	4.93	3.53	T3.4
litres/100km	14.50	14.73	13.98	10.93	14.21	10.18	
MJ per passenger-km car	3.52	3.12	3.45	2.62	3.03	2.12	T3.5
bus	2.52	1.64	1.61	1.32	0.84	0.74	
rail	0.74	1.12	0.51	0.49	0.16	0.24	
Cars per 1000 people	604	491	524	392	109		
km/person/year	16,045	10,797	9290	6601	2806		T3.8
% of total travelled distance on transit	3.1	7.7	10.2 (23.6)	22.6	48.7 (Hong Kong: 82.3)		T3.9
Indicator of urban form: Job and employment density							
Inner-area density People/ha	35.6	21.2	43.6 (60.0)	86.9	291.2		T3.12
Jobs/ha	27.2	26.2	44.6 (44.3)	84.5	203.2		
Outer-area density People/ha	11.8	11.6	25.9 (35.4)	39.3	133.3		
Jobs/ha	6.2	3.6	9.6 (16.3)	16.6	43.5		
Overall, jobs + people	23	18	(65)	82	251		T3.14
Indicators of socio-economic well-being related to urban transportation							
% of GRP spent on transportation Journey-to-work	6.9	6.3	(5.2)	5.4	6.3	7.7	F3.11
Total passenger	12.4	13.2	(7.4)	8.1	4.8	15.9	F3.16
Traffic deaths per 100,000 people per year	14.6	12.0	(6.5)	8.8	6.6	13.7	F13.3
Emissions of smog precursors (NO_x+SO_2 +CO+VOCs), kg/person/year	252	233	(216)	101	30	89	F13.5

Note: The cities used in the comparisons shown above are: US: Boston, Chicago, Denver, Detroit, Houston, Los Angeles, New York, Phoenix, Portland, Sacramento, San Francisco, San Diego, Washington; Australia: Adelaide, Brisbane, Canberra, Melbourne, Perth, Sydney Canadian cities: Calgary, Edmonton, Montreal, Ottawa, Toronto, Vancouver, Winnipeg; Europe: Amsterdam, Brussels, Copenhagen, Frankfurt, Hamburg, London, Munich, Paris, Stockholm, Zurich; Wealthy Asia: Hong Kong, Singapore, Tokyo; Poor Asia: Bangkok, Jakarta, Kuala Lumpur, Manila, Seoul, Surabaya. GRP = gross regional product.
Source: Tables and figures from Chapter 3 of Newman and Kenworthy (1999)

related to transportation energy use and urban form. US and Australian cities tend to have the largest per capita energy use, the lowest population and job densities, the lowest use of public transportation, the highest per capita expenditures on transportation, the greatest number of traffic-related deaths per capita and the highest emission of harmful pollutants per capita. Canadian cities (Toronto in particular) and European cities are distinctly better according to most or all of these social, economic and environmental indicators. Wealthy Asian cities also have high urban densities, high use of public transportation but low overall distance travelled per person, low overall expenditures on transportation and low traffic deaths and pollutant emissions per capita. Table 5.6 gives the proportion of trips made by car, public transportation, walking and bicycling in a selection of world cities in 2001. More compact cities generally have a larger percentage (50 per cent or more)

of trips made by public transportation, walking and bicycling than less compact cities.

The relationship between urban form and transportation energy use is strikingly illustrated by Figure 5.8, which shows the relationship between urban density and per capita gasoline or diesel energy use by private vehicles (after correcting for differences in vehicle fuel efficiencies and in income) across a sample of European, Asian and North American cities. High per capita energy use for private modes of transportation is associated with low urban density, with a factor of 25 difference between the two big extreme cities (Houston and Hong Kong). In terms of total energy use (including public transportation), there is only a factor of nine difference between wealthy Asian and typical US cities, as indicated in Table 5.5. Nevertheless, this factor of nine difference is much greater than the factor of two to four potential

Table 5.6 *Percentage of trips (not distance) made by various modes of travel in different cities in 2001*

City	Car	Public transportation	Foot	Bike	Foot + bike
Hong Kong	16	46	38	0	38
Moscow	26	49	22	2	24
Warsaw	29	52	19	0	19
Bangalore[a]	29	45	16	11	27
Budapest	33	44	22	1	23
São Paulo	34	29	37	0	37
Amsterdam	34	15	26	26	52
Prague	36	43	20	1	21
Vienna	36	34	27	3	30
Berlin	39	25	26	10	36
Helsinki	44	27	22	7	29
Singapore	45	41	10	4	14
Zurich	46	18	20	15	35
Paris	46	18	35	1	36
Stockholm	47	19	34	0	34
Copenhagen	49	12	19	20	39
London	50	19	30	1	31
Madrid	51	15	30	4	34
Rome	56	20	23	0	23
Brussels	59	11	23	8	31
New York[b]	61	30	–	–	9
Athens	64	28	7	2	9
Glasgow	66	11	23	1	24
Melbourne	76	6	17	1	18
Dubai	77	7	16	0	16
Washington[b]	77	16	–	–	7
Chicago	88	6	5	1	6
Atlanta[b]	95	5	0	0	0

Note: [a] From ADB (2001); car mode here consists of 11 per cent private automobile, 18 per cent motorcycle or three-wheeler.
[b] From L. R. Brown (2001), early 1990s data.
Source: Gilbert and Perl (2007), unless indicated otherwise

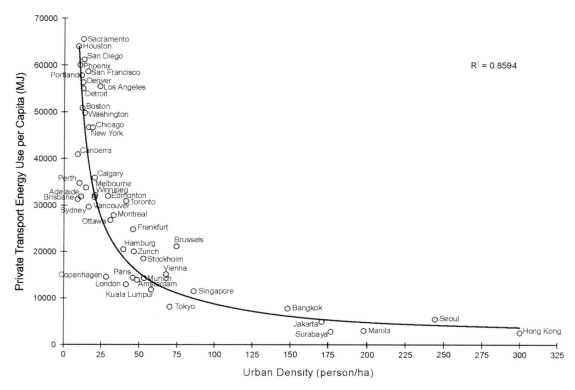

Source: Newman and Kenworthy (1999)

Figure 5.8 *Energy use per capita for private transportation versus urban density in global cities*

reduction in energy use through improvements in the fuel economy of private motor vehicles. However, in much of the existing urbanized landscape, the factor of two to four reduction through technical means will surely be easier to achieve.

In spite of the evidence presented in Table 5.5 and in Figure 5.8, some researchers have doubted whether urban population density plays a significant role in transportation energy use (Gordon and Richardson, 1997). Over the range of urban densities found in American cities, the relationship can be weak. Clearly, modest increases in urban density but with highly segregated land uses (so that considerable distances still need to be travelled) and with poor public transportation (so that cars will be the preferred choice) will have only a modest impact on urban transportation energy use. Conversely, there is no doubt that a combination of higher density, mixed land use (so that many daily destinations are within convenient public transportation, walking or cycling

distance of one another) and provision of high-quality rapid transit can achieve dramatic reductions in transportation energy use compared to alternative forms of urban development.

As seen in Table 5.1, the energy intensities of diesel and electric rail modes of travel are only about half and one third, respectively, that of typical cars. Hence, if urban *regions* grow in size, it will be difficult to reduce total energy use for travel even if accompanied by a shift from cars to rail. Thus, although it is important to build up the density of the existing urbanized area in urban centres poorly served by public transportation, it is also important to resist the tendency to allow sprawling urban regions with individual population centres linked together by commuter train if this makes longer distance commuting more viable. For example, a car commuter travelling 15km (the average commuting distance within the Greater Stockholm area) would use the same amount of energy as a rail commuter travelling 100km to Stockholm from satellite cities by rail (which can be done within 45–50 minutes) (Åkerman and Höjer, 2006). Thus, it is important to plan or redevelop cities so as to reduce the *distances* that are travelled, as well to provide or encourage efficient alternatives to single-occupancy vehicles.

5.2.2 Economic transportation energy intensity

Of particular interest with regard to projecting future transportation demand is the economic transportation energy intensity in different groups of cities, that is, the per capita transportation energy use divided by the per capita GRP (gross regional product). It is analogous to the concept of the energy intensity of the economy that was introduced in Chapter 1 as part of the Kaya identity, except that it is based on transportation energy use rather than total energy use. The transportation energy intensity is given in the third row of Table 5.5. It ranges from 2.40MJ/$ in US cities to 0.54MJ/$ in European cities and 0.34MJ/$ in wealthy Asian cities. However, the transportation energy use relative to income in poor Asian cities (2.2MJ/$) is comparable to that of US cities. The wealthy Asian cities of Table 5.5 have about eight times the per capita income as the poor Asian cities, but only 25 per cent greater transportation energy use per capita. Thus, if the poor Asian cities were to become like the wealthy Asian cities as they develop, the increase in per capita income would be almost completely offset by a decrease in transportation energy intensity. This would entail replacing an inefficient transportation system based on private motorized modes of transport, carrying relatively few people at slow speeds, with an efficient, rail-based public transportation system providing far higher levels of service with far less energy use per passenger-km. At the beginning, such systems could largely involve more affordable light-rail transit in dedicated surface rights-of-way on major roads. This option, coupled with supportive land-use planning, is discussed by Lefèvre (2009) in the context of Bangalore, India. Given the impending entry into the market of inexpensive cars such as the Nano, it is all the more urgent to begin construction of rail transit systems as soon as possible.

5.2.3 Constraints on growth in transportation energy use in developing countries

It is very clear that, however wealthy most developing world cities become during the next century, they will not be able to attain the intensity of car ownership and use prevalent in North America. This is because there simply isn't the space to provide enough roads and parking for per capita car use as large as found in North America. This, however, has not stopped planners in many cities (such as Beijing and Bangkok) from trying to emulate the North American approach to transportation by building networks of freeways. The results (congestion and pollution) were entirely predictable. The ostensible goal in owning an automobile is to increase personal mobility. However, if the majority of adults in most parts of the world were to own and use an automobile, the result would be a decrease in mobility through congestion.

As another indication of the impossibility of achieving high levels of car ownership in most of the world, consider China: its land area is 9.56 million km², of which only 13.6 per cent (1.30 million km²) is agricultural land, according to the 2006 *China Statistical Yearbook*. If a future peak population of 1.4 billion were to achieve American levels of car ownership (about 0.76 cars/person according to Figure 5.6), and if each car requires a land area (roads and parking) of 750m² (the American average, based on the per capita area of 573m² for roads and parking given by MacKay, 2007), then the total area devoted to cars would be 798,000km², or about 61 per cent of the agricultural land area in China and just over 8 per cent of the total land area of China. Conversely, achieving western European levels of car ownership (470 cars per person) with the UK allotment of land per person for roads and parking (72m² according to MacKay, 2007) would require a land area of about 101,000km² or 8 per cent of the agricultural land area. As most of China's population is concentrated in agricultural regions, most of the land devoted to cars would be in the same regions and therefore would subtract from the agricultural land base. Given China's precarious food balance and the risk of adverse effects of global warming, even an 8 per cent reduction in the agricultural land base could be critical.

The only viable development model is one oriented toward rail-based public transportation, supplemented with a strong reliance on bicycles as a natural complement to rail transit. This model is prevalent in large European cities such as Copenhagen and Amsterdam and in many smaller cities in northern Germany, The Netherlands and Denmark, where 20–30 per cent of all trips and sometimes more are made by bicycle (see subsection 5.2.5). Auto-sharing programmes (which began in Europe and are now developing in North America) provide an alternative to individual car ownership that will conserve precious space in highly populated cities (Gardner, 1999). In Switzerland, an average of 18 members of an auto-sharing cooperative share one car, and the average distance driven per member is one quarter that of privately owned cars (Strickler, 1997). The fact that it is simply not physically possible to provide the majority of citizens in the world with a car, however wealthy they become, imposes a strong constraint on future transportation energy demand – a constraint that is ignored in many projections of future transportation energy use. This is not to say, however, that large increases in transportation energy use in developing countries are not possible in the absence of policies to restrain the growth in demand, given the very low per capita transportation energy use in much of the world today.

5.2.4 Achieving compact urban form

Compact urban form, with appropriate mixes of land use and rapid-transit infrastructure, is most readily achieved in new urban developments. Once an urban area has been built in a low-density, car-dependent form, it is very difficult to convert the area to higher density, transit-friendly urban form. Thus, failure to plan new urbanized areas in such a way as to minimize transportation energy use represents a significant *lost opportunity*. Nevertheless, there are a number of ways in which the density of present low-density urbanized areas can be increased, including:

- development of housing on main streets (two to three storeys of residential housing above two storeys of commercial/retail floor space, as typifies main streets in most European cities);
- redevelopment of abandoned industrial lands as moderate-density mixed-used developments or as high-density development;

- allowing basement or back-alley apartments;
- wholesale demolition and reconstruction of areas that have fallen into disrepair.

The process of increasing the population density in the already-urbanized areas is referred to as *urban intensification*. For a diverse set of viewpoints concerning the sustainability of compact cities, see Jenks et al (1996).

5.2.5 Promoting bicycling

In cities that are already relatively compact, or as urban densities are increased and a greater mix of land uses are created in a given neighbourhood, bicycling is or becomes a viable transportation mode for many trips. Half of all urban trips in the UK and 40 per cent of those in the US are 3.2km (2 miles) or shorter (Gardner, 1998), and are therefore ideal candidates for bicycling when weather permits. Bicycles are also a natural complement to rail rapid transit, providing an easy way to reach rapid transit stations (which must, of course, be provided with adequate bicycle parking). Great strides were made in increasing the share of trip made by bicycles in West Germany between the 1970s and 1990s, as indicated in Table 5.7, and the same is true of other north European countries. A number of small cities in The Netherlands have bicycle shares greater than 30 per cent, with Groningen having the highest, at 37 per cent (ECF, 2009). In cities with high bicycle use, road space has been devoted to dedicated bicycle lanes and secure facilities for locking and storing bicycles have been provided.

Table 5.7 *Growth in the share of trips made by bicycle in West Germany between the mid-1970s and early 1990s*

City	Percentage of urban trips made by bicycle	
	Mid-1970s	Early 1990s
Munich	6	15
Nuremberg	4	10
Cologne	6	11
Freiburg	12	19
Essen	3	5
Bremen	16	22
Muenster	29	32
Average of all urban areas	8	12

Source: Pucher (1997)

5.3 Role of vehicle choice in passenger transportation energy use

There is a wide range in the fuel efficiency of vehicles that perform essentially the same function, namely, transporting people from one point to another safely and with a given cargo capacity. Of course, consumers base purchase decisions on a wide range of factors that have no real utilitarian value (such as the ability to drive at speeds that are far in excess of the legal speed limits, or for the sake of off-road capabilities that they will never use). However, consumer choices are influenced by the vehicle purchase price and by marketing. The former can be altered through taxes that increase with the fuel consumption of the vehicle being purchased, while the latter is determined by the automobile manufacturers and could be readily altered (particularly if manufacturers are required to increase the average fuel efficiency of the cars and light trucks that they sell).

5.3.1 Comparative fuel consumption and market trend in the US

Table 5.8 lists the fuel consumption for the vehicles with the lowest and highest fuel consumption within each of several vehicle categories for the 2009 model year. Within many of the vehicle categories there is a factor of two or more difference between the fuel consumption of the most and least fuel-efficient vehicles, and substantial differences between vehicle categories, with SUVs, minivans and light trucks having the highest fuel consumption. The average fuel consumption of cars and light trucks sold in the US rose from 10.6 litres/100km in 1987 to 11.3 litres/100km in 2004 (equivalent to a decrease from 22.2mpg to 20.8mpg). The most fuel-efficient vehicles in each of the two-seater, subcompact, compact and small station wagon categories have half the current average fuel consumption or less. Thus, the average fuel efficiency of the US fleet could be doubled simply by shifting consumer purchases (through taxes and

Table 5.8 *Fuel economy (litres/100km and mpg) of the most and least fuel-efficient vehicles in each US car and light truck size category for the 2009 model year*

Category	Litres/100km (city/highway)			Mpg (city/highway)		
	Worst	Best		Worst	Best	
		Regular	Hybrid		Regular	Hybrid
Two-seater	29.4/18.1	7.1/5.7		8/13	33/41	
Mini-compact	21.4/13.8	8.4/6.4		11/17	28/37	
Subcompact	21.4/13.8	8.1/6.5		11/17	29/36	
Compact	23.5/16.8	7.8/5.7	5.9/5.2	10/14	30/41	40/45
Mid-size	23.5/16.8	9.0/6.9	4.9/4.2	10/14	26/34	48/45
Large	26.1/15.7	10.7/7.4		9/15	22/32	
Small station wagon	14.7/10.2	7.8/5.7		16/23	30/41	
Mid-size station wagon	18.1/13.1	11.8/8.7		13/18	20/27	
Small pickup truck, 2WD	15.7/12.4	11.2/9.0		15/19	21/26	
Small pickup truck, 4WD	16.8/13.1	14.7/10.7		14/18	16/22	
Standard pickup truck, 2WD	26.1/18.1	15.7/11.2	11.2/10.7	9/13	15/21	21/22
Standard pickup truck, 4WD	26.1/19.6	16.8/12.4	11.8/11.8	9/12	14/19	20/20
Passenger van	26.1/19.6	18.1/14.7		9/12	13/16	
Minivan	21.4/14.7	10.7/8.4		11/16	22/28	
SUV, 2WD	23.5/16.8	10.7/7.8	6.9/7.6	10/14	22/30	34/31
SUV, 4WD	26.1/19.6	8.1/8.7	10.2/8.4	9/12	29/27	23/28

Note: 2WD = two-wheel drive; 4WD = four-wheeel drive.
Source: US DOE (2009)

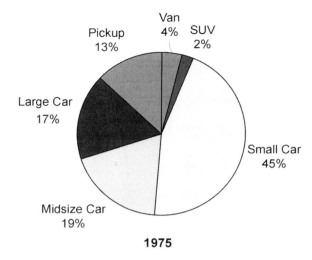

1975

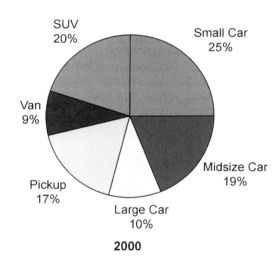

2000

Source: Friedman et al (2001)

Figure 5.9 *Mix of vehicle purchases in the US in 1975 and 2000*

advertising) away from SUVs, minivans and light trucks toward efficient vehicles as large as a small station wagon.

However, until mid-2008, recent trends in the US had been in the opposite direction. This is illustrated in Figure 5.9, which shows how the mix of vehicles purchased in the US for personal transportation changed between 1975 and 2000. In 1975, small cars accounted for 45 per cent of all vehicle sales and pickup trucks, vans and SUVs together accounted for 19 per cent.

By 2000, the proportions were 25 per cent and 46 per cent, respectively – almost a complete reversal. With the sudden increase in gasoline costs in 2008, sales of light trucks in the US sharply declined even before the financial crisis began.

5.3.2 Safety issues

The average fuel efficiency of the automobile fleet can be increased either through a shift to more fuel-efficient vehicles (as discussed above), or through technical measures that can improve fuel efficiency in all vehicle classes (discussed in section 5.4). However, the need to maintain safety is sometimes raised (especially by the automobile industry) as a reason why vehicle efficiency or fleet-average efficiency cannot be raised further. This argument is without merit, as shown below.

The only fuel-efficiency measure that could adversely affect vehicular safety is a reduction in vehicular weight and this only in two-vehicle collisions. However, the key safety factor here is the compatibility of the vehicles involved, not their absolute weights. Vehicle compatibility depends on differences in vehicle weight, differences in the height of the frame and differences in rigidity. Compatibility *within* the North American automobile fleet has increased over time if light trucks (which includes pickup trucks, SUVs and vans) are excluded, while light trucks – which accounted for half of all vehicle sales in the US in 2007 – violate all three principles of compatibility (Friedman et al, 2001). Forcing light trucks to meet the same fuel efficiency standards as cars, and across-the-board strengthening of these standards, would require a proportionately larger reduction in light truck weight, thereby increasing vehicular compatibility and improving safety (for example, in the efficiency scenarios of de Cicco et al, 2001, weight reductions of 10 per cent, 20 per cent and 30 per cent are assumed for small cars, large cars and SUVS and pickup trucks, respectively). Furthermore, enforcing a speed limit of 90–100 kilometres per hour (kph) (compared to typical highway speeds of 120kph and more) would increase safety (as well as reducing fuel consumption by another 10 per cent, as discussed later). As noted by Noland (2005) and Ross et al (2006), many other vehicle characteristics besides weight affect safety.

Figure 5.10 compares the risk to drivers of vehicles and the risk to drivers of other vehicles on the road for

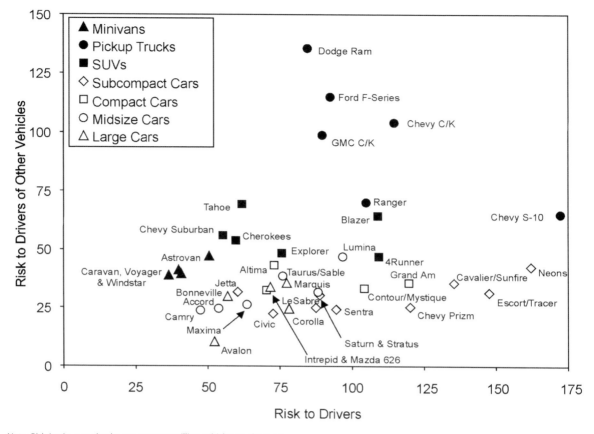

Note: Risk is given as deaths per year per million vehicles on the road.
Source: Ross and Wenzel (2002)

Figure 5.10 *Risk of death in the US to drivers of the most popular cars, SUVs and light trucks for 1995–1999, and risk of death to drivers of other vehicles on the road*

different kinds of cars and light trucks. Contrary to popular perception, gas-guzzling vehicles (SUVs and other light trucks) are not particularly safe for the drivers of these vehicles (compared to many compact and subcompact cars), and are several times more hazardous to the occupants of *other* vehicles on the road. SUVs and other light trucks also pose a much greater risk of death to pedestrians in the event that they hit pedestrians. This is because they strike pedestrians in the abdominal area rather than in the legs, due to the greater height of the vehicular frame above the road. Thus, measures to reduce fuel efficiency – by requiring weight reduction of light trucks, or their elimination altogether – would increase road safety.

However, as shown below, recent trends in vehicle weight in both the US and Europe have been in the opposite direction.

5.4 Technical options for reducing automobile and light truck energy use

In this section, technical options for reducing energy use by automobiles and light trucks at the point of use are discussed, beginning with conventional internal combustion engine (ICE) vehicles, then proceeding to ICE–electric hybrid vehicles, plug-in hybrid and all-electric vehicles, and finishing with fuel cell vehicles. Upstream energy use, associated with providing fuels to the consumer, is also discussed (energy losses during the conversion from crude petroleum to refined petroleum products are likely to increase over time, in spite of efforts to improve the conversion efficiency, for reasons discussed in Chapter 6, section 6.11). By including upstream energy use, the impact of alternative powertrain systems on primary energy use can be compared even when they involve different fuels. The technical options discussed here complement the substantial reductions in transportation energy use, discussed above, that are possible through reduced dependence on automobiles for personal transportation and through a shift in the mix of vehicles purchased.

5.4.1 Trends in automobile mass, power and fuel consumption

There have been substantial technological advances in automobile technology during the past 30 years, but these have gone into making engines more powerful, thereby reducing vehicle acceleration times at the same time that vehicles have become heavier. This is shown in Figure 5.11, which shows the trends in fuel economy (mpg) and fuel consumption (litres/100km) since 1976 in the US, Europe and Japan; and in Figure 5.12, which shows the trend in vehicle mass, acceleration and engine power in the US and Europe, as well as vehicle mass in Japan and the ratio of engine power to piston displacement in the US. In the US, fuel economy for cars and light trucks under test conditions has not changed since 1988, but the proportion of light trucks has increased, causing a gradual decrease in overall fuel economy (see Figure 5.11a). It is well known that fuel consumption under real-world driving conditions (with typical aggressive behaviour and use of air conditioning) is greater than under test conditions, and the gap has grown during the past two decades from 7 per cent in 1980 to 13 per cent in 1997 (Zachariadis, 2006). Thus, a downward trend of 6 per cent from 1980 to 1997 can be added to the variation seen in Figure 5.11a (upward in Figure 5.11b). From 1982 to 2004, US vehicle mass increased by 27 per cent (see Figure 5.12a) yet acceleration time decreased by 31 per cent (the rate of acceleration increased by 44 per cent) (see Figure 5.12b). This was accomplished through a doubling in engine power (see Figure 5.12c), which

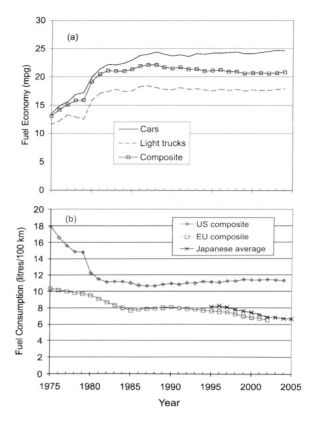

Source: Data files used in Zachariadis (2006), kindly provided by T. Zachariadis, except for Japanese fuel use, which is from Konuma (2007)

Figure 5.11 *Trend in (a) fuel economy (mpg) of US cars, light trucks and the weighted average of the two from 1975–2004, and (b) US weighted-average fuel consumption (litres/100km) and comparison with European and Japanese average fuel consumption*

also permitted an increase in maximum vehicle speed from 172kph to 217kph (see Figure 5.12b) – an absurdly high limit. The increase in engine power in turn was accomplished by increasing the power to displacement ratio, which almost exactly tracks the variation in engine power (see Figure 5.12c).

Reynolds and Kandlikar (2007) examined the relationships between automobile fuel consumption and mass and power. When both mass and power are used as predictors, only mass is significant, with each 100kg increase in vehicle mass causing an increase in fuel consumption of 0.7 litres/100km. This is a statistical

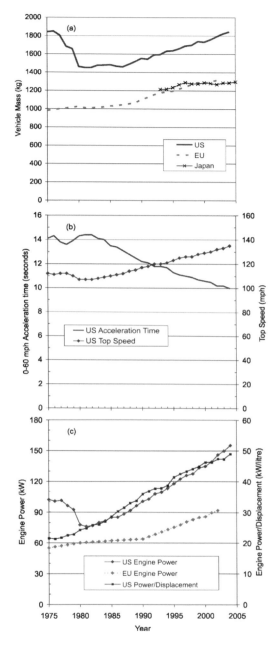

Source: Data files used in Zachariadis (2006), kindly provided by T. Zachariadis, except for Japanese vehicle mass, which is from Konuma (2007), and engine power/displacement ratio, which is from US EPA (2006)

Figure 5.12 *Trend in (a) average passenger-vehicle mass in the US and Europe, (b) acceleration time and top speed in the US, and (c) engine power in the US and Europe and the ratio of engine power to piston displacement in the US*

correlation, and could be affected by changes in other vehicle attributes that are correlated with vehicle mass and that affect fuel consumption. Simulations by Bandivadekar et al (2008) that leave vehicle acceleration and size unchanged indicate a smaller impact of vehicle mass, an increase of 0.40 litres/100km for a 100kg mass increase for cars and an increase of 0.49 litres/100km for light trucks (in both cases for combined city and highway driving). Figure 5.12a indicates an increase in average US vehicle mass of about 300kg, so if the 1980 vehicle mass (and associated power) had been held constant, the 2005 average fuel consumption would have been 9.2–10.1 litres/100km instead of 11.3 litres/100km – a saving of 10–20 per cent.

5.4.2 Kinds of automobile engines

There are two main kinds of ICE used in automobiles: the *spark ignition* engine and the *compression ignition* engine.

Spark ignition engines run on gasoline. Combustion occurs in a flame that proceeds from the spark through the mixture of vaporized fuel and air. The power output is reduced by reducing the flow of both fuel and air. The airflow is reduced by means of a *throttle*, which consists of vanes that restrict the flow of air into the intake manifold. This creates a partial vacuum as each piston is pulled out during its intake stroke, reducing the work that it can do and producing a major fuel-efficiency penalty.

Compression ignition engines run on diesel fuel. Fuel droplets are mixed with air and compressed to a high compression ratio. Diesel fuel is such that this causes spontaneous combustion, without a spark.[3] At typical power levels, there is excess air (the fuel:air mixture is lean). Power is reduced by reducing the amount of fuel injected, but the amount of air admitted is the same (there is no throttle). Due to the absence of throttling, the high compression ratio and the lean fuel mixture, diesel engines are substantially more efficient than spark ignition engines (details are provided later). In a side-by-side comparison of otherwise comparable diesel and gasoline vehicles, Schipper et al (2002) found that diesel vehicles use 24 per cent less fuel. As diesel fuel contains about 10 per cent more energy per litre, this translates into about 16 per cent less energy use.

Spark ignition engines use three-way catalytic converters to oxidize (add oxygen to) CO and hydrocarbons in the exhaust, while reducing (removing oxygen from) NO_x. For this to work, the air to fuel ratio introduced to the mixture must be stoichiometric (such that all the fuel and oxygen could combine to form CO_2 and H_2O, with neither fuel nor oxygen left over). Diesel engines use a lean fuel:air mixture, and until recently, three-way catalytic converts could not reduce NO_x because of the excess oxygen. However, technological advances based on the reaction of ammonia with NO_x have solved this problem with no fuel-economy penalty, although the release of unreacted ammonia is of some concern (and a small amount of energy would be required to produce the required ammonia). Particulate filters now exist that can be continuously regenerated, and these will lead to a dramatic reduction in particulate emissions in the next generation of diesel passenger vehicles, trucks and buses. Indeed, as shown in Chapter 2 (Figures 2.35 and 2.36) emission standards for diesel passenger vehicles will reach or exceed those of gasoline vehicles within a few years in Europe and Japan (and gasoline emission standards themselves have tightened considerably during the last few years). However, to achieve the strictest emission standards in both gasoline and diesel vehicles requires a much lower sulphur content in gasoline or diesel fuel than permitted in the vast majority countries (see Chapter 2, subsection 2.7.1).

This means that, as low-sulphur fuels become more widely available and vehicles are fitted with advanced control technologies, it will be possible to improve vehicle fuel economy without entailing an increase in pollutant emissions if diesel vehicles take an increasing share of the vehicle market.

5.4.3 Automobile energy flow and efficiency

The efficiency of an automobile engine, defined as the ratio of output energy to input (fuel) energy, is the product of two factors: the *thermal efficiency*, which is the fraction of the fuel energy supplied as energy to the pistons by the heated combustion gases, and the *mechanical efficiency*, which is the fraction of the piston energy that is converted to mechanical energy of the shaft.[4]

The thermal efficiency is less than 100 per cent because some of the thermal energy from combustion is unavoidably lost as waste heat. In a typical spark ignition internal combustion engine the thermal

efficiency is about 38 per cent (relative to the lower heating value[5] of the fuel) and is largely independent of the power output (Ross, 1994). In diesel engines the thermal efficiency is about 48 per cent (Weiss et al, 2000).

The mechanical efficiency is less than 100 per cent because some of the piston energy is used for pumping (moving air and vaporized fuel into the cylinders and the combustion products out through the exhaust system) and to overcome piston and crankshaft friction. The mechanical efficiency in typical US cars, averaged over urban and highway driving, is about 52 per cent (Ross, 1994). It ranges from zero when the engine is idling to 90 per cent near wide-open throttle (when there is minimal pumping resistance). The overall efficiency is then $0.38 \times 0.52 = 20$ per cent. The mechanical efficiency of diesel engines is higher due to the absence of throttling at part load, but this is partly offset by greater engine friction.

Some of the mechanical output from the engine (shaft energy) is used to drive vehicle accessories (primarily the air conditioner, alternator and hydraulic power steering and braking). The balance passes through the transmission to the wheels; a typical transmission efficiency is 85 per cent. The output from the transmission is energy delivered to the wheels and is used to satisfy four loads: overcoming rolling resistance (that is, friction between the tyres and road), overcoming aerodynamic resistance (that is, drag by air), accelerating and climbing hills. The mathematical dependence of these loads on vehicle mass, frontal area, velocity and the drag and aerodynamic resistance coefficients, is given in Table 5.9. The engine power required to overcome rolling resistance varies linearly

Table 5.9 *Automobile loads (W) and the formulae for their calculation*

Load	Representation
1 Tyre rolling resistance	$P_{tire} = C_R M g v$
2 Air drag	$P_{air} = 1/2 \varrho C_D A v^3$
3 Vehicle acceleration	$P_{inertia} = 1/2 M [\Delta v^2 / \Delta t]$
4 Hill climbing	$P_{grade} = M g v \sin \theta$

Note: C_R is a dimensionless coefficient of rolling resistance, M is the vehicle + payload mass, g is the acceleration due to gravity (9.8 m/s²), v is the vehicle speed (m/s), ϱ is the density of air (about 1.2 kg/m³), C_D is a dimensionless coefficient of aerodynamic drag, A is the frontal area and θ is the slope of the hill being climbed (tanθ is the grade – the ratio of height gained to horizontal distance travelled).

with velocity, while the power required to overcome aerodynamic resistance varies with v^3.[6]

Figure 5.13 shows the dependence of the rolling, aerodynamic and hill-climbing loads on vehicle speed for a typical US automobile (having a mass of 1600kg, rolling resistance coefficient of 0.009, aerodynamic resistance coefficient of 0.33 and a frontal area of $2m^2$) climbing a 6 per cent slope. At a speed of 100kph, the required power on level terrain is less than 13kW, and even climbing a 6 per cent grade at 100kph, the required power is only 39kW. Conversely, to accelerate from 0 to 100kph in ten seconds requires 61.8kW, but the average car sold today in the US has a power of 150kW (see Figure 5.12). This is far more than is needed and necessitates a large engine. Most of the time, however, the engine is operating at a small fraction of its capacity, which reduces it efficiency due to relatively large frictional losses from the large engine. As well, the greater weight of the unnecessarily large engine further increases fuel consumption.

Figure 5.14 shows how fuel use varies with vehicle speed. As the speed increases, the engine friction decreases relative to the energy input (that is, the mechanical efficiency increases). Energy used per unit time (power draw) by auxiliaries is fixed, so the auxiliary energy use per unit distance travelled also decreases with increasing speed. However, the power required to overcome air drag increases sharply (with v^3). The net result is that minimum fuel consumption per km travelled occurs at a speed of around 90kph (with the absolute value and exact location of the minimum and the rate of increase on either side varying from car model to model).

Figure 5.15 shows the energy flow in a vehicle with a fuel consumption of 8.92 litres/100km (equivalent to 26.4mpg). The rolling, aerodynamic and acceleration loads are computed from the equations given in Table 5.9 and converted to an energy use per km, assuming one stop every 4km and acceleration to a speed of 65kph in ten seconds after every stop. The thermal, mechanical and transmission efficiencies are the typical values given above, while the auxiliary load is taken from Ross (1994) but obviously depends on many variable factors. As seen from Figure 5.15, the energy output from the engine is only 20 per cent of the input, the energy to the wheels is only 15 per cent of the input and the overall useful energy output (energy to wheels plus energy to auxiliaries) is only 17 per cent. For the assumptions used here, the loads due to rolling and aerodynamic resistance and due to acceleration are comparable in magnitude.

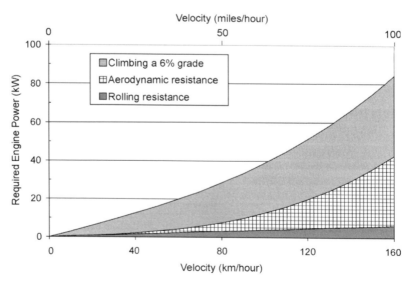

Note: See main text for assumptions.

Figure 5.13 *Variation of rolling, aerodynamic and hill climbing loads with vehicle speed for a typical US automobile climbing a 6 per cent slope, computed using the equations in Table 5.9*

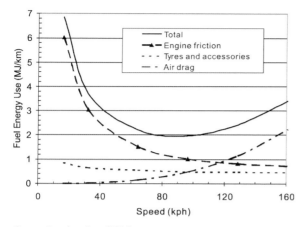

Source: Based on Ross (1994)

Figure 5.14 *Fuel consumption to meet different loads in an automobile, as a function of vehicle speed*

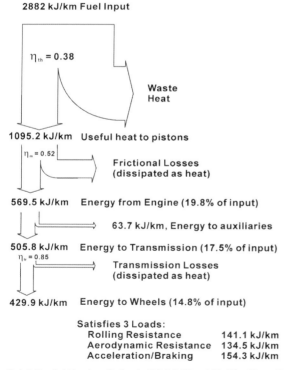

Figure 5.15 *Energy flow in a representative current-day automobile with a fuel consumption of 8.92 litres/100km (26.9mpg), computed as explained in the text*

5.4.4 Improving automobile fuel economy in conventional vehicles

Automobile fuel economy can be improved through incremental improvements to existing systems, or through whole new vehicle concepts. Incremental improvements in fuel economy can be achieved by: (1) increasing the engine thermal efficiency, (2) increasing the engine mechanical efficiency, (3) increasing the transmission efficiency, and (4) decreasing the loads that the engine must supply.

Increasing engine thermal efficiency

The present average thermal efficiency of 38 per cent for spark ignition (gasoline) engines could be increased to 45 per cent or even 50 per cent through the following options:

- *Leaner fuel:air mixtures.* A leaner mixture improves the thermal efficiency because a gas of simple molecules (such as air molecules) increases its pressure when heated more than a gas of complex molecules (such as vaporized gasoline) (but recall that lean-burn engines tend to have worse NO_x emissions). With complex molecules, much of the thermal energy goes into internal molecular motions rather than increasing pressure (for this reason, use of a fuel with simpler molecules – hydrogen or methane – also improves the thermal efficiency if the engine is designed for that fuel).
- *Variable compression ratio (VCR).* Current engines are limited to a compression ratio of about 10:1, due to the need to avoid engine knock at high loads. However, higher compression ratios without knock are possible at low load (which is the driving condition for most of time). VCR engines are currently under development and should give a fuel saving of 2–6 per cent, or 10–15 per cent if combined with a supercharged downsized engine (TRB, 2002).
- *Direct injection gasoline engines.* Cars in the 1970s used a carburettor to mix fuel and air before it went to the cylinder to be burned. Fuel injection – in which fuel is sprayed into the air just before it enters the cylinder – is now the standard. The next step will be direct-injection, in which the fuel will be sprayed directly into the cylinder at high pressure. This produces higher peak efficiency and a broader range

of conditions with relatively high efficiency (Friedman et al, 2001), and is already done for several vehicles. Fuel savings are 4–6 per cent (TRB, 2002).

- *Variable stroke.* It may be possible to design engines that can switch between four-stroke and two-stroke operation. Four-stroke engines carry out the stages of air intake, compression, combustion and exhaust in four strokes of a piston, whereas two-stroke engines carry out these operations in two strokes of the piston. Four-stroke operation is most efficient at high speeds, whereas two-stroke operation is most efficient during acceleration. Using each mode of operation when it is most efficient to do so could reduce fuel consumption by about 25 per cent (Graham-Rowe, 2008a).

A large fraction of the energy released by combustion of the fuel is lost as heat. BMW has developed a steam-based system that is said to recover 80 per cent of the exhaust heat to drive a small steam turbine linked to the crankshaft, giving an expected 15 per cent improvement in fuel economy (IEA, 2006).

Increasing engine mechanical efficiency

Five options for increasing the mechanical efficiency are:

- *Aggressive transmission management.* This amounts to reducing the engine speed (revolutions per minute) at a given power output, which can be done by building more gears and smaller gear ratios into the transmission and then, in driving, to gear up as quickly as possible. This in turn requires replacing four-speed automatic and five-speed manual transmissions with five-speed automatic and six-speed manual transmissions. There is a technology called *continuously variable transmission* (CVT) that goes even further by allowing an infinite number of variations in gear ratio between minimum and maximum gear ratios. This allows the engine speed and torque to be set so as to maximize efficiency over a wider range of operating conditions. Five-speed automatic transmissions should give a 2–3 per cent fuel saving compared to four-speed transmissions, and six-speed transmissions a further 1–2 per cent, while CVTs (which are available in Europe and Japan and in some Honda and Toyota vehicles in the US) give a 4–8 per cent saving (TRB, 2002).

- *Reduced engine displacement (piston volume).* This amounts to reducing the engine size for a given power output. As seen from Figure 5.12c, the specific power (ratio of power to engine size) more than doubled from 1976 to 2004. Specific power is increased through such measures as extra valves per cylinder, higher compression ratios, advanced fuel injection, sophisticated controls (i.e. ignition timing) and tuning of the intake and exhaust manifolds. The past increase in specific power has been used to increase engine power rather than reduce engine size. If, instead, engines were downsized by reducing power, fuel efficiency would increase because there would be fewer engine frictional losses and smaller vehicle weight (recall, present engines are significantly more powerful than legitimate power requirements). A 10 per cent reduction in engine size improves fuel economy by 6.6 per cent.

- *Variable valve control.* Variable early closing of the intake valve can be used instead of throttling (with fixed valve timing) in order to reduce the power output. This largely eliminates the pumping energy losses at part load, with up to a 10 per cent improvement in fuel economy. The technology also allows for greater peak power output, allowing further engine downsizing and associated fuel savings (Ross, 1994).

- *Reduced rubbing friction.* Fuel savings of 1–5 per cent are thought to be possible through improved lubricants (1 per cent saving) and better design, depending on the state of the baseline vehicle (TRB, 2002).

- *Idle-off when stopped.* Cars will soon be introduced whose engines shut off rather than idle when the car stops (or coasts) and start up as soon as the gas pedal is pressed. This will require a small motor/generator directly attached to the engine. It will serve as an integrated starter-generator that replaces the current starter motor and alternator, and will run on 42 volts instead of the 12-volt system of today's cars (Friedman et al, 2001). The 42-volt system will increase the efficiency of any accessory running on the current 12-volt system (since the required electrical current and associated resistance loses will be almost four times smaller). Overall fuel savings would be 1–2 per cent (TRB, 2002). Idling currently wastes 5–15 per cent of the fuel input over an average driving cycle.

There is plenty of scope for engine downsizing by reducing the excessive power of present automobiles. Recall from Figure 5.12b that acceleration times have fallen by 30 per cent since the 1980s while top speeds are far in excess of safe limits. Further downsizing (without reducing power) will be possible with turbochargers in diesel engines. These are commercially available and use hot exhaust gases in a turbine that drives a compressor to increase the pressure of the air entering the engine. This produces greater power for a given engine size, or allows a smaller engine for a given power. They also have the potential to become a key element of spark-ignition engines once variable valve control and advanced combustion become available (IEA, 2006).

Increasing the transmission efficiency

Transmission management, discussed above, can be used to improve the engine mechanical efficiency by allowing the engine to operate in the higher efficiency regions of the engine performance map. In addition, the efficiency of the transmission itself can be increased. Typical transmission efficiencies are as follows (Weiss et al, 2000): present five-speed manual: 94 per cent; present four-speed automatic: 70 per cent (city) to 80 per cent (highway); future five-speed automatically shifting clutched: 88 per cent; future continuously variable, 88 per cent; future single-gear ratio direct drive (in all-electric vehicles or series hybrids (see below)), 93 per cent.

Load reduction

A reduction in the load (power) that an engine must supply directly reduces the fuel consumption, with further, indirect savings due to the engine downsizing that can occur (recall: smaller engines save energy by reducing the vehicle weight and are more efficient if they run at a larger fraction of their peak power). A 10 per cent load reduction reduces fuel use by 4 per cent

in urban driving and by 5 per cent in highway driving through direct savings, and by a further 6.6 per cent through engine downsizing (Ross, 1994).

The engine load can be reduced by:

- reducing tyre rolling resistance through higher-pressure tyres;
- reducing aerodynamic resistance through changes in car shape;
- reducing vehicle weight;
- reducing vehicle accessory loads.

Heywood et al (2003) indicate that tyre rolling resistance could be reduced from 0.009 at present to 0.006. The average drag coefficient, C_D, dropped from 0.45 in the mid-1970s to 0.35 in 1990; Heywood et al (2003) assign a value of 0.33 for the 2001 baseline car, 0.27 to the 2020 baseline car and 0.22 to the 2020 advanced car. A value of 0.19 is conceivable according to Ross (1994).

Vehicle weight can be reduced in four ways:

1. through the use of advanced lightweight materials in place of heavier materials (such as steel);
2. through the redesign of vehicles so as to require less materials for a given vehicle size (for example, by changing from a body-on-frame to lighter unibody designs, which has not yet occurred in all vehicles);
3. through the downsizing of vehicle components (such as the engine) that is possible as a result of the first two weight reductions; and
4. through a market shift to smaller vehicles.

Bandivadekar et al (2008) compare the density, strength parameters (yield and tensile strength and elastic modulus) and cost of various alternative automotive materials. These properties are summarized in Table 5.10. Carbon fibre polymer composites have a fifth the density of mild steel, a comparable yield strength and elastic modulus and almost three times the

Table 5.10 *Properties and relative cost of alternative structural materials that can be used in cars*

Material	Density (gm/cm³)	Yield strength (MPa)	Tensile strength (MPa)	Elastic modulus (GPa)	Relative cost per part
Mild steel	7.86	200	300	200	1.0
High strength steel	7.87	345	483	205	1.0–1.5
Aluminium	2.71	275	295	70	1.3–2.0
Magnesium	1.77	124	228	45	1.5–2.5
Carbon fibre composite	1.57	200	810	190	2.0–10.0

Source: Bandivadekar et al (2008)

tensile strength. However, the cost per part would be two to ten times higher than using mild steel. Magnesium and aluminium have one quarter and one third the density of mild steel, respectively, and are less expensive than carbon fibre composites, but are also less favourable from a strength point of view. Through judicious use of all three alternative materials (aluminium, magnesium, carbon fibre), an acceptable balance between weight reduction and increased cost can be achieved. Weight reductions of 20–45 per cent have been achieved in a few concept vehicles (taking into account the secondary savings due to downsizing after the primary savings).

The major accessory loads are air conditioning (usually the largest and most amenable to reduction), the alternator (usually second) and hydraulic power steering (third). Energy use by automobile air conditioning systems (kWh of electricity per unit of heat removed) differs by a factor of two between the least efficient and the most efficient systems (IEA, 2006), with peak power draws of up to 5–6kW (Rugh et al, 2004). The production of 1kWh of electricity in turn consumes about 1 litre of gasoline, and at typical gasoline prices of $1–2/litre, this electricity is several times more expensive than even present-day PV power (40–80 cents/kWh). This provides a strong economic incentive to reduce air conditioning energy use, which can be done by (1) reducing thermal loads by up to 75 per cent through increasing use of recirculated air, improved glazings and improved spatial control of interior temperatures; and (2) doubling the air conditioner COP from 1.3 (typical of today) to an achievable 2.6 (Farrington and Rugh, 2000). The net result would be a reduction in cooling energy requirements by up to a factor of eight (and probably by at least a factor of four).

Further reductions in automobile air conditioning energy use in the US are possible through improved system design. As discussed by Clodic et al (2005), air conditioners in US cars work by producing more cooling than needed, then mixing the cold air with heated air as needed in order to produce the desired temperature. The control system is a simple on/off option. Variable capacity systems were introduced in Europe and Japan in the mid-1990s.

Figure 5.16 shows the energy flow in a spark-ignition (gasoline) automobile with the following improvements, which are based on the preceding discussion: an improvement in the engine thermal efficiency from 0.38

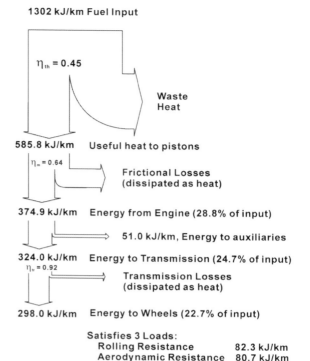

Figure 5.16 *Energy flow in an advanced future automobile with a fuel consumption of 4.03 litres/100km (58.4mpg), computed as explained in the text*

to 0.45 and in the mechanical efficiency from 0.52 to 0.64; an improvement in the transmission efficiency from 0.85 to 0.92; a reduction in vehicle + payload mass from 1600kg to 1400kg; a decrease in the auxiliary load by 20 per cent; a decrease in the rolling-resistance coefficient from 0.009 to 0.006; and a decrease in the drag coefficient from 0.33 to 0.22. The result is a fuel consumption of 4.03 litres/100km (equivalent to 58.3mpg) – a saving of 55 per cent compared to the current vehicle (featured in Figure 5.15). Of this 55 per cent saving, just over half (53 per cent) is due to reduced loads and the rest is due to improved thermal, mechanical and transmission efficiencies.

Further savings would be possible if the engine power to vehicle mass ratio were to return to the values of the 1980s. Yet greater fuel energy savings will be possible in vehicles with alternative drivetrains. Four alternative drivetrain systems for automobiles are currently under development: hybrid gasoline–electric vehicles, plug-in hybrid vehicles, battery electric vehicles and fuel cell-powered vehicles. These are discussed next.

5.4.5 Hybrid gasoline–electric and diesel–electric vehicles

Current hybrid electric vehicles (HEVs) use a gasoline engine to supply average power requirements and to charge a battery. During times of peak power (i.e. during acceleration or climbing a grade), power from the battery supplements the power from the engine, which can therefore be reduced in size. Hybrids can save energy in five ways:

- by downsizing the engine, which reduces engine friction losses;
- by optimizing the contribution from the battery so that the engine operates more of the time near conditions (moderate speed and moderate to high torque) that give peak efficiency. This is less important in diesel vehicles, which have little efficiency drop at part load; in diesels, the main saving is through engine downsizing;
- through *regenerative braking*, which is facilitated by automatically rewiring the motor to act as a generator when the brakes are applied, so some of the vehicle's kinetic energy can be used to recharge the battery. This gives the largest savings in stop-and-go traffic, but is less important than the saving due to engine downsizing and operation at peak efficiency;
- by eliminating engine idling, giving another 5–15 per cent energy savings;
- by shifting power steering and other accessories to (more efficient) electric operation.

The first modern hybrid vehicle was the Toyota Prius in Japan in 1997, followed by the Honda Insight. Both of these are gasoline–electric hybrids, although diesel–electric hybrids have also been produced. The 1998 Prius compact sedan attained 42mpg under US CAFE[7] test conditions with a 0–60 acceleration time of 14.5 seconds, while the larger (medium-size) 2004 Prius attained 55mpg and a 0–60 acceleration time of 10.5 seconds (Kobayashi et al, 2009). Fuel economy would have been better still if the previous acceleration time had been maintained.

Two basic hybrid configurations are possible: a *series hybrid*, in which the engine generates electricity that recharges the battery and drives a motor that drives the wheels, with or without supplemental electrical power to the motor from the battery; and a *parallel hybrid*, in which the engine powers the wheels directly through a transmission, with supplemental mechanical power to the transmission from the electric motor when needed. All hybrid vehicles currently use the parallel configuration. Within the parallel hybrid configuration, several options are possible, as outlined in Weiss et al (2000). For example, power from the motor can be added before or after engine power has passed through the transmission. Some hybrid vehicles are 'mild' hybrids in that there is minimal battery storage and so minimal driving on battery power; rather, the main features are regenerative braking and battery assistance to start the vehicle moving. Electric motors provide maximum output (torque) at zero or near-zero rates of rotation, which is when it is most needed (for starting and accelerating from low speeds).

The components of one possible parallel powertrain are illustrated in Figure 5.17, along with component efficiencies. The battery is charged both with mechanical power from the engine and from regenerative braking. The steps in the first case involve an electric generator (95–98 per cent efficiency) to charge the battery (95 per cent efficiency) and, later, withdrawal of electricity from the battery, passing through a DC/AC inverter (94 per cent efficiency), motor (92–95 per cent efficiency) and gears (92–98 per cent efficiency). Thus, only about 72–82 per cent of energy taken from the engine to recharge the battery is available later (as input to the transmission). Regenerative braking captures some of the kinetic energy of the vehicle and converts it to chemical energy in the battery. This energy then passes through the same steps as above, with the same losses. Thus (using the efficiencies from Figure 5.17), a minimum of $0.60 \times 0.90 \times 0.94 \times 0.92 \times 0.92 \times 0.88 = 0.38$ (38 per cent) and a maximum of $0.85 \times 0.95 \times 0.94 \times 0.95 \times 0.98 \times 0.88 = 0.62$ (62 per cent) of the kinetic energy taken from the vehicle is returned to the wheels. If 30 per cent of the engine output is used for acceleration, the output from the engine (and hence, fuel economy) is potentially enhanced by (0.38 to 0.62) × 30 per cent = 11–19 per cent through regenerative braking alone. This is in addition to the other efficiency advantages of hybrid vehicles mentioned above.

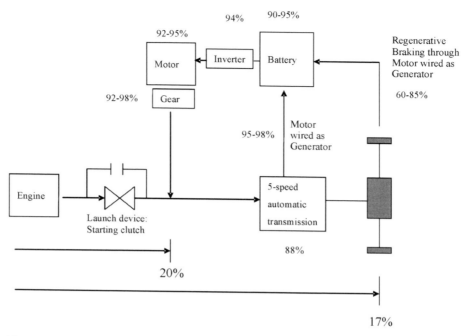

Source: Modified from GMC et al (2001)

Figure 5.17 *Components of the drive train in an ICE–battery hybrid vehicle (parallel configuration), and component efficiencies*

Table 5.11 *Characteristics of fuel-efficient, gasoline-powered vehicles*

	Family car		SUV	
	Evolutionary	Hybrid	Evolutionary	Hybrid
Engine System	GDI, VVC, four valves per cylinder, ISG	Advanced GDI	GDI, VVC, four valves per cylinder	Advanced GDI
Transmission	Continuously variable	Integrated gearset	Five-speed automatic	Integrated gearset
Vehicular properties				
– Drag coefficient, C_D	Down 10%	Down 10%	Down 10%	Down 10%
– Rolling resistance, C_r	Down 20%	Down 20%	Down 20%	Down 20%
– Weight	2660lbs	2782lbs	2633lbs	2749lbs
0–60mph acceleration time	10 seconds	10 seconds	11 seconds	11 seconds
Fuel economy, litres/100km (mpg)				
Baseline vehicle, CAFE test	8.97 (26.2)		11.57 (20.3)	
Advanced vehicles, CAFE test	5.13 (45.8)	3.96 (59.3)	5.86 (40.1)	4.40 (53.4)
Real world	6.42 (36.6)	4.96 (47.4)	7.32 (32.1)	5.50 (42.7)
Improvement (CAFE)	75%	125%	97%	163%
Baseline vehicle cost	$19,535		$29,915	
Price increase	$1292	$5089	$2087	$5472
Fuel savings over 10 years[a]	$4382	$5715	$6226	$8192
Lifetime cost savings	$3090	$626	$4139	$2720

Note: GDI = gasoline direct injection; VVC = variable valve control; ISG = integrated starter-generator. [a] Based on 19,000 miles (30,570km) driven per year and gasoline at $3/gallon, with annual cost savings discounted to the present at 6 per cent per year.
Source: de Cicco et al (2001)

Table 5.11 summarizes the characteristics and fuel economy of family cars and SUVs incorporating evolutionary changes (as described in subsection 5.4.4) and of the hybrid gasoline–electric drivetrains described here. These characteristics were calculated by de Cicco et al (2001). Fuel economy under CAFE test conditions could drop from 9.0 litres/100km to 5.13 litres/100km (45.8mpg) for an evolutionary family car and from 11.6 litres/100km to 5.86 litres/100km (40.1mpg) for an evolutionary SUV, representing savings of over 50 per cent in both cases (this is consistent with the simple analysis shown earlier in Figures 5.15 and 5.16). Forming a hybrid saves another 20–25 per cent, giving final fuel economies of 4.0 litres/100km and 4.4 litres/100km for the family car and SUV, respectively – savings of 56 per cent and 62 per cent compared to the present-day vehicles. However, this assumes that hybrid technology is used to improve fuel economy rather than to improve other measures of performance (such as acceleration). In the most recent hybrid vehicles, technological improvements have not been used to improve fuel economy, but rather, to provide faster acceleration.

5.4.6 All-electric and plug-in hybrid vehicles

The HEV described in the previous section relies entirely on the onboard fuel (gasoline or diesel fuel) to power the car. The battery is recharged solely with energy from the engine (either directly, or indirectly when the brakes are applied). All-electric vehicles (referred to as BEVs – battery-electric vehicles), in contrast, depend entirely on an external electricity source to charge the battery. They are thereby limited by the amount of energy that can be stored in the battery and in the long times (several hours) required to fully recharge the lead acid batteries that were used in the first all-electric vehicles. The increase in vehicle mass associated with heavy batteries imposes an energy efficiency penalty that further limits the driving range (to no more than 300km before recharging the battery).

An alternative to the HEV and BEV is the plug-in hybrid electric vehicle (PHEV) – an HEV with a battery that can be recharged from an external electricity source. Two powertrain configurations are possible in PHEVs. The first is like an HEV but with a larger battery, with traction provided by a combination of the electric motor and ICE based on the circumstances. The second, known as an *extended range electric vehicle*, would have an electric drive only, with the ICE operating only when the battery is nearing depletion and used to power an electric generator.

Types of batteries, performance and cost

Lead acid batteries are the batteries currently used in non-hybrid cars, and had been used in the first generation of all-electric vehicles. They provide low energy storage and power in proportion to their mass and volume, so large and heavy batteries were needed that in turn imposed a significant energy penalty. The most promising battery technologies are the nickel-metal hydride (NiMH), used in all hybrid vehicles today, and the lithium-ion battery. NiMH batteries are available in stores as rechargeable AAA, AA, C and D batteries, while lithium-ion batteries are used in cameras, cell phones and laptop computers. The key performance parameters are: the specific energy (Wh of energy storage per kg of battery mass), specific power (kW/kg), the number of times the battery can be discharged and recharged, and cost. Figure 5.18 plots specific power versus specific energy for a variety of NiMH and lithium-ion batteries available as of early 2004, in comparison with lead acid batteries. For both NiMH and lithium-ion batteries, higher energy storage per unit weight is associated with lower power per unit

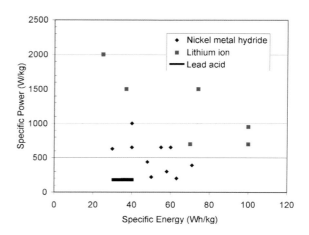

Source: Data from EPRI (2004)

Figure 5.18 *Specific power versus specific energy for various NiMH, lithium-ion and lead acid batteries*

weight. Lithium-ion batteries available in 2004 had a maximum specific energy of about 100Wh/kg (50 per cent better than the best NiMH batteries), while the most recent cells (as of 2009) had a specific energy of 200Wh/kg but a specific power of only 300W/kg. Developments in nanotechnology may push the performance of lithium-ion batteries to 3000W/kg with 100–150Wh/kg energy storage (Altairnano, 2009). Lithium-ion batteries may be more easily recycled than lead acid or NiMH batteries, an important consideration from both an environmental and cost point of view.

Battery lifespan and retention of capacity will be more challenging in PHEVs than in HEVs because of the deeper battery discharging in PHEVs. The number of discharge cycles possible during the life of a battery depends on how deeply the battery is discharged before recharging; the industry target is 2500 cycles at 80 per cent discharge and 3000 cycles at 70 per cent discharge, although the Electric Power Research Institute (EPRI) reports that some batteries manufactured with a high degree of quality control achieve 6000–8000 cycles with 60 per cent discharge (EPRI, 2004). In the first test of a lithium-ion battery pack by EPRI, the battery retained 90 per cent of its initial capacity after 2500 deep discharge cycles (Douglas, 2008).

NiMH and lithium-ion batteries are currently expensive, ranging from about $300/kWh at a low ratio of power to energy to $1400/kWh at a high ratio, compared to $50/kWh for lead acid batteries. However, costs are expected to fall substantially with further technical improvements and high volume (≥100,000 units/year) manufacturing. Figure 5.19 compares the current cost of lithium-ion batteries with projected future costs of NiMH or lithium-ion batteries. Projected costs range from $235 to 400/kWh (increasing with increasing power to energy ratio). IEA (2006) indicates a minimum long-term cost for lithium-ion batteries of $160/kWh, unless alternative materials can be used for some components of the battery. These are the costs of the modules and do not include the cost of the housing, connectors and dealer markup.

For a projected cost of $270/kWh achieved through high-volume production, the cost breakdown is: materials, $156/kWh; manufacturing cost,

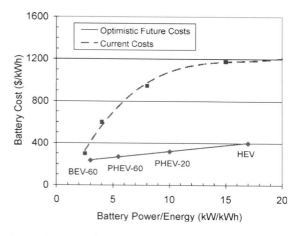

Source: Current cost from Kromer and Heywood (2007), optimistically projected cost from EPRI (2004) and Kromer and Heywood (2007)

Figure 5.19 *Current cost of lithium-ion batteries and optimistically projected cost of NiMH or lithium-ion batteries*

$47/kWh; profit margin, $68/kWh (Kromer and Heywood, 2007). Inasmuch as materials account for more than half the costs, future increases in material costs due to scarcity would have a significant impact on the battery cost. Lithium-ion batteries used in consumer products cost about $300/kWh, but this cost reflects the benefits of high-volume production without the cost of the additional safety features that would be needed for automotive applications.

Ultra-capacitors

A capacitor is a device that stores electrical energy. It consists of two conducting plates between a non-conducting material. An electric field is created between the plates that stores energy. Capacitors can be quickly charged and discharged with only 5 per cent energy loss, but their energy density is only 15–20Wh/kg (Kobayashi et al, 2009), compared to 100–200Wh/kg in recent vehicle-scale lithium-ion batteries (as discussed above). Nevertheless, they have been used in combination with batteries in tests of a diesel–electric hybrid bus (Lammert, 2008). Some people have speculated that developments in nanotechnology may eventually permit 'ultra' capacitors with energy densities of 30–60Wh/kg.

Fuel savings

The efficiency of an electric drivetrain originating from AC grid electricity will be given by the product of the battery charging efficiency, the discharge efficiency, and the inverter, motor and gear efficiencies. This will be in the order of 65–70 per cent, whereas the efficiency of an internal combustion engine is only about 20 per cent at 25 per cent of full load and 25 per cent at full load, and the transmission efficiency might be 92 per cent, giving an overall average drivetrain efficiency of about 22 per cent, or about 3 times smaller. The efficiency gain using electricity instead of diesel vehicles should be smaller, because of the higher efficiency to begin with of diesel engines compared to spark ignition engines. This expectation is confirmed by Figure 5.20, which plots grid electricity requirements per km driven versus fuel consumption per km for various gasoline and diesel PHEVs, as simulated in a study sponsored by the Electric Power Research Institute (based in California). A diesel vehicle with the same fuel economy as a gasoline vehicle is a heavier vehicle (compensated by the greater efficiency of the diesel engine), so more electricity is needed than for the gasoline vehicle. Also shown in Figure 5.20 is the amount of grid electricity needed to save one litre of fuel. This ratio is about 3kWh/litre saved for gasoline vehicles and about 4kWh/litre saved for diesel vehicles. Inasmuch as 1 litre of gasoline contains 32.2MJ and 1kWh of electricity contains 3.6MJ of energy, the efficiency using electricity is about 32.2/(3 × 3.6), or 3.3 times greater than using gasoline, which is consistent with the rough estimate given above. Inasmuch as electricity typically costs 10 cents/kWh, whereas gasoline costs $1–2/litre, one can also see that there will be a substantial saving in energy costs using electricity.

The greater the energy-storage capacity of the battery, the larger the fraction of total driving that would be powered with electricity from the grid, and the greater the savings in gasoline or diesel fuel. Figure 5.21 gives the gasoline savings as a function of the electric-only range for typical US commuting patterns (in places where commuters drive less per day, the relative savings using a battery with a given range would be larger because a larger fraction of the driving would be powered by the battery). Results are shown assuming either that the battery is used exclusively until the driving range is reached, or that the battery provides only half of the required energy but over twice the driving range. For the first case with a 95km (60-mile) range, the gasoline saving is almost 75 per cent.

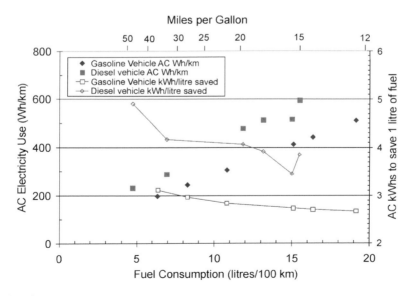

Figure 5.20 *Grid electricity requirements per km driven versus fuel consumption per km for various gasoline and diesel PHEVs, and the amount of grid electricity needed to save 1 litre of fuel*

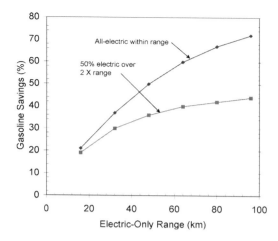

Note: In countries where daily driving distances our shorter, the relative savings would be larger.

Source: Data from Kliesch and Langer (2006)

Figure 5.21 *Savings in gasoline energy consumption using plug-in hybrid vehicles as a function of the electric-only range, for typical US driving patterns*

These expected savings will be reduced by aggressive driving behaviour and by cold conditions. Aggressive driving increases the energy consumption of HEVs and PHEVs by more (>30 per cent) than it increases the fuel consumption of conventional vehicles (5–10 per cent) (Sovacool and Hirsh, 2009). This is partly because less energy is recaptured through regenerative braking with aggressive driving. Cold conditions reduce the efficiency of electric drivetrains but not that of conventional vehicles, so the savings compared to conventional vehicles will be less in cold climates.

Resource implications of widespread use of NiMH and lithium-ion batteries

The metal in NiMH batteries is a mixture of abundant and rare-earth metals, the most common formula being AB_5, where A is a rare-earth mixture of lanthanum, cerium, neodymium and praseodymium, and B is nickel, cobalt, manganese and/or aluminium. Despite the name 'rare earth', these elements are relatively abundant in the earth's crust (9.5ppm for praseodymium, 32ppm for lanthanum, 38ppm for neodymium and 4600ppm for cerium) and most occur in a large number of different minerals. Lithium is the 33rd most abundant element in the earth's crust, with an average concentration of 20–70ppm. However, the relevant parameters are the amounts of the various elements that are concentrated to at least 10,000ppm (1.0 per cent) in ores, since the amount of rock that needs to be mined to yield one tonne of a mineral (assuming no losses during processing) is $1/x$ tonnes, where x is the ore grade ($x = 0.01$ for ores that are 1.0 per cent the mineral of interest). If one envisages a hypothetical future worldwide automobile fleet of 1–2 billion hybrid vehicles, it is likely that mineral supply (as reflected in costs) will become a limiting factor before a fleet this size is reached.

The common lithium-ion battery ($LiCoO_2$) consists of a cobalt oxide cathode, a graphite carbon anode and a $LiPF_6$ electrolyte, and has a cell specific energy of 110–190Wh/kg.[8] Cobalt is expensive, so alternative cathode materials are under development. There are also concerns about the safety (the battery tends to catch fire if overheated or overcharged), stability and longevity (they tend to lose capacity and power with time and with use) of lithium-ion batteries of the size needed for automotive applications and under realistic driving conditions, so a large number of alternative lithium-ion batteries are under development. Some of these, and their key advantages, are listed in Table 5.12. Current research focuses on developing non-flammable electrolytes and fire retardant additives.

The US Geological Survey identifies a world lithium reserve of about 4.1Mt, a reserve base (which includes the reserve and resources thought to have a reasonable potential of becoming economic in the future) of 11Mt and a resource base of 14Mt (USGS, 2009). Given a battery energy density of about 200Wh/kg using $LiCoO_2$ (see Table 5.12) and an energy storage requirement of 20kWh (see Table 5.13), a fleet of only 500 million hybrid vehicles would require 3.55Mt of lithium. This would largely exhaust the current lithium reserve and is almost a third of the entire estimated land-based resource. Although LiS batteries promise an energy density of at least 350Wh/kg, the lithium requirements would be larger still because lithium is 17.8 per cent of this battery and only 7.1 per cent of the $LiCoO_2$ battery by weight. Seawater is 0.11ppm lithium by weight and so is a potential source of lithium.[9]

Table 5.12 *Alternative lithium-ion batteries*

Battery type	Comment
Alternative cathodes	
$LiCoO_2$	Commonly used in consumer products. Best specific energy in 2006 of 190Wh/kg, about double the value when first introduced in 1991. Expensive
$Li[Ni_{0.8}Co_{0.2}]O_2$	Has started replacing the $LiCoO_2$ battery
$Li[Ni_{1/3}Co_{1/3}Mn_{1/3}]O_2$	This and the next two are used in power tools and medical equipment
$LiMnO_2$	Has a spinel structure, giving it permanently low resistance, high loading capability and inherent stability
$LiFePO_4$	Inexpensive
LiS	Specific energy of 350Wh/kg already achieved
Alternative anodes	
Nano-sized $LiTi_5O_{12}$ particles	Inherently safe, outstanding rate capability, but may not be cost-competitive
C fibre instead of graphitic C	
Alternative electrolytes	
$LiB(C_2O_4)_2$	
Solid state LiPON (lithium phosphorus oxynitride)	Specific energy of 300Wh/kg already achieved

Note: Specific energies are given in terms of the cell mass rather than in terms of the total module mass.
Source: Ritchie and Howard (2006) and Buchmann (2006)

Lifecycle vehicle costs

Lifecycle vehicle costs and lifecycle energy savings are intertwined, but it is more convenient to begin with lifecycle vehicle costs. Table 5.13 gives the gasoline and/or electric energy use for various vehicles, assuming a baseline HEV with a gasoline consumption of 5.6 litres/100km (42mpg) and an electricity consumption

Table 5.13 *Characteristics of various electric vehicles*

	Vehicle type					
	HEV	**PHEV-20**		**PHEV-60**		**BEV-40**
Gasoline use (litres/100 km)	5.6	6.0	6.1	6.5	6.9	6.2
Electricity use (kWh AC/km)	0.19	0.20	0.21	0.22	0.24	0.21
Electric range (miles)		20	20	60	60	40
Battery mass (kg)	33	122	163	262	349	175
Specific energy (Wh/kg)[a]	30	50	50	70	70	70
Battery capacity (kWh)[b]	1	6.1	8.2	18.3	24.5	12.2
Depth of discharge		0.8	0.6	0.8	0.6	0.8
Number of cycles[c]		2000	4000	2000	4000	2000
Lifetime electric miles[d]	0	40,000	80,000	120,000	240,000	80,000
Electric miles per year[e]	0	6000	6000	12,000	12,000	12,000
Years of battery use	10	6.7	13.3	10	20	6.7
Battery module cost ($/kWh)						
Near-term	2000	1600	1600	1400	1400	1400
Long-term[f]	400	320	320	270	270	235
Battery cost (total $)						
Near-term[g]	2693	13,010	17,120	33,084	43,885	22,282
Long-term[g]	1093	3226	4074	7170	9333	4472

Note: PHEV-20 = plug-in hybrid with a 20-mile range using the battery; PHEV-60 = plug-in hybrid with a 60-mile range using the battery; BEV-40 = electric vehicle with a 40-mile range. [a] From EPRI (2004, Table 2.1). [b] Given by electric driving range (km) × AC electricity use x charging efficiency (0.8) divided by discharge fraction. [c] From EPRI (2004, Table 4.7). Some battery manufacturers claim 6000–8000 cycles at 60 per cent discharge. [d] Given by number of cycles times the indicated electric driving range per cycle. [e] Computed assuming 300 cycles per year times the indicated electric driving range for each cycle. [f] From Figure 5.19 (taken from EPRI, 2004). [g] Includes balance-of-system and dealer markup cost of $680 + $13/kWh, taken from EPRI (2004, Table 4.6).

Source: Based on EPRI (2004) and Kliesch and Langer (2006)

by a PHEV of 0.19kWh AC/km if the fuel consumption is 5.6 litres/100 km (based on Figure 5.20). Actual gasoline and electricity consumption per km travelled have been adjusted to take into account the effect of the added mass of the battery. The battery mass and specific energy, energy storage capacity, depth of discharge, number of charge/discharge cycles over the life of the battery, the number of miles driven using the battery, and the near-term and long-term battery cost projections are also given in Table 5.13. The battery mass depends on the required battery capacity and battery specific energy; the required battery capacity in turn depends on the desired driving range, permitted battery discharge and kWh DC (that is, kWh in the battery) required per km driven. The given kWh AC/km reflects losses in charging the battery, so multiplying this number times the charging efficiency (assumed to be 0.8) gives kWh DC required per km. For the PHEV vehicles, results are given assuming a small battery but deeper discharge and fewer cycles over the lifetime of the battery, and for a large battery with shallower discharge and more cycles. Projected

long-term battery costs range from $1093 for the HEV to $9333 for the PHEV-60 with a large battery.

Table 5.14 shows the incremental cost of various electric vehicles compared to a conventional gasoline-powered vehicle, the annual cost savings and the NPV of the cost savings (defined below). The incremental vehicle cost excluding the battery is taken from EPRI (2004) and assumes high-volume production (100,000 units per year) of motors, controllers and electric drive components, while the long-term battery costs from Table 5.13 are used. This results in additional vehicle costs compared to the conventional gasoline vehicle of about $515 for the HEV, $2910–3589 for the PHEV-20, $6066–7797 for the PHEV-60 and $7991 for the BEV-40. Annual energy cost savings are computed assuming a gasoline cost of $4/gallon, an electricity cost of $0.10/kWh and an annual distance driven of 15,000 miles (24,140 km). The NPVs of the annual costs are then computed by multiplying the annual cost savings by the NPV factor (which is computed using Equation D.2 of Appendix D). In computing the NPV here, I assume ten years of savings for all vehicles except the

Table 5.14 *Economic analysis of conventional vehicles using gasoline (at 29mpg) and various electric vehicles of Table 5.13*

	Type of vehicle						
	CV	HEV	PHEV-20	PHEV-20	PHEV-60	PHEV-60	BEV-40
Vehicle cost excluding battery ($)	18,860	18,282	19,054	19,054	19,054	19,054	23,138
Battery cost ($)	0	1093	2716	3395	5872	7603	3713
Total vehicle cost ($)	18,860	19,375	21,770	22,449	24,929	26,657	26,851
Incremental vehicle cost		**515**	**2910**	**3589**	**6066**	**7797**	**7991**
Maintenance cost ($/mile)	0.063	0.056	0.048	0.048	0.048	0.048	0.035
Total miles/year	15,000	15,000	15,000	15,000	15,000	15,000	12,000
Gasoline miles/year	15,000	15,000	9000	9000	3000	3000	0
Electricity miles/year	0	0	6000	6000	12,000	12,000	12,000
Gasoline cost/year ($)	2069	1428	912	936	332	350	0
Electricity cost/year ($)	0	0	122	126	269	285	253
Total energy cost/year ($)	2069	1428	1033	1062	601	635	253
Maintenance cost/year ($)	945	840	720	720	720	720	420
Total cost/year ($)	**3014**	**2268**	**1753**	**1782**	**1321**	**1355**	**673**
Savings/Year ($/year)		746	1261	1232	1693	1659	2341
NPV factor		6.145	4.703	7.194	4.703	7.194	4.703
NPV of lifetime savings ($)		4581	5928	8862	10,401	14,126	11,008
NPV of lifetime net gain ($)		**4066**	**3017**	**5273**	**4335**	**6239**	**3016**
	Alternative incremental vehicle costs and resulting NPV						
Incremental vehicle cost ($)		3266		8436		13,289	
NPV of lifetime net gain ($)		1315		426		837	

Note: Energy costs are computed assuming gasoline at $4/gallon and electricity at $0.10/kWh. Vehicle costs and maintenance costs are from EPRI (2004), while electricity and gasoline use and battery costs are from Table 5.13. The NPV of the annual cost savings is computed assuming an interest rate of 10 per cent and the battery lifespans indicated in Table 5.13. NPV results using alternative and much higher incremental vehicle costs from Simpson (2006) are given in the last row.

all-electric vehicle, for which only seven years of savings are assumed because of the shorter lifetime of the battery when used as the sole source of power. A discount rate of 10 per cent is also assumed.

For the assumptions given above, the NPV of the lifecycle energy cost savings more than offsets the incremental cost of the electric vehicle in all cases (HEV, PHEV and BEV). This outcome is dependent on achieving the long-term projected battery costs given in Table 5.13, which are several times smaller than current costs, and would require medium-scale (100,000/year) production of batteries. The net cost saving is greater for the PHEV-20 and PHEV-60 with larger batteries, due to the longer lifetime resulting from shallower discharge after each recharge. As the HEV is closest to being economic today, growth of the HEV market could drive down the cost of batteries, thereby making PHEV more economic, increasing the demand and scale of manufacturing for batteries and further driving down the cost to the point where BEVs may become economic. In the long run, BEVs could be less expensive than PHEVs, as the engine and associated fuel system would be eliminated. One concept involves exchanging battery packs with recharged batteries in a matter of minutes with a robotic arm at exchange stations.

Others researchers have come to less optimistic conclusions concerning the extra cost of HEVs and PHEVs in the long run. Kromer and Heywood (2007) estimate an extra cost for the HEV of $1400 relative to a turbocharged gasoline vehicle, instead of $515, but comparable additional costs for the PHEV. Simpson

(2006) estimates consistently higher long-term costs: an extra $3266 for an HEV, an extra $8436 for a PHEV-20 and an extra $13,289 for a PHEV-60. However, even with these higher extra costs, the HEV and PHEVs would still yield lower discounted costs; the NPV of the net gain is shown using the higher incremental costs at the bottom of Table 5.14.

Even if PHEVs become economically attractive (in terms of entailing lower lifecycle costs at economically justifiable discount rates), consumers do not perform an NPV calculation before making purchase decisions. There may therefore be substantial resistance to purchasing PHEVs even when they result in lower costs over their lifespan. Sovacool and Hirsh (2009) provide a long list of other obstacles that will need to be addressed if PHEVs are to be successfully marketed once they represent the minimal lifecycle-cost choice. Shiau et al (2009) find, based on an analysis of the cost, weight and performance of PHEVs with different electric driving ranges, that PHEVs with a relatively small battery capacity and that can be recharged frequently will be most attractive economically. Thus, efforts to promote adoption of PHEVs could initially be targeted at this market segment.

Primary energy use and CO_2 emissions

Table 5.15 gives the primary energy use per km travelled when powered by gasoline and by grid electricity for the vehicles featured in Tables 5.13–5.14, computed here using the information from those tables and assuming:

Table 5.15 *Primary energy use and CO_2 emissions per km driven powered by gasoline and powered by electricity for the plug-in hybrid and all-electric vehicles featured in Table 5.14*

	Type of vehicle						
	CV	HEV	PHEV-20	PHEV-20	PHEV-60	PHEV-60	BEV-40
Primary energy, gasoline vehicle (MJ/km)	3.14	2.17	2.31	2.37	2.53	2.66	2.39
Primary energy use with electricity (MJ/km)	0.00	1.06	1.14	1.17	1.25	1.33	1.18
Average primary energy use (MJ/km)	3.14	2.17	1.84	1.89	1.51	1.59	1.18
CO_2 emissions (gmC/km), gasoline	62.9	43.4	46.2	47.4	50.5	53.2	47.8
CO_2 emissions (gmC/km), electricity	0.0	26.6	28.4	29.3	31.3	33.2	29.5
Average CO_2 emissions (gmC/km)	62.9	43.4	39.1	40.2	35.2	37.2	29.5

Note: Primary energy use for gasoline vehicles is computed assuming a heating value for gasoline of 32.2MJ/litre and an upstream energy use for exploration, extraction, refining and delivery equal to 20 per cent of the delivered gasoline energy, while primary energy use for electric vehicles is computed assuming an efficiency of 40 per cent in generating and transmitting electricity to the point of use (see Chapter 2, subsection 2.3.5 for justification of these markup factors). CO_2 emissions are computed assuming emission factors of 20gmC/MJ for crude petroleum and 25gmC/MJ for coal (assumed to be the source of the electricity). Average energy use and CO_2 emissions are for the gasoline to electric distance-driven ratios implied in Table 5.14. gmC = grams of carbon.

(1) grid electricity to be generated from coal with a generation × transmission efficiency of 0.40; (2) a heating value for gasoline of 32.2MJ/litre; and (3) an upstream energy use in supplying gasoline of 20 per cent of the delivered gasoline energy (see Chapter 2, subsection 2.3.5 for justification). With these assumptions, the primary energy required to drive a given distance is about 25 per cent less for the PHEV-60 than for the hybrid vehicle. The CO_2 emission, however, is only 15 per cent less because the emission factor for coal is 25 per cent greater than that for petroleum (25kgC/GJ instead of 20kgC/GJ). If the electricity comes from natural gas (having an emission factor of about 16kgC/GJ when small upstream emissions are included), emissions are reduced by about 35 per cent. However, natural gas supplies are limited. Thus, a significant reduction in CO_2 emissions from plug-in hybrid vehicles requires that they be charged largely at times when there is excess renewably based electricity available that could not otherwise be used, or that CO_2 emissions from the generation of electricity with coal be captured and sequestered (as discussed in Volume 2, Chapter 9) if this turns out to be practical and economically viable in the future.

Implications for national electricity demand

The impact of converting the entire automobile fleet to PHEVs and charging them with electricity from the grid would, in most if not all nations, have only a minor effect on total domestic electricity demand. Taking the US as an example, suppose that there is a fleet of 200 million PHEVs, driven on average 19,000km/yr (the current US average), with about two thirds – or 12,000km/yr – powered with grid electricity. With an electricity consumption of 0.22kWh AC/km, the total electricity demand would be 528TWh/yr. This is only 12 per cent of the 2005 electricity demand in the US of about 4300TWh. The power draw per vehicle during recharging would be 1.4–2.0kW, but recharging would occur at off-peak times and so would not require an increase in system capacity (200 million vehicles at 2kW would give a total demand of 200GW – about 20 per cent of the total US capacity of 1078GW in 2005).

Vehicle-to-grid (V2G) systems

Plug-in hybrid vehicles could eventually be designed to send power back to the grid when parked and when

extra power is needed. Utilities could draw power from those vehicles that are parked during times of peak load, while customers could recharge their vehicles at night, when utility rates are lower, thus generating a profit while the utility levelizes the electricity demand (some mechanism would need to be in place to prevent the utility from drawing so much power from the vehicle battery that there is insufficient energy for the homebound trip). Such a system is referred to as a 'vehicle-to-grid' or *V2G* system. As noted in Volume 2 (Chapter 2, subsection 2.2.6), the potential electricity production from 15 per cent efficient PV modules covering all parking lots in the US exceeds total US electricity demand, so parked bidirectional plug-in hybrids could play a significant role in storing and supplying PV electricity. A mere 10 million vehicles (about 2 per cent of the current US automobile fleet) with 10kW batteries could supply a standby power of 100GW (compared to a 2005 US powerplant capacity of 1078GW). In the meantime, tests of a V2G system by Google and Pacific Gas and Electric have started in California using a converted Toyota Prius (see www.google.org/recharge).

5.4.7 Impact of air conditioning in conventional and hybrid vehicles

Operating an air conditioner in a conventional (non-hybrid) vehicle makes the vehicle operate a little like a hybrid, in that when the brakes are applied, energy is directed toward the air conditioner compressor – a form of regenerative braking.[10] As well, the additional load on the engine when the engine is operating at part load (the normal situation) moves the engine from a lower-efficiency regime to a higher-efficiency regime. The energy requirements and source of energy for air conditioning under urban and highway driving for advanced HEVs and conventional vehicles (as estimated by Kromer and Heywood, 2007) are shown in Figure 5.22. For conventional vehicles under urban conditions, about a third of the energy required to operate the air conditioner comes from braking energy and a third of the required energy is offset by more efficient operation of the engine, so only a third of the energy required to operate the air conditioner results in increased fuel consumption. The energy required to operate an air conditioner in an HEV comes largely from increased fuel use (since regenerative braking already operates through the battery and the engine is

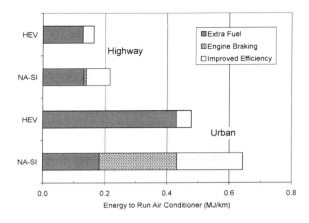

Note: NA-SI = naturally aspirated spark ignition.
Source: Kromer and Heywood (2007)

Figure 5.22 *Contributions to air conditioning energy requirements for advanced naturally aspirated spark ignition (gasoline) and hybrid gasoline–electric vehicles for urban and highway driving*

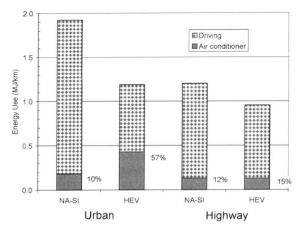

Note: Also given is the air conditioning energy use as a percentage of the non-air conditioning energy use.
Source: Kromer and Heywood (2007)

Figure 5.23 *Air conditioning and non-air conditioning energy use (MJ/km) under urban and highway driving conditions for advanced conventional and hybrid vehicles*

already in a more efficient operating regime), but the air conditioning energy requirement is smaller. The net result is that the impact of air conditioning on fuel consumption is twice as large in an HEV as in a conventional vehicle. Air conditioning requires less energy with highway driving, and there is essentially no difference in the additional fuel requirements by hybrid and conventional vehicles.

Figure 5.23 shows the air conditioning and non-air conditioning energy use under urban and highway driving conditions for the advanced vehicles considered above. Also indicated is the percentage increase in fuel consumption due to air conditioning; for advanced conventional vehicles in urban driving, 10 per cent more fuel is required, while for advanced HEVs in urban driving, almost 60 per cent more fuel is required. The greater relative impact in HEVs is because the absolute additional fuel use is almost 2.5 times greater for hybrid vehicles (for the reasons given above) and the fuel consumption for driving is about half as large for the vehicles illustrated in Figure 5.23. The information in Figure 5.23 can be used to compute the ratio of hybrid to conventional fuel consumption with and without air conditioning. This is shown in Figure 5.24, where it can be seen that (according to the calculations that underlie Figure 5.23), future HEVs will reduce fuel consumption by about 52 per cent compared to

future naturally aspirated (not turbocharged) gasoline vehicles under urban driving without air conditioning, by about 38 per cent with air conditioning, and by about 25 per cent under highway conditions with or without air conditioning (assuming that further technological improvements in hybrids are directed toward improving fuel efficiency rather than increasing power and acceleration).

As noted above (subsection 5.4.4), air conditioning energy requirements can be reduced by up to a factor of eight through a combination of up to a factor of four reduction in cooling requirements and a factor of two increase in the COP of automobile air conditioners.

The presence of air conditioners in road vehicles has a global warming effect through leakage of the refrigerant (primarily HFC-134a, a powerful GHG) that is many times that arising from the extra fuel use in conventional vehicles. According to Clodic et al (2005), the worldwide effect of leakage of mobile air conditioner refrigerant had a warming effect in 2003 about six times that arising from the extra fuel use. Applying this to the absolute fuel consumption impact shown in Figure 5.23 for conventional vehicles (to which the 6:1 ratio would apply), it can be seen that the combined effect of the fuel penalty and refrigerant

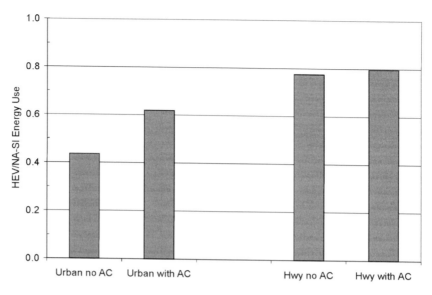

Source: Computed from the data in Figure 5.23

Figure 5.24 *Ratio of energy use by hybrid to energy use by conventional vehicles under urban and highway driving conditions with and without air conditioning*

leakage (at today's rate of about 15 per cent/year, given in Table 6.10 of Clodic et al, 2005) would be to almost triple the equivalent CO_2 emission from advanced HEVs.[11] Clearly, measures to reduce refrigerant leakage (a factor of two reduction is considered to be achievable) and the development of alternative but viable refrigerants with much lower GWP (defined in Chapter 2, subsection 2.6.5) are needed.

5.4.8 Fuel cell vehicles

Fuel cells are electrochemical devices for converting a fuel into electricity, which can then be used in an electric drivetrain that is similar to that used in hybrid electric vehicles. All fuel cells require a hydrogen-rich gas as a fuel input and produce water, electricity and waste heat by combining the hydrogen with oxygen from the air. The fuel cell would operate either on pure hydrogen (which would have to be stored on board the vehicle somehow), or on some hydrocarbon fuel (gasoline, diesel fuel, naphtha, compressed natural gas or methanol) that is converted on board the vehicle to a hydrogen-rich gas. Pure hydrogen could be produced from fossil fuels, either at central facilities or at

refuelling stations, or could be produced by electrolysis of water using fossil-based or renewably based electricity. The chemical formulae and energy related properties of some different fuels that could be used in fuel cell vehicles are given in Appendix B. Hydrogen contains three times more energy per unit mass than gasoline, but it is bulky (even if compressed to 1000 times atmospheric pressure, hydrogen would occupy three times the volume of gasoline with the same energy content). Systems for producing and distributing hydrogen are discussed in Volume 2, Chapter 10. Here, we discuss issues related to the use of fuel cells in automobiles and the onboard processing of hydrocarbon fuels or storage of hydrogen.

Fuel cells for transportation

A fuel cell power source consists of a series of individual *cells*, in the shape of a flat plate, that are joined together to form a fuel cell *stack*. Each cell consists of a flat plate called an anode (where hydrogen is added) and another plate called a cathode (where oxygen is added), separated by an electrolyte (an ion-conducting substance). Figure 5.25 shows an exploded view of a

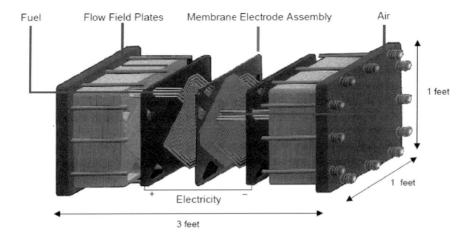

Fuel   Flow Field Plates   Membrane Electrode Assembly   Air

1 feet

1 feet

+ Electricity −

3 feet

Note: Each stack consists of 188 cells at 0.8V, and two stacks would be used in series, to give a total peak power of 50kW at 300V. The 50kW output is the net power, after accounting for the power requirements of the system itself (which amount to 6kW).
Source: Little (2000)

Figure 5.25 *A 25kW (net) fuel cell stack for automobile applications*

fuel cell stack for automotive applications. The first fully functional fuel cell vehicle (FCV) was built in the early 1990s, the tenth around 1996 and the hundredth shortly after the turn of the century. Each FCV built in 2003 cost around $1 million to build, by hand (Sperling and Cannon, 2004). It is thought that production levels would have to reach 100,000 vehicles per year before major economies of scale come into play. Prior to that, a number of significant issues related both to FCVs and to how to introduce a radically new technology will need to be resolved.

The different kinds of fuel cells were introduced in Chapter 3 (Tables 3.2 and 3.3). The solid polymer fuel cell (also known as the PEMFC) is the most promising fuel cell for transportation applications (whether in cars, buses or locomotives). An 85kW PEMFC for vehicular applications, manufactured by Ballard Power Systems, is shown in Figure 5.26. The PEMFC operates at a temperature of 50–90°C and has the highest efficiency, power density (W/litre) and specific power (W/kg) of any candidate fuel cell. It can also change its output rapidly. The electrolyte is solid (eliminating corrosion and safety concerns associated with liquid-electrolyte fuel cells), consisting of a perfluorosulphonic acid membrane. This is essentially Teflon, a substance built around CF_2 groups that has a very high conductivity for protons. The protons move

through the electrolyte as hydronium (H_3O^+), so water is needed and the reactant gases need to be humidified to replace the water that is lost through evaporation (Srinavasan et al, 1999). The electrolyte is very thin (10–100µm), which minimizes resistance losses. Like many types of fuel cells, it requires a Pt catalyst but can

Note: It measures 805 × 375 × 250mm, has a mass of 96kg and produces electricity at 284V.
Source: www.ballard.com

Figure 5.26 *An 85kW PEMFC manufactured by Ballard Power Systems of British Columbia, Canada*

start in seconds, whereas the PAFC requires no Pt catalyst but takes a few minutes to start.

Smoothing out the load placed on the fuel cell

Some sort of system to level the fluctuations in the load placed on the fuel cell and to reduce the peak fuel cell power requirement would probably be used in fuel cell vehicles. One possibility is to use a battery, and most of the existing fuel cell demonstration vehicles are hybrid fuel cell–battery vehicles. The battery is recharged by the fuel cell and stores energy for peak power. It also facilitates partial recapture of energy during regenerative braking. As in gasoline–electric hybrid vehicles, the battery used in a fuel cell vehicle would be an order of magnitude or more lighter than the battery in an all-electric vehicle, since the battery would not be used for storing energy (rather, the energy comes with the fuel carried by the vehicle). Flywheels and ultra-capacitors are other possibilities.

Figure 5.27 shows the components of the powertrain in a hydrogen-powered fuel cell–battery hybrid vehicle. By comparison with Figure 5.17, it can be seen that forming a fuel cell–battery hybrid is much simpler than forming an ICE–battery hybrid, since both the fuel cell

and battery output pass through the inverter, motor and gears, while the transmission is eliminated.

Fuel processing using hydrocarbon fuels

A fuel cell drivetrain system in a hydrocarbon-fuelled vehicle would consist of the following components:

- an onboard reformer to convert hydrocarbon fuels to hydrogen as part of a larger fuel-processing system;
- a compressor to compress the hydrogen-rich fuel that is produced from reforming of hydrocarbon fuels (this is eliminated if the vehicle is fuelled directly with hydrogen that is stored on board as a compressed gas);
- the fuel cell stack, which produces DC electricity;
- a system for separating and recycling unused hydrogen (about 15 per cent of the input) that passes through the fuel cell stack;
- an optional expander, to recapture some of the energy from pressurized exhaust gases;
- a DC to DC or AC to DC power conditioning unit;
- an AC or DC electric motor;
- gears, to convert the motor shaft rotation to wheel rotation at the desired rate.

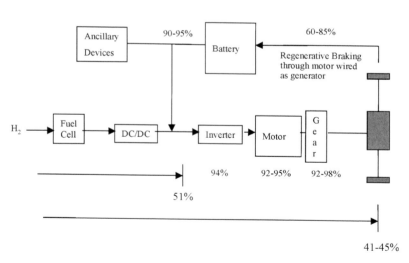

Source: Modified from GMC (2001)

Figure 5.27 *Components of the drivetrain in a hydrogen-powered fuel cell–battery hybrid vehicle, and component efficiencies*

Table 5.16 *Characteristics of different processing system/fuel combinations for processing of hydrocarbon fuels*

Process and fuel	Operating temperature (K)	Efficiency (%)	CO (volume %) prior to removal	Output (volume %) after removal of CO			
				H$_2$	CO$_2$	H$_2$O	N$_2$
Steam reforming of							
Methane	1000–1100	89	11.2	64.1	16.3	17.8	1.8
Methanol	500–560	87	0.8	61.8	21.1	14.1	3.0
Ethanol	800–1000	87	10–14	62.6	21.4	12.5	3.5
Gasoline	1000–1150	85	20	58.2	19.7	20.6	1.5
Partial oxidation of							
Methane	500–1600	85	20	43.8	15.4	5.4	35.4
Gasoline, diesel	1150–1900	74	25	32.0	19.4	5.8	42.8

Source: L. F. Brown (2001)

A hydrogen-rich fuel could be produced through onboard processing of hydrocarbon fuels in one of two ways: through steam reforming or through partial oxidation. Table 5.16 lists the fuels that can be used for each process. Steam reforming of ethanol at this stage is only a theoretical possibility but is of interest because of the growing use of ethanol made from biomass (see Volume 2, Chapter 4, subsections 4.3.5 and 4.3.6). In steam reforming, water is reacted with the hydrocarbon fuel to produce a mixture of CO, CO$_2$ and H$_2$. The reaction is endothermic, so heat must be added, either by burning some of the fuel or H$_2$. This is followed by a water shift reaction, in which water reacts with the CO to produce more CO$_2$ and H$_2$. In the case of steam reforming of methane, the reactions are:

$$CH_4 + H_2O \rightarrow CO + 3H_2 \text{ (steam reforming)} \quad (5.1)$$

and:

$$CO + H_2O \rightarrow CO_2 + H_2 \text{ (water shift reaction)} \quad (5.2)$$

The water shift reaction is really a two-stage reaction (with two separate reactor vessels). The first stage occurs at a temperature of around 400°C and uses a chromium catalyst, while the second stage occurs just above 200°C with a copper oxide-zinc oxide catalyst (L. F. Brown, 2001). In the case of steam reforming of methanol, only a single water shift stage is needed, and it occurs simultaneously with steam reforming. The water shift reaction is followed by a final, preferential partial oxidation stage that removes the residual CO (0.4–1.2 per cent). In partial oxidation, the fuel is

reacted with a quantity of O$_2$ insufficient for complete combustion. To prevent excessive temperatures when using partial oxidation, steam can be added, which reacts with the fuel, so that the net result is a combination of steam reforming and partial oxidation. Partial oxidation is also followed by a one- or two-stage water shift reaction.

Extra water must be carried on board, in addition to fuel, when hydrocarbon fuels are used. The water requirements are indicated by the net reforming reactions, which are:

$$CH_4 + 2H_2O \rightarrow 4H_2 + CO_2 \quad (5.3)$$

$$CH_3OH + H_2O \rightarrow 3H_2 + CO_2 \quad (5.4)$$

and:

$$C_2H_5OH + 3H_2O \rightarrow 6H_2 + 2CO_. \quad (5.5)$$

and:

$$CH_4 + 0.5O_2 + H_2O \rightarrow 3H_2 + CO_2 \quad (5.6)$$

for steam reforming of methane, methanol and ethanol, and partial oxidation of methane, respectively. The ratio of H$_2$ produced to water required is 2:1 for steam reforming of methane and ethanol and 3:1 for steam reforming of methanol and partial oxidation of methane.

The operating temperature, efficiency (hydrogen energy output/fuel energy input) and fuel composition for different processing systems and fuels are summarized in Table 5.16. Steam reforming produces more H$_2$ and is more efficient than partial oxidation, but is poor in responding to changing power requirements. Partial oxidation is good for transient operation, but is slightly less efficient than steam reforming and produces only 35–45 per cent H$_2$ (the balance being CO$_2$ and N$_2$). This is a significant drawback for automotive applications, as it reduces the fuel cell power output. Steam reforming requires temperatures of 500–1150K, whereas partial oxidation requires temperatures of 1500–1600K using methane and 1150–1900K for multi-carbon fuels. Higher temperature processing tends to be less efficient, requires more costly materials, needs finer controls and poses more safety concerns. The high temperature of the fuel processor undoes the safety advantage of using the low temperature (80°C) PEMFC.

As for fuels, methanol requires by far the lowest processing temperature (500–560K) with simultaneous steam reforming and a single water shift reaction. All other fuels produce substantially greater CO and so need a two-stage reaction. This is a significant drawback for all systems except steam reforming of methanol. However, methanol is toxic and soluble in water, so leakage and spillage would threaten drinking water supplies. From an energy efficiency point of view, methanol has the added disadvantage that it is

produced by processing some prior fuel, usually methane, so onboard processing to produce hydrogen would represent a second processing step. Some energy is unavoidably lost each time a fuel is converted from one form to another. However, methanol could in the future be produced from biomass carbon and hydrogen produced by electrolysis using electricity generated from carbon-free energy sources such as wind. Methanol in this case would serve as a chemical carrier of hydrogen, and could be used if current problems and safety concerns associated with other methods of onboard storage of hydrogen (discussed later) cannot be satisfactorily resolved.

Table 5.17 compares the estimated emissions of VOCs (volatile organic compounds), CO and NO$_x$ for fuel cell vehicles using gasoline, methanol and hydrogen, along with US and California emission standards. Hydrocarbon fuels use processors with catalysts that are poisoned by sulphur. This is a problem for gasoline, so a new grade of gasoline with a sulphur content below the latest North American standard of 15ppm might be necessary for gasoline-powered fuel cell vehicles. Fuel cell vehicles powered by gasoline would have essentially zero emissions of CO and NO$_x$, but would not meet US Tier II emission standards for volatile organic compounds, according to the emission estimates given in Table 5.17.

Table 5.17 *Estimated emissions of VOCs, CO and NO$_x$ (NO and NO$_2$) from conventional gasoline ICE vehicles and from future fuel cell vehicles*

	VOCs		CO		NO$_x$	
	Best	Probable	Best	Probable	Best	Probable
Gasoline ICEV		0.755		7.553		0.704
Gasoline FCV	0.268	0.371	0.004	0.005		0.001
Methanol FCV	0.020	0.023	0.003	0.004		0.001
Hydrogen FCV		0.004		0.003		0.001
			Emission standards			
Tier I		0.250		3.4		0.4
Tier II (full phase in by 2007)		0.125		1.7		0.2
ULEV		0.04		1.7		0.2
SULEV		0.01		1.0		0.02
EZEV		0.004		0.17		0.02

Note: Also given are the US Tier II and California emission ratings. VOC = volatile organic compound; FCV = fuel cell vehicle; ICEV = internal combustion engine vehicle; ULEV = ultra-low emission vehicle; SULEV = super ultra-low emission vehicle; EZEV = electric zero-emission vehicle.
Source: Thomas et al (2000)

In summary, using hydrocarbon fuels in a fuel cell vehicle adds considerable complexity and cost to the vehicle, requires high operating temperatures and entails an efficiency penalty. Neither gasoline nor methanol fuel cell vehicles are fully satisfactory from an environmental point of view, and methanol has the greatest water requirements of any hydrocarbon fuel (which could be a problem in dry regions). However, methanol is attractive because it is the easiest hydrocarbon to use in a fuel cell, it is an effective medium for storing hydrogen (as discussed below) and it could be produced sustainably from a combination of biomass (which would serve as a source of carbon) and H_2 produced electrolytically from renewably based electricity.

Onboard storage of hydrogen

An alternative to onboard processing of hydrocarbon fuels is to use hydrogen directly. However, the development of systems for the onboard storage of hydrogen has been a major challenge.

The weight and volume of different hydrogen storage systems are compared with each other and with gasoline storage in Table 5.18. Each of the hydrogen storage options has distinct advantages and disadvantages. Storage as compressed hydrogen is low-weight and requires about 10 per cent of the energy content of the fuel being stored for compression to

Table 5.18 *Mass and volume of alternative hydrogen storage systems for a mid-size hydrogen fuel cell automobile, with 3.9kg of stored hydrogen*

Fuel storage system	Fuel + storage system mass (kg)	Storage system volume (litres)
Compressed gas at 350atm (5144psi)	72.2 (5.5% H_2)	217
Compressed gas at 700atm (10,287psi)	89.8 (4.6% H_2)	150
Liquid hydrogen	31.0 (14% H_2)	116
Metal hydride ($FeTiH_{1.8}$)	232 (1.9% H_2)	41
Gasoline	64	50

Source: Based on Ogden (1999), Lasher et al (2009), Table 5.19 and author's estimate for a gasoline vehicle with the same range as for the hydrogen vehicle (610km) and a fuel consumption of 8 litres/100km

350atm (see Volume 2, Chapter 10, subsection 10.4.1), but is bulky compared to other storage options. Storage at 350atm (about 5000psi) has already been used in demonstration vehicles, and 700–840 atm (10,000–12,000psi) storage containers are currently under development (compression to 700atm would require about 14 per cent of the stored energy – much less than double that required for compression to 350atm). There are now some 140 hydrogen refuelling stations around the world, as part of hydrogen vehicle demonstration projects, and this experience indicates that routine refuelling of compressed-hydrogen vehicles in less than three minutes would not be a problem (Maus et al, 2008).

Storage as liquid hydrogen (LH_2) entails even less weight and is much less bulky than compressed hydrogen, but entails a significant energy penalty (about one third the energy content of the fuel at present, as little as 20 per cent in the future, as discussed in Volume 2, Chapter 10, subsection 10.4.1). Boiloff of less than 1 per cent of the stored hydrogen per day can be expected, adding to the energy penalty. Boiloff can be avoided if hydrogen is consumed at a sufficiently great rate; for heat absorption at a rate of 1W by a tank storing 5kg of H_2 (the current state-of-the art), an 80mpg (34km/litre) equivalent vehicle would need to be driven 25km per day, and the car could remain parked for only five days without venting of H_2. LH_2 has a large thermal expansion coefficient, so LH_2 tanks can be filled to only 85–90 per cent of their capacity so as to prevent LH_2 spills (Berry et al, 2003).

A third option is to store hydrogen as metal hydrides. These are compounds in which hydrogen is absorbed by a metal under pressure and released when heat is applied. Hydride storage systems may consist of H_2 absorbed by:

- a single metal (as in MgH_2 or TiH_2), or
- intermetallic hydrides of the form AB, A_2B, AB_2 or AB_5, where A is a metal that strongly absorbs hydrogen (such as Mg, Ti or La) and B is an element that does not (usually lighter transition metals from the first row of the periodic table, such as V, Cr, Fe, Co, Ni and Cu), or
- hydride complexes, in which a metal stabilizes an ion that contains hydrogen, an example being magnesium nickel hydride (Mg_2NiH_4), in which Mg stabilizes the $(NiH_4)^{2-}$ ion.

Table 5.19 *Hydrogen mass percentage and density for different metal hydrides, glass microspheres and carbon nanotubes (referred to here as the matrix), and derived matrix + H_2 density, specific energy and energy density (on a LHV basis)*

Hydride	H_2 mass (%)	H_2 density (kg/m³)	Matrix + H_2 density (kg/m³)	Specific energy (MJ/kg)	Energy density (GJ/m³)
$NaAlH_4$	7.5	93.6	1248	9.0	11.3
MgH_2	7.5	110	1470	9.0	13.2
$MgNiH_4$	3.7	93.0	2500	4.5	11.2
TiH_2	4.1	151	3690	4.9	18.2
$TiFeH_2$	1.9	105	5470	2.3	12.6
$LaNi_5H_6$	1.4	90	6380	1.7	10.8
Glass microspheres, now	4.5	36	800	5.4	4.3
Glass microspheres, target	10.0	84	840	12.0	10.1
Carbon nanotubes, low	4.0	40	1000	4.8	4.8
Carbon nanotubes, high	8.0	80	1000	9.6	9.6

Source: Hydrogen mass percentage and density for different metal hydrides from Berry et al (2003)

The density of hydrogen, percentage of the hydride mass that is hydrogen and the overall density of the hydride are given in Table 5.19 for a few hydrides. Many hydrides have hydrogen densities of 80–120kg/m³, compared to 71kg/m³ for LH_2, and equivalent to that of gaseous H_2 at over 1000atm. However, hydride systems usually achieve only 50 per cent volumetric efficiency because allowance must be made for the substantial expansion of the hydride as it absorbs H_2, so the final storage density will usually be somewhat less than that of LH_2. As well, not all of the H_2 mass can be extracted, so the usable storage capacity would be less (by 10 per cent or more). Furthermore, hydride storage systems are very heavy and refuelling would take 10–20 minutes, compared to a few minutes for liquid hydrogen and no more than three minutes for compressed hydrogen. Magnesium hydride (MgH_2) has the lowest density (about 1500kg/m³) among simple hydrides, and Mg is abundant and so would appear to be a good candidate, but the magnesium bond requires significant energy (75kJ/molH_2 at 550K) for separation (the exhaust from automotive fuel cells would be at only 380K). Complex hydrides of light metals (such as Li, Be, B, Na and Al) contain 5–14 weight per cent H_2 but usually do not absorb H_2 directly. Sodium alanate ($NaAlH_4$) would appear to be a promising candidate, as it has 7.5wt per cent H_2, a density of 1248kg/m³ (see Table 5.19) and the hydrogen can be released using heat at only 400K if catalysed by titanium or zirconium. However, only about half of the hydrogen is recoverable, and refuelling

generates temperatures above 455K, which in turn drive refuelling pressures as high as 100atm. These and other problems associated with various hydride systems are discussed more fully in Berry et al (2003) and Sakintuna et al (2007).

It is instructive to compare the hydrogen specific energy (energy content per unit weight) and energy density (energy content per unit volume) of the various hydride systems with the specific energy and energy density of advanced batteries. These quantities are given as the last two columns in Table 5.19 for hydrides (assuming 100 per cent volumetric efficiency). Specific energy ranges from 1.7 to 9.0MJ/kg, compared to 0.36–0.72MJ/kg (100–200Wh/kg) for lithium-ion batteries, while energy density ranges from 10–13GJ/m³, compared to 0.86–1.72GJ/m³ for lithium-ion batteries.[12] Thus, even heavy metal hydride storage would contain a higher specific energy and energy density than lithium-ion batteries.

Another option that has been suggested for transportation applications is to store H_2 in small (50μm diameter) glass spheres at pressures of up to 690atm (10,000psi). The microspheres would be filled with hydrogen by heating them (to increase their permeability) and applying the hydrogen under pressure. Hydrogen is released within less than one second by exposing the microspheres to intense infrared radiation. The glass is doped with 0.5wt per cent Fe_3O_4 to increase the permeability to hydrogen. Present storage density is 4.5wt per cent H_2 and 36kg H_2/m³, implying an overall system density of

$800kg/m^3$, which is exceptionally light. In the long term, a storage density of 10wt per cent H_2 is hoped for. However, the specific energy and energy density would not be attractive compared to lithium-ion batteries (see Table 5.19). As well, the refilling half life (the time required to fill half of the volume remaining at any given time) at a temperature of 350°C would range from 25 minutes to several hours (depending on the composition of the microsphere), increasing to a minimum of 18 days at a temperature of 50°C according to Kohli et al (2008). This makes glass microspheres unsuitable for transportation applications.

A final, speculative option is hydrogen storage in carbon nanotubes (having a diameter of about 10µm). Hydrogen in this case would be physically adsorbed onto the surface of the nanotube structure. There are indications that 4–8wt per cent H_2 is possible at 300K, which for a bulk nanotube density of $1000kg/m^3$ would correspond to a hydrogen density of $40–80kg/m^3$ (Berry et al, 2003). This is less than for metal hydrides ($80–140kg/m^3$), but has the advantage of lightness ($1000kg/m^3$ versus $1500–6400kg/m^3$ for the hydrides shown in Table 5.19). However, the specific energy and energy density would still not be attractive compared to lithium-ion batteries (see Table 5.19). The storage density achieved so far at ambient temperatures is less than 1wt per cent H_2 (W. C. Xu et al, 2007; Yürüm et al, 2009).

Sarkar and Banerjee (2005) have computed the energy required to make various storage systems capable of storing 3.75kg H_2, which would be adequate for 5000 cycles at 600km per cycle, or a total driving range of 300,000km. Table 5.20 summarizes the results of their calculations and compares the storage system embodied energy per km with other energy requirements. The extraction and refining of the metals in metal hydride systems requires an amount of energy, when spread over the total amount of hydrogen stored during the lifetime of the hydride system, equal to 70 per cent of that required for liquefaction in the case of FeTi hydride. As well, hydride systems increase the fuel consumption because of the greater weight of the vehicle. The energy penalty for the storage system is equal to the energy required

Table 5.20 *Characteristics of different energy storage systems and overall energy use to drive 1km*

	Compressed-H_2 tank (345atm)	Cryogenic-H_2 tank	FeTi hydride	Mg hydride
Storage mass (kg)	56.9	22.9	353	145
Tank embodied energy (GJ)	10.3	4.68	53.2	18.2
Compression or liquefaction (MJ/kg)	18.6	39		
Hydrogen and energy use				
H_2 consumption (gm/km)	6.24	6.40	8.04	9.70
Direct H_2 energy use (kJ/km)	750	769	966	1166
Upstream energy use, production (kJ/km)	188	192	242	292
Upstream energy use, transmission (kJ/km)	37.5	38.5	48.3	58.3
Compression or liquefaction (kJ/km)	116[a]	250[b]	0.0	0.0
Tank (kJ/km)	34.2	15.6	177.3	60.7
Total energy use (kJ/km)	1125	1265	1434	1576
Energy required for storage				
kJ/km	150	285	394	477
As a % of direct energy use	20	38	52	64
As a % of direct and upstream energy use	15	29	40	49

Note: The assumptions here are that H_2 is produced by steam reforming of methane at 80 per cent efficiency (LHV basis), total driving is 300,000km and energy to transport hydrogen is 5 per cent of the energy transported. [a] The compressional energy requirement here is 13.4 per cent of the direct + compressional energy use, which is somewhat higher than the 10 per cent value given in Volume 2 (Chapter 10, subsection 10.4.1). [b] The liquefaction energy requirement here is 24.5 per cent of the direct + liquefaction energy use, compared to 30–40 per cent at present and perhaps 20 per cent or less in the future.
Source: Sarkar and Banerjee (2005)

for compression or liquefaction of hydrogen, the energy required to make the storage system, and the additional direct and upstream energy consumption associated with the greater energy requirement to drive a given distance due to the mass of the storage system. This energy penalty as computed here is shown in Table 5.20 and is also given as a percentage of the direct + indirect H_2 energy use.[13] The energy cost of storage ranges from 15 per cent of the H_2 energy content for compressed-gas storage to 47 per cent for Mg hydride storage. In terms of the additional lifecycle energy requirement compared to compressed-gas storage, liquefaction requires an extra 12 per cent, FeTi hydride an extra 27 per cent and Mg hydride an extra 40 per cent.

Overall, the most promising method for storing hydrogen on board automobiles and trucks would appear to be as compressed hydrogen. Compressed hydrogen would be largely decompressed before being fed to the fuel cell, and perhaps 70 per cent of the energy used to compress it could be recovered with a microturbine at peak power and hydrogen flow rate (Cownden et al, 2001). Incoming air would be compressed to the same pressure as the hydrogen being fed to the fuel cell, which would range from 1.3atm at 10 per cent output to 3atm at peak output so as to maximize the net system efficiency. The relative bulkiness of compressed hydrogen (particularly compared to gasoline) can be mitigated through use in vehicles that would be about four times as efficient as present-day vehicles when powered by pure hydrogen. However, there are a number of concerns about the safety of compressed hydrogen that may result in significant public opposition or consumer resistance (see Volume 2, Chapter 10, subsection 10.8.1), although these issues are technically solvable.

Methanol (CH_3OH) also serves as a hydrogen carrier, in this case a liquid chemical carrier containing 12.5wt per cent H_2, a hydrogen density of 99kg H_2/m^3 and an overall density of only 792kg/m^3. As discussed above, an onboard reformer would be required to process the fuel prior to feeding it to the fuel cell, thereby increasing cost and complexity and reducing the overall efficiency of fuel use, but the operating temperature of the reformer would be far lower (500–560K) than for other hydrocarbon fuels (800–1150K), so methanol may still represent the best

option if the technical problems and/or safety concerns associated with onboard storage of pure hydrogen, or the difficult problems associated with the development of an infrastructure to supply hydrogen made from renewable energy (discussed in Volume 2, Chapter 10), cannot be resolved.

Fuel cell drivetrain efficiency

Figure 5.28 shows the variation with load in the stack and system efficiency of a PEMFC using pure hydrogen as the input fuel, as simulated by Kromer and Heywood (2007) for their base case assumptions (conservative assumptions produce efficiencies at most a few percentage points lower). Technical details on the determinants of the efficiency of a fuel cell are given in Box 5.2. Stack efficiency continuously increases as load decreases, while system efficiency peaks at about 60 per cent around 25 per cent of peak output. This variation is opposite to that of an ICE, where the combined thermal times mechanical efficiency is largest at peak power and decreases at partial power. Since a typical driving load is around 25 per cent of the peak load, this works to the advantage of fuel cells.

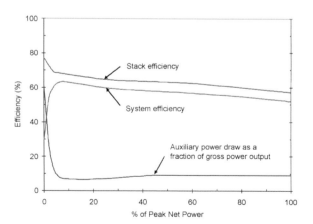

Note: Also given is the power use by auxiliaries as a fraction of the gross power production.
Source: Kromer and Heywood (2007)

Figure 5.28 *Variation in the stack and system efficiency of a PEMFC with net power output as a percentage of peak output*

Box 5.2 Efficiency of a fuel cell

The net reaction in a hydrogen-powered fuel cell ($H_2 + 1/2\ O_2 \rightarrow H_2O$) is the same as the combustion of hydrogen, a process that releases heat. The heat involved in a chemical reaction carried out at constant pressure is referred to as the change in the system's *enthalpy*, ΔH. When heat is released, ΔH is negative. In a fuel cell, some of the chemical energy is *directly* converted into electrical energy, without the intermediate steps of combustion and the production of mechanical energy, as in a turbine. The maximum amount of chemical energy that can be converted into electricity by the fuel cell reactions is equal to the *Gibbs free energy*, ΔG. Thus, the theoretical maximum efficiency of a fuel cell in producing electricity, referred to as the *reversible efficiency*, is given by:

$$\eta_r = \Delta G / \Delta H \tag{5.7}$$

The heat released from combustion of a fuel depends on whether one assumes that water vapour or liquid water is produced by the reaction of hydrogen with oxygen; in the latter case, the latent heat of condensation is added, giving the HHV of the fuel. If the water vapour is not condensed, one gets the LHV of the fuel. Efficiencies based on HHVs will be lower than efficiencies based on LHVs, since one is dividing by a larger number. Since it is possible to condense the water vapour and obtain additional heat, we have and will use efficiencies here that are based on HHVs. On this basis, the reversible efficiency for a hydrogen-powered fuel cell is 83 per cent (or 94.5 per cent if based on the LHV).

The efficiency that can be achieved in practice is lower for several reasons. First, crossover of some of the fuel from the anode to the cathode and galvanic currents within the oxygen electrode reduce the efficiency. At low current densities, limits on the rate of reaction of oxygen are important in reducing the efficiency. At intermediate current densities, resistance to the movement of the conducting ion across the electrolyte is important, while at high current densities, limits on the movement of the reactant through the porous gas diffusion materials are important. These three limitations lead to losses in efficiency, and are referred to as the activation, resistance and mass transport losses, respectively. Together, with the fuel crossover and galvanic current effects, they determine the *voltage efficiency*, η_v.

Limitations in the use of the fuel together produce a third efficiency, the *Faradaic efficiency*, η_f. These limitations involve flow of the fuel from the anode to the cathode, causing fuel to be consumed without generating an external current (a serious problem with direct methanol fuel cells), incomplete oxidation of the fuel (also a problem with methanol) or non-consumption of some of the fuel (which instead exits with the anode exhaust). The overall efficiency is given by the product of these three efficiencies:

$$\eta = \eta_r \eta_v \eta_f \tag{5.8}$$

The Faradaic efficiency is nearly always equal to unity. The voltage efficiency is larger if the fuel cell is operated at a higher voltage but lower current density; for typical operating conditions, η_v is around 50–60 per cent. Whatever portion of the total enthalpy of the reaction that is not converted to electricity is converted to heat and must be removed by a cooling system. In low-temperature fuel cells, the cooling demand requires a water cooling plate for every two cells.

Kim et al (1995) have shown that the voltage potential of a PEMFC can be computed as:

$$E = E_0 - b \log i - Ri - m \exp(ni) \tag{5.9}$$

where i is the current density (mA/cm^2) and E is in mW/cm^2. Kim et al (1995) have determined values of the constants E_0, b, R, m and n. The theoretical maximum voltage is the *reversible voltage E_r*, which is 1.229V for hydrogen-powered PEMFCs, and corresponds to the reversible efficiency of 83 per cent given above. The difference between E_r and E_o is due to fuel crossover and galvanic currents. The next three terms in Equation (5.9) represent the activation, resistance and mass transport losses. Figure 5.29a shows E for a fuel cell operating on air at 70°C and 5atm pressure, along with the contributions to the reduction in E from each of the terms in Equation (5.9).

The voltage efficiency η_v is given by:

$$\eta_v = \frac{E}{E_R} \tag{5.10}$$

and, neglecting the Faradaic efficiency, the overall efficiency is given by:

$$\eta = \eta_r \eta_v = 0.83 \frac{E}{E_R} = \frac{E}{E_{TN}} \tag{5.11}$$

where $E_{TN} = E_R/\eta_r = 1.481V$ is the *thermoneutral voltage*. It is the reversible voltage that would occur if $\Delta G = \Delta H$. The right-hand scale of Figure 5.29 shows the efficiency as given by Equation (5.11).

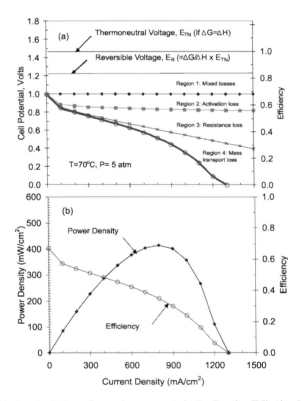

Note: The cell voltage potential is given by the heavy line, and was computed using Equation (5.9). Also shown are the thermoneutral voltage potential, the reversible voltage potential and the decrease in voltage potential from the reversible potential associated with each of the terms in Equation (5.9).

Figure 5.29 *(a) Variation of voltage potential, and corresponding efficiency, for a PEMFC operating on air at 70°C and 5atm pressure, and (b) variation in cell efficiency for the case shown in (a), along with the variation in power density*

The fuel cell power density (mW/cm^2) is given by the current density (in mA/cm^2) times the voltage. Figure 5.29b shows the variation in power density and efficiency for the case shown in Figure 5.29a. If one restricts the maximum current density to, say, $400mA/cm^2$, such that full load occurs at $400mA/cm^2$, and if one plots efficiency versus power as a fraction of peak power, one will obtain a plot similar to that shown in Figure 5.28 (except for the drop in efficiency at very lower power seen in Figure 5.28).

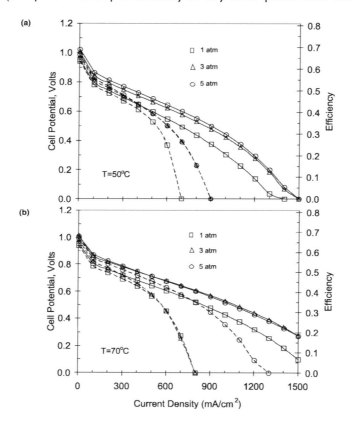

Note: Cell performance is given for operating temperatures of (a) 50°C and (b) 70°C.
Source: Computed using Equation (5.9) with coefficients given in Kim et al (1995)

Figure 5.30 *Variation of voltage potential and efficiency with current density for a PEMFC, using either air (dashed lines) or pure O_2 (solid lines) at pressures of 1atm (squares), 3atm (triangles) and 5atm (circles)*

Figure 5.30 shows the fuel cell voltage and efficiency for a variety of temperatures and operating pressures, using air and pure oxygen at the cathode, computed from Equation (5.9) using coefficients given in Kim et al (1995). Operation with pure oxygen, at higher pressure or at higher temperature increases the fuel cell efficiency. The benefits of using pure oxygen operating under pressure will be offset to some extent by the energy required to separate oxygen from air (although pure oxygen will be produced as a byproduct of the electrolysis of water to make hydrogen in the first place) and for compression of the input hydrogen and oxygen (although both may have been already compressed for storage, in which case the efficiency gain from higher pressure operation can be thought of as offsetting the energy losses associated with compression for storage). The curves shown in Figure 5.30 are based on data for fuel cell performance in the early to mid-1990s; the performance of current fuel cells will be better than shown here.

The losses in the energy flow from the fuel cell to the wheels can be seen in Figure 5.27. Losses occur in the power inverter (94 per cent efficiency), motor (92 per cent efficiency) and single-gear ratio transmission (93 per cent efficiency). There is a small additional loss at present due to utilization of only 96–98 per cent of the hydrogen supplied to the fuel cell (Nishikawa et al, 2008); a loss of 2 per cent will be assumed here.[14] Assuming all of these losses to be independent of load, and assuming a fuel cell system efficiency of 60 per cent at 25 per cent of peak load and 52 per cent at full load, the resulting fuel-to-wheels efficiencies are 47 per cent under typical driving conditions (25 per cent of peak load) and 40 per cent at peak load. By comparison, the thermal times mechanical times transmission efficiency in a conventional ICE vehicle is about 15 per cent ($0.38 \times 0.52 \times 0.75$) at 25 per cent of peak load and about 26 per cent ($0.38 \times 0.90 \times 0.75$) at full load (wide open throttle). Thus, a hydrogen fuel cell vehicle would be *more than three times* as efficient as present ICE vehicles (47 per cent versus 15 per cent) under normal driving conditions and about 1.5 times as efficient (40 per cent versus 26 per cent) at peak load.

The efficiencies of gasoline ICE and hydrogen fuel cell powertrains at 25 per cent and 100 per cent of full load, given above, are summarized in Table 5.21. Also shown are the engine-to-wheels efficiency for a conventional gasoline vehicle and the fuel cell-to-wheels efficiency for a comparable hydrogen-powered fuel cell vehicle, as measured by Villatico and Zuccari (2008) over a real urban driving cycle. The average efficiency of the fuel cell vehicle is 43 per cent, compared to 18 per cent for the gasoline vehicle.

Cost of automotive fuel cells

Carlson et al (2005) performed a detailed analysis of the cost of PEMFCs running on pure hydrogen that could be attained assuming high-volume production (500,000 units/yr) but with 2005 technology. The estimated fuel cell cost is \$108/kW, of which \$67/kW is the stack cost and \$41/kW is the balance-of-system cost. The target of the US Department of Energy is a stack cost of \$30/kW – about half the projected cost with current technology. Ford Motor Company expects stack costs of \$45/kW to be achieved for fuel cells with a peak net efficiency of 62 per cent, cold start capability at $-40°C$ and a 6000–8000 hour lifespan (Frenette and Forthoffer, 2009). These costs are factory costs, which include fixed and variable costs, but corporate charges (research and development costs, profit, corporate taxes, sales and marketing, warranties) are not included.

To reduce stack costs it is necessary to improve the areal power density (an improvement from 600mW/cm² to 800mW/cm² alone would reduce the stack cost to \$50/kW) and reduce Pt loading while improving catalyst stability and fuel cell performance (at present, the cost of Pt accounts for half the cost of a fuel cell stack).

Current Pt loadings are already about a factor of 100 less than in the first PEMFCs. The industry goal is to reduce Pt loading from 0.75 milligram (mg)/cm² today to 0.2mg/cm² in the near term and 0.1mg/cm² in the long term. All else being equal, a reduction in the Pt loading leads to lower power output per unit cell area, so larger fuel cells (which also happen to be more efficient) are required. The challenge then is to reduce Pt loadings while maintaining cell performance. With a reduction in Pt loading from 0.75mg/cm² to 0.1mg/cm² and the improvement in power density given above, the Pt loading would decrease from 1.25gm/kW in recent fuel cells to 0.125gm/kW. In the latest fuel cells the loading has been reduced to 0.7gm/kW (Frenette and Forthoffer, 2009). The price of Pt fluctuated between \$15–25/gm during the 1980s

Table 5.21 *Comparison of tank-to-wheels efficiencies for gasoline ICEs and for hydrogen-powered fuel cell vehicles*

Technology	Deduced in text (%)		Measured over a real urban driving cycle (%)
	At 25% of peak load	At full load	
Gasoline ICE	15	26	18
Hydrogen FCV	47	40	43

and 1990s, then rose to about \$40/gm (\$1200/oz) by 2004 and reached an all-time peak of \$74/gm (\$2100/oz) in June 2008 before dropping to about \$25/gm in November, 2008 and then rising to about \$50/gm by February 2010.[15] At \$50/gm, the cost of Pt at a Pt loading of 0.7gm/kW is \$35/kW, while at a plausible longer-term cost of \$100–200/gm combined with the long-term loading target of 0.125gm/kW, the cost of Pt is \$12.5–25/kW. Given that long-term Pt costs are likely to be high (for reasons discussed below), it is very unlikely that the cost of automotive fuel cells can fall below about \$60/kW. This is more than twice the typical \$25–30/kW cost of an ICE.

A more complete comparison of the cost of alternative automobile drivetrains, including the cost of storage tanks, transmission and controls, is presented in subsection 5.4.11.

Material constraints

Fuel cells intended for automotive applications will of necessity be low-temperature fuel cells (for safety reasons and because this permits fast start-up). Operation at low temperature, however, requires a precious metal catalyst. As noted above, a Pt catalyst is used in PEMFCs (the leading candidate for automotive applications), with a loading in the latest fuel cells of about 0.7gm/kW. Substantially greater catalyst loadings are needed in the fuel processors that would be used in vehicles powered by hydrocarbon fuels. As well, the fuel cells of hydrocarbon-powered vehicles require a ruthenium catalyst to prevent CO impurities in the hydrogen from poisoning the Pt catalyst. It is therefore pertinent to examine potential material constraints on the size of a worldwide fuel cell vehicle fleet. Råde and Anderson (2001) comprehensively review potential precious metal material constraints on fuel cell vehicles, and the key points from their work are summarized here.

Currently, about 170 tonnes of Pt are mined per year. The main uses of Pt are in jewellery (50 per cent), automotive catalytic converters (20 per cent) and industry (16 per cent). The major industrial uses are as a catalyst for the production of nitric acid, which is largely used to make fertilizers, as a catalyst in petroleum refining and as a trace element in computer hard discs and speciality glasses. The global Pt reserve is estimated at about 34,000 tonnes (34kt), and an

optimistic estimate of the exploitable resource is around 66kt, distributed as shown in Figure 5.31. Three quarters of the exploitable resource occurs in South Africa, almost all of that in the Bushveld complex north of Pretoria. This is a saucer-shaped geological deposit 370km across and 7km thick, with its rim exposed at the surface, discovered early in the 20th century. The amount exploitable is limited by the depth to which mines can be built. The main constraint on that in turn is the geothermal gradient of 20K/km. The temperature reaches 40°C at a depth of 1.2km, so expensive cooling is required. The deepest mine in 1999 was about 2000m deep, with a temperature of 63°C, and a mine to a depth of 2365m is under construction. Only one Pt deposit of any significance has been discovered anywhere in the world since 1960.

There is a tradeoff between the fuel cell Pt loading and efficiency; as discussed in Box 5.2, higher efficiency requires a lower current density, which in turn requires a greater cell area and hence a greater Pt loading for a given power. Given the Pt loadings of 0.7gm/kW at present and the long-term target of 0.125gm/kW mentioned earlier, a hybrid fuel cell–battery vehicle with a 64kW fuel cell (a reasonable power for a fuel cell–battery hybrid vehicle) would

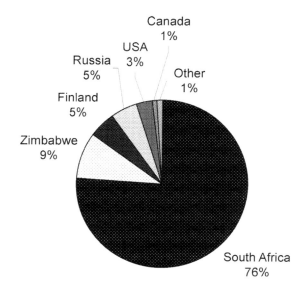

Source: Data from Råde and Andersson (2001)

Figure 5.31 *Distribution of the estimated exploitable platinum resource*

require 45gm of Pt at present and 8gm if the long-term target can be achieved. Here, we shall develop scenarios of Pt demand assuming a loading of 10gm per vehicle. If a hydrocarbon fuel is used, many times this amount (i.e. 50–150gm) would be required in the fuel processor, along with about 25gm of ruthenium (Ru), which is used to counter the poisoning of the Pt catalyst that otherwise occurs due to trace CO in the hydrogen produced from hydrocarbon fuels. A Pt/Ru/Co catalyst might reduce the required precious metal loading by a factor of eight compare to a Pt/Ru catalyst when using methanol as a fuel (Strasser et al, 2003). However, as noted earlier, Co is in demand for lithium-ion batteries but alternatives are being sought because it is an expensive mineral.

In a scenario considered by Råde and Anderson (2001), the annual Pt demand for jewellery and industrial uses grows to 200–250Mg/yr by 2100, and cumulative consumption for these purposes reaches 30Gg. The current global automobile population is about 700 million, and scenarios will be considered here in which the automobile population reaches 5 billion, 2 billion and 1 billion by 2100. It is assumed that, by 2050, 50 per cent of all new vehicles produced are hydrogen fuel cell vehicles. Details in the construction of the scenarios are provided in Box 5.3, along with the sigmoidal growth curves for the size of the vehicle fleet and the fraction of fuel cell vehicles in the fleet. Figure 5.32 shows the resulting annual demand for refined Pt from primary Pt (i.e. excluding recycled Pt) and the cumulative demand for raw Pt for the three vehicle-population scenarios. About 20–30 per cent of the mined Pt is lost between in situ resources and mill-head materials, and 10–20 per cent of the mill-head material is lost during the concentration, smelting and refining stages. In constructing the curve of cumulative extraction of raw Pt from the annual demand for refined Pt, losses of 20 per cent during extraction and 10 per cent during processing have been assumed. A fuel cell loading of 10gm/vehicle, a fuel cell lifetime of 10 years and 90 per cent recycling efficiency are also assumed here.

As seen from Figure 5.32, the Pt demand for automotive purposes peaks at about 950Mg/yr for the high vehicle-population scenario – about five times present world demand. The cumulative extraction of Pt for transportation applications reaches 62Gg by 2100 – about equal to the upper limit of the estimated

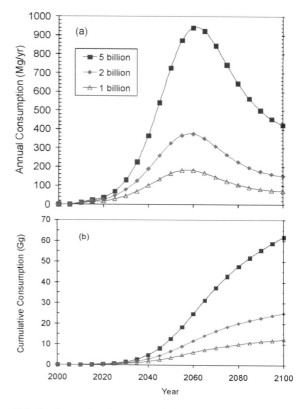

Note: Results are given for cases in which half of new vehicle production is as fuel cell vehicles by 2050. See text and Box 5.3 for the complete set of input assumptions.

Figure 5.32 *(a) Annual demand for new refined Pt (i.e. excluding recycled Pt), and (b) cumulative extraction of raw Pt, for scenarios in which the vehicle population reaches 5 billion (high), 2 billion (medium) or 1 billion (low) by 2100*

exploitable resource. To this would be added another 30Gg for other uses of Pt. Unless very large new discoveries of Pt are made – something that seems quite unlikely – creation of a fuel cell automobile fleet of 5 billion will not be possible even if the factor-of-ten reduction in Pt loading target is achieved. A fleet size of 2 billion fuel cell vehicles may also face material constraints, as cumulative consumption for transportation reaches 25Gg if the factor-of-ten reduction in loading can be achieved, bringing total consumption close (55Gg) to the most optimistic estimate (66Gg) of the exploitable resource. A fleet of 1 billion fuel cell vehicles running on pure hydrogen

Box 5.3 Constructing scenarios of future platinum demand

It is quite simple to construct illustrative scenarios of future Pt demand and cumulative Pt consumption by fuel cell vehicles, using an Excel spreadsheet. The following information needs to be specified: the rate of production of automobiles in the starting year (2000), P(2000), the average lifetime of automobiles, τ, and the population of automobiles when the population has stabilized, N_{eq}. The rate of production must plateau at $P_{eq} = N_{eq}/\tau$ if the population is to stabilize at N_{eq}. It is convenient to assume that the production rate follows a sigmoidal variation, initially growing rapidly but with ever slower growth as it approaches P_{eq}. This variation is described by the logistic equation:

$$\frac{dP}{dt} = aP - bP^2 \tag{5.12}$$

where a is the initial fractional growth rate and b causes the growth rate to decline as P gets larger (see also Box 2.1). It can be readily seen that the equilibrium production rate (i.e. when $dP/dt = 0$) is given by b/a, so $b = a/P_{eq}$. The solution to Equation (5.12) is:

$$P(t) = \frac{a}{b + ((a - bP_0)/P_0)e^{-a(t-t_o)}} \tag{5.13}$$

where t is time since the starting point (t_0, taken to be year 2000).

The growth of the vehicle population can be projected as follows: the population at the end of each year is given by the population at the start of the year plus the annual production (given above) minus the number of vehicles scrapped, which is taken as the number produced τ years before. Figure 5.33a shows the growth in vehicle production rate and in vehicle population for the following inputs: production rate in 2000 = 40 million/year; vehicle population in 2000 = 530 million; $a = 0.057$, $N_{eq} = 5.175$ billion and $\tau = 15$ years (so that $P_{eq} = 345$ million/year). This scenario is similar to many conventional scenarios, in which the global human population stabilizes at 10–11 billion by about 2100, and average car ownership rates reach the Western European level of 0.4–0.5 cars per capita.

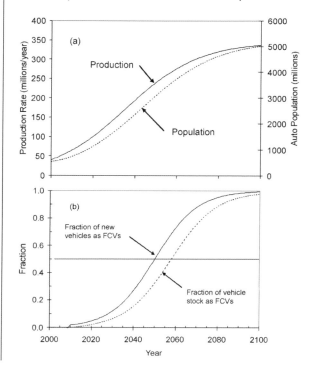

Figure 5.33 (a) Growth in the rate of production of new vehicles, and in the vehicle population, for a scenario in which the automobile population reaches 5 billion by 2100, and (b) growth in the fraction of vehicle production as fuel cell vehicles, and in the fraction of all vehicles as fuel cell vehicles for the scenario adopted here

A sigmoidal growth curve can also be used to describe the fraction F of vehicle production as fuel cell vehicles:

$$F(t) = \frac{e^{c(t-t50)}}{1+e^{c(t-t50)}} \qquad (5.14)$$

where t_{50} is the year in which fuel cells vehicles constitute 50 per cent of total annual production. Figure 5.33b shows the growth in fuel cell vehicle production fraction for the case where $t_{50} = 2050$ and $c = 0.1$. Also shown is the fraction of total vehicle stock that consists of fuel cell vehicles. The halfway point in the vehicle stock is reached about 8 years after fuel cell vehicles reach 50 per cent of annual vehicle production.

The annual Pt consumption is the annual production of fuel cell vehicles times the assumed loading per vehicle. The amount of Pt made available each year through recycling is the annual fuel cell vehicle production τ_{fc} years ago (τ_{fc} is the fuel cell lifespan, which can differ from the vehicle lifespan) times the assumed fraction f_r of Pt that is recovered by recycling. The primary Pt demand each year is the consumption minus the amount made available through recycling. The cumulative demand for raw Pt is obtained by adding the annual demand for refined Pt, taking into account losses during the mining and processing of raw Pt.

would require a minimum of 12Gg. By 2100, near 100 per cent recycling efficiency would be necessary in order for the fuel cell fleet to be sustainable. The Ru loading in hydrocarbon-powered fuel cell vehicles would be comparable to the Pt loading assumed here for hydrogen-powered vehicles, so the Ru required in a hydrocarbon-powered fleet of 1 billion fuel cell vehicles would also be around 12Gg. This is more than twice the estimated recoverable Ru resource according to Råde and Andersson (2001) and, as noted above, the Pt requirement for hydrocarbon-powered fuel cell vehicles is several times that of hydrogen-powered vehicles.

Thus, on the basis of the above, it can be concluded that:

- a hydrocarbon-powered fuel cell vehicle fleet of 1 billion is not feasible due to both Pt and Ru resource constraints;
- a hydrogen-powered fuel cell vehicle fleet of 1 billion up to the year 2100 is feasible if the upper estimates of the Pt resource are correct and a factor-of-ten reduction in Pt loadings can be achieved compared to today; and
- near total recycling of Pt from fuel cells will be necessary if a fuel cell vehicle fleet of 1 billion is to be sustained beyond 2100.

Given the material constraints on climate-friendly and pollution-free vehicles, it is clear that great emphasis

should be placed on urban development that eliminates the need for private automobile transportation altogether, as discussed at the beginning of this chapter.

5.4.9 Hydrogen internal combustion engine vehicles

An alternative to using onboard hydrogen in a fuel cell is to use the hydrogen in an ICE, in place of gasoline or diesel fuel. This may be an attractive option if fuel cells do not achieve the cost reduction needed to be viable in transportation applications, and it overcomes the constraint from limited supplies of platinum if the world's automobile fleet were to be converted to operation with fuel cells. However, the efficiency of an optimized gasoline ICE using hydrogen is only 10–20 per cent greater than that using gasoline (White et al, 2006), which in turn increases the problem of storing adequate hydrogen on board for an acceptable driving range (in fuel cells, hydrogen would be used about twice as efficiently). Hydrogen is more attractive in a diesel engine, where engine efficiencies (thermal × mechanical) of 38–45 per cent have been obtained in test rigs when running on hydrogen, compared to only 28 per cent using diesel fuel (Antunes et al, 2008, 2009). This is an efficiency gain of 35–50 per cent.[16] If the baseline vehicle is two to three times more fuel efficient than today's vehicles (as discussed above in subsection 5.4.5), the required storage volume for the same driving range as today would be comparable to that required today for gasoline. If hydrogen is used only as a range extender in plug-in hybrid vehicles, the required storage volume would be smaller still. The only non-negligible pollutant using hydrogen in an ICE would be NO_x, but even in this case the emissions would be very low. Hydrogen ICEs would have less power than a gasoline or diesel ICE, but this would be overcome by using a hydrogen plug-in hybrid electric vehicle.

5.4.10 Lifecycle energy use for alternative drivetrain and fuel combinations

As discussed in subsection 5.4.4, the fuel economy of gasoline-powered ICE vehicles could be doubled through the use of identifiable advanced technologies, including use of lightweight materials. Forming a gasoline–electric hybrid would lead to a further one third improvement in fuel efficiency under a mix of urban and highway driving conditions (see subsections 5.4.5 and 5.4.7). Another jump in fuel efficiency would occur through the direct use of hydrogen in fuel cells. However, in evaluating the primary energy requirements of different drivetrain–fuel combinations, one has to take into account the energy associated with producing the fuel and putting it into the vehicle storage system (the *well-to-tank* analysis), as well as the losses in going from the onboard storage to the wheels (the *tank-to-wheels* analysis). The overall energy use is given by a *well-to-wheels* (WTW) analysis. If vehicle fuel economy is improved through replacement of heavy materials with lightweight materials, then one must also take into account the different amounts of energy used in manufacturing the different materials that can be used in an automobile. Finally, one should take into account the energy associated with dismantling the automobile and energy recovered through recycling of used materials. Only then can one properly compare the impact on primary energy use of alternative vehicle systems. This is referred to as a *lifecycle* analysis.

Here, we discuss the Massachusetts Institute of Technology (MIT) analyses of potential future automobile energy use by Weiss et al (2000) and updated by Heywood et al (2003), which considered the full spectrum of measures identified above for improving automobile fuel economy, as well as differences in energy used to provide different fuels to the automobile and differences in the energy used to produce different vehicles. Another WTW analysis, carried out by General Motors in conjunction with the Argonne National Laboratory in the US, British Petroleum, ExxonMobil and Shell Oil (GMC et al, 2001), uses as its baseline vehicle a full-sized GM pickup truck. Fifteen different vehicle drivetrains and 13 fuels (from an initial list of 75 fuels) were analysed but, apart from changes in the vehicle drivetrain and fuel system, the other vehicle characteristics were left unchanged in their analysis, and so are of less interest here.

Tank-to-wheels energy use

The vehicle characteristics and energy use of most of the vehicles analysed in the MIT study are presented in Table 5.22. The baseline vehicle is a 2001 gasoline ICE vehicle with a fuel consumption of 7.67 litres/100km (31mpg) and an empty mass of 1408kg.[17] The authors of the MIT study caution that predicting the performance of future technologies is highly uncertain, such that an uncertainty of ±20 per cent should be attached to the results for advanced technologies. The baseline vehicle in 2020 is the result of incremental improvements to existing technologies. Vehicle mass decreases by about 200kg, a five-speed automatic clutched transmission is used in place of current four-speed automatic transmissions, and there are modest improvements in the drag coefficient, rolling resistance, engine thermal efficiency and friction, and transmission efficiency. Fuel consumption decreases by about 30 per cent (from 7.7 to 5.4 litres/100km). The advanced 2020 vehicle achieves a further 100kg mass reduction through greater use of lightweight materials, and achieves further reductions in the drag coefficient, rolling resistance and engine displacement, resulting in a further 10 per cent reduction in fuel consumption. The next step involves hybrid vehicles with parallel direct injection ICEs and storage batteries. The specific power (W/kg) of the battery assumed in this analysis is somewhat higher than current nickel-metal hydride batteries, but thought to be readily achievable. Fuel consumption drops by another 30 per cent compared to the advanced ICE vehicle. For both advanced ICE and hybrid vehicles, compression ignition (diesel) vehicles are about 15 per cent more efficient than spark ignition (gasoline) vehicles.

Table 5.22 *Characteristics and performance of current and possible future ICE, ICE–battery hybrid and fuel cell–battery hybrid automobiles*

Fuel	Gasoline					Diesel		H₂
Year	2001	2020	2020	2020	2020	2020	2020	
	Current	Baseline	Advanced	Advanced	Advanced	Advanced	Advanced	Advanced
Propulsion system	SI ICE	SI ICE	SI ICE	SI Hybrid	FC Hybrid	CI ICE	CI Hybrid	FC Hybrid
Transmission	Auto	ACT	ACT	CVT	Direct	ACT	CVT	Direct
Mass (kg)								
Body and chassis	930	845	746	750	776	757	758	754
Propulsion system	392	264	252	269	463	293	297	378
Total (includes 136kg payload)	1458	1245	1134	1155	1375	1186	1191	1268
Vehicle characteristics								
Rolling resistance coefficient	0.009	0.008	0.006	0.006	0.006	0.006	0.006	0.006
Drag coefficient	0.33	0.27	0.22	0.22	0.22	0.22	0.22	0.22
Frontal area (m²)	2	1.8	1.8	1.8	1.8	1.8	1.8	1.8
Auxiliary power (W)	700	1000	1000	1000	1000	1000	1000	1000
Propulsion system								
Engine displacement (litres)	2.5	1.79	1.65	1.11		1.75	1.16	
Engine thermal efficiency (%)	38	41	41	41		51	51	
Maximum engine power (kW)	110	93	85	58		89	59	
Maximum fuel cell power (kW)					69			63
Maximum motor power (kW)				29	103		30	95
Battery specific energy (Wh/kg)				50	50		50	50
Battery specific power (W/kg)				800	800		800	800
Battery power (kW)				29	34		29	32
Energy and fuel use during operation								
US urban MJ/km	2.82	2.00	1.78	1.20	1.16	1.53	1.03	0.66
US highway MJ/km	2.06	1.45	1.25	0.91	0.88	1.04	0.78	0.51
Litres gasoline eq/100km, urban	8.7	6.2	5.5	3.7	3.6	4.7	3.2	2.0
Litres gasoline eq/100km, highway	6.4	4.5	3.9	2.8	2.7	3.2	2.4	1.6
Miles per gallon, urban	27	38	43	63	65	50	74	115
Miles per gallon, highway	37	52	61	83	86	73	97	149

Lifecycle energy use (MJ/km)

Operation	2.48	1.75	1.07	1.03	1.31	0.92	0.59
Fuel production[a]	0.52	0.37	0.22	0.22	0.18	0.13	0.25 (1.51)
Vehicle manufacture[b]	0.29	0.25	0.26	0.28	0.26	0.26	0.28
Total	3.29	2.37	1.55	1.53	1.75	1.31	1.12 (2.38)
Percentage energy savings							
Operational, compared to current	29	38	57	58	47	63	76
Operational, compared to SI hybrid				3	−22	14	45
Operational, compared to CI hybrid							35
Lifecycle, compared to current	28	36	53	53	47	60	66
Lifecycle, compared to SI hybrid				1	−13	16	28
Lifecycle, compared to CI hybrid							14

Note: SI = spark ignition, CI = compression ignition, ACT = auto-clutch transmission, CVT = continuously variable transmission. Lifecycle energy use is computed assuming driving to be 55 per cent urban, 45 per cent highway. [a] Energy use during fuel production is assumed to be 21 per cent of the delivered fuel energy for gasoline and 14 per cent for diesel (as in Heywood et al, 2003). Energy use during the production of hydrogen given outside brackets is computed from Equation (5.16), assuming hydrogen to be produced by electrolysis at 80 per cent efficiency using renewably based energy, and to require 2 per cent of the hydrogen energy content (as electricity) for transmission 1000km at 30atm pressure and 10 per cent for compression to 350atm pressure at the point of use. Energy use in brackets is computed from Equation (5.16), assuming electricity to be generated with an efficiency of 40 per cent. [b] Assumes recycling of 90 per cent of vehicle metals and 50 per cent of vehicle plastics. Results are obtained from a process-based analysis for the energy directly used in producing materials and in assembling the vehicle and so cannot be directly compared with the vehicle manufacturing energy requirements given in Table 5.4, as the latter are based on input–output tables that include second order, third order and higher energy inputs.

Source: Heywood et al (2003)

Also shown in Table 5.22 are results for fuel cell–battery hybrid vehicles, using either gasoline or hydrogen as the fuel. The fuel cell–battery hybrid uses a simpler, single-gear ratio transmission that is lighter and more efficient than used in any of the other vehicles. Vehicles using gasoline require a substantially heavier fuel cell system than vehicles using hydrogen due to the need for a fuel processing system and for a more powerful fuel cell. Greater battery and motor power are also required for the gasoline fuel cell vehicle, however, the fuel storage system mass is greater in the hydrogen-fuelled vehicle. The net result is that the hydrogen-fuelled vehicle is 107kg lighter and is almost twice as fuel-efficient as the gasoline-powered fuel cell vehicle. Indeed, the fuel consumption of the gasoline fuel cell vehicle is no better than that of the advanced gasoline hybrid and 10 per cent worse than that of the advanced diesel hybrid.

Compared to the 2001 baseline vehicle, the advanced 2020 gasoline hybrid uses 57 per cent less energy, the advanced diesel hybrid uses 63 per cent less energy and the hydrogen fuel cell vehicle uses 76 per cent less energy for combined city and highway driving. Even compared to the advanced diesel hybrid vehicle, the hydrogen fuel cell vehicle reduces energy use by 36 per cent per unit distance driven, which means that it can be driven 56 per cent $(1/(1.0-0.36) \times 100$ per cent) further using the same amount of onboard energy.

The factor-of-four better fuel economy of advanced hydrogen-fuelled vehicles compared to current gasoline-fuelled vehicles reduces the amount of hydrogen energy that needs to be stored for a given driving range by the same factor, which is an important consideration given the greater bulk of hydrogen storage systems compared to gasoline. The approximate factor of 1.5 greater fuel economy (in terms of km/MJ) using hydrogen compared to advanced gasoline ICE vehicles means that the price of hydrogen (per unit energy) can be 1.5 times the future price of gasoline with no difference in the cost of fuel to drive a given distance. As discussed in Volume 2 (Chapter 10, subsection 10.7.4), the future cost of hydrogen produced from renewable energy sources and delivered to refuelling station is likely to be the equivalent of about $1–2/litre of gasoline ($4–8/gallon). Given the almost two times greater distance that can be travelled on a unit of hydrogen energy compared to a unit of gasoline energy, hydrogen-fuelled vehicles would be less costly to operate than gasoline-fuelled vehicles even if the cost per unit of energy is the same ($1–2/litre).

In an extension of the analysis that is summarized in Table 5.22, Bandivadekar et al (2008) compare the relative savings possible though advanced technology when applied to a typical American light truck, a typical American car and a typical European car (which already uses about 25 per cent less fuel than a typical American car). Their results are summarized in Table 5.23. Applied to typical American cars or light trucks, the potential energy savings through the application of

Table 5.23 *Comparison of gasoline-equivalent fuel use (litres/100km) for American cars and light trucks and for European cars using various current and future technologies*

	American vehicles			European
	Car	Car	Light truck	Car
	H2003	B2008	B2008	B2008
Current SI ICE	7.7	8.8	13.6	6.57
Current CI		7.4	10.1	5.48
Current SI hybrid		6.2	9.5	5.01
Current CI hybrid				4.51
Advanced SI	4.8	5.5	8.6	4.11
Advanced CI	4.1	4.7	6.8	3.48
Advanced SI hybrid	3.3	3.1	4.8	2.73
Advanced CI hybrid	2.8			2.45
Energy use relative to current SI ICE vehicles				
Advanced SI hybrid	0.43	0.35	0.35	0.42
Advanced CI hybrid	0.36			0.37

Source: Heywood et al (2003) for H2003; Bandivadekar et al (2008) for B2008

advanced technology while still using gasoline is 65 per cent, that is, *a factor-of-three reduction in energy intensity with no change in fuel*. As applied to the less energy-intensive European car, the projected reduction in fuel requirements is just over a factor of two (58 per cent savings).

Lifecycle energy analysis

Energy is required for the extraction, transportation and refining of crude oil, and for the transportation of petroleum products to the point of use. As discussed in Chapter 6 (section 6.11), the energy used for extraction and refining of conventional onshore oil in the US is about 10 per cent of the energy content of the delivered petroleum products. The fraction rises to about 16 per cent for offshore oil and to 38 per cent for the Canadian tar sands. The analysis of globally aggregated data presented in Chapter 2 (subsection 2.3.4) indicates that the average upstream energy use for extracting and refining of oil products is about 11 per cent of the delivered energy. The energy requirement for transporting crude oil by supertanker is 0.06–0.11MJ/tonne-km according to Table 5.32. Given that 1 tonne of crude oil contains 42.1GJ of energy (see Appendix B), the energy required to transport crude oil 10,000 km is only 1.4–2.6 per cent of the energy content of the fuel. The energy intensity for transport by heavy truck is about ten times that of supertanker, so if gasoline is transported on average 500km by truck, this would add another 1 per cent to the upstream energy use. Thus, the total upstream energy use will be 13–17 per cent of the delivered energy.

As discussed in Chapter 3 (subsection 3.2.10), the energy required to transport natural gas by pipeline is about 2.5 per cent of the gas energy content per 1000km transported. For an average transport distance of 3000km in North America, this amounts to an upstream energy use of 7.5 per cent of the delivered energy. For transport of Middle Eastern natural gas to North America or Japan by LNG supertanker, the upstream energy use is 10–20 per cent of the delivered energy.

The upstream energy use associated with the provision of hydrogen to refuelling stations depends on the efficiency in producing hydrogen, the energy required to transport it (if not produced on site) and the energy used for compression or liquefaction (the choice of which depends on the onboard storage system

used). All of these energy terms are discussed in Volume 2, Chapter 10. For production of hydrogen from natural gas, efficiencies range from 75–80 per cent on an LHV basis but could be increased to 85 per cent with better recovery of waste heat (see Volume 2, Chapter 10, subsection 10.3.1). The efficiency in producing hydrogen from coal is much smaller, about 55 per cent with current technology and perhaps 64–68 per cent in the future. Production of hydrogen by electrolysis using electricity generated by fossil fuels is not of interest, given that it will always be cheaper and more efficient to directly produce hydrogen from fossil fuels. However, production of hydrogen using electricity produced from wind or solar energy is of interest. An optimistic future electrolysis efficiency (accounting for electricity use by pumps and other auxiliary devices) is 80 per cent (see Volume 2, Chapter 10, subsection 10.3.3). Transport of hydrogen a distance of 1000km at 30atm pressure would require 2 per cent of the hydrogen energy content (on a LHV basis) as electricity, which is equivalent to a transmission efficiency of 0.98. Compression of hydrogen to 350atm pressure would require about 6 per cent of the energy content of the stored hydrogen and liquefaction would require 30–40 per cent of LHV of the hydrogen with present technology (see Volume 2, Chapter 10, subsection 10.4.1), corresponding to storage efficiencies of 0.94 and 0.6–0.7, respectively.

If η_{el}, η_{tr} and η_s are the efficiencies in producing hydrogen by electrolysis, transmitting hydrogen and in storing it as a compressed or liquefied gas, then the upstream energy use per unit of delivered stored hydrogen is:

$$E_{up} = \frac{1}{\eta_{el}\eta_{tr}\eta_s} - 1 \qquad (5.15)$$

For η_{el} = 0.8, η_{tr} = 0.98 and η_s = 0.94, E_{up} = 0.36 (i.e. 1.36 units of electricity are required to produce 1.0 unit of stored hydrogen). However, if there is fossil fuel electricity on the grid, the C-free electricity (assumed to be the only kind used to produce hydrogen) could instead be used to displace some of the fossil fuel-generated electricity. If η_{pp} is the efficiency of the fossil fuel powerplant, then the primary energy that could have been saved if hydrogen were not produced – the 'opportunity cost' of the hydrogen – minus the hydrogen energy supplied, per unit of supplied hydrogen, is given by:

$$E_{up} = \frac{1}{\eta_{pp}\eta_{el}\eta_{tr}\eta_{s}} - 1 \qquad (5.16)$$

For η_{pp} = 0.4, E_{up} = 2.54 (i.e. 3.54 units of fossil fuel could be displaced for each unit of stored hydrogen supplied using renewable-based electricity). Assuming that one unit of hydrogen fuel displaces 1.81 units of gasoline (based on Table 5.22) with an upstream energy use of 20 per cent, the savings in petroleum primary energy is only 2.17 units. Thus, in terms of primary energy savings, it is about 60 per cent more effective to use C-free electricity to displace coal-based (40 per cent efficiency) electricity than to use it to produce hydrogen for use in advanced fuel cell vehicles.

This assumes that there is fossil fuel electricity available to be displaced with C-free electricity. If hydrogen is produced during times of excess supply of wind or solar energy, then – *at the times when hydrogen is produced* – there is no displaceable fossil fuel electricity. Thus, even if there is some fossil fuel-generated electricity in the supply mix, it can still be appropriate to use Equation (5.15) rather than Equation (5.16) in computing the upstream energy use associated with providing hydrogen in a form ready to be used.

Table 5.22 gives the energy use in producing the amount of fuel required by the various vehicles. For gasoline and diesel fuel, the upstream energy use is assumed to be 21 per cent and 14 per cent, respectively, of the delivered fuel, while for hydrogen the upstream

energy use is computed using Equation (5.15) with the assumptions given above. Also given in Table 5.22 is the energy required to manufacture the various vehicles, divided by an assumed lifetime distance travelled of 300,000km. The manufacturing energy includes the energy used to produce the materials used in the vehicles (assuming 95 per cent of the metals and 50 per cent of the glass and plastics to be recycled materials), as well as the energy used at the manufacturing facility. Table 5.24 gives the mass of ferrous metals and aluminium assumed for the different vehicles. The mass of ferrous metals decreases from 886kg in the 2001 vehicle to 325kg in the advanced SI ICE vehicle, while the mass of aluminium increases from 81kg to 342kg. The fuel cell

Table 5.24 *Mass of ferrous metals and aluminium in the automobiles of Table 5.22*

Vehicle	Mass (kg)	
	Ferrous metal	Aluminium
Current SI ICE	886	81
Incremental SI ICE	667	97
Advanced SI ICE	325	342
Advanced CI ICE	379	337
Advanced SI hybrid	350	334
Advanced CI hybrid	387	330
Gasoline FC hybrid	640	305
H₂ FC hybrid	477	355

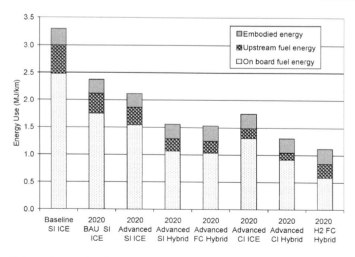

Source: Operational and embodied energies are taken from Heywood et al (2003) while energy associated with production of fuel is computed as explained in the main text

Figure 5.34 *Components of the lifecycle energy use of different vehicles*

vehicles have about 10 per cent less aluminium but 50–100 per cent greater mass of ferrous metals than in the advanced SI ICE vehicle. The amounts of other materials in the different vehicles are largely unchanged and have very little impact on the total vehicle embodied energy. The differences in total embodied energy between the different vehicles are very small compared to the savings in operational energy in moving to more fuel-efficient vehicles.

The lifecycle energy use per km travelled for the different vehicles is summarized in Figure 5.34. The hydrogen fuel cell vehicle has the lowest lifecycle energy use (about a third that of the current SI ICE vehicle), with the diesel hybrid a close second.

5.4.11 Comparative lifecycle cost of advanced conventional, hybrid, plug-in hybrid, battery electric and fuel cell vehicles

Table 5.25 lays out the calculation of the total costs of the drivetrain for an advanced but naturally aspirated gasoline vehicle (the NA-SI case) and for HEVs, PHEVs, BEVs and hydrogen FCVs. The engine, battery, motor and fuel cell sizes in the various vehicles, as well as unit costs of these components and the costs of other features (such as emission controls, transmission and wiring) are taken from Bandivadekar et al (2008), the only exception being the fuel cell cost, where a less optimistic cost of $75/kW is assumed here instead of $50/kW. Note that quite different energy:power ratios and costs per kWh are adopted for the batteries in HEVs, PHEVs and BEVs. Also given in Table 5.25 are the gasoline-equivalent fuel consumption for ICEs, HEVs, PHEVs and FCVs; the AC electricity consumption per km for the PHEV when running on the battery; and annual maintenance costs (taken from Table 5.14 or estimated) for the different vehicles. Energy costs are assumed to be $1.50/litre for gasoline (a value already found in some parts of the world), $30/GJ for hydrogen, and $0.10/kWh for electricity. The resulting annual costs and the net present value of ten years of the annual costs are shown, the latter assuming a 10 per cent discount rate. The last row of Table 5.25 gives the reduction in the NPV of the annual costs compared to

the NA-SI case (that is, the discounted ten-year savings) minus the extra vehicle cost. A positive value means that the discounted operational savings exceed the extra upfront purchase cost of the vehicle, and the larger this term, the more favourable the economics. For the various cost assumptions made here, the HEV is the most economical vehicle, following by the FCV and the PHEV vehicle, but the BEV entails a large net cost compared to the NA-SI vehicle.

The relative costs of the different vehicles are compared graphically in Figure 5.35. The core vehicle (that is, excluding the drivetrain and fuel storage system) is assumed to have the same cost for all drivetrains, although this will probably not be the case. Burns et al (2002) note that in advanced fuel cell vehicles, steering, braking and other vehicle systems would be controlled electronically rather than mechanically. All of the vehicle's propulsion and control systems would be contained in a 28cm thick 'skateboard-like' chassis at the bottom of the car. The car would have a flat floor and a spacious interior (due to the absence of an ICE, steering column and foot pedals). Interchangeable bodies could be offered for the same chassis, which could be mass-produced and sold to individual car companies, with the prospect of low cost. Thus, all else being the same, the FCV could be more competitive than indicated in Figure 5.35.

As noted earlier, the future costs of batteries and fuel cells under conditions of mass production are quite uncertain, as are the future costs of hydrogen and gasoline. Figure 5.36 plots the net discounted lifecycle savings (the last row of Table 5.25) for three different gasoline prices ($1/litre, $1.5/litre and $2/litre), for fuel cell costs of $50/kW and $100/kW, and for hydrogen costs of $20/GJ and $40/GJ. Unless battery costs are far below the cost assumed here (given in Table 5.25), BEVs will not be competitive with any of the other vehicles. Many different combinations of assumptions other than those shown here could be considered. The important point is that for cautious assumptions concerning the future costs of fuel cells, batteries and hydrogen, either PHEVs or FCVs could be the more competitive of the two, and for any reasonable projection of future gasoline costs (≥$1/litre), one or both could have lower lifecycle cost than advanced, fuel-efficient conventional or hybrid vehicles.

Table 5.25 *Comparison in the sizing and cost of the drivetrain components in different advanced vehicles (thought to be feasible by 2030), annual and discounted lifecycle cost, and discounted net savings*

	Vehicle type				
	NA-SI	HEV	PHEV-30	BEV	FCV
Drivetrain characteristics					
ICE (kW)	95	70	50		
Battery (kWh)		1	8.2	48	1.3
Battery (kW/kWh)		28	5.5	3	28
Battery (kW)		28	45.1	144	37.3
Fuel cell (kW)					60
Motor		25	40	85	90
Drivetrain component costs					
Engine ($/kW)	39	53	74		
Battery ($/kWh)		750	320	250	750
Fuel cell ($/kW)					75
Motor ($/kW)		15	15	15	15
Motor fixed ($)		200	200	200	200
Drivetrain cost ($)					
ICE	3700	3700	3700		
Battery		750	2624	12,000	1000
Fuel cell					4500
H$_2$ storage					1800
Motor		575	800	1475	1550
Transmission		300	300	200	200
Exhaust	300	300	300		
Wiring		200	200	200	200
Battery charger			400	400	
Weight reduction	700	700	700	700	700
Total	4700	6525	9024	14,975	9950
Incremental cost × 1.4		**2555**	**6054**	**14,385**	**7350**
Vehicle mass, fuel consumption and maintenance costs					
Vehicle mass (kg)	1284	1290	1338	1617	1320
Fuel consumption (L/100 km)	5.5	3.1	3.0		2.3
Fuel consumption (MJ/km)	1.77	1.00	0.97		0.74
Electricity use (kWh AC/km)			0.13	0.15	
Electricity use (MJ/km)			0.47	0.54	
Maintenance cost ($/km)	0.041	0.035	0.030	0.021	0.025
Annual and discounted lifecycle costs					
Annual energy cost	1650	930	580	300	444
Annual maintenance cost	820	700	600	420	500
Total annual cost	2470	1630	1180	720	944
NPV of annual costs	15,177	10,016	7251	4424	5800
NPV of savings		2606	1873	−3632	2027

Note: Calculation procedures are explained in the main text. The vehicles are assumed to be driven 20,000km/yr, with 50 per cent from gasoline and 50 per cent from grid electricity in the case of the PHEV.
Source: Input data largely from Bandivadekar et al (2008)

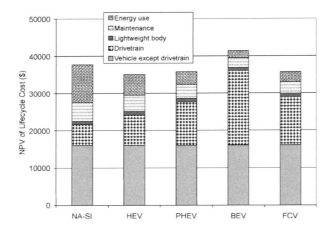

Figure 5.35 *The lifecycle cost components from Table 5.25*

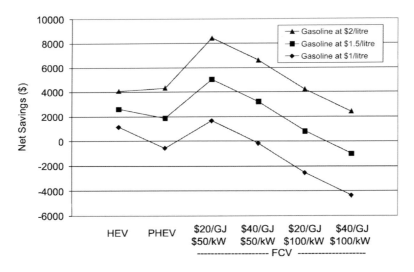

Note: Results are given for different costs of gasoline, different fuel cell costs and different costs of hydrogen at the refuelling station.

Figure 5.36 *The NPV of the savings in operational costs minus the extra purchase cost for various vehicles compared to an advanced non-hybrid gasoline vehicle*

5.4.12 Sunlight-to-wheels efficiency using renewable energy

Table 5.26 compares the sunlight-to-wheels efficiency for four different vehicles powered by solar energy: a methanol-powered FCV, where the methanol is produced from biomass energy; a methanol-powered FCV, where the methanol is produced from biomass energy and electrolytic hydrogen; a hydrogen-powered FCV with the hydrogen produced by PV electricity;

and a battery-powered vehicle with the batteries recharged using PV electricity. Photosynthesis has a very low efficiency (1 per cent) compared to that of commercially available PV modules (10–15 per cent) (see Volume 2, Chapter 4, subsection 4.9.1 and Chapter 2, subsection 2.2.2). The efficiency in producing methanol solely from biomass is no more than 67.5 per cent, so the efficiency from solar energy would be 0.675 per cent, whereas the efficiency in producing methanol from solar energy via a

Table 5.26 *Comparison of energy flows and overall efficiency in the utilization of renewable energy for hydrogen produced from biomass and used in a fuel cell, for hydrogen produced from solar electricity and used in a fuel cell, and for electricity produced from renewable energy and used in a battery electric vehicle*

FCV using methanol from biomass

Step	Sunlight	Photo-synthesis	Methanol production	Methanol transport	Fuel cell	Drivetrain	Solar energy-to-wheels
Efficiency		1%	67%[a]	98%	45%[b]	87%[c]	
Remaining energy	1.0000	0.0100	0.0067	0.0066	0.0030	0.0026	0.26%

FCV using methanol produced from biomass and solar electrolytic hydrogen

Step	Sunlight		Methanol production	Methanol transport	Fuel cell	Drivetrain	Solar energy-to-wheels
Efficiency			1.25%[a]	98%	45%[b]	87%[c]	
Remaining energy	1.0000		0.0125	0.123	0.0055	0.0048	0.48%

FCV with compressed H$_2$ produced using electricity from PV arrays

Step	Sunlight	PV array	Power conditioning	H$_2$ production	H$_2$ transport	H$_2$ compression	Fuel cell	Drivetrain	Solar energy-to-wheels
Efficiency		15%	85%	80%[d]	98%[c]	90%[f]	50%[g]	87%[c]	
Remaining energy	1.0000	0.1500	0.1275	0.1020	0.1000	0.0900	0.0450	0.0391	3.91%

Battery-powered vehicle using electricity from PV arrays

Step	Sunlight	PV array	Power conditioning	DC transmission	Battery recharger	Drivetrain	Solar energy-to-wheels
Efficiency		15%	85%	96%[h]	95%[i]	87%	
Remaining energy	1.0000	0.1500	0.1275	0.1224	0.1163	0.1012	10.12%

Note: [a] Optimistic future production efficiency, taken from Volume 2 (Chapter 10, subsection 10.6.2).
[b] 90% of the efficiency assumed for a fuel cell operating on pure hydrogen, based on the processing losses given in Table 5.16.
[c] Åhman (2001) gives the following efficiencies for the electric drivetrain components: electric motor and control system, 86 per cent today, 89 per cent potential; transmission, 98 per cent.
[d] A reasonable target for future system efficiency.
[e] Assuming transport a distance of 1000km at a pressure of 30atm, with an energy use equal to 2.1 per cent of the hydrogen energy, based on Volume 2 (Chapter 10, subsection 10.3.3).
[f] The additional energy required for compression from 30atm to 700atm is about 9.6% of the hydrogen energy, based on Volume 2, Figure 10.9.
[g] Net efficiency under typical driving conditions, after subtracting electricity use by the fuel cell system itself, based on Section 5.4.8.
[h] Based on an energy loss of 3.7 per cent for a transmission of 1000km as HVDC, as given in Volume 2, Figure 3.39.
[i] Åhman (2001) gives battery recharging efficiencies of 81 per cent for NiMH batteries and 95 per cent for lithium-ion batteries.

combination of biomass and electrolytic hydrogen would be 1.25 per cent (see Volume 2, Chapter 10, subsection 10.6.2 for the specific assumptions in deriving this efficiency). Given the efficiencies of the subsequent steps (given in Table 5.26), the hydrogen FCV ends up having a 16-fold greater sunlight-to-wheels conversion efficiency than the methanol FCV (about 4 per cent compared to 0.25 per cent) if methanol is produced solely from biomass, and a 9-fold greater sunlight-to-wheels efficiency if methanol is produced from biomass and electrolytic hydrogen. However, the battery-powered vehicle avoids the losses associated with production of methanol or the production and compression of hydrogen, and recharging the battery is about 60 per cent more efficient than the operation of a fuel cell. The net result is that the battery-powered vehicle is more than twice as efficient in its use of solar energy as the hydrogen FCV. The better efficiency of battery-powered vehicles compared to hydrogen FCVs would of course also apply to electricity produced from wind energy.

Given a fuel economy for a plug-in hybrid vehicle using gasoline of 5.6 litres/100km (42mpg) and an AC (grid) electricity energy use of 0.2kWh/km when using the battery (based on Figure 5.20), the corresponding energy use is 1.80MJ/km and 0.72MJ/km for gasoline and grid electricity, respectively. Given an efficiency of about 85 per cent in finding, extracting and converting crude petroleum to gasoline and delivering it to the vehicle, 1MJ of grid electricity displaces 3.0MJ of petroleum. If the grid electricity is renewably based and is used instead to replace coal-based electricity, then 1MJ of renewable grid electricity displaces 3MJ of coal at the current typical coal powerplant efficiency of 33 per cent (conversely, if the electricity is coal based at 33 per cent efficiency, there is no net saving in total primary energy use using BEVs or PHEVs). Given the 25 per cent higher CO_2 emission factor for coal compared to petroleum (see Table 2.10), the savings in CO_2 emission would be 25 per cent higher if the renewable energy is used to displace coal instead of petroleum. However, when a 50 per cent coal powerplant efficiency is achieved, 1MJ of renewable grid electricity would displace only 2MJ of coal, and 20 per cent more CO_2 emission would be avoided if renewable electricity is used to charge PHEVs than to displace coal-based electricity. If PHEVs are charged at times of excess renewable energy, then the net CO_2

saving is equal to the avoided petroleum emission even if there is coal on the grid at other times.

In light of the above, the best option for a car that is ultimately powered by renewable energy would be a PHEV that is charged with renewably based electricity, with either renewably produced hydrogen in a fuel cell or renewably based methanol in an ICE as a range extender.

5.4.13 Rebound effect from improved fuel economy

Improved fuel economy, by reducing the cost to drive a given distance and thereby reducing the effective cost of fuel, may lead to an increase in driving that offsets some of the savings due to improved fuel economy that would otherwise occur. This offset is referred to as the *rebound effect*. It is limited by that fact that increased driving also requires increased time and increases congestion, both of which serve to limit driving. The rebound effect is also an issue with regard to other forms of energy use, such as heating and lighting energy use in the residential sector, and is discussed more fully in Chapter 11 (section 11.7). For transportation, the effect appears to be smaller the larger the income (higher-income people are closer to saturation in terms of the number of hours that they are prepared to drive). The rebound is larger in the long term than in the short term. Small and van Dender (2005) have estimated the following rebound effects for the US: a short-term rebound of 4.7 per cent based on 1966–1997 data and 2.6 per cent based on 1997–2001 data, and a long-term rebound of 22.0 per cent based on 1966–1997 data and 12.1 per cent based on 1997–2001 data. These effects could be nullified by increasing the price of fuel (through taxes) in proportion to the increase in fleet-average fuel economy, so as to prevent the effective cost of fuel from falling.

In the absence of fuel cost increases through taxes, and assuming that there is no increase in driving due to the lower cost of driving, there will be additional money available to be spent elsewhere in the economy, with associated energy use and GHG emissions (this is an indirect rebound effect). However, to the extent that more fuel-efficient vehicles entail a greater upfront cost that needs to be financed over time, there will be a reduction in disposable income that will tend to reduce energy use and associated CO_2 emissions through the reduction in discretionary spending. The net effect depends on the relative importance of the decrease in fuel cost and of the reduction in disposable income due to the greater purchase cost of the vehicle.

5.5 Intercity passenger rail and urban and intercity buses

Summary information on the energy use per seat-km for intercity travel by cars, buses, diesel and electric trains and by air was presented in Table 5.3. In terms of onsite energy, diesel trains are not as efficient as electric trains, in part due to their greater structural mass so as to be able to hold fuel on board and due to the weight of the fuel itself. Buses have an energy use that is comparable to or slightly better than that of diesel trains, but much higher than for electric trains (due largely to the better efficiency of a motor compared to a diesel engine but also because of low aerodynamic and rolling resistance by trains). However, electric trains are subject to a markup factor of 2.5 (to account for losses in generating electricity at an assumed efficiency of 40 per cent). As a result, the energy intensity of a fully loaded intercity bus (about 0.26MJ/passenger-km) is comparable to the lower end of the range (0.22–0.41MJ/passenger-km) of fully loaded high-speed trains on a primary-energy basis.

A present-day fuel-efficient car (6 litres/100km in highway driving) with four persons would require modestly more primary energy per person-km than high-speed electric rail, but advanced hybrid gasoline vehicles would require much less than present electric trains and also possibly less than present diesel trains and buses (depending on the passenger loading). However, the energy intensities of buses and trains will also decrease over time, as discussed next.

5.5.1 Intercity trains

Table 5.27 gives the energy use for various low-speed (≤210km/hr) and high-speed (≥270km/hr) trains. Diesel trains are often used for low-speed, short-distance travel. Diesel–electric hybrid trains would reduce fuel use compared to purely diesel trains. In the hybrid train, the diesel engine would at times recharge a battery. Acceleration from a stop to 45kph would be accomplished entirely using the battery, after which the

diesel engine would come on. Some such trains are now operating in Japan. Fuel use is said to be reduced by 20–40 per cent (Graham-Rowe, 2008b).

The French TGV (Train à Grande Vitesse) and the Japanese Shinkansen have each been, at one time or the other, the fastest trains in the world, with the German Intercity Express taking up third spot. The Sanyo Shinkansen travels at 300kph but a 360kph train is under development, while the TGV routinely travels at 300kph, with short segments at 320kph and 350kph on some lines. High-speed trains exist or are under development in Korea (Seoul–Busan, using TGV technology), Taiwan (Taipei–Kaushsiung using Shinkansen technology), Germany (Cologne–Rhein/Main) and Spain (Barcelona–Madrid) (Takagi, 2006). Figure 5.37 shows the energy use per seat-km as a function of train speed for three generations of the German Intercity Express, while Figure 5.38 shows the progressive reduction in energy use per seat-km for the Japanese Shinkansen trains. The Intercity Express-3 is more efficient than earlier generations for speeds greater than 200km/hr, using about half the energy (0.14MJ/seat-km) of the Intercity Express-1 at 300km/hr. Shinkansen energy use fell from 0.126MJ/seat-km for the first Shinkansen in 1964, to 0.086MJ/seat-km for the N700 Series, introduced in 2005 – a reduction of 32 per cent. Magnetically levitated trains at 300kph have about twice the energy use per seat-km as the Series N700 Shinkansen at the 300km/hr Shinkansen typical speed and three times the energy use at 400km/hr, even though aerodynamic drag is almost the same for the two trains (see Kemp and Smith, 2007, Figure 9).

The very low energy use by the Shinkansen N700 is due in part to a very low train mass, a very aerodynamic design (necessitated to reduce track-side noise and to avoid sonic booms from compressed air at the far end of tunnels as the train enters them) and the use of 400m trainsets carrying 1323 passengers. Table 5.28 compares the number of cars and seats, train mass and mass per seat for the Japanese, French and German high-speed trains. The Japanese trains have the lowest mass per seat, due in part to a tighter packing of seats, and this contributes to lower energy use when accelerating and climbing a grade.

When travelling at a constant speed on horizontal terrain, energy is needed only to overcome resistance, which is overwhelmingly aerodynamic (the remainder

Table 5.27 *Onsite energy use by various intercity train systems*

	Speed (km/hr)	Energy use per car (MJ/km)	Passenger loading		Energy intensity (MJ/passenger-km)	
			Typical	All seats used	Typical	All seats used
British trains (low speed)						
Diesel[a]	≤200			120–600		0.28–0.42
Electric[a]	≤200			282–591		0.12–0.18
Japanese commuter trains[b]						
Low-power diesel						0.15
High-power diesel						0.29
Electric						0.079
Low-speed intercity electric trains						
Danish Rail	180	24.1	97	237	0.248	0.102
Swedish 2-car	200	21.3	63	180	0.338	0.118
Swedish Rail X2000	200	42.7	176	320	0.243	0.133
Flytoget (to Oslo airport)[a]	210			168		0.162
High-speed intercity electric trains						
Shinkansen Series 0 (1964)				1285		0.126
Shinkansen Series 300 (1992)				1321		0.115
Shinkansen Series 700 (1998)				1323		0.106
Shinkansen Series N700 (2005)						0.086
TGV Sud-est, 1st generation	270	63.72	294	368	0.217	0.173
TGV Duplex trainset	300	64.80	436	545	0.149	0.119
TGV Atlantique		47.52	291	485	0.163	0.098
German Intercity Express, 1st generation	280	86.72	290	645	0.299	0.134
Spanish AVE	300	51.17	266	313	0.192	0.163
Transrapid (MegLev)	300	74.40		440		0.169
Transrapid (MegLev)	400	104.5		440		0.238

Note: [a] From Kemp (2007); [b] From Takagi (2006)
Source: James Strictland, http://strickland.ca/efficiencjy.html, except where indicated otherwise

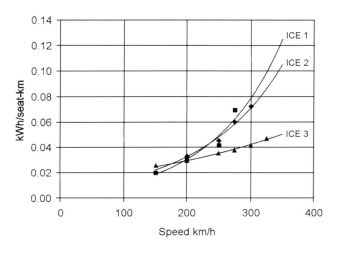

Source: Kemp (2007)

Figure 5.37 *Variation in energy use per seat-km with speed for three generations of the German Intercity Express high-speed train*

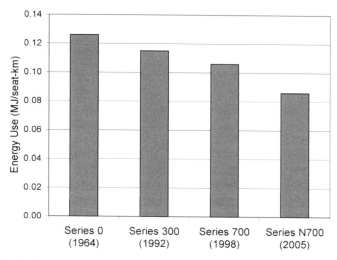

Source: Data from Takagi (2006) and other sources

Figure 5.38 *Energy use per seat-km by the Japanese Shinkansen train*

Table 5.28 *Characteristics of Japanese (Shinkansen), French (TGV) and German (Intercity Express) high-speed trains*

Train	Number of cars	Number of seats	Mass (t)	Length (m)	Mass/ length	Seats/ metre	Mass/seat (t/seat)
Shinkansen 0 (1964)	16	1285	967	400	2.42	3.21	0.75
Shinkansen 300 (1992)	16	1321	740	400	1.85	3.30	0.56
Shinkansen 700 (1998)	16	1323	634	400	1.59	3.31	0.48
TGV-PSE (1981)	10	384	380	200	1.90	1.92	0.99
TGV-PBKA (1990)	10	485	385	200	1.93	2.43	0.79
TGV Réseau-2N (1994)	10	545	384	200	1.92	2.73	0.70
Intercity Express	15	693	854	384	2.22	1.80	1.23
Intercity Express-3 (2000)	8	415	409	200	2.05	2.08	0.99

Source: Kemp (2007)

being rolling resistance). Aerodynamic resistance increases with speed squared, so the energy use per km tends to increase with speed squared (and the power that must be supplied to the train varies with speed cubed). However, the absolute time saving with equal increments of speed is subject to diminishing returns. Thus, increasing train speed from 300kph to 400kph increases energy use by 1.4 times the increase in going from 200kph to 300kph, while reducing travel time by only half as much. Thus, limits on further increases in the speed of high-speed trains can play an important role in reducing future energy use by high-speed trains.

The argument could be made that greater train speeds will increase their competitiveness against air travel, which is several times more energy intensive.

However, for short distances (300–600km), current high-speed trains are already competitive with air travel. Furthermore, a substantial fraction of the additional passengers taking current or faster high-speed trains are likely to be people who would not have otherwise travelled, as it is generally assumed that travel demand between two given population centres varies inversely with the square of the travel time (so cutting the travel time in half would increase travel demand by a factor of four). If a substantial fraction of the passengers on high-speed trains represents travel that is not displaced from inefficient or low-occupancy cars or from air travel, then introduction of faster trains will not decrease total transportation energy use.

5.5.2 Urban and intercity buses

Like trucks, the majority of buses are powered by diesel engines. The measures that are applicable to reducing the energy intensity of medium and heavy trucks (discussed in subsection 5.7.2) should be broadly applicable to buses. These include improvements in drivetrain efficiency (giving a 20 per cent fuel savings without forming a diesel–electric hybrid and a 30 per cent saving with hybridization), reduction in aerodynamic drag and rolling resistance, reduced energy use by auxiliaries (with air conditioning being particularly important for buses) and reduced idling (by installing auxiliary power units). Reductions in aerodynamic drag would be particularly beneficial for intercity buses, due to their greater speed, while forming a diesel–electric hybrid would be most useful for urban buses. By analogy with trucks (see subsection 5.7.2), it should be possible to reduce the energy intensity of intercity buses by at least a factor of two.

A variety of diesel–electric hybrid urban buses have been built and tested. Some of the reported increases in distance travelled per unit of fuel compared to diesel buses on comparable service runs are as follows: 8.5 per cent for buses in Long Beach, California (Lammert, 2008); 27 per cent for buses in the Seattle area (Chandler and Walkowicz, 2006); 50 per cent and 75 per cent for buses on test cycles that simulate conditions in Orange County and Manhattan, respectively (Chandler and Walkowicz, 2006); 29 per cent for buses on real routes in New York City (Barnitt, 2008); and 43 per cent for buses on the Chinese Transit Bus Driving Cycle, based on computer simulations for a bus with optimal switching between series and parallel hybrid configurations (Xiong et al, 2009).

Fuel cell buses have also been tested in several cities in the world. Axe et al (2008) present results from an evaluation of 27 fuel cell buses in nine European cities. They find that current fuel cell buses consume considerably more energy than diesel buses because the current generation of fuel cell buses has been designed for reliability rather than efficiency. Now that reliability has been demonstrated, measures to reduce energy use will be implemented, but are projected to be sufficient to return energy consumption to no more than that of diesel buses.

As noted above in subsection 5.4.4, the electricity used for air conditioning of automobiles (among other loads) is produced by the vehicle engine at a cost of $1–2/kWh (given a gasoline cost of $1–2/litre), which is several times the cost of PV electricity. The cost of electricity produced in buses is likely to be similar. Thus, rooftop PV systems could competitively meet at least some of the auxiliary loads, but at the same time there is a strong economic incentive to reduce electrical loads through improved efficiency.

In closing this section, it should be noted that the single largest factor in bus fuel consumption per passenger-km is the average number of passengers per bus. Thus, transit-friendly urban development or redevelopment – but combined with measures such as dedicated bus lanes to permit efficient operation of buses – is key in reducing the energy intensity of urban buses.

5.6 Passenger aircraft

As shown earlier, passenger air travel has been growing substantially faster than the overall growth in the transport of people, with passenger-km growing by an average of 4.4 per cent/year over the period 1990–2005 and no noticeable long-term effect of the September 2001 dip in air travel (see Figure 5.3). Recent trends in technology, aircraft operations and air-traffic management have resulted in energy intensity reductions per revenue passenger-km (RPK) of only 1.2–2.2 per cent per year (Lee et al, 2001). The net result is a growth in energy use by passenger aircraft of 2–3 per cent/year. Recent projections see further significant increases in RPK. For example, Berghof et al (2005) see global passenger-km travelled increasing from 4.09 trillion in 2005 to 21.2 trillion in 2050 under an 'unlimited skies' scenario, and increasing to 14.6 trillion in a 'regulatory push and pull' scenario. These are increases by about a factor of three to five. As discussed below, the climatic effect of aircraft energy use is disproportionately large (by at least a factor of two) compared to other energy uses, while the prospects for reductions in energy intensity are much smaller (probably only 25–50 per cent). Given the large projected increases in air travel, it is likely that restrictions in passenger air travel will be required if air travel is to contribute to reductions in GHG heating proportional to that of other sectors. This is likely to occur naturally to some extent as the price of oil increases in response to the imminent peaking in the world oil supply (see Chapter 2, subsection 2.5.1), as the cost of fuel is a large part of the total cost of air travel.

5.6.1 Climatic effects of air travel

Air travel influences climate through emissions of CO_2, NO_x, water vapour, SO_x and soot. NO_x affects climate through the production of ozone (a warming effect) and the removal of CH_4 (a cooling effect). Emissions of water vapour from the combustion of fossil fuels near the surface have a negligible effect on climate because they have a negligible direct effect on tropospheric water vapour concentration. However, water vapour produced (along with CO_2) from the combustion of jet fuel is released in the stratosphere (where long-haul aircraft fly), where it does noticeably increase the water vapour concentration. This in turn has a warming effect because water vapour is a GHG. Contrails (which consist of condensed water vapour from aircraft exhaust) form cirrus clouds that also have a warming effect. Finally, sulphate aerosols produced from SO_x emissions have a small cooling effect, while soot emissions have a small warming effect.

The net effect is a radiative forcing (net heat trapping) estimated to be about twice that arising from aircraft CO_2 emissions alone (Sausen et al, 2005). As emissions of CO_2, H_2O, NO_x, SO_x and soot depend on the amount of fuel consumed (with the emissions of the last three also depending on the fuel quality and/or the combustion characteristics), reductions in aircraft energy use will have a disproportionately large beneficial effect on radiative forcing and hence on future warming.

5.6.2 Factors influencing the energy intensity of air travel

The energy required to fly a given distance depends on the aircraft weight, drag and engine technology. The energy used per seat-km (energy intensity) for a trip between two cities depends on these three factors and the number of seats in the aircraft, the fraction of seats occupied (the capacity factor), the ratio of flying time to total operating time (which includes taxiing and waiting for take-off), the ratio of flying time (which includes waiting for permission to land) to minimum flying time needed (which accounts for flight paths that, as is often the case, do not represent the shortest distance between two points) and the total distance (the energy-intensive take-off is spread over a greater number of kilometres with longer flights, thereby reducing the average energy use per kilometre).

The first aircraft were propeller-driven using piston engines. *Turbojet* aircraft based on the gas turbine were introduced in the 1950s, followed by the bypass jet or *turbofan* engines in the 1970s. The public refers to both turbojet and turbofan aircraft as 'jet' aircraft, whereas the term 'jet' properly refers to a concentrated high-speed airflow. This air is thrown back, thereby propelling the aircraft forward. In turbojet aircraft, the air that is thrown back is air that has been combusted in the turbine. In turbofan aircraft, the turbine drives a large fan that is in a casing surrounding the turbine. Most of the air that is thrown back bypasses the turbine and is not used for combustion. The *bypass ratio* is the ratio of mass of air bypassing the turbine to mass of air used for combustion. Advanced engines have a bypass ratio of five to nine. Gas turbines can also be used to drive a propeller in *turboprop* aircraft, which are used for some short-haul flights where cruise speed is not critical. Turboprop, turbojet and turbofan engines are illustrated in Figure 5.39, where the compressor and turbine components can be seen.

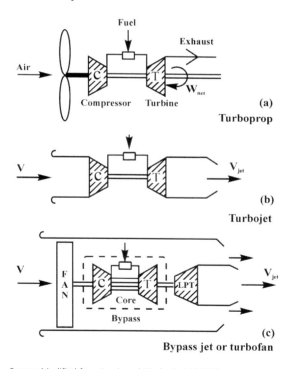

Source: Modified from Lewis and Niedzwiecki (1999)

Figure 5.39 *Gas-turbine aircraft engines: (a) turboprop, (b) simple turbojet and (c) bypass jet (turbofan) engine*

The efficiency of a jet engine is equal to the ratio of the power produced by the engine to the rate of consumption of fuel energy (engine power over fuel power). This in turn is the product of the thermal efficiency (the ratio of the rate of supply of kinetic energy to the airflow passing through the engine to fuel power, or more simply, the ratio of power supplied to the gas stream to fuel power) and the propulsion efficiency (the ratio of engine power to gas stream power). The thermal efficiency depends on the ratio of inlet temperature T_1 to outlet temperature T_2 and the ratio of peak pressure in the engine to atmospheric pressure, as well as on the compressor and turbofan efficiencies. Stoichiometric combustion (in which all of the oxygen is consumed) produces a temperature of about 2590K, but material constraints limited the allowed temperature to 1200–1600K in the mid-1970s and to 1800–2000K by year 2000 (Lewis and Niedzwiecki, 1999). The propulsion efficiency is related to the aircraft velocity V and the jet (gas stream) velocity V_j according to:

$$\eta_p = \frac{2V}{2V + V_j} \qquad (5.17)$$

Increasing the thermal efficiency requires increasing T_2/T_1, but this produces a larger V_j, which reduces η_p. For $T_2/T_1 = 5.6$ and a pressure ratio of 40 (typical of today's aircraft engines), $V_j = 817$m/s. At Mach[18] 0.85 at 10.7km altitude, $V = 252$m/s. This gives $\eta_p = 0.47$, and the corresponding thermal efficiency is 0.48, so the overall efficiency in this case is only 23 per cent (not much better than that of an automobile spark ignition engine) (Lewis and Niedzwiecki, 1999).

The thrust (force) imparted by the engine is equal to the change in the speed of air (as seen by an observer on the aircraft) as it passes through the engine times the rate of mass flow. If the rate of mass flow is doubled, the change in air speed can be halved while maintaining the same thrust. However, the amount of kinetic energy that must be added to the air decreases by a factor of four (because kinetic energy added varies with change in airspeed squared). Thus, less fuel needs to be consumed to produce the same thrust if the rate of mass flow is increased, which can be done through a larger bypass ratio. However, a larger bypass ratio requires a larger diameter engine casing, which increases the drag and engine weight, both of which would offset some or all of the benefit of the increased

bypass ratio. Efforts to reduce engine noise also tend to make them less efficient.

The mass of fuel consumed per unit of thrust supplied over a time of one second is referred to as the *thrust specific fuel consumption* (TSFC) and has units of kg/N-s. The TSFC is smaller the greater the efficiency of the engine (ratio of engine power to fuel power). The *specific air range* (SAR) is the distance travelled by the aircraft per MJ of fuel energy used. When the aircraft is cruising, the SAR for a jet aircraft depends on the TSFC, aircraft velocity (V), ratio of lift to drag (L/D) and aircraft weight (W) as follows (Babikian et al, 2002):

$$SAR = \frac{V}{TSFC \times h_F} \frac{L}{D} \frac{1}{W} \qquad (5.18)$$

where h_F is the heating value of the fuel (MJ/kg) and SAR is in km/MJ if V is in km/s. The ratio of lift to drag is called the *aerodynamic efficiency*, while the ratio of operating empty weight (OEW) to the maximum take-off weight (MTOW) is referred to as the *structural efficiency*. However, OEW/MTOW is really an inverse efficiency, as a smaller value means that a smaller structural mass is needed for a given payload and fuel mass, which in turn implies a smaller total aircraft weight W at each point in a flight with a given payload. The energy use per available seat-kilometre (ASK, the number of seats on the aircraft times the distance flown) is referred to as the *specific energy usage*, E_U. Under cruise conditions it is given by:

$$E_{UCR} = \frac{1}{SAR \times (Seating\_Capacity)} \qquad (5.19)$$

where the subscript CR indicates that $E_{U,CR}$ pertains to cruising conditions.

Figure 5.40 shows the trends in the TSFC and in the aerodynamic and structural efficiencies of new aircraft between 1959 and 1998, as well as the resulting tends in $E_{U,CR}$. The data are shown separately for long-haul turbofan, regional turbofan and (except for $E_{U,CR}$) for turboprop aircraft. Among long-haul aircraft, TSFC fell by almost 50 per cent from 1959 to 1998. The most recent regional turbofan aircraft have about a 20 per cent greater TSFC than recent long-haul turbofan aircraft due to the fact that the smaller engines in the regional aircraft have a smaller pressure ratio than in larger engines. Turboprop aircraft have about a 20 per cent smaller TSFC than long-haul aircraft. This

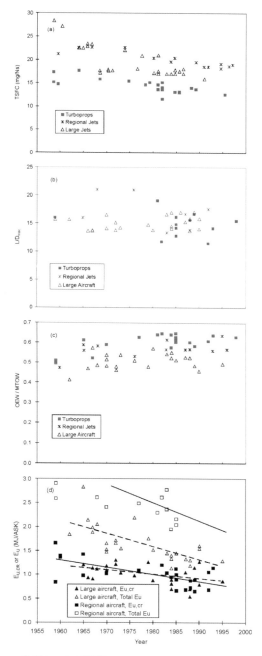

Source: Babikian et al (2002)

Figure 5.40 *Trend in (a) thrust specific fuel consumption, (b) aerodynamic efficiency (c) structural inverse efficiency, and (d) the resulting specific energy intensity while cruising for long-haul turbofan, regional turbofan and turboprop aircraft*

is due to the ability of propellers to accelerate large amounts of air at low speeds. As seen from Figure 5.40, there has been no trend in the lift/drag ratio – improvements in overall aerodynamics have offset the impact of fatter engines with larger bypass ratios – and there has been a slight upward trend in the ratio of empty to full weight, related in part to extra in-flight entertainment systems and to increases in the structural weight needed to enable better aerodynamics.

The energy used in flying between two points depends not only on $E_{U,CR}$ but also on the energy consumption during take-off and landing and on two additional efficiencies: the *airborne efficiency*, which is the ratio of the time required to fly between two points at cruising speed along the shortest path between the two points to the actual flying time, and the *ground efficiency*, which is the ratio of flying hours to total hours (including taxiing) that the plane is in operation. The energy use per km in flying between two points increases as the distance between them decreases because, with a shorter distance, the energy-intensive take-off is spread over fewer kilometres. Ground and airborne efficiencies also decrease as the distance decreases, because the time required for taxiing and delays at the airport are spread over fewer kilometres. This is seen in Figure 5.41, which shows the variation of the ground and airborne efficiencies with distance for turboprop, regional turbofan and long-haul turbofan aircraft. Turboprops have a greater airborne efficiency than regional turbofan aircraft for several reasons: they require about 30 per cent less runway than turbofan aircraft to take off, they climb to a lower altitude and they may experience less congestion-related delays because they fly at elevations not occupied by regional and long-haul jets and because they serve smaller airports on average. The greater ground efficiency of turboprops might also be related to the average size of the airports that they serve being smaller.

The net effect of all of the above factors gives the specific energy usage E_U averaged over the entire flight. Figure 5.40d compares E_U with $E_{U,CR}$ for regional and long-haul jets. Babikian et al (2002) indicate that regional turbofan E_U values are on average 3.2 times the calculated $E_{U,CR}$ values, regional turbofan E_U values are 2.5 times as large and long-haul jet E_U values are 1.6 times larger on average. Figure 5.42 shows E_U as a function of distance, separately for turboprop and

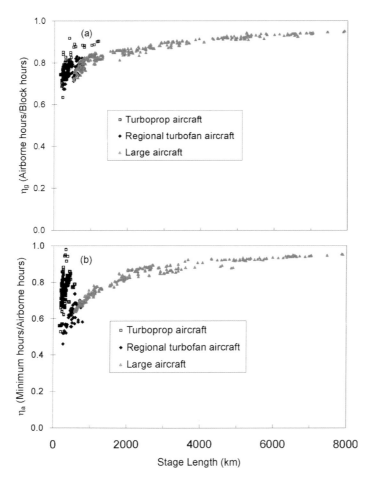

Source: Babikian et al (2002)

Figure 5.41 *Variation with distance travelled in (a) the ground efficiency, and (b) the airborne efficiency for regional turbofan and turboprop aircraft, and long-haul turbofan aircraft*

turbofan aircraft. The best E_U values for turboprop aircraft are significantly better (smaller) than the best E_U values for turbofan aircraft for distances of 500km or less.

Finally, the energy use per passenger-km is referred to as the *specific energy intensity*, E_I, and is given by E_U divided by the fraction of the seats occupied (the loading factor).

5.6.3 Prospects for further reductions in energy intensity

Despite the impressive 50 per cent reduction in TSFC achieved during the past 40 years, and the constraints

on further improvements in energy efficiency, there is still scope for further substantial improvement in aircraft energy intensity. Using the Boeing 777 as a baseline in 1995, J. J. Lee et al (2001) considered an average improvement in energy use per ASK of 1.0–2.0 per cent/year until 2025 to be feasible. With further improvements in capacity factor (which has increased from an average of 0.5 in 1970 to 0.75 in 2005 in the US) and operations, they projected a reduction in energy intensity per passenger-km of 30–50 per cent by 2025 for the most efficient aircraft on long-haul flights. Changes in fleet-average energy intensity could be larger or smaller than this, depending on a number of factors.

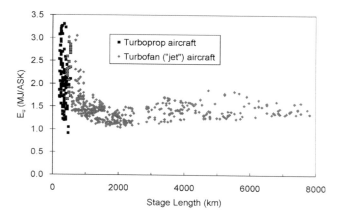

Source: Babikian et al (2002)

Figure 5.42 *Variation in overall energy intensity (MJ/seat-km) with distance travelled for turbofan and turboprop aircraft*

Lewis and Niedzwiecki (1999) believe that the thermal efficiency of aircraft engines can be improved by a further 10–20 per cent (from 50 per cent in high-bypass ratio engines to 60–70 per cent), that engine weight can be reduced by 20–40 per cent and that the propulsive efficiency can be improved by as much as 5 per cent. Each unit of reduction in engine weight reduces the overall aircraft weight by 1.5–4.0 units, amplifying the savings in fuel use.

One of the latest aircraft, the Boeing 787 Dreamliner (still under development in 2009), uses lightweight composites for 50 per cent of the primary structure, compared to only 12 per cent in the previous all-new Boeing aircraft. The engines are also lighter weight, and the aircraft is expected to be 20 per cent more fuel efficient than the B767 (introduced in 1978). The Airbus 380 (introduced in 2007) is 12 per cent more fuel efficient than the Boeing 747-400 (introduced in 1988), but could have been a further 11 per cent more efficient if the wing had been designed to minimize fuel use; however, operational constraints on wingspan at airports prevented optimization of the wing design. A solution with folding wing tips could have kept the A380 within the wingspan limit (80m) and also saved 11 per cent fuel (Peeters et al, 2005). Given these constraints, a 30–50 per cent improvement from 1995 to 2030 seems to be optimistic, however, a 20–25 per cent improvement from the current fleet average by 2030 does seem to be feasible, based on the improvements in the A380 and

B787 and the prospects for further improvements in engine efficiency and reductions in weight.

For short-haul flights, one option to reduce energy use is to switch from regional turbofan ('jet') aircraft to turboprop aircraft. As indicated in Figure 5.42, the energy intensity of the most efficient turboprop aircraft is 1.0–1.5MJ/seat-km for distances of 300–500km, whereas the energy intensity of turbofan aircraft over flights of this distance is 1.5–3.0MJ/seat-km. However, the cruising speed and range of turboprop aircraft are less than for turbofan aircraft with comparable passenger capacities, as indicated in Table 5.29. The latest Bombardier turboprop (the Q400) has a cruising speed of about 650kph and a range of 2400km, compared to 800–900kph and a 2700–3300km range for comparable turbojets. Noise and vibration have also been factors that have limited the appeal of turboprop aircraft, although Bombardier claims that the computer-controlled noise and vibration suppression system in its Q300 and Q400 models largely negates this disadvantage. According to Bombardier, the Q400 uses 30–40 per cent less fuel than older generation turboprops. However, even the most efficient turboprop aircraft are still many times more energy intensive than high-speed trains, some of which have energy intensities of 0.1MJ/seat-km or less (Table 5.27). Although train speeds are about half those of turboprop aircraft (300–350kph versus 500–650kph), they can provide comparable total travel times between city centres for journeys of up to 1000km.

Table 5.29 *Comparison of turboprop and small turbojet aircraft*

	Cruising speed (km/hr)	Range (km)	Number of seats
	Turboprop aircraft		
ATR-72-200	526 max, 460 economical	1195	64–74
Bombardier Q400	648	2400	70–78
	Turbofan aircraft		
Airbus 318	890	2780	107–117
Boeing 737-400	912 max 813 long range	3630	146
Embraer ERJ-170	890	3334	70

Source: www.airliner.net/aircraft-data

Another option for reducing CO_2 emissions from aircraft (but not energy use) is to use liquids produced from biomass using the Fischer-Tropsch process (see Volume 2, Chapter 4, subsection 4.3.8). Fischer-Tropsch liquids produced from coal have already been used in larger aircraft and have been observed to be of higher quality than conventional jet fuel (kerosene), with a factor of ten lower CO emission and 2–3 per cent less fuel consumption (IEA, 2006) (CO is a precursor for tropospheric ozone, a GHG).

5.7 Freight

The transport of goods accounts for about 30 per cent of transportation energy use in OECD countries (see Figure 5.2) and undoubtedly a much larger share in non-OECD countries. As noted earlier, freight movements worldwide increased by an average of 3.7 per cent/yr over the period 1990–2003 (see Figure 5.3). In 2003, ship, road and rail accounted for 73 per cent,

14 per cent and 13 per cent, respectively, of total international + domestic freight transport. For international trade, transport by water is even more dominant, accounting for 96.7 per cent of total tonne-km of freight movements in 2003, with road at 1.5 per cent, rail at 0.5 per cent, pipeline at 0.5 per cent and air at 0.3 per cent. However, goods traded by water accounted for only 49 per cent of the total monetary value of internationally traded goods, with road, rail, pipeline and air accounting for 11 per cent, 3 per cent, 2 per cent and 35 per cent, respectively. As for internal trade, Figure 5.43 shows the breakdown of freight tonne-km in Canada, Europe, Japan and the US by various modes (excluding pipelines). Rail constitutes 50–60 per cent of total freight transport in the Canada and the US due to the significant transport of coal and grains. Transport by water includes inland waters and coastal waterways, and is particularly important for Japan. Table 5.30 compares the value of transported goods per tonne for different modes of transport within the US, the cost per tonne transported 1000km and the cost of transporting goods 1000km as a fraction of the value of the transported goods. The cost of transport is generally a very small fraction of the value of the transported goods, so very large increases in the cost of energy (which is only part of the transport cost) would be required in order to have a significant effect on the final cost of delivered goods.

The energy used to transport goods can be reduced through: (1) shifting to less energy-intensive modes of transportation (particularly rail and ship) wherever possible; (2) improving the efficiency of each mode of transportation; and (3) reducing the movement of goods through greater reliance on locally produced goods. Each of these options is discussed below.

Table 5.30 *Value of transported freight per tonne of freight, cost of transporting freight, and transport cost as percentage of the value of the transported freight*

Mode	Freight value ($/tonne)	Transport cost ($/1000 tonne-km)	Transport cost as a percentage of value
Barge		5	
Road	2839	180	6.3
Rail	911	15	1.7
Pipeline	666	10	1.5
Air	86,816	551	0.6

Source: Gilbert and Perl (2007)

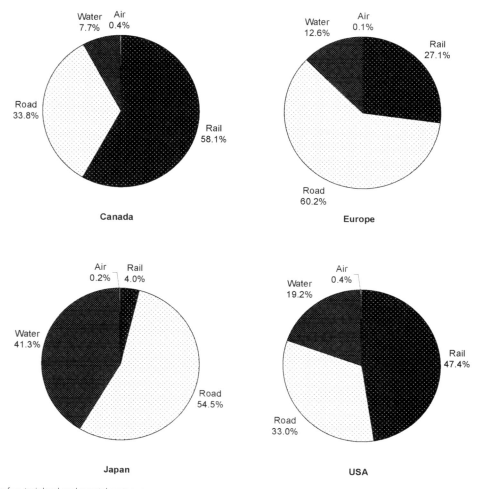

Note: Water refers to inland and coastal waterways.
Source: Victoria Transport Policy Institute, www.vtpi.org

Figure 5.43 *Relative amounts of freight (tonne-km) transported by different modes within Canada, Europe, Japan and the US*

5.7.1 Comparative energy intensities

Table 5.31 compares the energy intensity for different ways of transporting freight, based on international data collected for the International Maritime Organization and based on model simulations (Skjølsvik et al, 2000b). The model simulations take into account the energy use during different operations (loading and unloading, manoeuvring, transit), capacity factors (the load as a fraction of full load, given in Table 5.31) and transit speed. Rail is more efficient than truck for transport on land because of the extremely low rolling resistance of steel on steel and the lower average drag, especially of long trains. Marine transport by large cargo ship is even more efficient than transport by rail.

The dependence of the simulated energy intensities given in Table 5.31 on the distance transported is given in Figure 5.44, while the variation with capacity factor is given in Figure 5.45. The simulation results indicate that rail is about half as energy intensive as truck for transport distances throughout the entire range of 200–2000km, in spite of a lower average capacity factor for rail (0.65 versus 0.85). General cargo and container

Table 5.31 *Range of national average capacity factors and energy intensities of different methods of transporting freight*

Mode	National average data		Model simulations		
	Capacity factor	Energy intensity (MJ/tonne-km)	Capacity factor	Maximum speed (kph)	Energy intensity (MJ/tonne-km)
Air	n.a.	7–15			
Average road freight	0.2–0.4	1.8–4.5			
Heavy trucks	0.6–1.1[a]	0.6–1.0	0.85	88	0.65–0.9
Rail	0.5–0.8	0.4–1.0	0.65	80	0.27–0.52
Marine	0.5–0.75	0.1–0.4	0.65		
– General cargo			0.65	28	0.21–0.47
– Container cargo			0.65	37	0.21–0.32
– Bulk cargo			0.65	26	0.11–0.25
– Oil tanker			0.65	26	0.06–0.11

Note: Model-simulated energy intensities pertain to a transport distance of 200km (high values) to 2000km (low values). [a] A capacity factor greater than 1.0 indicates overloading.
Source: Skjølsvik et al (2000b)

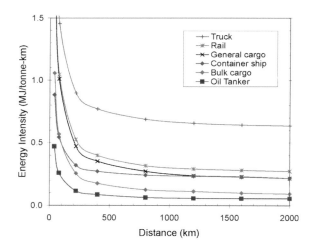

Note: Figure assumes a capacity factor of 0.85 for truck and 0.65 for rail and ship.
Source: Skjølsvik et al (2000b)

Figure 5.44 *Variation in energy intensity for different modes of freight transport as a function of the distance transported*

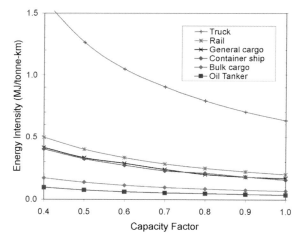

Source: Skjølsvik et al (2000b)

Figure 5.45 *Variation in energy intensity for different modes of freight transport as a function of the capacity factor, for a transport distance of 3218km*

ship transport is 30–40 per cent less energy intensive than rail at 200–400km transport distance but only 15 per cent less energy intensive at 2000km, while bulk transport by ship is 70 per cent less energy intensive than rail at a 2000km transport distance. However, the energy use by ship increases with the square of the transport speed, and ship speeds are considerably slower than rail speeds, as shown in Table 5.31.

5.7.2 Potential efficiency improvements in trucking

In the near term (until 2010), the fuel economy of trucks was expected to decrease by about 4 per cent according to Duleep (2007) and by 7–10 per cent according to Tokimatsu (2007) due to much more stringent pollutant emission standards by 2008–2010 for particulate matter and NO_x. Urea (with an embodied energy of 50MJ/kg according to Figure 7.10 and a density of 1.33kg/litre) is used to control NO_x (about 1–2 litres or 67–133MJ/100km), but the above energy penalty does not include the energy used to make urea. However, particulate matter is largely sooty carbon that strongly absorbs solar radiation, while NO_x contributes to the formation of tropospheric ozone, a GHG, so the sharp reduction in emissions of these pollutants will have a global warming benefit that probably greatly outweighs the additional CO_2 emission due to greater fuel consumption.

There are many commercially available technologies that could be implemented to reduce the fuel consumption of trucks in the medium term and more so in the long term. These technologies fall into the following categories: measures to reduce the loads that the engine must satisfy, improvements in the vehicle drivetrain, and measures to reduce energy use during idling. When a truck is being driven, the engine needs to provide power to overcome aerodynamic and rolling resistance, inertial resistance (related to vehicle mass), and to satisfy auxiliary loads (various pumps, fans and compressors). Collectively, these loads could be reduced by 17–34 per cent (based on the discussion in Vyas et al, 2002).

The best turbocharged diesel engines for heavy trucks achieve 45 per cent thermal efficiency (work output to fuel energy input ratio), although typical efficiencies are around 40 per cent. The US Department of Energy (DOE) Office of Transportation Technologies is developing a low-emission diesel engine with a near-term target efficiency of 55 per cent (the LE-55 engine). Recent work has identified technologies that should deliver 50 per cent efficiency while meeting 2010 pollutant emission standards for diesel-fuelled trucks (Easley et al, 2005). Hybrid diesel–electric technology can also offer large savings for trucks (and buses). According to Duleep (2007), overall savings of 25–45 per cent have been demonstrated in urban settings with demonstration vehicles. Cost is currently an impediment, but with production volumes of 2000–3000 hybrid vehicles per year, incremental costs will be in the $15,000–25,000 range, which can be justified in terms of fuel cost savings according to Tomić and van Amburg (2007).

Currently, freight trucks in the US idle an estimated average of 2400 hours per year so that the engine can provide power for heating or air conditioning the cabin while the driver sleeps, with a

fuel consumption of 0.6 gallons/hour without air conditioning in operation and 1.0 gallons/hour with the air conditioner. Direct-fired heaters for heating and auxiliary power units (APUs) for air conditioning would be far more efficient for these purposes. Solid oxide fuel cells could also serve as APUs, resulting in a 40 per cent reduction in fuel use during overnight idling (Lutsey et al, 2007), but at present they are prohibitively expensive (see Chapter 3, subsection 3.2.5). The air conditioning and heating loads that would need to be satisfied by APUs could themselves be reduced by more than 50 per cent through simple improvements in the thermal performance of the cabin enclosure (Proc et al, 2008).

Reductions in truck speed on highways can also yield significant energy savings – 6.0 per cent for reduction from 70mph to 65mph and another 7.6 per cent for a reduction from 65mph to 60mph according to Ang-Olson and Schroeer (2002), with another 3.8 per cent from improved driver training.

Calculation of the interactive effect of the various energy-saving measures requires a model of the energy flows and loads associated with various trucks, combined with simulation of realistic driving cycles (combining urban and highway driving). In the absence of such simulations (to the author's knowledge), a rough estimate will be made here using the following formula:

$$E_{final} = E_{driving} F_{load} F_{drive-train} F_{driver} + E_{idle} - S_{idle} \quad (5.20)$$

where $E_{driving}$ and E_{idle} are the portions of the total energy use of the baseline truck assumed to be used while the vehicle is being driven and during idling, respectively, F_{load} is the factor by which the total load on the engine is multiplied due to reductions in individual loads, $F_{drivetrain}$ is the factor by which the energy requirement while driving is multiplied due to improvements in the drivetrain efficiency, F_{driver} is a factor to account for changes in driving speed and driving habits, and S_{idle} is the savings in total energy use due to the elimination of idling with APUs and more efficient operation of the equipment powered by APUs. Here, it will be assumed that $E_{driving} = 0.88$, $E_{idle} = 0.12$ and $S_{idle} = 0.089$ (the last of these taken from Ang-Olson and Schroeer (2002), the first two being pure guesses). The savings due to reductions in individual loads are combined additively to give the overall savings, from which F_{load} can be calculated, while individual measures to improve the drivetrain efficiency are combined multiplicatively (this is of course a simplification of how the various measures would interact in reality).

Tables 5.32 and 5.33 lay out the calculation of the overall fuel savings using the above approach. Table 5.32 is based on near-term savings measures as identified by Ang-Olson and Schroeer (2002), while Table 5.33 is based on the work of Vyas et al (2002), which differentiates between different classes of trucks. In the latter study, hybrid diesel–electric drivetrains are thought to be viable for medium-weight trucks and to increase the distance travelled per litre of fuel by 40 per cent (which corresponds to a fuel saving of 29 per cent). For heavy trucks, a variety of other measures is expected to increase distance travelled per litre of fuel by 22 per cent (a fuel saving of 18 per cent – about the same as in going from an engine thermal efficiency of 45 per cent to 55 per cent). Savings due to improved driver operation of the truck are assumed to be applicable only to non-hybrid vehicles, and the savings due to reduced speed would undoubtedly be different for hybrid and non-hybrid vehicles, but the same factors are used here. Overall savings are 35 per cent according to Table 5.32 and 53–58 per cent according to Table 5.33.

Table 5.33 *Potential long-term energy savings in trucks of various classes (where Classes 2B and 3 are lighter medium-weight trucks, Classes 4–6 are heavier medium-weight trucks and Classes 7–8 are heavy trucks)*

Measure	Savings (per cent)		
	Class 2B, 3	Class 4–6	Class 7–8
Aerodynamics	2.5	7.5	10.8
Rolling resistance	2.5	2.5	7.2
Weight reduction	7.2	5	10
Auxiliaries (when not idling)	5	5	6
Cumulative loads	17.2	20	34
Loads multiplicative factor	0.828	0.800	0.660
Transmission	2	2	2
Diesel engine	29	29	18
Drivetrain multiplicative factor	0.696	0.696	0.804
Speed reduction (70 to 65mph)	6	6	6
Speed reduction (65 to 60mph)	7.6	7.6	7.6
Driver training	0	0	3.8
Driver multiplicative factor	0.869	0.869	0.836
Reduced idling	8.9	8.9	8.9
Overall savings	52.9%	54.4%	57.9%

Source: Component savings from Vyas et al (2002), except for effects of speed reduction and driver training, which are from Ang-Olson and Schroeer (2002)

Table 5.32 *Near-term potential to reduce energy use by heavy trucks*

Strategy	Fuel savings (%)	Multiplicative factor
Improved tractor aerodynamics	3.6	
Improved trailer aerodynamics	3.8	
Wide-base tyres (to replace dual tyres)	2.6	
Automatic tyre inflation	0.6	
Weight reduction	1.8	
Cumulative load reduction	12.4	0.876
Low-friction engine lubricants	1.5	
Low-friction drivetrain lubricants	1.5	
Speed reduction (70 to 65mph)	6.0	
Speed reduction (65 to 60mph)	7.6	
Driver training	3.8	
Drivetrain and driver, cumulative		0.811
Idling reduction	8.9	
Net result	34.4	0.656

Source: Ang-Olson and Schroeer (2002)

Table 5.34 shows the energy flow for base case and efficient, fully loaded heavy trucks travelling on flat terrain at 65mph as computed by Duleep (2007). Here,

Table 5.34 *Energy flow (kWh) for a fully loaded, heavy truck, travelling at 65mph on flat terrain for one hour*

	Base case	Efficient case	Reduction in total load (%)
Input energy	400	224.8	
Engine efficiency	0.4	0.5	
Total output	160	112.4	
Drivetrain losses	9	6.3	1.7
Aerodynamic losses	85	68	10.6
Rolling resistance losses	51	30.6	12.8
Auxiliaries loads	15	7.5	4.7
Overall reduction in losses			29.8

Source: Duleep (2007)

drivetrain losses after the engine are treated as an additional load. The required input energy for the efficient vehicle is 56 per cent that of the base case vehicle – a saving of 44 per cent. The savings would exceed 50 per cent if the driving speed is 70mph for the base case and 60mph for the efficient case.

Finally, capacity factors are an important determinant of energy use per tonne-km of transported material. For a fully laden truck on flat terrain, about 70 per cent of the fuel goes to moving the truck and 30 per cent to moving the load. Thus, a one-quarter-full large truck uses 2.5 times the fuel per tonne-km as a three-quarters-full truck (except in hilly country) (Richard Gilbert, personal communication, 2007).

5.7.3 Potential efficiency improvements in rail transport

A diesel locomotive uses a diesel engine to generate electricity that drives a motor that in turn drives the wheels (this is different from diesel trucks, cars and buses, where the diesel engine directly drives the wheel axle through a transmission). The steps involved in the transformation from input fuel to tractive force at the engine wheels for a diesel locomotive are shown in Figure 5.46. Two kinds of drive systems are shown in Figure 5.46: a DC drive system, which was used exclusively until recently, and an AC drive system,

which has been introduced recently. AC drive systems allow for more efficient adhesion control (adhesion control is the ability to reduce power to an axle when slippage of the wheel on the rail is detected) and give a larger power to volume ratio (Gavalos et al, 1995). The overall efficiency from fuel input to wheel traction is 32.6 per cent for a DC drive system and 33.4 per cent for an AC drive system. These efficiencies are for operation at full load, where the diesel engine efficiency is highest. The assumed diesel engine efficiency is 39 per cent after allowing for use of part of the power output from the engine to run auxiliaries.[19]

Rail freight energy efficiency in the US improved at an average rate of 2.8 per cent/year for the last 20 years of the 20th century (Brown et al, 1998). Further incremental improvements, probably at a slower rate, can be expected for some time through improvement in the diesel engine electrical efficiency, improved wheel traction and reduced auxiliary electricity use.

Some rail freight in Europe uses electric locomotives fed from overhead powerlines. For these, regenerative braking is readily available and can reduce energy requirements by 10–30 per cent, with another 20 per cent or greater savings through better driving style according to Vitins (2009). The efficiency in energy transfer from AC power lines to the wheel is

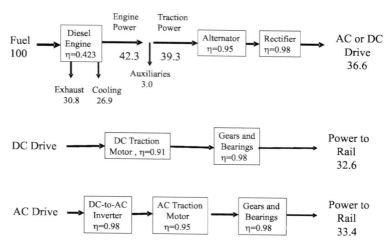

Note: The overall efficiency at 25 per cent of full power would be very close to that indicated here.
Source: Based on information in Gavalos et al (1995)

Figure 5.46 *Energy flow in a diesel locomotive operating at full power*

86 per cent. This can be increased to 90 per cent with DC power and even further with further technological development (Vitins, 2009).

Fuel cell locomotives using hydrogen

Locomotive engines are an ideal early application of hydrogen-powered fuel cells, which provide an opportunity for a large, non-incremental improvement in efficiency. Unlike automobiles, a large whole new infrastructure would not need to be built up for hydrogen fuel cells, because there would be only a limited number of sites where refuelling would occur. Furthermore, the safe onboard storage of hydrogen would be easier in locomotives than in automobiles, where the weight or bulk of the storage system would not be critical. Finally, locomotive diesel engines cost more than automobile engines ($300/kW versus $45/kW), so it will be easier for fuel cells to compete with diesel locomotive engines than with the automobile internal combustion engine. A global fleet of fuel cell locomotives, unlike a global fleet of fuel cell cars, would not exhaust the available supply of platinum even if low-temperature fuel cells (requiring platinum) were used.

Figure 5.47 shows the energy flow diagram for a fuel cell engine, based on Gavalos et al (1995) but with a fuel cell efficiency of 0.60, which is achievable at 25 per cent of full load (see Figure 5.28). The overall conversion efficiency from fuel to wheel traction is 50.6 per cent for DC drive and 53.7 per cent for AC drive. These efficiencies are 1.55 and 1.60 times the corresponding diesel engine efficiencies (however, as noted above, efforts are underway to eventually produce diesel engines for trucks with an efficiency of 55 per cent, so the advantage relative to diesel engines could be smaller in the future). The improvement in real operation would be even larger, since the locomotive would be running at part load much of the time, and the efficiency of a diesel engine decreases with decreasing load. The net result is at least a 55–60 per cent improvement in locomotive engine efficiency compared to current locomotives, which corresponds to a reduction in energy requirements of 35–38 per cent.

Another potential efficiency advantage is the use of regenerative braking to recharge a battery, which could be added as part of the fuel cell-motor system for peak power. The breaking system in present locomotives is a combination of rheostatic braking and frictional braking. Rheostatic braking involves a reverse-traction motor that

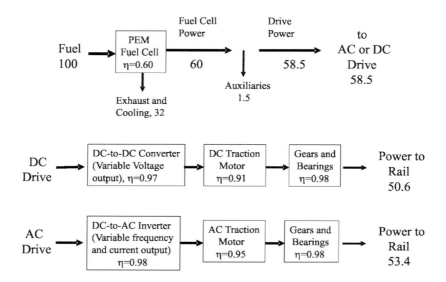

Source: Based on information in Gavalos et al (1995), except for an assumed fuel cell electrical efficiency of 60 per cent

Figure 5.47 *Energy flow in a fuel cell locomotive operating at 25 per cent of full power*

generates electricity. The electricity is dissipated as heat in resister elements that are placed on top of the locomotive. This heat in turn is dissipated to the atmosphere with a high-power fan (Scott et al, 1993).

5.7.4 Potential efficiency improvements in transport by ship

The International Maritime Organization assessed the potential for reductions in energy use and GHG emissions by maritime shipping, taking into account the normal rate of replacement of the existing stock and assuming implementation of energy efficiency measures in new ships (Skjølsvik et al, 2000a). They estimated a maximum reduction of 17.6 per cent by 2010 (if measures were to begin in 2000) and 28.2 per cent by 2020 compared to BAU energy use. A reduction in ship speed by 10 per cent would reduce the remaining energy use by 23.3 per cent (energy use per unit distance varies roughly with the square of the ship speed) and could be implemented immediately (as energy prices increase, trading off extra travelling time against reduced fuel costs will become increasingly attractive). The total reduction (relative to the baseline) would be 37 per cent over 20 years and 45 per cent over 30 years.

Small wind turbines could provide some of the required propulsive force for shipping. German wind turbine manufacturer Enercon has fitted four cylindrical-shaped Flettner rotors (named after the German inventor Anton Flettner, who developed the system in the 1920s) to the 130m × 22.5m barge used to transport its wind turbines to world markets. Each rotor is 4m in diameter and 27m high and is said to generate 10–14 times the propulsive force of a rectangular or trapezoidal sail with the same area. The four turbines are expected to reduce fuel use by 30–40 per cent (de Vries, 2007).

5.7.5 Potential efficiency improvements in transport by air

The potential to reduce the amount of energy used to transport goods by air is likely to be comparable to that for reducing the energy intensity of moving people by air, discussed in section 5.6. This would

appear to be no more than a 20–25 per cent reduction by 2030.

5.7.6 Reducing the need to transport goods

Globalization and free trade have led to a dramatic increase in the transport of agricultural and manufactured products between different countries. Part of this has been driven by low-cost labour in countries with lower standards of living (so wages are lower and benefits are often non-existent), a lack of safety, health and environmental standards in many countries, and artificially low valuation of currencies in major exporting countries such as China, all of which have led to an artificial stimulation of trade. Finally, energy costs have been comparatively low and have not included environmental and other externalities (such as military expenditures to protect oil interests), which also encourage artificially high levels of trade.

With the development of more equitable standards of living around the world and uniform health, safety and environmental standards, and with probably dramatic increases in the cost of petroleum products over the coming years, the growth in world trade volumes is likely to at least moderate and may even reverse in the future. This process can be accelerated through greater social awareness of the benefits of buying locally produced products.

There is a growing movement to buy locally produced food whenever it is available because such food tends to be of higher quality, and sourcing food locally provides a number of social and environmental benefits, as discussed at length by Halweil (2002). In many communities, high-end restaurants have been the catalyst behind this movement. As discussed by Halweil (2002, Figure 3a), the ingredients for a typical meal in Iowa are transported an average of 2577km, whereas most of the ingredients could be produced in the state with an average transport distance of 74km. A typical English Sunday meal (beef, potatoes, carrots, broccoli, beans, blueberries and strawberries) uses ingredients from around the world, with 650 times the transport-related CO_2 emissions as the same meal made from locally grown ingredients, most of which are available in England most of the year (Halweil, 2002, Figure 3b). Substitution of distant products with nutritionally equivalent local products, or consumption of some products only during the seasons when they are available locally (or greater use of preservation for out-of-season use) can also reduce energy use related to the transportation of food.

However, if the amount of land that is required to produce a given amount of food is much less for distant locations compared to local production, long-distance transport of certain products could be justified from an environmental point of view. The land saved by using the most productive land could be devoted to bioenergy crops (discussed at length in Volume 2, Chapter 4), offsetting or exceeding the transportation energy use, or could serve other environmental objectives (such as preservation of natural habitats). These and related issues are discussed more fully in Chapter 7.

5.7.7 Impact of e-commerce

The advent of the internet and e-commerce has a number of competing effects on transportation and overall energy use:

- it may have contributed to the trend of increasing freight energy use by supporting longer transport distances and greater delivery frequencies, although it can also contribute to a more efficient distribution system (Hesse, 2002);
- it may result in greater packaging, as found in a comparison of e-commerce versus conventional retailing in the Japanese book sector (Williams and Tagami, 2003);
- it can reduce the energy required to transport groceries from retail outlets to homes by permitting electronic shopping and home delivery services that replace trips by individual shoppers, as found for a case study in Finland (Siikavirta et al, 2003);
- it allows a reduction in warehouse building area (and associated energy use) by facilitating just-in-time delivery systems, which permits maintenance of less inventories (Matthews and Hendrickson, 2003).

Thus, although there has probably been some increase in transportation energy use due to e-commerce, there are other potentially offsetting impacts.

The effective use of e-commerce to reduce transportation energy use in the delivery of groceries

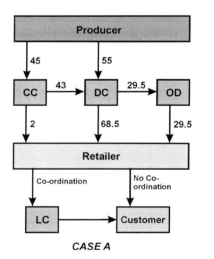
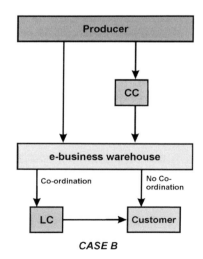

Note: The numbers in (a) represent the percentages of total flow in a district in Stockholm.
Source: Bratt and Persson (2001)

Figure 5.48 *Flow of goods from producers to consumers (a) at present, and (b) as might occur with an energy-efficient e-commerce arrangement*

and other household goods might require a reorganization of the relationships between producers and customers. At present, the relationship involves collection centres and distribution centres between producers and retailers, who then sell goods to customers, as illustrated in Figure 5.48a. Effective use of e-commerce might require simpler relationships that involve logistics centres in residential areas and direct involvement of wholesalers in online purchasing, as illustrated in Figure 5.48b. As envisaged by Bratt and Persson (2001), daily deliveries of all the household goods ordered over the internet would be made from retailers or e-business warehouses to the logistics centres. Customers could then pick up the goods themselves from the logistics centre (ideally on foot) or have them delivered during a certain time window. Standardized (and reusable) containers for groceries and other goods would be part of the package.

5.8 Summary

The most important points from this chapter are as follows:

- The single potentially most important factor in reducing future GHG emissions from personal transportation is urban form (in particular, density and the intermixing of different land uses) and transportation infrastructure (in particular, the provision of rail-based rapid transit). There is at present a factor of ten difference in per capita transportation energy use between wealthy Asian cities and American cities, and a factor of 25 difference in energy use by private motorized vehicles in Hong Kong and Houston, which represent the two extremes in terms of urban density and the type of transportation infrastructure available.

- There are substantial opportunities to improve vehicle fuel economy through technical measures – a factor of two without any change in the performance characteristics of vehicles. With vehicle downsizing and acceptance of slower acceleration, even larger savings could be achieved.

- If battery performance and cost continue to improve, direct use of grid electricity to recharge electric vehicles represents more than twice as effective a use of renewable energy as using it to make hydrogen for use in fuel cell vehicles; plug-in hybrid electric/hydrogen fuel cell vehicles rather than pure fuel cell vehicles may therefore represent the best option for renewable energy-based private vehicles with adequate driving range for intercity travel. Later, plug-in hybrids that are charged with excess PV electricity while parked could be designed to send power back to the grid when it is needed. A logical technology development sequence would thus be: hybrid gasoline–electric vehicles, followed by plug-in hybrid electric vehicles (PHEVs) using hydrocarbon fuels, followed by bidirectional PHEVs that serve to store intermittent renewably based electricity, followed by PHEVs using hydrogen or biomass- and hydrogen-derived methanol.

- There are significant constraints on the long-term viability of a global fleet of fuel cell vehicles at high (i.e. European and even more so American) rates of car ownership.

- The entire energy system needs to be considered when evaluating the merits of various alternative fuels in the transportation sector. In particular, biomass energy is much more effective in reducing CO_2 emissions if used to displace coal-fired electricity. Similarly, electricity from renewable energy sources such as wind and solar (PV or solar thermal) will be much more effective at reducing CO_2 emissions if used to displace coal-fired electricity than if used to displace petroleum used for transportation.

- If battery technology does not develop as hoped but the substantial obstacles to the use of H_2 in vehicles can be solved, then use of renewable energy to make H_2 for transportation makes sense once GHG emissions from electricity generation have been eliminated (either by displacing fossil fuel-based electricity with renewably based electricity, or through capture and sequestration of CO_2 from powerplants if this turns out to be feasible on the required scale). An important factor in the viability of renewably based hydrogen for transportation will be to reduce the cost of renewably based electricity and/or to substantially reduce the cost of fuel cells.

- With regard to intercity travel, a fully loaded bus or fully loaded high-speed electric train have comparable primary energy requirements per passenger-km, about four times less than that of a car with poor fuel economy (12 litres/100km) with four occupants. The primary energy requirement of high-speed rail per passenger-km is 5–10 times less than that of fully loaded passenger aircraft for distances of up to 1000 km. The energy requirements for rail increase sharply with speeds above 300km per hour, and when high-speed rail service is introduced some of the passengers will be people that would not have travelled otherwise, rather than passengers displaced from more energy-intensive modes of transportation.

- Substantial opportunities exist for reducing emissions per tonne-km for the transportation of goods, with potential savings in the order of 50 per cent. However, greater reliance on local production, so as to reduce the quantities and distances transported, could also contribute to large reductions in freight transport energy use.

- There are only modest opportunities for further improvements in the efficiency of air travel in the near term (a 20–25 per cent reduction in energy use per passenger-km on long-haul flights, relative to 1995), so restrictions on total air travel (through, for example, pricing mechanisms) will be required in order to achieve significant absolute reductions in energy use. Hydrogen-powered aircraft may be a feasible long-term solution to eliminate CO_2 emissions associated with air travel, as long as the hydrogen is produced without CO_2 emission, although NO_x emissions will have some warming effect. Biofuels are another possibility.

- Fuel cell railway locomotives represent a possible early niche application of renewably derived hydrogen, with a reduction in on-site energy requirements by about 35 per cent.

Notes

1   The term 'light truck' refers to pickup trucks, SUVs and passenger vans.
2   WBCSD (2004) gives energy use for all the modes of travel shown in Figure 5.2 except ships, so ship energy use was computed as the difference between the total transportation energy use given in Tables 2.1 and 2.2 and the total of all other modes given by WBCSD (2004).
3   Gasoline will combust in a diesel engine too, without a spark, but in a much less controllable manner.
4   In the automotive engineering literature one usually sees the terms 'indicated efficiency' and 'brake efficiency'. 'Indicated efficiency' and 'thermal efficiency' are the same thing, both being shortened forms of the full-term 'indicated thermal efficiency'. 'Brake efficiency' is sometimes also called 'brake thermal efficiency', but both terms mean the product of thermal efficiency and mechanical efficiency. The terms 'indicated' and 'brake' are used because they pertain to the experimental setups used to measure these efficiencies in test rigs.
5   Defined in Chapter 2, subsection 2.1.4.
6   Air drag (a force) varies with v^2, but the power required to overcome resistance varies with the speed of the vehicle times the drag, so the power required to overcome drag increases with the cube of vehicle speed. A similar relationship was noted between the rate of airflow in a building and the power required to maintain the airflow (Chapter 4, Equation (4.7)
7   CAFE = corporate automobile fuel economy, the regulated minimum required average fuel economy (mpg) for automobile fleets sold in the US.
8   In a device that provides power (such as a battery), the cathode is the positive electrode and the anode the negative electrode. The electrolyte is a conducting material between the cathode and anode.
9   As discussed in Volume 2 (Chapter 8, subsection 8.11.4), seawater has also been proposed as a source of uranium for nuclear reactors, an idea that appears to be impractical due to the vast volumes of water that would need to be processed. The concentration of lithium in seawater is 56 times that of uranium by weight, so its extraction would face less severe practical difficulties.
10 Conventional vehicles use a mechanical, belt-driven compressor, while hybrid vehicles use an electric compressor.
11 In current vehicles, with an average fuel consumption of 4.5MJ/km rather than 0.8MJ/km, as in the advanced HEV, the impact of refrigerant leakage would be equivalent to only about a 25 per cent increase in fuel-related CO_2 emissions (assuming an effect equal to 6×0.18MJ/km).
12 The conversion from MJ/kg to GJ/m^3 for lithium-ion batteries was done using a battery density of $2400kg/m^3$, which in turn is deduced from the statement in Kobayashi et al (2009) that the specific energy of the latest lithium-ion batteries is 130Wh/kg or 310Wh/litre. However, as noted earlier, 200Wh/kg has now been achieved.

13   The storage energy penalty given in Table 5.20 is based on the additional fuel consumption relative to the case with compressed-hydrogen storage, as the additional fuel consumption due to the relatively small mass of the storage tank is not available.

14   The lost hydrogen exits the fuel cell unutilized and is vented to the atmosphere along with water vapour as part of the vehicle exhaust. Presumably there will be a strong incentive to reduce these losses, perhaps by recycling unused H_2.

15   See   www.platinum.matthey.com/pgm-prices/price-charts/for up-to-date prices.

16   These results pertain to a large, stationary engine. The relative efficiency gain using hydrogen compared to diesel might be different for a smaller engine intended for mobile applications.

17   By comparison, a bicycle has a mass of about 13kg.

18   The Mach number is the ratio of the aircraft speed to the speed of sound, which depends on the air temperature and so varies with altitude.

19   Auxiliaries include the cooling system fans, the air compressor for the air brakes, blowers for traction motors, and electrical power for lights, air conditioning and computers.

6

Industrial Energy Use

Industrial energy use involves a large, heterogeneous mix of energy consumers. Even within a single industry, there is a tremendous variety in technologies and constraints (such as the quality of the raw materials) that affect energy use. It is therefore difficult to make reliable projections of the scope for reducing the energy intensity of key industrial products. However, a small number of industries account for the majority of industrial energy use. In this chapter we explain how key energy-intensive industrial products are produced today and how energy is used in their production. We then examine known or speculative ways of reducing industrial energy intensity, and from this, draw some useful generalizations. The products discussed in this chapter are iron and steel, aluminium, chromium, copper, zinc, cement, glass, pulp and paper, plastics and fresh water (from desalination of seawater). General principles concerning industrial cogeneration and heat management, motors and chemical reactions are presented. The energy used in making fertilizers and pesticides is discussed in Chapter 7 (sections 7.4 and 7.5) in the context of agricultural energy use.

6.1 Present-day patterns of industrial energy use

Figure 6.1 shows industrial energy use as a fraction of total energy use in different world regions in 2005. Worldwide, energy use by industry accounted for 37.5 per cent of total energy use in 2005, ranging from 27 per cent in North America to 49 per cent in Asia.

Figure 6.2 shows the breakdown of industrial primary energy use in OECD and non-OECD countries in 2005. The major industrial uses of energy are in the manufacture of iron and steel, chemicals and petrochemicals, pulp and paper, non-metallic minerals (mostly cement) and food and tobacco.

6.2 Overview of major industrial commodities

Metals that are produced from virgin ores and other raw materials are referred to as *primary metals*, whereas those that are produced through recycling of scrap are referred to as *secondary metals*. Energy is used to produce industrial commodities as *process energy* and, in some cases, as *feedstock energy*. Process energy is the fuel or electricity needed to power the conversion processes, while feedstock energy refers to energy commodities (such as natural gas or petroleum) that become part of the material product. Two important examples are the use of natural gas as a feedstock in the production ammonia (NH_3, the key ingredient in nitrogen fertilizer) and use of naphtha (from petroleum) in the production of ethylene (C_2H_4, a precursor to most plastics). Although these feedstocks are not used for energy purposes, they have energy value and represent a withdrawal from the available energy stock. The feedstock energy is equal to the heating value of the final products (the thermal energy released during combustion) and the process energy is equal to the total energy inputs minus the heating value of the final products. The *embodied energy* of a product is the total (feedstock + process) energy that went into making the product.

Table 6.1 provides summary information on the global production of the most important commodities from an energy point of view. Also given is the amount of primary energy used per tonne of material (averaged over primary and secondary production) and the total global primary energy used in the manufacture of each commodity. The primary energy requirements given in Table 6.1 were computed from the separate fuels and electricity uses given later in Table 6.3.[1] The global primary energy terms sum to 87.0EJ/yr, compared to a total worldwide industrial primary energy use in 2005 (excluding petroleum extraction and refining)

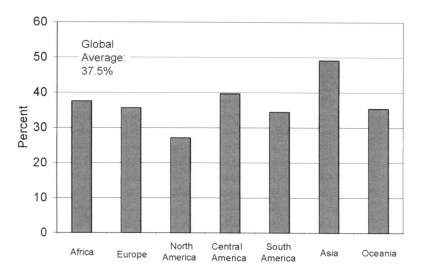

Source: Computed from data in IEA (2007b, 2007d)

Figure 6.1 *Industrial energy use in 2005 as a fraction of total energy use in various regions*

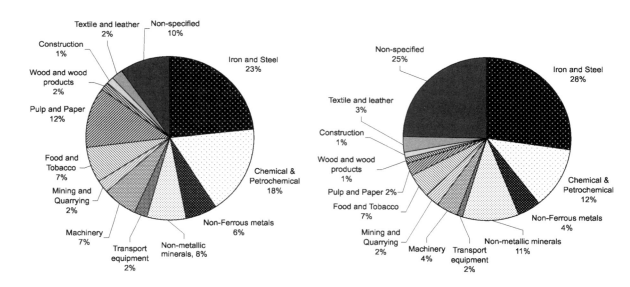

OECD Industrial Energy Use, 2005   Non-OECD Industrial Energy Use, 2005

Note: Fuel, electricity and heat consumption data were divided by the primary-to-secondary conversion efficiencies given in Table 2.6 to get the primary energy equivalents, and summed.

Source: Computed from data on fuel, electricity, and heat use given in IEA (2007b, 2007d) for individual industries, which are given aggregated over all industries in Tables 2.1 to 2.3

Figure 6.2 *Breakdown of industrial energy use in OECD and non-OECD countries in 2005*

Table 6.1 *Global mean primary energy use per tonne for the most important industrial materials in terms of total energy use, and total global energy use based on production of primary materials or based on the current estimated global average proportions of primary and secondary (recycled) materials (in which case the per cent secondary is indicated in brackets after the name of the commodity)*

Commodity	Production (Mt/yr)	Year	Energy intensity (GJ/t)	Global energy (EJ)	Reference for energy intensity
		Metals			
Steel (32.4%)	1320	2007	26.3	34.7	Table 6.6
Aluminium (18.7%)	38.0	2007	160.3	6.1	Table 6.9
Stainless steel (25%)	28.3	2006	69	2.0	Table 6.16
Chromium	20.0	2007	57	1.1	Table 6.16
Copper (12.9%)	15.6	2007	87.6	1.4	Table 6.13
Zinc (16%)	10.5	2007	42	0.44	Table 6.15
		Non-metallic minerals			
Cement	2600	2007	4.8	12.5	Table 6.21
Lime	277	2007	0.4	0.11	West and Marland (2002)
Container glass	57	2004	25.9	1.5	Table 6.25
Flat glass	38	2004	21.1	0.80	Table 6.25
Fibrous glass			58		Table 6.25
		Chemicals			
Ammonia	132	2003	39	5.2	IEA (2006)
Ethylene	110	2004	60[a]	6.6	Patel (2003)
Chlorine	44	2004	27	0.79	IEA (2006)
Pesticides	2.6	1999	60–580[b]	0.73	Table 7.5
N fertilizer	90.9	2005	42	3.8	Chapter 7
P fertilizer	36.8	2005	20	0.74	Chapter 7
K fertilizer	26.4	2005	5	0.13	Chapter 7
		Other			
Desalinated seawater	18,980	2007	0.265	5.0	Section 6.13
Paper and paper products	365	2006	26	9.5	Table 6.27
		Total			
Excluding stainless steel, N fertilizer and 50 per cent of Cl[c]				87.0	

Note: Primary energy is computed from the separate fuel and electricity uses given in Table 6.3, assuming any electricity used to be generated at 40% efficiency but with no markup for onsite fuels except for steel (where a 5% markup to coal is applied) and glass (see note to Table 6.25). [a] This is the energy intensity for ethylene from naphtha, which dominates global production. Energy intensities for ethylene from gas oil or light hydrocarbons are slightly greater. [b] This is the range for different pesticides, not an uncertainty. The average value is probably near 280GJ/t, the value used here in computing total energy use in making pesticides. [c] To avoid redundancy.
Source: 2008 US Geological Survey Mineral Commodity Summaries, Mineral Yearbooks (http://minerals.er.usgs.gov/minerals/pubs/commodity) for 2007. Other production data are from the references cited for energy intensity, except for glass, which is from Bernstein et al (2007)

computed in the same way from the data in Table 2.3 of 141EJ.[2] Thus, the industries listed in Table 6.1 account for about 62 per cent of total industrial primary energy use. Figure 6.3 gives the global primary energy use for the 12 most important commodities in terms of global energy use. The four most important commodities are steel, cement, ethylene and aluminium.

Figure 6.4 shows the variation from 1960 to 2006 in the global production of steel, aluminium, copper, zinc, cement, paper and plastics. The most dramatic increase has been in the production of plastics, which increased by a factor of 34 from 1960 to 2006 and by a factor of 2 from just 1993 to 2006. In the near term, the total consumption of metals and plastics can be expected to continue to grow, although in the long run, a decline in the consumption of primary metals can be expected as the world's material stock stabilizes and rates of recycling increase. In some cases, absolute scarcities and rising prices of metals and of feedstocks for plastics will force increases in the rate of recycling and decreases in total consumption.

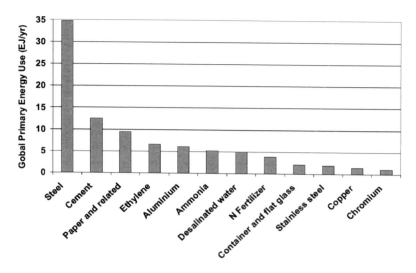

Source: Data from Table 6.1

Figure 6.3 *Estimated recent global primary energy use for the production of the 12 commodities using the most energy*

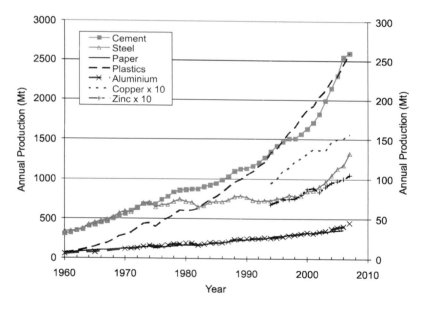

Note: Solid lines use the scale to the left, while dashed lines use the scale to the right. Copper and zinc production data have been multiplied by ten so as to be legible when plotted on the left scale.

Source: Data for steel and aluminium, WWI (2008); copper, USGS Minerals Yearbook (http://minerals.er.usgs.gov/minerals/upbs/commodity); cement, USGS Minerals Yearbook from 1994–2007, and inferred from data on cement-related emissions for 1960–2006 from Marland et al (2008); paper, Food and Agriculture Organization online Forestry data at http://faostat.fao.org; plastics, Plastics Europe (2008)

Figure 6.4 *Variation from 1960 to 2007 in the worldwide production of steel, aluminium, cement and plastics, from 1960 to 2006 for paper, and from 1994–2007 for copper and zinc*

Table 6.2 *Summary of chemical reactions that release* CO_2

Commodity	CO_2-producing reaction	CO_2 production (tC/t)
Iron	$2\,Fe_2O_3 + 3C \rightarrow 4Fe + 3CO_2$	0.161
Aluminium	$2Al_2O_3 + 3C \rightarrow 4Al + 3CO_2$ (net reaction)	0.334
Cement	$CaCO_3 \rightarrow CaO + CO_2$	0.134[a]
Glass	$nSiO_2 + mCaCO_3 + xNa_2CO_3 + \ldots \rightarrow Glass + CO_2$	0.046–0.054
Ammonia	$CH_4 + 2H_2O + N_2 \rightarrow H_2 + CO_2 + 2NH_3$ (net reaction)	0.353

Note: CO_2 production was computed from the stoichiometry of the reactions (6.1), (6.9) and (6.10). [a] For cement consisting of 95 per cent clinker (see subsection 6.6.1). For all materials except cement and glass, this CO_2 emission is accounted for in the fuel energy use, as the carbon on the left-hand side of the reactions is derived from the fuel input.

CO_2 emissions occur in association with the use of carbon-based process energy and, in many cases, through chemical reactions involving raw materials. These reactions and the C emission per tonne of commodity produced are summarized in Table 6.2 and stem from material later in this chapter. The important metals occur in nature as oxides or as sulphides (which are converted to oxides during processing). Metal oxides are converted to pure metals through removal of the oxygen – a process called *reduction*, which is the opposite of oxidation (the addition of oxygen). Reduction of metals occurs either through a chemical or an electrochemical reaction. An example of the former is the chemical reduction of iron using CO produced from coke, while an example of the latter is the electrochemical reduction of alumina (Al_2O_3) through electrolysis using a carbon anode that is consumed in the process. CO_2 is produced in both cases. However, this CO_2 is already accounted for in the energy use data because the C comes from the fossil fuel inputs (in the case of electrochemical reduction, the energy use data should include the energy content of the fossil fuels used to make the C anode). Similarly, the production of cement, glass, ethylene and ammonia entails emissions of CO_2 through the chemical reactions that form these materials. In the case of cement and glass, the CO_2 that is released is in addition to the CO_2 released through the combustion of fossil fuels because the C that appears as CO_2 comes from non-energy material inputs to the production process. Ethylene requires hydrogen as feedstock if made from heavy oils, which today is produced by steam reforming of natural gas, with attendant CO_2 emissions, but is accounted for in the energy-use data. Natural gas is normally used as a source of hydrogen in the production of ammonia as well. The hydrogen is combined with atmospheric nitrogen to make ammonia. The net reaction and CO_2 emission related to the use of natural gas as a source of H_2 for NH_3 (with production of excess H_2) are given in Table 6.2.

Emissions of perfluorocarbons (CF_2 and C_2F_6) occur in association with the production of primary aluminium. As discussed in subsection 6.4.1, these emissions have fallen dramatically since 1990, and in 2004 were equivalent (in terms of heat trapping averaged over a 100-year period) to an average CO_2 emission of 0.38tC/t of aluminium. Further reductions, by perhaps a factor of three, can be expected.

Table 6.3 provides summary information on the energy use (as fuels and as electricity) and CO_2 emissions per tonne of commodity produced, assuming any required electricity to be supplied (generated and transmitted) from coal at 40 per cent efficiency. Although much of the world's aluminium is produced in countries with a substantial supply of hydroelectric power, this renewable electricity could in most cases be used to displace the closest coal-based electricity if it were not used to produce aluminium. Thus, for accounting purposes, it is appropriate to assume the electricity used in making aluminium (and other commodities) comes from coal powerplants.[3] Provision of C-free electricity would eliminate over 75 per cent of the CO_2 emissions associated with the production of primary aluminium and copper. One quarter to one half of the remaining emissions are related to the reduction of the initial metal ores.

A factor of great importance in future energy use is the share of secondary (recycled) metals, glass, paper products and plastics as a fraction of the total production. The *recovery rate* is the fraction of discarded products that is recycled, while the *recycled fraction* is the fraction of total production that is made

Table 6.3 *Present average energy use and CO_2 emissions as tonnes of C per tonne of material produced for various primary materials or based on the current estimated global average proportions of primary and secondary (recycled) materials (in which case the per cent secondary is indicated in brackets after the name of the commodity)*

	Energy use		CO_2 emissions (tC/t)				For more
	Fuels (GJ/t)	Electricity (kWh/t)	Process-related	Fuels	Electricity	Total	detailed information
Steel (32.4%)	20.7	515	(0.16)	0.54	0.12	0.66	Table 6.6
Aluminium (18.7%)	36.6	12,930	(0.33)+0.38	0.96	2.91	4.24	Table 6.9
Chromium	28.1	4300	0.00	0.00	0.00	0.00	Table 6.16
Copper (12.9%)	34	5960	0.00	0.84	1.29	2.18	Table 6.13
Zinc (16%)	8.2	3780	0.00	0.08	1.04	1.11	Table 6.15
Cement	4.1	80	0.13	0.13	0.02	0.28	Table 6.20
Lime	0.15	117	0.00	0.003	0.026	0.030	Table 7.3
Container glass	9.6	1500	0.05	0.25	0.33	0.63	Table 6.25
Flat glass	12	650	0.05	0.32	0.15	0.52	Table 6.25
Fibrous glass	2.6	5800	0.04	0.07	1.30	1.42	Table 6.25
Newsprint	3	2060	0.00	0.00	0.46	0.46	Table 6.27
Writing paper	19	1600	0.00	0.00	0.36	0.36	Table 6.27
Sanitary paper	18	1700	0.00	0.00	0.39	0.39	Table 6.27
Ammonia	39.4		(0.35)	0.69	0.00	0.69	Sec. 7.4.2
Ethylene	60.4		0.00	0.61[a]	0.00	0.61	Fig. 6.3.5
Chlorine	2	2800	0.00	0.06	0.63	0.68	IEA (2006)
Pesticides	200	9000		3.96	2.03	5.99	Table 7.6
N fertilizer	40.2	200	(0.35)	0.70	0.05	0.75	Table 7.3
P fertilizer	17.4	290		0.30	0.07	0.37	Table 7.3
K fertilizer	3.0	223		0.05	0.05	0.10	Table 7.3

Note: Process-related emissions are from Table 6.2 and entries in brackets are accounted for in the energy-related emissions. Energy-related emissions were computed assuming the fuel to be coal for all materials except newsprint and paper (biomass fuel), ammonia and N fertilizer (natural gas fuel) and ethylene and pesticides (petroleum), while electricity in all cases is assumed to be generated from coal at 40 per cent efficiency. Coal, oil and natural gas amounts were multiplied by markup factors of 1.05, 1.2 and 1.25, respectively (see Chapter 2, subsection 2.3.5), and then by emission factors of 25kgC/GJ, 19kgC/GJ and 14kgC/GJ, respectively. The fuel for paper is assumed to be biomass, with an emission factor of zero. [a] Only process energy plus the markup component of feedstock energy is used here in computing CO_2 emission.

by recycling discarded materials. If the rate at which old products are being discarded is small compared to current production, the recycled fraction will be small even with a 100 per cent recovery rate. The difference between production and discarding gives the increase in the stock of material. As the stock of materials in existence reaches a constant value (assuming an eventual stable population and global economic development leading to saturation in the demand for end-use products), the discard and production rates will converge and the recycled fraction will equal the recovery rate. Table 6.4 provides summary information on the present-day recovery rate, recycled fraction, primary energy requirement per tonne of primary materials under best practice today and the energy savings with recycling compared to best practice today for primary materials.

In the remainder of this chapter, the manufacturing processes and current and potential future energy use in producing all of the commodities shown in Tables 6.1–6.4 except for ammonia, pesticides and fertilizers are presented. The discussion of these three commodities is deferred to the more general discussion of fertilizers and pesticides in Chapter 7 (sections 7.4 and 7.5).

6.3 Iron and steel

The growth from 1960 to 2006 in the worldwide production of steel in comparison with other energy-intensive commodities was illustrated in Figure 6.4. Figure 6.5 shows the trend in the production of steel from 1995 to 2007 in major world regions and globally, the percentage breakdown of production by

Table 6.4 *Recovery rate, recycled fraction, primary energy intensity for primary production, and primary energy savings with recycling for selected industrial commodities*

Commodity	Recovery rate	Recycled fraction	Average primary energy intensity of primary materials (GJ/t)	Primary energy savings (%)	Typical ore grade (%)
Steel	0.74	0.32	26	70	30–60
Aluminium	0.52	0.20	160	95	24
Copper	0.53	0.18	88	80	0.5–2
Zinc	0.41	0.16	42	75	3–9
Stainless steel		0.25	69	75	
Chromium	0.35[a]	0.15	57		
Nickel	0.53[a]	0.33	191		1.5–3
Glass			25	25	
Paper			26	30	
Plastics			60	60–80	

Note: [a] This is the recovery rate for recycling only, as recovery for downcycling to lower-value products does not contribute to sustainability and provides substantially smaller energy savings. Also given is the typical grade of ores for metals at present.

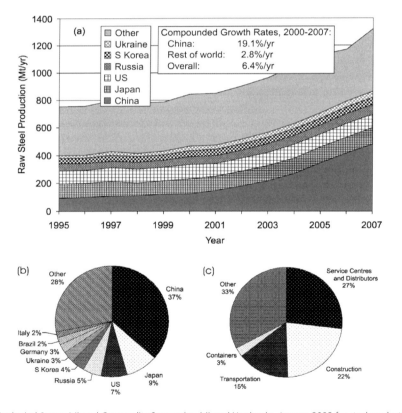

Source: Data from US Geological Survey Mineral Commodity Summaries, Mineral Yearbooks, January 2008 for steel production (http://minerals.er.usgs.gov/minerals/upbs/commodity), and US Geological Survey, Historical Statistics for Mineral and Material Commodities in the United States, September 2005 (http://minerals.usgs.gov/ds/2005/140) for end-use breakdown

Figure 6.5 *(a) Trends in the production of raw steel by various countries and regions, (b) breakdown of steel production by country and region in 2007, and (c) breakdown of the uses of steel consumed in the US in 2003*

various countries and regions in 2007, and the breakdown of end uses of new steel in the US in 2003. Production of raw steel in China grew very slowly between 1994 and 2000, then grew by an average compounded rate of 19.1 per cent/year between 2000 and 2007, following China's entry into the World Trade Organization. As a result, overall world growth was 6.4 per cent/year during this time period, reaching 1320Mt/yr by 2007. China alone accounted for 37 per cent of world steel production in 2007. In the US, 27 per cent of steel consumption is for service centres and distributors (an amalgam of different uses), 22 per cent is for construction and 15 per cent is for transportation equipment.

In 2005, world production of steel was 1130Mt and 441.5Mt of scrap steel were collected (IISI, 2007), representing 39 per cent of steel production. However, material losses during recycling are about 17 per cent of the collected scrap according to IEA (2006), so only 366Mt of secondary steel were produced, representing 32.4 per cent of total production. Figure 6.6 presents the global flows of iron as estimated for year 2000 by the Stocks and Flows (STAF) project of Yale University (Wang et al, 2007a). According to this figure, the rate of recovery of discarded iron was 267 teragram (Tg) out of a total scrap + discard flow of 361Tg, or 75 per cent, while the recycled fraction in new steel was 267Tg out of 840Tg, or 32 per cent, which happens to agree with the IEA-based estimate, although losses during recycling are not explicitly considered. Regional recovery

rates according to the STAF project are 73 per cent, 78 per cent and 81 per cent in North America, Europe and Asia, respectively.

The total secondary energy used for steel production worldwide amounted to 22.5EJ in 2005 (6.9 per cent of total world secondary energy demand), of which 68 per cent was coal, 14 per cent was electricity and 11 per cent was natural gas (based on tables in IEA 2007a, 2007c). Dividing by the 2005 steel production of 1130Mt gives a world average secondary energy intensity for steel of 19.9GJ/t.[4] Using the factors for conversion of different forms of secondary energy back to primary energy that are given in Chapter 2 (Table 2.6), this corresponds to a total primary requirement of 33.8EJ (7.0 per cent of total world primary energy demand) and a world average primary energy intensity for steel of 29.9GJ/t. However, this average primary energy intensity reflects the average electricity generation and delivery efficiencies of 37 per cent in OECD countries and only 21 per cent in non-OECD countries. Average efficiencies are unlikely to be applicable to the specific countries where most of the world's steel is produced. Assuming a standardized electrical supply efficiency of 40 per cent everywhere (as in de Beer et al, 1998a), as well as a standardized efficiency of 95 per cent in supplying coal to the steel plant, reduces the average primary energy intensity of steel to 26.3GJ/t. This in turn can be decomposed into average primary energy intensities for primary and secondary steel of 35.6GJ/t

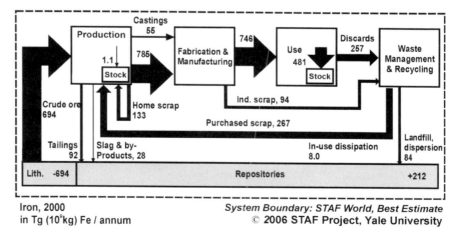

Iron, 2000
in Tg (10⁹kg) Fe / annum

System Boundary: STAF World, Best Estimate
© 2006 STAF Project, Yale University

Note: All flows are in units of Tg iron (Fe) per year.
Source: Wang et al (2007a)

Figure 6.6 *The global flow of anthropogenic iron in the year 2000*

and 7.1GJ/t, respectively, assuming a secondary fraction of 0.324.

The options to reduce energy use associated with the production of steel (and other metals) are:

1   to increase the efficiency of both primary (from virgin ores) and secondary (recycled) production;
2   to improve the efficiency in the onsite co-production of electricity in the case of primary iron;
3   to increase the proportion of new metals produced from recycled materials; and
4   to reduce the future demand for metals through reduced use of metals in consumer products (where this is feasible without adversely affecting quality), through lifestyle changes that result in less demand for material goods and by redesigning material products for greater longevity.

CO_2 emissions associated with the reduction of metal oxides can be eliminated through the use of hydrogen (if derived from non-carbon energy sources) as a reducing agent (as discussed in Volume 2, Chapter 10, subsection 10.6.7), while CO_2 produced from both onsite combustion of fuels and reduction of metal oxides could potentially be captured and sequestered underground (as discussed in Volume 2, Chapter 9, section 9.2).

6.3.1 Iron and steel production from raw ore

The major steps in the production of steel from iron ore are:

1   *beneficiation* of ore – the grinding of ore to a size small enough to separate iron mineral grains from other minerals, producing fine particles and iron ore concentrate;
2   *agglomeration* through heat treatment of fine particles into *sinter* (typically 20mm diameter) and of iron ore concentrate into *pellets* (9–13 mm diameter);
3   the *reduction* of iron ore agglomerations to produce crude iron;
4   the *refining* of crude iron to make steel by removing impurities and adding trace elements as required to give the desired steel properties; and
5   the *shaping* of the steel into products of the desired size and shape.[5]

These steps as carried out in conventional steel mills are described in more detail below, followed by a discussion of energy flows and use within steel plants today and a discussion of alternative technologies and their energy requirements.

Iron ore consists of various oxides of iron, primarily *haematite* (Fe_2O_3) but also *magnetite* (Fe_3O_4) and *wustite* ($Fe_{0.947}O$), mixed with other minerals. The typical iron content of iron ore is 60–70 per cent but can be as low as 30 per cent. The ore is ground to a fine powder to separate the grains of different minerals. Fine iron ore cannot be used in conventional blast furnaces, and so must be agglomerated to produce porous sinter and pellets (the proportions of which vary from country to country, a typical ratio being 70 per cent sinter and 30 per cent pellets).

The reduction of iron oxides can occur either chemically or electrochemically. The vast majority of iron ore today is reduced chemically using CO that is produced from gasification of *coke* in a blast furnace (coke is coal in which some of the volatile components have been driven off by heat). Iron ore, coke and limestone and/or dolomite are added to the blast furnace from the top, while pulverized coal along with hot compressed air is blown into the furnace from the bottom and combusted. The temperature of the blast furnace ranges from about 150°C at the top to about 2000°C at the bottom. Coke is gasified in the lower part of the blast furnace, producing CO that rises and reacts with (reduces) the iron in the middle and upper part of the blast furnace. Molten iron is produced, trickles down and collects in a well at the base of the blast furnace. Melting of pure iron ore occurs at a temperature of 1535°C, but the reduction reactions occur before melting occurs. Details are given in Box 6.1. The net reaction is, ideally:

$$2\,Fe_2O_3 + 3C \rightarrow 4Fe + 3CO_2$$
$$\Delta G° = 1.5 \text{ GJ/tonne of Fe} \qquad (6.1)$$

where $\Delta G°$ is the Gibbs free energy of the reaction (the difference in the energy between the products and reactants, which has to be added if positive). The energy value of pure carbon is 32.8MJ/kg and 161.1kg of carbon are required to produce 1 tonne of iron (based on the stoichiometry of reaction (6.1)), so another 5.3GJ/(tonne of Fe) are needed. The total minimum energy needed for the reduction of iron ore is thus 6.8GJ/(tonne of Fe) (de Beer et al, 1998a).

Box 6.1 Reactions in the reduction of iron ore in a blast furnace

Iron ore, coke and other ingredients are added to the top of a blast furnace. Temperature increases steadily from the top to bottom, with gasification of the coke occurring at the bottom and reduction of iron ore occurring in the middle and at the top of the blast furnace. The main reactions occurring in the various temperature zones are as follows (de Beer et al, 1998a):

150–600°C

$$3Fe_2O_3 + CO \rightarrow 2Fe_3O_4 + CO_2$$
$$2CO \rightarrow C + CO_2$$

600–1000°C

$$Fe_3O_4 + CO \rightarrow 3FeO + CO_2$$
$$FeO + CO \rightarrow Fe + CO_2$$

1000–1400°C

$$FeO + C \rightarrow Fe + CO$$
$$CO_2 + C \rightarrow 2CO$$

1400–2000°C

$$C + O_2 \rightarrow CO_2$$
$$2C + O_2 \rightarrow 2CO$$

Note that coke is oxidized to both CO and CO_2. The CO reacts with iron ore, converting Fe_2O_3 to Fe_3O_4 and then to FeO and finally to Fe, producing more CO_2 at each step. Some elemental C also reacts with FeO and CO_2 to produce CO.

In reality, agents such as $CaCO_3$ (from limestone) and SiO_2 (from impurities in the iron ore) must be added for slagging (removal of impurities) and fluxing,[6] respectively, and this increases the energy requirements because these added materials are also heated. The net reaction to produce reduced iron is thus closer to:

$$2Fe_2O_3 + 2CaCO_3 + SiO_2 + 3C$$
$$\rightarrow 4Fe + Ca_2SiO_4 + 5CO_2 \qquad (6.2)$$

where Ca_2SiO_4 is the slag (van Oss, personal communication, 2008). Energy is also lost in various ways in real blast furnaces. The net result is that the world average energy use in blast furnaces today is about 14.4GJ/t, the most efficient blast furnaces require 11.8GJ/t and a practical lower limit is about 10.4GJ/t, according to IEA (2006). Part or all of this energy can be supplied as coke, which serves both as a source of carbon for reaction (6.1) and as a source of the energy needed to drive the reaction (the remaining

energy is supplied by the pulverized coal that is introduced at the bottom of the blast furnace). Losses occur in the production of coke from coal (the conversion efficiency is about 65 per cent), so less coal will be used if more of the process energy is directly supplied from coal rather than from coke. Losses are reduced if the so-called *coke oven gases* are captured and used as a fuel in the blast furnace. Given a conversion efficiency from coal to coke of 80 per cent (applicable if all coke oven gas is captured and used), a minimum of 13GJ of coal are required per tonne of iron ore that is reduced if the required energy is supplied entirely as coke.

The second step is the refining of crude iron into steel, through the removal of carbon (which typically constitutes about 4 per cent of the crude iron) and other impurities (such as Si and Mn) by melting the iron. This can be done using an open-hearth furnace (first used in 1864), a basic oxygen furnace (BOF) (first used in 1952), or an electric arc furnace (EAF). The

BOF requires pure O_2, which could be produced in large quantities using industrial methods beginning in 1950. Heat from the oxidation of impurities in the BOF can be used to melt 20–30 per cent of additional scrap steel. The EAF was first introduced in the late 19th century but was restricted to speciality steels until 1965. Figure 6.7 shows the amount of world steel produced from crude iron using the open-hearth

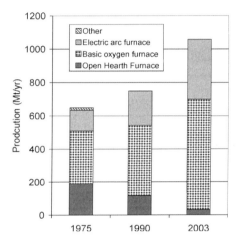

Source: de Beer et al (1998a) for 1975 and 1990, IEA (2006) for 2003

Figure 6.7 *Proportion of world steel produced from reduced iron using different refining techniques, in 1975, 1990 and 2003*

furnace, BOF and EAF in 1975, 1990 and 2003. Most scrap steel is processed using EAFs, and EAFs mostly process scrap, but some scrap goes into BOFs as 15–25 per cent of the feedstock, and EAFs can use direct-reduction iron (see below) and pig iron as well as scrap. Production using the open-hearth furnace has steadily decreased, while production by the BOF and EAF has increased (from 19 per cent of total steel production in 1975 to 27 per cent in 1990 and 34 per cent in 2003). Figure 6.8 shows the change in the amount of electricity used by EAFs per tonne of steel between 1965 and 1990; electrical energy use per tonne fell by almost half during this 25-year period. Energy use in the best facilities has fallen by another half since then. According to Kato et al (2005), old EAFs in Japan use 380kWh of electricity per tonne of steel produced, while new furnaces use 150kWh/t.[7] Heating of pig iron from 25°C to 1150°C (the melting point for iron with a carbon content of 4.3 per cent, which is typical) and subsequent melting requires 1.05GJ/t. In principle, almost all of this energy could be recovered when the refined iron cools, so the theoretical minimum energy in the refining step is essentially zero.

The third step is the shaping of the steel into the desired product, in a number of steps. Today, about 60 per cent of raw steel is cast directly into cubicle blocks, small bars or slabs in a *continuous caster*. These are then converted into various products using a *hot rolling mill*, *hot strip mill* and *hot plate mill* (depending on the product). Reheating and cooling of the steel

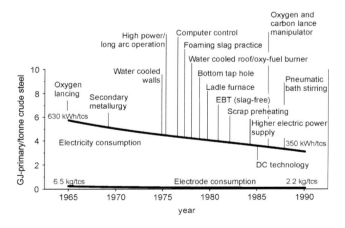

Source: de Beer et al (1998a)
Note: Direct electricity consumption is shown along with the embodied energy associated with the manufacture of the electrode material that is consumed during the heating process.

Figure 6.8 *Change in energy used by electric arc furnaces per tonne of crude steel, 1965–1990*

occurs between the various steps, resulting in large energy losses. The minimum theoretical energy required for shaping steel is essentially zero, however, assuming complete heat recovery.

The energy used in practice is much larger than the theoretical minimum required energy. This is due to energy inputs needed to heat non-iron ore inputs (such as lime), due to heat lost when material is heated and then partially cooled before the next heating step, and due to heat lost from the plant. Coal is converted into coke in a coke oven and iron ore is agglomerated in a sinter plant with temperatures of 700°C and 1000°C, respectively. Both coke and sinter cool to the ambient temperature before being fed to the blast furnace, where they are reheated. Crude iron leaves the blast furnace at a temperature of about 1400°C and cools by about 140K during transport to the BOF. Steel leaves the BOF at a temperature of 1650°C and enters a continuous caster, but cools to between ambient temperature and 500°C after continuous casting and then must be reheated to 1250°C before entering the hot rolling, strip or plate mill.

Considerable insight can be gained into the use of energy in the iron and steel industry by examining the energy balance of the Japanese iron and steel industry, which is presented in summary form in Table 6.5. The first column shows the inputs to the production of coke (primarily coal, with minor amounts of electricity, natural gas and petroleum products) as well as the outputs (indicated by negative entries). The outputs are coke, coke oven gas and tar. The efficiency in producing coke (that is, coke energy content over total energy inputs) is 65 per cent, while if coke oven gas and tar are included, the overall conversion efficiency is 80 per cent. The input to the blast furnace is 71 per cent coke and 23 per cent coal, but blast furnace gas with an energy content equal to 23 per cent of the total energy input is produced (along with crude steel) and is used in other processes. Coal and electricity each account for about 34 per cent of the input to the BOF, but BOF gas equal to 176 per cent of the total energy input is produced (the difference being supplied by heat released during the oxidation of C and other impurities in the crude iron). Thus, if this BOF gas is captured and put to use, the BOF is a net source of energy. The energy input to the EAF (used to melt scrap steel) is 83 per cent electricity, and gases equal to 1.4 per cent of the total energy input are produced. Forging, casting and rolling operations also obtain a substantial fraction of their input energy in the form of electricity. However, electricity constitutes only 7 per cent of the overall energy input to the Japanese iron and steel industry. Some electricity is generated on site, largely using coke oven, blast furnace and BOF gases, but the average efficiency of electricity generation is only 26.6 per cent. Substantial amounts of these gases are also used for rolling operations, but a substantial fraction of the gases go unutilized. Production of coke takes up 56 per cent of the overall energy inputs to the Japanese iron and steel industry and the blast furnace takes up 57 per cent (largely using coke).

The primary energy intensity of primary steel (made from virgin ores) ranges from 19GJ/t to 40GJ/t. Figure 6.9a shows the breakdown of primary energy use in the most efficient existing mills (at about 20GJ/t) using the blast furnace/BOF route.

6.3.2 More efficient production of steel from raw ore

The energy used to manufacture steel can potentially be reduced in seven general ways:

1   through improved conventional mills;
2   through direct use of coal in place of coke;
3   by using alternative reduction techniques;
4   by eliminating at least one heating and cooling step;
5   by using techniques that reduce the temperature required in different steps;
6   by using technologies that recover and use the heat available at high temperatures;
7   by making better use of residual blast furnace gases.

These techniques are briefly discussed below.

Improved conventional mills

Figure 6.9b shows the energy use for an improved steel mill, based on component technologies that existed in the late 1990s, as estimated by de Beer et al (1998a). The primary energy requirement is 16.9GJ/t (a reduction of about 15 per cent compared to best practice in the 1990s and a reduction of more than 50 per cent compared to the world average of about 36GJ/t), but this was not considered to be the minimum possible.[8]

Table 6.5 *Energy balance of the Japanese iron and steel industry in 1999*

	Production of coke	Sintering	Pelletizing	Blast furnace	BOF	EAF	Forging	Casting	Rolling	Other	Onsite power and cogeneration	Used by other industry	Total
Petroleum Products	19.58	0.17	0.08	18.35	3.16	4.41	4.21	1.44	41.22	9.29	24.68	4.04	130.63
Coal	1092.85	28.9	1.81	279.23	13.81	0.03	0	0	0	0.03	69.88	0	1486.54
Coke	−750.58	110.98	1.16	858.87	3.46	3.88	0	0	0	6.9	0	0.34	235.01
Tar	−23.78	0.09	0	2.36	0	0	0	0	0	0.1	0.69	0.01	−20.53
Coke oven gas	−162.57	2.4	0.86	24.81	5.75	0.49	0.55	0.04	65.75	5.88	29.55	1.23	−25.26
Blast furnace gas	47.01	0.17	0	−279.24	0.03	0	0.01	0	2.97	1.33	97.42	0.06	−130.24
BOF gas	1.95	0.23	0	9.6	−69.9	0.02	0.07	0	11.58	0.31	19.02	0.72	−26.4
EAF gas	0	0	0	0	0	−0.81	0	0	0	0	0	0.34	−0.47
LNG and town gas	0.06	0.06	0	0.26	0.34	1.17	1.35	0.24	30.08	8.47	7.82	2.21	52.06
Electricity	3.94	11.7	1.04	6.77	13.28	49.22	1.89	1.38	64.91	45.71	−66.17	9.42	143.09
Total inputs	1165.39	154.7	4.95	1200.25	39.83	59.22	8.08	3.1	216.51	78.02	249.06	18.37	2047.33
Inputs as a fraction of total energy use	0.560	0.074	0.002	0.576	0.019	0.028	0.004	0.001	0.104	0.054	0.120	0.009	1.000
Electricity fraction	0.003	0.076	0.210	0.006	0.333	0.831	0.234	0.445	0.300	0.486		0.513	0.073

Note: Positive entries are inputs, negative entries are outputs. Units are PJ (LHV basis) per year and correspond to a total steel production of 85.7Mt.

Source: Gielen and Moriguchi (2002)

(a) Best blast furnace reduction + BOF today

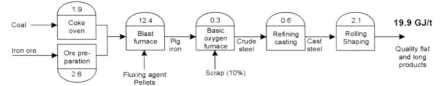

(b) Advanced blast furnace reduction + BOF

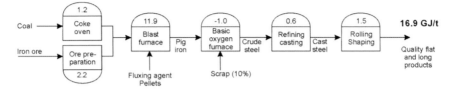

(c) Best direct reduction + EAF today

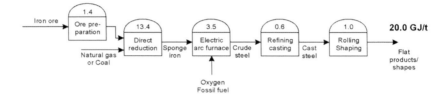

(d) Advanced direct reduction + EAF

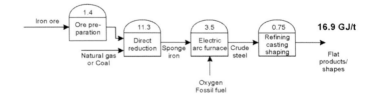

(e) Advanced smelting reduction + BOF

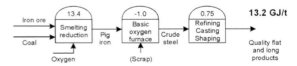

Source: Modified from de Beer et al (1998a) and Phylipson et al (2002)

Figure 6.9 *Primary energy use associated with the most efficient conventional (blast furnace) steel mills today, advanced conventional steel mills, current and advanced steel mills based on the direct-reduction process, and advanced steel mills based on the smelting-reduction process*

Larger mills tend to be more efficient than smaller mills. However, the trend in India and China – where much of the growth in steel production is occurring – has been to use mini-blast furnaces that feed electric arc furnaces. This allows small-scale steel production, which is better suited to local markets and requires less capital. Fuel requirements are 36 per cent larger for Chinese mini-blast furnaces than in advanced OECD coal plants, but only 15 per cent larger in large Chinese blast furnaces (IEA, 2006). Furthermore, the small mills use almost 100 per cent coke, whereas advanced blast furnaces use about 20 per cent coal and 80 per cent coke (direct use of coal reduces overall energy use). Thus, development of fewer but larger steel mills in China and India would tend to reduce energy use per unit of production.

Direct use of coal in place of coke

Given a conversion efficiency from coal to coke of only 65–80 per cent, the direct use of coal in place of some of the coke will save about 0.25 to 0.50 units of coal for every unit of coke that can be replaced. There has been some replacement of coke with direct use of coal in recent years.

Alternative techniques for reducing iron ores

As noted above, the primary method for reducing iron ores is through reaction of ore agglomerations with coke in a blast furnace. This requires 16–22GJ/t, of which 2–5GJ/t are losses during the production of coke from coal and another 2–3GJ/t are used in agglomerating the separated iron ore particles. Energy can be saved by eliminating the need for agglomeration and/or by eliminating the need for coke, and this can be done in two ways: through *direct reduction* and through *smelting reduction*.

Direct reduction is an alternative to the use of coke and blast furnaces in which DC electricity and solid or gaseous fuel are used in an electrochemical process at a temperature of 900–1000°C, without subsequent melting and heating to temperatures of 2000°C (as in blast furnaces). About 5–6 per cent of the 2004 world steel production was through direct reduction, two thirds of which was produced using the MIDREX process. According to Phylipson et al (2002), best-practice direct reduction requires 10.03GJ of natural gas and 0.36GJ of electricity per tonne of iron. For electricity and natural gas markup factors of 2.5 and 1.25, respectively, this translates into 13.4GJ/t of primary energy. Figure 6.9 (parts c and d) gives the overall energy use (including steps before and after the direct-reduction step) for best current and potential future direct-reduction routes. Overall energy use is 20.0GJ/t and 16.9GJ/t, respectively.

Direct reduction requires a hydrogen-rich gas, which can be provided either through gasification of coal, from decomposition of natural gas, or using hydrogen fuel itself. Orth et al (2007) describe a direct reduction process called Circofer that was developed by the German company Outotec GmbH,[9] in which iron ore and coal are injected into a preheater where the coal is partly combusted and partly gasified to H_2 and CO at 1050°C in oxygen. The mixture proceeds to a circulating fluidized bed reactor, where H_2 and CO achieve 85 per cent iron ore reduction at 950°C. The Circofer process produces a stream of 99 per cent purity CO_2 representing 70 per cent of the total onsite CO_2 emissions. This CO_2 could be captured and geologically sequestered (as discussed in Volume 2, Chapter 9). Alternatively, hydrogen can be directly used as a fuel and reducing agent, as in a pilot steel mill built in Trinidad in 1999 by Outokumpu (Nuber et al, 2006). Hydrogen in turn can be produced electrolytically from C-free electricity (as discussed in Volume 2, Chapter 10).

In smelting reduction, iron ore is directly reduced in a liquid state (rather than in the solid state, as in blast furnaces) using pulverized coal (rather than with coke that is made from coal). This allows reduction to occur more quickly. The reaction occurs in oxygen rather than air, so energy must be expended to separate oxygen from air. Gases and steam produced from the smelting furnace are used to generate electricity in a combined cycle powerplant. The most efficient smelting reduction process discussed by de Beer et al (1998a) is the cyclone converter furnace in The Netherlands, with an estimated energy use of 13.9GJ/t for the production of crude iron. This is a saving of about 18 per cent compared to the value of 16.9GJ given in Figure 6.9a for this part of the production process. De Beer et al (1998a) foresee this energy requirement dropping to 13.4GJ/t (see Figure 6.9e).

Alternative casting techniques

A number of progressively more difficult but more efficient casting techniques have been developed over time. These include *thin-slab casting, thin-strip casting* and *powder metallurgy*. These techniques eliminate at least one heating and cooling step. In thin-slab casting, a semi-finished product is cast with reduced thickness and sent directly to the hot rolling mill. In thin-strip casting, liquid steel is cast directly onto a belt or onto rollers. Conventional continuous casting produces

200–300mm thick slabs in a 500–800m long production line; thin-slab casting produces 50–60mm slabs in a 300–400m production line; and thin-strip casting produces 1–5mm strips in a 60m production line (Luiten and Blok, 2003) – an order of magnitude shorter than for conventional continuous casting. In Figure 6.9a (corresponding to best conventional practice), shaping of steel uses 2.7GJ/t. The energy requirement for strip casting based on data from pilot projects is 0.15–0.25GJ/t (de Beer et al, 1998a) – a reduction of 90–95 per cent. Capital costs for the casting step are expected to be 30–60 per cent lower due to the elimination of reheating furnaces and the shorter production line (IEA, 2006). Strip casting is used in a few pilot plants for producing stainless steel, but many problems remain to be solved before it can be applied to the mass production of high-quality grades of steel, such as those used in automobile manufacturing (EC, 2005).

Powder metallurgy entails producing refined steel in a powder form, pressing the powder into a mould with the desired shape and heating it to 1050–1250°C in order to bond the particles together. This replaces the multiple heating and cooling in the usual sequence of casting and shaping steps with a single heating step, saving perhaps another few tenths of a GJ/t. Powder metallurgy is used today to produce steels with very precise shapes and properties, but de Beer et al (1998a) do not see it as being commercially viable for bulk steel production. It has existed for more than 100 years and comprises several different technologies for fabricating semi-dense and fully dense components.

Greater recovery and use of waste heat

Heat is lost to the surroundings during steel production in hot gas streams (from the coke oven, blast furnace and BOF or EAF) and in flows of hot solids (coke, blast furnace and BOF slag and hot steel). Heat recovery requires the use of heat exchangers, and is easiest in the lower part of the temperature range used in steel making. Recovery of heat from the higher temperature steps in steel making is a major challenge. In the US, only 7 per cent of steel mills employ any form of waste heat recovery (US DOE, 2002). Maruoka and Akiyama (2006) discuss the idea of using a nickel-copper phase-change material for capturing waste heat at temperatures warmer than 1000°C.

More efficient onsite generation of electricity

As noted above, coke oven gas (COG), blast furnace gas (BFG) and BOF gas are produced during conventional steel making. The COG has a high heating value (19.3MJ/Nm^3 LHV basis) and is largely put to use elsewhere in the steel mill (as seen in Table 6.5). BFG and BOF gas have lower heating values: 3MJ/Nm^3 and 9MJ/Nm^3 respectively. BFG is produced during the gasification of coke and coal within the blast furnace and consists of a mixture of about 21 per cent CO, 20 per cent CO_2, 3 per cent H_2 and 56 per cent N_2 (from the original air). CO is largely used as a reducing agent. The residual gas mixture contains CO and H_2, about half of which has to be combusted to preheat the incoming air, while the balance can be used to generate electricity. However, the gas heating value of 3MJ/m^3 is less than 10 per cent that of natural gas due to dilution with CO_2 and N_2, so natural gas or COG has to be added to the residual blast furnace gas in order to produce a gas that can be used in gas turbines or combined cycle plants. The effective efficiency of the added natural gas is only 14–18 per cent (Gielen, 2003), compared to 55–60 per cent in state-of-the-art combined cycle (gas + steam turbine) plants using pure natural gas (NGCC) (see Chapter 3, subsection 3.2.3 'Natural gas turbines and combined cycle powerplants'). In the US, flue gases are used almost exclusively to produce useful heat only, rather than to produce heat and electricity; only 2 per cent of US steel mills employ any type of cogeneration (US DOE, 2002).

Blast furnaces can operate using oxygen that has been separated from air rather than using air, and a pilot oxygen blast furnace was built in Japan in 1986. Less coke and more pulverized coal can be used, and the incoming oxygen does not need to be preheated (saving about half of the residual gas) (Jianwei et al, 2003). Due to the absence of N_2, the residual gases after reduction of iron ore have a large enough energy value that they can be used to generate electricity in a combined cycle powerplant with an efficiency in the order of 40–45 per cent. An oxygen blast furnace would be similar to an IGCC powerplant, in which coal is also gasified with oxygen rather than air and used to generate electricity in a combined cycle powerplant (see Chapter 3, subsection 3.2.2), the major difference being that part of the gasified coal would be used first for iron reduction. Additional energy is required compared to a conventional blast furnace, for pressurization of the gases to 20atm and for CO_2 capture, but about half of this could be supplied by the additional electricity that can be produced with the residual gases. This integrated process can be thought of as a form of cogeneration, producing electricity and reduced iron rather than electricity and useful heat (as in Chapter 3, section 3.3).

As with IGCC, conversion of the residual CO to additional H_2 and CO_2 through a water shift reaction (with combustion of H_2 rather than H_2 and CO) would create additional CO_2 that could be readily captured (however, the CO_2 concentration prior to capture would be about 41 per cent, compared to 64 per cent in IGCC powerplants according to Gielen, 2003). A single steel plant may have two dozen or more blast furnaces and will constitute one of the largest single sources of CO_2 in the world (around 5MtC/yr). Gielen (2003) argues that carbon capture and sequestration should be preferentially applied to steel production rather than to continued coal-based electricity generation, because there are comparatively inexpensive renewable energy alternatives to coal-based electricity (particularly baseload wind energy systems, as discussed in Volume 2, Chapter 3, section 3.14 but not for steel production.

Impact of declining quality of ore

The quality of the iron ore affects energy use. For mini-blast furnaces, a reduction in the iron content of the ore from 55 per cent to 50 per cent increases the fuel requirement by 25 per cent (from 600kg/t to 750kg/t) (IEA, 2006). Over time, the quality of the remaining iron ores will decline, which will therefore tend to increase the energy intensity of primary iron (at least using blast furnaces).

Synthesis

The world average primary energy requirement for the production of primary steel is estimated to have been about 36GJ/t in 2005. Energy use for best current technology and for expected future technologies is summarized in Figure 6.9. Best current blast furnace and direct reduction processes each require about 20GJ/t but future energy use in new plants of about 17GJ/t or less is thought to be possible. Advanced smelting reduction is thought to be possible with an energy use of about 13GJ/t – a reduction by about 65 per cent compared to the 2005 world average. The energy required to provide pure oxygen (separated from air cryogenically, and amounting to 1.2GJ primary energy per tonne of steel according to de Beer et al, 1998a) is not included in this energy intensity, but in a future hydrogen economy (see Volume 2, Chapter 10), pure oxygen would be produced at no additional energy cost (except for storage and transportation) as a byproduct of the electrolysis of water to make hydrogen using renewable energy.

6.3.3 Use of wastes as an energy source in blast furnaces

Wastes such as plastics, used tyres and carpets have been used as a partial fuel for blast furnaces. As these materials are currently made from fossil fuel feedstocks, their combustion represents a fossil fuel CO_2 emission. More energy can be saved by using wastes in blast furnaces than can be saved through incineration with heat recovery, but this is partly because the effectiveness of heat recovery from incinerators tends to be low (see Chapter 8, subsection 8.3.4). In the long run, these materials could be produced from biomass feedstocks (see Volume 2, Chapter 4, section 4.7), so they would be CO_2-neutral (or close to it) if the biomass is grown sustainably. As in the use of wastes in cement kilns (see subsection 6.7.2), emissions of toxic materials can be a problem unless stringent advanced control systems are implemented. As well, the use of wastes may add impurities that require additional slagging elements to remove and could lead to the excessive buildup of coatings on the brick linings of the blast furnace (van Oss, personal communication, 2008).

6.3.4 Production of new steel from scrap metal

In steel production from iron ore, up to 90 per cent of the required primary energy is used to prepare the raw materials and to chemically reduce iron ore. This energy can be saved through recycling and reprocessing of scrap steel, but melting is still required. Melting can be done using: (a) a BOF, (b) an EAF, (c) a scrap melter using both electricity and fossil fuel, or (d) an all-fossil fuel melter. During recycling, the quality of the scrap can be upgraded by chemical and mechanical separation processes, which require 0.5–2GJ/t. The energy needed to heat scrap from 25°C to the melting point, and subsequent melting, is 1.05–1.36GJ/t (depending on the carbon content of the steel, which determines the melting point). The most efficient EAFs require 1.5GJ/t of electrical energy (de Beer et al, 1998a), so EAFs are efficient melters. However, for an electricity generation + transmission efficiency of 40 per cent, the primary energy requirement increases to 3.75GJ/t (note, however, that the present average efficiency in generating and transmitting electricity in non-OECD countries is only 21 per cent, as shown in Table 2.6). Altogether, the average worldwide primary energy use in scrap-based mills is about 7GJ/t, with the most efficient mills requiring about 5GJ/t, as shown in Figure 6.10. This is about 15 per cent of the average energy requirement (36GJ/t) today for producing primary steel.

It should be noted that the energy required for the production of speciality steels can be substantially greater than for the production of typical steel, but there are also large energy savings through recycling. For example, Johnson et al (2008) calculate a primary energy requirement of 80GJ/t for the production of stainless steel

(containing 74 per cent Fe, 18 per cent Cr and 8 per cent Ni) using primary materials, but only 26GJ/t (three times less) using recycled materials. More details concerning stainless steel are provided in subsection 6.6.2.

Apart from increasing the efficiency of electricity generation (which is discussed in Chapter 3), the major opportunity for reducing the amount of primary energy used for recycling steel lies in recovering and using heat from waste gases and from processed scrap and slag as they cool. Systems for recovering heat from waste gases produced by heating the scrap, and its use in preheating incoming scrap, are already commercially available. Figure 6.10b shows the primary energy requirement for an advanced scrap-based mill, assuming electricity to be generated with an efficiency of 60 per cent (which could be the norm in the future). The total primary energy requirement is 3.5GJ/t – half the world average today. This includes energy used to make oxygen, which, as noted above, could be produced as a byproduct of hydrogen production in a hydrogen economy. Over time, as the stock of steel products builds up and is in need of replacement, recycling as a share of total production can be expected to increase.

(a) Present efficient mill using scrap steel

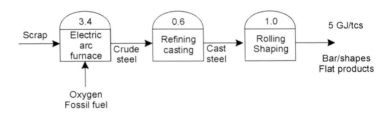

(b) Advanced mill using scrap steel

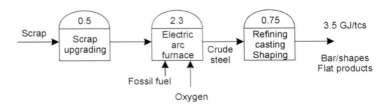

Source: de Beer et al (1998a)

Figure 6.10 *Primary energy use associated with existing efficient scrap-based mills (assuming 40 per cent electricity generation efficiency) and advanced scrap-based mills (assuming 60 per cent electricity generation efficiency)*

6.3.5 Steel energy use – summary and future scenarios

Table 6.6 gives the present-day breakdown between fuels and electricity for the production of primary and secondary steel. Also shown is the breakdown between fuels and electricity under current best practice and for anticipated future technologies (assumed to be smelting reduction for primary steel), based on the information reviewed here. Fuel and electricity use as well as primary energy use are given for average steel based on the present-day secondary fraction (0.324) and for the hypothetical future secondary fraction (0.90). The breakdown between fuels and electricity is relevant to the construction of global energy-supply scenarios, which are presented in Volume 2 (Chapter 12). The key results are:

- If steel production in the future is 10 per cent primary and 90 per cent secondary using current best practice, then the average primary energy intensity of steel production would decrease from 26.3GJ/t at present to 6.9GJ/t – a reduction by a factor of 3.8.
- If the secondary fraction increases to 90 per cent and future steel production occurs at the lowest potential energy intensity identified here, the average primary energy intensity would be 5.9GJ/t – a reduction by a factor of 4.5.
- If, in addition, the remaining required electricity is supplied with an efficiency of 60 per cent rather than 40 per cent, the average primary energy intensity would drop to 4.5GJ/t – a reduction by a factor of 5.8 compared to today. Inasmuch as the assumed baseline energy efficiency of 40 per cent for today is higher than in reality, the true potential for reducing primary energy use is even larger.
- The energy intensity ratios (best future energy intensity over current energy intensity) for fuels and electricity for average steel are 0.11 and 0.71, respectively, indicating that the anticipated savings in steel energy use occur disproportionately through reduced fuel use.

As economies develop, a larger fraction of annual steel production will be secondary steel. By 2100, with the human population probably stabilized or falling and the required infrastructures and buildings presumably already built throughout the world, steel would be largely needed only to replace aging structures. A recycled fraction of 90 per cent in a future world with a constant stock of steel is therefore reasonable. In 1993, 95 per cent of steel scrap produced in the US was recycled (Sibley and Butterman, 1995), while for the world as a whole, about 75 per cent was recovered in 2000 (Figure 6.6). However, there might always be a need for some small portion of steel made from raw materials due to problems of contamination of scrap steel or of incompatible ratios of trace elements when different alloys of scrap steel are mixed. And as noted in the introduction to this section, about 17 per cent of collected scrap steel is lost at present during the production of new steel.

Table 6.6 *Breakdown of energy use as fuels and as electricity for present-day average primary and secondary steel and for the weighted mean of the two using the present-day recycled content of new steel and the weighted mean energy use with 90 per cent recycled content and (1) present average and best foreseen future energy intensities with 40 per cent electricity generation efficiency, and (2) best foreseen future energy intensities with 60 per cent electricity generation efficiency*

Onsite energy intensity	Electricity efficiency	Recycled fraction	Primary steel		Secondary steel		Mean steel			PE reduction factor
			Fuels (GJ/t)	Electricity (kWh/t)	Fuels (GJ/t)	Electricity (kWh/t)	Fuels (GJ/t)	Electricity (kWh/t)	PE (GJ/t)	
Average today	0.4	0.324	30.0	450	1.15	650	20.65	515	26.3	
Best practice	0.4	0.9	16.4	290	1.15	475	2.68	457	6.9	3.8
Future best	0.4	0.9	11.8	115	1.15	395	2.22	367	5.9	4.5
Future best	0.6	0.9	11.8	115	1.15	395	2.22	367	4.5	5.8
Energy intensity ratio for mean steel, future best versus average today							**0.107**	**0.713**		

Note: Also given are energy intensity ratios for future steel with 90 per cent recycled content. PE = primary energy.

In conclusion, the technical potential to reduce the average primary energy intensity of steel is a factor of four to six reduction compared to today.

6.4 Aluminium

Aluminium metal is produced from bauxite ore, which contains oxides of aluminium mixed with reactive silica; there are at least 18 different types of bauxite, differing in the kinds of aluminium oxides present and the reactive silica content. The dominant minerals are gibbsite ($Al(OH)_3$) and boehmite and diaspore ($AlO(OH)$). Bauxite forms either in limestone regions in Mediterranean-type climates or in tropical regions in association with deep weathering, with 98 per cent of bauxite occurring close enough to the surface that strip

mining is employed. In a survey of bauxite mines accounting for 70 per cent of the world's total bauxite extraction, the International Aluminium Institute (IAI, 2004) reports that the average thickness of bauxite deposits is about 5m (typically 5–30m) and the average thickness of overlying material is about 2m (typically 1–10m). About 40 per cent of mined bauxite is from Mediterranean climates and 60 per cent from tropical climates. Native hardwood covered 41 per cent of the land disturbed by bauxite mining, tropical rainforest covered 15 per cent and native pasture covered 8 per cent. Formal written rehabilitation plans are in place at 97 per cent of the surveyed mines but cannot be expected to restore the landscape and its ecosystems to their former state. As discussed below, enormous (factor of 5–23) reductions in the energy intensity of

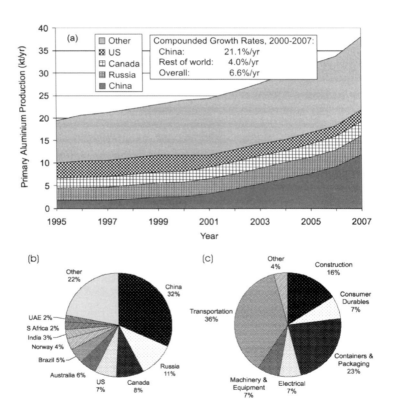

Source: Data from US Geological Survey Mineral Commodity Summaries, Mineral Yearbooks, January 2008 (http://minerals.er.usgs.gov/minerals/upbs/commodity), and US Geological Survey, Historical Statistics for Mineral and Material Commodities in the United States, September 2005 (http://minerals.usgs.gov/ds/2005/140)

Figure 6.11 *(a) Trends in the production of primary aluminium by various countries and regions, (b) breakdown of aluminium production by country and region in 2007, and (c) breakdown of the uses of aluminium consumed in the US in 2003*

new aluminium are possible through recycling. Given the prevalence of strip mining of forested landscapes to extract the raw material for aluminium, increased recycling of used aluminium will provide significant non-energy-related environmental benefits.

The growth from 1960 to 2006 in the worldwide production of aluminium in comparison with other energy-intensive commodities was illustrated in Figure 6.4. Figure 6.11 shows the trend in the production of primary aluminium from 1995 to 2007 in major world regions and globally, the percentage breakdown of production by various countries and regions in 2007 and the breakdown of end uses of new aluminium in the US in 2003. Production of primary aluminium in China grew very slowly between 1994 and 2000, then grew by an average compounded rate of 21.1 per cent/year between 2000 and 2007, thereby contributing strongly to overall world growth of 6.6 per cent/year during this time period. China accounted for 32 per cent of world aluminium production in 2007. About one third of the aluminium consumed annually in the US is for transportation equipment and about one quarter is for packaging.

6.4.1 Production of aluminium and associated GHG emissions

The three main steps in the production of primary aluminium are *mining*, *refining* and *smelting*.

Refining transforms bauxite into alumina (Al_2O_3) through the following steps in what is known as the *Bayer process*:

- Digestion, in which the bauxite is ground in mills and mixed with caustic soda (sodium hydroxide, NaOH) at a temperature of 175°C to dissolve the aluminium minerals in the ore, forming aluminium hydroxide. The reactions for the gibbsite and boehmite/diaspore components of the bauxite ore are:

$$Al(OH)_3(solid) + NaOH \rightarrow$$
$$Al(OH)_4^-(aqueous) + Na^+ \quad (6.3a)$$

and:

$$AlO(OH)(solid) + NaOH + H_2O \rightarrow$$
$$Al(OH)_4^-(aqueous) + Na^+ \quad (6.3b)$$

respectively.

- Clarification, in which the caustic soda and alumina solution pass into tanks where solid impurities sink to the bottom. The remaining alumina trihydrate is filtered.
- Precipitation of the alumina trihydroxide solution:

$$Al(OH)_4^- + Na^+ \rightarrow Al(OH)_3 \downarrow + NaOH \quad (6.4)$$

- Calcination, in which the crystals are washed, then heated to warmer than 1100°C in a kiln (normally heated with natural gas) to remove water molecules. The reaction is:

$$2Al(OH)_3 \rightarrow Al_2O_3 \downarrow + 3H_2O \quad (6.5)$$

The resulting fine white powder is alumina.

Smelting transforms alumina into aluminium. Alumina has a high melting point (2000°C), so separating the aluminium from alumina by melting the alumina is not practical. Instead, alumina is dissolved into cryolite (Na_3AlF_6), which lowers the melting point to about 900°C. As the mineral cryolite is extremely rare, synthetic cryolite made from fluorite (CaF_2) is used instead. The liquid cryolite serves as an electrolyte, and the alumina is reduced to elemental aluminium through electrolysis. This is referred to as the *Hall-Heroult process*. The reactions at the cathode and anode, both of which are made of carbon, are:

$$Al^{3+} + 3e^- \rightarrow Al \quad (6.6)$$

and:

$$2O^{2-} \rightarrow O_2 + 4e^- \quad (6.7)$$

$$O_2 + C \rightarrow CO_2 \quad (6.8)$$

respectively. The balanced net reaction is:

$$2Al_2O_3 + 3C \rightarrow 4Al + 3CO_2 \quad (6.9)$$

The anode is consumed by the anode reactions, and three quarters of a mole of CO_2 is produced for every mole of aluminium that is reduced (this CO_2 is in addition to the CO_2 released through the combustion of any fossil fuels used to supply energy, although most aluminium smelters are dependent on electricity from hydropower). Molten aluminium is deposited at the bottom of the electrolytic cell, siphoned off and taken to a holding furnace where it can be blended with other metals to produce a given alloy, then cleaned and cast.

Global mass flow of aluminium

Figure 6.12 gives the global aluminium mass flow in 2005 as estimated by the International Aluminium Institute. From this figure it can be seen that about 2.15 tonnes of dry bauxite are required per tonne of alumina produced (so the average grade of bauxite is 46 per cent alumina or 24 per cent aluminium), while the alumina to aluminium ratio is 1.94, so the dry bauxite to aluminium ratio is about 4.2. About half of the input to aluminium ingots (blocks of aluminium prior to shaping into final products) is primary aluminium and half is secondary (recycled) aluminium. However, much of the recycled aluminium is recycled internally within aluminium fabrication plants. The real fraction of secondary aluminium is the flow of post-consumer or 'old' scrap (7.4Mt in 2005) divided by the flow of finished products (37.6Mt). This gives a secondary fraction of 20 per cent. It can also be seen from Figure 6.12 that 3.5Mt were not recycled and the final disposition of another 3.3Mt is uncertain. Along with metal losses of 1.3Mt, potentially 57 per cent of the aluminium taken out of service in 2005 was not

recycled. The potential secondary aluminium share of total production was (7.4+3.5+3.3+1.3)/37.6 or 41 per cent – about twice the actual share.

Figure 6.12 also shows the breakdown of the existing stock of aluminium among different products. Aluminium in buildings accounts for 32 per cent of the existing stock, transportation equipment 28 per cent, engineering and cables 28 per cent, packaging 1 per cent and other materials 11 per cent. The packaging share of the cumulative aluminium stock is much smaller than the share of annual production of aluminium that goes into packaging (23 per cent for the US in 2003 according to Figure 6.8b) because of the very short lifespan of packaging.

Energy use in making primary aluminium

Table 6.7 shows the electricity and fuel energy used in the production of 1 tonne of product for various steps in the production of primary aluminium. Total energy use per tonne of aluminium is summed assuming 4.2 tonnes of ore per tonne of aluminium produced and 1.95 tonnes of alumina per tonne of aluminium produced. For smelting, the range and average energy use today as given by IEA (2006) is shown, as well as the lowest energy use foreseen by Norgate and Rankin (2001) assuming continued use of C anodes and with

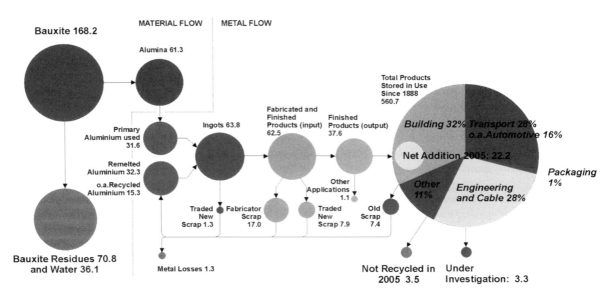

Source: International Aluminium Institute (www.world-aluminium.org)

Figure 6.12 *Estimated global mass flow (Mt) of bauxite, alumina and aluminium in 2005*

Table 6.7 *Range and global average energy use in the production of primary and secondary aluminium, and potential future energy use*

Step	Electricity (kWh/t)	Fuel (GJ/t)	Total primary energy (GJ/t)	
			40% efficiency	60% efficiency
Primary aluminium				
Mining (per tonne of ore)	10–20[a]	0.05	0.15–0.24	0.12–0.18
Refining (per tonne of Al_2O_3)	200–300	7–18	10–24	10–23
Smelting (per tonne of Al)				
Range today	14,000–18,000	18[b]	146–184	104–130
Average today	16,000	18	159	113
Best foreseen with C anodes	12,000	18	130	94
Best foreseen with inert anodes	11,000	0	99	66
Total (4.2 × (mining energy) + 1.95 × (refining energy) + smelting energy)				
Range today	14,230–18,670	32–53	166–232	124–176
Average today	15,816	42.5	193	146
Best foreseen with C anodes	12,430	31.8	150	113
Best foreseen with inert anodes	11,430	13.8	119	85
Secondary aluminium				
Scrap preparation	1	2.6	3.1	3.1
Melting	302	4.6	8.3	7.4
Refining	0	1.8	2.1	2.1
Slag treatment	76	2.2	3.3	3.1
Total	379	11.2	16.8	15.7

Note: [a] Schwarz et al (2001) give mining electricity use as simply <100kWh/t. The range here is chosen to bracket that for mining of copper (see Table 6.10). [b] This is the amount of energy (as fossil fuel feedstocks and process energy) given in IEA (2006) for production of anodes from pitch and petroleum coke, but it can be derived approximately as follows: from reaction (6.9), 333kg of C are consumed per tonne of aluminium produced. The emission factor for combustion of coal is 24kgC/GJ, so if the C in coal is turned into anode C with no losses and process energy, 13.9GJ of coal would be required. Thekdi (2003) gives the efficiency in producing C anodes from carbonaceous fuels as 71 per cent, so if the anode were to be produced from coal at this efficiency, coal with an energy value of 19.5GJ would be required. Primary energy use is computed assuming a fuel markup factor of 1.2 and assuming electricity to be supplied (generation + transmission) at 40 per cent or 60 per cent efficiency.
Source: Primary aluminium, IEA (2006), Norgate and Rankin (2001), Schwarz et al (2001); secondary aluminium, Schlesinger (2007)

the use of inert anodes (discussed later). The electricity requirement (about 14,000–19,000kWh/t) is larger by far than that of any other metal considered here. Total onsite energy use is 83–120GJ/t of aluminium, while the primary energy requirement is about 170–230GJ/t assuming that electricity is supplied (generated + transmitted) at 40 per cent efficiency. Even if electricity could be produced from primary energy without any energy losses, it can be seen that the production of aluminium requires *several times more primary energy* (per tonne) than the production of steel. Because so much electricity is required to produce aluminium, much of the world's production occurs in regions with access to inexpensive hydroelectric power. However, substantial amounts of aluminium are made in China, South Africa and Mozambique using coal-fired electricity, requiring about three to four units of primary energy

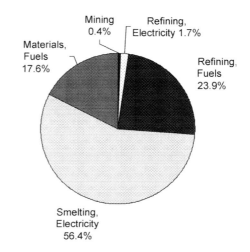

Source: Data from Schwarz et al (2001)

Figure 6.13 *Relative global energy use in the production of primary aluminium*

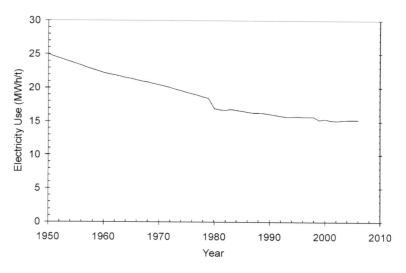

Source: Data from WWI (2008)

Figure 6.14 *Variation in the estimated world average electricity use per tonne of aluminium produced from 1950 to 2006*

for every unit of electrical energy supplied to the aluminium plant.

Figure 6.13 gives the breakdown in the average amount of energy used to make aluminium in terms of electricity and fuels for refining, smelting and the production of anodes. Figure 6.14 gives an estimate of the variation in world average electricity use per tonne of aluminium produced from 1950 to 2006. Average electricity intensity fell by about 40 per cent from 1950 to 2002 (a compounded rate of decline of 1.0 per cent/year), but increased slightly since then.

Energy use in making secondary aluminium

The recycling of scrap aluminium into new aluminium products (secondary aluminium) requires about ten times less primary energy than the production of primary aluminium. The lower part of Table 6.7 shows one breakdown of electricity and fuel use for scrap preparation, melting, refining and treatment of slag. Total primary energy use (at 40 per cent electricity supply efficiency) amounts to about 17GJ/t (compared to an average of about 200GJ/t for primary aluminium). The production of solid wastes and atmospheric emissions is also reduced by about 90 per

cent, while the capital cost of an aluminium recycling facility is 80–85 per cent that of a plant to produce primary aluminium (Schlesinger, 2007).

Emissions of perfluorinated hydrocarbons

GHGs – primary CO_2 – are emitted in association with the energy used to heat the kiln used for alumina calcination, the generation of electricity used for electrolysis, the production of caustic soda and the chemical reaction at the anode of the electrolytic cell (see Equation (6.8)). Perfluorinated hydrocarbons (PFCs) (CF_4 and C_2F_6), which are very strong GHGs, are also produced at the anode.

On a molecule-per-molecule basis, CF_4 and C_2F_6 are about 7100 and 18,600 times stronger than CO_2, respectively, and they have average lifespans in the atmosphere of about 50,000 and 10,000 years, respectively (Forster et al, 2007). Integrated over a 100-year time period, their GWPs are 7390 and 12,200 times that of CO_2, respectively.

PFCs are not generated during normal smelting operating conditions. Rather, they are produced during brief upset conditions known as 'anode effects' that occur when the level of the dissolved alumina in the cell drops too low and the electrolytic bath itself begins to undergo electrolysis. Measures to reduce the frequency and duration of anode effects not only reduce PFC emissions but improve energy and mass utilization. According to Bernstein et al (2007), average CF_4 emissions fell from 0.60kg/t of aluminium in 1990 to 0.16kg/t of aluminium in 2004, while the best-practice emission is 0.05kg/t of aluminium. Emissions of C_2F_6 fell from 0.058 to 0.016kg/t of aluminium over the same time period. These emissions can be multiplied by the GWPs given above and divided by 3.67 to give the equivalent CO_2 emission in terms of mass of C emitted. The result using average emissions in 2004 is a CO_2-equivalent emission of 0.375tC/t of aluminium. This is comparable to the chemical emission associated with the electrolysis of alumina (see Table 6.2) and about 10 per cent of the emissions that would be produced if the electricity used for electrolysis is supplied from coal at 40 per cent efficiency (see Table 6.3). PFC emissions are avoided altogether when secondary aluminium is produced through recycling of primary aluminium.

6.4.2 Reductions in energy use and GHG emissions

Energy savings in the production of primary aluminium

Thekdi (2003) discusses the potential for reducing energy use in the production of primary aluminium. As an indication of the potential energy-saving opportunities, Figure 6.15 gives the ratio of the theoretical minimum energy use for each step in the production of aluminium to the actual energy use. With the exception of anode production, these efficiencies are all very low (3–30 per cent). Although the theoretical minimum energy use can never be achieved in practice, the low efficiencies do indicate that there are still substantial opportunities to reduce energy use. Among the areas identified by Thekdi (2003) where energy use can be reduced are:

*   compressed air systems, which currently have a wire-to-work efficiency of about 10 per cent, and where optimization can provide improvements of

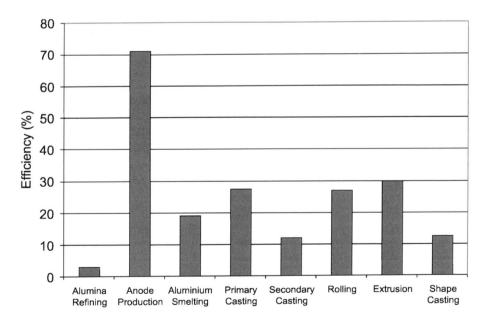

Figure 6.15 *The efficiency of energy use in various steps in the production of aluminium, computed as the ratio of the theoretical minimum required energy to actual energy use*

20–50 per cent, often without major capital investments;

- electric motors, where reductions in electricity use by 10–25 per cent are possible through more efficient motors and better controls;
- pumping systems, where 10–20 per cent savings are easily achievable;
- process heating, the largest single energy use, where fuel savings of 15–25 per cent are possible with low capital investment.

The development of inert anodes would eliminate emissions of CO_2 (and PFCs, a group of powerful GHGs) from the electrolysis process, reduce the electrolysis electricity requirement from 13kWh/kg (the current state-of-the-art) to 11kWh/kg, and save 18GJ of coal and oil energy used to manufacture the anode material (IEA, 2006). However, the eventual feasibility of inert anodes is still not proven, despite 25 years of research.

Energy savings in the production of secondary aluminium

Energy is used primarily in two steps during the production of secondary aluminium: (1) cleaning the scrap aluminium by heating it in a kiln to a temperature (150–230°C) hot enough to vaporize or carbonize any organic materials including coatings (about 30 per cent of total energy use), and (2) melting the cleaned scrap and casting it into new materials (about 70 per cent of total energy use). New techniques for cleaning can reduce the energy use at this step by 50 per cent, thereby reducing total energy use by 15 per cent (Swanson et al, 2005).

Energy savings through increased recycling of aluminium

The possible energy savings through improvements in the production of primary and secondary aluminium are dwarfed by the savings that are possible in shifting production from primary to secondary aluminium, through increased recycling of aluminium. As noted above, the production of secondary aluminium requires about ten times less primary energy than the production of primary aluminium. Recycling is therefore the key to major reductions in the energy required to produce new aluminium. Figure 6.16 shows the world production of primary and secondary (recycled) aluminium from 1971 to 2006, as well as the share of total aluminium production that was

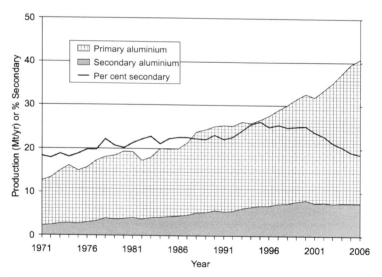

Source: Data from WWI (2008)

Figure 6.16 *Variation from 1961 to 2006 in the worldwide primary and secondary production of aluminium, and in the secondary share of total aluminium production*

secondary.[10] The secondary share peaked at 26 per cent in 1995, and has fallen since, especially after 2000.[11] The secondary share depends on the amount of aluminium taken out of use, the fraction of this aluminium that is recycled and the total aluminium production. Aluminium production grew at a slightly faster rate after 2000 than before, while secondary production remained constant, leading to the fall in the secondary share. If the global stock of aluminium in use eventually stabilizes, the rate at which aluminium is taken out of service will approach the demand for new aluminium and the potential secondary share will rise. If, by 2100, 90 per cent of all the aluminium being produced is secondary, then average energy intensity of aluminium production will decrease by a factor of 4.6 compared to today even with no change in the average energy intensity of primary or secondary aluminium.

Recycling of aluminium is constrained by the vast range of aluminium alloys in existence. There are over 200 alloys, which differ in the content of additives such as Si, Fe, Cu, Mn and Mg, most with over 95 per cent aluminium (Schlesinger, 2007). If a mixture of different alloys is melted down, the resulting average concentrations of the various additives will in general not be suitable for any given alloy, so fresh primary aluminium needs to be added in order to dilute the additives that are in excess. Thus, a transition to 100 per cent recycling (or close) will require some rationalization of the current mix of alloys, or extensive sorting of different aluminium products. However, something as simple as a beverage can uses three different alloys (one for the body of the can, one for the lid and one for the pull tab). The alloy produced by melting the can has too much Mg to be used for the body and too much Mn to be used as the lid. However, one producer has used a single alloy for the entire can, and this may have to become the norm.

The mass flow model of the International Aluminium Institute, shown in Figure 6.12, has been used to project that the amount of post-consumer scrap aluminium that is recycled globally will double between 2005 and 2020 (Martchek, 2006). Table 6.8 gives the fractions of aluminium in different product categories that are estimated to be recycled worldwide today. In the US, 60 per cent of scrap aluminium is new scrap (offcuts produced during the manufacture of aluminium products) and 40 per cent is old scrap (discarded products). Of the old scrap, 50 per cent comes from used

Table 6.8 *Estimated world average rate of collection for recycling of aluminium in different products in 2000, and rates of recovery during melting*

Product	Collection (%)	Melting recovery (%)
Buildings	70	96
Autos and light trucks	75	96
Aerospace	75	96
Other transport	75	96
Containers	59	85
Foil packaging	16	30
Machinery	44	96
Electrical cables	51	96
Other electrical	33	96
Consumer durables	21	96

Source: Martchek (2006)

beverage cans. However, the rate of recycling of beverage cans in the US decreased from a peak of 65 per cent in 1992 to only 45 per cent by 2004 (Schlesinger, 2007).

6.4.3 Aluminium energy use – summary and future scenarios

Table 6.9 gives the present-day breakdown between fuels and electricity for the production of primary and secondary aluminium. Also shown is the breakdown between fuels and electricity under current best practice and for anticipated future technologies, based on the information reviewed here. Fuel and electricity use as well as primary energy use are given for average aluminium based on the present-day secondary fraction (0.187) and for the hypothetical future secondary fraction (0.90). The key results are:

- If aluminium production in the future is 10 per cent primary and 90 per cent secondary using current average energy use for primary and secondary aluminium, then the average primary energy intensity of aluminium production would decrease from 160.3GJ/t at present to 34.5GJ/t – a reduction by a factor of 4.6.
- If the secondary fraction increases to 90 per cent and future aluminium production occurs at the lowest potential energy intensities identified here, the average primary energy intensity would be 23.3GJ/t – a reduction by a factor of 6.9.

Table 6.9 *Breakdown of energy use as fuels and as electricity for present-day average primary and secondary aluminium and for the weighted mean of the two using the present-day recycled content of new aluminium and the weighted mean energy use with 90 per cent recycled content, and (1) present average and best foreseen future energy intensities with 40 per cent electricity generation efficiency, and (2) best foreseen future energy intensities with 60 per cent electricity generation efficiency*

Onsite energy intensity	Elec Effic	Recycled Fraction	Primary aluminium		Secondary aluminium		Mean aluminium			PE reduction factor
			Fuels (GJ/t)	Electricity (kWh/t)	Fuels (GJ/t)	Electricity (kWh/t)	Fuels (GJ/t)	Electricity (kWh/t)	PE (GJ/t)	
Average today	0.4	0.187	42.5	15,816	11.2	380	36.6	12,929	160.3	
Average today	0.4	0.9	42.5	15,900	11.2	380	14.3	1924	34.5	4.6
Future best	0.4	0.9	13.8	11,430	8.4	285	8.9	1400	23.3	6.9
Future best	0.6	0.9	13.8	11,430	8.4	285	8.9	1400	19.1	8.4
Energy intensity ratio for mean aluminium, future best versus average today							**0.244**	**0.108**		

Note: Also given are energy intensity ratios for future aluminium with 90 per cent recycled content. A markup factor of 1.2 is applied in converting fuel energy to primary energy.

- If, in addition, the remaining required electricity is supplied with an efficiency of 60 per cent rather than 40 per cent, the average primary energy intensity would drop to 19.1GJ/t – a reduction by a factor of 8.4 compared to today.
- The energy intensity ratios (best future energy intensity over current energy intensity) for fuels and electricity for average aluminium are 0.244 and 0.108, respectively, indicating that the anticipated savings in aluminium energy use occur disproportionately through reduced electricity use.

Thus, it is concluded that the technical potential to reduce the average primary energy intensity of aluminium is a factor of five to eight reduction compared to today.

6.5 Copper

Copper is the third most widely used metal, after iron and aluminium. It occurs both as sulphide minerals and as oxide minerals (the latter as Cu-carbonates and Cu-silicates). Figure 6.17 shows the trend in copper mining from 1995 to 2007 in major world regions and globally, the percentage breakdown of mining by various countries and regions in 2007 and the breakdown of end uses of new copper in the US in 2003. The global mining of copper grew at an average rate of 5.6 per cent/year between 1994 and 2000 but slowed to an average of

2.4 per cent/year between 2000 and 2007. Chile alone accounted for 36 per cent of global copper mining in 2007. In the US in 2003, building construction accounted for almost half (48 per cent) of total copper consumption, while electrical and electronic equipment accounted for another 20 per cent.

Figure 6.18 presents the global flows of copper as estimated for year 2000 by the aforementioned STAF project of Yale University (Graedel et al, 2004). According to this figure, the rate of recovery of discarded copper was 2040Gg out of a discard flow of 3850Gg, or 53 per cent, while the recycled fraction in new copper was 2040Gg out of 11,650Tg, or 18 per cent.

6.5.1 Production of primary copper

Copper can be extracted from metal-containing ores through two broad sets of processes, which are discussed in textbooks on extractive metallurgy (such as Davenport et al, 2002). These are *pyrometallurgy* and *hydrometallurgy*. Pyrometallurgy is applied to most of the sulphide copper ores (including chalcopyrite ($CuFeS_2$) and bornite (Cu_5FeS_4)), while hydrometallurgy is applied to oxide ores and to the sulphide ore chalcocite (CuS). According to Davenport et al (2002), about 20 per cent of world copper production is through hydrometallurgical processing and the rest by pyrometallurgical processing.

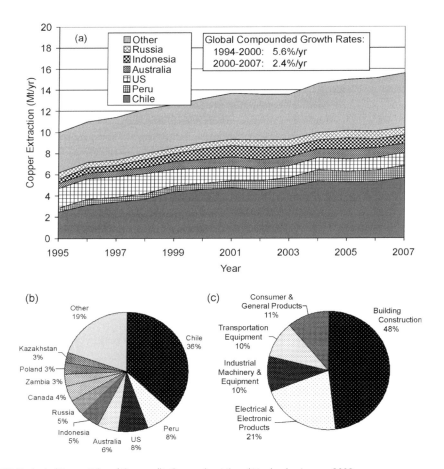

Source: Data from US Geological Survey Mineral Commodity Summaries, Mineral Yearbooks, January 2008 (http://minerals.er.usgs.gov/minerals/upbs/commodity), and US Geological Survey, Historical Statistics for Mineral and Material Commodities in the United States, September 2005 (http://minerals.usgs.gov/ds/2005/140)

Figure 6.17 *(a) Trends in the mining of copper by various countries and regions, (b) breakdown of copper mining by country and region in 2007, and (c) breakdown of the uses of copper consumed in the US in 2003*

Pyrometallurgy consists of the following steps (Davenport et al, 2002):

- *Beneficiation* and *froth flotation.* This involves grinding the ore so as to separate Cu-containing mineral grains from other mineral grains and mixing the ground ores with water and chemical foaming reagents. These chemicals make the Cu-containing minerals water repellent. Air blown through the mixture forms bubbles to which the copper minerals adhere, causing them to float to the top of the slurry, where they are removed

with a skimmer. This produces a concentrate (such as $CuFeS_2$) consisting of 25–30 per cent copper. The remaining debris (the tailings) is dewatered and disposed of in tailings ponds, while the water is recovered and recycled. About 10 per cent of the copper is lost during the concentration step.

- *Smelting.* This involves heating the copper concentrate to 1200–1250°C, thereby oxidizing the Fe and S in the concentrate and producing three products: a molten copper matte, a molten slag and gases. The overall reaction is:

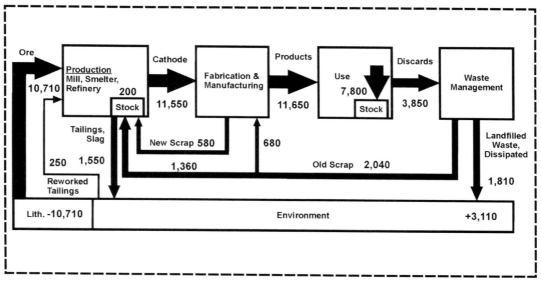

Figure 6.18 *The estimated global flow of anthropogenic copper, circa 1994*

$CuFeS_2 + O_2 + SiO_2 \rightarrow$

$Cu\text{-}Fe\text{-}S \quad + \quad FeO\cdot SiO_2 + SO_2 + heat \quad\quad (6.10)$
molten matte   molten slag

The oxygen is usually supplied as oxygen-enriched air. The oxidation of Fe and S is exothermic, with sufficient heat released to provide all the heat required for melting once the process has started. The matte contains 50–70 per cent copper (largely as Cu_2S), the balance being largely FeS. The slag contains oxides of various metals (such as oxides of Fe, Si, Al, Ca and Mg), in addition to FeO. Silica (SiO_2) is added to the smelter along with copper concentrate so as to inhibit the reaction of FeO with Cu_2S in the matte, which would otherwise produce some Cu_2O that would end up in the slag. The addition of SiO_2 increases the energy requirement (since it must be heated along with the copper concentrate) and increases the viscosity of the slag. Particles of matte are supposed to settle through the slag and collect at the bottom of the smelter, but some copper dissolves into the slag and is lost if it is not later separated from the slag.

Dissolution losses would increase with a more viscous slag, so there is an optimal silica input that minimizes the overall loss of copper from the matte. The greater the iron content of the concentrate, the more slag that is generated and the more copper that is lost.

- *Converting.* The copper matte is largely $Cu_2S\cdot FeS$. Converting is the process of separating the FeS from the Cu_2S and oxidizing the S in both components using oxygen-enriched air to produce molten *blister* copper (99 per cent Cu) and further iron slag. The reactions are:

$2FeS + 3O_2 + 2SiO_2 \rightarrow$
$\quad\quad 2FeO\cdot SiO_2 + 2SO_2 + heat \quad\quad (6.11)$

$Cu_2S + O_2 \rightarrow 2Cu° + SO_2 + heat \quad\quad (6.12)$

The heat required for converting is supplied entirely as the heat released by reactions (6.11) and (6.12) once the process starts. The product of converting contains less then 1 per cent Fe, 0.1–0.8 per cent O and 0.001–0.03 per cent S. Nevertheless, if copper were cast with these O and S concentrations, the S and O would form SO_2 bubbles or blisters, hence the name blister copper. The traditional converting process is a batch process (it starts and stops), which introduces energy losses and makes capture of SO_2 more difficult. Modern converting processes are continuous with the heat supplied entirely by the exothermic reactions.

- *Fire refining and casting of anodes.* Fire refining is a process for removing most of the remaining O and S and, like previous steps, is carried out at a temperature of about 1200°C. The O is removed as CO_2 through reaction with a hydrocarbon reducing agent (typically 5–7kg per tonne of copper), while S is removed as SO_2 through reaction with atmospheric oxygen. Fire refining is carried out in special rotating furnaces that are heated by combusting hydrocarbon fuels. The liquid product has 0.15 per cent O and 0.002 per cent S, and is directly cast into thin anodes that are interleaved with cathodes in electrorefining cells.

- *Electro-refining.* This is an electrolysis process that involves electrochemically dissolving copper from impure copper anodes into a $CuSO_2$-H_2SO_4-H_2O electrolyte and electroplating pure copper from the electrolyte onto a cathode without the impurities. After 7–14 days the cathodes are removed from the cell and the pure metal is scrapped off. The reactions are:

$$Cu°_{anode} \rightarrow Cu^{2+} + 2e^- \qquad (6.13)$$

at the anode, and:

$$Cu^{2+} + 2e^- \rightarrow Cu° \qquad (6.14)$$

at the cathode.

In the smelting of $CuFeS_2$ concentrate, about 2 tonnes of SO_2 equivalent are produced per tonne of copper (Alvarado et al, 2002). The sulphur dioxide, if sufficiently concentrated, is captured and converted to sulphuric acid (H_2SO_4). This is now mandatory in many parts of the world, so as to avoid emissions into the atmosphere (SO_2 itself is as an air pollutant with adverse effects on human health, and is eventually converted to H_2SO_4, a component of acid rain). Sulphuric acid is a useful product that can be sold (but is expensive to store or transport) or used in the hydrometallurgical process (as explained below).

There are four major types of copper smelter in use today, described in Ayres et al (2003) and Davenport et al (2002): reverberatory furnaces, electric furnaces, flash smelters and Noranda smelters. The first two are batch smelters and so are inherently inefficient due to heat losses after each batch is finished. They do not make use of the heat value of the sulphur, do not produce SO_2 gases in concentrations high enough for capture, and have largely been phased out. The flash and Noranda smelters are suitable for continuous processing without interruptions. There are two kinds of flash smelters, the Outokumpu smelter (about 30 in operation worldwide and accounting for over half of all copper smelting today) and the Inco smelter (about 5 in operation worldwide). The Outokumpu smelter produces a gas stream with only 10–15 per cent SO_2, while the Inco smelter uses pure oxygen and yields a gas stream of 80 per cent SO_2. The Noranda smelter has the lowest energy consumption but produces a slag with 10 per cent copper, so the slag must be reprocessed by froth flotation.

As with smelters, there has been a transition from batch converters toward continuous converters, which improves efficiency.

Hydrometallurgy consists of the following steps:

- *Leaching.* This involves first crushing the ore from the mine and then dissolving the ores in an acid (typically sulphuric acid). It can be done by piling the crushed ore in a heap or placing it in a large vat, or (rarely) by drilling holes into the ore body, pumping in the leaching solution, then pumping out the leachate. The leaching reactions for representative oxide and sulphide ores are:

$$CuO + H_2SO_4 \rightarrow Cu^{2+} + SO_4^{2-} + H_2O \qquad (6.15)$$

and:

$$Cu_2S + 5/2\ O_2 + H_2SO_4 \rightarrow$$
$$2Cu^{2+} + 2SO_4^{2-} + H_2O \qquad (6.16)$$

respectively. Bacteria that occur naturally in the sulphide ore provide enzyme catalysts that speed up the leaching process of sulphide copper by a million-fold. Leaching occurs over a period of years, during which time sulphur dioxide is emitted to the atmosphere.

- *Concentration.* This involves organic solvents that selectively absorb the copper from the solution, but also possibly removing any undesirable metals that may have been taken into solution during the leaching process. After the organic solvent has absorbed copper, it is separated from the solution and the copper is stripped from it with a fresh acidic solution. This produces a clean, concentrated solution.

- *Refining.* This is done through an electrolytic process called *electro-winning*. Electro-winning is similar to electro-refining (discussed above under pyrometallurgy), except that the anode consists of an inert Pb-Sn-Ca mixture. The reactions are:

$$Cu^{2+} + 2e^- \rightarrow Cu^\circ \qquad (6.17)$$

at the cathode (as in electro-refining), and:

$$H_2O \rightarrow H^+ + OH^- \rightarrow 1/2O_2 + 2H^+ + 2e^- \quad (6.18)$$

at the anode.

The combination of leaching, extraction of the mineral from solution with a solvent and electro-winning, is referred to as the *solvent-extraction electro-winning* (SX-EW) process. The SX-EW process was initially used on ores that were uneconomic to process with pyrometallurgy (Bartos, 2002). In the mid-1980s, many mines integrated SX-EW into their operations in an optimized fashion, with pyrometallurgy applied to ores above some minimum grade and SX-EW applied to ores below that grade but still worth processing. A consideration in choosing the extent to which SX-EW is used is the need to balance, as much as possible, the supply of sulphuric acid from pyroprocessing and the demand for sulphuric acid for leaching. Leaching of copper ores has been aided by bacteria during the past two decades, although bacteria have probably always played a role that was not originally recognized.

There are two disadvantages to SX-EW: (1) any gold, silver and molybdenum in the copper or zinc ore is lost, as the leaching of these metals requires alkaline conditions while the leaching of copper requires acidic conditions, and (2) the fraction of the copper in copper ore that can be extracted is much less for SX-EW than for pyrometallurgy. The copper oxide minerals malachite, azurite and atacamaite yield 70–90 per cent of the contained copper over a leaching period of weeks

to months, but the oxide cuprite yields only 50 per cent over a period of one year (Bartos, 2002). The sulphide mineral chalcopyrite yields only 10–15 per cent per year over the first few years, with a cumulative yield of only 50 per cent over a six- to ten-year period. This is uneconomic. Thus, hydrometallurgy tends to be applied to oxide minerals and pyrometallurgy to sulphide minerals. A given ore deposit may contain some of both, as the upper zone will contain oxide minerals while deeper materials that have always been below the water table will contain sulphide minerals.

Hydrometallurgy is the clear choice for the upper zone and pyrometallurgy for the lower zone, but either technique could be chosen for ores in the transition zone, with the preferred technique depending on the mineralogy, permeability and fracturing ability. With time, the range of materials that hydrometallurgy (SX-EW) can handle has been expanding – both to ores that formerly could not be processed at all and to ores that had previously been processed through pyrometallurgy. In the US, the fraction of mined copper subject to SX-EW rose from about 1 per cent in 1970 to 38 per cent in 2000 (Bartos, 2002). In Chile the fraction reached 32 per cent by 2001 (based on data in Bartos, 2002). As previously noted, about 20 per cent of total copper production worldwide is through hydrometallurgy. Because a smaller fraction of the available copper is extracted using SX-EW, the trend toward increasing use of SX-EW to process ores that would otherwise have been processed through pyrometallurgy will accelerate the depletion of the copper ore resource. This in turn increases the energy required for mining and processing the remaining ores, as discussed below. Hydrometallurgy requires a considerable amount of water, which can be a limiting factor in dry areas (such as some parts of Chile).

The vast majority (95 per cent in the US) of refined copper is produced as a cathode, either through electro-refining or electro-winning. This cathodic copper must be melted and then cast in various ways to produce the various copper end products. Melting requires 1.9–2.6GJ per tonne of melted copper (Davenport et al, 2002). Casting machines are classified as billet casting (for extrusion and drawing to a tube), bar casting (rolling to a rod or drawing to a wire) and strip casting (for rolling to a sheet or forming a welded tube). All casters use metals with an input temperature of 1110–1130°C, while exit temperatures are 940–1015°C. As in the steel industry, there has been a trend toward more continuous casting. Efforts are underway to produce thinner strips (5–12mm instead of 15–20mm) so as to avoid the need for rolling.

6.5.2 Energy use in producing primary copper

Table 6.10 summarizes estimates of the amount of energy used in various steps during the production of copper products from virgin ores. Crushing to a size of 10–13cm is done in the mine in order to permit transport of the ore on a conveyor. Grinding is required down to the typical size of the mineral grains (which varies from ore to ore) so that copper-bearing and copper-free minerals can be separated. This typically requires grinding to the 100–200µm size. Typical electricity requirements are 0.3kWh per tonne of ore for crushing and 15–35kWh/t for grinding, plus 2kg of fuel

oil per tonne for mining operations (Norgate and Rankin, 2000; Davenport et al, 2002). These energy requirements need to be divided by the grade of the ore (the amount of copper as a fraction of the total ore) times the recovery fraction (the fraction of copper in the original ore that makes it into final products) in order to get energy use per tonne of final product. The recovery fraction is smaller for hydrometallurgical processing (0.3–0.7) than for pyrometallurgical processing (0.90–0.95), but normally only crushing but not grinding is required for hydrometallurgical processing.

Ayres et al (2003) report the following chemical requirements per tonne of ore for froth flotation of sulphide ores: 1.1kg of lime and 140gm of various chemicals. Assuming an embodied energy of lime of 0.4MJ/kg (as in Table 6.1) and a typical chemical embodied energy of 100MJ/kg, the energy input for the chemicals used in froth flotation is in the order of 15MJ/tonne of ore.

Table 6.10 *Energy used during individual steps in the production of copper*

	Fuels	Electricity	Total (GJ/t)	
	GJ/t	kWh/t	On site	Primary
Mining				
Mine operations (per tonne of ore)	0.086	13	0.133	0.219
Crushing (per tonne of ore)		0.3–2	0.001–0.007	0.003–0.018
Grinding (per tonne of ore)		15–35	0.054–0.126	0.135–0.315
Hydrometallurgy				
Leaching and solvent extraction		2500	9	22.5
Electro-winning	0.05	1800–2800	6.5–10.1	16.3–25.3
Pyrometallurgy				
Froth flotation	0.15		0.15	0.015
Smelting	0–10	120–670	0.4–12.4	1–16
Converting	0–5[a]		0–5	0–5
Fire refining	2–3		2–3	2–3
Electro-refining	1	300–400	2.1–2.4	3.7–4.6
Cathode melting	1.9–2.6		1.9–2.6	1.9–2.6
Casting				
Rolled copper tubes	11.5		11.5	11.5
Totals at 1.0% copper in ores				
Hydrometallurgy	26.7–27.4	6300–7600	50–55	87–99
Pyrometallurgy	25.9–42.6	3500–6600	39–66	60–104
Totals at 0.5% copper in ores				
Hydrometallurgy	39.9–40.6	8400–9900	70–76	121–135
Pyrometallurgy	35.3–51.0	6600–12,100	59–95	99–164

Note: Energy use is per tonne of copper, except for mining energy use, which is per tonne of ore. Primary energies are computed using markup factors of 2.5 and 1.2 for electricity and for fuels, respectively. [a] Upper limit is a pure guess, assumed to be half that of smelting.
Source: Based on estimates given in Norgate and Rankin (2000), Davenport et al (2002) and Ayres et al (2003)

As noted earlier, there is substantial energy value in the unoxidized S in copper ores, and in modern continuous smelting and converting the heat released during oxidation of S provides essentially all of the energy needed for the process. Thus, the lower limit for the fuel energy requirement given in Table 6.10 for these steps is zero, while the upper limit represents fuel use in older batch furnaces.

Fire refining fuel consumption is 2–3GJ/t of copper, while electro-refining requires 300–400kWh/t for electrolysis and a hydrocarbon fuel for heating the electrolyte and for melting anode scrap (Davenport et al, 2002). Electrolysis energy use can be reduced by (1) using a lower current density (which slows the rate of refining), and (2) minimizing the amount of unused current. At present, 2–7 per cent of the current is unused, due to short circuits (which occur when an anode and cathode touch each other), stray current to the ground (when electrolyte is spilled), or through reoxidation of cathode copper (when oxygen is absorbed by the electrolyte due to uneven electrolyte flow). The electrolyte heating is minimized by insulating the tanks and pipes and by covering the electrolyte cells with canvas sheets. The need for melting anode scrap is minimized by casting thick,

equal-mass anodes and by equalizing the current between all anode-cathode pairs. In five plants with a current efficiency of 90–95 per cent documented by Davenport et al (2002), electro-winning requires 1840–2000kWh per tonne of copper shipped.

The overall primary energy use, assuming an ore grade of 1 per cent, 50 per cent recovery with hydrometallurgy, 95 per cent recovery with pyrometallurgy and a 40 per cent electricity supply (generation + transmission) efficiency, is about 90–100GJ/t using hydrometallurgy and 60–100GJ/t using pyrometallurgy. This assumes casting of rolled copper tubes with a casting energy requirement of 11.5GJ/t. Casting of copper wire would require somewhat more energy.

Figure 6.19 gives the variation in the total primary energy requirement for copper tubes with the ore grade as computed from the data in Table 6.10. Lower grade ores require proportionately more energy for mining, crushing and grinding per unit of final product, but the energy use for subsequent steps is unaltered. The total primary energy requirement increases sharply with decreasing ore grade below about 1 per cent copper.

Table 6.11 tabulates various published estimates of the primary energy requirement for the production of copper

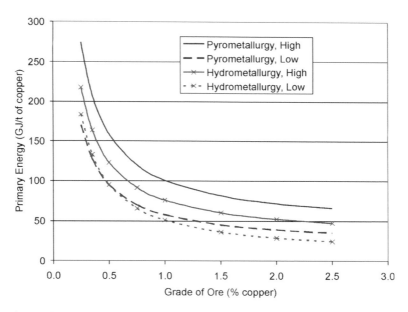

Source: Computed from data in Table 6.10

Figure 6.19 *Variation with the grade of copper ore in the amount of primary energy required to produce primary copper*

Table 6.11 *Published estimates of the amount of primary energy required to produce primary copper*

Reference	Conditions and comments	Energy intensity (GJ/t)
Norgate et al (2007)	Based on Australian LCA model for 3% ore grade, 35% electricity generation efficiency	33GJ/t for pyrometallurgy 64GJ/t for hydrometallurgy
Schleisner (2000)	Based on Danish LCA model with energy supply system appropriate for Denmark	78GJ/t, of which 3GJ/t are coke and 45GJ/t coal
Voorspools et al (2000)	Based on a Belgian LCA model with energy supply system appropriate for Belgium	113GJ/t
Ayres et al (2003)	Average energy consumption in German plants, 0.5% ore grade	42GJ/t fuels, 57GJ/t utilities
Ayres et al (2003)	Swedish smelters	32GJ/t for mining, concentration and transport 15GJ/t for smelting at 100Mt/yr (1995) 10GJ/t for smelting at 130Mt/yr (1996) 7.3GJ/t projected for Outokumpu flash smelting
Ayres et al (2003)	For production of brass (63% Cu, 37% Zn) and bronze (6% Sn) from 'average' ores.	52.2GJ/t brass 43.5GJ/t bronze

Note: LCA = lifecycle assessment.

products. Details (including the assumed grade of ore) are not available for most of these estimates, but the published estimates are broadly consistent with the energy intensities calculated here and shown in Figure 6.19.

6.5.3 Production of secondary copper and associated energy use

The first step in recycling old copper (after its collection and transport to the recycling facility) is the separation of copper from other materials. Scrap wires and cables are chopped into pieces of about 6mm in length using electric *granulators* and then directed to a *specific gravity separator* using air currents so as to separate the cable from the insulation. The major sources of copper from cars are the radiator, which is separated from discarded cars along with other major components before the car is shredded, and the wiring, which is extracted in various ways from shredded cars. Processing of discarded electronic equipment also involves initial disassembly to recover the larger items, followed by shredding and various separation techniques.

The methods for processing recovered scrap copper and the associated energy use strongly depend on the extent to which the scrap copper is contaminated with other materials. The purest copper scrap can be directly melted and recast. Electric furnaces are well suited for melting scrap. Less pure scrap is remelted and cast as anodes, followed by electro-refining to remove impurities (as in pyrometallurgical production of primary copper). Impure copper scrap must be smelted and converted. This can be done either in a dedicated smelter for secondary copper or in a smelter for primary copper (blended with concentrate from froth flotation). Primary ores contain Fe and S that are oxidized to provide most or all of the heat required for smelting and converting in the most efficient plants, but this energy source is not available for the smelting of contaminated secondary copper, so coke is used as an energy source. Coke is also used as a source of CO for reducing the oxides of the various metals (including copper oxide) that are in the scrap that is fed to the blast furnace, the reactions being (Davenport et al, 2002):

$$CO + Cu_2O \rightarrow CO_2 + 2Cu°(l) \qquad (6.19)$$

$$CO + ZnO \rightarrow CO_2 + Zn°(g) \qquad (6.20)$$

$$CO + PbO \rightarrow CO_2 + Pb°(l,g) \qquad (6.21)$$

$$CO + NiO \rightarrow CO_2 + Ni°(l) \qquad (6.22)$$

$$CO + SnO_2 \rightarrow CO_2 + SnO(l,g) \qquad (6.23a)$$

$$CO + SnO \rightarrow CO_2 + Sn°(l) \qquad (6.23b)$$

Iron in the scrap will do some of the reduction, especially of the easily reduced copper oxide, but then this iron would subsequently need to be reduced if it were also to be recycled.

Figure 6.20 is a flowchart for the smelting and refining of contaminated scrap copper. Contamination is a particular problem when recycling copper from electronic equipment (as seen in Figure 6.17c, electrical and electronic equipment account for about a fifth of copper use in the US). Electronic scrap consists of about 30 per cent plastic, 30 per cent refractory (heat-resistant) oxides and 40 per cent metals, with half the metal content being copper (Davenport et al, 2002). Electronic scrap also contains significant amounts of gold and silver, which are already recovered in primary copper smelters. High temperature oxygen smelting is required to insure complete combustion of the plastic contaminants. The metal content of electronics has been declining over time (the gold content having fallen from 0.1 per cent in 1991 to 0.01 per cent in 2000 according to Davenport et al, 2002), which makes

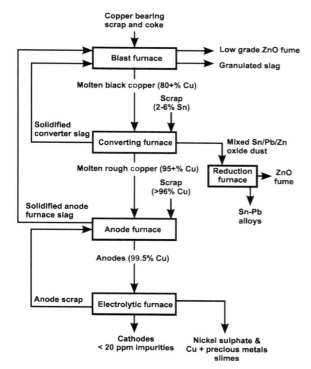

Source: Davenport et al (2002)

Figure 6.20 *Flowchart for the production of secondary copper from contaminated scrap copper*

conventional recycling less profitable. Clearly, there is a need to develop production systems that will permit the easy disassembly of products containing metals and the separation of the metals from other materials.

According to the Copper Development Association (www.copper.org), the production of secondary copper requires about 7–13GJ/t using clean scrap and 20–66GJ/t using scrap that needs to be refined. Ayres et al (2003) give the following energy requirements for recycling various kinds of copper scrap:

- Scrap grade 1 (wire, cable, copper tubing) – 4.45GJ/t for melting in a reverberatory furnace (additional energy, in the order of 10GJ/t for copper tubing, would be required for casting).
- Scrap grade 2 (unalloyed scrap, ≥94 per cent copper) 18.4GJ/t for melting plus 1.75GJ/t for space heating and pollution control (again, with additional energy for casting).
- Scrap grade 3 (25–35 per cent copper) 46GJ/t + 3.2GJ/t for ancillaries. Remelting of brass and bronze requires 6.7GJ/t in process energy plus 1.4GJ/t in ancillaries.

Compared to the amount of energy used to produce primary copper tubing (50–100GJ/t) from 1 per cent grade ore, there is about a factor of three to seven reduction in energy use for scrap grade 1 and up to a factor of three using scrap grade 2. However, there is potentially no savings in the recycling of scrap grade 3, due in part to the lack of energy from oxidation of Fe and S. Better sorting of waste streams will reduce the need for refining of melted scrap, and hence reduce the energy required for recycling.

6.5.4 Strategies for reducing energy use in making copper

As noted earlier, the use of energy to produce metals can be reduced through: (1) improved efficiency in the production of primary and secondary metals, (2) increased recycling of scrap metal, and (3) reduced demand. We discuss the prospects for each of these below.

Improved production efficiency

The transition from batch smelting and converting to continuous processing, with heat requirements provided through oxidation of S and Fe, would largely eliminate the fuel energy requirements for pyrometallurgy. However, a shift toward more production via hydrometallurgy will increase electricity requirements for refining but eliminate the use of electricity for grinding (which is not needed with hydrometallurgy). Research directed at more energy-efficient ways of making secondary copper from scrap copper is likely to be more effective than further research involving production of primary copper, especially since recycling will have to play an increasingly important role in satisfying future copper demand. One option to reduce the energy required for recycling of scrap copper is better sorting to reduce the amount of impurities.

Increased recycling of copper

The potential for a high rate of recycling of copper is good, as most of the world's copper is used in applications (such as wires and pipes, electric motors, generators and transformers, heat exchangers in air conditioners, and as a roofing material) from which it can be readily collected and separated (see Figure 6.17c for the US breakdown). Brass (an alloy of copper and zinc) is used in industrial valves and fittings and decorative hardware (such as locks and latches).

Some estimates of the fraction of copper in discarded materials that is recycled today are:

- Germany: >80 per cent overall recycling but only 40 per cent from vehicles (35 per cent is transferred to combustible waste and ends up in ash, and 25 per cent remains with ferrous scrap, where it poses a problem to the steel industry) (Ayres et al, 2003).
- Japan: 17 per cent from TVs, 35 per cent from fax machines and PCs, 50 per cent from residential household wiring, 80 per cent from industrial machinery, ships, railway equipment and wiring in non-residential buildings; 90 per cent from automobiles that are dismantled prior to shredding but only 10 per cent for automobiles that are shredded without dismantling; 100 per cent from heavy electrical machinery and equipment (Ayres et al, 2003).
- US: about 30 per cent overall (Ayres et al, 2003).
- Western Europe, about 70 per cent overall (Ruhrberg, 2006).

- Overall recovery rates of 50 per cent, 60 per cent and 40–75 per cent in North America, Europe and Asia, respectively according to the STAF project (the results of which are shown in Figure 6.18 at the global scale) (the recovery fraction in Asia is unclear because the reconstructed flows for Asia do not balance due to poor data).

Estimates of the overall rate of recycling of copper are based on observed present-day scrap flows, estimates of the average lifetime τ of copper products and comparison with the rate of production of copper products τ years ago. The average lifetime is uncertain, and modest changes in the assumed lifetime can have a large effect on the calculated recycling rate. Metals that are dissipated to the environment cannot be recycled, but dissipative losses in Sweden and the US seem to be in the order of 1 per cent of annual consumption (Ayres et al, 2003).[12]

Figure 6.21a shows the variation from 1966 to 2005 in the worldwide production of primary copper (from virgin ores) and of secondary copper (from scrap), as well as the supply of old scrap available each year.[13] The annual supply of old scrap has risen dramatically but the production of secondary copper has remained roughly constant during the 1966–2005 time period, such that the proportion of old scrap worldwide that is recycled had fallen from about 40 per cent to 20 per cent, as shown in Figure 6.21b. This decline in recycling is closely correlated with the fall in the price of copper from about $7000/t in 1966 to $2000/t by 2001 (in constant 2005$), as also shown in Figure 6.21b. Prices have since risen sharply, having fluctuated between $5000 and 8000/t between mid-2006 and mid-2008. Secondary copper constituted about 13 per cent of total production in 2005 according Figure 6.21. The recovery rate in 2000 is substantially smaller according Figure 6.21 than in Figure 6.18 (recovery rates of 25 per cent versus 53 per cent), while the recycled fractions are similar (14 per cent versus 18 per cent).

Reduced use

As the income of a country increases, the stock of copper (and other metals) will increase due to the buildup of copper-containing infrastructure, buildings and equipment. The current stock of copper in the US is estimated by Gordon et al (2006) to be 238kg per

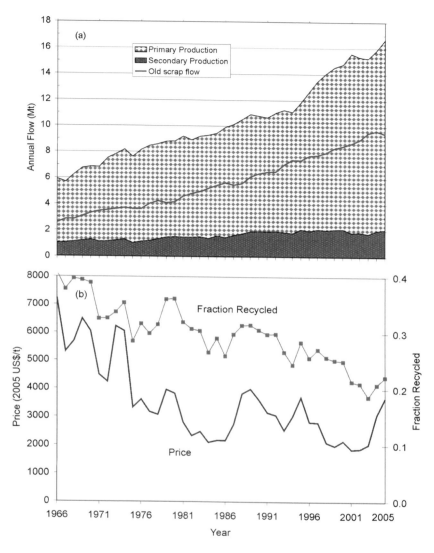

Source: Data from Gómez et al (2007)

Figure 6.21 *Variation from 1966 to 2005 in (a) the worldwide production of primary and secondary copper and the supply of old scrap available each year, and (b) the price of copper and the fraction of old scrap that is recycled*

person, broken down as shown in Figure 6.22. If a future world population of 9 billion were to achieve the same per capita stock, the total stock of copper would reach 2.1 billion tonnes. If 80 per cent of the mineable copper is recovered and all discarded copper is recycled without losses as this stock is built up (an unrealistic assumption), the required resource base is about 2.5 billion tonnes. The total worldwide extraction of copper to date is about 400 million tonnes but perhaps

only half of this is still in use. There is disagreement concerning the size of the available long-run resource (Gordon et al, 2007; Tilton and Lagos, 2007), with estimates for land-based resources ranging from 1.6 to 3.0 billion tonnes. Even if the upper estimate is more accurate than the lower estimate, resource limitations are very likely to prevent attainment of US levels of copper stock as world income increases, even if there is 100 per cent recycling of discarded copper. Large price

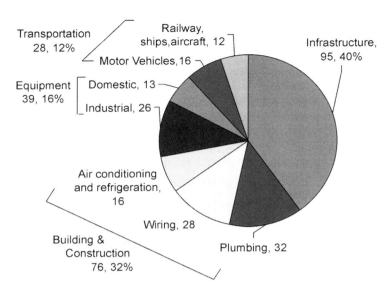

Source: Data from Gordon et al (2006)

Figure 6.22 *Distribution of the total copper stock of 238kg per person in the US*

increases as the resource is depleted will provide the incentive for very high rates of recycling, but will also limit the future buildup of copper stock.

A greater emphasis on compact housing that makes more efficient use of interior space (as discussed in Chapter 4, subsection 4.2.7) can yield significant savings in the use of copper (and other resources); according to Holtzclaw (2004), a suburban house in Davis, California uses five times the amount of copper pipe as a typical Nob Hill apartment in San Francisco. Lower levels of per capita car ownership than in the US and Europe (which, it is argued in Chapter 5, subsection 5.2.3, will be necessary due to space limitations) will reduce the required per capita stock in automobiles, although greater reliance on tethered electric vehicles (trolley buses and electrified intercity rail) would tend to increase copper demand. There are also a number of energy efficiency measures that would tend to increase the demand for copper beyond what it would otherwise be, as discussed below.

First, doubling the diameter of copper wire quadruples its cross-section and reduces electrical resistance losses by a factor of four. Overall resistance losses account for 4–8 per cent of all electricity generated (but these losses can be reduced through

more decentralized power generation). Second, frictional losses in pipes vary inversely with pipe diameter to the fifth power, so wider pipes translate into reduced pumping energy use. Third, doubling the area of a heat exchanger decreases the temperature differential across the exchanger by a factor of four to eight (so that incoming air can more closely approach the initial temperature of the outgoing air, meaning that more heat is recovered).

The demand for copper and other metals is also affected by substitutions with other materials that in some cases reduce energy use. In telecommunications, copper has been replaced with glass fibre, which has less volume and weight than copper: 18 optical fibre pairs weighing 117kg/km are equivalent to 2400 copper wire pairs weighing 7225kg/km. Energy consumption per voice channel is 0.1MJ instead of 23MJ, and the energy to produce the cables is 15.9GJ/km instead of 827.4GJ/km (Ayres et al, 2003). As existing telecom lines are converted to fibre optic cables, the telecom sector will probably become a significant source of high-quality scrap copper. However, fibre-optic cables utilize electric power, which cannot be carried through glass fibres, so there will be an increased peripheral demand for electrical wiring. High-temperature superconducting (HTS) cables using quaternary ceramic superconductors could replace copper wires. These would be cooled with liquid nitrogen, which would be separated from air in large quantities as a byproduct of separating oxygen for use in a variety of energy transformation processes (such as gasification of coal and later of biomass for use in combined cycle electricity generation).

In plumbing, polyvinyl chloride (PVC) pipes are replacing copper pipes. The energy intensity of PVC is 50–80GJ/t while that of copper is up to 180GJ/t for production from low grade ores. The energy use of PVC compared to copper per metre of pipe would be lower still due to the smaller density of PVC (1380kg/m³) compared to copper (8960kg/m³). Thus, replacement of copper pipes with PVC pipes represents a significant reduction in energy use for a given length of pipe.

New markets can drive up demand for metals. For example, a new market for copper is in automotive brake linings, where copper or brass powder in a plastic matrix has largely replaced asbestos. Copper in this application cannot be recycled, as it is dissipated to the environment. However, a global society that is less automobile dependent than current American or European societies – as is required on practical if not climatic grounds (see Chapter 5, subsection 5.2.3) – will moderate this and other demands for metals. Moderating the demand for electrical power will also moderate the demand for copper.

6.5.5 Implications of resource depletion (use of lower grades of ore) for energy use in mining and concentrating copper

Most elements that are mined from the earth are heavier than the average density of the earth's crust (which is largely Si and O), and so are very scarce in the crust (having been concentrated, instead, in the earth's mantle and core). A number of very scarce elements occur in two forms in the earth's crust: at concentrations several hundred to a thousand times greater than the crustal average, and as random substitutions for other elements in common minerals at the atomic level. From this, it is likely that many geochemically scarce elements (such as copper, which constitutes 0.132 per cent of the crustal mass) have a bimodal distribution of concentrations, as illustrated in Figure 6.23: one mode corresponds to deposits where the element was geochemically concentrated by chance, and the second, larger mode corresponds to diffuse substitution for other elements in common minerals. This implies that there is not an endlessly increasing quantity of available element as lower grades of ore are considered, but rather, that there is a dropoff in quantity with decreasing grade below some grade. This is referred to as the 'mineralogical barrier'. When copper is extracted from ores, for example, one is simply *physically separating grains of copper minerals from other minerals*. To extract copper

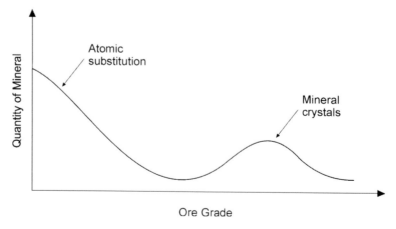

Source: Ayres et al (2003)

Figure 6.23 *Likely distribution of the quantity of metals with grade, for many metals*

Table 6.12 *Estimated average grade of the remaining copper ores in various world regions*

Region	Grade (% copper)
North America	0.47
Latin America	1.00
Europe	1.50
Oceania	1.56
East Asia	1.13
Central Asia	1.51
Africa	3.00

Source: Giurco (2005)

from the silicate matrix of common minerals requires separating copper and non-copper elements at the atomic level and processing vastly greater quantities of material, which would require 100–1000 times more energy as well as enormous (and unavailable) amounts of water (Ayres et al, 2003).

The average ore grade of US copper reserves has declined from 2–3 per cent copper from the first mines, in the late 1800s, to about 0.5 per cent by the late 1990s (Ayres et al, 2003). Table 6.12 presents estimates of the average grade of remaining copper ores in major world regions. As shown in Figure 6.19, the energy use in producing copper from virgin ores increases sharply as the ore grade drops below 1 per cent.

6.5.6 Copper scenarios

Ayres et al (2003) have developed a number of alternative scenarios for the future production and recycling of copper. The main results are:

- the fraction of discarded copper that is recycled increases from 51 per cent to 56–57 per cent globally by 2100, or from 72 per cent to 78–80 per cent globally (it being unclear what the current recycling fraction is);
- the fraction of new copper made from recycled copper increases from 20 per cent at present to 36–62 per cent by 2100;
- global consumption of refined copper increases from 12Mt/yr in 2000 to a peak of around 62Mt by 2065, or 90Mt/yr in 2100 and still rising (depending on the scenario);
- global rates of mining of copper peak anywhere between 50 and 66Mt/yr, some time during the 21st century.

These scenarios do not consider prices or resource constraints but are driven by projected growth of population, GDP/person and the demand for goods containing copper, and by an assumed modest increase in recycling rate. However, Ayres et al (2003) strongly doubt that the availability of copper of sufficiently high grade will permit rates of extraction anywhere close to that projected by their model (cumulative extraction by 2100 ranges from 8 to 10 times the present reserve base and 11 to 14 times the cumulative extraction to present). They believe that the maximum mine output will not be more than twice and possibly three times current output, and that the peak will occur within two decades.

Giurco and Petrie (2007) have considered scenarios whereby CO_2 emissions associated with the production of refined copper in the US could be reduced by 60 per cent by 2050. The extent to which demand grows over the coming decades and the extent of recycling are key factors. They ask, 'What changes to enterprises, regulations and prices would be required to transform the industry from "make and sell" to a "service" industry that rents metals to users and returns them at the end of their use as proposed by Ayres?'. They find that expanded use of hydrometallurgical processing coupled to C-free electricity sources does not play a significant role in achieving stringent emission reduction targets because of the smaller yield of copper with hydrometallurgical processing, which accelerates the decline in the grade of the available copper ores due to faster depletion, which in turn increases the mining and processing energy requirements. Instead, an increased focus on recycling and demand management is required.

Table 6.13 gives an estimate of the present-day breakdown between fuels and electricity for the production of copper, assuming that 80 per cent of primary copper is produced through pyrometallurgy and the rest through hydrometallurgy, that the average ore grade is 1 per cent, and that 13 per cent of total copper production is secondary copper. Also shown is estimated fuel and electricity use assuming a decrease in the average ore grade to 0.5 per cent and with either current average practice and secondary fraction, best practice, or best practice with a 90 per cent secondary fraction. For secondary copper, half of the required energy is assumed to be supplied as fuels and half as electricity. All of the input numbers are chosen from within the ranges given in Table 6.10, with values near the lower limits chosen for future production. The

Table 6.13 *Breakdown of energy use as fuels and as electricity for primary and secondary copper and for the weighted mean of the two assuming an electricity supply efficiency of 40 per cent and using the present-day ore grade, hydrometallurgical fraction, recycled fraction and energy intensities, and for combinations of possible future values of these parameters, including current best practice, future energy intensities and a decline in average ore grade*

Case	Electricity efficiency	Recycled fraction	Ore grade (%)	Hydro fraction	Primary copper					
					Pyrometallurgy			Hydrometallurgy		
					Fuels (GJ/t)	Electricity (kWh/t)	PE (GJ/t)	Fuels (GJ/t)	Elect (kWh/t)	PE (kWh/t)
Average today	0.4	0.129	1.0	0.2	35	5500	85	30	7800	100
DG + PI	0.4	0.129	0.5	0.2	45	9000	126	48	10,500	143
DG + BP	0.4	0.129	0.5	0.3	35	6500	94	48	9800	136
DG + BP + IR	0.4	0.900	0.5	0.3	35	6500	94	48	9800	136
DG + BP + IR	0.6	0.900	0.5	0.3	35	6500	74	48	9800	107

Case	Mean primary copper			Secondary copper			Mean copper			PE reduction factor
	Fuels (GJ/t)	Electricity (kWh/t)	PE (GJ/t)	Fuels (GJ/t)	Electricity (kWh/t)	PE (GJ/t)	Fuels (GJ/t)	Electricity (kWh/t)	PE (GJ/t)	
Average today	34	5960	88	15	4167	53	34	5960	87.6	
DG + PI	45.6	9300	129	10	2778	35	41	8460	117	0.7
DG + IP	38.9	7490	106	10	2778	35	35	6883	97	0.9
DG + IP + IR	38.9	7490	106	10	2778	35	13	3249	42	2.0
DG + IP + IR	38.9	7490	84	10	2778	27	13	3249	32	2.6
Energy intensity ratio for mean copper, future best versus average today							0.409	0.567		

Note: PI = present energy intensities, IR = increased recycling, DG = decreased ore grade, BP = best practice, IP = improved practice. Present energy intensities are computed from the data in Table 6.10. A markup factor of 1.2 is applied in converting fuel energy to primary energy.

resulting energy intensity factors for fuels and electricity for future production (0.5 per cent ore grade, 30 per cent hydrometallurgy for primary copper, 90 per cent secondary copper) are 0.41 and 0.57, respectively.

6.6 Zinc, stainless steel and titanium

Zinc, stainless steel and titanium are all very energy-intensive metals. The global production of stainless steel (at 28.3Mt in 2006) rivals that of aluminium (38Mt in 2007), and it is over twice as energy-intensive as regular steel. The total energy use worldwide in producing zinc is not large because global production is not large compared to that of other commodities (about 10Mt worldwide in 2006), but zinc warrants brief consideration here because the energy required in producing it is comparatively large (as discussed below) but it can be used in very small quantities to extend the lifetime of copper and steel products, thereby

leveraging much larger energy savings. Titanium is produced in very small quantities but has an exceptionally high energy intensity and can be used in ways that reduce other energy uses, a tradeoff that is also briefly examined here.

6.6.1 Zinc

Zinc ores tend to be of higher grade than copper ores (1–15 per cent zinc, typically 3–9 per cent, compared to 0.5–3 per cent for copper), so less energy is expended during the mining, crushing and grinding stages. Zinc ores occur either as oxide ores mixed with sulphur or as sulphide ores, and can be processed using hydrometallurgy or pyrometallurgy, which are similar to the procedures for the processing of copper.

The initial steps involved are (Ayres et al, 2003) production of zinc concentrates (containing about 50 per cent zinc); and optional roasting of zinc concentrates to drive off the sulphur (as SO_2, converting Zn to ZnO in the process and producing a material called *calcine* with a typical residual S content of 2 per cent.

These steps are followed by either hydrometallurgy or pyrometallurgy. The steps under hydrometallurgy (which accounts for 81 per cent of world primary refining capacity) are:

- leaching of calcine with sulphuric acid, producing a solution of copper, zinc, cadmium and other impurities;
- precipitation of copper and cadmium from the solution through successive additions of zinc dust, leaving a pure zinc sulphate solution;
- production of pure zinc in a cathode through electro-winning. This is analogous to the SX-EW process for copper;
- casting.

The steps under pyrometallurgy (19 per cent of primary refining capacity, applied to lead zinc ores) are:

- sintering, the process of roasting the lead-zinc concentrate (13 per cent of zinc production) or zinc-only concentrates (6 per cent) to obtain a material suitable for subsequent reduction in a blast furnace (metal sulphides and sulphates are transformed into oxides);
- smelting in a blast furnace, where oxides are reduced using CO from coke (as in a steel blast furnace);
- final refining and casting.

The energy released during the oxidation of sulphur exceeds the energy required to transform zinc ores into zinc slabs. Nevertheless, the typical primary energy requirement for the production of zinc slabs in the 1970s was about 60GJ/t from utilities (presumably electricity) and about 40GJ/t as coking coal (Ayres et al, 2003). More recent smelters make substantial use of the energy released from the oxidation of sulphur, significantly reducing the need for coal and reducing

Table 6.14 *Energy used during individual steps in the production of zinc*

	Fuels (GJ/t)	Electricity (kWh/t)	Total (GJ/t)	
			On site	Primary
Mining (energy use per tonne of ore)				
Mine operations	0.1[a]	13	0.1	0.2
Grinding		31	0.1	0.3
Hydrometallurgy (energy use per tonne of zinc)				
Leaching	1.4		1.4	1.4
Electrolysis, leaching, roasting, acid plant		4000	14.4	36.0
Lime	0.1		0.1	0.1
Oxygen		32	0.1	0.3
Total	2	4032	16	38
Pyrometallurgy (energy use per tonne of zinc)				
Sintering	0.2	532	2.1	5.0
Smelting furnace	24.2	174	24.8	25.8
Refining	5.5	25	5.5	5.7
Total	30	731	32.5	36.4
Totals at 9% zinc in ores				
Hydrometallurgy	2.9	4643	19.6	44.7
Pyrometallurgy	31.3	1342	36.1	43.3
Totals at 3% zinc in ores				
Hydrometallurgy	5.7	5865	26.8	58.6
Pyrometallurgy	34.1	2564	43.3	57.2

Note: Energy use is per tonne of zinc, except for mining energy use, which is per tonne of ore. Primary energies are computed using markup factors of 2.5 and 1.2 for electricity and for fuels, respectively. [a] Fthenakis et al (2009) cite various sources giving an onsite mining energy requirement of 0.08–0.8GJ per tonne of ore mined.
Source: Based on estimates given in Norgate and Rankin (2002)

overall energy requirements to about 40–50GJ/t. The energy inputs for recent hydrometallurgical and pyrometallurgical processing are given in Table 6.14.

Zinc is far more difficult to recycle than copper, due to the fact that 60–70 per cent of zinc produced worldwide is used to make galvanized (corrosion-resistant) steel. The next largest uses are in zinc-based alloys of steel, to make brass (an alloy of copper and zinc) and to make die castings. Some of the zinc coating on galvanized steel is lost due to corrosion (this is in place of corrosion of the steel, which is the intended effect). It is difficult to remove the remaining zinc coating before the steel is remelted for recycling, although it can be removed and collected through leaching of the scrap steel followed by electrolysis. Some of the zinc could be recovered from the ash that accumulates in electric arc furnaces, but at present most steel makers find it cheaper to dispose of the ash in hazardous-waste landfills (Gordon et al, 2003). In the US, zinc from EAF ash could supply 10 per cent of the country's zinc demand if it were recovered. According to the global zinc flow diagram presented in Graedel et al (2005), the global mean recovery rate for zinc in 2000 was 41 per cent and the recycled fraction was 16 per cent. As for the energy use in recycling zinc, this will be largely the energy required to refine discarded zinc. Martchek (2000) indicates a saving of 75 per cent.

Table 6.15 gives an estimate of the present-day breakdown between fuels and electricity for the production of primary and secondary zinc, assuming that 80 per cent of primary zinc is produced through pyrometallurgy and the rest through hydrometallurgy and that the average ore grade is 6 per cent. Also shown is estimated fuel and electricity use assuming a decrease in the average ore grade to 4 per cent but improved practice. Assuming a shift from 16 per cent secondary zinc to 80 per cent secondary zinc, energy intensity factors for fuels and electricity for future production are 0.39 and 0.42, respectively.

Table 6.15 *Breakdown of energy use as fuels and as electricity for primary and secondary zinc and for the weighted mean of the two assuming an electricity supply efficiency of 40 per cent and using the present-day ore grade, hydrometallurgical fraction, recycled fraction and energy intensities, and for combinations of possible future values of these parameters, including 10 per cent better future energy intensities for primary zinc and 20 per cent better for secondary zinc improved practice and a decline in average ore grade*

Case	Electricity efficiency	Recycled fraction	Ore grade %	Hydro fraction	Primary Zinc					
					Pyrometallurgy			Hydrometallurgy		
					Fuels (GJ/t)	Electricity (kWh/t)	PE (GJ/t)	Fuels (GJ/t)	Electricity (kWh/t)	PE (kWh/t)
Average today	0.4	0.16	6	0.80	32	1650	47	3.6	4950	48
DG + PI	0.4	0.16	4	0.85	33	2100	52	4.7	5400	53
DG + IP	0.4	0.16	4	0.85	30	1890	47	4.2	4860	48
DG + IP + IR	0.4	0.80	4	0.85	30	1890	47	4.2	4860	48
DG + IP + IR	0.6	0.80	4	0.85	30	1890	41	4.2	4860	33

Case	Mean primary zinc			Secondary zinc			Mean zinc			PE reduction factor
	Fuels (GJ/t)	Electricity (kWh/t)	PE (GJ/t)	Fuels (GJ/t)	Electricity (kWh/t)	PE (GJ/t)	Fuels (GJ/t)	Electricity (kWh/t)	PE (GJ/t)	
Average today	9.3	4290	48	2.5	1100	12	8.2	3780	42	
DG + PI	10.4	4740	53	2.5	1100	12	9.1	4155	47	0.9
DG + IP	8.1	4415	48	2.0	880	10	7.1	3849	42	1.0
DG + IP + IR	8.1	4415	48	2.0	880	10	3.2	1587	18	2.4
DG + IP + IR	8.1	4415	35	2.0	880	7.3	3.2	1587	13	3.3
Energy intensity ratio for mean zinc, future best versus average today							**0.392**	**0.420**		

Note: PI = Present energy intensity, IR = increased recycling, DG = decreased one grade, IP = improved practice. Present energy intensities are computed from the data in Table 6.14. A markup factor of 1.2 is applied in converting fuel energy to primary energy.

As noted above, the major use of zinc is as a corrosion-resistant coating for ferrous metals. In energy terms, this is a highly favourable tradeoff: the application of 30–70kg of zinc (requiring 125–300kWh of energy to make) can extend the life of 1 tonne of steel products (requiring 2500kWh to make) by a factor of three to five (Gordon et al, 2003). The established method of galvanizing is to dip the steel in molten zinc. However, electrolytic deposition has been used and permits thinner coatings. This may be a method to significantly reduce the demand for zinc.

6.6.2 Stainless steel and titanium

Stainless steel is an alloy of steel that can be divided into categories: austenitic stainless steel, consisting of iron, chromium and nickel; and ferritic stainless steel, containing Cr without nickel. Austenitic stainless steel commonly contains 18 per cent Cr and 8 per cent Ni, with much of the remainder being Fe (the Ni and Cr content may vary by a few per cent depending on the grade, and some of the Fe may be displaced with Mn). The use of chromium in stainless steel accounts for 86 per cent of all the chromium that is mined worldwide (Johnson et al, 2006), while the use of nickel in stainless steel accounts for 68 per cent of the

nickel that is mined worldwide (Reck et al, 2008).[14] Chromium is obtained from ores containing the mineral chromite, a mixture of chromium and iron oxides ($FeO \cdot Cr_2O_3$). Processed nickel is produced in three forms: as ferronickel, nickel oxide sinter and refined nickel. The iron content of stainless steel comes in part from ferrochromium, ferronickel and from dedicated iron. The chromium, nickel and iron inputs are processed in an EAF to produce stainless steel. The energy used to make the various inputs to stainless steel and the overall energy use as estimated by Johnson et al (2008) are given in Table 6.16. The overall primary energy intensity of stainless steel is 83GJ/t. This is comparable to the 75GJ/t energy intensity calculated by Norgate and Jahanshahi (2004) (but includes a markup factor of 1.1 for fuels), and over twice the 34GJ/t world average for regular steel given earlier. Also shown in Table 6.16 is the estimated energy requirement for recycling stainless steel. The estimated primary energy intensity for recycled (secondary) stainless steel is 26GJ/t – less than one third that of primary stainless steel.

According to the global chromium flow diagram presented in Johnson et al (2006), 670Gg of chromium were recycled back to stainless steel in 2000 out of a total scrap, discard and post-manufacturing waste flow

Table 6.16 *Energy used in producing the various inputs to stainless steel and in the final production of stainless steel in an EAF, and the overall energy use taking into account the amounts of the various inputs*

	Energy use per tonne of input			Input per tonne of stainless steel produced
	Fuels (GJ)	Electricity (kWh)	Primary energy (GJ)	
Primary stainless steel (from virgin materials)				
FeCr	20	3830	57	183kg Cr, 132kg Fe
FeNi	120	9630	218	24kg Ni, 49kg Fe
NiO	22	128,570	1181	4kg Ni
Ref Ni	49	15,200	191	52kg Ni
Direct reduced iron	17	106	20	573kg Fe
Energy use per tonne of output				
Electric arc furnace	7	1740	23	
Overall	34	5010	83	
Secondary stainless steel (recycled)				
Total	9.09	1743	26	
Average stainless steel				
25% secondary	28.1	4299	68.8	
80% secondary	14.2	2400	37.2	
Energy intensity factors	0.504	0.571		

Note: The disaggregation into separate fuels and electricity use given here is based on data files kindly provided by Jeremiah Johnson. Primary energies are computed using markup factors of 2.5 and 1.1 for electricity and for fuels, respectively.
Source: Johnson et al (2008)

of 1920Gg, and another 670Gg were recycled into lower-value products (such as C steel) that do not take advantage of the particular properties of chromium. Thus, the true recovery rate for recycling (as opposed to downcycling) was 35 per cent. The recycled portion of new production was 15 per cent. For nickel the recovery rates for recycling and downcycling were 58 per cent and 13 per cent, respectively, while the recycled fraction was 33 per cent, based on the nickel flow diagram in Reck et al (2008). This suggests a recycled fraction of around 25 per cent for stainless steel. As seen from Table 6.16, this leads to an average primary energy intensity today for stainless steel of 69GJ/t, which would decrease to 37GJ/t if the recycled fraction rose to 80 per cent. Energy intensities for future production are given in the last row of Table 6.16, assuming an increase in the recycled fraction from 25 to 80 per cent but no efficiency improvements (for lack of information).

When stainless steel is downcycled rather than recycled in a closed loop, the high embodied energy of the original chromium and nickel inputs is lost. Closed-loop recycling requires separating stainless steel scrap from regular steel scrap. This is easy for austenitic stainless steel because it is non-magnetic, but ferritic steel – like regular steel – is magnetic and so cannot be separated from regular steel by magnetic means. Thus, in Japan, 95 per cent of austenitic stainless steel is recycled as stainless steel but only 2 per cent of ferritic stainless steel is recycled as such (Igarashi et al, 2007) (just under a third of all stainless steel produced in Japan is ferritic, and is used in the same applications as austenitic stainless steel and in similar proportions). A larger price differential between the cost of nickel and iron would justify use of more costly, non-magnetic separation techniques.

Although stainless steel has higher energy intensity than regular steel, it is corrosion resistant, which leads to energy (and cost) savings by increasing the longevity of steel products. Condensing furnaces and boilers require the use of stainless steel so as to be able to withstand the effects of the condensate, but achieve an extra 10–20 per cent greater combustion efficiency at full load and permit modulation of the fuel flow at part load, giving an even greater efficiency gain (see Chapter 4, subsection 4.3.3).

Even greater corrosion resistance is provided by titanium, which is also lightweight and strong, making it a good candidate for partially replacing steel in transportation applications. The energy intensity of titanium today is about 360GJ/t (90GJ/t fuels and 29,000kWh/t electricity at an assumed supply efficiency of 40 per cent). Assuming a vehicle lifespan of 200,000km, Norgate et al (2004) estimate that substitution of titanium for steel in cars does not pay back in energy terms (through reduced fuel use), although substitution of aluminium and especially of magnesium for steel does pay back in energy terms. Not accounted for in this estimate is that titanium components could be thinner (due to their greater strength per unit thickness), or the possibility of end-of-life recycling. In any case, Norgate et al (2004) find that the use of titanium in place of stainless steel in chemical machinery in environments where corrosion is a problem does pay back in energy terms. At present the use of titanium metal is very limited – only 35,400 tonnes in the US in 2007 (Gambogi, 2008).

6.7 Cement

The growth from 1960 to 2006 in the worldwide production of cement in comparison with other energy-intensive commodities was illustrated in Figure 6.4. Figure 6.24 shows the trend in the production of cement from 1995 to 2007 in major world regions and globally, the percentage breakdown of production by various countries and regions in 2007 and the breakdown of end uses of cement production in the US in 2003. Production of cement in China grew very slowly between 1994 and 2000, then grew by an average compounded rate of 11.5 per cent/year between 2000 and 2007, thereby contributing strongly to overall world growth of 6.9 per cent/year during this time period. China accounted for 49 per cent of world cement production in 2007.

'Cement' in common parlance usually refers to 'Portland cement', the generic term for a class of fine manmade mineral powders that harden by reaction with water. Portland cement is usually manufactured from limestone or chalk, clay and smaller amounts of various other minerals that are added as needed to give the desired final mineralogical composition. Limestone, chalk and clay are very abundant and widespread, so cement can be produced in most parts of the world, but because material transportation costs are a major factor in the onsite cost of cement, cement plants tend to be located close to large sources of those materials that are themselves reasonably close to the points of use or to convenient shipping points.

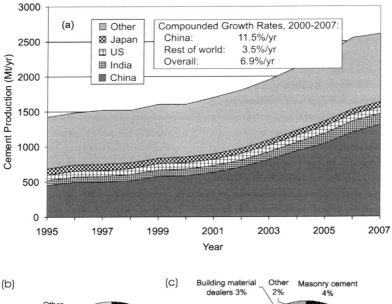

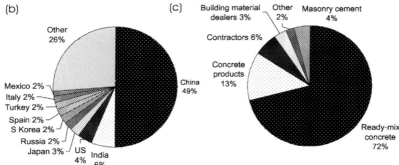

Source: Data from US Geological Survey Mineral Commodity Summaries, Mineral Yearbooks, January 2008 (http://minerals.er.usgs.gov/minerals/upbs/commodity), and US Geological Survey, Historical Statistics for Mineral and Material Commodities in the United States, September 2005 (http://minerals.usgs.gov/ds/2005/140)

Figure 6.24 *(a) Trends in the production of cement by various countries and regions, (b) breakdown of cement production by country and region in 2007, and (c) disposition of cement produced in the US in 2003*

Cement is mainly used as a binder for concrete, which is by far the most widely used of all construction materials (about 10km³ per year).[15] Concrete is a mixture of cement (typically 8–23 per cent by mass) and aggregates (gravel, crushed stone and sand) (typically 60–70 per cent by mass), with the balance as water. Higher-strength concrete usually has a higher cement content per unit volume (and a lower water to cement ratio) and so has a higher embodied energy, as cement is usually the most energy intensive of the major components of concrete.

Most structural applications of concrete include steel reinforcement, which is necessary to provide tensile (pulling) strength. Unreinforced concrete has a high compressive strength but is brittle and cannot support high flexural (bending) or tensile loads, so it is mainly used for making simple blocks and bricks. While steel is needed to give the concrete tensile strength, the concrete matrix provides both corrosion protection and fire protection to the steel, as well as reducing the amount of steel needed for any given structure and providing most of the compressive strength. As noted earlier, virgin steel from best-practice steel mills has an embodied energy of about 18–20GJ/t, whereas concrete has an embodied energy of only 0.5–1.0GJ/t (as will be seen later). Thus, the use of reinforced concrete as opposed to pure steel is preferable from an environmental impact viewpoint and also from a cost viewpoint in almost all cases. Steel corrodes in the presence of air and moisture, and many of the problems referred to as 'concrete durability' are in fact problems related to the corrosion of the steel in the concrete.

6.7.1 Cement production process

The main steps involved in the production of Portland cement are:

* crushing, grinding and blending the raw materials into a homogeneous powder;
* heating the blended raw materials to over 1400°C in a kiln to produce *clinker*;
* grinding the clinker to a fine powder and mixing it with additives, producing cement.

Table 6.17 gives a typical chemical composition for a Portland cement clinker. The components are given in their oxide form because, in the final product, the elements are mainly present in their most highly oxidized common states. Limestone and chalk are calcareous (Ca-containing) materials, but often also contain substantial amounts of the SiO_2, Al_2O_3 and Fe_2O_3 that are needed in the production of clinker. Any deficiencies in the key elements are made up by adding other materials that are mined nearby or imported. Portland cement is so-called because the prototypical product (patented in 1824) resembled the natural limestone from the peninsula of Portland, in England, which was very fashionable for construction use at the time. The original Portland cement consists of about

Table 6.17 *Typical elemental composition of clinker*

Chemical formula	Shorthand notation	Amount (%)
CaO	C	65.0
SiO_2	S	22.0
Al_2O_3	A	6.0
Fe_2O_3	F	3.0
MgO	M	1.0
K_2O+Na_2O	K+N	0.8
Other (incl. SO_3)	...($\hat{S}$)	2.2

Source: van Oss and Padovani (2002)

95 per cent clinker and 5 per cent gypsum, but many modern Portland cements also contain other ingredients, as will be explained later.

All modern cement plants use rotary kilns to produce clinker. Clinker can also be produced in vertical shaft kilns, which are still used in China and India (accounting for about 85 per cent and 10 per cent of total cement output, respectively, in the mid-1990s according to Worrell, 2004). These kilns are smaller (about a tenth the size of a typical US rotary kiln) and less capital intensive, so they lend themselves to local cement production in villages, but they are less efficient than larger kilns and very hard to control. A modern rotary kiln consists of a long (typically 50–200m) steel tube lined with heat-resistant (refractory) material (giving a typical internal diameter of 3.5–5.5m) that is gently inclined and slowly rotated. A fuel-air mixture is burned at the lower end of the kiln and the hot combustion gases move up the kiln. The raw mix, usually in the form of a fine powder, enters the upper part of the kiln and gradually works its way to the bottom end, forming itself into nodules (typically one or two cm in diameter) due to partial melting of the raw mix at 1200–1300°C. The hottest zone of the kiln, close to the flame, usually heats these clinker nodules to over 1400°C. At the bottom end of the kiln, the hot clinker nodules pass into the 'clinker cooler', in which most of their sensible heat is used to preheat the incoming combustion air, improving fuel efficiency and giving very high flame temperatures (which is also one reason why cement kilns can burn a very wide range of fuels, including many waste materials).

There are two main classes of rotary kiln process: wet and dry. In the wet process, the raw materials are ground wet to make an aqueous slurry, which facilitates uniform mixing. The slurry is then introduced into the back end of a very long rotary kiln. In the dry process, raw materials are dried (usually using the kiln exit gases) and ground and blended as a dry powder, which is then introduced to the kiln as such. In modern dry process plants the powder first passes through a series of heat-exchanger cyclones (known as a 'preheater tower') which use the heat in the kiln exit gases to preheat the powder before it enters the rotary kiln, thus giving much higher energy efficiency and also allowing the rotary kiln itself to be much shorter. The water content of the powder entering dry process kilns is typically less than 1 per cent, compared to a third or more for the slurry entering a wet kiln. All water must be evaporated before clinker formation, so wet kilns require far more energy than dry kilns and so are slowly being phased out.

Drying is followed by *calcination* at temperatures of 750–1200°C, *nodulization* at temperatures of 1200–1350°C and *clinkering* at temperatures of 1350–1450°C, followed by cooling of the clinker back to near ambient temperatures in the clinker cooler. Calcination is the reaction:

$$CaCO_3 \rightarrow CaO + CO_2 \qquad (6.24)$$

while clinkering is a sintering (partial fusion) process. It involves the reaction of CaO with the other raw material inputs containing silica, aluminium and iron. The overall reaction in a typical case is approximately (van Oss and Padovani, 2002):

$$29C + 8S + 2A + F \rightarrow 2C_2S + 6C_3S + C_3A + C_4AF \qquad (6.25)$$

where C, S, A and F are the shorthand notations given in Table 6.17 and C_2S, C_3S, C_3A and C_4AF are shorthand for the main clinker phases, whose chemical composition, name and typical proportions as found in Portland cement are given (along with gypsum) in Table 6.18.

Given that one mole of CO_2 is emitted for every mole of CaO that is input to the production of clinker, and given that CaO forms about 65.8 per cent of the average clinker, it follows that the chemically produced CO_2 is about 0.141tC/t of clinker (or 0.134tC/t of a Portland cement containing 95 per cent clinker).[16] A few per cent of this chemical CO_2 is reabsorbed from

the atmosphere during a typical 50-year cement lifespan, but much more can be absorbed when the concrete is broken up during demolition (as this greatly increases the surface area for absorption of CO_2).

Figure 6.25 shows the layout of the various zones in various kinds of kilns. Table 6.19 gives the proportion of cement production from different kinds of kilns in different countries and regions, along with the fuels used (which are discussed later). In most dry kilns, drying, preheating and most of the calcination occur in a specialized tower (a 'cyclone preheater') ahead of the kiln. Once the clinker has been cooled, it is ground and blended with a few per cent of gypsum ($CaSO_4 \cdot 2H_2O$) and/or anhydrite ($CaSO_4$) plus, in many cases, other additives such as limestone, granulated blast furnace slag, fly ash or natural pozzolans (discussed later).

When cement is mixed with water, the water reacts chemically with all of the mineral phases, but it is the hydration of the silicate phases C_3S (alite) and C_2S (belite) that accounts for most of the strength development. Alite hydrates more quickly than belite and so imparts early setting and strength to the cement, while belite is a major contributor to long-term strength. The typical hydration reactions for alite and belite are (van Oss, 2005):

$$2C_3S + 6H (water) \rightarrow C_3S_2H_3 (C - S - H \; gel)$$
$$+3CH (hydrated \; lime \; or \; 'portlandite') \qquad (6.26)$$

and:

$$2C_2S + 4H (water) \rightarrow C_3S_2H_3 + CH \qquad (6.27)$$

respectively.

Table 6.18 *Information on the components of Portland cement and the typical range of mineralogical compositions of pure Portland cement*

Chemical formula	Shorthand notation	Description	Amount (%)	Comment
Ca_3SiO_5	C_3S	Tricalcium silicate ('alite')	50–70	Imparts early strength
Ca_2SiO_4	C_2S	Dicalcium silicate ('belite')	10–30	Imparts long-term strength
$Ca_3Al_2O_6$	C_3A	Tricalcium aluminate	0–15	Acts as a flux and contributes to early strength
$Ca_4Al_2Fe_2O_{10}$	C_4AF	Tetracalcium aluminoferrite	0–15	Acts as a flux; contributes to long-term strength and imparts grey colour
$CaSO_4 \cdot 2H_2O$	$C\hat{S}H_2$	Calcium sulphate dihydrate (gypsum)	3–7	Controls early setting

Source: van Oss and Padovani, 2002 and Gartner, personal communication, 2008

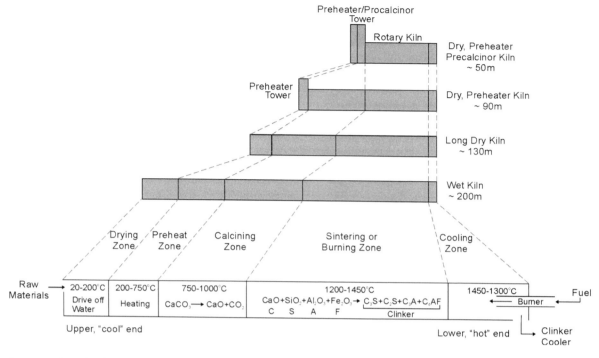

Source: van Oss and Padovani (2002)

Figure 6.25 *Layout of the various zones of a cement kiln in various types of rotary kilns*

The C-S-H gel is the main binder in the hydrated cement paste and, in the case of a pure Portland cement, constitutes 55–60 per cent of the hardened paste, while portlandite constitutes 20–25 per cent and calcium sulphoaluminate compounds constitute 15–25 per cent. Some of the lime released in reactions (6.26) and (6.27) is taken up in the hydration of C_3A and C_4AF, but most remains available to activate any slag or pozzolans (discussed later) that are added to the finished cement. The hydration of C_3A is very rapid and highly exothermic, and so must be controlled, usually by the addition of gypsum. C_4AF acts to lower the temperature at which clinker can form, but imparts the grey colour to cement and concrete. It is therefore avoided in the production of white cements (used for architectural applications), but such cements require more expensive (purer) raw materials and greater energy to manufacture due to the higher temperature that is required to produce a good quality clinker without C_4AF.

6.7.2 Energy use and fuels

Table 6.20 shows the range in the amount of energy used per tonne of cement. Electricity is used for crushing and grinding the raw materials, operating the kiln and grinding the clinker to make the final cement product. The energy used in the crushing and grinding of raw materials depends on their hardness, while the finish-grinding energy depends largely on the desired fineness of the cement. About 70–80 per cent of the energy used to manufacture cement is the thermal energy used in the manufacture of clinker. The theoretical minimum energy required to produce clinker is 1.76GJ/t of clinker.[17] Actual energy use is greater since energy is needed to evaporate water (much more so in wet kilns, which require up to twice the energy of dry kilns) and to compensate for heat lost directly from the kiln (by radiation and convection) and in hot materials leaving the kiln (dust, gases and clinker). Clinkering energy requirements thus range

Table 6.19 *Kiln types, fuel mixes and proportion of clinker used in cement in various countries and regions*

Region or country	Kiln type (% of output)				Fuel share (%)					Clinker fraction
	Dry	Semi-dry/wet	Wet	Vertical	Coal	Oil	Natural Gas	HFO	AFR	
North and South America										
Canada	71	6	23	0	52	6	23	4	15	0.88
US	65	2	33	0	58	2	13	0	26	0.88
Latin America	67	9	23	1	20	36	24	8	12	0.84
Europe, Middle East and Africa										
Western Europe	58	23	13	6	48	4	2	4	42	0.81
FSU	12	3	78	7	7	1	68	24	< 1	0.83
Other Eastern Europe	54	7	39	0	52	34	14	0	< 1	0.83
Middle East	82	3	15	0	0	52	30	14	4	0.89
Africa	66	9	24	0	29	36	29	2	5	0.87
Asia, Australia and New Zealand										
Australia and New Zealand	24	3	72	0	58	< 1	38	0	4	0.84
China	5	0	2	93	94	6	< 1	0	0	0.83
India	50	9	25	16	96	1	1	0	2	0.89
Japan	100	0	0	0	94	1	0	<1	3	0.80
Korea	93	0	7	0	87	11	0	0	2	0.96
Southeast Asia	80	9	10	1	82	9	8	0	1	0.91

Note: HFO = heavy fuel oil; AFR = alternative fuels and resources.
Source: Humphreys and Mahasenan (2002)

Table 6.20 *Energy used in making Portland cement (95 per cent clinker, 5 per cent gypsum)*

Step	Fuel (GJ/t)	Electricity		Total primary energy (GJ/t)
		On site (kWh/t)	Primary (GJ/t)	
Crushing and grinding		16–22	0.14–0.20	0.14–0.20
Clinker kiln, Wet	5.9–7.0	25	0.23	6.4–7.6
Dry	2.8–4.2	22–26	0.20–0.23	3.1–4.6
Theoretical minimum	1.67			
Finish grinding[a]		28–55	0.25–0.50	0.25–0.50
Total	2.8–7.0	66–103	0.59–0.93	3.5–8.2

Note: Primary energies are computed using markup factors of 2.5 and 1.05 for electricity and for fuels, respectively.
[a] Applies to grinding to a specific fineness (4000 Blaine, cm² of surface area per gm of material).
Source: Worrell et al (2001)

from 2.9–4.2GJ/t of cement (dry kilns) to 5.9–7.0GJ/t of cement (wet kilns). Lower energy use is achieved if waste heat from the kiln is used for drying and pre-heating the raw materials, as in all modern and some old kilns. The clinker cooler also plays an important role in heat recovery from the clinker.

Table 6.21 compares the estimated energy intensity of cement production in different countries. The present average energy intensity of cement ranges from 3.1GJ/t (in Japan) to 6.1GJ/t (in Columbia). Moving the world-average energy intensity (4.8GJ/t) to the existing Japanese intensity

Table 6.21 *Primary-energy intensity of cement production (excluding any markup factor for fuels) and share of total world cement production for different regions and countries*

Country	Intensity	Share	Country	Intensity	Share
	GJ/t	%		GJ/t	%
China	5.0	33.0	North America	5.4	7.0
Europe	4.1	11.5	– Canada	5.1	0.8
– Italy	4.5	2.2	– US	5.5	6.2
– France	4.1	1.2	Eastern Europe/FSU	5.5	7.1
– Germany	3.8	2.2	– Poland	5.6	1.2
– Spain	3.9	1.8	– Ukraine	6.0	0.8
– Rest of Europe	4.2	4.1	– Russia	6.0	2.5
OECD–Pacific	3.5	9.3	– Rest of Eastern Europe/FSU	4.9	2.5
– Japan	3.1	5.1	Latin America	4.7	6.3
– Korea	4.3	3.7	– Brazil	4.1	1.4
– Rest of OECD–Pacific	4.2	0.5	– Mexico	4.5	2.0
Other Asia	4.9	9.3	– Columbia	6.1	0.6
– Thailand	4.8	2.4	– Venezuela	5.7	0.5
– Taiwan	4.9	1.8	– Argentina	5.3	0.4
– Indonesia	5.3	1.7	– Rest of Latin America	4.9	1.4
– Rest of other Asia	4.9	3.4	India	5.0	5.1
Middle East	5.1	8.4	Africa	4.9	2.9
– Saudi Arabia	4.7	1.1	– Morocco	4.8	0.5
– Egypt	5.8	1.3	– South Africa	4.9	0.6
– Iran	5.3	1.2	– Rest of Africa	4.9	1.8
– Turkey	4.9	2.6			
– Rest of Middle East	4.9	2.2	World total	4.8	100.0

Source: Worrell et al (2001)

would result in a 35 per cent reduction in the average energy intensity of cement production. However, the low average energy use in Japan is due in part to the blending of 14 million tonnes of blast furnace slag and fly ash with clinker to make 70 million tonnes of cement in 2007, but this high proportion is unlikely to be achievable everywhere in the long run (as discussed below) unless natural supplementary cementitious materials can be used as well (as also discussed below). Nevertheless, a thermal energy requirement of less than 3.5GJ/t for clinkering should be achievable in almost all locations and with almost all raw materials if the best available clinkering technologies are fully implemented.

A wide variety of different fuels, including wastes, can be burned to provide the heat needed for clinkering. Coal is the dominant fuel used in kilns,

especially in Asia, although natural gas and oil are significant energy sources in some regions. Among the alternative materials that have been used as fuel for cement kilns are used tyres (53 million per year in the US alone, amounting to 41 per cent of all tyres that are burned), used solvents, sewage sludge, municipal solid waste, petroleum coke and biomass residues. Only biomass residues would entail zero net emissions on a lifecycle basis at present, although a number of waste materials that presently are derived from fossil fuel feedstocks (such as plastics and tyres) could eventually be produced from biomass feedstocks (see Volume 2, Chapter 4, section 4.7). Another usable kiln fuel is 'Profuel', a fuel made from paper that cannot be recycled and from offcuts from carpet manufacturers (Placet and Fowler, 2002).

The use of wastes as fuel for cement kilns can be an effective way of breaking down toxic organic compounds, and the incorporation of the residues into clinker can be an effective way of immobilizing these materials. However, toxic emissions (particularly Hg, Cd, As, Pb and chlorinated dioxins) can increase by a factor of 10–100 when wastes (including sewage sludge) are used as a fuel (Reijnders, 2007). Through a combination of measures, Hg emissions from burning municipal wastes in cement kilns can be reduced by 97 per cent compared to uncontrolled emissions (Reijnders, 2007). The burning of chlorinated organic materials can increase vaporization of toxic metals in the kiln and also leads to kiln operating problems, so total chlorine contents must be tightly controlled. The kiln destroys any dioxins in the raw materials, but trace amounts of VOCs in the exhaust can be catalysed by dust particles at the dust collector into dioxins if the exhaust is warmer than about 230°C, so it is necessary to cool the exhaust to below 230°C before it reaches the dust collector. This is now done in modern cement plants (van Oss, personal communication, 2008). There are also still some unresolved questions about the leaching of toxic metals from hardened cement products, which could be a particular problem after demolition of concrete structures at the end of their life.

6.7.3 Reducing the energy intensity of conventional cement

Since most of the energy used to make cement is in making clinker in kilns, and the majority of CO_2 emissions occur in association with clinker, the greatest opportunities for reducing cement energy and CO_2 intensity lie in either reducing the energy intensity of clinker or reducing the amount of clinker in the final cement product. These options are:

- to shift from wet to dry kiln production and to upgrade existing dry kilns to more advanced forms;
- to use alternative materials in the production of clinker (requiring less heat and reducing the chemical production of CO_2);
- to increase the use of non-clinker additives in blended cement;
- to maximize the use of waste heat.

Shifting from wet to dry kilns

As indicated in Table 6.19, a substantial (but decreasing) fraction of cement is still produced in wet kiln systems, with the largest proportion (78 per cent) being in the Former Soviet Union. As noted above, the minimum theoretical energy required to produce clinker (the enthalpy of formation) is 1.76GJ/t, but the feed entering wet kilns is typically one third water by mass. Since the ratio of kiln feed to clinker is usually very close to 1.5 (the difference being mainly due to the release of CO_2 from limestone, as shown in Equation (6.24)), this implies an additional minimal energy requirement of about 500kg/t × 2.5MJ/kg (the latent heat of evaporation of water), or 1.25GJ/t. Shifting from a wet kiln feed with 33 per cent water to a dry kiln feed with only 1 per cent water therefore reduces that theoretical minimum energy use by about 1.25/(1.76+1.25) or 40 per cent, with a comparable saving in actual energy use.

Use of alternative inputs for the production of clinker

The lime input required for the production of clinker is normally produced by the calcination of carbonate minerals in limestone, with the emission of CO_2 (Equation (6.24)), but slag produced from electric arc steel furnaces and bottom ash are also sources of lime that can be used in producing clinker. Slag melts easily, requires only course grinding before being fed to the kiln, and can substitute for 3–10 per cent of the limestone input (van Oss, 2005).

Use of blended cements

In blended cements, some of the energy-intensive clinker is replaced with some combination of glassy blast furnace slag, fly ash from pulverized coal-fired powerplants and ashes from other industrial sources, as well as natural volcanic ashes, fired clays, silica fumes and some limestones. These materials are referred to generically as *supplementary cementitious materials* (SCMs), but they are very different from one another in their chemical and physical nature and in the extent to which they can substitute for clinker. One major class of SCMs is amorphous (and therefore more reactive) alumino-silicates, known generically as *pozzolans*, because the original representative of this class was a volcanic ash from Pozzuoli, near Mount Vesuvius, which was used to great effect in Roman concrete. Coal fly ashes and burnt clays are also pozzolans.

The major SCMs are discussed in the following paragraphs. Since they are already available as byproducts of other industrial processes, or are naturally occurring materials, the additional energy required in using them in place of energy-intensive clinker is negligible. The critical issues are, rather, the strength and durability of the cements produced when clinker is diluted with SCMs.

Pozzolans react with water in the presence of portlandite (hydrated lime) to form calcium silicate hydrates (C-S-Hs) and calcium aluminate hydrates of the same types as form during ordinary Portland cement hydration. These additional hydrates contribute to the total binding capacity of the cement. The pozzolan particles fill some of the voids in cement, making the hardened cement matrix less permeable and thus less susceptible to certain types of chemical attack (Phair, 2006). On the negative side, by diluting the clinker and consuming some of the portlandite that would otherwise be present in the hardened cement paste, the total alkalinity of the cement is reduced, which can reduce the protection that the cement provides to corrosion of steel reinforcing elements (Gartner, personal communication, 2008). Another drawback of the use of pozzolans is that their reactions are usually rather slow compared to the hydration of alite and belite, so long periods of moist curing (often several months, especially in colder climates) are needed to obtain the desired ultimate strength and durability properties. The early-age strengths of concretes made with pozzolans are thus usually much lower than those of concretes made with 'pure' Portland cement. For this reason and also because of the fact that Portland cement clinkers only hydrate to provide about 20–25 per cent of their mass in the form of the portlandite required to activate pozzolans, it is unusual to find more than about 35 per cent pozzolan in a modern blended cement (Gartner, personal communication, 2008).

Rice husk ash, which is almost pure silica, has long been touted as a very reactive pozzolan, and has been used as such in concretes and mortars in those areas where it is readily available (Mehta, 1977). In Malaysia, ash from palm oil plants (the ash is produced from the burning of palm oil husks and shells during palm oil milling) has also been tested as a SCM. However, the global quantities of such materials are relatively small compared to potential market needs for reactive pozzolans, and their generalized use would require centralized collection and treatment facilities.

Blast furnace slag is not technically considered to be a pozzolan because it is far richer in lime (CaO), but it imparts similar benefits to concrete. If allowed to cool slowly, the molten slag from a blast furnace crystallizes and becomes totally unreactive (suitable only for use as an aggregate), but if it is quenched rapidly with water, a glassy slag is produced that has the properties of cement. The quenching is usually done with excess water in a device called a *granulator*, and the product is called *granulated blast furnace slag* (GBS). GBS, if finely ground, can hydrate on its own (albeit rather slowly), but it is usually used in combination with Portland cement clinker to make blended cements. The proportion of GBS in such cements can be as high as 90 per cent, but is usually much less because higher clinker contents are needed to get sufficient early-age concrete strengths for use in modern concrete applications. GBS usually decreases the permeability of concrete and provides chemical resistance to corrosion of steel (Placet and Fowler, 2002).

However, if steel is eventually produced without blast furnaces and the use of coal is phased out, two of the major industrial byproduct SCMs that can currently be blended with clinker (blast furnace slag and fly ash) will have been eliminated. Even now, with the increasing use of low-NO_x burners at coal powerplants, less ash suitable for blending with cement will be available in the coming years. Figure 6.26 compares the estimated regional supply of fly ash and blast furnace slag with the regional demand for cement in 2020, as estimated by the Battelle Corporation (Humphreys and Mahasenan, 2002). In most regions, and especially in developing countries, the supply of fly ash and slag is estimated to be only a few per cent of the cement demand, even without considering restrictions on the use of coal. Thus, efforts will be needed to find natural substitutes for clinker.

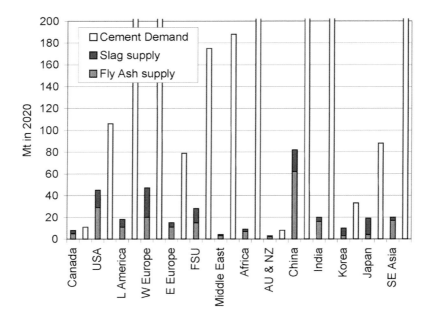

Source: Humphreys and Mahasenan (2002)

Figure 6.26 *Comparison of the regional supply of fly ash and blast furnace slag with the regional demand for cement in 2020*

Limestone is one possibility. The European cement industry has led the world in the use of limestone as a SCM, and cements with up to 35 per cent limestone are permitted under the European norms (Gartner, personal communication, 2008). Only about 5 per cent limestone per mass of clinker can react (with the aluminate phases), but limestone also catalyses the hydration of the calcium silicates by providing nucleation sites for C-S-H gel formation. It is thus useful as a cement extender, especially for applications that do not require very high strengths or very high degrees of impermeability and resistance to corrosive chemicals. Limestone also has advantages in being very easy to grind compared to clinker, is almost totally innocuous from a health and safety viewpoint and is readily available to most cement plants.

Ordinary quartz sand is another possibility, in combination with energetically modified cement (EMC) technology. This technology involves intensively grinding ordinary Portland cement, creating a network of micro-cracks and defects that allows a higher percentage of the potential binding capacity of Portland cement to be used. This in turn permits blends of up to 50 per cent fly ash or silica sand and 50 per cent Portland cement (in the case of 50 per cent fly ash, use of EMC results in 40 per cent higher strength than Portland cement after 24 hours, although the two have equal strength after 28 days (Placet and Fowler, 2002)). However, respirable (i.e. finely ground) quartz is a potential health hazard that must be tightly monitored, and the grinding energy requirements for quartz are rather high (Gartner, personal communication, 2008).

Several cement companies are working on the use of waste sludge and soils from contaminated clean-up sites, or dredged harbour sediment as a raw material to the cement kiln. The high temperatures in the kiln destroy PCBs (polychlorinated biphenyls), insecticides, dioxins, furans and other toxic materials, while heavy metals are immobilized (Placet and Fowler, 2002). One example is 'Ecocement' (from the Taiheiyo Cement Company of Japan), in which up to 50 per cent of cement kiln feed is replaced with ash from the incineration of municipal solid waste and sewage sludge. However, in order to control toxic emissions, the cement plant must be very sophisticated and capital costs are much higher than for an ordinary cement plant, so this type of cement manufacturing process can currently only be justified in locations where other methods of waste disposal (usually landfills) are exceedingly expensive, as is currently the case in parts of Japan.

Net energy savings using blended cements

In reinforced concretes, the durability of the steel is usually more important than the durability of the cement paste itself. The high pH buffering capacity of a Portland cement matrix, due to the presence of excess solid portlandite ($Ca(OH)_2$), helps keep the steel in a passive (non-corroding) state over long periods of time, even though the pH tends to decrease slowly due to atmospheric carbonation. Almost anything that dilutes the Portland clinker also tends to reduce this buffering capacity, which means that a greater depth of concrete cover over the steel and/or a higher cement content per unit volume of concrete is needed when pozzolanic cements are used. This will reduce the net energy savings, but diluting the clinker with, say, 50 per cent pozzolans would not require doubling the thickness of cement (Gartner, personal communication, 2008).

Reducing electricity use

The electrical energy used for crushing and grinding can be reduced through the use of high-efficiency motors, process-control systems and inherently more efficient grinding techniques. The most commonly used grinding machine is the ball mill, in which materials are broken by the shock of metal balls falling onto the material. The efficiency of this technique is in the order of 5 per cent, as most of the kinetic energy is dissipated as heat, but this is a problem for all fine grinding techniques. For modern high early-strength cements, which tend to have high specific surface areas in order to promote rapid hydration, ball mill efficiencies can be improved significantly by running them in a closed circuit with high-efficiency particle separators and by mixing very small amounts (100–300ppm) of 'grinding aids' (usually cheap glycols) with the material to be

ground. Grinding aids work by dispersing dry powder agglomerates, allowing the separator to better do its job of returning coarse particles to the ball mill for regrinding while efficiently sending the fine particles out of the system as product (Gartner, personal communication, 2008).

There are also a number of alternative grinding processes, most of which involve compression of a bed of particles between two hardened metal surfaces. One such technique, developed in France, is the Horomill. This involves a roller that grinds raw materials on the inner surface of a rotating tube. Other grinding techniques, such as vertical roller mills, use rollers to grind the raw material on a bed under moderate to high pressure. Compared to ball grinding, energy savings are reported to be around 45–57 per cent for the grinding of cement raw materials, 30–36 per cent for grinding ordinary Portland cement and 38–62 per cent for grinding various cement blends (Placet and Fowler, 2002). Vertical roller-type grinding mills are especially useful for grinding kiln feed, as they can easily permit circulation of hot gases (usually exhaust gases from the kiln system) for concurrent drying of the material, and they are now widely used in such applications. They are less used for clinker grinding, due partly to higher maintenance costs, but their use is increasing. However, the less efficient ball mills are still preferred when grinding cements to very high finenesses, as is often necessary for the higher early-strength cement classes.

Cement grinding energy consumption increases strongly with the desired fineness of the product, which in turn depends on market demand for high early strength cement. Granulated blast furnace slags are very hard to grind and so consume appreciably more grinding energy than Portland cements. However, as the primary energy used for finish grinding during the manufacture of Portland cement is only about 10 per cent of the thermal energy used (see Table 6.20), the partial substitution of blast furnace slag for clinker reduces the overall energy use.

6.7.4 Development of new cements

A variety of alternative cements have been developed that would be used completely in place of Portland cement clinker. One such group is called *geopolymers*, although the cements in this group are not polymers in the usual sense. They are formed through the reaction of certain alumino-silicates with water-soluble alkali hydroxides or silicates (such as NaOH or Na_2SiO_3), forming amorphous alkali alumino-silicates as the binding phase (instead of calcium silicate hydrates). Any pozzolanic compound or source of alumino-silicates that readily dissolves in an alkaline solution can serve as the raw material for the production of geopolymer. This includes materials such as metakaolin, produced by heating clays to relatively low temperatures (600–800°C), and glassy coal combustion ashes (i.e. fly ashes). Although much has been written about the potential of such cements, there have been almost no serious studies of their performance and energy requirements on an industrial scale. Thus, figures touted in the literature showing enormous CO_2 and energy reductions relative to Portland cements are not credible, because (1) nowadays it is already quite common to introduce significant percentages of pozzolans and other low-energy SCMs with Portland cements in concrete, and (2) most such studies ignore the high energy costs required to make the chemical activators (such as NaOH), which involve substantial energy inputs (Gartner, personal communication, 2008). In any case, the use of geopolymer has been limited by an inability to maintain a consistent mix and the need to cure it at a temperature >30°C if appreciable strength is to be attained in a reasonable length of time (Phair, 2006).

A second class of alternative cement is *calcium sulphoaluminate* (CSA) cements (Gartner, 2004). These are usually produced from limestone, clay, bauxite and gypsum or anhydrite. The formation of CSA clinkers requires about 20–40 per cent less energy and emits 20–40 per cent less CO_2 than Portland cement clinker, due mainly to the lower $CaCO_3$ content of the kiln feed. Less energy is usually required for grinding the clinker because it is softer, partly as a result of its formation at a lower temperature (1200–1300°C rather than over 1400°C). Greater amounts of gypsum, fly ash and metallurgical slags could be used in producing CSA-based clinkers than in making Portland cement clinkers. This could save a considerable amount of energy and could make use of low-grade coal combustion ashes and slags that are not suitable for use

as SCMs in ordinary Portland cement concretes. The manufacture of CSA-based cements is more widespread in China than elsewhere because the technology was mainly developed there, but at present, manufacturing costs are too high to justify its use in conventional concrete applications, so it is essentially limited to speciality applications. However, recent developments in this area lead to the hope that it may one day provide a useful alternative to Portland cements (Gartner and Quillin, 2007).

6.7.5 Integrated production of cement and electricity

A technology called the *Global New Energy Process* has been tested in the US, China and Canada. It involves burning a mixture of coal and a proprietary mixture of other materials in a coal powerplant, rather the coal alone. The resulting fly ash and bottom ash can be converted to a form of cement clinker (rich in belite) with no additional production of CO_2 compared to the production of electricity alone and with very little impact on the combustion efficiency of the powerplant. There are also no CO_2 emissions from chemical reactions to produce the clinker; thus, the effective CO_2 emission in producing the clinker is close to zero (Humphreys and Mahasenan, 2002). The process also captures some of the SO_x that would otherwise be emitted. Another technology, not yet tested, involves a specially designed fluidized bed combustor (FBC) in a coal powerplant that is integrated with a cement kiln. The feed materials are crushed coal, limestone and additives, and the outputs are electricity, cement and clean flue gas. Sulphur and trace elements such as mercury are isolated in a concentrated solid stream, while all other solid waste products are incorporated into the cement product. Both technologies represent a form of cogeneration, producing electricity and cement rather than electricity and useful heat.

Another option is cogeneration using an Organic Rankine Cycle, in which low temperature (300°C) exhaust air from the clinker cooler is used to generate electricity at an efficiency of about 10 per cent. At a test cement plant in Germany, the generated electricity was able to satisfy 25 per cent of the electricity needs of the plant (Humphreys and Mahasenan, 2002, Appendix B).

However, if the cement plant is located in an industrial park having industries that can use heat at less than 300°C, the waste heat could undoubtedly be used directly with greater efficiency. Conversely, the waste heat could be supplied to a district heating system if one is nearby. Although this does not reduce the amount of energy used to make cement, the waste heat that would otherwise be discarded can displace the other fuels otherwise used to generate heat.

6.7.6 Use of solar energy in the manufacture of cement

Work at the Paul Scherrer Institute in Switzerland has examined the feasibility of using concentrated solar energy for the calcination reaction. Up to 85 per cent of the energy for calcination would be supplied by solar energy, giving an overall saving in the production of cement of 20 per cent (Placet and Fowler, 2002). A 60kW solar calciner has been built and operated at the Paul Scherrer Institute. Solar production of cement would have to be restricted to arid regions, having sunshine almost every day.

6.7.7 Durability of cement and energy use

Efforts to improve the durability of concrete could reduce long-term energy use by reducing the need to replace concrete structures and to produce fresh clinker. However, as pointed out earlier, it is often the corrosion of reinforcing steel that limits concrete service life. This suggests that the development of inexpensive stainless steels (preferably without the need for high chromium contents), or of other inexpensive materials that could substitute for steel reinforcement, would be an excellent avenue of research. Not only could it lead to better concrete durability, but it could also allow the use of alternative binders that cannot currently be used simply because they carbonate too rapidly in air and thus do not sufficiently protect mild steel.

6.7.8 Long-term energy savings potential

Worrell et al (2000) estimate that the average intensity of cement production by existing plants in the US could be reduced by 2.22GJ/t (a 40 per cent saving) through implementation of all measures with an internal rate of return of 7 per cent or better.[18] This would reduce average energy use to 3.3GJ/t – slightly more than the current Japanese average. A draft report by the TATA Energy Research Institute in India (cited in Worrell et al, 2001) concludes that future technologies could reduce the Indian cement energy intensity (currently at 5.0GJ/t) by almost 48 per cent, giving an energy intensity of 2.5GJ/t. This seems reasonable, as it is only 20 per cent less than present Japanese practice, which surely leaves room for improvement.

Assuming that, over the next 50 years, we can find ways to reduce the energy intensity of conventional Portland cement by 20 per cent compared to current best practice (in Japan) and bring the whole world to this level, the average energy intensity for cement production would fall by a factor of two to about 2.5GJ/t.

Further savings will probably be possible through increased use of pozzolans to make blended cement or through the development of alternative cements. Additional reductions in the use of fossil fuels can be achieved through use of biomass as an energy source for the kiln. As discussed in Volume 2 (Chapter 9, section 9.2), cement plants also lend themselves to direct capture of the CO_2, whether produced from burning fossil fuels or biomass.

6.8 Glass

There are four major types of glass: container glass, flat or 'float' glass (used for windows, mirrors, windshields, tabletops), fibrous glass used in insulation and textile fibreglass (used for reinforcing rubber tyres, plastics and roof shingles, as well as for electrical insulation tapes). These glasses are made from varying mixtures of sand, limestone and (depending on the type of glass) soda ash, borate, feldspar and clay, as shown in Table 6.22. During glass manufacture, calcination of the limestone input occurs (reaction 6.24), releasing CO_2. The mass of CO_2 released is equal to the difference between the sum of the input mass given in Table 6.22 and the output mass (1 tonne), and amounts to 15–20 per cent of the mass of glass produced. Also given in Table 6.22 are the proportions of different oxides in the various glass products, the dominant oxides being silica (SiO_2,

Table 6.22 *Typical inputs used per tonne of various kinds of glass produced*

Material	Type of glass			
	Container glass	Flat (float) glass	Fibreglass insulation	Textile fibreglass
	Inputs (tonnes)			
Sand (SiO_2)	0.65	0.73	0.54	0.54
Limestone ($CaCO_3$ or $CaMg(CO_3)_2$)	0.19	0.24	0.19	0.12
Soda ash (Na_2CO_3)	0.22	0.23	0.22	0.00
Borate	0.00	0.00	0.10	0.15
Feldspar	0.11	0.00	0.11	0.00
Clay	0.00	0.00	0.00	0.34
Total	1.17	1.20	1.16	1.15
	Chemical CO_2 emission			
(tCO_2/t glass)	0.17	0.20	0.16	0.15
(tC/t glass)	0.046	0.054	0.044	0.041
	Composition of glass (%)			
SiO_2	72.5	70.7	57.0	55.0
CaO	10.0	9.5	8.1	20.5
Na_2O	14.5	13.3	14.5	1.0
Al_2O_3	1.5	0.7	8.0	14.8
B_2O_3	0.5	1.1	5.2	7.0
MgO	0.5	3.8	4.2	0.5
K_2O	0.5	0.9	2.1	1.0

Source: Ruth and Dell'Anno (1997) for inputs; Ross (2005) for glass composition

55–73 per cent), CaO (8–21 per cent) and alumina (Al_2O_3, 1–15 per cent).

6.8.1 Energy use and CO_2 emissions

The processes involved in the production of glass are: (1) batch preparation (mixing the inputs in the required proportions), (2) melting of the raw materials and removal of bubbles (refining), and (3) shaping the molten glass into the desired shapes. Melting can be carried out in one of four ways, using regenerative, recuperative, direct-fired or electric furnaces.

In *regenerative furnaces*, burners are fired along one side of the furnace and hot exhaust gases pass through a chamber on the other side containing refractory materials such as bricks that absorb heat.[19] Air that is fed to the burner (the combustion air) is preheated to up to 1400°C by passing through the heated chamber, then boosted by the burners to above the melting point of sand (1650±75°C). After about 20 minutes, firing switches to the other side. This results in an efficiency of up to 55 per cent in large furnaces (less in smaller furnaces); that is, 55 per cent of the heat from combustion of the fuel is used to melt the raw materials, with the balance lost in exhaust gases and from the furnace structure. In *recuperative furnaces*, a heat exchanger is used to transfer heat to the combustion air, but it is preheated to only 800°C, resulting in lower efficiency but at less capital cost. Recuperative furnaces are generally smaller than regenerative furnaces, which also makes them less efficient (the surface area, through which heat is lost, is larger relative to the volume and hence capacity of the furnace). In *direct-fired furnaces*, combustion of the fuel occurs in oxygen rather than in air, and the combustion gases directly heat the raw materials without the use of a heat exchanger. In *electric furnaces*, electrodes heat the materials through resistance heating.

Table 6.23 compares the efficiencies of different furnaces. Melting of raw materials in pure oxygen rather than in air reduces the energy requirements by 45 per cent in small furnaces and by 15 per cent in larger furnaces, although much of the latter savings are offset

Table 6.23 *Efficiency of different kinds of furnaces used in melting the raw materials to make glass*

Type of furnace	Efficiency (%)
Recuperative	20–40
Regenerative	35–55
Regenerative with batch/cullet preheater	50–65
Oxy-fuel	40–55
Oxy-fuel with batch/cullet preheater	50–65
Electric	70–90

Note: The energy required to produce oxygen for oxy-fuel furnaces is not accounted for here.
Source: Sinton (2004)

by the energy required to separate oxygen from air (Levine and Jamison, 2001).[20] The efficiency of electric furnaces is potentially twice that of regenerative furnaces and more than twice that of recuperative furnaces, but the efficiency in generating electricity also has to be taken into account. If the electric furnace requires half the energy but the electricity generation and transmission efficiency is only 30 per cent then the primary energy requirement increases. At the current average electricity generation and transmission efficiencies of 21 per cent and 38 per cent in non-OECD and OECD countries, respectively (see Table 2.6), use of electric furnaces entails substantially greater use of primary energy.

Beerkens (2006) carried out an extensive survey and analysis of the energy requirements used for melting the raw materials for glass. They ranked the plants in terms of increasing energy use. Based on data from 90 container-glass plants worldwide but mostly in Europe, the energy requirements at the 10th and 90th percentiles[21] are as follows:

- for recuperative furnaces, 4.8GJ/t and 9.0GJ/t, respectively;
- for regenerative furnaces, 3.7GJ/t and 8.3GJ/t, respectively;
- for direct-fired furnaces, 3.7GJ/t and 4.9GJ/t, respectively (including the energy required to separate oxygen from air).

The typical energy consumption of 40 float-glass plants, adjusted to 25 per cent recycled fraction, is 5.7GJ/t to 8.7GJ/t. For both container- and float-glass

plants, larger plants tend to have smaller energy intensities. The energy intensity of recuperative fibreglass plants (there being no regenerative plants in the survey) ranges from 8GJ/t to 25GJ/t, with an average energy intensity of 12GJ/t and 20 per cent less with oxygen-firing (including the energy required to separate oxygen from air).

Table 6.24 gives the average energy used for melting and refining per tonne of different kinds of glass produced in the US and the EU. Although potentially requiring less energy at the point of use, electric melters on average require greater energy per tonne of glass produced than regenerative furnaces for some types of glass. This is because electric melters tend to be used with small furnaces, perhaps with less insulation than in bigger furnaces, and potentially with discontinuous operation.

Table 6.25 gives estimates of the overall primary energy use and CO_2 emission in making glass in the US, assuming either that electricity is generated from coal at 40 per cent efficiency or natural gas at 60 per cent efficiency, or is C-free. Onsite energy use ranges from 14 to 23GJ/t, but primary energy requirements

Table 6.24 *Energy consumption (GJ per tonne of glass) used in melting and refining the raw materials for glass*

Glass and furnace type	US	EU
Flat glass		
Regenerate side-port	8.4	6.3
Electric boost side-port	6.2	
Container glass		
Large regenerative	7.5	4.2
Electric boost	5.3	
Oxy-fuel	4.5	3.3
Small electric	2.7	
Pressed and blown glass		
Regenerative	5.3	
Recuperative		6.7
Oxy-fuel	3.5	
Electric	9.9	
Insulation fibreglass		
Recuperative	6.7	4.3
Oxy-fuel	5.4	
Electric	7.2	
Textile fibreglass		
Recuperative	10.1	
Oxy-fuel	5.4	

Source: Sinton (2004)

Table 6.25 *Energy used in various stages in the production of flat glass, container glass and fibrous glass*

Step	Mass flow (t/t glass)	Total onsite energy use		Energy (% or absolute)		Primary energy use (GJ/t glass)		CO₂ emissions (tC/t glass)		
		Per unit mass flow (GJ/t)	Per unit glass produced (GJ/t)	Electricity	Natural gas	Electricity from coal at 40%	Electricity from natural gas at 60%	Electricity from coal at 40%	Electricity from natural gas at 60%	C-free
Flat glass										
Batch preparation	1.32	0.26	0.34	100%		0.9	0.7	0.02	0.01	0.00
Melting and refining	1.26	7.75	9.77		100%	12.2	12.2	0.16	0.16	0.16
Forming	1.26	1.39	1.75	100%		4.6	3.6	0.11	0.05	0.00
Post forming	1.2	2.11	2.53	10%	90%	3.5	3.4	0.05	0.05	0.04
Total (GJ/t)		11.50	14.38	2.34	12.04	21.2	19.9	0.36	0.27	0.20
Total (kWh/t)				650						
Container glass										
Batch preparation	1.32	0.51	0.67	100%		1.8	1.4	0.04	0.02	0.00
Melting and refining	1.26	6.07	7.65		100%	9.6	9.6	0.13	0.13	0.13
Forming	1.2	3.84	4.61	100%		12.1	9.6	0.30	0.13	0.00
Post forming	1.13	1.76	1.99		100%	2.5	2.5	0.03	0.03	0.03
Total (GJ/t)		12.17	14.91	5.28	9.64	25.9	23.0	0.51	0.31	0.16
Total (kWh/t)				1470						
Fibrous glass										
Batch preparation	1.3	1.10	1.43	100%		3.8	3.0	0.09	0.04	0.00
Melting and refining	1.24	9.47	11.74	100%		30.8	24.5	0.77	0.33	0.00
Forming	1.11	6.93	7.69	100%		20.2	16.0	0.50	0.22	0.00
Post forming	1.0	2.62	2.62		100%	3.3	3.3	0.04	0.04	0.04
Total (GJ/t)		20.12	23.48	20.86	2.62	58.0	46.7	1.41	0.63	0.04
Total (kWh/t)				5790						

Note: Primary energy use has been computed from onsite use assuming markup factors for the delivery of coal and natural gas to the glass factory or powerplant of 1.05 and 1.25, respectively, and powerplant efficiencies of 0.40 and 0.60 for coal- and natural gas-based electricity. CO₂ emissions have been computed assuming emission factors of 13.5kgC/GJ and 25kgC/GJ for natural gas and coal, respectively.

Source: Mass flow, onsite energy and proportions of onsite energy as electricity and natural gas are from Ruth and Dell'Anno (1997)

are in the order of 20GJ/t for flat glass, 24GJ/t for container glass and 45–60GJ/t for fibrous glass.

6.8.2 Energy saving opportunities

The maximum theoretical efficiency of a regenerative furnace is 80 per cent rather than 100 per cent, because the mass of the exhaust air times its heat capacity exceeds that of the incoming combustion air, so it will never be possible to fully capture the heat in the exhaust gas. The efficiency can be increased from the typical value of 50 per cent through the use of more refractory bricks and better designs of the heat transfer chamber. Losses from the furnace structure can be reduced through increased insulation, but addition of insulation is limited by the fact that the furnace itself becomes hotter the more it is insulated. Melting energy use can also be reduced by reducing the amount of time that the raw materials need to spend in the furnace. This in turn can be accomplished by ensuring the optimal grain size of the input materials and through pelletizing or pre-reacting some of the inputs (Sinton, 2004). The industry goal is to reduce the current 24- to 48-hour melting time to an hour. Sinton (2004) mentions a number of concepts for completely re-engineered methods of melting glass that would improve efficiency by up to 25 per cent while also reducing NO_x emissions. According to EC (2001a), it is unlikely that there could be a cost-effective regenerator design with an efficiency greater than 70–75 per cent, but this still represents a 30–35 per cent saving in energy use compared to current furnaces.

Further energy savings are possible (in the case of fossil-fuel furnaces) if the exhaust gases (which will be at a temperature of 270–300°C after preheating the combustion air) are used as a heat source for other industrial processes that can utilize heat at this temperature. This will be possible if industries that can use this heat are located next to a glass factory. Conversely, the waste heat could be supplied to a district heating grid. If half of the exhaust heat is recovered, the overall efficiency would be improved by another 10 per cent.

6.8.3 Impact of recycling

Recycled glass is referred to as *cullet*. Due to the presence of sodium and other elements in glass, it has a melting

point lower than that of silica (sand), the main raw material used to make glass. As a result, about 25–30 per cent less energy is needed to melt and recycle cullet than to make glass from raw materials (EC, 2001b). As well, CO_2 emission associated with the calcination of limestone is avoided. Table 6.26 gives a breakdown of the amounts and kinds of energy used in the manufacture of low-cullet (25 per cent) and high-cullet (59 per cent) glass. Excluding grid electricity and transport energy use, the remaining energy use (which is probably almost entirely thermal energy) decreases by 16 per cent for an increase in the cullet fraction by 34 per cent.

Figure 6.27 gives rates of recycling of glass containers in various countries. The recycling rate varies from a high of 88 per cent in Austria to a low of 21 per cent in the US. Long-term high rates of recycling will be inhibited by impurities in the glass (such as bottle caps and foils), which are acceptable in small quantities but will build up after several cycles if not diluted with fresh glass each time glass is recycled. Metals mixed with glass also damage the furnace by sinking to the bottom of the melted mixture and boring holes through the bottom of the furnace through chemical reactions. The high-quality requirements of flat glass and textile fibreglass preclude the use of post-consumer cullet, while container and other fibreglass are limited to 85 per cent cullet and 50 per cent cullet respectively due to the wide range of cullet compositions, which need to be balanced with fresh material (Ruth and Dell'Anno, 1997). The use of small amounts of virgin raw materials also aids in the refining process, such as the elimination of gaseous bubbles (Ross, 2005).

Table 6.26 *Energy inputs (MJ/tonne) used in the production of low-cullet and high-cullet glass*

Input	Low-cullet glass	High-cullet glass
Cullet	25%	59%
Diesel for transport	514	282
Grid electricity	411	469
Coke oven gas	0	327
Oil	1957	828
Natural gas	2872	3265
Unspecified energy	1226	672
Total	**6980**	**5843**

Source: EC (2001b)

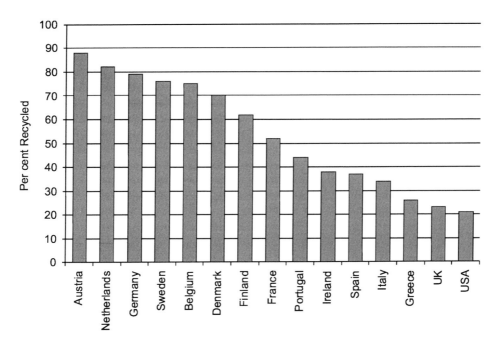

Note: European recycling rates are for 1997, the US rate for 2001.
Source: Data from EC (2001b) for European countries and US EPA (2003) for the US

Figure 6.27 *Rates of recycling of glass containers in various countries*

6.8.4 Synthesis

Improved non-electric furnaces should be able to give savings of 30–35 per cent in the energy required to make glass from raw materials, although higher energy prices than at present will be needed to make such improvements economically justifiable. It might be possible to save another 10 per cent through recycling of hot exhaust gases, and another 15 per cent by increasing the cullet (recycled fraction) from 25 per cent (roughly the present world average) to 60 per cent. Thus, a reasonable but rough range for the potential reduction in the energy intensity of glass is 45–60 per cent.

6.9 Pulp and paper

In 2006, 365 million tonnes of paper and paper products were produced, an increase by a factor of 4.7 since 1961 and by a factor of 2.0 since just 1984. Figure 6.28 shows the trend in paper production from 1961 to 2006 by major world region and by major groups of paper

products. About half of the paper products produced are packaging and wrapping paper. Figure 6.29 shows the trends in annual per capita paper consumption in various countries and regions. Per capita paper consumption in the US is about twice that of Europe but has been falling since 1999. Figure 6.30 gives rates of paper recovery for recycling in 27 European countries, the US, Canada and Japan. Paper recycling rates are higher than 80 per cent in several countries and greater than 90 per cent in Ireland, with an overall recycling rate in the EU27+2 (defined in Appendix C) of 67 per cent in 2008.

6.9.1 Processes for making paper and kinds of paper products

The production of paper involves four steps: (1) acquisition of fibres, (2) pulping, (3) bleaching (in some cases), and (4) manufacture of paper from pulp. Fibres used in the manufacture of paper can be obtained from roundwood (wood removed from forests or other areas), sawmill residue or wastepaper.

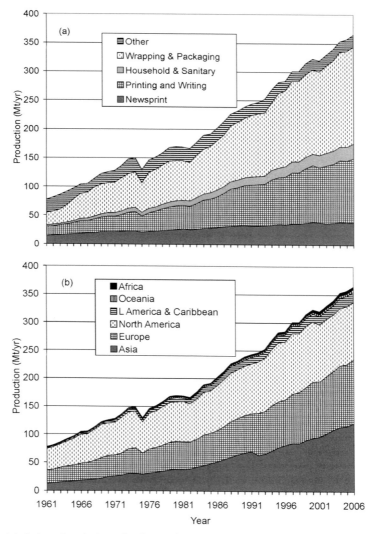

Source: Data from Food and Agriculture Organization online forestry data at http://faostat.fao.org

Figure 6.28 *Variation from 1961 to 2006 in the annual production of paper and paper products by (a) product, and (b) region*

Pulping

Fibres obtained from roundwood or sawmill residues can generally be used in any of the following pulping processes:

- mechanical – groundwood;
- mechanical – refiner process;

- chemi-thermomechanical;
- semi-chemical;
- chemical – kraft;
- chemical – sulphite.

Wastepaper can be processed using either mechanical pulping without de-inking, or mechanical and chemical pulping with de-inking.

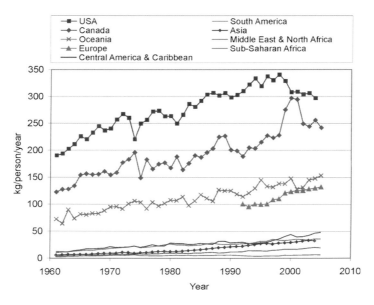

Source: Data from www.earthtrends.wri.org

Figure 6.29 *Annual per capita paper use in major world regions, 1961–2005*

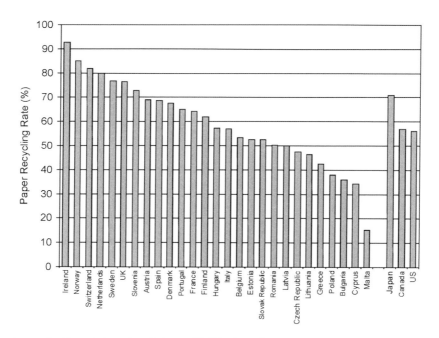

Note: CEPI data have been audited by an independent third party, PriceWaterhouseCoopers.

Source: Data from Confederation of European Paper Industries (CEPI, www.cepi.org) for European countries, American Forest and Paper Association for the US, Canadian Pulp and Paper Products Council for Canada, Highbeam Research (www.highbeam.com) for Japan.

Figure 6.30 *Rates of recovery of paper for recycling in various European countries in 2008, in the US in 2008, Canada in 2007 and Japan in 2004*

Figure 6.31 shows the variation in the worldwide trade of mechanical, chemical, semi-chemical and other pulps from 1961–2006. This is only a rough indicator of the relative amounts of different kinds of pulp produced, since it does not include pulp that is produced in integrated pulp and paper mills (as this pulp is not traded). Chemical pulps account for about half of the total traded pulp.

In the groundwood mechanical pulping process, the log is pressed against a rotating grinder stone and water is applied, whereas in the refiner process the wood fibres are broken apart between discs. In both cases, both lignin and fibre (cellulose and hemi-cellulose) become part of the pulp, so there is a high pulp yield (85 per cent of the wood is used). A large amount of heat is generated that can be recovered through *clean steam heat recovery* and supplied to the paper-making stage. This is now standard practice in new mills. The two fully chemical processes – kraft and sulphite – use only the fibre (40–55 per cent of the wood) to make pulp (because the lignin is dissolved) but the pulp is of high quality. The process begins by softening the wood chips with steam. Next, the softened chips are mixed in a digester with a highly alkaline solution called white liquor, which contains sodium hydroxide (NaOH) and sodium sulphide

(Na$_2$S). The mixture is heated to 160–170°C under pressure for several hours, dissolving the lignin. The spent liquor and dissolved materials, now called black liquor, are washed out and the fibres sent to the bleaching phase. The black liquor and bark residues are burned to produce heat and electricity in separate boilers, but with low (typically 10 per cent) electrical efficiency. Two thirds of the pulp in the world is produced using the kraft chemical pulping process.

Bleaching

Bleaching is used to remove residual lignin in the pulp, which otherwise causes the pulp to be dark brown. Bleaching requires the use of a chemical (Cl$_2$, ClO$_2$, or H$_2$O$_2$ and O$_3$) that oxidizes the lignin but not the cellulose and hemi-cellulose in the pulp; oxidation causes the lignin to break apart and become water soluble, allowing easy separation from the cellulose. The various kinds of pulp can be bleached, semi-bleached or left unbleached, and are then blended in different proportions to make various products: newsprint, sanitary paper, container board, construction paper, writing paper, packaging and industrial paper and boxboard. Newsprint is 65 per cent groundwood pulp, container board is close to 80 per cent unbleached kraft

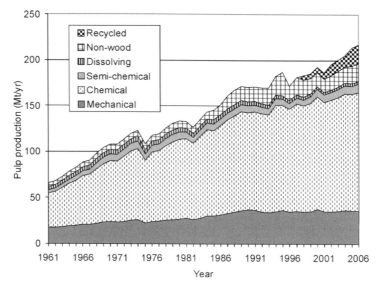

Source: Data from Food and Agriculture Organization online forestry data at http://faostat.fao.org

Figure 6.31 *Variation from 1961 to 2006 in the annual production of different kinds of pulp used to make paper and paper products*

pulp, and sanitary paper is close to 80 per cent fully bleached kraft pulp. Sulphite pulp is a minor (<20 per cent) constituent of sanitary paper, packaging and industrial paper.

Paper making from pulp

At the paper mill, the pulp fibres are mixed with 100 times their weight in water, which is subsequently removed as paper is made. The initial fibre fraction or *consistency* is thus about 1 per cent. The steps in making paper from pulp are:

1 *forming*, in which the stock is spread over a wire screen to form a sheet, with sufficient water removed by gravity and suction to increase the consistency to 20 per cent;
2 *pressing*, in which the consistency is increased to 40–45 per cent by passing the sheet, supported by felt, through three to four pairs of press cylinders; and
3 *drying*, by passing the sheet through 40–50 steam-heated cylinders to produce a consistency of 90–95 per cent.

Pressing normally occurs at a temperature of 40–50°C, with a 10K increase in temperature giving a minimum increase in consistency of 1 per cent. Drying occurs at a temperature of 100°C.

6.9.2 Energy use in making paper

Pulp can be produced in one location and then transported elsewhere to make paper, or pulp and paper can be manufactured in a single, integrated mill. In non-integrated production, the pulp has to be dried for transportation to the paper mill. At the paper mill, water is added back to the pulp and then gradually removed (largely through the addition of heat) after the pulp-water mixture has been spread into sheets, as explained above. In Europe, most paper production is not integrated with pulp production.

The minimum energy requirement (E_{min}) for heating water from 50°C to 100°C, followed by evaporation, is 2.46GJ per tonne of water evaporated. An additional 0.07GJ/t of water is required to heat the pulp fibres if the initial water content is 55 per cent and a further 0.02GJ/t is needed for desorption of water from the fibres, which brings E_{min} to 2.55GJ/t of water evaporated (de Beer et al, 1998b). If w_f is the initial

water fraction, then there are $(1-w_f)$ units of dry matter per unit of paper produced. As the amount of dry matter is conserved during the drying process, it follows that the energy required to evaporate water per tonne of paper produced is given by:

$$E_{min}^* = E_{min} \left(\left(\frac{1-w_f}{1-w_i} \right) w_i - w_f \right) \qquad (6.28)$$

where w_i is the initial water fraction.

The energy required for drying per tonne of paper produced depends strongly on the initial water content of the pulp: for $w_i = 0.55$ and $w_f = 0.05$, $E_{min}^* = 2.83$ GJ/t, but for $w_i = 0.60$, $E_{min}^* = 3.51$ GJ/t. Increasing the consistency from 40 per cent to 45 per cent after the forming and pressing steps thus reduces drying energy use by 20 per cent. The actual energy use at these consistencies will be greater still, due to heat losses to the surroundings during the drying process. Steam energy use based on mid-1990s or earlier data was 3.5–6.7GJ/t for newsprint dryers in Canada and 2.4–5.5GJ/t in Sweden (de Beer et al, 1998b), with additional heat from electricity. Onsite fuel energy use will be greater still, due to losses in the production of steam from fuel or biomass.

Table 6.27 gives the energy used for various pulping and paper-making processes, for ozone bleaching and wastewater treatment, and the totals. Mechanical and semi-chemical pulping are sources of heat but use about twice the electricity of chemical pulping. Pulping of pre-existing paper for recycling requires only a quarter to a third the energy for pulping of wood, but the energy use in subsequent steps is the same, so recycling reduces the overall energy requirement by only 20–40 per cent.

Of particular importance is the fact that the main required energy inputs are electricity and steam in a roughly 1:2 to 1:3 ratio. This makes the pulp and paper industry ideal for cogeneration, in which steam and electricity are co-produced. Moreover, in modern pulp mills and in modern integrated pulp and paper mills, there is sufficient biomass waste – either wood waste or pulping liquor (if chemical pulping is used) – to supply all of the energy needed by the cogeneration plant, although most mills in most countries are not energy self-sufficient. In the US, for example, only half of the steam requirements are met from biomass. Further, the electricity to heat production ratio from cogeneration in the pulp and paper industry is low, around 1:5 (Khrushch et al, 2001), so electricity must be supplied

Table 6.27 *Energy use (GJ/t) for pulp and paper production in the mid-1990s*

Process	Fuel/heat	Electricity	Primary
Pulping			
Mechanical	−2.1	5.3	11.2
Chemical	7.8–14.0	2.1–2.7	15.3–16.8
Semi-chemical	−3	6	12.0
Recycled	0.4	1.4	3.9
Bleaching			
Using H_2O_2 and ozone[a]	0.57	0.72	2.37
Paper making			
Newsprint	2.3–5.3	1.4	5.8–8.8
Printing/writing paper	5.5–8.9	1.9–3.2	10.3–16.9
Sanitary paper	5.3–7.0	2.4–3.6	11.3–16.0
Packing paper	5.0–7.0	1.3–2.9	6.3–14.3
Effluent treatment			
Mechanical or chemical[b]	0.0	0.14	0.35
Total (with % of world production in 2006 by type)			
UB mechanical + packing (46%)	3–5	6.7	20–22
B mechanical + newsprint (11%)	0.8–3.8	7.4	19–22
B chemical + writing paper (30%)	14–24	4.8–6.7	26–41[c]
B chemical + sanitary paper (7%)	14–22	5.3–7.1	27–40[c]
Rough mean	7–12	6.1–6.8	22–29

Note: Where a single value is given, it is usually a best-practice value. Primary energy is computed assuming an electricity generation efficiency of 40 per cent. UB = unbleached, B = bleached.[a] These are partial totals based on Tables 2.21, 2.28, and 2.30 of EC (2001c). [b] From Tables 4.11 (mechanical mill) and 4.12 (chemical mill) of EC (2001c). [c] In forming the totals, I have combined low heat use with low electricity use, and vice versa. This might not be right – processes that have low heat use might have high electricity use, and vice versa, in which case the range shown here is too large but the mid-value will be right.
Source: Farla et al (1997)

by the power grid. The low electricity to steam ratio is a result of using a simple back-pressure steam turbine for cogeneration (see Chapter 3, subsection 3.3.1).

Table 6.28 gives the energy requirements for two model pulp mills using best existing technology, in which significant use is made of heat flows that would otherwise be wasted. One model mill was developed by the Canadian pulp industry, the other by the Swedish pulp industry. Also given in Table 6.28 is the average energy use in 23 Canadian mills and the range of energy use in 6 Swedish and Finnish mills. There is a large variation in energy use among the Scandinavian mills that is not related to mill age or capacity, but is attributed to a lack of interest in energy efficiency due to low energy costs. The average energy use in the Canadian mills is near the high end of the range reported for Scandinavian mills. According to Table 6.28, heat and electrical energy

requirements in Scandinavian mills could be reduced by up to one third and 10 per cent, respectively, while heat and electrical energy use in the Canadian mills could be reduced on average by 44 per cent and 30 per cent, respectively. Many existing mills are self-sufficient in energy, but have the potential to be substantial net exporters of energy. Gross primary energy use (i.e. excluding energy from biomass wastes) is about 16–24GJ/t in Sweden and about 30GJ/t in Canada according to Table 6.28. In China, IEA (2006) reports an average energy use of 45GJ/t and even higher (but unspecified) energy use in India.

Proper accounting of the energy used in making paper requires that the energy used to produce the chemicals used in the production of paper also be counted, most of which is chlorine. Throughout most of the last century, elementary chlorine (Cl_2) was used

Table 6.28 *Comparison of energy use and energy production per tonne of pulp in existing Canadian and Scandinavian pulp mills, and in model mills using best existing technology developed by the Canadian and Swedish pulp industries*

| | Existing mills | | Model mills | |
	Canadian, average of 23 mills kWh/t (GJ/t)	Scandinavian, range of 6 mills kWh/t (GJ/t)	Canadian kWh/t (GJ/t)	Swedish kWh/t (GJ/t)
Heat use	6000 (21.6)	3000–5000 (10.8–16.5)	3390 (12.2)	3000 (10.6)
Electricity use	900 (3.2)	600–830 (2.2–3.0)	638 (2.3)	742 (2.7)
Heat production		5084 (18.3)	4389 (15.8)	5521 (19.9)
Electricity production		600–1190 (2.2–4.3)	655 (2.4)	1390 (5.0)

Note: Heat and electricity are produced solely from wood wastes generated by the mill.
Source: Klugman et al (2007)

worldwide for bleaching, but it produces severe water pollution. In much of the developed world, elemental chlorine has been replaced with chlorine dioxide (ClO_2), which produces fewer chlorinated pollutants. An alternative is *oxygen bleaching*, which uses hydrogen peroxide (H_2O_2) and oxygen for bleaching (this is also referred to as *totally chlorine-free* (TCF) bleaching, as opposed to *elemental chlorine-free* (ECF) bleaching using ClO_2). At a pulp mill in Louisiana, Bevilacqua-Knight Inc. (2000) report a bleaching steam energy use of 2.97GJ/Adt (air-dried tonne) of pulp in 1993, prior to TCF; 2.47GJ/Adt in 1995 using TCF bleaching; and 1.42GJ/Adt using TCF bleaching in 2000. Thus, TCF bleaching reduces the amount of energy used for bleaching by a factor of two compared to old (pre-1990) techniques, while eliminating chlorine pollution. This represents about a 10 per cent saving in total energy use for Swedish mills.

6.9.3 New or emerging technologies for drying pulp

The energy required to dry pulp can account for a third to a half of the total amount of energy required to make paper. New or emerging technologies have the potential to reduce the energy use for this step by 75–90 per cent. These technologies are outlined here.

Dry sheet forming entails making paper without the addition of water, a process that was commercialized in the 1970s but used only for the production of speciality papers. Dry fibres are dispersed over a flat sheet, then bonded either by adding resins or spraying a polymer

latex on the web. If followed by thermal bonding by passing the paper through electrically heated hot presses, the electrical energy for this step would be about 0.6–1.0GJ/t according to de Beer et al (1998b). The thermal energy required is reduced by 50–100 per cent. Research would be required to determine whether this technology could be applied to other types of paper and paper products on a large scale.

Press drying involves pressing a sheet between two hot (100–250°C) surfaces, followed by conventional drying with a reduced number of cylinders (15–30 instead of 40–50). A lot of research was done on this process, but it was never commercialized in spite of greater drying rates. De Beer et al (1998b) estimated a reduction in thermal energy for drying of 5–30 per cent.

Electric impulse drying involves feeding the paper onto an electrically heated metal roll. The surface water flashes into steam, which expels water from within the web. The consistency increases to about 55 per cent for paper board, 60–65 per cent for newsprint and 78 per cent for lightweight paper, so conventional drying is required to finish the drying with 10–20 cylinders. Steam requirements are reduced by 50–75 per cent, with the additional electricity use equal to 10–15 per cent of the savings in steam energy. Electric impulse dryers were developed in the 1980s and can be retrofitted into existing paper-making plants, but there are still many technical obstacles to their full-scale implementation.

Air impingement drying involves blowing hot (300°C) air at high velocity against the wet paper sheet. Some of the heat in the exhaust air can be recovered. All of the conventional drying rollers can be replaced with this process. The thermal energy requirement ranges from 1.9 to 4.3GJ/t, with lower values at low air velocity and temperature and a low drying rate. This is a 10–40 per cent saving.

Steam impingement drying involves superheated steam impinging on the wet paper rather than air. Superheated steam at 300°C and 1.1 bar pressure is blown onto the sheet, cooling to 150°C and releasing about 0.3GJ per tonne of steam. Also, 15–20 tonnes of steam are required to produce 1 tonne of dried paper, corresponding to 4.5–6.0GJ/t. However, all of the evaporated water is condensed in a condenser, releasing 3.0GJ per tonne of paper. The remaining heat requirement in a paper mill is only 0.5–1.0GJ/t of paper.

Airless drying involves steam-heated cylinders, as in conventional drying, but the drying section occurs inside an airtight and well-insulated hood. The required steam is produced by compressing the air into which the water has evaporated to 4atm pressure. Heating energy required for drying is reduced by 70–90 per cent, but the electrical energy required for compression is equal to 15–25 per cent of the savings in heating energy. As of the late 1990s, this technology was still in an early stage of development and had been applied only to batch drying of ceramics and laundry.

In the long term, de Beer et al (1998b) estimate that the thermal energy used for drying can be reduced by 75–90 per cent through a combination of impulse drying, a latent heat recovery system (such as steam impingement drying) and a number of small improvements. All of the technologies except airless drying potentially entail smaller investment costs than conventional drying. Another possibility is to use some substance other than water as the medium for forming paper from pulp fibres. Possibilities that have been suggested are ethanol and supercritical CO_2.

6.9.4 More efficient cogeneration of electricity

Some to all of the steam and electricity needs of pulp and paper mills are generated on site through cogeneration using biomass wastes produced by the plant. These wastes could be bark and wood scraps, or black liquor remaining after chemical pulping. The efficiency of electricity generation is low in most existing pulp and paper cogeneration facilities – typically around 10 per cent using existing boilers, 14–15 per cent with new boilers (Möllersten et al, 2003). However, electrical efficiencies of 27–30 per cent (and 72 per cent overall LHV efficiency) can be obtained through gasification of black liquor and use of the gasified liquor in combined gas and steam turbines (black liquor gasification combined cycle or BLGCC cogeneration). This process is likely to be commercially available by 2015–2020 according to Eriksson and Harvey (2004). There has been a long line of biomass gasification development and demonstration plants in Sweden going back at least to 1987 (see www.chemrec.se), but only for the production of heat. The main challenge in combined cycle cogeneration applications of biomass gasification is to produce a gas sufficiently clean to avoid disturbing the gas turbine. Gasification would occur at about 950°C and 25 bar in pure oxygen. Solid biomass wastes such as bark and wood chips can be gasified at the same time as black liquor, also providing energy at a greater efficiency than alternative processes. Gasification of black liquor and solid wastes would lend itself to relatively easy capture of CO_2 for sequestration (see Volume 2, Chapter 9, subsection 9.3.7).

A decrease in the heat load of a mill leads to a decrease in the amount of electricity cogenerated at the mill (some of which could have been exported if it is in excess of the mill needs) but increases the amount of excess biomass that can be sold. If the saved biomass can be supplied to a nearby district heating system with biomass gasification combined cycle cogeneration, then the marginal efficiency of electricity generation in the district heating system is likely to be greater than the marginal efficiency of electricity generation at the mill, so the total production of electricity will increase (see Chapter 3, subsection 3.3.3, on the concept of marginal efficiency). As sustainably grown biomass entails near-zero net CO_2 emission, total CO_2 emission is reduced if generation of electricity from fossil fuels elsewhere can be decreased. Ådahl et al (2006) have analysed the economics (in Sweden) of four measures that reduce heat loads in pulp mills (increased use of heat exchangers, use of heat pumps in combination with heat exchangers, and partial or total integration of the evaporation plant with increased heat exchange). They find that the measures that save the greatest amount of biofuel are also the most profitable for a wide range of assumptions concerning the future evolution of the Nordic electricity system.

6.9.5 Integrated production of heat, electricity and dimethyl ether

An alternative use of gasified black liquor is the production of dimethyl ether (DME) to serve as a transportation fuel. Joelsson and Gustavsson (2008) find that, after a pulp and paper mill's internal electricity and heat needs are met, using excess gasified biomass to produce electricity for export is more effective in reducing CO_2 emissions if the displaced external electricity is generated from coal, but that using excess biomass to produce DME is more effective if the external electricity is generated from natural gas. Altogether, a third of Sweden's present diesel fuel use could be displaced in this way. Joelsson and Gustavsson (2008) also compared the biomass requirements for the following cases that yield equivalent CO_2 emission reductions: generation of electricity beyond the mill's needs using surplus gasified mill biomass with standalone production of DME from biomass, or production of DME beyond the mill's fuel needs with standalone production of electricity from biomass. Production of both electricity and DME requires less biomass when integrated into the pulp and paper mill than if both are produced for separately, but less biomass is needed if production of DME is integrated into the pulp and paper mill and electricity is produced separately, then vice versa.

6.9.6 Development of completely closed mills

A Swedish project entitled 'Ecocyclic pulp mill' aims to develop a completely closed pulp mill process, with the near-total elimination of organic and inorganic material flows to water. Although modern mills can now be completely self-sufficient in energy (through the use of biomass wastes to generate heat and electricity, as discussed above), large amounts of low-grade heat are still rejected to the environment. Closure entails recycling liquid flows back through the mill process, rather than rejecting liquid with heat (and pollutants) to the environment. There is a synergy between pollution control and energy savings in that advances in pollution treatment will facilitate increased system closure that in turn makes thermal energy available. For example, membrane techniques are showing promise in separating harmful organic and inorganic substances from process streams. With the reduction of aqueous discharges from $40m^3/t$ of pulp to $10m^3/t$, about 15GJ heat/t would theoretically become available at a high enough temperature for external use (EC, 2001c). However, at the moment there is no technology that could facilitate the use of more than a small fraction of this energy. Another synergy with reduction of pollution and energy savings is related to gasification of black liquor, which has the potential to double the electricity production from biomass wastes. The gasification process would be less complicated and costly if the initial cooking of the biomass could be done without sulphur, and sulphur pollution would be reduced.

6.9.7 Energy savings through recycling of paper

An accounting of the energy savings in recycling paper should account for:

- the net energy required in making paper from virgin fibres, subtracting the energy that can be produced from black liquor (in the case of chemical pulping) and from forestry residues (but adding the energy needed to produce fertilizers to replace the nutrients removed with residues);
- the energy that can be supplied through the incineration of wastepaper if there is no recycling;
- the energy input required during the recycling of wastepaper, including the energy required to collect and transport the wastepaper to the recycling plant;
- the energy that could be supplied from the biomass that is saved due to the recycling of wastepaper.

These issues are discussed at some length in Farahani et al (2004) and Byström and Lönnstedt (1997). Plant fibres can be recycled only about five times, due to damage to the fibre with repeated recycling, so with the recycling option, account would also have to be taken of the portion of the paper flow that is removed from the wastepaper stream to accommodate the addition of virgin fibres. Here, we present a simplified set of calculations – in Table 6.29 – in order to provide insight into these tradeoffs. Key input assumptions are listed at the top of the table, followed by data on the heat and electricity requirements for advanced primary and secondary production processes and data on the heat and electricity that can be produced through gasification of black liquor and bark residues. This is followed by the calculated net energy balance for various cases. The production of paper from virgin fibres in advanced mills has a small heat requirement but provides a substantial electricity surplus that could be exported to the electricity grid. The production of new paper from wastepaper requires a significant net heat and electricity input, as there are no biomass residues that can be used to supply the internal energy needs. The energy required in recycling paper waste is less than the gross energy required in producing paper from virgin fibres, but is greater than the net energy requirement (gross energy use minus biomass energy supply). In effect, production of paper from wood entails a use of biomass for energy purposes comparable to the amount used as a feedstock, whereas recycling of paper waste does not. If de-inking of recycled pulp is required, this adds another 5GJ/t, and if the pulp is processed elsewhere, it must be dried prior to transport, adding another 15GJ/t (Counsell and Allwood, 2007). Altogether, recycling requires an energy input of 20–40GJ/t, whereas production of virgin fibres with use of biomass residues or black liquor waste (equal in mass to the amount of wood converted to pulp) yields a small net energy surplus.

However, recycling 1 tonne of wastepaper saves 2 tonnes of biomass that could be used to generate electricity and heat at high efficiency through biomass integrated gasification combined cycle power generation (see Volume 2, Chapter 4, subsection 4.4.3). This supplies or saves about 52GJ/t, resulting in a net energy saving with recycling of 12–32GJ/t. Conversely, wastepaper can also be gasified and combusted to produce heat and electricity at high efficiency. This supplies or saves 26GJ/t, which can be added to the energy surplus from the production of paper from virgin fibres. Incineration with production of virgin paper to replace the incinerated waste generates an overall saving of about 28GJ/t. Neither transportation energy use nor the production of fertilizers to replace nutrients withdrawn in forest residue have been included in this analysis but these are expected to be small based on the studies reviewed in Volume 2 (Chapter 4, subsection

4.6.3). Also not accounted for in Table 6.29 is the fact that 15–20 per cent of the incoming wastepaper that is recycled ends up as sludge. Its disposal as a landfill poses problems in many communities and will lead to methane emissions, while its incineration requires fossil fuels since it cannot support combustion on its own. It would seem that either virgin paper + incineration with energy recovery, or recycling with use of saved biomass for energy, could supply or save more energy, depending on local circumstances.

6.9.8 Reduced demand for paper

Aggregate energy use in making paper can also be reduced by reducing the demand for paper. Hekkert et al (2002) estimated the potential reduction in paper use in Europe resulting from the following measures:

- reducing the thickness of printing and writing paper from 80gm/m^2 to 70gm/m^2 and reducing the thickness of magazine and book paper by 10 per cent, something that will not noticeably affect the appearance;
- reducing office paper use by 10 per cent through good housekeeping programmes, something already shown to be achievable in The Netherlands;
- increasing the average fraction of duplexing in photocopying from 30 to 60 per cent;
- reducing the printing of materials that are never used through printing-on-demand technology (40 per cent of all commercial printing in The Netherlands is not used at all, as it is outdated before it is distributed, while 30–50 per cent of all published books are returned to the publisher rather than sold in bookstores).

The estimated aggregate effect of these measures is a 37 per cent reduction in paper demand by 2015 compared to the baseline scenario.

As shown in Figure 6.29, per capita paper use in the US is over twice that of Europe, with Canada not far below the US, in spite of a recent downward trend in both the US and Canada. This implies that there is an enormous potential to reduce paper demand in the US and Canada. However, per capita paper use in most of

Table 6.29 *Energy balance for production of paper from wood with landfilling, incineration with energy recovery or recycling of the paper at the end of its life*[a]

Assumptions	
Tonnes of wood per tonne of paper produced	2
Tonnes of black liquor per tonne of paper produced	1
Tonnes of biomass residue per tonne of paper produced	0.2
Biomass saved per tonne of paper recycled (including residues)	2.2
Fraction of virgin fibres in recycled paper	0.2
Heating value of recycled paper	17GJ/t
Heating value of biomass	19GJ/t
Efficiency in generating electricity from wastepaper or biomass	0.5
Efficiency in producing useful heat from wastepaper or biomass	0.3
Efficiency of central electrical powerplant	0.4

Energy balances				
	Heat (GJ/t)	Electricity		Total (GJ/t)[b]
		(kWh/t)	(GJ/t)	
Energy used to make pulp from wood	10.4	588	2.1	15.7
Energy used to make paper from pulp	6.9	760	2.7	13.7
Energy used to make paper from wood (sum of the above)	17.3	1348	4.9	29.4
Energy produced from black liquor + bark	14.8	1797	6.5	31.0
Net energy balance making paper from wood	**2.5**	**−449**	**−1.6**	**−1.5**
Energy used to make pulp from wastepaper	0.4	390	1.4	3.9
Energy used to make pulp from wood (from above)	10.4	588	2.1	15.7
Energy used to make paper from pulp (from above)	6.9	760	2.7	13.7
Energy to make paper from wastepaper[c]	**9.3**	**1190**	**4.3**	**20.0**
Energy from incineration of 1 tonne of wastepaper	5.1		8.5	26.4
Energy from BIGCC-cogen per tonne of forest biomass	5.7		9.5	29.5
Energy from BIGCC-cogen per tonne of secondary paper[d]	10.0		16.7	51.8
Net balance, all virgin paper plus landfilling of waste	2.5		−1.6	−1.5
Net balance, all virgin paper plus incineration of waste	**−2.6**		**−10.1**	**−27.9**
Net balance, recycling + saved forest biomass for energy	–0.7		−12.4	−31.8
Net balance, recycling + partial incineration + forest biomass[e]	**−1.8**		**−14.1**	**−37.1**

Note: [a] The energy requirements for making paper from wood or for wastepaper given in the table pertain to advanced mills and depend on the specific pulping process used and mix of paper products produced, while the energy supplied by gasification of black liquor pertains to technology that could be available by 2015–2020. A negative energy balance means that excess energy is available that could be used elsewhere. [b]This is the primary energy requirement (or savings) assuming that electricity supplied to the paper mill (or displaced by exported electricity) is generated at an efficiency of 40 per cent, and assuming that any heat required beyond that supplied through biomass cogeneration is generated at an efficiency of 85 per cent. [c] Equal to 0.2 × (energy for pulp from wood) + 0.8 × (energy for recycled pulp) + (energy for making paper from pulp), the assumption being that 20 per cent of used fibres must be replaced with fresh fibres. [d] This is the energy supplied from the amount of biomass that is saved per tonne of wastepaper that is recycled. It is equal to (2.2t biomass/t paper recycled) × (1 − fraction of virgin fibres) × (Energy from BIGCC/t biomass). [e]Equal to above line minus (Energy from Incineration) × (Fraction of virgin fibres in recycled paper). This assumes that the amount of used paper that is incinerated for energy equals the amount of virgin fibres added for the next cycle.
Source: Energy requirements are taken from Farahani et al (2004)

the world is five to ten times less than that in the US, so there is an enormous potential for increasing demand – demand that may outstrip the capacity of the world's forests to supply fibres on a sustainable basis even with maximum feasible recycling rates, and that will surely outstrip the sustainable supply without significant recycling.

6.9.9 Paper energy use – synthesis

In summary, the main options for reducing the energy intensity of paper production involve:

1 recovering and using waste heat as much as possible (as in clean steam heat recovery);

2 reducing the energy required for bleaching;
3 reducing the energy used for drying;
4 increasing the use of recycled fibres; and
5 optimizing the co-production of heat, electricity and transportation fuels (with any energy exports serving as an energy credit).

Currently available technologies could reduce the heat and electrical energy requirements in Scandinavian mills by up to one third and 10 per cent, respectively, while heat and electrical energy use in Canadian mills could be reduced on average by 44 per cent and 30 per cent, respectively. Several other technologies are under development that could lead to yet further substantial reductions in energy intensity – perhaps another factor of two reduction.

Other available options do not reduce the energy intensity of the production of paper per se, but involve maximizing the amount of useful energy that can be extracted from the biomass wastes associated with paper production. This entails increasing the overall efficiency of cogeneration. With clean steam heat recovery, the overall steam requirement that needs to be met by cogeneration is reduced (or perhaps eliminated altogether), so that the electricity to steam production ratio will need to be increased. As discussed in Chapter 3, a ratio as large as 1.25:1 can be achieved through combined cycle cogeneration, rather than through a simple back-pressure steam turbine (with a ratio of 0.25:1). However, this requires first gasifying the biomass waste (since the first step in combined cycle cogeneration is a gas turbine). This topic is discussed further in Volume 2, Chapter 4, subsection 4.4.3). If no steam is required from cogeneration, the combined cycle plant could be run to maximize electricity production from biomass waste. The electricity so produced would probably exceed the electricity needed for paper production, and so could be exported from an integrated pulp and paper mill (via the power grid) for use elsewhere. The amount of electricity that could be produced for export at a given mill would depend on the details of the steam and electricity requirements of the mill and on the amount of biomass waste (i.e. whether pulping liquor is produced or not), which in turn depends on the mix of paper products produced at the mill.

If the marginal efficiency of electricity generation using excess biomass residues outside the pulp and paper mill is greater than the marginal efficiency with which additional electricity can be generated at the plant, then the excess biomass should be exported. This will often be the case when there is a biomass power/cogeneration/district heating facility nearby.

6.10 Plastics

The dramatic growth in the annual production of plastics was illustrated in Figure 6.4. Figure 6.32 shows the breakdown of production by various countries and regions in 2007, and the breakdown of end uses of plastic production in Europe (EU27+2, defined in Appendix C) in 2007. Europe and North America each accounted for about 25 per cent of world plastic production in 2007. Packaging accounted of 37 per cent of plastics use in Europe in 2007, followed by building and construction (21 per cent) and automotive uses (8 per cent).

Plastics are produced today by steam cracking hydrocarbons to make a variety of intermediate products that are then combined in various ways to produce a range of different plastics. Steam cracking is the process of reacting steam with hydrocarbons at temperatures as high as 1100°C (typically 750–900°C) so as to produce a variety of bulk chemicals by breaking the carbon-carbon bonds in the hydrocarbon precursors. This is followed by a series of fractionation steps to separate the various products, and involves compression to 15–20atm and cooling to −150°C (Ren et al, 2006). Table 6.30 gives the mix of chemical products produced from the cracking of naphtha (an intermediate product of petroleum refining), natural gas and a mixture of light hydrocarbons. These products consist of methane, olefins (compounds containing at least one carbon-to-carbon double bond) and aromatics (particularly stable ring compounds). The major olefins and aromatics, along with intermediate and final products of methane and the various olefins and aromatics, are shown in Figure 6.33. The two most important products by mass of steam cracking are ethylene and propylene, which are the basic building blocks of most plastics and resins. Most plastics consist of polymers, or long chains consisting of repeating units called monomers. These are the compounds shown in Figure 6.33 beginning with

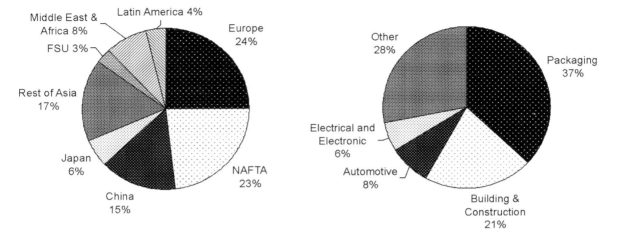

Figure 6.32 *(a) Breakdown of the global production of plastics in 2006, and (b) end uses of plastic in Europe*

the prefix 'poly'. The four most important plastics are polyethylene, polypropylene, polystyrene and polyvinyl chloride (PVC). About half of the ethylene that is

Table 6.30 *Product yields and energy inputs for steam cracking of various hydrocarbon feedstocks*

	Feedstock		
	Naphtha	**Gas oil**	**Light hydrocarbons[a]**
Product yields (kg product/tonne feedstock)			
Ethylene	299	252	568
Propylene	174	157	106
C4 chemicals[b]	106	108	58
Fuel oil	36	150	5
Pyrolysis gasoline and aromatics[c]	237	219	55
Methane	130	101	177
Hydrogen and losses	18	13	31
Process energy input (GJ) per tonne of feedstock			
Fuels	6.0	6.9	7.8
Electricity	0.1	0.1	0.1
Process energy input (GJ) per tonne of ethylene produced			
Fuels	20.1	27.4	13.7
Electricity	0.3	0.4	0.2

Note: [a] Outputs are for a feedstock composition of three sevenths ethane, one-seventh propane and three sevenths butane. Yields vary strongly with the mix of feedstocks used. [b] Includes butane, butadiene and butylenes. [c] Includes benzene, toluene and xylene.
Source: Patel (2003)

produced worldwide is used to make polyethylene. Figure 6.34 shows the relative amounts of different types of plastic produced in Europe. Polyethylene is the dominant type of plastic (at 29 per cent of the total), followed by polypropylene (18 per cent), polyvinyl chloride (12 per cent) and polystyrene (8 per cent).

6.10.1 Energy use in making plastics

The energy used to make plastics occurs as feedstock energy (petroleum or natural gas used as the raw material for the plastic) and as process energy (largely supplying the heat needed to drive the chemical transformation and to cool liquid plastic products after their shaping). Figure 6.35 gives one estimate of the feedstock energy input and the process energy input for a variety of final products as manufactured in Germany in the mid-1990s. The process energy input is computed as the total primary energy input minus the heating value of the plastic, where the primary energy input takes into account losses in providing the hydrocarbon inputs to the plastics plant as well as losses in the generation of electricity used by the plastics plant, and the heating value is the heat released when the plastic is combusted. Also given, where available, is the range of estimates from different studies; there is often a factor of two difference between the lowest and highest estimate of the energy requirement for a given product. Total energy inputs range from 50 to 160GJ/t,

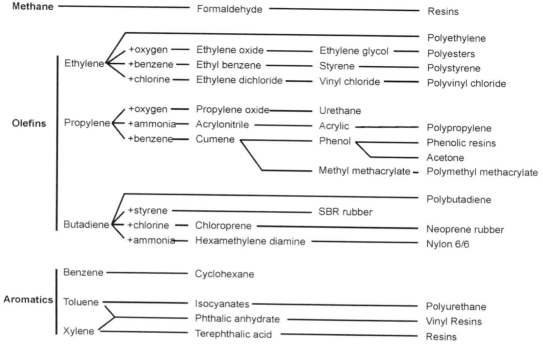

Source: Geiser (2001)

Figure 6.33 *The principle petrochemicals produced from hydrocarbon fuels*

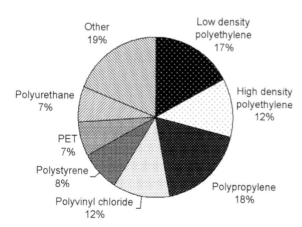

Note: PET = polyethylene terephthalate.

Source: Data from Plastics Europe (2008)

Figure 6.34 *Relative amounts of different types of plastic produced in Europe (EU-27 plus Norway and Switzerland)*

with the feedstock energy accounting for 18–43GJ/t. Potential energy savings apply only to the process energy, as the feedstock energy is embedded in the products. However, the substitution of biomass feedstocks for fossil fuel feedstocks is a means of reducing this component of the fossil fuel energy input.

6.10.2 Reducing process energy requirements

Process energy can be saved through better use of waste heat and more effective cogeneration. Typical crackers use 25–30GJ per tonne of ethylene produced, while state-of-the-art crackers use 20–25GJ/t. Further advances (including high-temperature cogeneration) should reduce this to 17–22GJ/t (IEA, 2006) – a saving of 25 per cent or more. At the other end of the production stream,

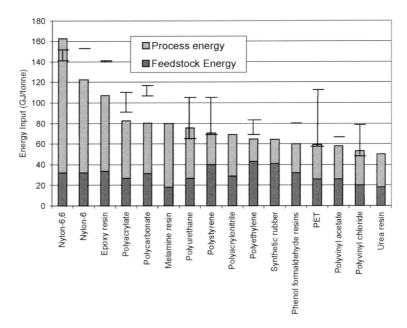

Source: Patel (2003) for bar data; Patel and Mutha (2004) for ranges

Figure 6.35 *Process primary energy and feedstock energy used to produce major plastics in Germany (bars) in the mid-1990s and a range of estimates (where available) for production in Germany and elsewhere*

where plastics are shaped into final products, improved temperature controls can save cooling and/or fan energy use, typically by more than 30 per cent and sometimes by up to 85 per cent, while also improving product quality and/or increasing throughput, as documented by several examples in Rappa and Lawas (2005). Plastics are produced by extruding molten plastic through a die, then cooling it to below its melting point. Plastic sheets use water-cooled rollers, but the water is generally at a temperature (8–14°C) much colder than needed, which reduces the efficiency of the chiller (see Chapter 4, subsection 4.4.4). Indeed, water provided by a cooling tower would be sufficiently cold, eliminating the need for chillers altogether. Sometimes, air-cooled chillers are produced that dump the heat from the hot plastics into the air-conditioned factory space, which must then be removed by the air conditioning system! Direct use of cooling tower water reduces the air conditioning load and the indoor temperatures, creating a more comfortable working environment. Over time, as a

plastics factory expands, the network of pipes in the cooling system can become quite complicated and difficult to understand without detailed modelling and measurements. Rappa and Lawas (2005) document a case where adjustments in the piping system, incorporation of variable speed pumping and chiller controls and increasing the temperature of the water leaving the chillers resulted in a saving in electricity use of almost 50 per cent with a 1.6-year simple payback, while resolving previous problems of inadequate temperature control.

6.10.3 Recycling of plastics

Plastics can be recycled mechanically or chemically. *Mechanical recycling* involves recycling plastic without changing the basic structure of the material. Chemical recycling – also known as *feedstock recycling* – involves breaking the plastic polymers into their constituent monomers or into other chemicals. Inasmuch as plastics are made from fossil fuels that could instead have been used as a source of energy, the combustion of plastics with energy recovery is sometimes referred to as *thermal recycling*. Figure 6.36 shows the portions of plastic waste in the EU-15 (prior to 2005) and in the EU-27 (2005 onward) that were subject to feedstock or mechanical recycling, incinerated with energy recovery or sent to landfill. In 2007, mechanical recycling and incineration accounted for 20 per cent and 28 per cent of the plastic waste generated in Europe, respectively, with Germany having the top recycling rate, at about 35 per cent.

Mechanical recycling requires separating different kinds of plastic into plastic streams that are at least 96 per cent pure. Mixing of different plastics reduces the strength of the resulting product, due to formation of separate immiscible phases having little adhesion. Thus, different types of plastic need to be separated from each other. For intimately intermixed plastics (as in many consumer goods), separation of different plastics requires first shredding the materials to a size such that they can be separated through a combination of electrical and gravimetric means. For

example, polystyrene can be made to acquire a positive electric charge and polyethylene and PVC can be made to acquire a negative charge, and air flow can be used to separate polyethylene from PVC based on their difference in density. The shredding energy requirement depends on the initial and final sizes (the electricity use in kWh/t is given by $W = 10W_b(1/\sqrt{(D_f)} - 1/\sqrt{(D_i)})$, where $W_b = 13.8$ kWh/t, D_f is the final size (typically 0.005m) and D_i is the initial size (typically 0.0026m)). This gives an energy use of 0.76kWh/kg of 6.8GJ primary energy per tonne). The combination of electric and gravimetric separation applied to a polystyrene-ployethylene-PVC mixture requires 0.74kWh of electricity per kg (Dodbiba et al, 2008), or 6.7GJ primary energy per tonne (assuming 40 per cent electricity supply efficiency). Froth flotation has been used in laboratory tests to separate PVC, polycarbonate, polyacetal and PPE (polyphenylene ether) from each other, and to separate PVC and PET (polyethylene terephthalate) from each other (Borchardt, 2006). Infrared spectroscopy has been proposed to identify

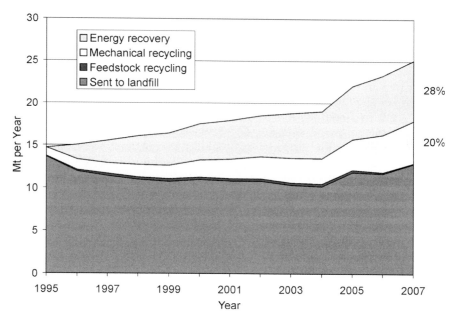

Note: The discontinuity between 2004 and 2005 is due to a change for EU15+2 data to EU27+2 data.
Source: Plastics Europe (2008)

Figure 6.36 *Disposition of plastic wastes in Europe: landfilling, feedstock recycling, mechanical recycling and incineration with energy recovery*

different plastics passing along a conveyor belt, permitting automated sorting. According to Geiser (2001), remelting of plastics for reforming tends to damage the polymer chains, so most plastics cannot be recycled back to the original products. Instead, they are 'downcycled' to less demanding products. For example, plastic lumber can be made from commingled plastic wastes. This is not sustainable in the long run, as it does not form a closed loop.

Recycling of plastics requires only 10–15 per cent of the energy required to refine petroleum and manufacture virgin resins according to Borchardt (2006). Summation of the fuels needed to produce a mixture of primary and secondary plastics (polystyrene, PVC and polyethylene from discarded TVs) as given in Dodbiba et al (2008) indicates that mechanical recycling requires about 43 per cent of the primary energy used in making primary plastics (20.0GJ/t versus 46.4GJ/t). The saving through recycling in this case (26.4GJ/t) is less than the mean heating value of the plastics (33.4GJ/t), but the amount of heat that can be recovered and put to use through combustion would be less than the heating value.

An alternative to downcycling is the high-temperature, high-pressure conversion of mixed plastics back to the more elementary chemical feedstocks (chemical recycling). Steam cracking of commingled plastics can produce higher yields of ethylene (35 per cent versus 28 per cent) and butylenes (12 per cent versus 7 per cent) than thermal cracking of naphtha (the original fossil fuel feedstock), with comparable yields of propylene. According to Boustead (2008), production of ethylene (a monomer) requires about 20GJ/t of process energy, whereas production of polyethylene requires about 30GJ/t, so less process energy is saved with recycling to monomers. However, the entire feedstock energy of about 45GJ/t is saved. The reduced savings in process energy (and possibly greater recycling energy) has to be balanced against the possibly greater fraction of materials that can be recycled to monomers than can be recycled as polymers.

Originally, PET was used almost exclusively to package soft drinks, resulting in a rather homogeneous recycling stream consisting of unpigmented 1- and 2-litre bottles with a small percentage of green bottles. PET is now used for a much wider variety of containers containing various pigments, label adhesives and resins, which serve as impurities and make recycling more difficult. PET is recycled either by breaking it down to its component ethylene glycol and terephthalate monomers using methanol, or by partial breakdown (Borchardt, 2006).

6.10.4 Replacing fossil fuel feedstocks

Another major opportunity to reduce fossil fuel energy use in plastics is through the use of biomass feedstocks instead of petroleum feedstocks in the manufacture of plastics, or to replace plastics with natural fibres wherever feasible. These options are discussed in Volume 2 (Chapter 4, section 4.7). When bioplastics are made from biomass feedstocks, other products are sometimes produced as well. These co-products can also be used in place of materials made from fossil fuels, and sometimes the avoided CO_2 emissions from the co-products alone exceed the CO_2 emission associated with the production of bioplastics and co-products, so the net savings in CO_2 emissions is greater than 100 per cent.

6.11 Petroleum refining

The most effective means to reduce energy use for petroleum refining will be to dramatically reduce the demand for petroleum products, primarily transportation fuels, using the mix of transportation efficiency measures discussed in Chapter 5. In the long run, fossil fuels as an energy source for transportation will have to be completely replaced. In the meantime, energy use associated with the refining of crude petroleum into secondary energy forms (such as gasoline, diesel fuel and jet fuel) should be reduced as much as possible. Worrell (2005) indicates that, in the US, refinery energy use per unit of production can typically be reduced 20 per cent.

However, two factors will tend to drive up the overall energy use associated with the extraction and refining of petroleum during the transition to a fossil fuel-free future: (1) increasingly stringent limits on the sulphur content of gasoline; and (2) the decreasing quality of and ease of extracting the remaining petroleum resources. Reducing the permitted sulphur level from 30ppm to 1ppm increases refinery energy use by about 10 per cent. More significant is the increasing energy cost of extracting and processing the remaining oil. This is illustrated in Table 6.31, which compares the energy required for extraction and refining of conventional onshore US oil, offshore oil, heavy oils and the Canadian tar sands. Processing the least energy-intensive resource listed (conventional onshore oil in the US) requires an energy input equal to 10 per cent of the energy content of the final products, while processing the most-energy intensive resource listed (the Canadian tar sands) requires about 40 per cent of the energy content of the final products according to Table 6.31.

More detailed information on the energy requirements associated with processing of tar sands is given in Table 6.32. The tar sands consist of highly viscous bitumen that can be extracted either through surface mining or through injection of steam into the ground, thereby reducing the viscosity of the bitumen to the point that it can be withdrawn from the ground. The required steam is at present produced by burning natural gas – about $0.25m^3$ per litre of fuel products that are ultimately produced. Surface mining requires substantially less energy than steam extraction, but only about 20 per cent of the tar sands are amenable to surface mining (which, of course, poses other environmental problems). Once extracted, the tar sands require an energy-intensive upgrading process that entails heating the bitumen to 500°C and converting about 75 per cent of it to a liquid called syncrude. CO_2 emissions during upgrading amount to 6–9kgC/GJ syncrude, compared to 3kgC/GJ for conventional oil refining. This is followed by refining of the syncrude to a mix of fuel products. About 1.2 litres of syncrude are required per litre of fuel products

Table 6.31 *Energy required for extracting and refining various forms of oil*

	Energy use (GJ/tonne oil)			Energy use as a % of energy in products
	Extraction	Refining	Total	
US conventional onshore oil	1.3	2.9	4.2	10.0
Canadian conventional onshore oil	2.4	3.0	5.4	12.8
Conventional offshore oil	3.9	3.0	6.9	16.4
Heavy oils	2.9	3.0	5.9	14.1
Canadian tar sands	12.9	3.0	15.9	37.8

Source: (S&T)² Consultants (2005)

Table 6.32 *Energy used during the production of a 50:50 mixture of gasoline and diesel fuel from the Canadian tar sands*

Step	m³ NG per litre of crude	Energy per litre of fuels		Energy (MJ, except Wh for electricity) per MJ of fuels produced			
		MJ NG	Wh electricity	NG	Diesel	Electricity	Primary energy
Alternative bitumen extraction techniques							
In situ	0.190	8.78	91.5	0.250		2.60	0.273
Surface mining	0.045	2.07		0.058			0.058
Other steps							
Upgrading	0.089	4.10		0.117			0.117
Transmission			7.8			0.22	0.002
Refining		0.004	81.6			2.32	0.021
Delivery					0.006		0.008
Total							
In situ route	0.279[a]	12.88	180.9	0.336	0.006	5.14	0.420
Surface mining	0.134[a]	6.17	89.4	0.176	0.006	2.54	0.206

Note: Primary energy requirements have been computed assuming markup factors of 2.5 and 1.25 for electricity and diesel fuel, respectively. The given natural gas requirements include energy required to transmit the gas to the tar sands sites, so no markup factor is applied for natural gas.
[a] Jordaan et al (2009, Figure 6) give almost the same total but a different breakdown between natural gas use for extraction and upgrading
Source: McCulloch et al (2006)

consisting of 50 per cent gasoline and 50 per cent diesel. According to Table 6.32, the total upstream energy requirement is about 42 per cent of the energy content of the delivered fuels if the bitumen is extracted through steam injections and about 21 per cent if extracted through surface mining.

With regard to in situ extraction, upgrading and refining of oil from US oil shales, Brandt (2008) calculates an external energy ratio (i.e. the ratio of the energy in refined products to the sum of all energy inputs outside of the oil shale itself) of 2.4–15.8. This corresponds to an upstream energy input of 6–42 per cent of the energy content of the final products. For surface processing of US oil shales, Brandt (2009) calculates an external energy ratio of 2.6–6.9, corresponding to an upstream energy input of 15–38 per cent.

6.12 Chemical industries

Energy use per unit of output in the chemical industry has fallen dramatically over the last two decades, but the possibilities for further energy intensity reductions are far from exhausted. To drive a chemical reaction, energy must be provided to break existing chemical bonds and form new ones. Existing processes generally use far more energy than is theoretically required; the excess energy is used for such things as heating the reactants to high temperatures so that the desired reaction will occur or energy is wasted exciting the wrong bonds. Two generic areas where substantial further energy savings could occur are: (1) *improved catalysts* that speed up particular reactions and, indirectly, minimize the production of undesired products (the introduction of catalysts reduced the energy required to produce ammonia fertilizer by a factor of three); and (2) new or improved techniques for *separation* and *concentration*, including membrane technology (already penetrating the chlor-alkali industry), use of *supercritical fluids* (fluids at temperature and pressure combinations such that there is no distinction between liquid and gaseous states) and use of *freeze crystallization* (which can give energy savings of 75–90 per cent compared to distillation).

About 40 per cent of all energy used by the chemical industry is used for separation processes, much of this being used for distillation. Even in the most efficient plants, basic operations such as the separation of oxygen from air or the production of petrochemical feedstocks use four to six times the

thermodynamic minimum amount of energy needed. Membrane separation processes, illustrated in Figure 6.37, can greatly reduce the energy used for separating materials. A typical energy saving is between 20–60 per cent (IEA, 2002). However, suitable membranes do not exist at present for many important separations. Membrane reactors (combining chemical conversions and separation) are another area that needs considerable research and development.

In nature, remarkable chemical compounds (such as the shell of an egg) are produced at ambient temperatures. Production of chemicals and compounds

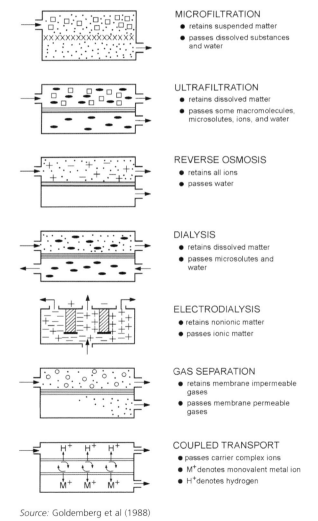

Source: Goldemberg et al (1988)

Figure 6.37 *Types of membrane separation processes and key characteristics*

by humans usually entails very high temperatures, through more of a brute-force approach, resulting in large energy use. There is every reason to believe that, as we learn better how to replicate natural processes, the energy requirements in the chemical manufacturing industry will fall dramatically.

In the manufacture of a chemical compound, the energy input under ideal conditions (with no losses) would exactly equal the heating value of the products minus the heating value of the inputs. If this difference is positive, the reaction is endothermic and energy must be added. If the difference is negative, the reaction is exothermic but, in principle, the heat released could be captured and used somewhere. Greater energy is used in practice due to: (1) non-selectivity of the reactions, (2) heat losses to the surroundings, and (3) inability to recover and use all of the heat released from exothermic reactions. Energy losses due to non-selectivity involve either over-oxidation of the feedstock, or the production of substances that are not wanted. In the first case, energy becomes available as process heat and could be recovered or used in the process itself or supplied to other processes, while in the latter case the unwanted materials could be burned for their energy content. With the exception of steam cracking to produce olefins and chlorine production, most processes in the chemical industry are exothermic. Neelis et al (2007) estimate that the total worldwide exothermic heat release is 1191PJ, which is equal to 60 per cent of the process energy input used in the manufacture of products produced by exothermic reactions. This is the upper limit to the amount of process energy that could be saved by capturing the heat released by exothermic reactions.

6.13 Desalination of seawater

The worldwide desalination capacity in 2008 stood at around 52 million m³ per day, or 18,880Mt/year. The need for desalination of salt water is likely to grow substantially during the coming decades, both due to high population growth rates in many arid coastal regions of the world (as indicated by the map of fertility rates in Chapter 11, Figure 11.1) and due to global warming. García-Rodríquez (2003), Tonner and Tonner (2004) and Kalogirou (2005) provide recent comprehensive reviews of seawater desalination. Here, we outline the different processes for desalinated

seawater, their past and current energy requirements and the prospects for further reductions in the required energy. In Volume 2 (Chapters 2 and 3), we examine issues and costs related to using solar and wind energy as the energy source for desalination.

Seawater can be desalinated through a variety of processes, which can be classified as either thermal (phase-change) or mechanical (single-phase) processes. Among the thermal processes are multi-stage flash (MSF) and multi-effect boiling (MEB) distillation. Both consist of a series of stages to evaporate water at successively smaller temperatures and pressures, and are driven by steam – at about 100°C for MSF and about 70°C for MEB.

Mechanical desalination uses membranes. The most common mechanical distillation process is reverse osmosis (RO), in which a pump increases the pressure of the seawater to about 70atm on the saline side of the membrane. Water but not salt passes through the membrane, leaving behind salt-enriched brine. Impulse turbines or Pelton wheels (see Volume 2, Figure 6.2c) can be installed in the brine stream to recapture some of the energy content of the high-pressure brine and used to assist in pressurizing the incoming seawater, thereby reducing the overall energy requirement. Another mechanical method is electrodialysis (ED), which uses electricity to ionize water. The dominant processes for desalination are MSF and RO, accounting for 44 per cent and 42 per cent, respectively, of worldwide capacity.

Table 6.33 gives the thermal and electrical energy requirements for the major desalination processes. The minimum theoretical energy requirement for desalination of seawater is 0.65kWh/t at 0°C and 0.90kWh/t at 100°C (Tonner and Tonner, 2004). MSF requires about 300MJ/m³ thermal energy, or about 470MJ/m³ primary energy if heat is generated by a boiler at 80 per cent efficiency and we apply a markup factor of 1.2 to the heating fuel. In contrast, the current typical energy requirement with RO is 2–3 kWh/m³, which translates into a primary energy requirement of 27–36MJ/m³ if the electricity is generated at 40 per cent efficiency. The best energy use achieved with RO is 1.58kWh/m³.

If we approximate the world's desalination as 50 per cent MSF at 470MJ/m³ and 50 per cent RO at 32MJ/m³, the current mean primary energy intensity for desalination is 251MJ/m³. However, MSF also

Table 6.33 *Seawater desalination processes and energy use*

Process	Energy use			
	Thermal (MJ/m³)	Electrical (kWh/m³)	Primary (MJ/m³)	Investment (€/(m³/day))
Theoretical minimum				
At 0°C		0.65	5.9	
At 100°C		0.90	8.1	
Thermal or phase-change processes				
MSF	294	2.5–4.0	~470	950–1900
MEB	123	2.2	~200	900–1700
Vapour compression		8–16	~70–140	1500–2500
Membrane or single-phase processes				
RO				
– early practice		10–15	90–135	900–2500 for
– current typical practice		3–4	27–36	membranes (replaced
– best		1.58	14	every 4–5 years)
ED		12	~110	

Note: Primary energy requirements were computed assuming heat generation at 80 per cent efficiency and electricity and fuel markup factors of 2.5 and 1.2, respectively.
Source: Kalogirou (2005) for thermal process and ED energy use and all investment costs; Service (2006) and von Medeazza and Moreau (2007) for RO energy use; Tonner and Tonner (2004) for the theoretical minimum energy requirement.

requires an amount of electricity for auxiliary equipment (such as pumps) comparable to that of RO (McGinnis and Elimelech, 2008), which brings the total primary energy requirement to about 265MJ/m³. With an eventual replacement of the world's entire desalination capacity with the current state-of-the-art RO technology, the primary energy intensity would be 14.2MJ/m³ at 40 per cent electricity generation efficiency (a factor of 19 reduction) and 9.5MJ/m³ at 60 per cent electricity generation efficiency (a factor of 28 reduction).

The steam temperatures required for MSF (100°C) and MEB (70°C) are such that, were the steam taken from a steam turbine as a cogeneration scheme, there would be a penalty in terms of reduced electricity production that would add to the direct electricity requirements of these methods (see Chapter 3, subsection 3.3.1). However, forward osmosis (FO) is an alternative process under development that could use 40°C heat while requiring an order of magnitude less electricity then RO. Extraction of heat from a steam turbine at this temperature would not reduce the electricity production, so the effective energy use would be an order of magnitude less than that of RO.

6.14 Cogeneration and heat management

There is an enormous potential to utilize high-temperature exhaust streams that are currently vented to the atmosphere, to utilize high pressures in natural gas transmission systems, to provide more efficient cogeneration of heat and electricity than where practised at present, to implement more widespread cogeneration and to better integrate heating and cooling requirements within industrial processes. These options are briefly discussed here.

6.14.1 Capturing the energy in hot exhaust flows and pressurized gases

Casten and Ayres (2007) discuss the potential for generating electricity from hot exhaust gases produced by high-temperature industrial processes. Hot (≥300°C) exhaust gases from coke ovens, glass furnaces, silicon production, refineries, petrochemical processes and many processes in the metals industry are available that could be used to produce steam to generate electricity with a turbine. About 10GW of electricity are generated in this way in the US, but the technical potential is estimated to be 95GW, 64GW of which should be achievable in practice. As it would run continuously, this capacity would supply 13 per cent of the 2005 US electricity supply without any increase in the use of fossil fuels. Capital costs range from $300/kW for large (160MW) back-pressure turbines to $1800/kW for small (40kW) steam turbines, but there are no fuel costs. One 95MW powerplant producing electricity and extra steam from hot exhaust at a steel mill near Chicago generated as much electricity with no incremental fossil fuel use as *all* of the grid-connected PV panels in the world in 2004. The world's silicon smelters produce enough hot exhaust gas to generate 6.5 billion kWh/yr of electricity, which is comparable to the worldwide generation of electricity from solar power in 2006 (which can be estimated as 6000MW peak capacity × 8760 hours/year × a likely average capacity factor of about 0.16, or 8.4 billion kWh).

Natural gas is transmitted under pressure (at an energy cost of 2.5 per cent/1000km), but the pressure is reduced at each city gate with valves. The potential energy of the pressure drop is wasted, but could be used to provide 6500MW of electrical power in the US (0.65 per cent of total capacity) using back-pressure turbines according to Casten and Ayres (2007).

6.14.2 Improving the efficiency of existing industrial cogeneration

We have already discussed cogeneration (the co-production of electricity and useful heat) in the production of iron and steel, cement and pulp and paper. Petroleum refining also lends itself to cogeneration. However, existing industrial cogeneration systems are not particularly efficient, both in terms of electricity efficiency and in terms of overall efficiency (based on the combined electricity and useful heat output).

Figure 6.38 shows the flows of energy for cogeneration involving various combinations of boilers, gas turbines and/or steam turbines, and heat recovery devices. Ozalp and Hyman (2006) used US national data to deduce average efficiencies of this equipment

when used for cogeneration in the pulp and paper, petrochemical and chemical industries. Results are given in Table 6.34. Efficiencies in generating electricity are 22–35 per cent using ICEs, 24–29 per cent using gas turbines and only 9–13 per cent using steam turbines. Even the fraction of fuel input converted to useful heat is rather small, ranging from 16–21 per cent for ICEs, 14–37 per cent for gas turbines and 57–65 per cent for steam turbines. The overall efficiencies (electricity + useful heat output divided by the total fuel input), as calculated here using the component efficiencies shown in Figure 6.38, are also given in Table 6.34. For combined cycle power generation, the overall efficiencies depend on the proportion of fuel input that is directly supplied to the gas turbine (which has a relatively high efficiency) versus to the boiler and thence to the steam turbine (which has a relatively low efficiency). Overall efficiencies are only 61–68 per cent for steam turbine cogeneration and 58–59 per cent for combined cycle cogeneration with no use of a supplemental boiler (efficiencies are smaller otherwise). These are rather low efficiencies. In the long run, molten carbonate and solid oxide fuel cells (which operate at 650–1000°C) may offer opportunities for cogeneration in industries with heat requirements in this temperature range (see Chapter 3, section 3.4). Hydrogen produced as a byproduct of the manufacture of chlorine can already be used in low-temperature (120°C) proton-exchange membrane fuel cells.

6.14.3 Integrating heating and cooling loads

Many industrial processes have simultaneous heating and cooling requirements (for example, heating ingredients for processed foods, then cooling the product at the end of the processing). To the extent that hot process streams can transfer heat to cold process streams when the hot streams need to be cooled and the cold streams heated, both heating and cooling energy requirements can be reduced. *Pinch analysis* is a technique for determining the arrangement of heat exchangers that will minimize heating + cooling energy requirements (Trivedi, 1997). It entails plotting the temperature of the hot and cold process streams against their heat content (relative to a common reference state), as illustrated in Figure 6.39. The region of overlap (on

a)

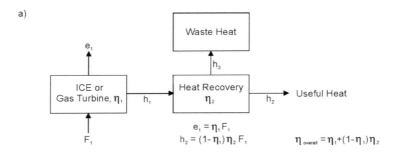

$e_1 = \eta_1 F_1$

$h_2 = (1-\eta_1)\eta_2 F_1$

$\eta_{overall} = \eta_1 + (1-\eta_1)\eta_2$

b)

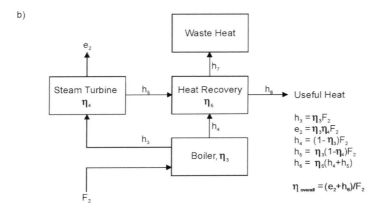

$h_3 = \eta_3 F_2$

$e_2 = \eta_3 \eta_4 F_2$

$h_4 = (1-\eta_3)F_2$

$h_5 = \eta_3(1-\eta_4)F_2$

$h_6 = \eta_5(h_4 + h_5)$

$\eta_{overall} = (e_2 + h_6)/F_2$

c)

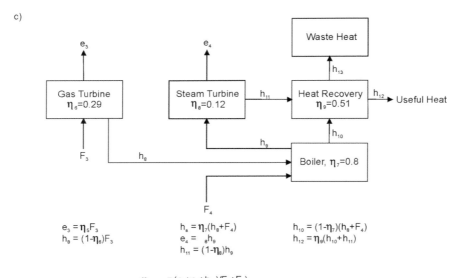

$e_3 = \eta_5 F_3$

$h_8 = (1-\eta_6)F_3$

$h_8 = \eta_7(h_8 + F_4)$

$e_4 = {}_8 h_9$

$h_{11} = (1-\eta_8)h_9$

$h_{10} = (1-\eta_7)(h_8 + F_4)$

$h_{12} = \eta_9(h_{10} + h_{11})$

$\eta_{overall} = (e_3 + e_4 + h_{12})/F_3 + F_4)$

Source: Based on Ozalp and Hyman (2006)

Figure 6.38 *Energy flows in industrial cogeneration systems: (a) using ICEs or simple cycle gas turbines, (b) using steam turbines and (c) with combined cycle cogeneration*

Table 6.34 *Efficiencies of ICE, gas turbines, steam turbines and heat recovery devices in industrial cogeneration systems in the US*

Energy conversion step		Industry		
		Pulp and paper	Petroleum refining	Chemical
ICE or simple cycle gas turbine cogeneration (Figure 6.38a)				
ICE	η_1	0.260	0.220	0.350
Heat recovery	η_2	0.210	0.000	0.160
Overall efficiency		0.415	0.220	0.454
Gas turbine	η_1'	0.290	0.240	0.270
Heat recovery	η_2'	0.140	0.330	0.370
Overall efficiency		0.389	0.491	0.540
Steam turbine cogeneration (Figure 6.38b)				
Boiler	η_3	0.800	0.800	0.800
Steam turbine	η_4	0.120	0.090	0.130
Heat recovery	η_5	0.650	0.580	0.570
Overall efficiency		0.684	0.610	0.615
Combined cycle cogeneration (Figure 6.38c)				
Gas turbine	$\eta_6 (=\eta_1')$	0.290	0.240	0.270
Boiler	$\eta_7 (=\eta_3)$	0.800	0.800	0.800
Steam turbine	$\eta_8 (=\eta_4)$	0.120	0.090	0.130
Heat recovery	η_9	0.510	0.640	0.360
Overall efficiencies when F_4:F_3=0				
Overall electrical efficiency		0.358	0.295	0.346
Overall thermal efficiency		0.327	0.451	0.235
Overall efficiency		0.685	0.746	0.581
Overall efficiencies when F_4:F_3=1				
Overall electrical efficiency		0.227	0.183	0.225
Overall thermal efficiency		0.394	0.523	0.279
Overall efficiency		0.621	0.706	0.504
Overall efficiencies when F_4:F_3=2				
Overall electrical efficiency		0.183	0.146	0.185
Overall thermal efficiency		0.416	0.546	0.294
Overall efficiency		0.600	0.693	0.478

Note: The boiler efficiency of 0.8 is an assumed value. Overall efficiencies have been computed here from the component efficiencies (and are not the same as the prime mover efficiencies given in Ozalp and Hyman, 2006). For combined cycle cogeneration, the overall efficiencies depend on the ratio of fuel supplied to the boiler (F_4) and to the gas turbine (F_3).
Source: Ozalp and Hyman (2006)

the heat-content axis) indicates the maximum possible extent of heat recovery. The point where the two curves come closest together is called the pinch point. The principles used in a designing a thermal system are:

- heating should only take place above the pinch temperature;
- cooling should only take place below the pinch temperature;
- no heat should be allowed to flow from above the pinch point to below the pinch point.

Ideally, heating is required only for the region where the cold stream extends beyond the hot stream, and cooling is required only for the region where the hot stream extends beyond the cold stream (as indicated in Figure 6.39). Through the application of pinch analysis, heating and cooling energy savings of 50 per cent or more can sometimes be achieved.

Many industrial processes have heating demands with temperatures of 100–120°C, while waste heat is typically available at 30–50°C. Heat pumps would be ideal for upgrading waste heat to the required

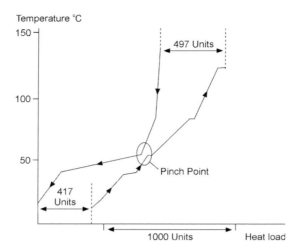

Source: Mercer (1991)

Figure 6.39 *Illustration of temperature versus heat content for the hot (upper) and cold (bottom) process flows in an industrial process*

100–120°C, but heat pumps normally cannot supply heat in this temperature range (purely vapour compression heat pumps using ammonia can achieve a maximum condenser temperature of 78°C). However, a hybrid absorption/compression heat pump recently developed in Norway can upgrade heat from 50°C to 110–115°C with a COP[22] of over 3.5 (Grandum and Horntvedt, 2000), while vapour compression heat pumps using water as a working fluid typically achieve temperatures of 80–150°C (but 300°C was achieved in a test plant in Japan) (HPC, 2000). Heat pumps with water as a working fluid require large and expensive compressors due to the low volumetric heat capacity of water compared to conventional working fluids. However, such heat pumps are another option for low-temperature industrial applications.

6.15 Motors and motor-drive systems

In the mid-1990s, it was estimated that electric motors used 58–68 per cent of all the electricity generated in the US (de Almeida and Greenberg, 1997). Motors are found in homes in the compressors of refrigerators and air conditioners and in furnace blowers; in commercial buildings in air conditioning and refrigeration equipment and in ventilation fans; in municipal water works; and in industry in pumps, compressors, blowers and fans and machine tools (see Table 6.35).

Motors in buildings and in industrial facilities are often part of a much larger system, involving pipes, ducts, fans or pumps and even the electrical transmission grid itself.[23] The energy required in a pumping or ventilation system depends on (1) the *efficiency* of the motor and fans or pumps, and of the coupling between the two, and (2) the *load* that the system has to supply. The load is the amount of power that needs to be supplied to the air or fluid in order to keep it moving against the force of friction. This depends on the pipe system: straighter, fatter and smoother pipes will entail less frictional losses. For a given flow, the frictional losses along straight segments decrease with increasing pipe diameter to the *fifth power*, so a mere 20 per cent increase in the diameter of the pipes will reduce the frictional loss along straight segments by a factor of $(1.2)^5 = 2.5$. The load also depends on the flow rate: for a given piping system, the required power varies with the flow rate to the *third power*. Thus, if and when the required flow can be cut in half, the power that needs to be supplied to the fluid will fall by a factor of eight. As discussed in Chapter 4 (subsection 4.6.3), measures such as displacement ventilation can reduce the required airflow in buildings and hence the work required by a motor by a factor of

Table 6.35 *Proportion of total electricity use that is due to motors in various end-use applications, and the potential savings in motor electricity use through the use of variable speed drives*

Sector	End use	Electricity use as a % of US total	Potential savings with VSDs as a % of each end use
Residential	Refrigeration	6.9	20
	Space heating	1.4	25
	Air conditioning	3.8	20
	Other	0.9	0
	Total	13.1	19
Commercial	Air conditioning	7.5	15
	Ventilation	2.2	25
	Refrigeration	1.3	25
	Other	0.1	20
	Total	11.3	18
Industrial	Pumps	7.7	25
	Fans	4.5	25
	Compressors	3.9	25
	Machine tools	2.1	10
	DC drives	2.5	15
	HVAC	0.6	15
	Other	4.9	0
	Total	25.4	18
Public and miscellaneous	Water works	1.7	15
	Electric utilities	5.4	25
	Other	0.2	0
	Total	7.3	22
Total		57.1	19

Source: de Almeida and Greenberg (1997)

eight if the ducts are not downsized, although a slight decrease in the device (motor) efficiency at lower flow occurs, such that system energy use is reduced by a factor of 6–7 if ducts or pipes are not downsized. Here, we discuss potential savings in motor system efficiencies in the industrial sector.

If P_{fluid} is the power that needs to be supplied to the fluid (that is, the load), then, as discussed in Chapter 4 (subsection 4.1.2), the required electrical power is:

$$P_{electric} = \frac{P_{fluid}}{\eta_m \eta_p} \qquad (6.29)$$

where η_m and η_p are the motor and pump efficiencies, respectively. Motor and pump efficiencies are largest at

full load, and decrease with decreasing load. This decrease is greater in motors or pumps that have smaller full-load efficiencies to begin with so, at part load, the relative difference between efficient and inefficient pumps or motors is greater than at full load. The combined motor and pump efficiency might fall from 80 per cent at full load to 50 per cent at 30 per cent of full load for conventional devices, and from 90 per cent at full load to 70 per cent at 30 per cent of full load in high-efficiency devices. However, P_{fluid} would decrease to only $(0.3)^3 = 0.027$ of the full-load power for flow at 30 per cent of the peak flow. Thus, in spite of the significant drop in efficiency at part load, the required electrical power still drops dramatically because P_{fluid} drops with the flow rate to the third power. This is illustrated in Figure 4.6 of Chapter 4.

This, however, assumes that the pump is only attempting to provide P_{fluid}. Conventional pumps operate at a fixed speed, irrespective of the load. Throttling is used to reduce the flow when required; that is, the flow is partially obstructed. This wastes a lot of energy, so there is very little reduction in the energy use by the pump as the flow rate decreases (this is also illustrated in Figure 4.6).

The components in motor systems are systematically oversized, since it is difficult to predict system flow rates and friction in advance, and the cost (in terms of lost productivity) of a system failure is greater than the extra cost of an oversized system. However, when each successive component in a system is sized to handle the load presented by the previous component plus a safety margin, the oversizing of the motor can become very large. With an oversized motor, greater throttling is required at all times than if the motor were not oversized, thereby compounding the energy wasted through throttling.

To change the speed of a motor requires a VSD, which converts the input AC electricity from 50 or 60Hz to some other frequency, thereby altering the motor rotation rate.

Examples of energy savings obtained with VSDs in pilot projects in industrial facilities are:

- a 50 per cent reduction in electricity used to produce compressed air for use in a glass wool manufacturing process (Maaløe and Johansson, 1995);
- a 35 per cent saving in fan energy in an aluminium works;

- a 45 per cent saving of exhaust fan energy in a sludge treatment plant;
- a 25–30 per cent water pumping energy saving in a brewery; and
- a 41 per cent pump energy saving in a whey plant (Dutch National Team, 1995).

VSDs can also provide significant capital and maintenance cost savings (Baldwin, 1989). Given the greater operational flexibility of variable speed operation, it might be possible to alter design rules that currently lead to extreme oversizing, thereby reducing capital costs (as well as reducing power factor losses, since the motor would operate closer to peak power if not oversized as much). VSDs permit slow, controlled starts, thereby reducing electrical stresses on motors, transformers and switchgear, and reducing mechanical stress on motors, gears and related equipment. Equipment lifetimes would also be increased by avoiding the constant back-pressure of throttled systems, or the intermittent bearing lubrication of on–off systems (on–off cycling is another way to achieve part-load operation).

Even in systems with constant flow, VSDs might be able to provide savings. The past 100 years have been spent optimizing pumps, fans and compressors to run with a fixed motor speed of either 3600rpm (in North America) or 3000rpm (elsewhere). However, for a given flow rate and pressure head, pump efficiency increases with the shaft speed (from 53 per cent at 3600rpm, to 66 per cent at 7200rpm and 70 per cent at 10,800rpm in one illustrative case discussed by Baldwin, 1989). VSDs may therefore allow re-optimization of pumps and fans for greater efficiency.

Table 6.36 shows the net effect of a variety of motor system improvements on the overall efficiency of a hypothetical industrial pumping system. In the standard system, the overall efficiency in converting fossil fuel energy at a powerplant into fluid energy in a piping system is 5.4 per cent. In the efficient system, the overall efficiency is 14.7 per cent – almost *three times* greater. This is achieved largely through avoiding throttling and by 25 per cent larger pipes (which reduce pipe friction by 59 per cent if the flow is non-turbulent), with substantial additional contributions from more efficient motors and pumps.

A more extreme example, discussed by Amory Lovins, is the design of a carpet factory in Shanghai. The manufacture of carpet involves melting of bitumen in a hot-oil pumping loop. A Dutch engineer, designing the plant to minimize total lifecycle costs, found that fatter pipes laid out as straight as possible reduced the

Table 6.36 *Energy flow in a hypothetical motor-pump system, where the average flow rate is 80 per cent of the peak flow*

Step	Standard system		Efficient system	
	Efficiency	Remaining energy	Efficiency	Remaining energy
Input to powerplant		100.0		100.0
Powerplant	40	40.0	40	40.0
Transmission	92	36.8	94	37.6
Transformer	96	35.3	96	36.1
VSD	–	35.3	95	34.3
Motor	88	31.1	92	31.5
Shaft coupling	98	30.5	100	31.5
Pump	77	23.5	83	26.2
Pump throttle	66	15.5	–	26.2
Piping system	35	5.4	56	14.7

Note: In the standard system, throttling is used to give reduced flow, while in the efficient system, lower motor power output is achieved through a VSD. The efficient system also has a more efficient motor, pump and motor-pump shaft coupling; a higher power factor, which reduces losses in transmitting electricity from the powerplant; and 25 per cent larger pipes in order to reduce pipe friction. In real systems, the energy losses in the standard system (and the energy savings in the efficient system) could be much larger or much less, depending on the variability of the flow and the details of the piping system
Source: Baldwin (1989), except that the powerplant efficiency has been assumed to be 40 per cent rather than 30 per cent, to reflect more modern conditions.

expected pumping energy by 92 per cent compared to the conventional design (Anonymous, 1997).

Compressed air systems are a commonly found motor-driven system used in manufacturing. The September 1999 issue of the *CADDET Energy Efficiency Newsletter* contains many case studies of the energy savings that can be achieved through reconfiguration and upgrading of existing systems. Savings of 30 per cent are the rule rather than the exception, with savings of 50–60 per cent in some cases and savings of 10 per cent sometimes from nothing more than repairing leaks.

6.16 Summary

Industrial energy use accounts for about one third of global energy use and energy-related CO_2 emissions, with the industrial share smaller (25–30 per cent) in regions (such as North America and Europe) that industrialized first, and larger (40–50 per cent) in regions that are rapidly industrializing now (such as Asia). The major energy-using industries in the world are the iron and steel, aluminium, copper, cement, glass, pulp and paper, fertilizer, and chemical and petrochemical industries. The present-day global average energy intensity for the production of these industrial products and future energy intensities (probably achievable long before 2100) are summarized in Table 6.37.

Iron, aluminium, copper and zinc occur in nature as oxides (and also as sulphides in the case of copper and zinc), and the first step in transforming metal-containing minerals is to drive off the oxygen (and sulphur, if present). This can be done in a variety of ways: in a blast furnace at 2000°C (95 per cent of world production) or through direct reduction at 900–1000°C (5 per cent of world production) in the case of iron; refining at 1100°C followed by electrolysis at 900°C in the case of aluminium; or by smelting and fire refining at temperatures of 1200–1250°C in the case of the pyrometallurgical processing of copper ores. These high temperatures are largely supplied by combusting coal or, in some cases, solid wastes and coal. The majority (80 per cent) of the energy used in the transformation of aluminium ores into aluminium, however, is electricity used for electrolysis. The hydrometallurgical processing of copper ores also uses significant amounts of electricity for electrolysis, rather than heat. High temperatures are required in the

transformation of limestone and other ingredients into cement, in the transformation of sand and other ingredients into glass, and in most chemical industries so as to drive the required chemical reactions. Large amounts of heat are used in the chemical production of pulp from wood and in drying pulp to make paper, whereas large amounts of electricity are used in the mechanical production of pulp, but significant amounts of waste heat are generated that can be used in the subsequent paper-making step if pulp and paper production are combined at a single facility.

The opportunities for reducing the energy intensity of iron and steel involve alternative methods for reducing iron ores that require lower temperatures, a shift from the use of coke to the direct use of coal for reducing iron ore, alternative casting techniques that reduce the number of heating and cooling steps required, and greater recovery and use of waste heat. Hydrogen produced from renewable energy could be used as an alternative reducing agent in place of coke or coal, eliminating the chemical production of CO_2. In the cement industry, a shift from wet to dry kilns and the use of shorter and better-integrated kilns for the production of clinker, the use of natural alternatives to clinker (pozzolans), and possibly the development of alternative cements promise significant reductions in energy use. Energy use in the production of chemicals can be reduced through the use of catalysts and membranes (for separation processes) and through better control of reactions. The production of many industrial commodities involves significant uses of electricity for pumps and motors, and in this case, energy savings are possible through more efficient pumps and motors. In some cases, significant savings are possible through redesign of entire systems to greatly reduce the load placed on the pumps or motors (through better layout of pipes and through thicker pipes) and through the use of VSDs.

In many industries there are significant opportunities to save energy through cogeneration of heat and electricity. This is especially true for the pulp and paper industry which, with the application of advanced concepts such as biomass gasification combined cycle, could be self-sufficient in heat and a net exporter of electricity. Other forms of cogeneration are possible too: the cogeneration or integrated production of steel and electricity in the steel industry (through the gasification of coal primarily for the

Table 6.37 *Summary of potential energy-intensity reductions for key industrial products*

| | Production of primary materials | | | | | | Production of secondary (recycled) materials | | | | | | Intensity factors[a] | |
| | Present | | | Future | | | Present | | | Future | | | | |
	Fuels (GJ/t)	Electricity (kWh/t)	PE (GJ/t)	Fuels (GJ/t)	Electricity (kWh/t)	PE (GJ/t)	Fuels (GJ/t)	Electricity (kWh/t)	PE (GJ/t)	Fuels (GJ/t)	Electricity (kWh/t)	PE (GJ/t)	Fuels	Electricity
Iron and steel	30	450	36	12	115	13	1.2	650	7.1	1.2	395	3.6	0.107	0.713
Aluminium	43	15,820	193	14	11,430	85	11.2	380	17	8.4	285	12	0.244	0.108
Copper	34	5960	94	39	7490	92	15	4170	53	10	2780	29	0.409	0.567
Zinc	9.3	4300	48	8.1	4415	36	2.5	1100	12	2	880	7.7	0.392	0.420
Stainless steel	34	5010	83	34	5010	68	9.1	1740	26	9.1	1740	21	0.482	0.550
Cement	5	80	6.0	2.5	64	3.0							0.5	0.8
Flat glass	12	650	21.2				9.6	650	18[b]				0.5	0.5
Container glass	9.6	1470	25.9				8.1	1470	22[b]				0.5	0.5
Fibrous glass	2.6	5790	58										0.5	0.5
Pulp and paper	7–12	610–680	26	2.5	−450	−0.2	9.3	1200	20	9.3	1200	17	0.34	0.34
Plastics			50–160[c]						10–30					
Fresh water, MSF[d]	350	2.5–4.0	470				0.7–2.9	0.2–0.8	0.7–2.9					
Fresh water, RO[e]	0	3.5	0.032	0	1.2	0.007	0	< 0	< 0				0.34	0.34

Note: Primary energy (PE) use is given as onsite fuel energy use plus electricity use assuming an average efficiency for electricity generation of 40 per cent at present and 60 per cent in the future. MSF = multi-stage flash, RO = reverse osmosis. [a] Assuming 90 per cent recycled content of metals, 60 per cent recycled content of glass. [b] Based on an assumed 15% savings compared to present. [c] 20–40GJ/t as feedstock energy, 30–120 GJ/t as process energy. [d] Energy use given for recycled water is the energy use of a conventional water treatment plant with heating and combustion of sewage sludge. [e] Energy use given for recycled water assumes anaerobic digestion of sewage sludge to produce methane that is combusted to produce heat and electricity in excess of the water treatment plant energy needs. The intensity factor is for future versus present reverse osmosis.

chemical reduction of the iron ore but with use of unreacted gases for combined cycle power generation in an integrated system), and the cogeneration or integrated production of cement and electricity in the cement industry (by combusting the coal for power production but with the integrated production of belite from the fly ash and bottom ash with a minimal reduction in power output and no chemical CO_2 emissions).

Other efficiency gains are possible through recovery and use of heat that would otherwise be dissipated to the surroundings. This requires the use of heat exchangers and the cascading of waste heat from a high-temperature process to processes that can use successively lower temperature heat. The opportunities for using heat that would otherwise be wasted can be increased further by locating industries that can use low-temperature heat next to industries (such as the metal, cement and petrochemical industries) that produce high-temperature waste heat.

Significant energy intensity reductions are possible through recycling of primary materials (particularly for iron and steel, aluminium, copper and plastics). Recycling of primary metals eliminates the energy required in mining and concentrating the metal (which is a significant fraction of total processing energy use in the case of copper) and in reducing (removing oxygen from) the metal, and eliminates the chemical production of CO_2. The combination of advances in the production of primary and secondary steel and aluminium, and 90 per cent recycling, would reduce the average energy intensity of steel by almost a factor of four to six and that of aluminium by a factor of five to seven. Production of pulp from discarded paper products requires only a third to a quarter the energy required to produce pulp from raw wood, although this saving is partly offset by the energy needed for de-inking (if carried out) and for treating the sludge waste producing by recycling paper products. However, the picture with regard to recycling of paper products is not straightforward, as advanced production of paper from virgin fibres with full use of wood residues and biomass wastes is a small net source of energy, while recycling of paper is a net consumer of energy. If paper waste is gasified and used to produce heat and electricity through combined cycle power generation, instead of recycled, significant additional energy savings would accrue. Conversely, recycling saves large amounts of wood that could be used in the efficient production of electricity (through integrated biomass gasification and combined cycle power generation), such that either continuous efficient production of fresh paper products followed by efficient gasification of wastepaper with energy recovery, or recycling with efficient use of the saved wood, could save or supply a greater amount of energy. In the case of plastics, recycling consistently saves more energy than incineration with energy recovery, but there are limits at present in the extent to which plastics can be recycled (instead of downcycled). Calculation of the energy benefits of recycling should take into account the energy required to collect and separate recyclable materials, as well as consider alternative waste management options. Such analyses are presented in Chapter 8 (section 8.3). In the long run, plastics comparable to those available today could be produced from biomass feedstocks.

Notes

1   In computing primary energy from electricity and fuel energy, markup factors of 2.5, 1.05, 1.2 and 1.25 were applied to electricity, coal, oil and natural gas, respectively, based on Chapter 2, subsection 2.3.5.
2   Excluded from this sum are the energy used to produce stainless steel and N fertilizer, as the former is largely redundant with the portion of the chromium and steel energy uses given in Table 6.1 (these being two major inputs to stainless steel), while the N fertilizer energy use is largely redundant with a portion of the energy used to make ammonia (ammonia being the major input to the manufacture of N fertilizer).
3   In economic jargon, the coal powerplant emissions that could be but are not displaced represent the opportunity cost of producing the aluminium.
4   IEA (2006) gives an average 'final' energy use of 18.8GJ/t based on 2003 data.
5   Crude iron that is allowed to solidify is called *pig iron*, but now most crude iron is transported while still molten to the steel furnace.
6   'Fluxes' are materials that have a lower melting point than the primary materials and that, by melting to a liquid sooner, facilitate heat transfer throughout the mixture.
7   EAFs also use a small amount of oxygen, increasing from $33Nm^3$/tonne in old Japanese EAFs to $45Nm^3$/tonne in new EAFs, but the extra $12Nm^3$ of oxygen requires only 6kWh through cryogenic separation.

8   De Beer et al (1998a) compute primary energy as the onsite fuel energy use plus electricity use times a markup factor of 2.5, that is, with no markup factor to account for the losses in extracting and delivering fuels to the point of use. All primary energies given in this chapter have been adjusted to include, where applicable, markup factors of 1.05, 1.2 and 1.25 for coal, petroleum products and natural gas, respectively.

9   Outotec GmbH is the German subsidiary of the Finnish company Outotec Oyj, which was originally called Outokumpu Technology. As discussed later, Outokumpu was also involved in the development of alternative methods (including flash smelting) for processing of copper.

10 The secondary share in this figure corresponds to the 'Old Scrap' term in Figure 6.12, while the total corresponds to 'Finished Products'. The estimates for 2005 in Figure 6.16 are slightly higher than the corresponding terms in Figure 6.12.

11 In 1998, about 37 per cent of the aluminium supply in the US was from recycled aluminium.

12 The major routes by which copper is dissipated into the environment are through human food (humans require 1.5–3mg/day with food, but this intake does not accumulate in the body, so a comparable flux is lost in faeces and ends up in sewage, but could eventually be taken up and concentrated by plants); through chemical uses (in wood preservatives, fungicides, pigments and dyes and anti-fouling paint for ships); and through corrosion of copper roofs and pipes (average runoff from copper roofs in Stockholm is about 1–1.5gm/m²/yr). Rates of corrosion of copper pipes can be reduced by a factor of seven by controlling the alkalinity (as bicarbonate ion), pH, and sulphate and chloride concentrations of tap water (Landner and Lindström, 1999, cited in Ayres et al, 2003). The major benefit in terms of copper demand would be to extend the lifetime of pipes.

13 Scrap copper, as with other metals, is divided into new scrap and old scrap. New scrap is scrap metal generated during the production process, the vast majority of which is recycled at the production facility. Old scrap refers to metal in discarded consumer products.

14 The other major uses of chromium are as a pigment and as an ingredient in preservatives for wood, while the other major uses of nickel are in alloys of steel and copper, and in batteries, catalysts, dyes and chemicals.

15 This can be derived from the world cement production in 2007 of 2600Mt (see Figure 6.24a), assuming a density for cement and concrete of about 2.6t/m³ (i.e. comparable to that of typical rocks) and assuming concrete to be 10 per cent cement by volume.

16 This emission factor is based on the assumption that all of the CaO input to clinker comes from carbonate, which is not quite right, but also ignores the CO_2 emission associated with the small amount of MgO input to clinker (not shown in Equation (6.25)), which also comes from carbonate. The two small errors largely cancel out.

17 This is the enthalpy of formation, given by Worrell et al (2000).

18 This is for present-day energy prices. With higher energy prices to reflect climatic and non-climatic externalities associated with energy use, the rate of return on energy-saving measures would be greater.

19 Refractory materials are materials that are strong at high temperature, are chemically inert and have a low thermal conductivity and coefficient of expansion.

20 A minimum estimate of the amount of oxygen needed can be obtained from stochiometry for the combustion of natural gas. Using a heating value for natural gas of 50MJ/kg, an electricity consumption of 0.25kWh per kg of O_2 separated from air and an electricity generation efficiency of 40 per cent, the primary energy used to separate O_2 from air is equal to 18 per cent that of the combusted natural gas.

21 That is, the energy use not exceeded by 10 per cent and 90 per cent of the plants.

22 See Chapter 4 (subsection 4.4.2) for a definition of COP.

23 Motors introduce a phase lag between the voltage and current in the electric grid, which increases the energy

7

Agricultural and Food System Energy Use

7.1 Introduction

Energy is used in the agricultural sector both directly – through the use of electricity for pumps and motors, fuel for tractors and natural gas for heating buildings and sometimes for drying crops – and indirectly, through the energy used to manufacture fertilizers and pesticides, both of which are energy-intensive products. Figure 7.1 shows the breakdown of secondary energy use in the US agricultural sector in 2002. Fertilizers and pesticides account for about one third (34 per cent) of agricultural energy use, fuels for farm vehicles account for another third (36 per cent) and electricity and natural or liquid petroleum gas account for the remaining third (30 per cent). Direct energy use on US farms per unit of agricultural output fell by about 40 per cent from the 1960s to late 1990s and indirect use fell by about 20 per cent, with most of the decrease occurring during the late 1970s and early 1980s (Schnepf, 2004). Figure 7.2 places agricultural energy use in the broader context of the US food system: agricultural energy use accounts for about 20 per cent of total food system energy use, alongside household storage and preparation (at 31 per cent), processing (16 per cent), transport (14 per cent), retail and commercial food services (11 per cent) and packaging (7 per cent). Carlsson-Kanyama (2007) gives a similar breakdown for Sweden (~20 per cent of total food system energy use agricultural, ~30 per cent household). The total energy input into the US food system was about 10.1EJ/yr (of which only 2.2EJ/yr are related to the production of food), while the available food energy output was 1.4EJ/yr in the mid-1990s according to Heller and Keoleian (2000). Thus, about seven units of non-solar energy are used to produce one unit of food energy in the US according to this analysis. The total food-related energy use is about 10 per cent of total US energy demand.

In this chapter we discuss broad strategies to reduce agricultural energy use. These strategies are:

- to reduce the energy required to produce a given amount of fertilizer or pesticide by improving the production process;
- to reduce the amounts of these and other inputs that are required by avoiding waste or ineffective application, or by shifting to low-input agricultural systems; and
- to promote changes in diet toward less energy-intensive foods (in particular, a shift away from meat in general and beef in particular).

Pesticides and fertilizers have a host of adverse environmental side-effects, so efforts to reduce their use in order to reduce energy use will yield a number of significant additional environmental benefits. As well, N fertilizer entails the production of N_2O (a powerful GHG) independently of the emissions associated with providing the energy and feedstock needed for its manufacture, while the supply of mineable phosphorus may peak within two to three decades, leading to concerns about the long-term availability of P fertilizer. Runoff of both N and P fertilizers to the oceans contributes to eutrophication of coastal oceanic regions and a growing number of oceanic dead zones. Thus, we may soon have no choice but to adopt measures to reduce the use of fertilizers. In the next two sections, non-energy issues related to N and P fertilizers are briefly discussed, followed by a comprehensive discussion of the options to reduce agricultural energy use.

7.2 Nitrogen fertilizer and the N cycle

Nitrogen cycles between different forms as part of a natural biogeochemical cycle, illustrated in Figure 7.3. Nitrogen in the atmosphere occurs as N_2 but is not

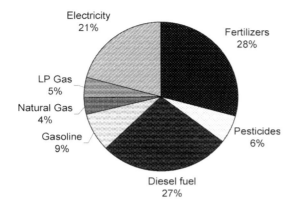

Source: Redrafted from Schnepf (2004)

Figure 7.1 *Breakdown of agricultural energy use in the US in 2002*

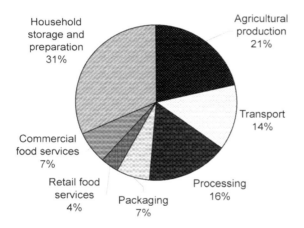

Source: Data from Center for Sustainable Systems Fact Sheet, University of Michigan, http://css.snre.umich.edu/css_doc/CSS01-06.pdf

Figure 7.2 *Energy use in the US food system*

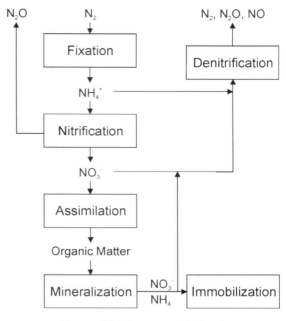

Figure 7.3 *The terrestrial N cycle*

useful to plants in this form. Instead, N_2 has to be converted to forms that can be used by plants – a process called nitrogen *fixation*. Fixation is carried out by bacteria that live in the roots of certain plants in a symbiotic relationship. Among these plants are legumes such as pulses, soybean and groundnut, and trees such as *Acacia*. These plants have a higher N content in their tissues, and this N is added to the soil when the plant parts fall to the ground and decay. Nitrogen fixation largely produces NH_4^+ (ammonium), which is oxidized to NO_3^- (nitrate) in a process called *nitrification*. Nitrogen in the form of NH_4^+ and NO_3^- can be absorbed by plants and incorporated into organic matter,

a process called *assimilation*. When plant matter falls to the ground and decays, the N is released back to the soil as NH_4^+ and NO_3^-, a process called *mineralization*. The conversion of NH_4^+ and NO_3^- back to N_2 or to other gases (N_2O, NO, NO_2) is the process of *denitrification*.

Humans have massively perturbed the natural N cycle by fixing atmospheric N at a rate (160MtN/yr) comparable to the natural rate of N fixation. Human-induced fixation occurs through the production of N fertilizer (at a rate of 95MtN/yr), through land use changes that have enhanced biological N fixation (an extra 35MtN/yr), and through high-temperature combustion of fossil fuels and biomass (producing 30MtN/yr as N_2O, NO and NO_2). This greatly accelerated conversion of N_2 to other forms has a number of adverse environmental effects:

*   emissions of N_2O (nitrous oxide) contribute to the greenhouse effect, with an N_2O molecule having 298 times the heat-trapping effect of CO_2 and an average atmospheric lifespan of 120 years;
*   emissions of NO contribute to destruction of stratospheric ozone;
*   emissions of NO and NO_2 (collectively referred to as NO_x) contribute to the formation of tropospheric ozone and acid rain, and react to form aerosol particles that penetrate deep into the lungs;

- nitrite ions (NO_2^-) in drinking water are a threat to human and especially infant health, as they can enter the bloodstream and inactivate haemoglobin;
- nitrate ions (NO_3^{2-}) in aquatic ecosystems can lead to eutrophication (over-fertilization, leading to algae growth and subsequent decomposition in deep water, depleting the deep water of oxygen and therefore prohibiting other organisms from living in the water).

Emissions of N_2O occur directly from agricultural soils (through both nitrification and denitrification when soils are supplied with N by either chemical fertilizer or biological fixation), from animal waste (urine and faeces) and indirectly through atmospheric deposition of NO_x and NH_3 on soils, which fertilize soils and surface waters and ultimately lead to increased N_2O emissions. Mosier et al (1998) estimate all three sources to be roughly comparable in importance at a global scale.

A striking example of eutrophication from nitrogen fertilizer is a zone of about 13,000km² in the Gulf of Mexico next to the Mississippi delta, which is rendered anoxic (oxygen depleted) each summer. However, such dead zones are found along coastlines worldwide, with a total area of 245,000km² and serious consequences for marine ecosystems (Diaz and Rosenberg, 2008).

Under conventional practice, large additional increases in the use of N fertilizers, and in water and air pollution, would be required in many parts of the world in order to continue increasing food yields so as to keep up with a growing population and an increasing shift to meat consumption (which requires more food as animal feed than if food crops were directly consumed by humans).

7.3 Phosphorus fertilizer and potential supply constraints

Phosphorus fertilizer is manufactured from phosphorus-containing rocks (phosphates), so the availability of P fertilizer in the future will be limited by the supply of mineable phosphates and our ability to recycle it. It thus differs from nitrogen fertilizers, which can be manufactured from atmospheric N_2 as long as there is a supply of energy and H_2, or can be created by nitrogen-fixing micro-organisms. Estimated global annual flows of phosphorus in the food system are shown in Figure 7.4. About 18MtP/yr are mined and processed into 15MtP/yr as inorganic P fertilizers. An additional input to the food system of about 12MtP/yr occurs through the uptake of P by animals grazing on pastures, giving a

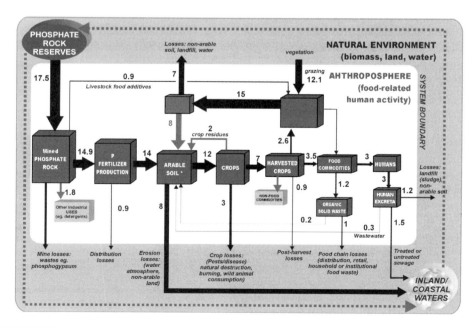

Source: Cordell et al (2009a)

Figure 7.4 *Global annual flows of P in the food system, in units of MtP/yr*

total input of about 27MtP/yr. Animal wastes (manure) are a large source of recyclable P, with flows of 16–20MtP/yr, about half of which is lost and half of which is returned to the soil (particularly large losses are associated with feedlot operations housing thousands of animals). The P added to arable soils from manure amounts to about 50 per cent of the amount of P added as inorganic fertilizer, or a third of total applications. Crop residues not removed from the field recycle another 1–2MtP/yr back into the soil, and the return of human and food wastes back to the soil adds another 0.5MtP/yr. Altogether, inputs to arable soils amount to about 24MtP/yr, of which about half (12MtP/yr) is incorporated into food crops and one third (8MtP/yr) is lost. The balance is accumulating in soils, and will largely be available for uptake in future years. In many parts of the world (particularly the US and Western Europe), soils are saturated with respect to P, so only an amount of P equal to annual losses need be added each year. Only 3MtP/yr ends up in food consumed by humans, and this P is then excreted in faeces and urine. Sludge from sewage treatment plants is thus a potential source of a portion of the 3MtP/yr and could be returned to the soil, but in most cases this cannot be done because of contamination of the sludge with heavy metals. Most (60–70 per cent) of the P in most cereal and leguminous grains is in forms that are not digestible by non-ruminant mammals (such as pigs), so inorganic P is added to the animal diet, resulting in large losses of P in excreted manure (Smil, 2000).

The current reserve of phosphate is estimated to be about 2.4GtP (Jasinski, 2009), which implies a lifespan of about 135 years at the current global rate of extraction.[1] The P reserve is highly concentrated geographically, with about 70 per cent in China and Morocco and Western Sahara alone, as shown in Figure 7.5. More critical are (1) the decline in the amount of remaining high-quality (uncontaminated) phosphate resources, and (2) the timing of peak extraction rate. Sedimentary phosphates contain low levels of As, Cd, Cr, Hg, Pb, U and V (among other elements), with cadmium being the most enriched element. The World Health Organization recommends a maximum daily intake of Cd of 1µg/kg body weight or 50–70µg/day for most adults. Intake rates in Europe are around 40µg/day, and a number of European countries have enacted Cd limits on P fertilizers (Smil, 2000). A universal ban on high-Cd phosphate ores would eliminate a large fraction of the available phosphate according to Smil (2000), particularly the

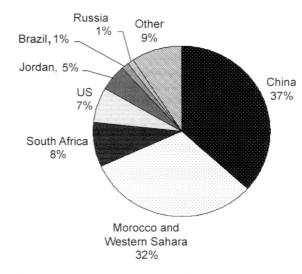

Source: Based on estimates given in Jasinski (2009)

Figure 7.5 *Distribution of phosphate reserves*

Moroccan phosphates. Removal of Cd and other contaminants would be costly and energy intensive, and would generate hazardous wastes. However, the whole array of heavy metals in P fertilizer contributes to terrestrial and aquatic ecotoxicity.

Figure 7.6 shows annual P extraction from 1900 to 2006, alongside the variation in annual extraction assuming that extraction rate varies logistically with an ultimate cumulative extraction of 3212MtP (this is the cumulative extraction to 2007 of 854MtP plus the estimated remaining reserve of 2358MtP). The extraction rate is projected in this way to peak around 2035. This is a highly uncertain projection, as one would normally expect the remaining ultimate extraction to exceed current reserve estimates, but the usable P will probably by severely constrained by contaminants in the phosphate ores. The important point is that supplies of a critical fertilizer may peak in the near future, and when the peak occurs, major price increases will in turn prompt more efficient use and greater recycling of nutrients (a scenario for balancing P needs with P supply after the peak is presented in Cordell et al, 2009b). This in turn will reduce the energy use associated with producing fertilizers. Conversely, measures to use fertilizers more effectively as part of a broader strategy to reduce agricultural energy use would delay the peak in P supply. According to Smil (2000), ten of the world's most populous low-income countries, with more than

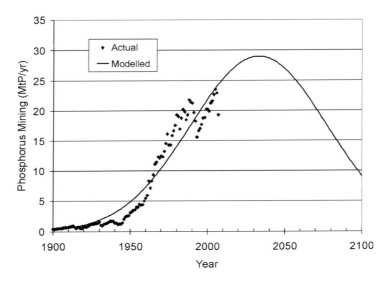

Source: Cordell et al (2009a)

Figure 7.6 *Annual global mining of phosphate (MtP/yr) from 1900 to 2007 and variation in annual P mining from 1900 to 2100 assuming an ultimate cumulative extraction of 3212MtP*

50 per cent of the world's population and an enormous potential need for P fertilizer, have only 5 per cent of all high-quality reserves.

7.4 Reducing fertilizer energy use

Fertilizers can be classified as organic and inorganic fertilizer, the former supplied from organic materials (plant residues and animal and human waste), the latter produced by industrial processes. The production of inorganic fertilizers is an energy-intensive process, and the process of applying them to agricultural fields accounts for a significant fraction of the liquid fuels used by farm vehicles. However, given that the use of inorganic fertilizers results in greatly increased agricultural yields, which in turn spares forests (a carbon store and habitat for innumerable species of life) and frees land that could be used for biomass energy (which is discussed in Volume 2, Chapter 4), the net effect on the demand for non-renewable energy of using inorganic fertilizers (even when made from non-renewable energy sources) is generally favourable. Furthermore, elementary calculations show that present world food production could not be sustained without the continuous input of fertilizers. In principle, organic fertilizers would be sufficient if there were 100 per cent recycling of nutrients taken from the soil back into the

soil (by returning all human and animal wastes, food-processing wastes and plant residues back to the soil). However, 100 per cent recycling is not achievable in practice, and some of the applied organic fertilizer would be lost rather than incorporated into new crop biomass. The rate of release of new nutrients from chemical weathering of soil parent materials is too slow to make up for losses at the present global scale of food production. Thus, some production of inorganic fertilizers would probably be required in order to sustain present rates of food production even in an ideal world with maximum achievable recycling of nutrients.

The addition of N fertilizer causes a decrease in soil pH, which in turn increases the solubility of aluminium and magnesium and decreases the availability of phosphorus, which decreases yields. The reduction in pH is counteracted by the addition of lime (CaO) in quantities such that additional indirect energy inputs equal to about 3 per cent of the fertilizer energy use are required in the case of US corn (maize; see Volume 2, online supplemental Table S4.1).

The addition of N fertilizer thus affects the GHG balance of agriculture in four ways:

1   through emissions associated with its manufacture;
2   through emissions associated with the production of lime;

3   through increased crop yield (freeing up land for other purposes, including bioenergy crops); and
4   through emissions of N_2O.

The amount of energy used in the production of fertilizer can be reduced in two ways: (1) by reducing the energy required per unit of fertilizer manufactured; and (2) by reducing the amount of fertilizer used per unit of crop produced. Energy requirements per unit of fertilizer can be reduced through technical improvements in the manufacturing process and by shifting from energy-intensive forms of a given fertilizer to less energy-intensive forms of that fertilizer. Fertilizer use per unit of crop can be reduced by reducing the amount of applied fertilizer that is lost from the soil–plant system rather than taken up by the plant, and by making better use of agricultural residues, animal wastes and (in the case of N fertilizer) biological nitrogen fixation. Improvements in crop yield (through development of improved varieties, irrigation or conservation of soil moisture) reduce the loss of fertilizer by increasing plant demand for fertilizer. The potential in each of these areas is briefly discussed below for the major fertilizer groups, but first we present background information on trends in fertilizer use and on the amounts and kinds of energy used to make fertilizers at present.

7.4.1 Trends in fertilizer use

Figure 7.7 shows the growth in fertilizer use from 1970 to 2005 by OECD countries, the FSU and Eastern-bloc countries, and other non-OECD countries. Fertilizer use in OECD countries has been relatively constant or declined slightly since the 1980s, while fertilizer use in the FSU and Eastern bloc fell precipitously between the mid-1980s and mid-1990s. Fertilizer use in non-OECD countries has grown dramatically during the past 10–15 years. Figure 7.8 gives the worldwide consumption by different N, P and potassium fertilizers in 2001. Table 7.1 gives the average rates of application of N fertilizer for different crops in different world regions. Vegetables require the most N fertilizer in most regions, with corn and sugarcane taking top spot in one region each. Wheat and especially fodder have relatively low N fertilizer demands.

The total primary energy use for the manufacture of fertilizer in 2001 amounted to 3660PJ (or 3.66EJ,

which is 0.8 per cent of the worldwide primary energy demand in 2001 of 455EJ). The distribution of this energy use among the major fertilizer groups is shown in Figure 7.9, where it can be seen that N fertilizer dominates, at 72 per cent of total fertilizer energy use.

7.4.2 Increasing the efficiency in manufacturing fertilizer

All N fertilizers begin with the production of ammonia (NH_3). Ammonia can be used directly or can be transformed to nitric acid (HNO_3) through partial oxidation or transformed to urea (NH_2CONH_2) by reaction with CO_2. Nitric acid can be reacted with more ammonia to produce ammonium nitrate (NH_4NO_3), and ammonium nitrate can be reacted with calcium carbonate ($CaCO_3$) to produce calcium nitrate ($Ca(NO_3)_2$). Ammonia can also be reacted with sulphuric acid to produce ammonium sulphate (see Ramírez and Worrell, 2006, for the major reactions).

The major energy-consuming step is the production of ammonia, which is produced by reacting H_2 with atmospheric N_2. Over 99 per cent of the H_2 today is supplied from a hydrocarbon feedstock (natural gas, coal or naphtha) that could otherwise be used to supply energy. Additional hydrocarbons are burned to supply the heat that is required to drive the reaction processes. Table 7.2 lists the feedstocks, production processes, feedstock plus process energy use and CO_2 emissions for the most efficient plants today, and the share of global ammonia production in 1994. Most of the world's ammonia is produced by steam reforming of natural gas, in which 30 per cent of the natural gas supplies process energy and 70 per cent serves as the feedstock. The reactions involved are:

$$CH_4 + H_2O \rightarrow CO + 3H_2 \qquad (7.1)$$

$$CO + H_2O \rightarrow CO_2 + H_2 \qquad (7.2)$$

and:

$$3H_2 + N_2 \rightarrow 2NH_3 \qquad (7.3)$$

The first two reactions together constitute steam methane reforming with a water–gas shift. The third reaction is carried out under high pressure as part of the

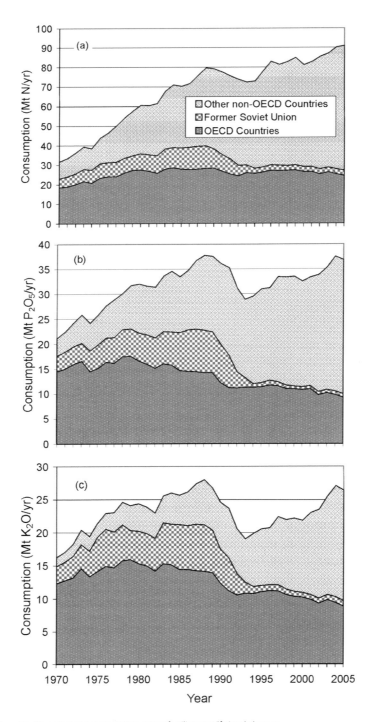

Source: Data from International Fertilizer Industry Association, www.fertilizer.org/ifa/statistics.asp

Figure 7.7 *Variation in fertilizer consumption from 1970 to 2005 in OECD countries except Eastern Europe and Central Asia (taken to be the FSU), the FSU and other non-OECD countries for (a) nitrogen, (b) phosphorus and (c) potassium*

(a)

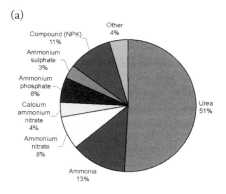

Total = 82.5 million tonnes N

(b)

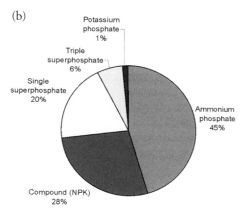

Total = 31.9 million tonnes P₂O₅

(c)

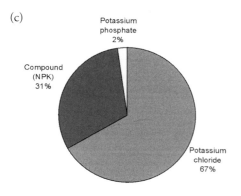

Total = 22.2 million tonnes K₂O

Note: Compound is a mixture of ammonia, nitric acid and ammonium sulphate for nitrogen; phosphoric acid and TSP for phosphorus; and potassium chloride for potassium (K).
Source: Data from Ramírez and Worrell (2006)

Figure 7.8 *Breakdown of worldwide N, P and potassium fertilizer consumption in 2001 by type of fertilizer*

Haber-Bosch process. The net reaction in producing ammonia from natural gas is:

$$CH_4 + 2H_2O + N_2 \rightarrow CO_2 + H_2 + 2NH_3 \quad (7.4)$$

The energy required to produce ammonia from natural gas fell from about 80GJ/t in 1930 to 28GJ/t in the most efficient plants in operation today, while the theoretical minimum energy requirement is 19.4GJ/t (Rafiqul et al, 2005).[2] However, energy use in much of the world was still in the 36–44GJ/t range in 2003, with a global mean energy intensity of 39.4GJ/t (IEA, 2006). The best identified future technologies would require 23.6GJ/t according to Blok (2005) – a saving of 40 per cent compared to the current world average. Partial oxidation requires more energy and is more expensive than steam reforming, but is flexible in terms of the feedstock: it can be applied to gaseous, liquid or solid hydrocarbons (including biomass). Partial oxidation of coal requires 42GJ/t in the most efficient plants, but Rafiqul et al (2005) give a mid-1990s energy use of 166GJ/t in India. Steam reforming and partial oxidation produce a mixture of H_2 and CO_2, but the CO_2 has to be separated from the H_2 before the H_2 can react with N_2 to form NH_3. Much of the separated CO_2 is used for the production of urea (CH_4N_2O), a popular fertilizer, but the balance constitutes a ready supply of separated CO_2 that could be sequestered geologically (see Volume 2, Chapter 9, subsection 9.3.9).

The reactions to produce nitric acid and ammonium nitrate are exothermic, so there is net energy production in the form of heat, and zero or close to zero net energy use in theory. According to Blok (2005), production of nitric acid requires 26.8GJ/t with best practice, best identified future technology would require 15.3GJ/t and the theoretical minimum is 3.2GJ/t.

Figure 7.10 compares the energy intensity (GJ primary energy used per tonne of fertilizer produced) for various fertilizers in 1960 and 2000. The energy intensity of most fertilizer products fell by about 40–60 per cent during this 40-year period but, as noted above, the worldwide use of synthetic fertilizers increased several-fold during that time and there has been a shift from less energy-intensive to more energy-intensive forms of phosphorus fertilizer. Nitrogen fertilizer is the most energy-intensive fertilizer (40–50GJ/tonne at present) but differs little among the major types because they all begin with the production of ammonia, which dominates overall energy use. The energy intensity of

Table 7.1 *Average rates of application of N fertilizer (kgN/ha/yr) for different crops in different regions*

Region	Crop						
	Wheat	Corn	Rice	Vegetables	Sugarcane	Cotton	Fodder/pasture
Asia-Pacific	65	82	63	**132**	111	97	50
Middle East	101	177	89	**218**	144	99	59
Africa	78	82	71	118	**181**	60	97
Europe	120	**144**	139	132	139		98
North America	97	130	123	**138**	116	**138**	81
Latin America	63	96	85	**142**	111	111	36

Note: The most N-intensive crop in each region is indicated in bold. Averages pertain only to cases where fertilizers are applied.
Source: Peoples et al (1999)

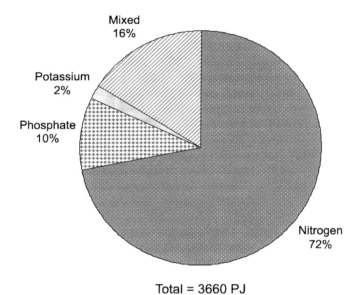

Total = 3660 PJ

Source: Data from Ramírez and Worrell (2006), who computed primary energy use as onsite fuel use + electricity use divided by an unspecified efficiency in generating electricity.

Figure 7.9 *Worldwide fertilizer primary energy use in 2001*

phosphorus fertilizer ranges from about 8GJ/tonne for single superphosphate (SSP), to 10GJ/tonne for triple superphosphate (TSP), 20 GJ/tonne for ammonium phosphate and 32GJ/tonne for PK 22-22. The shift toward the more energy-intensive phosphate fertilizers contributed about three quarters as much as has increased use of phosphate fertilizer to increased worldwide energy use for phosphate fertilizer during the period 1961–2001. Potassium is the least energy intensive of the three fertilizers (5–12GJ/tonne).

If the rest of the world can be brought up to the current best practice in the production of ammonia

(28GJ/t), the implied reduction in world average energy intensity for the production of ammonia is about 30 per cent but this probably requires shifting to natural gas where coal is used. For urea, the world average direct energy use (i.e. not counting the energy used to make the input ammonia) is 9.3GJ/tonne N but the best-practice energy intensity is only 5.6GJ/tonne N – 40 per cent less.

Table 7.3 shows the different energy sources used for the production of fertilizer (and lime, which is added to the soil to prevent acidification by the fertilizers) in the US as given by West and Marland (2002) and by Graboski (2002). There are significant differences

Table 7.2 *Processes for producing ammonia, feedstock plus process primary energy use, CO_2 emissions and global production share in 1994*

Feedstock	Process	Energy use (GJ/t NH_3)	CO_2 emission (tC/t NH_3)	Global production share (%)
Natural gas	Steam reforming	28.5	0.44	77
Coal	Partial oxidation	42	1.06	13.5
Naphtha, LPG, refinery gas	Steam reforming			6
Heavy hydrocarbons	Partial oxidation	34.5	0.68	3
Water	Electrolysis			0.5

Source: Neelis et al (2005)

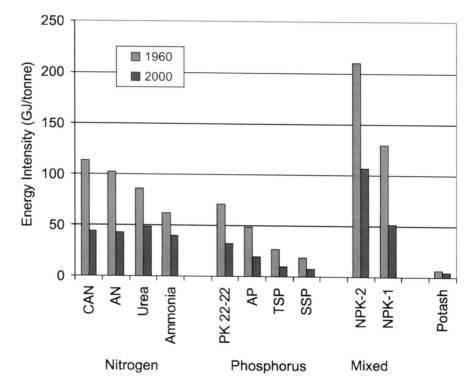

Note: CAN = calcium ammonium nitrate, AN = ammonium nitrate, SSP = single superphosphate, TSP = triple superphosphate.
Source: Data from Ramírez and Worrell (2006)

Figure 7.10 *Comparison of world average fertilizer primary energy intensities in 1960 and 2000*

between the two sources, and neither source is consistent with the total primary energies given in Figure 7.10. The lower part of Table 7.3 contains fuel and energy uses consistent with the primary energies given in Figure 7.10 and adopted in Chapter 6 (Table 6.3). Not included in Table 7.3 (or in Figure 7.10) is energy use for transporting the inputs to or outputs from the fertilizer plant. In terms of the primary energy required to produce it, electricity accounts for only 4 per cent of the N fertilizer energy input, about 40 per cent of the potassium energy input, about 80 per cent of the P energy input and about 90 per cent of the lime energy input.

In the long run, the hydrogen needed for reaction with N_2 to produce ammonia could be produced by

Table **7.3** *Energy inputs for the production of fertilizers in the US*

	N	P	K	Lime
From West and Marland (2002)				
Natural gas (GJ/t)	51.81	0.63	2.69	0.006
Distillate (GJ/t)	0.01	0.40	0.00	0.033
Steam (GJ/t)	0.91			
Coal (GJ/t)				0.003
Gasoline (GJ/t)				0.004
Electricity (kWh/t)	263	510	200	36
Total primary energy (GJ/t)	55.32	5.62	4.49	0.37
% electricity PE	4.3	81.7	40.1	87.6
From Graboski (2002)				
Natural gas (GJ/t)	37	12	2.4	0.0
Electricity (kWh/t)	567	200	167	36
Total primary energy (GJ/t)	42.1	13.8	3.9	0.3
% electricity PE	12.1	13.0	38.5	100.0
Values adopted here				
Fuels (GJ/t)	40.2	17.4	3.0	0.15
Electricity (kWh/t)	200	290	223	117
Total PE (GJ/t)	42	20	5	1.2
% electricity PE (kWh/t)	4.3	13	40	88

Note: Primary energy has been computed assuming steam to be produced at 80 per cent efficiency and electricity to be produced at 40 per cent efficiency.

electrolysis of water using renewable energy, rather than supplied from natural gas or coal. Hydrogen could also be used as a fuel to supply process energy, thereby eliminating emissions altogether. Hydrogen produced using electricity from wind or solar energy would, however, be much more expensive than the cost of natural gas at present ($20–40/GJ versus $5–10/GJ). Conversely, nitrogen fertilizer plants represent a source of CO_2 that could be easily captured and geologically sequestered (see Volume 2, Chapter 9), as it must be separated from H_2 in any case as part of the ammonia production process.

7.4.3 Increasing the efficiency of fertilizer use

The effectiveness with which fertilizers are used can be increased by reducing losses, so that more of the applied fertilizer is used by the plants. Fertilizer loss can be reduced through more accurate application of fertilizer in terms of the amount, timing and depth of application, and in the physical form in which the fertilizer is added. Table 7.4 presents data on the fraction of applied N that is taken up by crops during the growing season, based on field measurements. The fertilizer-use efficiency is generally well below 50 per cent. Results for India are for two different years: one with low crop yields and low fertilizer efficiency (18 per cent) and another with high crop yields (due to better weather) and high fertilizer efficiency (49 per cent). Two cases are shown for a sample of Asian countries: with and without guidance to farmers concerning the appropriate fertilization level. Farmers in most parts of Asia have few guidelines for adjusting the amount of fertilizer to account for the large differences in the indigenous nutrient supply; provision of guidance based on soil characteristics and applying the N in as many as four applications increased the efficiency from 30 per cent to 40 per cent. This corresponds to a 25 per cent reduction in fertilizer use for the same benefit.

Nitrogen losses occur through leaching (the penetration of fertilizer deep into the soil, below the rooting zone, by rainwater), runoff (into rivers, lakes and eventually the ocean), gaseous emission (volatilization), denitrification and erosion. In The Netherlands, the breakdown of losses is estimated to be as follows: volatilization, 21 per cent; leaching,

Table 7.4 *Nitrogen fertilizer uptake efficiency for major food crops in major food regions, based on measurements at farms*

Crop	Region	Number of farms sampled	N fertilization rate (kgN/ha/yr)	Uptake (%)
Maize	North central US	55	103	37
Rice	China, India, Indonesia, Phillipines, Thailand, Vietnam	179	117	31[a]
Rice	China, India, Indonesia, Phillipines, Thailand, Vietnam	179	112	40[b]
Wheat	India, poor year	23	145	18
Wheat	India, good year	21	123	49

Note: [a] Results without guidance given to farmers. [b] Results when fertilizer application is adjusted to balance soil nutrient supply and crop demand.
Source: Cassman et al (2002)

18 per cent; accumulation in the soil, 11 per cent; and unaccounted for, 20 per cent. Losses in the form of N_2O emission depend on the type of fertilizer used, ranging from about 0.03 per cent for nitrate, 0.11 per cent for urea, 0.4 per cent for ammonium nitrate and 1.6 per cent for anhydrous ammonia (Worrell et al, 1995). For non-OECD countries, Wohlmeyer (1998) estimates the breakdown of N losses as follows: denitrification, 22 per cent; leaching, 14 per cent; volatilization of NH_3, 7 per cent; and erosion losses, 57 per cent. Erosion, which dominates N loss in non-OECD countries, is the only significant loss mechanism for P.

Worrell et al (1995) have assessed the potential savings in fertilizer use in The Netherlands from the following measures:

- use of recommended amounts of fertilizers (farmers often apply more fertilizer than recommended in order to be sure of the highest possible crop production);
- proper maintenance of machines for spreading fertilizer, so that the fertilizer is applied uniformly (this is estimated to save 22 per cent);
- implementation of a 50cm-wide unfertilized zone next to ditches so as to reduce fertilizer losses into ditches (a 4 per cent saving);
- adjustment of distribution geometries on fertilizer spreaders to prevent fertilizer losses into ditches;
- application of fertilizer in rows during seeding, rather than over the entire field before seeding (10–30 per cent savings);

- analysis of soil conditions and application of just the required amount of fertilizer two or more times during the growing season, rather than one single application (fertilizer requirements depend on weather and growth conditions, and so vary from year to year) (10–20 per cent savings);
- choice of alternative types of fertilizer (replacement of calcium nitrate by urea can reduce volatilization losses by 12–35 per cent);
- use of higher-quality fertilizer granules, which release the fertilizer more gradually.

Altogether, Worrell et al (1995) estimated an overall potential saving relative to 1988 of about 30–45 per cent (excluding the last two items in the above list, due to uncertainties). Only the application of fertilizer in rows could lead to an increase in non-fertilizer energy use, but this impact would be small.

Nitrate accumulates in the soil during fallow periods between cropping seasons and is lost through leaching. Cover crops planted during fallow periods will assimilate nitrogen and then release it slowly to the soil during the next season after the cover crop has been ploughed into the soil. This reduces pollution of groundwater by nitrate and reduces N fertilizer requirements.

Figure 7.11 shows the variation in average US corn yields and in fertilizer use per tonne of corn produced from 1970 to 2005. Yields (tonnes per hectare per year) have almost doubled, while N fertilizer use per tonne of crop has decreased by about 40 per cent and P and potassium use per tonne of crop have decreased by about

60 per cent. The trend of decreasing fertilizer use per unit of production shows no sign of slowing down, and a consideration of current practice in the US indicates that there is substantial room for further reductions.

First, many fields receive both manure and N fertilizer. Better accounting of manure additions and even application of manure could lead to more optimal use of N fertilizer. About half of the soils in North America could benefit from additional P and potassium fertilizer, which in turn would improve the effectiveness of N fertilizer. In soils with high potassium levels, 80 per cent of the applied N fertilizer is taken up by plants, while in soils with low potassium, less than 45 per cent of applied N is taken up (Fixen and West, 2002). (Soils in China appear to be developing an increasing potassium deficit, and N utilization rates for the 1990s are estimated to be around 35 per cent.) For US maize, 28 per cent of N fertilizer is applied in the autumn (with large losses) because suppliers offer a discount for N purchased in the

autumn (Cassman et al, 2002), 45 per cent is applied in the spring before or during planting and 25 per cent after planting. Many farmers do not split the fertilizer application in two but instead have one large application.

Precision farming is a system of adjusting N and other inputs to allow for variations in soil conditions across a field. These make use of satellites to accurately position fertilizer spreaders as they move across a field, with programmed variations in the rate of fertilizer applications. Systems are under development to adjust N fertilization rates based on measurements of chlorophyll content as the spreader moves across a crop (Jenkinson, 2001).

Worldwide, there is a wide variation in the effectiveness with which nitrogen fertilizer is used to produce rice, ranging from 20–45kg rice/kgN in southern Asia and China to 72kg rice/kgN in Japan (Roy et al, 2002). Thus, there is almost a factor of four difference between lowest and highest N effectiveness.

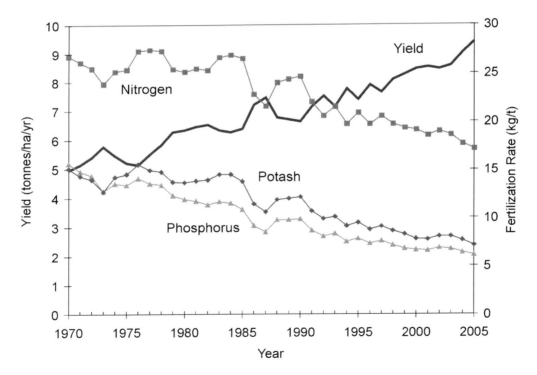

Source: Data from Wang et al (2007b)

Figure 7.11 *Variation in average US corn yields and in fertilizer use per tonne of corn produced, 1970–2005*

Trends in the amount of organic matter in soils will affect the apparent efficiency of nitrogen utilization (Fixen and West, 2002). During the first half of the 20th century, soil carbon declined by 40–50 per cent in much of the central US and Canada due to intensive cultivation. This released N that in turn contributed to the N requirements of crops, creating the appearance of high uptake of N fertilizers. More recently, conservation tillage (which is practised on over 35 per cent of US cropland) has led to a buildup of soil carbon, but also a sequestration of some of the applied N fertilizer. In the US cornbelt, soil carbon has recovered from a low of about 53 per cent of the 1907 carbon content in the 1960s to about 61 per cent of the 1907 content by the early 1990s. This corresponds to a sequestration of 250kgC/ha/yr and 22kgN/ha/yr.

7.4.4 Enhancing biological nitrogen fixation

In the case of nitrogen fertilizers (which account for three quarters of total fertilizer energy use), the alternation of N-fixing plants with non-N-fixing plants, or simultaneous cropping of the two in alternate rows, can be used to build up soil N. The effectiveness of leguminous plants can be enhanced through artificial inoculation of the plants with N-fixing bacteria. Grain legumes currently occupy 11 per cent of the world's arable lands and fix about 25 per cent of all crop-fixed N (Roy et al, 2002).

7.4.5 Recycling nutrients back into the soil

Fertilizers that are not lost from the soil end up in plant material. This material consists of organic residues (the uneaten parts of the plants) and food processing, human and animal waste. If all of these materials were returned to the soil that produces the food, only those fertilizers lost from the plant–soil system would need to be replenished through artificial fertilizers. The N flux in manure worldwide (about 80MtN/yr) is comparable to the total industrial production of N fertilizer (82MtN/yr), so full utilization of manure already represents a significant N fertilizer resource.

With an increasing reliance on feedlot livestock operations rather than pasture operations, and the often considerable distance between the food sources and the feedlot, there are significant logistical difficulties and financial costs in recycling manure nitrogen back to the soil that produced the animal feed. Smaller-scale, mixed farming systems are more conducive to recycling of nutrients. Such systems allow one to feed agricultural residues to animals that otherwise could not be made use of in this way, with the nutrients in the residues eventually recycled back to the soil as animal waste. An increase in consumer demand for meat from grass-fed animals would increase the opportunities for recycling of manure back into the soil by facilitating more mixed farming operations.[3]

Nitrogen in animal manure is less effective (typically by about 50 per cent) than nitrogen in chemical fertilizers, so more manure nitrogen needs to be added than chemical nitrogen displaced. This is one reason why manure fertilizer requires extra labour and equipment compared to chemical fertilizers. The extra on-farm fuel use is quite small compared to the savings in fertilizer embodied energy, but if the manure has to be transported a long distance by truck, the energy savings could easily be eliminated.

Manure also contains phosphorus, and the rate at which manure can be added to the soil is limited by the maximum permissible rate of addition of phosphorus, which depends on the crop. If this limitation is too low, an inadequate amount of nitrogen will be supplied by manure.

It is more difficult to synchronize the supply of nitrogen from plant residues and manure with crop demand, than using inorganic fertilizers. This is because nitrogen in organic materials can be mineralized (released) at times when the crop is not growing. Increased leaching from green manures and animal manures has been observed compared to inorganic fertilizer. One approach, currently being investigated, is to modify the organic material by adding on-farm and off-farm waste products (such as primary fibre sludge from paper mills) or to manipulate the C to N ratio of green manures by varying the phenol content or the types and concentrations of carbohydrates (Bergström and Goulding, 2005). Phosphorus in manure can also be lost because it is water soluble, but the losses can be reduced by at least 50 per cent without affecting the long-term availability of P by adding aluminium sulphate or byproducts from coal powerplants (Dou et al, 2003).

Another source of nutrients is human urine (faecal matter has very low nutrient content). Effective recycling of urine requires replacing existing toilets with toilets designed to separate urine from faecal matter, and directing the urine to separate underground containers (perhaps one serving many residential units), from which the urine would be periodically collected. Apart from creating a supply of nutrients for food production, diversion of urine in this way would reduce the pollution of water bodies with nutrients (which cause eutrophication) as well as with endocrine disruptors and pharmaceutical residues (both of which are found predominantly in urine rather than in faecal matter and are able to pass through wastewater treatment systems largely unaffected, but are subject to breakdown in soils) (Cordell, 2006). Separation of nutrients (only a small portion of which ends up in sewage sludge) from bulk wastewater is an expensive and energy-intensive process, although, as discussed in Chapter 8 (subsection 8.2.2), the lifecycle energy costs of urine collection systems may not be much less.

7.4.6 Shifting to a diet with less meat

The production of one unit of animal products in a form suitable for human consumption typically requires 10–100 units of grains as animal feed. Much of this animal feed is grown with the aid of fertilizers, so direct consumption of plant products and less consumption of meat will reduce overall fertilizer requirements. This strategy is discussed more fully in subsection 7.9.1 in the context of the overall savings in the energy required to produce food that can be achieved through a shift to low-meat, vegetarian or vegan diets.

7.5 Pesticides

Pesticides can be divided into herbicides (for controlling weeds), insecticides and fungicides, as well as a number of other smaller categories. Table 7.5 gives the annual worldwide and US use of pesticide active ingredients averaged over the period 1998–1999, along with the energy intensities reported for the range of products within each category and an estimate of global energy use. Herbicides constitute the single largest category (about 30 per cent of the total pesticide use). The estimated global energy use for all pesticides is 0.72EJ/yr, which is just under 1 per cent of total industrial primary energy use (95EJ/yr). Table 7.6 gives a representative breakdown of energy inputs for the manufacture of pesticides in the US. Naphtha is the dominate feedstock, while the primary energy used to supply electricity constitutes about 25–30 per cent of the total primary energy requirements.

7.5.1 Reducing the energy intensity of pesticides

There appears to be no published information on the opportunities to reduce the energy intensity of specific pesticides. However, different pesticides differ greatly in the energy requirements for their manufacture, so changes in the proportions of different pesticides used will change the overall pesticide energy intensity. For example, with the development of corn varieties that have been genetically engineered to be resistant to glyphosate, this energy-intensive (454MJ/kg) herbicide has begun to replace atrazine (190MJ/kg) and

Table 7.5 *Worldwide and US pesticide use during 1998–1999 and energy intensities*

Pesticide	Pesticide use (million kg AI/yr)		Energy intensity (MJ/kgAI)	Energy use (EJ/yr)	
	World	US		World	US
Herbicides	948	246	80–450	0.251	0.065
Insecticides	643	52	70–580	0.209	0.017
Fungicides	251	37	60–400	0.058	0.009
Other	721	219	280[a]	0.203	0.062
Total or mean	2563	554	280	0.721	0.152

Note: AI = active ingredient. [a] This is the weighted average of the midpoints of the above intensity ranges.
Source: Pretty (2005) for use and Helser (2006) for intensities

Table 7.6 *Energy inputs for the production of pesticides in the US*

	Herbicide	Insecticide	Fungicide
Production			
Naphtha (GJ/t)	72.0	63.3	92.2
Natural gas (GJ/t)	43.0	49.8	31.1
Coke (GJ/t)	0.3	0.8	0.0
Distillate fuel (GJ/t)	12.3	7.9	11.1
Steam (GJ/t)	44.2	48.0	53.3
Electricity (kWh/t)	6830	8770	7443
Formulation			
Natural gas (GJ/t)	20	20	20
Total			
Primary energy (GJ/t)	264	281	288
% electricity PE	23	28	23

Note: Primary energy has been computed assuming steam to be produced at 80 per cent efficiency and electricity to be produced at 40 per cent efficiency.
Source: West and Marland (2002)

metolachlor (276MJ/kg). However, the last two were used together in quantities such that the energy use per hectare is 25 per cent lower using glyphosate (Helser, 2006). Another example is the use of fluazifop-butyl (518MJ/kg) in place of metolachlor. Although it is almost twice as energy intensive per kg, it requires only two thirds of the energy on a per hectare basis.

7.5.2 Reducing the use of pesticides

Pesticide use can be greatly reduced through systems of integrated pest management (IPM). Pesticides are of concern not only because of the energy required to produce them, but more so because of their adverse health impacts (on farm labourers) and environmental impacts (on wildlife and through the pollution of rivers and lakes). Although the use of pesticides may increase income (by increasing crop yields) by more than the cost of the pesticides, the use pesticides for rice farming in the Philippines (for example) is not economical when health costs are included (Pretty, 2005). A number of countries or jurisdictions within countries have set aggressive targets for reductions in pesticide use. For example, the Canadian provinces of Ontario and Quebec have targets for a 50 per cent reduction in pesticide use. Pretty (2005) reviews the relationship between pesticide use and crop yields in 62 IPM projects from 26 countries. In 60 per cent of the projects, pesticide use declined and yields increased – contrary to what one would expect. One explanation is that increased yields are due to improvements in other

management practices (involving soil, water and nutrients) that resulted from the general training in IPM given to the farmers participating in the studies. Another explanation is that money saved on pesticides was invested instead in other inputs such as higher quality seeds. For the production of pork, the use of pesticides per unit of meat produced can be reduced by over 50 per cent through a more ecological feed production involving, among other things, the integration of protein crops and grain crops (Cederberg et al, 2005).

7.6 Direct energy use on farms

Energy is used directly on farms as electricity for motors and pumps (largely for irrigation), lighting, ventilation and cooling; usually as fuels for heating of buildings and water and for drying and curing; and as fuels to operate machinery for tilling (if performed), applying fertilizer, and for planting seeds and harvesting crops. Brown and Elliott (2005) have assessed the potential for direct energy efficiency savings in the US agricultural sector and estimated an overall savings potential (among end uses examined) of 34 per cent (arising from 25 per cent savings of gasoline, 48 per cent savings of diesel fuel, 30 per cent savings of natural gas and 31 per cent savings of electricity).

In Flanders (Belgium), the 5 per cent most energy-efficient farms producing milk had an energy productivity (litres of milk produced per MJ of energy input) 70 per cent greater than the average (Meul et al, 2007). These farms also had yields (litres of milk per hectare of land area devoted to the milk system) 25 per cent greater than the average. The greater energy efficiency of the top farms is attributed to a lower use of mineral fertilizers (-38 per cent) and of concentrates (-31 per cent), but this is not due to weather or soil differences, but rather, better management. For farms specializing in fattening pigs, the top performing farms had 62 per cent greater energy productivity than the average (including indirect energy inputs through fertilizers and pesticides). These results imply that there is room for substantial reductions in agricultural energy use by bringing the performance of the average dairy or pig farm up to that of the top performers.

The net energy use on farms with animals can be significantly affected by the processing of wastes. Manure can be processed to produce biogas that can then be used to generate electricity, while creating a more concentrated residue that can make it easier and less costly to return nutrients to distant fields (see Volume 2, Chapter 4, subsection 4.3.3). However, a more common practice is simply to dispose of it as a waste.

7.7 Comparative energy use and yield of conventional and low-input systems

Conventional farming systems in developed countries today involve significant inputs of fertilizers and pesticides, and tilling the fields (breaking up and overturning the soil) with machinery prior to planting in order to control weeds. Lower-input systems range from no- or minimum-till practices (but often accompanied by increased use of fertilizer and herbicides) to organic farming, in which animal and crop wastes are recycled back into the soil and no chemical fertilizers or pesticides are used at all. Less extreme systems seek to optimize (according to some criteria) the mix of organic and chemical fertilizers and will allow some use of chemical pesticides. Critical issues concern the yield (food production per unit of land area), cost, labour requirements, and energy and GHG balance of various low-input farming systems. Differences in yield with different degrees or kinds of fertilization should take into account any differences in crop quality; the protein content of wheat, for example, increases with increasing

fertilization (up to some point), which amplifies the benefit of increasing yield (Charles et al, 2006), but this qualitative change is generally not taken into account when comparing different systems.

7.7.1 No-till agriculture

No-till farming is now practised on about a third of US farmland, has allowed the partial restoration of soil organic matter that had been lost under previous practices and provides significant savings in fuel and machinery. West and Marland (2002) and Marland et al (2003) examined the energy and GHG balances of no-till farming averaged over all cropland in the US, compared to conventional practice with tilling, taking into account accumulation of soil carbon, changes in the use of N and other fertilizers (affecting both energy-related and N_2O emissions) and accounting for reduced energy use by and embodied energy in farm machinery. In the case of corn and soybeans in the US, no-till systems require substantially more fertilizers, while winter wheat requires less fertilizer under no-till. An analysis of their data indicates that reduced energy requirements to make and operate machinery are more than offset by increased use of energy for extra fertilizer and herbicide in the case of corn, resulting in a small (4.5 per cent) increase in total energy inputs. For soybeans and winter wheat, there is a small (<2 per cent) saving in total energy use. Given uncertainties in the data, we can regard the difference in energy use between conventional and no-till systems as zero. Zentner et al (1998) also find, for spring wheat grown in Saskatchewan, that energy savings in fuel and machinery are offset by the extra energy required for greater fertilizer and herbicide inputs in no-till systems. However, for hay grown in Canada, Dyer and Dejardins (2005) calculate a 50 per cent reduction in total diesel and gasoline use for field operations in no-till systems compared to conventional systems, with only a temporary increase in the required application of fertilizer.

An additional consideration with regard to GHG emissions is the change in N_2O emission associated with changes in the use of N fertilizer under no-till systems. As noted above, no-till agriculture requires increased N fertilizer for corn and soybeans but decreased N fertilizer for winter wheat in the US. However, the buildup of soil carbon over the first 40 years provides a significant net GHG benefit, estimated by Marland et al (2003) at 6–14 tC/ha averaged over all

US cropland. After 40 years there would be no further increase in soil carbon and a small ongoing gain or loss in net GHG emission depending on the balance between reduced fossil fuel use and, in the case of corn and soybeans, increased N_2O emissions.

7.7.2 Organic farming

The impact of organic farming on crop yields depends on the type of crop, soil conditions and other variables. Badgley et al (2007) reviewed 293 published studies of the difference in yield between conventional and organic farming systems and determined the changes in yield averaged separately over all the studies pertaining to different crop groups (such as grains, starchy roots, sugars, oily crops, legumes, vegetables and fruits) and animal products (meat, milk and eggs), and separately for developed and developing countries. In developed countries, organic farming on average resulted in 10–20 per cent lower yield for three food categories (starchy roots, legumes and vegetables), 0–10 per cent reduction for three categories (grains, oily crops and fruits), and essentially no change for one category (sugars). In developing countries, organic systems multiplied yields by a factor of 1.5 to 4 compared to current practice. However, yields under current practice are very low due to lack of suitable inputs. A more meaningful comparison would be between best-practice conventional and organic systems in developing countries. Nevertheless, these findings indicate that *present* world food production could be increased while switching entirely to organic methods, using the same amount of land as at present. Badgley et al (2007) also conclude that leguminous cover crops could fix enough nitrogen to replace the amount of inorganic fertilizer currently in use. These conclusions are not without controversy, however, as exemplified by the contrary viewpoints expressed in the same issue of the journal as the Badgley paper. In particular, a switch to plant rotations that involve more frequent years with leguminous cover crops to build soil N reduces yield averaged over the entire rotation cycle.

With regard to energy, fertilizer and direct energy requirements were reduced by up to 34–53 per cent and pesticide use by 97 per cent in one 21-year comparison of conventional and organic farming systems in Central Europe, but yields were reduced by 20 per cent (Mäder et al, 2002). Soil carbon content is greater, and soil fungi and insects are two to three times more abundant in soils subjected to organic farming. The lower inputs are compensated by greater labour requirements. In a comparison of organic and inorganic apricot farming in Turkey, Gündoğmuş (2006) found that yields were 9 per cent lower with organic methods but the energy input per unit mass of apricot was 34 per cent lower. The organic method involved no chemical fertilizer, three times as much manure as the conventional method and 10 per cent more fungicide (and so was not entirely organic).

Energy inputs and land area requirements for producing rye bread and milk using conventional and organic farming methods in Finland are compared in Table 7.7. The organic systems require much more land than the conventional systems because crop years alternate with fallow years using clover to fix atmospheric nitrogen in place of chemical N fertilizer. Energy inputs are required for the fallow fields even though they are not producing cereals, and this reduces the energy efficiency of the organic system. Pre- plus on-farm energy use with the organic system is about half that of the conventional system, but land requirements are almost twice as large. However, the relative savings in total energy use to the point of retail sale is much smaller – about 15 per cent for bread and 30 per cent for milk. In the case of bread, this is because the energy used in the bakery (including the flour mill) is a significant fraction of the total energy embodied in rye bread, being several times the energy embodied in the rye itself. Much of the bakery energy is in the form of heat that could be recovered if the bakery were part of a district heating system (see Chapter 9, section 9.2). In the case of milk, packaging is a significant (~20 per cent) fraction of the total energy embodied in milk. The total primary energy use to make bread is 13–15MJ/kg (13–15GJ/tonne), which is a third to three quarters of that of steel (19GJ/tonne for the most efficient mills, as shown in Chapter 6, section 6.3)).

Table 7.8 compares the energy inputs per unit output for conventional and organic farming methods for the production of barley and milk in Denmark. The near-elimination of energy use for fertilizers and pesticides is partly offset by a small increase in the embodied energy in machinery and in fuel energy use. In this case, the difference in energy use per unit of production between the two is smaller than the uncertainties in the estimates (with organic methods being slightly less energy intensive). Table 7.9 gives other data concerning typical energy use and yields for a variety of conventionally grown and organically grown food crops in Denmark. The greatest yield reduction with organic farming is for cereals (about 30 per cent at

Table 7.7 *Comparison of energy inputs (unspecified units) for conventional and organic farming in Finland*

	Rye		Milk	
	Conventional	Organic	Conventional	Organic
Pre-farm total	2.38	0.10	3.02	0.81
– Electricity	0.40	0.07	0.09	0.05
– Purchased fodder			0.19	0.13
– Fertilizer	1.98	0.03	2.74	0.63
On-farm total	1.26	1.51	1.12	1.33
– Electricity	0.08	0.08	0.74	0.73
– Fuels	1.18	1.43	0.39	0.59
Pre- and on-farm total	3.64	1.61	4.14	2.14
Post-farm total	11.02	11.02	1.99	1.99
– Bakery or dairy electricity	6.76	6.76	0.59	0.59
– Bakery or dairy fuels	3.81	3.81	0.40	0.40
– Packaging	0.45	0.45	1.00	1.00
Transportation of inputs and outputs up to the point of retail sale	0.67	0.71	0.25	0.29
Grand Total	15.33	13.34	6.39	4.41
Land area required (ha/FU):	0.188	0.319	0.240	0.434

Note: Required land areas are given as hectares per functional units (FUs) of either 1000kg bread or 1000 litres milk.
Source: Grönroos et al (2006)

Table 7.8 *Comparison of energy inputs (GJ/ha/yr) for conventional and organic systems of farming for two case studies in Denmark*

	Spring barley on irrigated sandy soil			Milk	
	Conventional	Organic		Conventional	Organic
	Direct energy use				
Fuel	3.4	5.0	Grazing fodder	3.6	2.3
Lubricants	0.3	0.4	Grass silage	2.4	1.5
Irrigation	1.5	1.5	Whole crop silage	1.0	0.8
Drying	0.5	0.4	Grain cereals	2.7	3.3
Subtotal	5.7	7.3	Concentrates	7.4	6.7
	Indirect energy use		Straw bedding	0.4	0.4
Machinery	1.1	1.6	Milking, milk cooling	8.0	8.0
Fertilizers and Lime	6.7	0.1	Farm buildings	2.5	2.5
Pesticides	0.3	0.0			
Subtotal	8.1	1.7			
Total energy use	13.8	9.0	Total energy use	28.0	25.6
Yield (kg/ha)	5000	3600	Yield (kg)	9000	9000
Energy use (MJ/kg)	2.8	2.5	Energy use (MJ/kg)	3.1	2.8

Source: Jørgensen et al (2005)

present, 20 per cent projected in the future), while energy use per unit of metabolizable energy output is about 25–70 per cent less. Case study results with 30–50 per cent less energy use under organic farming and negligible differences in yield are presented in

Tables 7.10 (for corn in Canada) and 7.11 (for a barley–wheat–sugar beet rotation in Germany). The energy inputs and yields for conventional and organic systems of olive production in Greece, as determined by Kaltsas et al (2007), are as follows: conventional,

Table 7.9 *Comparison of measured energy inputs and yields for current conventional and organic farming in Denmark, and as expected for future organic farming*

	Energy use (GJ/ha)	Yield (GJ/ha)	Energy input/output	Energy output/input
	Grass/clover			
Conventional	15.2	81	0.187	5.3
Organic now	4.0	65	0.062	16.3
Organic future	4.2	71	0.059	17.0
	Cereals			
Conventional	12.7	63	0.203	4.9
Organic now	6.3	43	0.148	6.7
Organic future	6.4	49	0.131	7.6
	Row crops			
Conventional	21.9	130	0.169	5.9
Organic now	15.2	121	0.125	8.0
Organic future	15.6	121	0.129	7.8
	Permanent grass			
Conventional	1.8	25	0.072	13.9
Organic now	1.1	23	0.049	20.5
Organic future	1.1	23	0.049	20.5

Source: Dalgaard et al (2002)

Table 7.10 *Comparison of energy inputs (MJ/kg) and yield (t/ha) for corn in south-western Ontario, Canada, using chemical fertilizers or swine manure*

	Course soil		Medium soil		Fine soil	
	Chemical	Manure	Chemical	Manure	Chemical	Manure
Seeds	0.31	0.33	0.35	0.33	0.40	0.40
Starter fertilizer	0.12	0.06	0.06	0.06	0.52	0.53
Fertilizer	0.75	0.00	1.58	0.00	0.92	0.00
Grain drying	0.59	0.59	0.61	0.59	0.59	0.58
Herbicides	0.28	0.29	0.20	0.18	0.24	0.24
Fuel	0.11	0.15	0.12	0.14	0.14	0.16
Total (MJ/kg)	2.16	1.41	2.92	1.29	2.81	1.91
Yield (t/ha)	8.84	8.44	7.90	8.34	6.78	6.82

Source: McLaughlin et al (2000)

20.7MJ/kg and a yield of 3.7t/ha/yr; organic, 17.5MJ/kg (a saving of 15 per cent) and a yield of 2.4t/ha/yr (a reduction of 35 per cent).

As discussed by Mason and Spaner (2006), modern crop varieties have been bred for conditions of high fertilizer input with weed and pest control through herbicides and pesticides. Breeding programmes may now have to re-optimize crops in order to maximize output in organic or other low-input farming systems, but such re-optimization would reduce the differences in yield between conventional and low-input systems.

If there were a widespread shift to organic agriculture with lower yields, this could in principle be compensated by a shift to less meat in the average diet (as explained below). As a quantitative example, suppose that organic methods reduce crop yields on average by 10 per cent. With about 50 per cent of global cropland devoted to providing food directly for humans and 50 per cent for animals, the consumption of animal products supported by croplands would need to be reduced 40 per cent (in fact, by slightly more than 40 per cent because the need for food directly consumed by humans would increase slightly). However, as not all

Table 7.11 *Comparison of energy inputs (GJ/ha) during the last 5-year rotation in a 32-year field experiment involving winter barley, winter wheat and sugar beet in Germany on relatively fertile soil using either chemical fertilizers or manure*

	Chemical	Manure
Seeds	1.12	1.12
Fertilizer	9.43	4.13
Herbicides	0.93	0.93
Fuel	4.22	4.68
Machines	1.81	2.54
Total (GJ/ha)	17.51	13.40
Dry yield (t/ha/yr)	10.80	10.85

Source: Hülsbergen et al (2001)

animals are fed with food from croplands, the required overall reduction in meat consumption would be considerably less.

7.7.3 Urban agriculture

Urban agriculture is agriculture within an urban area, in one of three locations: on residential plots (backyard or frontyard gardens), in community gardens and on small or medium-sized commercial farms in and around cities. According to Halweil (2002), there are 800 million urban farmers in the world, providing 60 per cent or more of the vegetables consumed in some cities. Urban agriculture is labour intensive but – in residential and community gardens – entails no energy inputs for farm machinery, need not require the use of pesticides and entails little or no energy in transporting the food to the consumer. Some forms of urban agriculture are referred to as *permaculture* (Mollison, 1990). Urban agriculture is an especially attractive option in low-density suburbs, which are difficult or impossible to retrofit to substantially higher density, and many of which are likely to otherwise become unviable if the cost of transportation increases substantially.

7.8 Energy use by fisheries

The collection of fish is generally energy intensive due to the fuel used by fishing vessels. Large vessels require a greater amount of fuel per tonne of fish caught than smaller vessels because they need to travel a greater distance and on average have a greater load than smaller vessels. For example, consider ten 10-tonne vessels versus

one 100-tonne vessel, each with the same energy requirement per tonne-km of cargo transported. Suppose that the vessel must travel x km per tonne of fish collected. The ten small vessels and the one large vessel each travel a total distance of $100x$ km during the fishing operation (plus an additional distance to and from the fishing area), but the average load will be ten times larger for the larger vessel. Thus (neglecting the empty weight of the vessels), the total tonne-km transported will be ten times larger. Although the energy intensity per tonne-km might be somewhat smaller for the larger vessel, it is evident that the single large vessel will require substantially more energy than the ten smaller vessels. More specifically, Pimentel and Pimentel (2008) state that twenty-two 15-tonne vessels use 44 per cent less energy than one 330-tonne vessel with the same fish yield. Near-shore fisheries require substantially less fuel than long-distance fisheries; Pimentel and Pimentel (2008) indicate almost a factor of six greater energy use per kg of fish for the Japanese fleet compared to vessels from Washington state fishing in the Alaska region.

Table 7.12 gives the ratio of fossil fuel energy input to protein energy output for various US fisheries. Herring requires the least energy input (2MJ/MJ) and king salmon the greatest energy input (40MJ/MJ) among fish species, with a large jump to the energy inputs required for shrimp (150MJ/MJ) and lobster (190MJ/MJ). Table 7.13 gives the ratio of fossil fuel energy input to protein energy output for various aquaculture systems. In the case of aquaculture, the largest energy input is often the energy required to provide the feed supplies to the fish. In many cases, fertilizer is applied in order to stimulate growth of phytomass for fish consumption. Input to output ratios range from 10:1 to 189:1.

Table 7.12 *Ratio of fossil fuel energy input to protein energy output for various US fisheries*

Fishery	Input/output ratio
Herring	2:1
Perch (oceanic)	4:1
Salmon (pink)	8:1
Cod	20:1
Tuna	20:1
Haddock	23:1
Halibut	23:1
Salmon (king)	40:1
Shrimp	150:1
Lobster	192:1

Source: Rawitscher and Mayer (1977)

Table 7.13 *Ratio of fossil fuel energy input to protein energy output for various aquaculture fisheries*

Fishery	Input/output ratio
Pond polyculture, Israel	10:1
Catfish, Louisiana	34:1
Sea bass, Thailand	65:1
Shrimp, Thailand	70:1
Oyster on land, Hawaii	89:1
Prawn, Malaysia	130:1
Lake Perch, Wisconsin	189:1

Source: Pimentel and Pimentel (2008)

Currently, most of the major fished stocks in the world have been overfished and if current trends continue, the total extermination of the world's commercial fisheries is projected to occur by about 2050 (Worm et al, 2006) (algae would take over). If and as commercial fisheries become further depleted, more energy will be required per unit of fish harvested – in the same way that more energy is required per unit of metal recovered during the mining and concentration stages of metal production as the grade of the ore declines (see Chapter 6, subsection 6.5.5). Mitchell and Cleveland (1993) report that the fuel requirements for the New Bedford fishery in Massachusetts per unit of fish protein increased by a factor of seven between 1966 and 1986, due in part to a decline in fish stocks and due in part to more international fishing competition.

7.9 Impact of a consumer shift to less energy-intensive diets

The proportion of meat in the human diet and specific choices within meat and vegetable categories strongly influence the total energy required in providing a meal with a given nutritional value. Daily protein requirements are roughly 2.2gm per kg of non-fat body mass (equal to about 85 per cent of total body mass for people who are not overweight) or 85140gm/day for adults. Average per capita protein consumption in many countries is greatly in excess of the required amount, and most of that is in the form of meat. However, the daily protein requirements can be easily supplied in a balanced way with various high-protein plant foods, some of which are listed in Table 7.14. The ready availability of plant foods with more than adequate protein content means that shifts to

Table 7.14 *Protein content of high-protein plant foods on a dry-weight basis and comparison with dairy products and meat*

Food product	Protein content (%)	Food product	Protein content (%)
Vegetables		*Nuts and seeds*	
Mushrooms	43	Hemp seeds	27
Spinach	39	Pumpkin seeds	23
Asparagus	34	Flax seeds	17
Hemp bread	30	Peanut butter	15
Broccoli	27	Almonds	14
Squash	24	Walnuts	14
Legumes		Pistachios	14
Soy (veggie) burgers	50	*Diary products*	
Tofu	40	Eggs	33
Tempeh	34	Cheese	33
Edamame	30	Milk	27
Lentils	30	Ice cream	7
Peas	26	*Meat*	
Refried beans	24	Lean beef	45
Humus	18	Salmon	44

Note: Differences in crop yield (mass produced per hectare per year) need to be taken into account (along with inverse feed efficiencies for animal products) when comparing different protein sources in terms of land area requirements to meet human protein needs.

Source: Precision Nutrition (*www.precisionnutrition.com*)

low-meat or vegetarian or vegan diets are feasible options (from a nutritional point of view) in reducing the energy requirements of the food system and in freeing up land that can either be used for bioenergy crops, allowed to revert to nature or used to produce additional food for direct human consumption in the event that real (physical) food shortages reappear in the future.

7.9.1 Role of meat consumption

Table 7.15 gives global mean ratios of food energy input to edible animal product output (in terms of energy) for various animal products, as well as the average values in different regions. In systems where animals are largely pasture-fed, much of the food energy consumed is used in moving (and perhaps also in keeping warm), so large feed energy inputs are required for a given increase in animal bulk. Average energy input to output ratios range from 7.7 for milk to 117 for lean beef. Global mean input to output ratios, weighted by total consumption of each commodity, are 7.7 for diary products (milk and eggs), 24 for all fatty meats (excluding seafood) and 52 for all non-fatty meats (excluding seafood). Inasmuch as the land used to produce feed for animals could be used instead to produce bioenergy crops, the energy value of the feed given to animals will be counted here as a real energy cost of animal-based food for humans.[4] This energy cost is in addition to energy inputs required for producing animal feed (such as fertilizers, pesticides and fuel for farm machinery) and in transporting and processing animals and animal products.

Figure 7.12 gives an overview of the flows of phytomass energy in the world food system, based on the same data as used in preparing Table 7.15.[5] There is almost 2.5 times as much pastureland as cropland in the world, and about half of the world's croplands are devoted to producing food for animals rather than to directly producing food for humans. Cropland produces about 72EJ/yr of energy in the form of food crops for human or animal consumption, and another 61EJ/yr as humanly inedible crop byproducts. About 28EJ/yr of phytomass energy (as grains, fruits and vegetables) flows directly into the food-processing system but only 17.7EJ/yr of processed foods are produced – an efficiency of 62 per cent. Only 4.8EJ/yr are consumed as unprocessed food. About 109EJ/yr of phytomass energy flows into the animal-production system, but only 4.1EJ/yr is supplied as meat – an extraordinarily low efficiency of only 3.8 per cent. The low efficiency of the meat production system in meeting food requirements means not only that much land is required, but also that more fertilizers, pesticides, herbicides and fuels are required to provide a given amount of food energy to people. As well, significant emissions of methane and N_2O are associated with livestock (Steinfeld and Wassenaar, 2007). Finally, a further 30 per cent of the combined plant plus animal food supply is wasted (giving a final consumption of only 19.2EJ/yr out of 27.1EJ/yr delivered at the wholesale level).

The overall global average inverse efficiencies (phytomass energy input over food energy output at the wholesale level) consistent with the flows shown in Figure 7.12 are as follows: for plant products (vegetables, fruits,

Table 7.15 *Ratio of phytomass energy input to the metabolizable energy of animal products consumed by humans (MJ/MJ)*

Region	Animal product							
	Beef	Pork	Poultry	Fatty beef	Fatty pork	Fatty poultry	Milk	Eggs
East Asia	145	22	20	67	8.3	9.1	7.7	7.7
Eastern Europe	71	21	18	36	7.7	7.3	6.7	7.7
Latin America and Caribbean	125	36	17	59	12	7.7	9.1	7.1
North Africa and Middle East	133	22	20	59	8.3	9.1	10	7.7
North America and Oceania	59	16	13	31	6.3	6.7	4.8	5.9
South and Central Asia	227	31	24	104	11	11	10	9.1
Sub-Saharan Africa	172	33	26	77	11	11	19	10
Western Europe	56	16	12	29	6.3	5.9	5.3	5.3
Weighted world average	117	21	17	55	7.9	8.0	7.7	7.4

Source: Computed from data in Wirsenius (2000)

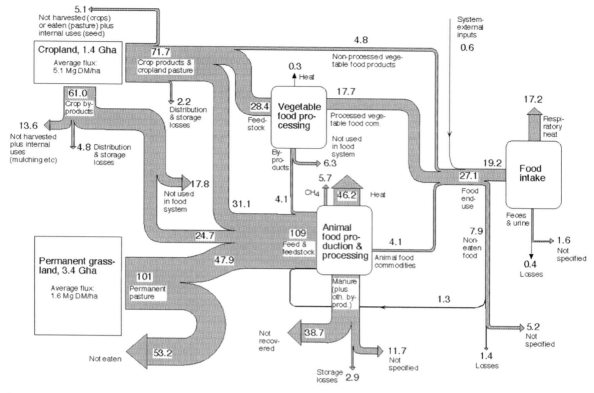

Source: Wirsenius (2003)

Figure 7.12 *Phytomass energy flows (EJ gross energy/yr) in the world food system estimated for the period 1992–1994 (with a global population of 5.54 billion)*

grains), 1.5; diary products, 7.7; and meat products, 44.6. However, only 71 per cent of the food delivered at the wholesale level is consumed by people (19.2EJ/yr out of 27.1EJ/yr). Thus, the inverse efficiencies based on food consumed are 2.3 for plant products, 11 for dairy products and 60 for meat products.

The large phytomass requirement per unit of meat food energy, compared to food energy from vegetables, implies that diets high in meat, especially beef, require greater demands for fertilizer (and other agricultural inputs) than low-meat diets, except where animals are fed on unfertilized pastures (in which case large land areas are required). Even with recycling of all animal wastes (manure and urine) back to the soil, large amounts of N fed to animals are lost from the soil–animal system because of the ready volatilization of NH_3 (about 25 per cent of total N excreted in open cattle feedlots is lost in this way) and leaching of nitrate

from animal wastes. Thus, Bleken and Bakken (1997) calculate that 21gm of N must be added to the soil in order to supply 1gm of N to humans in a mixture of milk and several kinds of meat, whereas only 3gm of N have to be supplied to the soil in order to supply 1gm of N to humans in wheat flour. The fertilizer requirements of a high-meat diet are such that Donner (2008) finds, using a model of nitrogen cycling in the US Mississippi basin, that a hypothetical shift from meat production to production of grains with the same protein content for direct consumption by humans would be sufficient to eliminate the anoxic dead zone in the Gulf of Mexico (which is related to runoff of N fertilizers, as explained earlier), and that other measures will almost certainly need to be accompanied by a partial shift away from meat production if the dead zone is to be eliminated.

To underline the importance of the amount of animal products in the human diet, Table 7.16 gives the

Table 7.16 *Food energy consumption (including losses by wholesalers and beyond) and phytomass energy requirements for different diets assuming inverse efficiencies of 1.5 for plant food, 7.7 for dairy products and 44.6 for land meat products[a]*

	Current average	Vegan	Vegetarian	Moderate	Affluent
			Diet scenarios		
Wholesale food energy supply (MJ/person/day)					
Plant energy	10.95	13.10	11.90	9.44	6.84
Land meat energy	1.04	0.00	0.00	0.60	2.20
Seafood energy	0.12	0.00	0.00	0.06	0.06
Dairy energy	0.99	0.00	1.20	3.00	4.00
Total	13.10	13.10	13.10	13.10	13.10
Required phytomass (MJ/person/day)					
Related to plant food	16.4	19.7	17.9	14.2	10.3
Related to meat food	46.3	0.0	0.0	26.8	98.1
Related to dairy	7.6	0.0	9.2	23.1	30.8
Total	70.3	19.7	27.1	64.0	139.2

Note: [a] Per capita wholesale food energy supplies are global mean gross energy consumptions computed from data in Wirsenius (2000), and are consistent with the regional per capita metabolizable energy values given in Figure 7.13. Global mean inverse feed efficiencies are from the same data source, and are consistent with the inverse efficiencies given for individual animal products in Table 7.15 and the proportions of meat and dairy consumption given here.

current global average per capita food energy supply (at the wholesale level) from plants, dairy products, land meat and seafood, as well as hypothetical breakdowns for a vegan diet (no meat or dairy products), a vegetarian diet, an affluent diet (where much of the world moves to North American and European levels of meat consumption) and a diet intermediate between vegetarian and affluent.[6] These times the inverse efficiencies given above (at the wholesale level) give the average land phytomass energy requirements for the different diets. As seen from Table 7.16, there is a factor of seven difference between the phytomass energy requirement of the vegan and affluent diets, and a factor of five difference between the vegetarian and affluent diets. Assuming comparable yields per unit of land area for phytomass grown for human and animal consumption, this translates into a five to seven times greater land area requirement for the affluent diet. This rough global average analysis is consistent with the detailed analysis of Peters et al (2007), who examined the land area requirements when producing mixtures of specific food products in New York state for 42 different diets. They found almost a factor of five difference between the diets with the greatest and smallest land area requirement.

It should also be noted that diets high in meat (especially beef) require substantially greater water inputs (often through the extraction of groundwater at unsustainable rates in order to irrigate the crops fed to animals). Thus, Pimentel et al (1997) indicates water requirements of 500 litres/kg for potatoes, 900 litres/kg for wheat and 1400 litres/kg for corn, but 3500 litres/kg for poultry and 43,000 litres/kg for beef.

7.9.2 Patterns and trends in meat consumption

Figure 7.13 compares the per capita supply of food energy from meat, dairy products and plant products, as well as the per capita losses, in different world regions. The world average supply of metabolizable food energy at the wholesale level is 13.1MJ/person/day, but ranges from 15.1MJ/person/day to 9.1MJ/person/day at the regional scale. The amount of food actually consumed per person, and relative losses, are largest for North America (9.3MJ/person/day) and smallest for sub-Saharan Africa (7.9MJ/person/day).

Figure 7.14 gives the trend in worldwide production of meat from land animals, the harvesting of fish and the production of seafood meat through aquaculture, as well as total meat consumption and world average consumption per capita from 1960 to 2005. Total per capita meat consumption (land animals plus seafood) has almost doubled over the past 40 years.

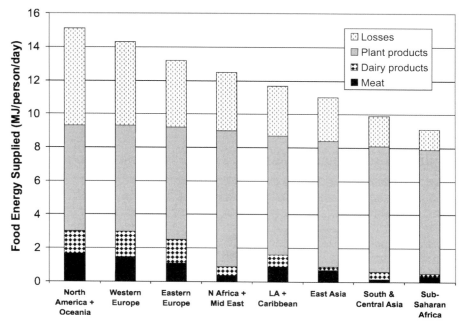

Source: Data from Wirsenius (2000)

Figure 7.13 *Comparison of per capita food energy consumption (MJ metabolizable energy/person/day) in the form of meat, dairy products and plant products in different world regions, as well as per capita waste of food after delivery to wholesalers*

Figure 7.15 compares per capita land meat and seafood consumption in Japan, the US, the EU, China and India according to one source. Table 7.17 compares estimates of annual average per capita meat consumption in 14 European countries; the average meat consumption in this table (about 85kg/person/yr) is larger than the EU average (about 70kg/person/year) given in Figure 7.15. Whatever the correct EU average, average per capita meat consumption in Japan and the US is substantially greater than in the EU – and, according to Pimentel, 2004, average US protein consumption (75gm/person/day from animals and 37gm/person/day from plants) is about twice the recommended daily allowance (56gm/person/day). As noted by Halweil and Nierenberg (2008), the world's fisheries receive about $30–40 billion/year in subsidies (in the form of low-interest loans, port development and fuel subsidies), while direct subsidies for livestock in the US alone amounted to a total of $2.9 billion between 1995 and 2005. Feedlot operators are also subsidized. Elimination of subsidies would be a first step in restraining or even reversing the growth in meat consumption.

A low-meat diet does not necessarily represent lower total energy use than a high-meat diet if the meats (especially poultry and pork) are replaced with vegetables flown a great distance. This tradeoff is discussed in the next section.

7.9.3 Embodied energy in different food products

The embodied energy of a food product involves the energy used in the production of the initial food crop (direct farm energy use and indirect farm energy use through fertilizers and pesticides) and the energy used in transporting, processing, packaging and storing the product. Table 7.18 is a compilation of estimates of the direct and indirect farm-level energy inputs for the production of various foods in different countries. The energy value of the food output is generally several times the fossil fuel energy input for production, but energy used for the production of food is generally a small fraction of the lifecycle energy use. The required farm-level energy input for most food crops is 2–7MJ/kg.

Table 7.19 gives the embodied energy of food products in Sweden, supplemented by additional

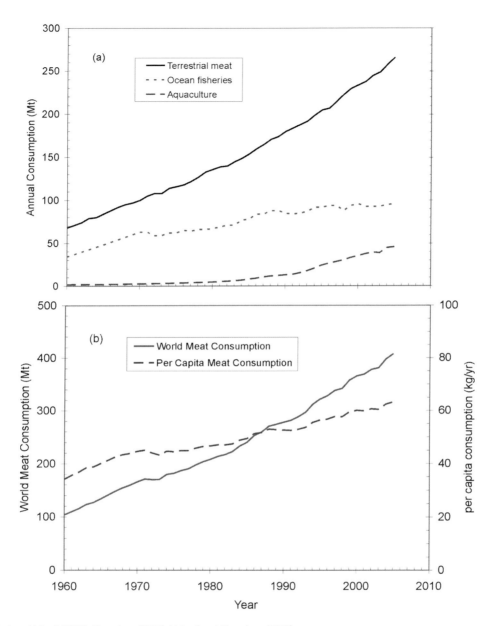

Source: Data from Halweil (2005), Nierenberg (2005), Halweil and Nierenberg (2008)

Figure 7.14 *Annual animal production for human consumption and catches of wild and aquaculture fish from 1960 to 2005*

estimates for beverages in Norway, as estimated by Carlsson-Kanyama et al (2003). Included in this analysis is the energy used on farms, in making pesticides and fertilizers, in drying, processing and transporting food products to the retailer, and in storing and cooking the food in households. Not included is the energy used to make machinery and buildings, for packaging, treating waste, transporting the food from the retailer to the consumer and washing dishes. Also not included is the energy content of the

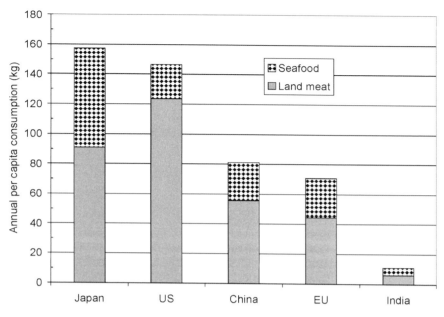

Source: Data from Halweil and Nierenberg (2008)

Figure 7.15 *Per capita meat consumption in 2005*

Table 7.17 *Comparison of annual per capita meat consumption in different European countries*

	Annual meat consumption (kg)					Daily mean meat consumption (grams)
	Beef	**Pork**	**Poultry**	**Other**	**Total**	
Spain	12.7	55.3	25.5	6.6	100.1	274
Denmark	17.6	64.2	15.3	1.2	98.3	269
Austria	19.6	56.9	15.3	1.2	93	255
Belgium	21.2	46.6	23.1	2.1	93	255
France	28.1	35.9	22.6	5.3	91.9	252
Ireland	14.5	37.9	30.9	7.2	90.5	248
Netherlands	19.8	46.3	20.1	1.3	87.5	240
Germany	16.6	55	13.4	1.1	86.1	236
Portugal	17.6	34.7	23	3.6	78.9	216
Italy	25.9	33.1	18.4	1.7	79.1	217
Greece	19.6	24.8	17.7	13.6	75.7	207
UK	17.5	23.1	25.1	6	71.7	196
Sweden	18.2	36.1	7.9	0.7	62.9	172
Finland	19.1	32.2	8.8	0.5	60.6	166

Source: Gerbens-Leenes and Nonhebel (2002)

Table 7.18 *Energy inputs at the farm for the production of various crops, and ratio of food energy produced to energy input*

Crop, region	Output/input (MJ/MJ)	Input (MJ per kg)	Yield (t/ha)	Reference
Grains and legumes				
Corn, Mexico, human	10.7	1.4	1.94	PP2008(1951)
Corn, Nigeria, human	6.4	2.3	1.00	PP2008(1971)
Corn, Guatemala, human	4.8	3.1	1.07	PP2008(1940)
Corn, Philippines, oxen	5.1	2.9	0.94	PP2008(1960)
Corn, Mexico, oxen	4.3	3.4	0.94	PP2008(1951)
Corn, Guatemala, oxen	3.1	4.8	1.07	PP2008(1940)
Corn, Indonesia, bullock	1.1	14.2	1.20	PP2008(1988)
Corn, US, modern	3.8	3.9	8.66	PP2008(2005)
Wheat, India	2.1–4.1	4.6–9.0		Singh et al (2007)
Wheat, Canada	3.7	5.2		Piringer and Steinberg (2006)
Wheat, US	3.9–6.1	3.1–4.9		Piringer and Steinberg (2006)
Winter wheat, Japan	7.2	2.6		Koga (2008)
Wheat, Europe	7.9–12.7	1.5–2.4		Piringer and Steinberg (2006)
Wheat, India, bullocks	1.0	14.4	0.82	PP2008(1966)
Wheat, US, modern	2.1	6.6	2.67	PP2008(2006)
Wheat, Kenya	3.3	4.3	1.79	PP2008(1989)
Oats, US, modern	5.1	2.8	3.22	PP2008(1980)
Rice, India	6.6–8.7	2.2–2.9		Baruah and Dutta (2007)
Rice, Borneo, human	7.1	2.1	2.02	PP2008(1955)
Rice, Philippines, carabao	3.3	4.6	1.65	PP2008(1965)
Rice, Japan, modern	2.8	5.4	6.33	PP2008(1991)
Rice, US, modern	2.2	6.7	7.37	PP2008(2006)
Sorghum, Sudan, human	14.4	1.0	0.90	PP2008(1965)
Sorghum, US, modern	2.0	7.4	3.03	PP2008(1980)
Soybeans, Illinois	5.0	3.4	3.00	PP2008(1999)
Soybeans, US	3.2	4.7	2.67	PP2008(2006)
Soybeans, Philippines	1.5	11.5	0.99	PP2008(2001)
Dry beans, US	1.8	7.9	1.46	PP2008(1976)
Cowpeas, Nigeria	6.5	2.2	1.53	PP2008(1977)
Adzuki bean, Japan	4.2	2.29		Koga (2008)
Peanuts, Thailand	2.6	6.3	1.28	PP2008(1977)
Peanuts, Georgia (US)	1.4	12.3	3.72	PP2008(1980)
Fruits and vegetables				
Apples, Greece	1.0	2.5		Strapatsa et al (2006)
Apples, US	0.61	3.8[a]	55.0	PP2008(2006)
Oranges, Florida	1.0	2.1[b]	46.0	PP2008(2006)
Potato, Japan	14.7	1.24		Koga (2008)
Potatoes, US	1.3	1.8	40.7	PP2008(2006)
Potatoes, UK	1.6	1.4	26.3	PP2008(1976)
Potatoes, Philippines	0.42	5.8	5.5	PP2008(2001)
Sweet potatoes, Vietnam	2.0	2.1	11.9	PP2008(2001)
Cassava, Tanga region, Africa	22.9	0.6	5.8	PP2008(1968)
Cassava, various	3.6	4.4	12.4	PP2008(1996)
Cabbage, US	1.8	1.2	38.4	PP2008(2001)
Cabbage, India	0.53	4.0	11.4	PP2008(2001)
Spinach, US	0.23	4.8[c]	11.2	PP2008(1977)
Tomatoes, US	0.26	3.2[d]	41.8	PP2008(2006)
Tomatoes, Pakistan	0.94	0.9	14.8	PP2008(2001)
Tomatoes, Turkey	0.8	1.0		Çetin and Vardar (2008)
Brussels sprouts, US	0.69	2.7[e]	12.3	PP2008(1976)
Sugar beet, Japan	29.2	0.54		Koga (2008)
Sugar beet, UK	3.6	0.8	35.5	PP2008(1976)

Note: Solar energy inputs are not included, but the energy input for rice includes human energy. Output to input ratios in some cases were deduced from, rather than given in, the indicated reference. PP2008(xxxx) = Pimentel and Pimentel (2008), where xxxx is the date of the source that is cited in PP2008. [a] 34 per cent as human labour. [b] 41 per cent as human labour. [c] 50 per cent for N fertilizer alone. [d] 45 per cent as human labour. [e] 33 per cent as N fertilizer.

Table 7.19 *Secondary energy inputs used for the production and delivery of food to consumers in Sweden and drinks in Norway*

Category	Food type and origin	Embodied energy (MJ/kg)
	Food and drinks in Sweden	
Meat from land animals	Beef, frozen, Central Europe	75
	Beef, fresh, Sweden	70
	Lamb, frozen, overseas	53
	Lamb, fresh, Sweden	43
	Pork, frozen, Central Europe	44
	Pork, fresh, Sweden	40
	Chicken, frozen, Central Europe	41
	Chicken, fresh, Sweden	35
	Pork sausage, fresh, Sweden	34
	Lamb sausage, fresh, Sweden	30
	Chicken sausage, fresh, Sweden	20
	Beef stew, Sweden	24
	Lamb stew, Sweden	18
	Pork stew, Sweden	17
	Chicken stew, Sweden	13
Meat from fish and crustaceans	Shrimps, without shells, Sweden	220
	Cod, fresh, Sweden	105
	Salmon, farmed, Sweden	84
	Canned tuna, overseas	44
	Mackerel, fresh, Sweden	37
	Herring, fresh, Sweden	22
	Clams, tinned, Sweden	19
Milk, cheese, and eggs	Cheese, Southern Europe	65
	Cheese, Sweden	60
	Milk powder, Sweden	58
	Cream (40% fat), Sweden	19
	Eggs	18
	Yoghurt, Central Europe	12
	Milk (4% fat), Sweden	5.9
	Milk (1.5% fat), Sweden	5.0
Legumes	Beans, canned, overseas	20
	Brown beans, overseas	11
	Brown beans, Sweden	8.9
	Soya beans, overseas	7.9
	Yellow peas, Sweden	5.0
Fruits	Tropical, fresh, overseas by plane	115
	Apples, dried with commercial energy, overseas	38
	Apples, sun dried, overseas	18
	Apples, fresh, overseas	8.6
	Apples, fresh (in-season), Central Europe	4.8
	Raisins, sun dried, overseas	23
	Bananas, fresh, overseas	12
	Grapes, fresh, overseas	9.7
	Cherries, fresh, overseas	9.6
	Oranges, fresh, overseas	9.4
	Oranges, fresh (in-season), Southern Europe	6.8
Vegetables	French fries, Sweden, cooked as one portion	60
	French fries, Sweden, cooked as four portions	30
	Potatoes, Sweden, baked	29
	Potatoes, Sweden, boiled	4.6
	Potatoes, Sweden, mashed powder	5.6
	Tomatoes, fresh, greenhouse, Sweden	66
	Tomatoes, canned, Southern Europe	15
	Tomatoes, fresh, Southern Europe	5.4
	Broccoli, frozen, Europe	18
	Vegetables, canned, overseas	18

Category	Food type and origin	Embodied energy (MJ/kg)
	Olives, canned, Southern Europe	15
	Carrots, canned, Central Europe	11
Berries	Strawberries, fresh, Middle East by plane	29
	Strawberries, fresh, Central Europe	16
	Strawberries, fresh, Sweden	6.2
	Raspberries, frozen, Central Europe	16
	Raspberries, fresh, Central Europe	7.5
	Blueberries, frozen, Central Europe	9.0
	Raspberry jam, 55% fruit, Sweden	16.0
Breakfast cereals	Baked cereal, Sweden	37
	Muesli with sundried raisins, Sweden	17
	Oat flakes	11
	Oat flake porridge	2.5
Cereals	Pasta, Southern Europe	7.5
	Couscous, Central Europe	5.3
	Rye flour, Sweden	5.2
	Wheat flour, Sweden	5.0
	Barley, Sweden	2.0
Bread and Pastries	Biscuits with butter, Central Europe	26.0
	Sweet bread with butter, Central Europe	21
	Sweet bread with margarine, Central Europe	18
	Sweet bread with margarine, Sweden	15
	Bread, frozen, bakery far away	13
	Bread, frozen, local bakery	12
	Bread, fresh, local bakery	8.9
Spices	Herbs, commercially dried, Southern Europe	36
	Herbs, sun dried, Southern Europe	16
	Herbs, sun dried, overseas	23
Oil and fat	Butter, Sweden	40
	Margarine, 80% fat, Sweden	17
	Olive oil, Southern Europe	24
	Sunflower oil, overseas	20
	Rapeseed oil, Central Europe	15
	Soya oil, overseas	14
Sugar and candies	Chocolate, Central Europe	44
	Candies, Sweden	18
	Ice cream, Sweden	15
	Honey, overseas	5.6
	Honey, Sweden	1.3
Drinks	Wine, overseas	14
	Wine, Southern Europe	12
	Beer, Sweden	12
	Orange juice, overseas	10
	Apple juice, Central Europe	7.1
	Soft drinks, Central Europe	7.1
	Soft drinks, Sweden	5.9
	Water from bottle, Central Europe	2[a]
	Drinks in Norway	
Drinks	Beer	8.6
	Juice	4.4
	Coffee	4.1
	Milk	3.8
	Soft drinks	2.5
	Bottled water	1.8[a]
	Tap water	0.0012

Note: Not included is the energy content of the phytomass inputs to the food production system.
[a] Much higher values (5.6–10.2MJ/litre) have been computed for bottled water delivered to Los Angeles (see Chapter 8, Table 8.1)
Source: For food in Sweden, Carlsson-Kanyama et al (2003); for drinks in Norway, Hanssen et al (2007)

phytomass fed to animals or used as input for the processing of plant foods, nor the energy used in transporting animal feed to the animals (which can involve distances of thousands of km). Only secondary energy is considered, such as electricity but not the energy lost in producing electricity. Food grown in greenhouses is rather energy intensive (so tomatoes grown in greenhouses in Sweden require an estimated 66MJ/kg, compared to 5.4MJ/kg for fresh tomatoes imported from Southern Europe), but if greenhouses with high-performance glazing were used this high energy intensity would be reduced. Also note the particularly high energy intensity for shrimp (220MJ/kg), tropical fresh fruit imported by plane (115MJ/kg), cheese (60–65MJ/kg), chocolate (44MJ/kg) and butter (40MJ/kg). Soybeans (which can be used to produce alternatives to milk and other dairy products) have an energy intensity comparable to milk (8MJ/kg versus 6MJ/kg), although additional energy would be required in processing soybeans into dairy substitutes (but milk entails substantial non-energy related emissions of methane and nitrous oxide, as discussed later). When daily diets are composed with comparable nutritional levels, a factor of four difference is found in daily food energy inputs between the lowest energy diet (13MJ/day) and the highest energy diet (51MJ/day) (both diets contain equal amounts of meat). Up to one third of food embodied energy is for sweets and drinks that have no nutritional value.

The absolute energy use indicated in Table 7.19 for comparable drinks is quite different according to the Swedish and Norwegian analyses. This largely reflects uncertainties in energy accounting. However, the relative differences between different drinks within a given analysis are probably reliable. On this basis, beer and wine are the most energy-intensive drinks, followed by juice, coffee and milk, then soft drinks, bottled water and tap water. The very small embodied energy in tap water is related to water purification and treatment.

Table 7.20 gives the embodied energy of food items consumed in the UK, as compiled by Coley et al (1998) from Dutch data (included is energy used for the production, transport and retailing of the food items, but not for household preservation and preparation). The energy embodied in British diets, based on a survey of 2197 real adult diets, ranges from less than 15MJ/person/day to more than 100MJ//person/day, with an average embodied energy

of 50MJ/person/day (comparable to the high embodied energy Swedish diet as estimated by Carlsson-Kanyama et al, 2003).

Figure 7.16 shows the energy used in different stages from the production to consumption of a can of corn in the US; energy used directly or indirectly on the farm is only 15 per cent of the total energy use, while packaging in this case accounts for 33 per cent of the total energy use – about four times that for the US food system as a whole, as given in Figure 7.2. The ratio of embodied energy to food energy is 8.1 – slightly higher than the ratio of seven for the US food system as a whole.

Table 7.21 compares the range of embodied energies for various food groups from Tables 7.19 and 7.20 with the range as estimated for The Netherlands. The most notable differences are the much lower energy inputs given for fruit and vegetables in The Netherlands. Note that the energy inputs required for fresh fruit transported by air can exceed the energy inputs required for consumption of land meat in Northern Europe, but will often be less than the embodied energy in seafood. Fresh fruits transported by ship generally entail substantially less energy inputs than land or sea meat.

The above assessments of the energy embodied in animal products and the comparison with fresh fruits from distant locations do not account for the energy content of the phytomass that is fed to animals. Given our interest, in Volume 2, in the prospects of developing an entirely renewably based energy system – one component of which would be biomass energy that could be grown on land presently used to support animals for human consumption – the phytomass energy input, although renewable, should be considered alongside the fossil fuel energy inputs. The calculation of the phytomass energy inputs and their summation with fossil fuel energy inputs is presented in Table 7.22, based on data from Pimentel (2004) for US products. The derived inverse feed efficiencies (MJ phytomass per MJ of animal product) are roughly comparable to those given in Table 7.15 for most products.[7] Per kg of animal product, phytomass energy inputs range from 13MJ/kg for milk from grain-fed dairy cattle to 3800MJ/kg for beef grazing on poor pasture. Fossil fuel energy inputs range from 11MJ/kg (milk) to 301MJ/kg (lamb). For beef, the fossil fuel input is about 90–180MJ/kg. This range is higher than the energy intensity range given for beef in Table 7.20 for the UK (45–109MJ/kg, including transportation

Table 7.20 *Embodied energy of food items consumed in the UK*

Food item	Energy (MJ/kg)	Food item	Energy (MJ/kg)
		Meat	
Whitefish	151	Coated chicken	88
Sausage	109	Shellfish	76
Pork	107	Oily fish	75
Meat pie and pastry	103	Liver	60
Beef	101	Bacon and ham	64
Burgers and kebabs	88	Chicken and turkey	45
		Dairy products	
Cottage cheese	75	Polyunsaturated margarine	26
Eggs	58	Skimmed milk	8.3
Butter	55	Whole milk	7.4
Milk and cream	34	Soft margarine	4.0
Yoghurt	30	Ice cream	2.7
		Fruits and nuts	
Unsalted nut and fruit mix	65	Apples and pears	23
Preserves	34	Oranges and tangerines	14
Canned fruit	31	Bananas	9.3
		Vegetables	
Fresh tomatoes	35	Peas	22
Baked beans	33	Green leafy	10
Green beans	32	Carrots (cooked)	9.3
		Grains and grain products	
Buns, cakes, pastries	61	Pasta	14
Biscuits	49	Wholewheat bread	9.5
Whole grain, high-fibre cereals	25	White bread	7.8
Rice	17		
Savoury snacks	188	Sugar confectionary	13
Chocolate confectionary	43	Potato chips	6.9
Sugar	15		
		Drinks	
Spirits and liqueurs	121	Fruit juice	15
Wine	62	Milk (from above)	7–8
Cider and perry	38	Coffee	3.7
Soft drinks	18	Tea	0.64

Note: Energy for storage and preparation of food at home is not included here, nor is the energy content of biomass fed to animals or intended for consumption by humans.
Source: Estimated by Coley et al (1998) from Dutch data

and processing energy use), and the value of 70MJ/kg given in Table 7.19 for fresh beef in Sweden.

The important result in Table 7.22 is the remarkably high total energy input for animal products, including both fossil and phytomass energy – about 21–26MJ/kg for dairy products except eggs, almost 300MJ/kg for eggs from grain-fed chickens, almost 150MJ/kg for grain-fed pork, about 350MJ/kg for gain-fed beef and over 3600MJ/kg for beef grazing on poor pasture. Inasmuch as the land used by grazing animals could almost always be used to produce bioenergy crops that could displace fossil fuels (an exception being alpine meadows grazed

during the summer), these energy costs are properly regarded as fossil fuel energy costs. By comparison, the energy cost of fresh fruit sent by air from the tropics to Northern Europe is approximately 100MJ/kg, while that of strawberries sent from the Middle East to Europe by air is about 30MJ/kg (see Table 7.19).

7.9.4 Comparative greenhouse gas emissions

Emissions of methane (CH_4) and nitrous oxide (N_2O) occur in association with livestock. Nitrous oxide

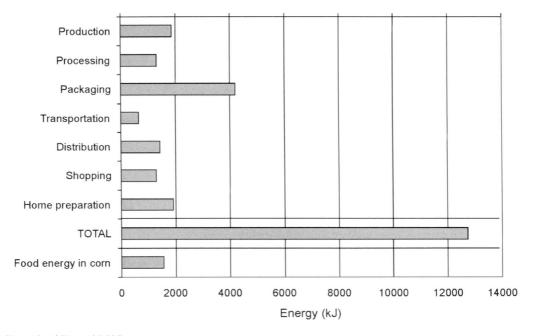

Source: Pimentel and Pimentel (1996)

Figure 7.16 *Energy in a 455gm can of sweetcorn*

Table 7.21 *Summary of the range in food lifecycle energy inputs from Tables 7.18 and 7.19 for food consumed in Sweden and the UK, and comparison where possible with estimates for food consumed in The Netherlands as reported by Dutilh and Kramer (2000)*

Food group	Lifecycle energy (MJ/kg)		
	Sweden	UK	The Netherlands
Meat from land	13–75	45–109	30–70
Meat from oceans	19–220	75–150	20–40
Cheese	60–65	75	50–60
Eggs	18	58	20
Milk	5–6	7–8	10
Legumes	5–20	22–33	
Other vegetables	11–60	9–35	1–4
Fruit, unspecified		9–31	2–5
Fruit, air freight	29–115		
Fruit, ship	9–10		
Fruit, local	5		
Breakfast cereals	2.5–37	25	
Cereals	2–7.5	10–25	
Breads	9–21	10	
Sweets	1–44	13–190	

Note: The lifecycle energy inputs given here do not include phytomass inputs.

Table 7.22 *Energy, protein and water content of different animal products; feed requirements per kg of animal product when fed grain or allowed to graze on pasture, and phytomass energy, fossil fuel energy and total energy inputs per unit of animal product*

Animal Product	Quantities per kg of animal product			Feed (kg) per kg of animal product		Phytomass (MJ) per MJ of animal product		Phytomass (MJ) per kg of animal product		Fossil energy input (MJ)			Energy input (MJ) per MJ of animal product		Energy input (MJ) per kg of animal product	
	Energy (MJ)[c]	Protein (gm)	Water (%)	Grain	Forage	Grain	Forage	Grain	Forage	per MJ of animal protein	per MJ of animal product[a]	per kg of animal product[b]	Grain	Forage	Grain	Forage
Lamb	10.6	220	47	21	30	36	51	378	540	57	21	223	57	72	601	763
Beef	10.7	186	49	13	30	22	50	234	540	40	12	132	34	63	366	672
	10.7	186	49		200		335		3600	20	6.2	66		341		3666
Turkey	5.0	123	55	3.8		14		68		10	4.4	22	18		90	
Eggs	6.1	116	74	11		32		198		39	13	81	45		279	
Pork	9.8	134	57	5.9		11		106		14	3.4	33	14		140	
Dairy	2.7	34	87	0.7	1	4.7	6.6	13	18	14	3.1	8	7.8	9.8	21	26
Poultry	5.7	238	71	2.3		7.3		41		4	3.0	17	10.3		58	

Note: Two forage entries are given for beef, the first for high-quality pasture and the second for low-quality pasture. [a] Computed as (MJ fossil energy input)/(MJ protein) × (MJ protein)/(kg animal product) ÷ (MJ animal product)/kg animal product), where (MJ protein)/(kg animal product) is computed from the given (gm protein)/kg times a protein energy content of 24MJ/kg. [b] Computed as (MJ fossil energy input)/(MJ protein) × (MJ protein)/(kg animal product). [c] Human-metabolizable energy per kg of product including water.

Source: First five data columns and column ten, Pimentel (2004); other columns, computed from first five columns assuming an energy value of 18MJ/kg for phytomass fed to animals and a human-matabolizable energy value of 17.8MJ/kg for protein (see Blaxter, 1989, Table 8.3).

emissions are also associated with the use of N fertilizer that is applied to plant crops and, as noted earlier, 5–30 times more plant mass is typically required to supply a MJ of food to humans in the form of animal products (depending on the product) than as processed plant products. This in turn increases the requirement for N fertilizer and associated N_2O emissions. Methane is produced through anaerobic fermentation of food inside the gut of ruminant animals (such as cattle and sheep), and both methane and nitrous oxide are produced during the handling and processing of animal manure (Steinfeld and Wassenaar, 2007).

As discussed in Chapter 2 (subsection 2.6.5), emissions of different GHGs can be converted to an equivalent CO_2 emission and summed using the global warming potential index. Figure 7.17a gives an estimate of the equivalent CO_2 emissions due to fossil fuel energy inputs (assuming the fossil fuel to be coal), methane emissions from enteric fermentation and disposal of manure, and N_2O emissions due to disposal of manure associated with various animal products (emission due to fertilization of crops fed to animals are not included). As seen from Figure 7.17a, CH_4 and N_2O emissions are more significant than CO_2

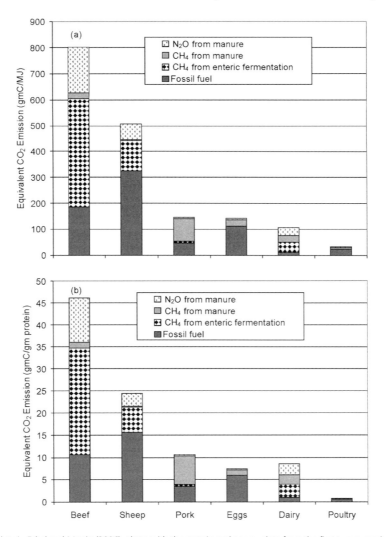

Source: Computed from data in Eshel and Martin (2006), along with the protein and energy data from the first two numeric columns of Table 7.22

Figure 7.17 *CO$_2$-equivalent GHG emissions (a) per MJ of animal product, and (b) per gm of protein in animal products*

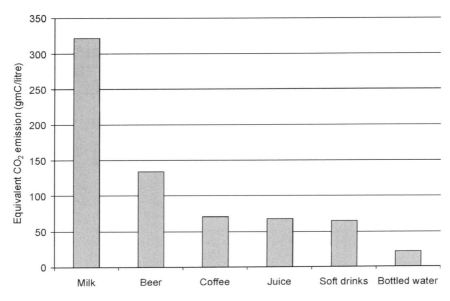

Source: Data from Hanssen et al (2007)

Figure 7.18 *CO_2-equivalent GHG emissions per litre of various beverages*

emissions associated with the use of fossil fuels, with the largest emissions associated with beef and the smallest associated with poultry. Figure 7.17b compares the emission per gm of protein provided by the various animal products.

Figure 7.18 compares the total equivalent CO_2 emissions per litre of various beverages. Milk has by far the largest emission per unit volume, but it also has a much higher protein content than the other beverages (some of which have no protein). However, the difference in emissions between milk and other beverages could be mitigated through measures to reduce methane and N_2O emissions associated with cattle (Johnson et al, 2007; Smith et al, 2007).

7.9.5 Comparison of energy consumption related to food and other consumer expenditures

A number of studies have estimated the total primary energy use related to consumer expenditures in different areas, such as for home heating and cooling, local vehicular travel, holiday travel, clothing, food and recreation. The usual procedure to estimate the energy use related to expenditures in different areas is to construct a large input–output matrix for the economy

of a given country. The matrix will have 50–400 sectors and gives the amounts of the output of each sector required as input to all the sectors. From data on the economic value-added for each sector and its total energy use, the average energy use associated with an average dollar (or currency unit) of value-added in that sector can be computed. By combining the input–output matrix with energy intensities (MJ/$, for example) and data on how much households spend in different aggregate sectors, the total energy use associated with each aggregate sector can be estimated. These estimates include all the upstream energy uses associated with a given activity. For example, the energy used to produce transportation vehicles includes the energy used at the manufacturing facility, the energy input at the steel plant that produced the steel that goes into the vehicle, the energy use to make the machinery at the steel plant, and so on. The resulting energy amounts are primary energy but are subject to errors because they are based on the average value-added and energy use in broad sectors, rather than the amount of energy associated with physical quantities of specific physical materials.

Hertwich (2005) summarizes the results of 18 such studies from the 1980s to the present. Figure 7.19 compares the estimated annual per capita primary energy use associated with different categories of household expenditures in the US, Australia, UK and Sweden. The definitions of the various categories may be slightly different for the different studies, and they apply to years ranging from 1993 to 2002. Nevertheless, they consistently show household energy consumption (electricity fuels for heating and hot water) to account for the single largest share, followed by either food or vehicle fuel. Figure 7.20 compares the estimated annual per capita primary energy use for food up to the point of purchase in different countries (so energy spent travelling to the retail food outlet, and in storing and cooking food and washing dishes is not

included in the food category). Per capita food energy use ranges from about 35GJ/yr in the US to about 3GJ/yr in India – more than a factor of ten difference.[8]

As a consistency check, it is instructive to compare the energy use of the low- and high-energy diets, discussed earlier, with the average per capita food energy uses given in Figure 7.20. The annual per capita food-related energy consumption for the energy-efficient and energy-intensive diets considered by Carlsson-Kanyama et al (2003) for Sweden (13MJ/person/day and 51MJ/person/day) amounts to 4.7GJ/yr and 18.6GJ/yr, respectively, of secondary energy. If this energy is 75 per cent fuels with a markup factor of 1.25 and

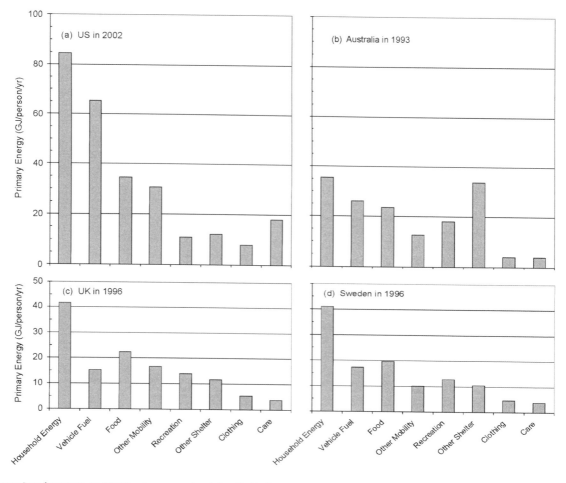

Source: Data from Hertwich (2005), using an associated data file kindly provided by Edgar Hertwich

Figure 7.19 *Comparison of annual per capita primary energy use associated with various household activities in different countries, as estimated in different published studies*

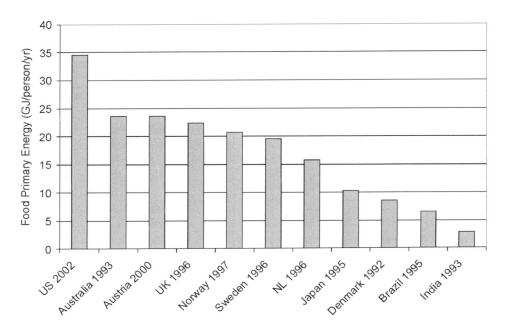

Source : Data from Hertwich (2005), using an associated data file kindly provided by Edgar Hertwich

Figure 7.20 *Comparison of annual per capita primary energy use associated with food consumption in different countries, as estimated in different published studies*

25 per cent electricity with a markup factor of 2.5, then the average markup factor is about 1.5 and the corresponding primary energy amounts are about 7GJ/yr and 28GJ/yr.[9] These bracket the average per capita primary energy use for food in Sweden of 19.6GJ/year given in Figure 7.20.

However, this does not include phytomass energy inputs. Adding the vegan and affluent-diet phytomass energy inputs from Table 7.16 to the low- and high-energy inputs given above results in extreme energy inputs to the food system of about 35MJ/person/day and 190MJ/person/day. Applying a markup factor of 1.5 to convert non-biomass inputs to primary energy, the resulting minimum and maximum food primary energy requirements for a family of four are about 60GJ/yr and 315GJ/yr, respectively.[10,11] By comparison, driving a car 15,000km/year with a fuel consumption of 8 litres/100km represents an annual energy use of 46GJ,[12] taking a family of four on a long-haul flight (10,000km return) requires 24–36GJ,[13] and heating a 200m² house with a heat requirement of 15kWh/m²/yr (the Passive House Standard) or 75kWh/m²/yr (typical of new construction today in many countries) corresponds to annual energy use of 14GJ/yr or 65GJ/yr, respectively.[14]

Thus, the difference in annual energy use between an energy-intensive and an energy-efficient diet for a family of four excluding phytomass energy inputs (83GJ/yr) is greater than the typical annual energy use by one car (about 45GJ/yr), the energy required for a long-haul return flight by the family members (about 24–36GJ), or the difference in annual heating energy use between typical and highly efficient houses (also about 50GJ/yr).[15] When phytomass energy inputs are included, the difference in the energy requirements for feeding a family of four greatly exceeds the sum of the energy requirements for a poorly insulated house, a not particularly efficient car and an annual long-haul flight by all the family members. Results are summarized in Table 7.23.

7.10 Impact of globalized versus localized agriculture

There is a growing interest in promoting greater reliance on locally produced food. Peters et al (2009) find, through a geographic information system-based optimization tool, that it is possible to meet 34 per cent of the food needs of New York state within a distance

Table 7.23 *Comparison of annual energy use by a family of four*

	Secondary energy use	Annual primary energy use (GJ)		Difference (GJ/yr)
		Low	High	
Food, excluding phytomass input	Low: 12MJ/person/day High: 50MJ/person/day	29	112	83
Food, including phytomass input	Low: 29MJ/person/day High: 182MJ/person/day	60	315	255
Housing heating (200m² floor space)	Low: 15kWh/m²/yr High: 75kWh/m²/yr	14	65	51
Transport	15,000km/yr by car at 8 litres/100km	–	61	61
Vacations	Long-haul flight (10,000km return)	–	24–36	24–36

of just 49km from the points of consumption. Part of the high energy use in the energy-intensive diets discussed above is due to long-distance transport (sometimes by air) of food grown a considerable distance from the final consumer. However, long-distance transport of food can be justified in energy terms if it permits food to be grown on much more productive land than if grown locally. There is readily a factor of two to three difference in land productivity in different regions. Producing food on the most productive land, wherever it occurs, in turn frees up land that can be used either for bioenergy crops or for increased supply of forest products (which can substitute to some extent for more energy-intensive building materials such as concrete), or which can be set aside for reversion to nature.

Of course, there are other considerations besides minimizing land area requirements and net energy use in producing food locally instead of importing it, including food security, providing diversity of employment opportunities (rather than overly concentrating jobs in the provision of the goods and services that would need to be exported in order to pay for food imports from more productive lands elsewhere), and permitting the recycling of at least some of the nutrients in human waste back to the land that produces the food (which is undoubtedly required for long-term sustainability).

A number of factors need to be considered in comparing the merits of local versus imported food from an energy point of view:

- differences in energy requirements during the production stage (particularly if locally produced food is grown in greenhouses outside the natural growing season);

- differences in the energy required for transport, including local distribution within the imported country and including refrigeration energy use;
- differences in the energy used during storage (particularly if locally produced fruits and vegetables are stored for off-season use);
- food losses during transport and storage;
- differences in the packaging of local and imported food.

These terms and tradeoffs are compared in Table 7.24 for apples grown and consumed locally (250km transport distance) in Europe and imported from New Zealand by ship. Transport by ship accounts for about half of the total energy in supplying apples to the retail outlet. If locally consumed apples are consumed after only ten days storage, the energy required to supply local apples is less than half that of the New Zealand apples. However, if local apples are consumed just before the next season's harvest (after, say, 350 days of storage), the total energy input is comparable to that of New Zealand apples. The typical storage loss after ten months storage is 25 per cent (but potentially as high as 40 per cent), making the energy input per unit of apple delivered to the consumer greater for the local than for the New Zealand apples (more detailed comparisons, for different seasons and taking into account storage in New Zealand prior to transport during some seasons, is given in Canals et al, 2007). Similarly, Saunders et al (2006) find that UK storage of locally produced onions entails greater CO_2 emission than transport by ship from New Zealand.

The data in Table 7.24 indicate the same energy for packaging of local and imported apples. However, Kooijman (1993) estimated a packing energy use of 8.5MJ/kg for fresh imported peas and 0.8MJ/kg for locally sourced peas.

Table 7.24 *Comparison of the energy inputs (embodied energy) in providing locally produced apples in Europe and apples imported from New Zealand by ship*

	Apples grown in Europe		Apples imported from New Zealand	
	Input data	MJ/t	Input data	MJ/t
Initial cultivation	1070MJ/t	1070	470MJ/t	470
Transport to coldstore	3.47MJ/t/km × 40km	138.8	3.47MJ/t/km × 40km	138.8
Initial cooling	86.3MJ/t	86.3	86.3MJ/t	86.3
Cooling in storage	(5.4MJ/t/day) × (10–350) days	54–1890	5.4MJ/t/day × 10 days	54
Transport to Europe by ship		0	0.11MJ/t/km × 23,000km	2530
Cooling during transport		0	10.8MJ/t/day[a] × 28 days	302.4
Transport to regional distribution centre	1.2MJ/t/km × 100km	120	1.2MJ/t/km × 250km	300
Transport to retail outlet	1.2MJ/t/km × 150km	180	1.2MJ/t/km × 150km	180
Packaging	650MJ/t	650	650MJ/t	650
Total (MJ/t)		2299–4135		4702

Note: Losses during transport and storage are neglected here. [a] Carlsson (1997) estimated a cooling energy use by ship of 0.035MJ/tonne-km, which would correspond to 28MJ/t/day for the ship speed of 800km/day that is implied here.
Source: Canals et al (2007)

For tomatoes in Sweden, Carlsson (1997) estimates an energy consumption in heated greenhouses of 57.3MJ/kg versus 1.5MJ/kg for unprotected production in Spain. For production in heated greenhouses in Belgium, van Hauwermeiren et al (2007) calculate an energy requirement of about 27MJ/kg for tomatoes and 23MJ/kg for lettuce. Transport to Sweden from Spain by truck requires 3.9MJ/kg, whereas transport from the Canary Islands by ship requires 2.2MJ/kg.

Wallgren (2006) compared the energy used in transporting food to farmers' markets in Sweden with the transport energy used in the conventional food system. Producers supply food to a local network called the Farmers Own Market (FOM). The distances transported range from 15–250km. In the conventional system, distances transported range from 750–24,000km, with 40 per cent of the food imported. Transport energy intensities range from 0.45–50 MJ/kg, the latter for air freight of tomatoes. Average truck loading factors in the conventional system are estimated to be about 70 per cent for long-distance transport and 50 per cent for local distribution. Excluding air freight, the greatest transport energy intensity in the conventional system is 10.1MJ/kg, and there is no significant difference in the transport energy intensity between the FOM and conventional systems. The

unexpectedly high energy intensity of the FOM system (compared to an intensity that might be expected to be lower than that of the conventional system) is due to most of the produce being taken in small quantity by producers in their own cars. There is therefore the potential to greatly reduce the transport energy intensity of the FOM system (and similar systems elsewhere) by creating a more organized delivery system, and through direct marketing to restaurants, schools and institutional systems. This would particularly be the case if the size of the FOM increases. Van Hauwermeiren et al (2007) also compared the transport energy intensities of local food systems in Belgium (comparable to the Swedish FOM) with conventional (i.e. supermarket-based) systems, and found that the local system often entails two to four times greater transport energy use.

Loading factors in the conventional system could perhaps also be improved. In the UK, on approximately one third of load journeys and a fifth of the total distance travelled (including the return trip), food transport vehicles are less than half full (Garnett, 2003). For around 15 per cent of the working time of refrigerated vehicles, they are loaded and ready to go but not moving, so cooling energy increases because the COP (see Chapter 4, subsection 4.4.2) of mobile refrigeration units is less than that of warehouse-based units.

Long-distance transport of fruit and vegetables is often needed in order to meet nutritional requirements. The World Health Organization recommends a daily fruit and vegetable intake of 400gm/day per person according to Saunders and Hayes (2007) but, within the EU, only Greece (at 587gm/day/person) and Italy (at 417gm/day/person) meet this target, with consumption in northern countries such as Sweden, the UK and Ireland averaging only about 250gm/day/person. As noted above, supplying fruits and vegetables locally through cultivation in greenhouses tends to be more energy intensive than importing fruits and vegetables. Production of food for export is a significant source of employment for many developing countries (for over 715,000 farmers in sub-Saharan Africa according to Saunders and Hayes, 2007), although there are also issues related to how much farmers are paid for their work, their ability to feed and support their own families, and the ability of developing countries to feed their own people while exporting food to developed countries. However, potential agriculture yields – with appropriate inputs – are far higher in many countries than at present. Reliance on products that can be transported by sea (such as apples, bananas and oranges) rather than by air freight will significantly reduce the transportation-related energy use.

Finally, Vringer and Blok (2000) estimate the energy intensity of cut flowers produced and sold in Holland to range from 3 to 195MJ per flower, with flowers grown in greenhouses requiring about four times the energy of those grown outdoors. In wintertime, each carnation requires about 50MJ of energy (equal to the energy content of 1.5 litres of gasoline). Purchase of flowers (on average 250 per household per year) in Holland amounts to 1 per cent of total household primary energy use. The energy required to import flowers by air from Kenya is quite small (1–3MJ per flower), so substantial energy savings are possible in some cases by importing flowers. Other techniques to reduce energy use are to preferentially buy flowers in summer, to buy less energy-intensive flowers, to extend the lifetime of the flowers by proper treatment and to buy bulbs instead of bulbous cut flowers. Altogether, these are estimated to be able to reduce the energy requirement for flowers by 50 per cent in Holland.

7.11 Summary

Energy is required in the food system for the production of fertilizers and pesticides, as direct energy use on farms in the form of fuels and electricity, and for the transport, processing, storage and/or packaging and cooking of food.

Four general ways to reduce the amount of energy used to produce fertilizers are:

- to reduce the need for artificial fertilizers through maximum recycling of nutrients back into the soil;
- to reduce the need for artificial fertilizers by minimizing the loss of applied fertilizer (in runoff or elsewhere);
- to reduce the need for fertilizer by promoting diets (such as low-met diets) with less fertilizer requirements
- to further reduce the amount of energy needed to produce a given amount of fertilizer.

With regard to the last of these, a minimum further saving of 20 per cent is reasonable for the most efficient production facilities, but more on a world-average basis, as many current facilities are old and inefficient. The main ingredient used in the manufacture of nitrogen fertilizer is ammonia, which is produced from natural gas, oil or coal feedstocks, but could be produced from hydrogen that in turn is produced electrolytically from C-free energy sources. More efficient use of fertilizer can reasonably be expected to reduce demand by 50 per cent compared to current practice in most industrialized countries. Reduced use of nitrogen (and, to a lesser extent, phosphorus) fertilizer would provide many environmental benefits in addition to reduced energy use, and reduced use of phosphorus fertilizer would delay the peak (which may occur as early as 2035) and subsequent decline in the availability of phosphate rocks.

Adoption of no-till agriculture in place of conventional agriculture has very little effect on overall energy use, as savings in fuel energy and machinery embodied energy are roughly offset by the increased use of energy to make the extra fertilizer and herbicides that are typically needed. Organic farming (no or minimal chemical fertilizers and pesticides) can readily reduce overall farming energy use per unit of output by 30–50 per cent, usually but not always with a 10–20 per cent reduction in yield. However, present crop varieties have been bred to produce maximum yields under conventional farming systems; re-optimization to account for organic methods could reduce or eliminate the difference in yield.

The ratio of harvested phytomass energy to food energy is typically 1.5:1 for processed plant foods, about 8:1 for eggs, milk, fatty poultry and fatty pork, 55:1 for fatty beef and over 100:1 for lean beef. Thus, a vegan or vegetarian diet requires substantially less phytomass energy than a high-meat diet, thereby freeing up land that could be used to produce bioenergy crops or allowed to revert to nature. A shift to diets with no or less meat would also significantly reduce the need for fertilizers and other energy inputs to the agricultural system. There is a factor of four difference in the total amount of fossil fuel energy used to produce and deliver food for the least and most energy-intensive typical meals in Northern Europe with comparable nutritional quality. The difference in total annual energy use (fossil fuel and phytomass) between low- and high-energy diets for a family of four is greater than the sum of the typical annual energy use by a car, the energy required for all four family members to take a long-haul (10,000km) trip, and the difference in heating energy use between conventional and highly insulated houses in cold climates.

Notes

1   Jasinski (2009) gives world reserves as 8Gt phosphate rock, of which is only 29–34 per cent is P_2O_5 according to Cordell et al (2009a), which in turn is only 43 per cent P.

2   According to Ramírez and Worrell (2006), the theoretical minimum energy use is 23.3GJ/tonne, while according to Blok (2005), the theoretical minimum requirement is 19.8GJ/t. The differences are probably due to differences in the assumed end states or other assumptions. IEA (2006) gives a theoretical minimum energy requirement of 21.2GJ/t, based on the hydrogen content of ammonia and the LHV for H_2 of 120.0GJ/t.

3   Such a shift would confer substantial health benefits to humans and animals, as cattle that are fed grain must also be given antibiotics to avoid vomiting, but these antibiotics persist in the meat consumed by humans (so it is advisable that pregnant women not eat grain-fed cattle).

4   This assumes, for simplicity, that the land used to produce animal feed could have produced bioenergy crops with the same energy content as the animal feed.

5   A similar figure for the US, but involving mass flows only and without giving land areas, is found in Heller and Keoleian (2000).

6   Average per capita seafood consumption for the moderate and affluent diets is assumed to decrease because the human population continues to grow, but it is questionable whether even the current rate of harvest of seafood is sustainable.

7   For example, for pork the inverse feed efficiency is 11MJ/MJ in Table 7.22 and 8MJ/MJ in Table 7.15, for chickens it is 7.3 and 8.0MJ/MJ, respectively, and for milk it is 5–7MJ/MJ in Table 7.22 and 7.7MJ/MJ in Table 7.15. The largest discrepancy is for eggs, where the inverse feed efficiency is 32MJ/MJ in Table 7.22 but only 7.4MJ/MJ in Table 7.15.

8   Because the estimated average per capita energy use for each consumption category in a given nation is constructed from input–output tables for that nation, the energy consumption for food (for example) assumes that all of the consumed food, including imported food, is produced with the same energy intensities (including the transportation component) as for corresponding products produced in the country in question. Thus, the energy consumption values shown in Figure 7.20 are more a comparison of the relative energy intensity of different national food systems, rather than a comparison of the embodied energy of the diets in different countries.

9   See Chapter 2, subsection 2.3.5, for an explanation of markup factors.

10   The actual phytomass energy inputs depend strongly on the inverse feed efficiency and the types of animals eaten, as indicated in Table 7.15.

11   It may or may not be correct to combine the low phytomass energy input (17MJ/person/day) in the vegan diet with the diet having the minimal fossil fuel input, as the use of dairy substitutes could increase the fossil fuel energy input.

12   Given an LHV for gasoline of 42.5MJ/kg, a gasoline density of 0.75kg/litre and a 20 per cent markup to account for upstream energy use.

13   Given the energy intensity of the Airbus 321/100 of 0.5–0.75MJ/seat-km (Table 5.29) and a 20 per cent markup to account for production of jet fuel from crude petroleum.

14   Assuming heating by natural gas with a markup factor of 1.25.

15   Although, as explained in Chapter 5 (subsection 5.6.1), the climatic effect of long-haul air travel is about twice as large as implied by its energy use.

8

Municipal Services

This chapter briefly discusses the energy use and GHG emissions related to the supply of clean water, the treatment of wastewater, the management of solid wastes and the provision of various municipal recreational facilities.

8.1 Supply of water

Energy is used to supply water through the energy required to build the supply infrastructure, through the energy used to make chemicals (such as chlorine) for treatment, and through the energy uses at water treatment facilities and pumping stations. In most parts of the world, water is pumped from the ground or is taken from rivers, lakes or reservoirs and pumped some distance to its point of use, all of which require energy.

8.1.1 Energy requirements

Table 8.1 summarizes estimates of the amount of energy used in the various steps in supplying water through municipal systems in California. Also given, for comparative purposes, is the embodied energy of bottled water supplied to Los Angeles; bottled water requires 500 to 2600 times the energy of municipal water. Several options are available for the source of municipal water, ranging from recycling of wastewater and pumping of fresh groundwater to desalination of brackish (slightly salty) groundwater, desalination of seawater, large-scale diversion from central to southern California by aqueducts, and towing of fresh water at sea in bags. Several options are given for treatment of wastewater, with and without recovery and use of the biogas energy

that is produced at the wastewater treatment plant. The total embodied energy ranges from a minimum of 0.6kWh/m^3 (if water can be extracted from an adjacent river with no use of energy) to a maximum of about 5.5kWh/m^3, almost all of it as electricity. This estimate does not include the energy used to produce the infrastructure that delivers the water or the plants that treat the water and wastewater, nor does it include the energy used to produce the chemicals used in the water treatment plant. For household water consumption ranging from 0.5 to $1.0\text{m}^3/\text{day}$, the energy used to supply household water amounts from about 100kWh/yr to 2000kWh/yr in California. This is significant compared to the major direct uses of electricity in a household, such as the air handling system (100–1000kWh/yr), refrigerator (500–1000kWh/yr) and consumer goods (1000–2000kWh/yr).

The energy requirements to supply water in California are large because water has to be provided from great distance. Table 8.2 compares the energy requirements for the collection, treatment and distribution of municipal water, and subsequent treatment of wastewater, in three cities or regions (Toronto, Sydney, and Wallonia in Belgium). The energy required to construct the water supply and wastewater collection infrastructure is not included, but Lundie et al (2004) estimate that this amounts to only 4 per cent of the total lifecycle energy use. Total operational energy requirements are $0.5–1.3\text{kWh/m}^3$ which, given a typical household water consumption of up to $300\text{m}^3/\text{yr}$, implies up to 150–400kWh/yr energy use. Thus, water conservation efforts are important even in cities located adjacent to their water supply.

Table 8.1 *Energy use to deliver and treat water supplied to urban consumers in California, given as kWh/m³ or its equivalent in MJ/m³*

Process	Energy use (kWh/m³)		Energy use (MJ/m³)	
	Minimum	Maximum	Minimum	Maximum
Conveyance to urban centres				
Diversion		2.43		8.75
Pumping groundwater	0.38	0.60	1.36	2.16
Desalinization of seawater by reverse osmosis	3.57	3.89	12.84	14.01
Desalinization of brackish groundwater by reverse osmosis		1.38		4.96
Water bags (towed 600km)		0.96		3.44
Treatment and local distribution				
Water treatment	0.03	0.06	0.12	0.20
Local distribution	0.14	0.76	0.50	2.74
Wastewater treatment				
Trickling filter	0.18 (0.11)	0.47	0.66	1.69
Activated sludge treatment	0.28 (0.18)	0.61	0.99	2.19
Advanced treatment	0.32 (0.23)	0.70	1.17	2.52
Advanced treatment with nitrification	0.42 (0.32)	0.79	1.52	2.86
Representative total				
All steps	0.6	5.5	2.16	19.8
Provision of bottled water to households in Los Angeles				
Manufacture of bottles			4000	4000
Transportation			1400	5800
All steps			5600	10,200

Note: For wastewater treatment, minimum energy use pertains to a 400 million litre/day facility, maximum energy use pertains to a 4 million litre/day facility, and the values in brackets are with energy recovery.
Source: Cohen et al (2004) except for bottled water energy use, which is from Gleick and Cooley (2009)

Table 8.2 *Energy used (kWh/m³) in various stages in the municipal water cycle*

Step	Region		
	Wallonia	Sydney	Toronto
Water collection	0.18	0.00	0.00
Water treatment	0.04	0.09	0.28
Water distribution	0.18	0.28	0.50
Wastewater treatment	0.10	0.40	0.50
Total	0.49	0.76	1.28

Source: Racoviceanu et al (2007) for Wallonia; Lassaux et al (2007) for Sydney; Lundie et al (2004) for Toronto

8.1.2 Energy savings opportunities

Energy savings opportunities related to water include: (1) reducing losses of water due to leaks in the delivery system, (2) improving the operation of pumps, and (3) reducing waste of both hot and cold water by end-users.

Reducing leakage

Leaks are a particular problem in developing countries, where 30–50 per cent of the water that enters the supply system is typically lost before it reaches the consumer (Barry, 2007). The 20 per cent largest leaks are generally responsible for 80 per cent of the total water losses, so targeted repair of the largest leaks will be the most cost-effective strategy. Assessing the overall level of leakage requires accurate monitoring of the system, including measurements of the amount of water supplied and billed and comparison of daytime and minimum night-time flow rates. Sophisticated equipment such as acoustic loggers can be placed on surface fittings such as fire hydrants to aid in determining the location of the largest leaks.

Improving the efficiency of the pumping system

Measures to improve the efficiency of the supply system include the use of automation to control pressure and output in the networks and to optimize the operation of pumping equipment, installing VSDs and implementing a system of regular, preventative maintenance. Matching pumps to system requirements saves 10–30 per cent, use of VSDs typically saves 10 per cent, and pump system optimization can save 20–40 per cent (Barry, 2007).

Optimization of the pumping system requires an adequate array of sensors to measure pressure, water level and flow rate at various points in the system and to convey the information to a central, computer-based control system. Automated control systems provide a number of benefits according to Barry (2007):

- water savings of up to 10 per cent (by avoiding excess pressure, thereby reducing leaks);
- electricity savings of 10–30 per cent;
- reduced maintenance costs.

Reducing waste by end-users

Frequently, the biggest opportunity to reduce the energy associated with municipal water is to avoid waste of both cold and hot water (with the latter also providing savings in the energy used to heat water). Opportunities to reduce the use of hot water are discussed in Chapter 4 (subsection 4.7.1), where savings potential of up to 50 per cent have been identified. The use of cold water can also be reduced by 50 per cent or more through, for example, more water-efficient clothes washers used for cold-water washing (see Chapter 4, Table 4.14). In order to encourage non-wasteful behaviour, it is essential that water consumption be metered so that users can be billed in proportion to their consumption. According to Barry (2007), only about 35 per cent of water consumption around the world is metered.

Improved desalination (where applicable)

In some parts of the world, fresh water is produced through desalination of seawater. Energy requirements for various desalination processes are discussed in Chapter 6 (section 6.13). About half of the world's desalination occurs through multi-stage flash distillation, a highly inefficient process. As discussed in Chapter 6, the eventual replacement of the world's entire existing desalination capacity with state-of-the-art RO technology would reduce the average primary energy requirement for desalination (assuming 40 per cent electricity generation efficiency) by about a factor of 20.

8.1.3 Discouraging the use of bottled water

The most egregious energy waste is the provision of water in bottles rather than through existing municipal

water supply systems. As noted earlier, provision of bottled water requires 500 to 2600 times as much energy as municipal water in Los Angeles. Bottled water is also expensive compared to the cost of upgrading water treatment facilities (where necessary). A number of municipalities in North America have banned the sale of bottled water in municipally owned facilities, the motivation in some cases being to reduce the unnecessary production of plastic waste.

8.2 Wastewater treatment

Energy is used to treat wastewater through the sewer infrastructure built to transport wastewater to treatment facilities, and through the construction and operation of the wastewater treatment plant. The infrastructure consists of cement and steel, whose energy requirements are discussed in Chapter 6 (sections 6.3 and 6.7). The main operational energy use will be various pumps and motors used to transport and aerate the wastewater. The global warming impact of emissions of CH_4 (from anaerobic decomposition of sewage sludge) and N_2O (from denitrification of nitrogen compounds, as discussed in Chapter 7, section 7.2) can potentially rival the impact of CO_2 emissions associated with energy used in treating liquid wastes.

8.2.1 Wastewater treatment techniques and associated energy use

Treatment of wastewater involves pretreatment, primary treatment and secondary treatment. The products of sewage treatment are *effluent* and *sludge*. Effluent is the decontaminated and purified water that is discharged into a nearby water body, while sludge consists of wet solids. Sludge can be incinerated (with use of additional fuel due to its high moisture content), anaerobically digested with recovery of methane for energy, or landfilled. Table 8.1 gives the onsite energy use for various wastewater treatment processes (excluding possible sludge incineration).

Water treatment steps

Pretreatment involves removing large particles and fats, oils and greases. During primary treatment, sewage flows through tanks that are large enough that sludge can settle and floating material such as further grease and oils can rise to the surface and be skimmed off. This produces a homogeneous liquid capable of being treated biologically and a sludge that can be separately treated or processed. The primary tanks usually have mechanically driven scrapers that continually scrape the collected sludge towards a hopper in the base of the tank, from where it is pumped to the sludge treatment facility.

Secondary treatment involves bacteria and protozoa that digest and respire various organic compounds in the sewage. Aerobic digestion of sewage requires mechanical aeration of the water so as to provide adequate oxygen to the microbes. Energy-saving measures include: use of VSDs, avoiding excessive aeration of aeration basins (that is, delivering the optimal amount of oxygen to microbes) and optimal thickening of the process sludge so as to reduce the quantity of water pumped with the solids (thereby reducing overall pumping rates). Water conservation methods (such as use of low-flush toilets and minimization of storm water runoff into sewage water flows), while not reducing the mass of solids that needs to be treated, will reduce pumping energy use by reducing the dilution of materials needing treatment.

Removal and recovery of nutrients

Although secondary treatment removes biological contaminants, about 80 per cent of the original nutrients remain in the effluent (with the balance in the sludge). This can cause eutrophication of the river or water body receiving the effluent. This is a problem along the Rhine River in Europe, for example, so nutrient loadings should be reduced. Ideally, this would be accompanied by recovery of the nutrients and their reuse as a fertilizer for food production. Maurer et al (2003) present an overview of various processes for nutrient removal and the associated energy requirements.

Various biological denitrification techniques can be used to convert effluent N to N_2-gas (which is then lost to the atmosphere), and these require about 20–110MJ/kgN (for aeration and internal recycling of water). Alternatively, ammonia can be stripped from effluent by reaction with sulphuric acid (with addition of CaO to control pH). The result is to produce ammonium sulphate, which can be used as a fertilizer (and constitutes about 3 per cent of world N fertilizer production today, as shown in Figure 7.8), although it is not clear to what extent ammonium sulphate can

replace other fertilizers without overloading soils with sulphate. As well, there are substantial material requirements with this technique: 2.4kg of CaO and 7.0kg of sulphuric acid per kg of recovered N. The energy requirement (including the energy to produce CaO and sulphuric acid) amounts to about 90MJ/kgN, which is about twice the energy intensity of N fertilizer plants today (see Chapter 7, Figure 7.10).

Removal of P from sewage effluent requires either chemical precipitation with iron(II) sulphate (at an energy cost of 49MJ/kgP) or enhanced biological uptake and incorporation into sludge (at an energy cost of 28MJ/kgP). By comparison, various P fertilizers today require 10–30MJ/kgP (see Figure 7.10), although saving energy need not be the only motivation for recovering P as a fertilizer, given that the global availability of mineable P will probably peak some time around the middle of this century (as discussed in Chapter 7, Section 7.3).

An alternative system for capture and reuse of nutrients involves decentralized collection of urine that is minimally diluted with water (using appropriate toilets), storage in tanks, collection by truck twice per year, concentration by a factor of ten in a centralized evaporation plant, and use as a urea fertilizer. An analysis of the lifecycle energy use of this scheme reported in Maurer et al (2003), taking into account the energy in pipes and tanks, and assuming average transport distances of 60km transport from local storage tanks to centralized evaporation plants and 100km from there to farms, indicates an energy requirement of 65MJ/kgN. This is somewhat more than the energy required to produce N fertilizer today. However, this figure is based on allocating all of the required energy to N recovery, whereas some P and K fertilizer would also be produced. The alternative, to recover all three nutrients from a conventional wastewater treatment plant, would require 153MJ/kgN.

The pending shortage of P fertilizer combined with the high energy cost of recovering P and other fertilizers from conventional sewage treatment plants, and the difficulty in retrofitting existing communities, makes a strong case for designing new communities now so as to be able to collect minimally diluted urine for use as a source of nutrients for the production of fertilizer. As a minimum, new housing developments can be built with toilets designed to separately collect urine, along with the required plumbing, so that a nutrient recycling system can be put in place later.

Fertilization of bioenergy plantations

A simpler and less energy-intensive method of removing and recycling nutrients from wastewater is to use sewage water to fertilize bioenergy plantations. A community where this is being done now is Enköping, Sweden. An 80ha willow plantation is irrigated and fertilized with sewage water. The willow is used to generate heat and electricity as part of a district heating system, and the ash is returned to the willow plantation. Further information is found in Volume 2 (Section 4.5.8).

Energy recovery

The most significant opportunity to reduce the net energy consumption by sewage treatment plants is likely to be the recovery of energy in the biogas produced from anaerobic digestion of sludge. Analyses by Hydromantis (2006) indicate that very large water treatment plants can become self-sufficient in heat and electricity in this way. In cold climate countries such as Canada, adequate insulation around the digester tank is essential. Otherwise, the biogas will need to be combusted along with supplemental natural gas in order to supply the heat required to maintain the target digestion temperature of 35°C or 55°C. Sewage treatment plants that are more sophisticated than the norm will produce less sludge and hence biogas, but a substantial fraction of the wastewater energy requirements can still be met.

8.2.2 Comparative energy use of alternative sewage treatment systems

Tidåker et al (2006) have assessed the lifecycle energy use and savings associated with three methods of treating household organic waste and wastewater sludge in combination with farming in Sweden. The three methods are:

1   to install food waste disposers in houses, which mill the food waste at the kitchen sink and discharge it to the sewage waste flow, then to digest the sewage waste, use the biogas in a gas engine at 30 per cent efficiency to generate electricity, and prepare a soil conditioner with the residue for use in parks, while composting food waste not handled by disposers and using the product as a soil conditioner;

2   to compost all household kitchen organic waste and digest the remaining sewage sludge to produce a soil conditioner that is applied to farm fields, displacing some chemical fertilizers (22 per cent of N fertilizer in the first year, 50 per cent thereafter);

3   to use a blackwater system, in which food waste, urine and faeces are collected separately from grey water and stored in underground 30m³ tanks that serve 50 households each and are emptied once per month by truck. This waste is digested as before, and the residue used as an agricultural fertilizer.

The blackwater system is similar to the urine source separation system considered by Maurer et al (2003) and discussed above as a strategy to recover nutrients, but includes solid wastes as well. The blackwater system makes maximal use of sewage sludge, displacing the most chemical fertilizer, but does not save energy overall because of the energy required to build the infrastructure. However, it serves to conserve non-renewable phosphate deposits used for P fertilizer.

8.3 Solid wastes

There are a number of complex issues related to the net impact on energy use and GHG emissions of alternative methods of treating solid wastes. These include the energy used by different treatment options, the energy that can be produced as a byproduct of various treatment options, the kinds of energy displaced by byproduct energy, and emissions of CH_4 from landfills or anaerobic digestion. These and other issues are discussed in some detail in the following sections.

8.3.1 Solid waste treatment options

The following options are available for the treatment of municipal solid waste (MSW):

*   *Landfilling.* Bulk untreated MSW is deposited in landfills, which in developed countries are of the 'sanitary' variety consisting of a liner of clay followed by layers of MSW alternating with thin inert layers. The clay is intended to prevent contamination of the regional groundwater system. Anaerobic conditions arise that are conducive to the production of CH_4 from the decomposable organic fraction in the waste.

- *Incineration, pyrolysis or gasification.* Incineration is the combustion of MSW in air with or without energy recovery; energy, if recovered, is in the form of hot steam that can be used to generate electricity or used for heating, or a combination of the two. Pyrolysis involves heating the waste in the near absence of air so as to drive off volatile materials as gases (CO, CH_4 and H_2) that can be used as a fuel, while gasification involves heating the waste in the presence of air or pure oxygen, converting the vast majority of carbon materials into CO, CH_4 and H_2 followed by combustion to generate electricity and useful heat.
- *Anaerobic digestion.* This is similar to composting, except performed under anaerobic conditions with the express purpose of generating methane under controlled conditions so that it can be entirely captured and used as an energy source. The organic residue is an excellent fertilizer and contains stabilized organic material that can be used as a soil conditioner. However, if the digestate is applied in excess, emissions of nitrogen and phosphorus can exceed those from chemical fertilizers (Sundqvist, 2006).
- *Composting.* Good quality garden and food wastes are separated from other wastes and composted, an aerobic process in which easily decomposable organic matter is oxidized to CO_2 (rather than to methane, as in landfills). The residue can be added to the soil, where some of the carbon will be oxidized but some will add to the long-term soil carbon reservoir. However, more nitrogen is lost to the atmosphere as ammonia during composting than is normally the case with anaerobic digestion, so the compost has a lower nitrogen content than the digestate produced by anaerobic digestion. Composting requires an energy input rather than producing energy.
- *Mechanical biological treatment.* Mechanical biological treatment (MBT) entails separating easily recyclable materials (metals, glass, plastics) from the waste stream and digesting or composting the remainder, producing a relatively inert residue that is then sent to a landfill, but with far less production of methane during the landfill stage than from MSW in a landfill.

The treatment of solid wastes is of relevance to climate both in terms of the energy used, saved or supplied through various treatment options – which has implications for net CO_2 emissions – and through the emissions of non-CO_2 GHGs (methane, nitrous oxide and halocarbons from refrigeration equipment and various foams). Energy and GHG balances are discussed in the following two subsections.

8.3.2 Considerations in lifecycle analysis of the net energy balance of alternative waste treatment options

The recycling of industrial materials into new materials of the same kind (secondary production) generally requires far less energy than production from raw inputs (primary production), as discussed in Chapter 6 for iron, aluminium, copper, zinc, glass, plastics and paper products. The alternatives to recycling of these commodities are incineration with other wastes, with or without energy recovery, or disposal in a landfill. For wet organic wastes, the options are composting, anaerobic digestion, pyrolysis, gasification, incineration and landfilling.

A full assessment of the energy savings from recycling and supplied by incineration with energy recovery, compared to landfilling, requires a consideration of the entire lifecycle of recycled and non-recycled products. This entails consideration of:

- the energy required to supply raw materials to primary production facilities;
- the energy required to collect, sort and possibly clean materials that are used by secondary production facilities;
- the energy used at the primary or secondary production facilities (the subject of Chapter 6), distinguishing between thermal and electrical energy and the ratios of these two forms of energy for different products;
- the energy value of co-products produced in primary or secondary production;
- the energy costs of disposal of the wastes associated with primary and secondary production;
- the electrical and/or useful thermal energy produced through incineration, if this is the alternative to recycling;
- the efficiency of methods that would be used to generate the heat and electricity not supplied by incineration if there is no incineration;

- the efficiency with which the wood saved from recycling of paper could be used to generate heat or electricity;
- the ratio with which recycled fibres can substitute for virgin fibres in the production of paper;
- any methane that is captured from landfill gases (LFGs) and used for energy purposes, if this is the alternative to recycling.

Both primary and secondary production facilities will use a mix of electrical and thermal energy. The primary energy entailed in either case depends on the efficiency with which the electricity that is supplied to the production facilities is generated. The efficiency that should be used is the efficiency of the next increment of electricity generation, rather than the average efficiency of the system that supplies the electricity. That is, marginal rather than average efficiencies should be used. Similarly, in determining the savings in primary energy by producing heat and/or electricity through incineration, one should consider how the next increment of heat or electricity would otherwise be produced, rather than looking at the average existing production by alternative facilities.

Another issue with regard to landfills is the climatic effect of methane emissions to the atmosphere. Apart from representing lost energy, these emissions have a disproportionately large effect on climate (see Chapter 2, subsection 2.6.5), so avoiding these emissions by recycling rather than landfilling those wastes that produce methane could easily exceed the climatic benefit of the CO_2 emissions avoided due to energy savings from recycling. That is, when methane emissions from landfills are considered, the climatic benefit of recycling rather than landfilling can be easily doubled.

A distinction needs to be made between closed-loop and open-loop recycling. Closed-loop recycling is recycling the used products back to the original products. An example of open-loop recycling is the use of discarded glass to make building materials rather than reconstituting the original products, recycling plastics into lower-grade products, or recycling high-grade paper to make lower-grade paper (such as packaging paper). This open-loop recycling is more accurately referred to as *downcycling*.

8.3.3 Mass flows and associated energy use

The mass flows associated with the recycling or non-recycling of a single material commodity and the associated process energy requirements per unit mass are shown in Figure 8.1. The mass of material used by the consumer and then discarded is m. With no recycling, this is the mass of raw material that must be fed into the system (assuming that any mass losses during the manufacturing process are internally recycled). Energy E_1 is used during initial processing of the raw material, E_2 during final manufacturing and E_3 for disposal. If a fraction f is recycled, then the required supply of raw materials is $m(1-f)$. Energy E_4 is used to collect and process the discarded material to the point where it is equivalent to raw materials after initial processing, with raw and recycled materials subsequently both requiring energy E_2 for final manufacturing, as indicated in Figure 8.1b. The total energy requirements are given by:

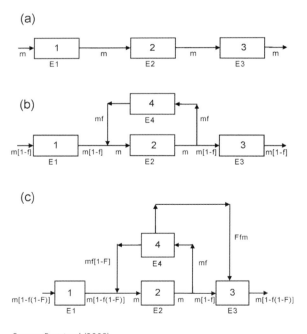

Source: Boustead (2008)

Figure 8.1 *Material flows and energy use associated with the consumption of a mass* m *of materials in consumer products and its ultimate disposal, for cases with (a) no recycling, (b) recycling with no material losses during the recycling process, and (c) recycling with material losses*

$$E_{total} = m(E_1 + E_2 + E_3) \qquad (8.1)$$

for the case without recycling, and by:

$$E_{total} = m(1-f)(E_1 + E_3) + mE_2 + mfE_4$$

$$= m(E_1 + E_2 + E_3) + mf(E_4 - E_1 - E_3) \qquad (8.2)$$

for the case with recycling. The difference in energy use is given by

$$\Delta E = mf(E_4 - E_1 - E_3) \qquad (8.3)$$

which is <0 if energy is saved.

In practice, some fraction F of the material that is collected for recycling is lost during the recycling process; the mass flows for this case are shown in Figure 8.1c. Assuming that this loss occurs before the processing energy E_4 is applied, and that material lost during recycling has to be processed as waste with the same energy expenditure E_3 as directly disposed material, the energy requirement is now given by:

$$E_{total} = m(1 - f(1-F))E_1 + mE_2$$

$$+ m(1 - f(1-F))E_3 + mf(1-F)E_4$$

$$= m(E_1 + E_2 + E_3) + mf(1-F)(E_4 - E_1 - E_3) \qquad (8.4)$$

for the case with recycling. The difference in energy use is given by:

$$\Delta E = mf(1-F)(E_4 - E_1 - E_3) \qquad (8.5)$$

That is, the energy savings given by Equation (8.3) is multiplied by the factor $(1-F)$.

Figure 8.2 illustrates the case when there are two material flows, m and M, with the recycling of m serving solely as an input to M. This is an example of *open-loop recycling*, but it would be downcycling if M is a lower-value commodity than m or if the energy chain $E_4 + E_5 + E_6$ is less than the chain $E_1 + E_2 + E_3$. The energy use is given by:

$$E_{total} = m(E_1 + E_2) + m(1 - f(1-F))E_3$$

$$+ mf(1-F)E_7 + (M - mf(1-f))E_4$$

$$+ M(E_5 + E_6) \qquad (8.6)$$

The difference in energy use compared to separate processing of both material flows with no recycling is given by:

$$\Delta E = mf(1-F)(E_7 - E_3 - E_4) \qquad (8.7)$$

which is similar in form to Equation (8.5).

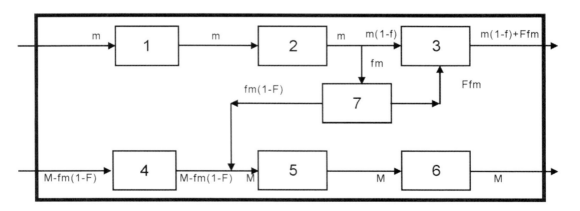

Source: Boustead (2008)

Figure 8.2 *Material flows and energy use associated with the recycling of two commodities with consumer consumption* m *and* M, *with a fraction* f *of* m *collected for recycling into* M *but with loss of a fraction* F *of the collected material*

As discussed by Boustead (2008), there are two important implications of Equation (8.7). First, the energy savings depends on the energy associated with both the first production process (E_3) and the second production process (E_4), as well as on the energy associated with the recycling operation (E_7). This means that there is no single way of allocating the energy savings between the two products. The second point is that maximizing the efficiency for the production of the individual products may not maximize the efficiency for the overall system involving the two products. The example given by Boustead (2008) is that where the sequence 1–2–3 involves the production of PET bottles and the sequence 4–5–6 involves the production of PET fibre. For a given production M, the energy savings for the system will be larger if m is larger (up to the limit for the fraction of recycled material that can be used), but this requires producing thicker bottles, which means a less efficient (in a broad sense) production sequence 1–2–3. However, this will not be necessary if the total mass flow M is substantially less than m. In this case, M cannot absorb all of the potential secondary flow $mf(1 - F)$. If M could take more secondary material, the solution is to increase the recovery rate of m rather than to make unnecessarily thick bottles.

Recycling will always save feedstock energy (and conserve raw materials) and reduce the volume of waste produced but may or may not save process energy, depending on the difference between E_7 and $E_3 + E_4$.

8.3.4 Energy balance of recycling, incineration and landfilling

WRAP (2006) comprehensively reviewed published lifecycle analyses of the energy benefits of recycling for the major product categories – paper/cardboard, plastics, aluminium, steel, glass, wood and aggregates. Many of the reviewed studies omit one or more of the considerations listed above, and consider average rather than marginal fuel mixes. When correcting for these deficiencies, WRAP (2006) concludes that, for the UK:

- recycling of paper and cardboard is clearly superior to incineration with energy recovery;
- recycling glass is preferable to incineration (a saving of $0.6 kgCO_2$/kg glass recycled) or to landfilling (a saving of $0.43 kgCO_2$/kg of glass recycled);

- recycling of plastics is preferable to incineration when cleaning is not necessary (a saving of $1.5–2.0 kgCO_{2-eq}$/kg of plastic recycled), but when cleaning is necessary, incineration can be preferable because of the hot water used in cleaning, the energy required to treat organic materials in municipal wastewater, and the fact that the organic materials are an added energy source if incineration occurs;
- recycling of steel and aluminium is strongly preferable to any alternative;
- recycling of aggregates is preferable to landfilling.

In the case of plastics, 100 per cent substitution of virgin materials with recycled plastic is not always possible because of a decrease in the quality of the plastic during recycling. In general, WRAP (2006) found that the energy used in collecting materials used for recycling is a very small part of the overall energy recycling budget in the UK.

Table 8.3 gives the savings in primary energy per tonne of various recycled waste materials, along with the heating value of the waste, which sets an upper limit to the amount of useful energy that can be obtained through incineration. It also sets a lower limit to the savings from recycling, since the heating value is equal to the feedstock energy, which is always saved with recycling. Some portion of the process energy will also be saved, as the processing of materials collected for recycling invariably requires less energy than production from virgin materials.

There are a number of reasons why incineration, even with energy recovery, is not a particularly attractive option from an energy point of view. First, many elements of the waste stream yield little or no energy when combusted. This is seen in Table 8.4, which gives the gross and net heating value for various wastes. The gross heating value is the heat released if all of the water vapour is condensed to a liquid and all the products of combustion are hypothetically cooled to the same temperature as the inputs and the excess heat utilized, while the net heating value neglects the latent heat of water vapour produced by combustion (as in the lower heating value) and also subtracts the energy required to vaporize any water initially present in the waste stream and to heat the combustible and non-combustible products from the initial temperature to the final temperature. Metal and mineral wastes do not yield energy when incinerated because they are not

Table 8.3 *Primary energy savings associated with recycling of various materials compared with manufacture from virgin materials*

	Savings (GJ/t)	Comments
Combustible materials		
Tyres	60	
Plastics, recycled to same	22 (PVC) to 53 (PUR)	Heating value of PVC is 20GJ/t and that of PU is 26GJ/t according to Figure 6.35, while the process energies are 35GJ/t and 50GJ/t, respectively
Plastics, back to feedstock	16	
Paper	21	According to Table 6.29, the energy balance of producing virgin materials with incineration is a net saving of 28GJ/t, compared to a net savings of 37GJ/t with recycling
Chipboard	16	
Non-combustible materials		
Aluminium	188	152GJ/t according to Table 6.4
Ferrous scrap	20	12GJ/t according to Table 6.4
Glass	2.3	1.2GJ/t savings in fuels, 0.06GJ/t extra electricity use when recycled content increases from 25 per cent to 59 per cent according to Table 6.26

Note: Compared with savings given in Chapter 7 and with the heating value of the waste.
Source: Summarized from various studies by Dornburg et al (2006)

Table 8.4 *Gross and net heating value of the waste components in MSW and in refuse-derived fuel, and the weighted average heating value in the EU*

Component	Per cent in MSW or RDF in the EU	Gross heating value (GJ/t)	Net heating value (GJ/t)
Paper/cardboard	29	13.1	11.5
Putrescibles	31	5.9	4.0
Plastic	8	33.5	31.5
Glass	11	0	0
Metals	5	0	0
Textiles	2	16.1	14.6
Other	12	10.0	8.4
MSW	**100**	**10.0**	**8.5**
Paper/cardboard	74	13.1	11.5
Plastic	21	33.5	31.5
Textiles	5	16.1	14.6
RDF	**100**	**17.6**	**15.9**

Note: RDF = refuse-derived fuel.
Source: Tables A3.36 and A3.37 of EC (2001d)

combustible, but would instead contribute to the ash residue after the incineration of the combustible components of the waste stream (largely organic wastes, paper products, plastics). This ash could be collected and recycled if it is impractical to separate the non-combustible wastes from combustible wastes prior to incineration. Wet organic wastes have a low heating value due to their water content, and if the energy required to vaporize the water in this portion of the waste is less than the energy content of the organic waste, the combustion of organic wastes would reduce the energy potentially available from incineration of the remaining combustible waste.

The other major reason why incineration even with energy recovery is generally unattractive is that the efficiency of energy recovery from incineration is low. Table 8.5 gives data on the performance of modern incineration facilities with energy recovery in the form of electricity or electricity and useful heat. Results are given for a small facility (serving 200,000 people,

Table 8.5 *Energy flow related to incineration with production of electricity only at large and small scales, and for incineration with cogeneration of electricity and heat for a district heating system*

	Small Electricity	Large Electricity	Large Cogeneration	Large Cogeneration
Feedstock input (MW$_{LHV}$)	25.4	152.1	152.1	152.1
Gross electric output (MW)	6.7	49.3	42.9	36.5
Net electric output (MW)	5.3	43.7	37.8	31.8
Useful thermal output (MW)	0.0	0.0	34.4	68.8
Electric efficiency (%)	**20.9**	**28.7**	**24.9**	**20.9**
Thermal efficiency (%)	0.0	0.0	22.6	45.2
Overall efficiency (%)	20.9	28.7	47.5	66.1
Thermal COP	0	0	5.9	5.8

Note: The two cogeneration options differ in the amount of steam sent to the district heating network.
Source: Consonni et al (2005)

65kt/yr waste for incineration) and a large facility (serving 1.2 million people, 360kt/yr waste). The efficiency of electricity generation is not particularly high: only 21 per cent at the small scale and 29 per cent at a large scale when no useful heat is withdrawn. In cogeneration mode, it is assumed that steam is withdrawn at a pressure of 2.6atm and used to make pressurized hot water for a district heating system at 115–120°C. In one case, 30 per cent of the steam flowing through the turbine is withdrawn and in the other case, 60 per cent is withdrawn. The amount of useful heat extracted is 22.6 and 45.2 per cent of the waste energy input, respectively, while the electricity production drops from 29 to 25 and 21 per cent, respectively. Overall efficiencies are 48 and 66 per cent for the two cogeneration cases. The ratio of useful heat gained to electricity production sacrificed can be thought of as a thermal COP. Its value is about 5.9, much better than can normally be achieved with an electric heat pump in heating mode. However, the electric and overall efficiencies are small compared to utility-scale powerplants because of the need to limit the temperature and pressure of combustion due to impurities and irregularities in the waste stream, which in turn limits the Carnot Cycle efficiency (see Chapter 3, Equation (3.16)). Temperatures also need to be limited (to less than 400°C) if the waste has a high chlorine content, as otherwise corrosion becomes a serious problem (Frankenhaeuser et al, 2008).

Because efficiencies are larger (and costs smaller) at larger scales, there is a preference for larger-scale waste-to-energy plants, which in turn requires

collecting waste from a larger area so as to have an adequate supply. Once built, there will be a strong disincentive to implement waste diversion (through recycling, composting or to biogas plants). Waste incineration to generate electricity is costly as well as inefficient – investment costs are $4000–4500/kW (Monni et al, 2006) and up to $7000/kW according to IEA (2005).

Higher overall efficiencies can be obtained if more of the waste heat can be utilized, which requires being able to use heat at lower temperatures than otherwise. This in turn requires that the district heating system serve well-insulated buildings that can be kept warm with relatively low-temperature heat. In the above example, steam is withdrawn from a steam turbine under pressure at 115–120°C, which entails a loss of electricity output. Distribution of heat with hot water from the steam turbine condenser at a temperature of 50–60°C would entail no loss in electricity production and would be adequate if the buildings that are served are well insulated. Additional heat from the combustion flue gases would also be available at about this temperature. According to Eriksson et al (2007), incineration with cogeneration for district heating in Sweden has an overall efficiency of 90 per cent. The heat to power ratio is about 5:1 using municipal waste (so mostly heat with very little electricity is produced), but improves to 2.2:1 using biomass and 0.9:1 using natural gas.

IGCC power generation is expected to yield much greater efficiencies in generating electricity from waste – in the order of 36–56 per cent, depending on the scale of the facility. BIGCC, in which only the woody materials from the waste stream would be gasified, yields efficiencies of 45–62 per cent. The variation of efficiency with scale in both cases is shown in Figure 8.3, along with the efficiency of conventional biomass combustion and of the incinerators profiled in Table 8.5.

With disposal in a landfill, one obtains neither the benefits of recycling nor (in the case of combustible materials) any of the energy released by incineration, although methane produced by the anaerobic decomposition of organic wastes in landfills can be captured and used as an energy source. However, the global warming impact of emissions of just part of the methane could offset or exceed the global warming benefit of saving energy by using the methane that is not emitted. Instead, dedicated anaerobic digesters to produce biogas are preferable so as to avoid or at least minimize leakage of methane to the atmosphere.

Table 8.6 gives the carbon content and heating value of different plastics, while Table 8.4 gives the gross and net heating value of the waste components in MSW and in refuse-derived fuel (RDF), and the weighted average heating value in the EU. The heating value of MSW with the EU average composition is quite low – about 8GJ/tonne – which is a reason for the low combustion efficiency noted above. RDF consists of only the high-energy-value waste products (paper, cardboard, plastics, textiles) and, with a typical net heating value in the order of 16GJ/tonne, can be used as a substitute for coal in cement kilns; the ash residue in this case becomes incorporated into the cement clinker, so there is a double energy-saving benefit: fossil fuel use is reduced and the amount of energy-intensive clinker that needs to be produced for a given amount of cement is reduced. Furthermore, all of the heat released from the combustion of waste in cement kilns can be used during clinkerization, compared to only 50–60 per cent in a typical municipal incinerator with heat recovery (EC, 2001b, Table A3.38).

Sathre and Gustavsson (2006) examined the energy and carbon balances of several different options involving

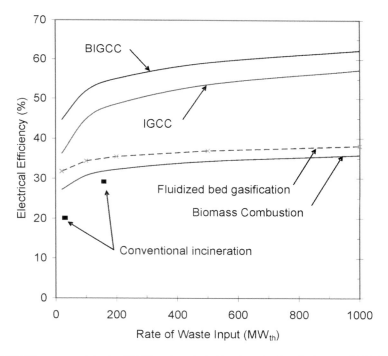

Note: A waste input of 1000MW$_{th}$ corresponds to about 2Mt/yr, using a typical heating value for mixed waste of 16GJ/t.

Source: Dornburg et al (2006); data points in the case of conventional incineration are from Table 8.3

Figure 8.3 *Variation in the efficiency of electricity generation from wastes with the scale of the waste treatment facility, for IGCC treatment of general waste and woody waste only, for biomass combustion and for conventional incineration*

Table 8.6 *Carbon content and heating value of different plastics, and embodied energy (the sum of feedstock energy plus process energy used in manufacturing the plastic)*

Type of Plastic	% of EU plastic waste	% carbon content	Higher heating value (MJ/kg)	Embodied energy (MJ/kg)
LDPE	21	86	45	68–136
HDPE	18	86	45	66–132
PP	20	86	46	55
PS	11	92	41	126
PVC	9	38	18	69–105
PET	9	63	22	45–79
PU	3	58	25	59–113
Others	9	75	35	65–105
Weighted average	100	78	35	

Note: LDPE and HDPE = low- and high-density polyethylene, respectively; PP = polypropylene; PS = polystyrene; PVC = polyvinyl chloride; PET = polyethylene terephthalate; PU = polyurethane.
Source: Carbon content and heating value from Table A3.34 of EC (2001d); embodied energy from Patel and Mutha (2004)

the use wood framing materials recovered from building deconstruction: reuse as lumber, reprocessing as particle boards, reprocessing as pulp and burning to provide energy. Forest biomass that is saved through various options for reusing framing materials is assumed to be used for energy. It is almost always better to process the recovered wood into some other product and use the saved forest wood for energy, rather than burning the recovered wood for energy and making new products from virgin wood. The only exception is for the production of particle board from recovered wood, where burning the recovered wood provides a greater net energy gain than processing the recovered wood. The greatest net energy savings (82GJ/t of recovered lumber) occurs if recovered wood is reused as lumber for 100 years, then transformed into particle board, then transformed into pulp after 140 years, then burned for energy the next year

8.3.5 GHG balance of different waste treatment options

Figures 8.4 and 8.5 give the flows of organic and fossil carbon, respectively, associated with different processes for the treatment of municipal solid waste. Organic (biomass-derived) carbon is removed from the atmosphere during the growth of plants (through photosynthesis), and is returned to the atmosphere in part during composting (as CO_2) and anaerobic digestion (as CO_2 and CH_4) and later as CO_2 after residues from these processes are added to the soil, and in whole through incineration. Carbon also returns to the atmosphere from landfills as CO_2 and CH_4.

Some carbon from landfill and from composting and anaerobic digestion residues remains in the soil indefinitely (this is referred to below as sequestered carbon).

Sanitary landfills are a significant source of methane emission to the atmosphere due to the anaerobic conditions under which the organic matter decomposes and the presence of cracks in the intermixed clay layers. As shown in Table 2.13, methane emissions from landfills and waste treatment are estimated to constitute 10–20 per cent of total anthropogenic emissions (35–69Tg (CH_4)/year). As the global warming potential of methane is 25 times that of CO_2 on a mass basis over a 100-year time horizon (see Chapter 2, subsection 2.6.5), this is equivalent to a CO_2 emission of 875–1725Tg (CO_2)/yr or 0.28–0.63GtC/yr. This in turn is equal to 3.5–8.0 per cent of the total fossil fuel plus cement CO_2 emission of 8.0Gt C in 2005. Furthermore, methane emissions from landfills are expected to grow dramatically over the coming decades under BAU scenarios, due to the shift from open to sanitary landfills in developing countries and increased generation of waste with increasing affluence. For example, Monni et al (2006) foresee an eight-fold increase in landfill methane emissions from 2000 to 2050 under BAU, with only a 27 per cent reduction in this growth with various mitigation options. Clearly, nearly 100 per cent diversion of degradable organic wastes from landfills into alternative management systems that maximize the energy recovery and eliminate methane emissions would be an important objective. Table 8.7 gives estimates of the fraction of landfills that

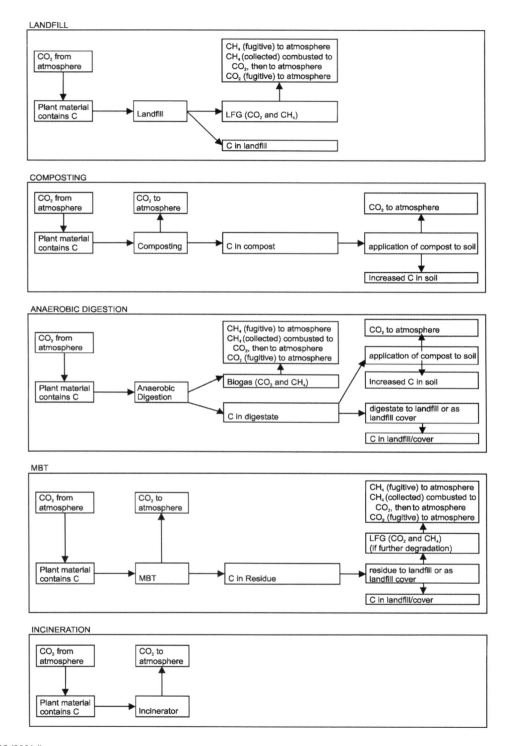

Source: EC (2001d)

Figure 8.4 *Flows of organic carbon associated with different processes for the treatment of municipal solid waste*

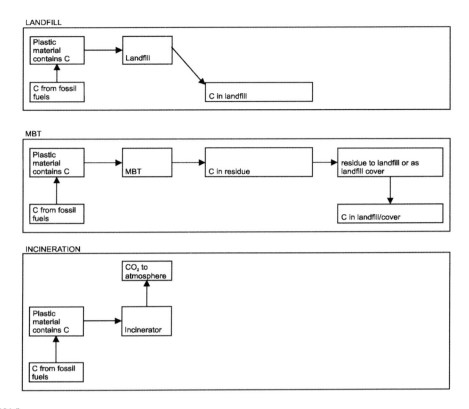

Source: EC (2001d)

Figure 8.5 *Flows of fossil carbon associated with different processes for the treatment of municipal solid waste*

Table 8.7 *Current methane recovery from landfills*

Region	Share of landfills with methane recovery (%)	Average fraction of methane recovered (%)	Fraction of generated methane recovered (%)
OECD and EIT	70	80	56
Other	50	50	25

Note: OECD are EIT country groups are defined in Appendix C.
Source: Monni et al (2006)

currently have some form of methane capture and the fraction of methane that is captured on average at those landfills.

In the following paragraphs, the GHG balance of different waste treatment options is briefly discussed.

GHG balance of landfills

Table 8.8 shows the estimated average CO_2-equivalent emissions associated with landfills in Europe. Included

are methane emissions (converted to CO_2-equivalent emissions assuming a methane 100-year global warming potential of 21), carbon still in the landfill after 100 years, emissions during collection of the waste and operation of the landfill, and avoided emissions assuming that some of the landfill methane is used to generate electricity at 30 per cent efficiency and to displace European average electricity having an emission factor of $0.45 \text{kgCO}_{2\text{-eq}}/\text{kWh}$. The total production of methane from MSW depends on the

Table 8.8 *CO_2-equivalent C fluxes (kgC per tonne of waste) associated with operation of landfills in Europe, excluding emissions associated with the replacement of materials disposed of in landfills*

	Paper		Putrescible		Raw MSW	
	Low	**High**	**Low**	**High**	**Low**	**High**
	With current landfill practice in Europe					
CH_4 emission	46.6	116.2	66.3	109.6	137.7	250.9
C sequestration in the ground	−76.1	−47.7	−32.2	−14.5	−124.4	−78.0
C emissions during transport	0.5	0.5	0.5	0.5	1.9	1.9
Avoided electricity emissions	−1.4	−3.5	−1.9	−3.3	−4.4	−7.6
C emissions from energy use	0.0	0.0	0.0	0.0	0.3	0.3
Net emission	**−30.5**	**65.7**	**32.7**	**92.7**	**11.5**	**167.5**
	With collection and use of 3 × the fraction of landfill methane as at present					
CH_4 emission	23.3	58.1	33.1	54.8	68.9	125.5
C sequestration in the ground	−76.1	−47.7	−32.2	−14.5	−124.4	−78.0
C emissions during transport	0.5	0.5	0.5	0.5	1.9	1.9
Avoided energy and materials	−4.1	−10.6	−5.7	−9.8	−13.1	−22.9
C emissions from energy use	0.0	0.0	0.0	0.0	0.3	0.3
Net emission	**−56.3**	**0.3**	**−4.2**	**31.1**	**−66.4**	**26.7**

Note: Results are given for a low fraction of dissimilable degradable organic carbon in the waste stream and for a high fraction (within the uncertainty range).
Source: EC (2001d, Figure 8

carbon content of the waste, the fraction of the carbon content that is degradable, the fraction of the degradable carbon that dissimilates to CO_2 or CH_4 (the balance does not degrade to gaseous products within the 100-year time horizon considered), and the proportion of the landfill gas so produced that is CH_4. The rate of production of CH_4 depends on temperature and moisture conditions. The dissimilable fraction is particularly uncertain and depends on the material. In calculating the methane emissions and the amount of energy used to generate electricity, the disposition of landfill methane in Europe was assumed, on the basis of the limited available data, to be as follows (EC, 2001b):

- Fraction of landfill waste with some system of methane collection: 68 per cent.
- Average fraction of methane collected from these sites: 54 per cent (so a total of 36.6 per cent of the methane generated in landfills is collected).
- Of the 36.6 per cent collected, 10 per cent is vented to the atmosphere, 30 per cent is flared and 60 per cent is used for energy (so a total of 22 per cent of the methane generated in landfills is used for energy).
- Of the 63.4 per cent not collected, 10 per cent is assumed to be oxidized in the soil before escaping to the atmosphere and 90 per cent escapes (so 60.6 per cent of the generated methane is emitted).

As only about a third of the generated methane is thought to be collected in Europe, there would appear to be scope for a significant reduction in landfill methane emissions and for significantly more production of energy from landfill methane. With 80 per cent collection efficiency from 83 per cent of the landfill waste and no venting or flaring, three times as much methane would be used for energy and CH_4 emissions would be cut in half. Net CO_2-equivalent emissions from raw MSW would be reduced by about a factor of six, or may even become negative, as shown in the lower part of Table 8.8. The use of a 'restoration' layer to increase the rate of oxidation of methane escaping from the landfill would reduce net emissions further (oxidation and combustion convert CH_4 to CO_2, but the resulting CO_2 has a much smaller warming effect, molecule per molecule). This is not to imply that landfills are a preferred method for treating waste; rather, this example serves to illustrate the potential GHG emission reduction from *existing* sanitary landfills.

The undesirability of landfill sites with decomposable wastes, even with maximal collection of methane, is illustrated in Figure 8.6, which shows the net CO_2-equivalent GHG emission as a function of the fraction of the methane that would otherwise escape to the atmosphere that is instead captured and used to produce electricity that displaces coal-fired electricity. Details of the calculation are provided in Box 8.1. For

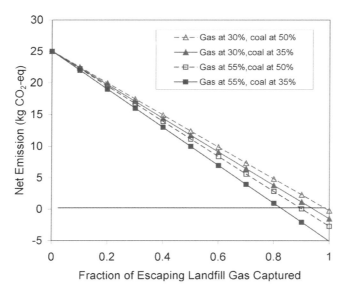

Note: Results are given for different combinations of efficiency in generating electricity from landfill gas and from coal. See Box 8.1 for calculation details.

Figure 8.6 *Net CO$_2$-equivalent emission associated with the partial emission to the atmosphere of methane from landfills and the capture and use of the remaining methane to generate electricity that is assumed to displace coal-fired electricity*

Box 8.1 Net GHG emission from landfills with partial capture of landfill gas

Using an updated methane GWP[1] of 25 (Forster et al, 2007), an energy content of methane (LHV_{CH4}) of 54.4MJ/kg, efficiencies in generating electricity from landfill gas and from the coal that would otherwise be used of η_{gas} and η_{coal}, respectively, and a coal CO$_2$ emission factor (E_{coal}) of 0.09167kgCO$_2$/MJ (25kgC/GJ), then the release of 1kg of landfill gas is equivalent to a CO$_2$ emission of 25kg. Also, the capture and combustion of 1kg of landfill gas saves LHV_{CH4} × (η_{gas}/η_{coal}) MJ of coal and avoids an emission of LHV_{CH4} (η_{gas}/η_{coal}) E_{coal} kgCO$_2$. For η_{gas}= 0.55 and η_{coal} = 0.35, 7.84kgCO$_2$ from the use of coal are avoided, but the combustion of 1kg of landfill gas produces 2.75kgCO$_2$, for a net reduction of 5.09kgCO$_2$/kg landfill gas.

For these assumptions, 83 per cent of the escaping landfill gas must be recovered and used to generate electricity in order for the net emission to be zero. For more efficient generation of electricity from coal than assumed above, a larger fraction of the escaping methane must be captured in order to yield zero net emission, and if the alternative is NGCC at the same efficiency with which the landfill gas is used, all of it must be captured and used. If allowance is made for some of the landfill carbon to be permanently sequestered in the landfill, these results are hardly changed. For example, suppose that 10 per cent of the landfill carbon that had been assumed to produce methane is instead permanently sequestered in the landfill. For the above efficiency assumptions, the capture and use of the remaining methane yields a net emission reduction of (0.9 × 5.09 + 0.1 × 2.75) kgCO$_2$ = 4.86kgCO$_2$. This is less than if no carbon is assumed to be permanently sequestered because less coal is displaced.

---

a wide range of reasonable assumptions concerning the efficiency with which the captured gas is used to generate electricity and concerning the efficiency with which coal would otherwise be used, the fraction of escaping methane that must be captured is far greater than current practice – 80 per cent or more. This reiterates the point that decomposable organic wastes should be diverted from landfills and either composted (producing CO_2 and resistant carbon that can be added to soils) or anaerobically digested to produce CH_4 with near-total capture and use. For more detailed calculations of the GHG balance of landfills, taking into account the chemical content of different decomposable waste streams and their relative proportions and the variation over time in methane emissions, see Lombardi et al (2006). They foresee a minimum CO_2-equivalent emission of about 400kg/tonne of MSW.

GHG balance of incineration

Table 8.9 compares the onsite emissions associated with the incineration of various solid wastes with the emissions that can be avoided through energy recovery of various forms (recycling of ash only, or recycling of ash along with electricity generation, cogeneration of electricity and heat, or use of high heating-value wastes as fuel in cement kilns). The last of these entails prior separation of putrescible and non-combustible waste. According to Table 8.9, incineration of paper or putrescible waste to produce electricity at only 18 per cent efficiency and displacing coal yields a net reduction of CO_2-equivalent GHG emissions based on the difference between avoided powerplant emissions and emissions at the incineration facility. With electricity generation at 18 per cent efficiency and heat recovery at 50 per cent efficiency, incineration of all wastes except glass yields a net emission reduction. However, this is only a partial picture, as it does not take into account the greater energy required to produce replacements for the incinerated materials compared to recycling. The analysis presented in Chapter 6 (Table 6.29) for paper indicates that recycling combined with partial incineration and biomass energy plantations on the saved forest land yields a 50 per cent greater net savings in energy compared to incineration alone.[2]

GHG balance of composting

The carbon balance for composting involves emissions during the collection and transport of waste from households to the composting plant and during the transport of compost to market; emissions during treatment, due to decomposition of the waste and from energy used for turning and processing the compost;

the buildup of soil carbon due to the application of compost to the soil; and avoided emissions due to peat not being mined as a compost. Table 8.10 gives an estimate of the carbon balance of composting a variety of different organic materials and compares that with the carbon balance of landfilling. Composting is consistently more favourable than landfilling (it entails large negative net emissions), even when 80 per cent of the methane generated in landfills is assumed to be captured.

Composting can be applied to food and garden waste mixed with 10 per cent paper, which makes the compost more fibrous. The end product consists of lignin, polymers formed from the remains of the biomass, and humus[3] – all of which are resistant to further decomposition. Very rough estimates indicate that, of 100 units of carbon in the initial putrescible waste, 40 will remain in the compost and 60 are lost to the air as CO_2 during composting. When the compost is applied to the soil, decomposition continues. After 100 years, only a few per cent of the carbon in the compost will remain in the soil. The amount of soil carbon will increase until the annual addition from new compost just equals the decomposition in that year of compost added during all previous years. If application of compost were to cease, the amount of carbon in the soil would gradually decrease.

The alternative to using compost from municipal putrescible waste as a soil conditioner is to add peat that has been mined. Most peat has accumulated over a period of several thousand years in water-logged conditions; when it is removed and added to the soil, decomposition begins, adding CO_2 to the atmosphere. High-quality compost applied to agricultural soils can displace some of the chemical fertilizers that would otherwise be required, with associated energy and emissions savings. However, insufficiently stabilized compost, with a high C to N ratio, can lock up the nitrogen in the soil due to stimulation of microbial activity (microbes require nutrients and outcompete plant roots in the uptake of nutrients).

GHG balance of anaerobic digestion

In anaerobic digestion, food and garden waste mixed with up to 40 per cent paper is shredded and mixed in a liquid medium, and kept at a temperature above 30°C (through heat generated from decomposition). Unlike

Table 8.9 Carbon balance ($CO_{2\text{-eq}}$ kgC/tonne of material or per tonne of MSW) for various incineration options: no energy recovery but recovery of metals from ash (displacing primary metals); energy recovery as electricity at 18 per cent efficiency and displacing coal; electricity and heat recovery at 18 per cent and 50 per cent efficiency respectively; and use of some waste materials as RDF for a cement kiln (displacing coal)

Waste material	% C	Emissions		Avoided emissions				Net fossil emissions			
		Biomass	Fossil	Ash	Electricity	CHP	RDF	Ash	Electricity	CHP	RDF
Paper	33	330	0		−150	−274	−298	2	−148	−272	−265
Putrescible	19	189	0		−52	−94		2	−50	−92	
Plastic	61	0	610		−410	−748	−813	612	203	−136	−171
Glass	0	0.0	0.0		0.0	0.0		2.2	2.2	2.2	
Metal	0	0	0	−373	−373	−373		−371	−371	−371	
Textiles	39	196	196		−190	−346	−377	198	8	−148	−149
Other	24	147	70		−109	−199		72	−37	−127	72
Mean MSW	**24**	178	63	−20	−130	−223	−160	45	−65	−158	−94

Note: Net emissions are based on fossil fuel onsite emissions, avoided emissions, an emission of 2.2$CO_{2\text{-eq}}$ kgC per tonne of material transported and process energy (for shredding, compressing and other operations) for RDF of 30.2$CO_{2\text{-eq}}$ kgC per tonne of refuse used (11.7$CO_{2\text{-eq}}$ kgC per tonne of MSW received).

Source: EC (2001d)

Table 8.10 Comparison of direct CH_4 and CO_2 emissions, carbon sequestration when organic wastes are disposed of through landfilling and composting, avoided CO_2 emissions due to avoided use of peat or fertilizers with composting, and emissions associated with energy used for the landfilling or composting process (all as kgC/tonne of waste)

Waste material	Disposal method	Direct emissions		C sequestration	Avoided emission		Energy use emission	CO_2-eq net emission (kgC/tonne)	
		CH_4	CO_2		Peat	Fertilizer		0% capture	80% capture
Paper	Landfilling	58	58	214				227	−109
	Composting		312	17			4	−114	−114
Food	Landfilling	55	55	37				381	63
	Composting		144	3	99	2	4	−100	−100
Garden waste	Landfilling	60	60	120				338	−11
	Composting		230	10	99	2	4	−107	−107
Textiles	Landfilling	28	28	37				175	13

Note: Given are net emissions (excluding direct CO_2 emissions, which are offset by the photosynthesis that produced the organic material) assuming no capture of methane and assuming 80 per cent capture and combustion but with no displaced CO_2 emissions.

Source: Tables A2.26 and A5.46 of EC (2001d) and inferences therefrom

generation of methane in landfills, leakage of methane is, or has the potential to be, negligibly small. Use of the recovered methane for CHP generation generates the most favourable GHG balance of any waste management option if the generated electricity displaces coal-fired electricity (EC, 2001b). The small amount of residue remaining after the anaerobic digestion process is relatively resistant to decomposition and so will serve as a C sink when added as a fertilizer to soil.

8.4 Recreational facilities

Energy use by recreational facilities is particularly important in the context of this book because, as discussed in Chapter 11 (subsection 11.2.2), one strategy to limit GHG emissions is to limit the growth in material consumption by, in turn, promoting a gradual decrease in working hours and more time for (among other things) leisure activities. As well, investments in recreational facilities provide activities for adolescents that will keep those that might otherwise get into trouble, out of trouble, and would undoubtedly serve as part of a broader strategy to deal with the growing incidence of obesity and the (at least partially) related over-dependence on TVs for entertainment. Here, a few brief comments will be made concerning the energy of

year-round indoor ice rinks, swimming pools and gymnasia and of ways to reduce these energy uses.

8.4.1 Indoor year-round ice rinks

Indoor artificial ice rinks are popular in many northern countries. A recent study identified over 2300 skating arenas and over 1300 curling rinks in Canada, for example, and most of these operate year round. These would appear to be energy-intensive luxuries, but the potential exists to reduce the energy requirements of such rinks to the point where, with addition of PV on their extensive roofs, they could become zero-net-energy facilities. Figure 8.7 gives the breakdown of annual energy use for a typical Canadian skating arena (with a 26m × 61m ice rink, a 3000m² floor area and total energy use of 1.3 million kWh/yr).[4] Refrigeration accounts for about 50 per cent of the onsite energy use, but heating accounts for another 25 per cent. Heat is required for viewing stands, washrooms and changing rooms, and for heating the subsoil beneath the concrete rink slab in arenas that operate year round so as to prevent freezing and heaving of the ground beneath the rink. Many arenas keep temperatures in the stands at 10–15°C, with some maintaining temperatures as high as 18°C when the building is occupied. Hot water is used for showers, taps and (at 65°C) for ice resurfacing.

One key to minimizing the use of energy is to supply the heating requirements by extracting heat from the condenser of the chiller that is used for cooling the ice surface (NRCan, 2003). If spectator stands are heated, this should be done with a low temperature (<32°C) radiant floor heating system, using heat that can be easily recovered from the chiller condenser (in some arenas in Canada, the stands are heated with overhead electric resistance heaters). Increasing the thickness of insulation below the concrete slab on which the ice occurs will minimize the downward transfer of coldness into the ground, thereby minimizing the required compensating heating of the subsurface. Greater amounts of external insulation will reduce conduction of heat from the outside during the summer. Enthalpy heat exchangers (exchanging both heat and humidity between incoming fresh air and outgoing stale air) can be employed.

In most arenas, the rink ice is cooled by a brine (or secondary coolant) that circulates through a network of pipes embedded in a concrete slab. Usually a

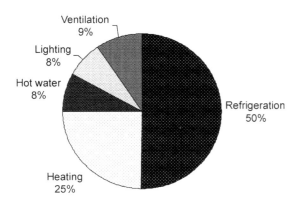

Total = 1.33 million kWh/yr

Source: Data from NRCan (2003)

Figure 8.7 *Breakdown of energy use at a typical Canadian indoor skating arena*

constant-speed brine pump is used, and the brine pump accounts for over 15 per cent of the refrigeration system energy use. Implementation of variable speed motors and optimization of the brine circuit (with use of two or more pumps) can reduce the total refrigeration system energy use by 15 per cent – in part through a reduction in pumping energy use, and in part indirectly through a reduction in the refrigeration load itself (NRCan, 2003).

Infrared radiation from the relatively warm arena ceiling to the ice represents up to 30 per cent of the ice surface heat gain. Use of low-emissivity materials for the ceiling (such as a suspended aluminized cloth or low-emissivity aluminized paint) can reduce the refrigeration system energy requirement by about 15 per cent. The higher visible reflectivity of low-emissivity ceilings allows for lower lighting power with the same brightness. A non-energy side-benefit of low emissivity ceilings is a reduction in condensation (because the ceiling remains warmer in response to less radiant heat transfer to the ice surface), which is a particular problem during the summer due to higher absolute indoor air humidity.

Other measures to reduce energy use include:

- use of displacement ventilation (providing ventilation air to the spectator area through floor vents below the seats, with air extraction at the ceiling level);
- lowering the temperature in the spectator stands;
- maintaining thinner ice (5cm thick ice requires about 10–15 per cent more refrigeration energy than 2.5cm thick ice according to Accent Refrigeration Systems (2009)); and
- use of a secondary cooling loop to melt snow collected by the zamboni (the ice-resurfacing machine), thereby extracting an amount of heat equivalent to the latent heat of melting snow and reducing overall refrigeration requirements by up to 30 per cent (Accent Refrigeration Systems, 2009).

Inasmuch as arenas are often located next to parks, they lend themselves to the use of ground-source heat pumps (GSHPs). With GSHPs, heat removed from the arena during the summer would be added to the ground (rather than rejected to the air) and extracted from the ground during the winter for heating purposes. This would improve the chiller performance during the summer and reduce primary energy requirements for heating arena air during the winter. Detailed simulations would be required to determine whether this strategy would be better from an energy use point of view than extracting heat from chiller condensers for heating during the winter, or if some combination of the two is optimal.

The cumulative effect of the various measures discussed here and other minor adjustments is a saving in annual energy use of well in excess of 50 per cent compared to conventional arena designs (to 500,000kWh/yr or so for the arena profiled in Figure 8.7). Given that arenas have a large roof area suitable for the installation of PV panels, a substantial fraction of the remaining energy use could be provided on site through rooftop PV. Given an energy use of 500,000kWh/yr after energy-saving measures, an arena roof area of 3000m², a solar irradiance of 1600kWh/m²/yr and coverage of 80 per cent of the roof area with modules having 10 per cent sunlight to AC conversion efficiency, the annual electricity generation would be 384,000kWh/yr. With 15 per cent efficient modules, 522,000kWh/yr would be generated – about equal to the target energy use. Warmer climates would have a greater energy requirement (for refrigeration) but also greater available solar energy. Alternatively, the large roof area could be covered with solar thermal collectors to collect heat during the summer that is stored underground and used in the winter as a heat source for a small district heating system that serves nearby buildings (as discussed in Volume 2, Chapter 11, section 11.2). Better still, combined PV/thermal collectors could be used to simultaneously collect heat and generate electricity (see Volume 2, Chapter 2, section 2.6).

8.4.2 Indoor swimming pools and related recreation complexes

The thermal conditions and loads for indoor swimming pools are substantially different than for other buildings. Air temperatures are typically 30–35°C, but the interior temperature of the outside walls can be substantially colder. This, combined with high absolute humidities due to evaporation of water from the swimming pool, can create problems of condensation unless relative humidity is kept quite low. A low relative humidity, however, increases the evaporation rate, which tends to cool the pool water. Low relative humidities are maintained through greater exchange of indoor air with outdoor air than is required on the basis of air quality, further increasing the energy requirements.

With high levels of insulation, the interior wall temperatures will be warmer in the winter. However, if this is to be translated into higher permitted indoor humidity, triple-glazed windows are required in cold climates so as to maintain high window inner surface temperatures, as the permitted relative humidity is set by the coldest surface element. A higher relative humidity reduces the evaporation rate and the required exchange of indoor air with outdoor air. The warmer inner surface temperatures associated especially with triple-glazed windows leads to a perception of warmer air temperature, improving occupant comfort. An indoor swimming facility in Germany recently built to the Passive House Standard (described in Chapter 4, subsection 4.11.1) is expected to achieve a 60–70 per cent reduction in total energy use compared to pools meeting the current German standard (Schulz, 2009).

Recreation centres lend themselves particularly well to the use of heat pumps for integrating heating and cooling needs. Waste heat from exhaust air, wastewater and refrigeration equipment can be collected, stored and upgraded to be used as needed for space heating and hot water production. Substantial energy savings can also be achieved through judicious use of heat exchangers even without heat pumps. For example, Herrera et al (2003) have shown how the use of just four heat exchangers in a complex in Aguascalientes, Mexico, involving a hospital, laundry centre, sports centre with a swimming pool and a family health centre could save 38 per cent of the thermal energy. Here, the pinch analysis technique (explained in Chapter 6, subsection 6.14.3) was used to identify the maximum amount of energy that can be saved through heat exchangers, and to identify where within the various flows to place the heat exchangers.

About 20 gymnasia have been built to the Passive House Standard in Germany (Kah et al, 2009). The insulation level must be such that heating the amount of air required for ventilation purposes alone is sufficient to keep the building warm when in use. This requires heating an air exchange of 0.7/hr to a temperature of up to 50°C (on the coldest days). In conventional gymnasia, the air temperature increases about 1K per metre of height, resulting in warm air next to the ceiling, thereby increasing heat loss. In Passive House gymnasia, the total temperature difference over a 6m height is about 2K. In conventional gymnasia, there are separate ventilation systems for the gym and for the changing room and shower areas. In the Passive House gymnasia, the changing room/shower area is ventilated with air that has first passed through the gym but is reheated first, because of the different temperature requirements of the two areas. The use of a single ventilation system for the two zones is facilitated by the high standard for the thermal envelope (insulation, windows, airtightness) but reduces the cost of the mechanical system, thereby offsetting at least part of the extra cost of the high-performance thermal envelope.

8.5 Summary

The provision of clean water and the treatment of wastewater represent uses of energy comparable, on a per capita basis, to that of major home appliances. The main options for reducing the amount of energy required to supply clean water are to reduce the amount that needs to be supplied and to improve the efficiency of pumping systems. The amount of water required can be reduced through (1) more efficient end use of water and (2) reducing leakage in water distribution systems. An end-use saving potential of up to 50 per cent is possible (depending on current practices), while 30–50 per cent of the water that enters the water supply system in developing countries is typically lost through leakage. The use of bottled water is a particularly energy-intensive way of supplying water and should be strongly discouraged (although this may require upgrading existing supply systems in many parts of the world in order to ensure that tap water is safe to drink).

Measures to reduce the volume of water that is mixed with wastes in the wastewater stream will serve to reduce the energy requirements at wastewater treatment plants. An important initiative at wastewater treatment plants themselves is to maximize the energy recovery from sewage sludge, through anaerobic digestion and capture of methane. Sludge methane has the potential to completely supply the heating and electricity needs of sewage treatment plants, making them energy self-sufficient. Decentralized systems for capturing urine alone, or urine and faeces, and using the nutrients as a fertilizer in place of chemical fertilizers may or may not save energy compared to conventional sewage treatment and the production of fertilizers. However, in the long term such systems will be needed in order to at least partly offset the anticipated decline the supply of usable phosphate, and so should be actively considered in new developments.

With regard to solid waste, there is a clear preference for recycling of paper, metals and plastics, along with anaerobic digestion of organic waste and capture and use of the methane so produced for CHP generation. Composting of organic wastes is another alternative for that part of the waste stream. For wastes that cannot be recycled (such as paper products after already having been recycled once), incineration with energy recovery is preferable to landfilling. Landfilling should be limited to the very small volumes of materials that cannot be recycled or incinerated.

Greater investment in recreational facilities is seen here as an important part of the strategy for the needed shift from a work- and consumption-dominated lifestyle to a more sustainable lifestyle. As with other building facilities discussed in Chapter 4, there is the potential to achieve reductions in energy use in recreational facilities such as indoor skating arenas, swimming pools and gymnasia by at least 50 per cent. As many of these have large roof areas, they have the potential to be equipped with sufficient PV capacity to serve as net zero-energy buildings. The economics of doing so are likely to be more favourable than for buildings with smaller roof areas, due to economies of scale.

Notes

1   See Chapter 2, subsection 2.6.5.
2   The Chapter 6 study assumes advanced incineration with 50 per cent electricity generation efficiency, but recycling still yields a larger net savings.
3   Highly condensed aromatic structures of high molecular mass.

9

Community-Integrated Energy Systems

9.1 Introduction

This chapter discusses the savings in energy use that are possible when the heating, cooling and electricity needs of a collection of buildings are linked together in an integrated system. This is accomplished through a network of underground pipes to distribute heat and another network to distribute chilled water. The heat is ideally provided through cogeneration (the simultaneous production of electricity and useful heat) and by tapping non-electrical sources of waste heat that would otherwise be discarded. The cooling can be provided through dedicated central electric chillers, or through central absorption chillers that use waste heat from electricity generation. Such integrated systems have been referred to as 'community-integrated energy systems'.

The potential to link buildings in an integrated system has implications for the design and operation of individual buildings, and especially for the planning of developments involving more than one building. Since energy losses occur during the distribution of heat and coldness through underground pipes, these losses can be reduced – and the overall efficiency of the system increased – if the distances that hot and cold water need to be distributed are minimized. This in turn provides another reason for building compact cities rather than sprawling, low-density suburbs (the first and stronger reason being to make public transportation, cycling and walking viable alternatives to the private automobile). Losses will also be reduced if heat is distributed at the lowest possible temperature and coldness at the warmest possible temperature, which in turn implies buildings designed to have minimal heating and cooling loads.

9.2 District heating

District energy pipes in modern systems are pre-insulated and installed in trenches, as illustrated in Figure 9.1. In modern systems using hot water, heat can be transported several tens of kilometres with only a few per cent loss (Karvountzi et al, 2002). However, in many older steam-based systems in the former Soviet Union and Eastern Europe, losses in the order of 20–30 per cent can occur.

Table 9.1 names the cities and gives the annual heat load for the ten largest district heating systems in the world. Figure 9.2 gives the share of the total space heating load that is met by district heating in Western Europe, Canada and the US. In cities in Northern Europe, the heat for district heating systems is typically supplied through cogeneration or from waste incineration, but elsewhere it is more common to have a dedicated central heating plant. Many smaller and medium-sized communities, especially in Sweden and Austria, use biomass as the energy source for district heating.

The easiest heat source to capture for district heating is the waste heat from electricity generation, through cogeneration (see Chapter 3, Section 3.3). However, if the electricity is already supplied by a C-free power source (such as wind, solar or hydropower), cogeneration powered by fossil fuels to produce electricity while providing heating would increase CO_2 emissions. Alternatively, biomass-powered cogeneration would be close to carbon neutral and would represent an efficient overall use of biomass as long as the majority of the waste heat can be put to use. A key question is whether or not centralized district heating systems without cogeneration can save energy compared to individual boilers in each building.

Figure 9.1 *District heating pipes about to be installed in a trench*

Table 9.1 *The world's ten largest district heating systems*

Location	Annual heat delivery	
	PJ	GWh
St Petersburg	237	66,000
Moscow	150	42,000
Prague	54	15,000
Warsaw	38	10,600
Bucharest	37	10,200
Seoul	36	10,000
Berlin	33	9247
Copenhagen	30	8000
New York	28	7800
Stockholm	27	7500

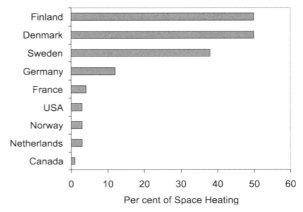

Figure 9.2 *Share of total space heating load supplied by district heating in selected countries*

9.2.1 Energy saving through district heating without cogeneration

Until recently, centralized production of heat for a group of buildings could be done more efficiently than producing heat with a boiler in each individual building. This is because a building will typically have one boiler (often with an identical backup boiler), sized to meet the peak load. Most of the time, the boiler operates at a small fraction of peak load, which for conventional non-condensing boilers, means at much lower efficiency. When buildings are clustered together into a district heating system with a number of separate boilers to serve the entire system, only the boilers that are needed will be used, which means that no more than one boiler will be operating at part load at any given time. On an annual basis, this can reduce energy use by 10–20 per cent.

However, condensing boilers have fallen in price (to no more than twice the cost of conventional boilers), and these boilers have greater efficiency at part load than at full load (see Figure 4.22). Furthermore, a small boiler could be installed in a given building and used for the summer domestic hot water load, with the main boiler shut done. Thus, in new developments, there is no longer an efficiency advantage in consolidating the heating load and operating only one boiler at part load at any given time. There could still be some efficiency gains in a centralized facility, in that boilers in a centralized facility are more likely to be properly maintained and operated in an optimal fashion (as the extra personnel costs can be more easily justified). However, the small efficiency gain (5–10 per cent) will be more easily offset by heat losses from the distribution network.

Tapping sources of heat other than from electricity generation

There could still be substantial energy savings if a district heating network collects heat from scattered sources that would otherwise be discarded. Examples include sewage treatment plants, bakeries, some manufacturing facilities and electrical transformer stations. As these heat sources are often at a lower temperature than in the district heating network, heat pumps will be needed to transfer heat from these heat sources to the heat distribution grid. In Tokyo, sewage has a stable temperature of about 16°C in February and is used as a heat source for production of hot water at 47°C with a COP[1] of 3.9 (Yoshikawa, 1997), while over 20 per cent of the heat for district heating in Gothenberg, Sweden, is extracted from wastewater (Balmér, 1997). In some cities in Sweden, steam produced by incineration plants is used to drive 6MW absorption heat pumps that are used to extract heat from the exhaust gas of the incineration plants; the exhaust gas is cooled from 70°C to 30°C in the process, and the extracted heat is used for district heating (Berntsson, 2002).

Low-grade geothermal heat, left over after higher-temperature heat has been used for heating or hot water purposes, is another example of a heat source that can be put to use with a heat pump if a district heating system is present. In Tianjin, China, geothermal 'waste' heat is available at 40°C but should not be discharged to waterways at a temperature above 30°C. A heat pump with an appropriate mixture of working fluids can be used to upgrade the return-water of the district heating system from 65°C to the supply temperature of 80°C while cooling the discarded geothermal water to 30°C (Zhao et al, 2003).

Distribution of heat as hot water or as steam

Heat can be distributed as either steam or hot water. Many older systems use steam, whereas most modern systems use hot water. Steam has a much higher thermal energy density than hot water, so much less mass needs to be circulated in a steam system than in a hot water system. However, hot water provides a number of efficiency advantages compared to steam: less electricity production is sacrificed if hot water rather than steam is extracted from a turbine, lower-temperature heat sources can be used and distribution losses are smaller due to the smaller temperature difference between the pipe and the surroundings. The supply temperature can be adjusted with the seasons, or there may be more than one piping network, carrying water at different temperatures, thereby allowing both low-grade (low-temperature) and high-grade (high-temperature) heat to be captured and utilized. Separate and optimal temperatures can be used for the space heating and tap water systems. The space heating system can be closed down during the summer and only the smaller tap water system maintained, with consequent energy savings.

Accommodating low distribution temperatures

In order to utilize low-temperature heat sources, a relatively low distribution temperature (<70°C) is needed. The heat flow from a radiator depends on the radiator temperature, size and hot water flow rate. In new developments, the building envelope can be designed to require smaller peak heating rates. However, if a high-temperature district heating system is converted to a low-temperature system in order to utilize low-grade heat sources, special provisions will need to be made to assure adequate heating of existing buildings. Wherever possible, the thermal envelope in these buildings should be upgraded prior to conversion so that the amount and temperature of the required heat can be reduced. A larger or more effective heat exchanger between the district heat water and building heating water would also help by reducing the temperature drop between the two. Other options are:

- to increase the flow rate (necessitating larger pumps and, in rebuilt systems, larger-diameter pipes); or
- to install larger radiators; or
- to add a peaking plant to boost the temperature during the coldest conditions.

All of these options entail added cost, so the split between the different options should be optimized, as discussed by Kilkis (1998). For a system in which the supply temperature is reduced from 90°C to 60°C, the radiators could be oversized by 60 per cent and the distribution temperature boosted to 76°C during peak conditions. Flow oversizing will not be the prime method of restoring equipment capacity, but can be used for fine tuning.

Another option is to use onsite heat pumps to boost the internal distribution temperature for those buildings that require higher temperatures (Curti et al, 2000). This avoids the need to provide heating water at the temperature required by the most demanding building. In buildings where a low internal distribution temperature is permitted, heat pumps can be used to extract heat from the return branch of the district heating system rather than from the supply branch. This will increase the efficiency of the central heating plant, and also reduce the required water flow in the district heating loop because the temperature differential between supply and return water will be increased.

Embodied energy in district heating networks

Energy is used in the manufacture of district heating pipes and insulation, in the excavation and installation of the pipes, and the replacement of pavement that may have been removed prior to excavation. Fröling and Svanström (2005) and Perzon et al (2007) present lifecycle assessments of the energy cost of district heating networks, including heat losses during transmission. Overall losses are dominated by the loss of heat during transmission, with the next largest term (the energy used to make the insulation) being about ten times smaller, so high levels of insulation are justified. Both assessments assume the use of insulation produced without the use of halocarbon blowing agents; if halocarbon blowing agents are used, the global warming effect of leakage from the insulation would probably be important compared to the energy used in producing the district heating network. Minimizing the energy used during excavation (by removing and replacing no more material than necessary) and the amount of pavement that needs to be replaced are also important. Fröling and Svanström (2005) recommend the use of a twin-pipe system (two pipes inside a single round casing) rather than two single pipes, so as to minimize the size of the required trench.

9.3 District cooling

District cooling can save energy due to the more efficient operation of large centralized facilities that is possible, and through the opportunities that district cooling provides to make use of low-temperature heat sinks.

9.3.1 Efficient operation of centralized chilling facilities

Commercial chillers with fixed-speed compressors are most efficient operating at full load, while those with variable speed compressors are most efficient at 50–80 per cent of full load. If a building is cooled with its own chiller, sized to meet the peak cooling load, the chiller will be operating at a small fraction of its peak load most of the time. If it has a fixed-speed compressor, it will not be operating at its peak efficiency most of the time, and this will also be true much of time even if it has a variable speed compressor.

A central cooling plant serving several nearby buildings can be built from many chillers that would otherwise go into individual buildings. The system can be designed with the number of chillers and size mix chosen so as to give the greatest operational flexibility, with the number of chillers in operation and the distribution of the cooling load among the chillers at any given time chosen so as to maximize the overall efficiency. For example, three chillers meeting 20 per cent, 30 per cent and 50 per cent of the peak load could be installed. When the cooling load is less than 20 per cent of the peak load, the smallest chiller would be used but at a relatively large fraction of its peak load. As the cooling load increases, the cooling requirement would be met by a changing combination of the three chillers. Variable speed drives, which entail about a 5 per cent efficiency loss even at full load, could be replaced with fixed speed drives. Another saving arises from the fact that larger chillers tend to be more efficient than smaller chillers, and larger chillers can be used in a centralized facility. Dharmadhikari et al (2000) present the example of the Expo '98 project in Lisbon, where a central plant reduced chilling-energy use by 45 per cent compared to meeting the same loads with chillers in each building. Thus, unlike district heating and hot water, there is a large potential efficiency gain through district cooling simply by centralizing the production of chilled water.

A large efficiency gain is also possible if a centralized chilling facility is used in place of individual, small air conditioners in multi-unit residential buildings, as air conditioners are less efficient than commercial chillers (with typical COPs of 2.5–3.5, compared to 5.0–7.5 for large commercial chillers). Of course, a relatively large chiller could be used in each building, but this requires a cooling tower on the roof, cold water piping and radiators. Connection to a district cooling system provides a more convenient alternative to air conditioners by avoiding the need for cooling towers, although an internal cold water distribution system and radiators are still needed. There is a small additional benefit from district cooling in that waste heat from a centralized cooling facility is concentrated in a thermal plume that rises high into the atmosphere without contributing to the urban heat island (Chen et al, 2008).

Utilizing dispersed heat sinks in district cooling systems

A chiller will operate more efficiently if it can operate with a cooler condenser (see Chapter 4, subsection 4.4.2). This will be the case if the condenser is in contact with a cooler medium than the outside air (if it is air cooled) or water from a cooling tower. Potential media for receiving the heat from a cooling system include sewage water, lake water or the ground itself. However, the majority of buildings in a region will probably not be situated so as to be able to use these heat sinks. By linking the buildings in a district cooling network, it may become possible to utilize heat sinks that could not otherwise be used. Even when this is not possible, it might be possible to utilize a centralized cooling tower for a group of linked buildings that, on an individual basis, would have to rely on air-cooled condensers due to site constraints.

In Japan, sewage water is used as a heat sink for cooling purposes in summer (and as a heat source, with heat pumps, for the heating network in winter). During extraordinary conditions in Tokyo in 1995, with a peak outside temperature of 37°C, the sewage temperature rose to 29.4°C, and the monthly average COP was 4.3 for the production of chilled water at 7°C (Yoshikawa, 1997). In Stockholm, water from the Baltic Sea is directly used for cooling through heat exchangers (that is, it replaces the central chillers

altogether, rather than merely cooling the condensers). When the water is not cold enough, it is cooled with a heat pump, with the ejected heat added to the district heating network. In Toronto, cold water (at 4°C) from below the 80m depth of Lake Ontario is also used in place of electric chillers in a small, pilot district cooling network. In other parts of the world, where sea or lake water is not cold enough for cooling purposes, the sea or lake water can be used to cool the condensers in a district cooling plant. This option has been actively considered for a possible district cooling system in Hong Kong (Yik et al, 2001; Parsons Brinckerhoff Asia, 2003; Ove Arup and Partners Hong Kong, 2003). Lake water is used with heat pumps as a heat sink for a district cooling system in a city in Hunan province, south China (Chen et al, 2006).

Opportunities for centralized daily storage of coldness

The benefits of diurnal storage of coldness in individual buildings, using chilled water or an ice/water slurry, were discussed in Chapter 4 (subsection 4.5.7). Coldness can also be stored centrally in the same way, as part of a district cooling system. The Chicago Trigen Peoples District Energy system has a large welded steel tank storage system. At Yale University, reinforced concrete tanks are located beneath parking lots and playing fields, and have a total capacity of 11 million litres (11,000m³). Large-scale, centralized storage is advantageous compared to storage for individual buildings because, as the scale of the storage system increases, the volume to surface area ratio increases. Since energy losses depend on the surface area, this tends to reduce the loss per unit of thermal storage.

9.4 Advantages and economics of district energy systems

From the perspective of the owner or operator of a building, there are a number of reasons why hooking up to a district heating or cooling system might be preferable to onsite production of hot and chiller water:

• there are savings in the upfront capital cost to the building owner since boilers and chillers do not need to be purchased as part of the building;

- there are savings in space, maintenance costs and insurance by virtue of having no onsite heating or cooling equipment;
- noise and vibration associated with heating and/or cooling equipment are eliminated;
- in the case of district cooling, there is no need for onsite cooling towers and condenser piping;
- the absence of roof-mounted cooling towers eliminates the risk of legionnaires disease, particularly if air intakes are also located on the roof;
- the absence of roof-mounted cooling towers increases flexibility in the design of the building, including increased opportunity for incorporation of PV, solar thermal collectors or rooftop gardens.

Centralized district heating and cooling systems entail a substantial upfront capital cost to install the pipe network. Offsetting this are a number of factors that tend to reduce the total capital cost of the heating and cooling system:

- total heating and cooling capacities do not need to be as large as the sum of the capacities that would occur in individual buildings, because peak demands would not all occur at the same time;
- less total backup capacity is needed (in buildings that are individually heated or cooled, it is not uncommon to have two identical boilers, one of which is backup, whereas in a central facility with, say, ten boilers, one backup would be sufficient);
- unit costs of electrical generators, heat exchanges, boilers and chillers all tend to fall with increasing capacity.

Boiler operation and maintenance costs will also be lower due to economies of scale, while the energy operating cost of the cooling plant will be less due to the greater efficiency of large chillers compared to small chillers, the better opportunity to schedule multiple chillers so as to maximize overall efficiency at part-load operation, and the relatively smaller cost of maintenance and supervisory personnel.

Representative capital costs of various methods of cooling buildings are compared in Table 9.2. The first method is the conventional system with electric chillers in each building. In the next two methods, the existing district heating network is used to provide heat to either absorption chillers or desiccant cooling systems (Chapter 4, subsection 4.5.4) in each building (the

network cost is zero in this case). In the last two cases, chilled water is produced centrally, either with electric or absorption chillers, and then distributed to each building with a district cooling network (which exists alongside the pre-existing district heating network). The savings in the cost of the chilling equipment with a district cooling system more than offsets the cost of the network for the network costs presented in Table 9.2, resulting in a smaller total capital cost. Centralized chilling using electric chillers is the least expensive of the three district cooling options and, as will be seen below, it also the most efficient.

The economics of district energy networks are less favourable where building densities are low or heating or cooling loads are low (as in high-performance buildings). Nilsson et al (2008) discuss techniques being developed to improve the economics of installing district heating in existing sparse developments. Connection to low-energy single-family residential buildings is hard to justify, while connection to low-energy multi-unit residential buildings is easier to justify (and as discussed in Chapter 4, subsection 4.2.8, achieving low energy use per unit floor area is inherently easier in multi-unit buildings).

The difficulties with district heating and cooling occur at the societal scale: district energy projects are typically complex and involve a large number of institutional, technological, legal and financial issues (MacRae, 1992). The required feasibility studies are expensive and time consuming. District energy systems involve a single large upfront investment rather than many smaller investments by individual building developers. Strong and determined leadership is required to bring about these systems.

9.5 Cogeneration and trigeneration

Here, we briefly analyse whether and to what extent cogeneration and trigeneration in district energy systems can reduce the use of primary energy in meeting a given set of electric, heating and cooling loads. Cogeneration in district heating systems occurs at a scale of several tens to hundreds of MW of electrical and thermal output. Cogeneration can also be done at the scale of individual buildings (5–500kW electrical or thermal output) using microturbines, reciprocating engines or fuel cells, as discussed in Chapter 4 (subsection 4.3.7). Waste heat would be used during

Table 9.2 *Comparison of capital costs, energy use and water use for electric vapour-compression chillers in individual buildings, for various systems driven by heat from a district heating system, and for centralized chillers serving a district cooling system that serves a 25 per cent larger total load due to individual peak loads not exactly coinciding*

Cost component	System				
	500kW$_c$ VCCs in individual buildings, peak COP = 4.2	District heat supplied to:		District cooling chilled water from:	
		500kW$_c$ absorption chillers	500kW$_c$ desiccant cooling units[a]	5MW$_c$ of VCCs serving 6.25MW$_c$ load	Four 1.25MW$_c$ Li/Br absorption chillers serving 6.25MW$_c$ load
Chiller, $	76,470	214,120	94,830	413,000	416,000
Chiller, $/kW$_c$	153	428	190	83	83
Cooling tower, $	41,650	26,120	0	110,000	196,500
Mechanical installation, $	60,060	82,350	included	275,000	508,700
Electrical installation, $	35,700	39,000	included	137,500	126,500
Building, $	57,060	57,060	26,470	76,000	258,800
Other, $	101,120	181,830	59,880	308,000	531,600
Network, low[b], $	0	0	0	674,100	674,100
Network, high[c], $	0	0	0	2,195,300	2,195,300
Total, $	372,000	600,350	181,180	1,993,600–3,514,800	2,754,700–4,275,900
Total, $/kW$_c$ of load served	**744**	**1200**	**362**	**319–562**	**441–685**
EFLH/yr	800	800	800	1000	1000
Chilling provided (MWh/yr)	400	400	400	5000	5000
Electricity Use (MWh/yr)	140	84	46.8	1220	737.5
Heat use (MWh/yr)	0	635	334	0	736
Average COP	2.9	0.63	0.84	4.1[d]	
Water use (m³/yr)	1098	2300	800	1020[e]	22,032
Water use (m³/GJ)	0.73	1.60	0.56	0.71	1.22

Note: VCC = vapour-compression chiller; EFLH = equivalent full-load hours. [a] Providing chilled and dehumidified air only. [b] 100m of network per MWh/yr of cooling load. [c] 400m of network per MWh/yr of cooling load. [d] Assumed here. [e] Reduced from first case in proportion to the reduction in required heat rejection.

Source: IEA (1999) for all cases except the district cooling system with VCCs, where all costs except network costs are taken from Spurr and Larsson (1996) and multiplied by 1.1 in order to be roughly comparable with the more recent costs in IEA (1999)

winter months for space heating requirements and year round for hot water requirements, but would be in excess during the summer. Trigeneration would make use of excess waste heat during the summer.

9.5.1 Maximizing the efficiency of cogeneration

As discussed in Chapter 3 (subsection 3.3.3), the effective efficiency in generating electricity through cogeneration is maximized if the overall efficiency is maximized, which in turn requires maximizing the amount of waste heat that can be used. This in turn requires being able to use waste heat at the lowest temperature at which it is produced. As also discussed in Chapter 3 (subsection 3.3.1), the amount of electricity that is sacrificed when heat is withdrawn is minimized (in combined cycle cogeneration) the lower the temperature at which the warmest heat is withdrawn. As well, heat losses in distributing heat through the pipe network will be smaller the lower the temperature of the heat in the network. Thus, the overall efficiency and possibly the electricity to heat

production ratio will be greater the lower the temperature at which heat needs to be supplied to the buildings in the district heating system.

This in turn depends on the performance of the building envelope (insulation levels and window U-values): with a better thermal envelope, the rate of heat loss is smaller, so heating radiators in the buildings do not need to be as warm, so the heating water supplied by the district heating system does not need to be as warm. As well, if the supply temperature is lower, the temperature of the water returning to the central cogeneration powerplant will be lower too, which means that more heat can be extracted from the flue gas exhaust by condensing water vapour in the exhaust and added to the return water as a first step in reheating it. This also increases the overall efficiency. Figure 9.3 shows the increase in electrical efficiency, overall efficiency (including useful heat) and the power to useful heat ratio for a combined cycle cogeneration system, as a function of the temperature at which heat is supplied to a district heating system.

The above discussion and much of the literature on cogeneration applies to steady, full-load operation.

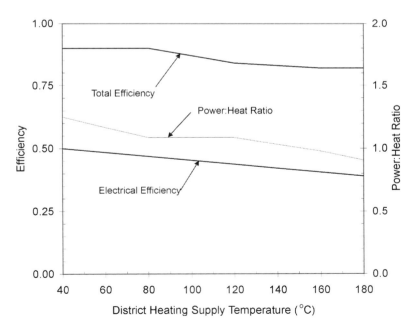

Source: Spurr and Larsson (1996)

Figure 9.3 *Variation in electricity generation efficiency, overall efficiency and electricity to useful heat ratio in a gas combined cycle system as the temperature of heat supplied to a district heating system varies*

Most cogeneration units supply heat in an on–off mode of operation, which means that electricity will also be generated in an on–off mode. This in turn requires the central powerplant to operate with a larger spinning reserve, or to frequently fluctuate in its power output. If the installation of a given amount of cogeneration prevents the installation of an equal amount of combined cycle power generation at a central plant, then other, less efficient generating units will need to be used when the cogeneration unit is not operating. In either case, the efficiency of the central powerplant will be reduced. These effects were investigated by Voorspools and D'haeseleer (2006) using a model that simulates a central powerplant on an hourly basis. They considered four different strategies for sizing the cogeneration unit, which are illustrated in Figure 9.4.

Shown are heat duration curves, where the x axis is the number of hours with a heat load greater than or equal to that shown on the y axis. The classical approach to sizing would be to size the cogeneration unit such that the heat output (Q in Figure 9.4a) and the hours of operation (U) maximize the area of the rectangle shown below the load duration curve. A second option (Figure 9.4b) is to choose a smaller cogeneration unit that can operate more hours per year. A third option (Figure 9.4c) is to discard some heat when the heat load is less than the heat supply, thereby allowing more hours of operation. The fourth option (Figure 9.4d) is to use multiple smaller units. Also considered was the possibility of heat storage, which allows the cogeneration unit to operate when electricity but not heat is needed, without having to discard the heat.

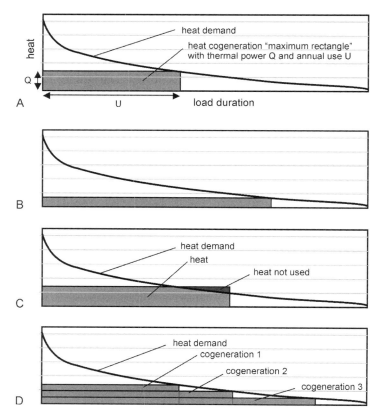

Figure 9.4 *Different strategies for sizing cogeneration: (a) maximizing the area of the rectangle under the load duration curve; (b) reducing the size of the cogeneration unit but operating it more hours per year; (c) sizing the unit as in (a), but with longer operation and some wasted heat; and (d) using multiple smaller units*

For the particular heat and electricity demand variations considered by Voorspools and D'haeseller (2006) (pertaining to a commercial building in Belgium), the energy saving is maximized if the cogeneration unit is about half the size that maximizes the area under the load duration curve. If the unit is sized according to the 'maximum-rectangle' rule, but then operated for more hours, total energy saving initially increases as the hours of operation increase, then decreases. For modest increases in the number of operating hours, net energy saving increases even though some heat is thrown away, because more less-efficiently generated central electricity is displaced. However, as more heat is thrown away, the effective electrical generation efficiency of the cogeneration plant decreases, and when the effective efficiency equals that of the central powerplant (times transmission losses), the maximum saving is reached.

9.5.2 Does trigeneration save energy?

The term trigeneration refers to the production of electricity, heat and chilled water as part of a single, *integrated* process. This can occur using waste heat from a gas or steam turbine to drive an absorption chiller. In the former case, the electricity production is low whether or not waste heat is used for absorption chilling, while in the latter case some electricity production is sacrificed. The alternative is to maximize the production of electricity and use the extra electricity to drive an electric vapour-compression chiller. This is a *sequential* process.

An analysis of the relative energy use for chilling using electric chillers compared to absorption chillers using waste heat from microturbines, fuel cells and combined cycle cogeneration was presented in Chapter 4 (subsection 4.5.4 and Table 4.8). Assuming an electric chiller COP of 4.6, the microturbine plus absorption chiller option was found to increase primary energy use if the alternative is a near state-of-the-art (50 per cent efficiency) powerplant, while the combined cycle plus absorption chiller option decreases primary energy use even with state-of-the-art alternatives.

Here, the electric chiller COP at which there is no net saving in primary energy use is computed for single-, double- and triple-effect absorption chillers. We use the absorption-chiller COP versus input temperature information given in Figure 4.40 and the ratio of electricity production lost to thermal energy gained versus temperature given in Figure 3.14. The choices are to extract some thermal energy at a high enough temperature to be used in an absorption chiller, or to maximize electricity production with no useful thermal energy, and use the extra electricity in an electric chiller. Equations for the calculation of the equivalent electrical COP of an absorption chiller are given in Box 9.1. For single-, double- and triple-effect absorption chiller COPs of 0.7, 1.2 and 1.7 (respectively), input temperatures of 80°C, 120°C and 180°C, and electricity loss to useful heat ratios of 0.11, 0.185 and 0.28, the break-even electric chiller COPs are 6.4, 6.5 and 6.1, respectively. However, this does not take into account the fact that absorption chillers also require some electricity (to operate pumps), and they require larger cooling towers than electric chillers, with greater fan and pump electricity use. Accounting for this, the break-even electric chiller COP is 5.5–5.8.[2] If the electric chiller COP is less than this break-even value, absorption chillers require less primary energy. Large centrifugal chillers (which can be used in district cooling systems or in large buildings) have COPs in the range of 5.3–7.0, so trigeneration with absorption chillers in place of dedicated electricity production and large chillers will usually *increase* total primary energy use compared to the best electric chiller alternative. That is, it is better to maximize electricity production even if the waste heat cannot be used for any other purpose, and to use the extra electricity so-produced in a large, highly efficient electric chiller.

Lin et al (2001) have analysed the prospect of using hot water from the district heating system in Beijing for cooling using an absorption chiller. Addition of cogeneration to the district heating system is being contemplated. At present, the Beijing system supplies 100MW of domestic hot water in summer using supply and return temperatures of 75°C and 50°C, respectively. Supplying water at this temperature through cogeneration would entail no penalty in terms of electricity production. However, to supply water at 95°C (needed for absorption chillers) would entail an electricity production penalty, and this penalty would apply to *all* of the hot water produced, whether for absorption chilling or for domestic hot water. When this interaction between the production of chilled water and hot water is taken into account, the absorption chiller option is decidedly less attractive than using an electric chiller.

Box 9.1 Electricity-equivalent COP of an absorption chiller

If ΔH is the heat extracted from a steam turbine and ΔE is the electricity production that is lost, a COP for heat extraction (COP_{ext}) can be defined as $\Delta H/\Delta E$. As ΔH times the absorption chiller COP is the amount of chilling produced, the electrical-equivalent COP of an absorption chiller can be defined as:

$$COP_{e\_abs} = COP_{ext}COP_{abs} = \frac{\Delta H}{\Delta E}COP_{abs} = \frac{chilling\ provided}{electricity\ lost} \tag{9.1}$$

This is analogous to the COP of an electric chiller, given by (chilling provided)/(electricity used). In both cases, there is additional electricity use by the cooling tower pumps and fans and the chilled water pump – so-called auxiliaries. If q_{aux} is the electricity used by auxiliaries per unit of cooling, then the total electricity used, q_{tot}, by absorption and electric chillers per unit of cooling is:

$$q_{tot} = \frac{1}{COP_{e\_abs}} + q_{aux-abs} = q_{abs} + q_{aux-abs} \tag{9.2}$$

and:

$$q_{tot} = \frac{1}{COP_e} + q_{aux-e} \tag{9.3}$$

respectively, where $q_{abs} = 1/COP_{e-abs}$. The difference in auxiliary energy use is significant, as $q_{aux-abs} \approx 0.04$–0.05 while $q_{aux-e} \approx 0.02$. The final effective electric COP of an absorption chiller is then given by:

$$COP_{e-abs}^f = \frac{1}{q_{tot}} = \frac{chilling\ provided}{electricity\ lost + electricity\ used} \tag{9.4}$$

This is the break-even electric chiller COP; electric chillers with a COP above this value use less primary energy than the absorption chiller alternative. If the choice is between onsite electric chillers and absorption chillers in a district cooling system, then the chilling-provided term used in Equation 9.4 should be the chilling delivered by the district heating system to the buildings themselves and the electricity-used term should include the electricity used by the district cooling pumps. If both electric and absorption chillers are being considered for a district cooling system, then energy use and loss by the district cooling system does not need to be included in Equation 9.4.

9.5.3 Impact of cogeneration on pollutant emissions

To the extent that cogeneration reduces overall fuel use for heating and the generation of electricity, it will tend to reduce total pollutant emissions (along with CO_2 emissions). However, the emission factors (pollutant emission per unit of fuel used) can be quite different for a central powerplant, onsite boilers or furnaces, and the cogeneration powerplant. Here, we assess the net effect on pollutant emissions of various cogeneration systems, given the relevant efficiencies and emission factors.

Table 9.3 gives SO_x, NO_x, PM_{10} (particulate matter smaller than 10 microns in diameter), CO and hydrocarbon emissions per unit of fuel use for furnaces, boilers, centralized NGCC and coal powerplants, and for cogeneration plants using gas engines, gas turbines, steam turbines powered by coal and NGCC. Note that there is a wide range of emission factors reported for a given technology. Table 9.4 gives representative

Table 9.3 *Pollutant emissions (gm/GJ fuel) as computed from the raw data given in the indicated reference and/or converted to common units*

	Pollutant					Reference
	SO_x	NO_x	PM_{10}	CO	HC	
	Stand-alone options					
Furnaces		11		4.7		Table 2.19
Boilers	0.5	9				Gulli (2006)
		3–9		2–6		Lazzarin & Noro (2006)
	2.8	61	2.8	33	4	Strachan & Farrel (2006)
		0.01–0.02		0.04		Table 2.19
Central NGCC	0.3	3.6	2.4			Gulli (2006)
	5.3	54	11	19	14	Strachan & Farrel (2006)
	0.3–0.4	5–37	4.6–5.8			DSS (2005)
Central Coal	1567	472	38	19	14	Strachan & Farrel (2006)
	91–513	29–107	1–18			DSS (2005)
	Cogeneration options					
Gas Engine	0.3	100	1.4			Gulli (2006)
	8.9	139	3.9	500	150	Strachan & Farrel (2006)
Gas Turbine	0.2	39	2.7			Gulli (2006)
	8.9	81	11	117	117	Strachan & Farrel (2006)
		19–45		5–15		Lazzarin & Noro (2006)
Steam Turbine		20–70		8.5–17		Lazzarin & Noro (2006)
NGCC		39–96		11–34		Lazzarin & Noro (2006)

Table 9.4 *Efficiencies assumed in computing the amount of fuel required to supply a given amount of heat and electricity separately or through cogeneration*

	Electrical	Thermal	Overall
Boiler	0.00	0.92	0.92
Central NGCC	0.55	0.00	0.55
Central coal	0.33	0.00	0.33
	Cogeneration options		
Gas engine	0.38	0.50	0.88
Gas turbine	0.25	0.40	0.65
Steam turbine	0.45	0.30	0.75
NGCC	0.45	0.40	0.85

efficiencies. Given these efficiencies, the amount of fuel required to produce the same amount of heat and electricity as in various cogeneration systems, but through separate heating and electricity production, can be computed. The fuel use can then be combined with the emission factors to compute the emissions for cogeneration and for separate production. The resulting ratios of emissions from cogeneration to emissions through separate production of heat and electricity are given in Table 9.5. If cogeneration displaces central electricity production using natural gas/combined cycle, emissions are increased unless the cogeneration system is also based on natural gas/combined cycle. When cogeneration displaces coal, emissions are not always decreased, depending on the pollutant considered and the type of cogeneration system.

The results given here are for a fixed production of heat and electricity that matches the output ratio of a given cogeneration system. If more heat is needed, then supplemental boilers will be required. The greater the heat to electricity demand ratio, the closer the emissions with cogeneration will be to those for separate production, because overall emissions will be increasingly

Table 9.5 *Ratio of emissions produced through cogeneration to emissions produced through separate production of heat and electricity*

	Cogeneration system that is matched with separate production of heat and electricity							
	Gas engine		Gas turbine		Steam turbine		NGCC	
Source of central electricity	NGCC	Coal	NGCC	Coal	NGCC	Coal	NGCC	Coal
Electricity produced	38	38	25	25	45	45	45	45
Heat produced	50	50	40	40	30	30	40	40
Total fuel use with stand-alone production of heat and electricity	123.4	169.5	88.9	119.2	114.4	169.0	125.3	179.8
Ratio of cogeneration emissions to emissions with separate production								
SO_x, min	0.7–5.0	0.00	0.7–6.5	0.00–0.01			0.2–0.6	0.00
NO_x, min	2–24	0.2–2.8	1.6–6.4	0.2–0.8	1.1–5.1	0.1–0.5	0.8–0.9	0.08–0.09
PM_{10}, min	0.4–0.7	0.1–0.5	1.2–3.0	0.4–1.4			0.8–1.0	0.1–1.0
CO, min	16–35	12–21	0.5–5	0.3–4	0.5–0.6	0.3–0.5	0.6–1.2	0.5–0.7
HC, min	12.8	8.3	14.6	9.6			1.1	0.7

Note: The cogeneration system uses 100 units of fuel to produce the amounts of heat and electricity indicated here, and can be compared with the total fuel used for separate production of heat and electricity.

Source: Computed from data in Table 9.3

dominated by those from supplemental heating (as illustrated in Strachan and Farrell, 2006). Conversely, if the cogeneration system is sized to the maximum heat output and operated as a baseload electric powerplant, excess heat will be discarded and excess electricity sold to the grid (this is a common practice). In the limiting case, the cogeneration unit operates as an electric generator only, but invariably with less efficiency and higher emissions than the central powerplant. Annual relative emissions and energy use then depend on the fraction of heat that is thrown away, but can be worse than for separate heating and electricity generation. The only attractive cogeneration option if the alternative is central NGCC, is NGCC cogeneration. Most district heating cogeneration facilities use steam or simple cycle gas turbines, but these are not attractive compared to central NGCC powerplants and highly efficient onsite boilers.

Another factor that must be considered is the human exposure to pollutants. Cogeneration will entail emissions closer to human populations, and in the case of cogeneration in individual residential buildings (as advocated by some), will be emitted closer to ground level. Thus, a smaller pollutant emission may entail greater exposure. In centralized cogeneration and district heating, the exhaust stack will be higher, thereby reducing human exposure compared to the same emission from residential furnaces and boilers. However, careful attention to local pollutant concentrations is needed in the design of district heating cogeneration systems, as local concentrations may not otherwise decrease (Genon et al, 2009).

9.5.4 Cost savings with decentralized power production

The generation of electricity through cogeneration involves relatively small generation units (typically less than 200MW$_e$), sized to the local matching thermal loads and next to the electrical demand centres. Compared to large, centralized power stations, costs are reduced in the following ways:

• transmission and distribution capital costs are reduced, from an average of $1300/kW for new systems in the US to $100–200/kW according to Ayres et al (2007);

- less powerplant capacity needs to be built because less electricity is lost through transmission and distribution;
- less backup capacity is needed: 4–5 per cent rather than the standard 15 per cent for centralized powerplants, due to the fact the distributed cogeneration systems consist of many independent generators subject to random failure.

If L_t is the fractional transmission loss, C_{pp} and C_T the powerplant and transmission capital costs ($/kW), respectively, and R the reserve margin as a fraction of the peak demand, then the capital cost of central power is given by:

$$C_{cap} = \frac{C_{PP}(1+R)}{1-L_t} + C_T \qquad (9.5)$$

If C_{pp}= \$1200/kW, R = 0.15, L_t = 0.20 (which can be the case during times of peak demand on hot days), the required powerplant investment is \$1725 per kW of peak demand – 44 per cent greater than the nominal cost per kW. With average transmission costs, the overall cost is around \$3000/kW. For the same unit powerplant cost of \$1200/kW, the net cost of a decentralized powerplant (including distribution costs) would be \$1400–1500/kW – about half that of a centralized powerplant. As shown in Chapter 3 (subsection 3.4.3), these differences result in electricity from cogeneration using \$10/GJ natural gas being no more expensive than electricity from \$2/GJ coal in a central powerplant.

In spite of the favourable economics for decentralized power generation, there are a number of non-technical barriers (mostly regulatory), listed by Ayres et al (2007), that have inhibited the widespread development of decentralized power systems.

9.5.5 Use of district energy systems today and potential future use

In the US there are around 6000 district energy systems (approximately 2000 at universities, 2000 at medical facilities and 2000 at other facilities, including district energy utilities, private industry and airports), serving more than 10 per cent of the country's commercial floor space. Only 10 per cent of these are combined with cogeneration of electricity, however, representing

3.5GW of generation capacity (Kaarsberg et al, 1999) out of a total of 1076GW in 2005. In both the US and Europe (as well as elsewhere), there is a huge potential for cogeneration, even without building more district heating systems.

District heating provides 60 per cent of the heating and hot water requirements in the countries of the former Soviet Union and Eastern Europe. However, distribution systems in these countries lose up to 30 per cent of the heat that they carry, the systems are over-sized and so tend to operate at partial capacity (with lower efficiency), and the buildings often lack thermostatic controls and are poorly insulated (IEA, 2004b). Thus, in these countries the large potential energy savings is not in the expansion of district heating with cogeneration, but in the improvement of the efficiency of the existing system and of the buildings that they serve.

9.6 Summary

Community-integrated energy systems involve centralized production of heat and possibly chilled water that are distributed to individual buildings through district heating and cooling networks.

District heating provides an energy saving if it makes use of heat that would otherwise be wasted. The most common source of waste heat is the heat produced from the generation of electricity in thermal powerplants. Dedicated production of heat for a district heating network using centralized boilers will provide little if any energy saving compared to the use of modern, condensing boilers in individual buildings, particularly when heat losses during distribution are taken into account. Hot water systems with separate pipes for space heating and domestic hot water, but in a single insulated casing, are preferred to steam systems. When heat is supplied through cogeneration of electricity, the heat should be supplied at the lowest possible temperature so as to minimize the reduction in electricity generation, maximize the fraction of waste heat that is used, and minimize heat losses during distribution.

District cooling from large, centralized chillers can provide large (up to 45 per cent) savings compared to the use of separate chillers in individual buildings. This saving is due to the larger full-load efficiency of large chillers than of small chillers, and the ability to operate each chiller in a centralized system at or close to its maximum efficiency. Further savings are possible if the centralized system can make use of heat sinks (such as sewage or lake, river or seawater) that would not be available to chillers in individual buildings. Due to small unit costs for large chillers, the need for less total capacity in centralized systems (because the peak cooling loads in individual buildings do not all occur at the same time) and the need for less backup capacity, the total cost of district cooling systems can be less than the total cost of equipping each building with its own chillers. District cooling systems also eliminate the need for rooftop cooling towers, thereby freeing up roof space for other purposes (such as rooftop gardens or solar panels).

Cogeneration (the simultaneous production of electricity and useful heat) coupled to district heating will save energy compared to separate production of heat and electricity, as long as the system is sized to avoid many hours of operation as an electric powerplant only with waste heat discarded or to avoid excessive operation at part-load efficiency. However, with dedicated electricity production now at 55–60 per cent efficiency using natural gas combined cycle, and boiler efficiencies of 95 per cent or more, the energy saving through cogeneration is not as large as it used to be. Unlike

cogeneration, trigeneration using absorption chillers will normally increase energy use compared to 55–60 per cent efficient electricity generation and large, efficient (COP > 6) centralized electric chillers. This is because absorption chillers are inefficient (COP = 0.6–1.2) but require extracting heat from the electric powerplant at a temperature of 120–160°C, which reduces the electricity output. The sacrificed electricity would provide more chilling with electric chillers than is provided by the heat extracted from the powerplant.

Cogeneration using NGCC should reduce pollutant emissions compared to separate NGCC electricity generation and heating using natural gas, but other forms of cogeneration (gas engine, simple gas turbine or simple steam turbine) probably increase pollutant emissions. Even compared to advanced coal, simple cycle gas cogeneration does not necessarily reduce pollutant emissions.

Notes

1   See Chapter 4, subsection 4.4.2.
2   Other turbines may have a different electricity loss to useful heat ratio than used here, in which case the break-even COP will be different.

10

Energy Demand Scenarios

10.1 Introduction

This book and Volume 2 are motivated by the need to stabilize atmospheric CO_2 at a concentration of no more than 450ppmv. A CO_2 concentration of 450ppmv, when combined with the heating effect of other GHGs, would correspond to an equivalent concentration (in terms of changes in climate) of 560ppmv – that is, a doubling of the pre-industrial concentration of 280ppmv CO_2. This assumes that there are *stringent* reductions in the emissions of other GHGs (such as methane, N_2O and the halocarbons) or of gases (such as nitrogen oxides, carbon monoxide and reactive hydrocarbons) that react to form ozone, another GHG. If emissions of these other GHGs or GHG precursors are not sharply reduced, a mere 450ppmv CO_2 concentration would correspond to more than a doubling in terms of eventual climate warming.

As discussed in Chapter 1 (subsection 1.2.1), there is a very wide body of evidence indicating that a CO_2 doubling would eventually warm the climate by 1.5–4.5°C on average – that is, that the *climate sensitivity* is likely to fall between 1.5°C and 4.5°C. As also discussed in Chapter 1 (subsection 1.4.1), there is a risk of increasingly severe impacts with increasing warming of the climate, beginning with the worldwide collapse of coral reef ecosystems with somewhere between 1°C and 2°C global mean warming relative to pre-industrial temperatures, increasingly severe impacts on global agriculture with a warming in excess of 2°C, and melting of the Greenland ice sheet and collapse of the West Antarctic ice sheet for a sustained warming of as little as 1°C and near certainty of loss of these two ice bodies with a 4°C global mean warming. The result would be an eventual sea level rise in excess of 10m. Ecological assessments indicate that between a sixth and a third of land animal species will probably be committed to extinction if the climate warms by 2°C by 2050.

Independently of whatever changes in climate and related impacts occur as a result of increasing CO_2 concentration, the absorption of anthropogenic CO_2 by the oceans causes an acidification of the ocean surface water. Even for an atmospheric CO_2 concentration stabilized at 450ppmv, severe ecological impacts are expected, and if fossil fuel emissions were to cease altogether, the time required for the ocean acidity to return to pre-industrial levels would be in the order of 100,000 years.

Thus, it is apparent that even a CO_2 concentration of 450ppmv is dangerously high, as it runs a high risk of severe impacts. As discussed in Chapter 1, stabilization of atmospheric CO_2 at 450ppmv requires rapid (within ten years) elimination of net emissions from deforestation, reduction of fossil fuel emissions from about 8GtC/yr at present to about 3GtC/yr by mid-century and their elimination altogether before the end of the century (see Figure 1.9b). By the time we bring about the changes needed to limit the CO_2 concentration to 450ppmv (if we do that), it will almost certainly turn out that widely unacceptable consequences will have already occurred and that many more will be judged likely if the CO_2 concentration is sustained at that concentration, so the decision will probably be made by some subset of the world's nations to draw down the CO_2 concentration by creating negative emissions through the sequestration of CO_2 released from the use of biomass energy, or through the concurrent harvesting of bioenergy crops with the buildup of soil carbon (as discussed in Volume 2, Chapter 4, section 4.8).

Chapter 1 introduced the Kaya identity, which represents CO_2 emissions as the product of human population, GDP per person, energy use per unit of GDP (the energy intensity) and C emission per unit of energy (the C intensity). The chapter then explored the tradeoff between the rate of improvement in energy intensity and the rate of deployment of C-free energy supplies needed in order to stabilize atmospheric CO_2 at 450ppmv. For a middle population scenario and assuming growth in world average GDP per person of 1.6 per cent/year throughout this century, the amount of C-free primary power needed by 2050 ranges from 21TW at 1 per cent/year improvement in energy intensity to 4.6TW with 3 per cent/year improvement (1TW = 10^{12} watts, and primary power refers to the rate of supply of primary energy of all forms). By comparison, total world primary power demand in 2005 was 15.3TW and the total C-free power supply was 3.3TW. However, alternative assumptions concerning future population and the growth of average GDP per capita have a significant effect on the energy intensity/C-free power combinations needed to limit atmospheric CO_2 to 450ppmv.

In this chapter, I examine in more detail the combinations of trends in the terms of the Kaya identity at the global scale that lead to reductions in CO_2 emissions consistent with CO_2 concentration peaking at 450ppmv. This will provide additional insight concerning the impact of plausible variations in the different terms in the Kaya identity. I then summarize the findings of this book concerning the potential for reducing the physical energy intensity in different sectors. These findings are then used to construct scenarios of fuel and electricity demand in distinctive world regions, and are summed to give global demand scenarios taking into account likely shifts in the relative importance of industry, transport and other sectors of the economy as the global economy grows. The final result – fulfilling one of the underlying purposes of this book – will be an estimate of the maximum feasible compounded rate of reduction in economic energy intensity over the period 2005–2050. The demand scenarios are combined in Volume 2 with conclusions concerning feasible rates of deployment of C-free energy to develop scenarios of industrial CO_2 emissions. These emission scenarios are then used – with the aid of a simple coupled climate–carbon cycle model – to assess the implications for global mean temperature change and for ocean acidification.

10.2 Drivers of future CO_2 emissions

Under the Kaya identity, CO_2 emission (E) is decomposed as the product of four factors:

$$E = P \times \text{GDP}/P \times \text{energy intensity} \times \text{carbon intensity} \quad (10.1)$$

where P is the world population, GDP represents national, regional or total world economic output (depending on the scale of analysis), energy intensity is an *economic* energy intensity – the primary energy used per unit of economic output – and *carbon intensity* is carbon emission per unit of primary energy. The economic energy intensity differs from the physical energy intensity (the energy use per unit of physical output), which is what was reviewed in this book. For now, I continue to use the economic energy intensity, but later I relate it to physical energy intensity.

The baseline or BAU scenario, used in constructing Figure 1.9, assumes that world population peaks at 9.0 billion in 2070 and then gradually declines, that world average GDP/person grows at 1.6 per cent per year during the entire period 2000–2100, that economic energy intensity decreases by 1 per cent/yr and that the amount of C-free power grows by 0.5 per cent/yr. The combination of population and exponential GDP/P growth leads to growth in world GDP by a factor of six (from \$55.7 trillion in 2005 to \$331.0 trillion in 2100). This combined with the 1 per cent/yr decline in energy intensity causes primary power demand to grow by more than a factor of two (from 15.3TW in 2005 to 35.2TW in 2100). As oil and gas are depleted, an increasing fraction of the fossil fuel portion of the total energy supply is provided by coal, which has a particularly high CO_2 emission factor (a world average of 25.2kgC/GJ, compared to 21.4kgC/GJ for oil and 14.8kgC/GJ for natural gas). As a result of this shift, and because the relative growth in C-free power is slightly less than the relative growth in total primary power demand, the carbon intensity increases modestly (from 0.501kgC/yr/W in 2005 to a peak of 0.576kgC/yr/W in the 2070s). The net result of all of the above is that fossil fuel CO_2 emission increases from 7.7GtC in 2005 to 20.3GtC in 2100 in the BAU scenario – almost a factor of three increase.

In the following section we examine the impact on the required C-free power of a broad range of assumptions concerning growth in world population and world average GDP/P, and of the rate of decline in energy intensity.

10.2.1 Tradeoffs between population, GDP per person, energy intensity and carbon intensity

Figure 10.1a shows three scenarios of population growth and two scenarios of growth in world average GDP per

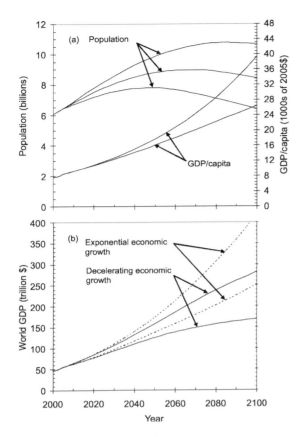

Note: For each set of GDP curves, the lower, middle and upper curves correspond to the low, middle and high population scenarios.
Source: The three population scenarios from Lutz et al (2001)

Figure 10.1 *(a) The three population scenarios and two GDP/P scenarios; (b) growth in world GDP for decelerating growth in GDP/P (solid lines) and for exponential growth in GDP/P (dashed lines), with the lower curve for each pair corresponding to the low population scenario from (a) and the upper curve corresponding to the high population scenario*

capita. The three population scenarios are based on the work of Lutz et al (2001), who generated thousands of scenarios of future population by randomly altering the values of the many factors that determine future population in their calculations. This allows one to develop a histogram or probability distribution for future population trajectories. The low, medium and high scenarios shown in Figure 10.1a are such that 20 per cent,

50 per cent and 80 per cent, respectively, of the population trajectories generated by Lutz et al (2001) fell on or below the trajectories shown in Figure 10.1a (that is, these three curves correspond to the 20th, 50th and 80th percentiles of the set of population scenarios generated). Global population peaks at 7.8 billion near 2050, at 9.0 billion near 2070 and at 10.8 billion near 2085 for the low, middle and high scenarios, respectively.

With regard to global average GDP per capita, consider two scenarios: one with growth of 1.6 per cent per year for the entire century (as in the BAU scenario of Chapter 1) and another with growth that decreases linearly from 1.6 per cent/year in 2000 to 0.8 per cent/year in 2100. The low-growth scenario should not be construed as representing a poorer global society or a lower standard of living. As discussed in Chapter 11, lower growth could arise as a result of an increasing labour productivity partly channelled into an increase in leisure time and other positive developments, rather than solely into an increase in per capita consumption. In any case, there is little difference in GDP/P between the two scenarios prior to 2050. Figure 10.1b shows the resulting growth of world GDP for the two GDP/P scenarios for the low and high population scenarios.

Figure 10.2a shows the variation in economic energy intensity over the period 2000–2100 assuming historical intensities during 2000–2005 and either a 1 per cent/year or 2 per cent/year decrease thereafter. Figure 10.2b shows the resulting variation in world primary power demand for the 1 per cent/year and 2 per cent/year rates of improvement in energy intensity for high population and GDP/P or low population and GDP/P combinations. Primary power demand increases by a factor of three compared to 2000 in the highest case and decreases by about 50 per cent in the lowest case. The result is a factor of six difference in primary power demand in 2100 between the lowest and highest cases.

The final step in computing CO_2 emissions is to make some assumption concerning the rate of growth of C-free power. Consider again two cases: one in which C-free power grows by 0.5 per cent/year and another in which it grows by 2.0 per cent/year. With three population scenarios, two GDP/P scenarios, two energy intensity (EI) scenarios and two C-free power scenarios, a total of 24 different scenario combinations

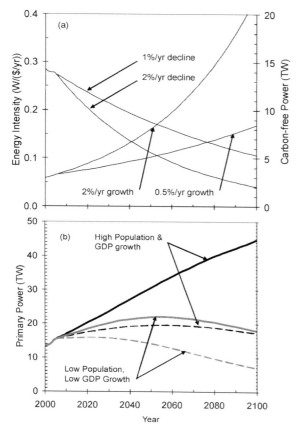

Note: For each population-GDP/P pair, the upper curve is for a rate of reduction of energy intensity of 1 per cent/year and the lower curve is for a 2 per cent/year rate of reduction.

Figure 10.2 *(a) Variation in energy intensity (EI) when EI decreases by 1 per cent/year or 2 per cent/year, and growth in C-free power when C-free power increases by 0.5 per cent/year or 2.0 per cent/year; (b) variation of world primary power demand for various combinations of population and GDP/P growth*

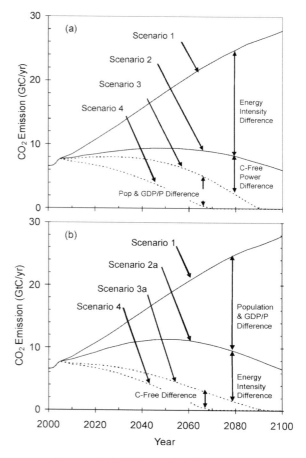

Figure 10.3 *Variation in CO_2 emissions for (a) scenarios 1, 2, 3 and 4, and (b) scenarios 1, 2a, 3a and 4*

is possible. Results for the following four scenarios are shown in Figure 10.3a:

- Scenario 1: High population, high GDP/P, low EI improvement, low C-free power;
- Scenario 2: High population, high GDP/P, high EI improvement, low C-free power;
- Scenario 3: High population, high GDP/P, high EI improvement, high C-free power;
- Scenario 4: Low population, low GDP/P, high EI improvement, high C-free power.

In Scenario 1, CO_2 emissions rise to 28GtC/yr and in Scenario 2 they peak at 9.7GtC/yr around 2050, while in Scenarios 3 and 4, emissions are eliminated by 2095 and 2070, respectively. Scenario 3 is consistent with the emission reductions needed to stabilize atmospheric CO_2 at 450ppmv (see Figure 1.9), while Scenario 4 provides a margin of safety for stabilization at 450ppmv.

From Figure 10.3a, it appears that the single most important factor in reducing CO_2 emissions is to increase the rate of improvement of EI from 1 per cent/year to 2 per cent/year; the impact of quadrupling the rate of growth in C-free power, or the combined effect of lower population and GDP/P growth, is substantially less. However, the relative impact of the different changes depends on the order in which they are implemented. This is illustrated in Figure 10.3b, which shows CO_2 emissions for Scenarios 1 and 4 as well as for the following:

- Scenario 2a: Low population, low GDP/P, low EI improvement, low C-free power;
- Scenario 3a: Low population, low GDP/P, high EI improvement, low C-free power.

Now, the combined effect of low population and low GDP/P, when implemented first, is *comparable* to the effect of the higher rate of improvement in EI when it is implemented first. Thus, reducing the product of population and GDP/P and accelerating the rate of improvement in energy intensity are *both* important to the long-term reduction of CO_2 emissions. However, policy discussion focuses almost exclusively on the latter option.

Significant economic growth is of course needed in the developing countries in order to improve the material standard of living, but economic growth should not be the goal in itself. Rather, the underlying goal should be human happiness which, once basic human needs are satisfied, is related to many non-material factors, including the diversity and richness of human relationships (family, friends, community), and not the accumulation of material wealth. Indeed, there is substantial evidence that, in the US and other developed countries, economic growth has been *negatively* correlated with various measures of human well-being and happiness for the past two to three decades (Daly and Cobb, 1994). Issues related to the current growth paradigm and how we might shift away from it are discussed in Chapter 11.

Figure 10.4a shows the total primary power demand in 2050 as a function of the rate of reduction in energy intensity, in comparison to the fossil fuel power permitted in 2050 if atmospheric CO_2 is to be stabilized at 450ppmv, assuming either high population combined with high GDP/P, or low population combined with low GDP/P. The difference between the primary power demand and the permitted fossil fuel primary power gives the required C-free primary power, which is shown in Figure 10.4b for 2050 in comparison to the global primary power supply in 2005. Average global primary power supply in 2005 was 15.3TW, of which 12.0TW were from fossil fuels and 3.3TW from C-free sources (of which 0.78TW were nuclear and not likely to be wholly replaced as existing facilities are retired). If energy intensity continues to decrease at 1 per cent/year, global

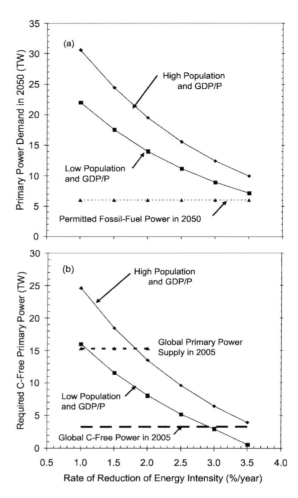

Figure 10.4 (a) *Primary power demand in 2050, as a function of the rate of decrease of global energy intensity for the high population and GDP/person or low population and GDP/person scenarios, and the approximate permitted fossil fuel power in 2050 if atmospheric CO_2 is to be stabilized at 450ppmv; (b) amount of C-free power required in 2050 for the same conditions as in (a), and global primary power and C-free primary power supply in 2005*

primary power demand in 2050 would be 22TW and 31TW for the low and high GDP scenarios, respectively. The permitted fossil fuel use supplies 6.0TW of primary power, so the required C-free primary power in 2050 is 16–25TW and probably impossible on this timeframe. If global average energy intensity decreases at 2.5 per cent/year, the required C-free power in 2050 is

5.2–9.6TW, or two to four times the present non-nuclear world C-free power supply of 2.45TW. The required C-free power under the low GDP scenario is only 54 per cent that of the high GDP scenario, which underlines the importance of low population growth and a moderation in the rate of growth of GDP/person.

10.3 Summary of the potential for physical energy intensity reductions

Physical energy intensity – the subject of Chapters 3 to 8 of this book – is the energy use per unit of physical activity or output. Building floor area is often used as a proxy for activity levels in buildings, while distances travelled by people, the product of distance and mass of goods transported, and the physical outputs of various manufacturing sectors represent other activity levels. The physical energy intensity in terms of primary energy depends on the intensity of each form of secondary energy used (electricity, refined petroleum products, coal and natural gas ready for use) divided by the efficiency in converting from primary energy to secondary energy. As incomes and populations grow, the rate of change in economic energy intensity will depend on: (1) changes in activity levels (which may increase faster or more slowly than GDP); (2) changes in the mix of activities (so the relative importance of less energy-intensive and more energy-intensive activities in generating GDP will change); (3) changes in physical secondary energy intensity levels; and (4) changes in the efficiency in converting from primary to secondary energy. In this section, the potential for improving the efficiency in producing electricity from primary energy and in reducing physical energy intensities for the major energy uses is summarized based on the detailed information presented in Chapters 3 to 8. This will then be translated into changes in economic energy intensity, which in turn can be related to the analysis of the Kaya terms presented in the previous section.

Overall *energy intensity factors* for fuels and for electricity will be computed here for buildings, transportation and industry. These are the ratios, for each sector, of the future fuel or electricity use to the present use, with the future energy uses pertaining to the time when all of the average efficiency gains identified here have been achieved but assuming no change in activity levels or in the relative proportions of different activities. Later, we will consider the impact of a shift in the mix of physical activities (including changes in the mix of industrial outputs) in response to saturation of demand, congestion and price increases as the global economy grows and non-renewable resources are depleted.

10.3.1 Electricity generation from fossil fuels

The current average efficiency in supplying electricity to end-users from fossil fuels and biomass is about 37 per cent in OECD countries and 21 per cent in non-OECD countries (Table 2.6). The particularly low efficiency in non-OECD countries is a result of lower generation efficiencies, substantial power consumption by powerplants themselves and significant distribution losses (some portion of which is stolen rather than dissipated as heat during transmission).

The efficiency and capital cost of current and advanced systems for generating electricity from fossil fuels were summarized in Table 3.11. As existing powerplants are replaced with new powerplants, we have the potential for:

- a 50 per cent increase in the efficiency of electricity generation from coal (from a world average efficiency of 34 per cent to 50 per cent);
- a doubling in efficiency if typical coal powerplants (at 34 per cent) are replaced with fuel cell/turbine hybrid powerplants powered by natural gas (at 70–75 per cent);
- up to a tripling in the efficiency if older coal powerplants (at 25–30 per cent efficiency) are replaced with cogeneration (having an 80–90 per cent marginal efficiency).

The high marginal efficiency for cogeneration is applicable only if the waste heat produced from electricity generation can be fully used. This will not always be possible. Here, we shall assume an increase in the average efficiency in supplying electricity to 65 per cent in both OECD and non-OECD countries.

10.3.2 Buildings

The evidence reviewed in Chapter 4 indicates that, in climates where heating is required in winter, high-performance residential buildings can be readily constructed with a heat requirement of no more than 15kWh/m²/yr for climates as cold as that of southern Finland. This is the Passive House Standard, and it can be achieved in both single-family and multi-unit housing, although it is easier and less costly to achieve in multi-unit housing. As typical heat requirements for new construction are 60–75kWh/m²/yr, this standard represents a factor of four to five reduction compared to current conventional practice in most parts of the world. A number of buildings – especially high-rise multi-unit housing – have been renovated to or close to this standard. As the typical heating requirement in much of the older building stock in Europe, the former Soviet Union, North America and the cold regions of China is 200–400kWh/m²/yr, renovation to the Passive House Standard represents up to a factor of 15–25 reduction compared to the average of existing buildings.

With regard to residential cooling loads in hot climates, the evidence reviewed in Chapter 4 indicates that a factor of two to four reduction compared to current practice for new buildings is feasible in many regions – up to a factor of two through windows and shading that greatly reduce heat gain, and a factor of two or more through use of passive cooling techniques (especially night cooling combined with thermal mass and external insulation) and/or more efficient cooling equipment (including a near doubling in the real operating efficiency of air conditioners). For existing buildings with large expanses of unshaded glazing, solar heat gain can be reduced by a factor of five or more through some combination of addition of external shading and replacing existing windows with windows with minimal solar heat gain. The impact that this has on overall cooling load depends on the relative importance of solar heat gain and other sources of heat. Other measures to modestly (20–30 per cent) reduce cooling loads include more efficient cooling equipment (including fans), more effective operation of the equipment and greater reliance on passive ventilation and cooling whenever outdoor conditions permit.

For a given level of service demand, a factor of two reduction in the energy requirements for appliances, residential lighting and consumer electronics is clearly feasible. However, service demand is growing rapidly in developing countries and even in already-affluent developed countries through, for example, use of ever larger TVs and the proliferation of consumer gadgets.

With regard to commercial buildings, the evidence and examples reviewed in Chapter 4 indicate that a maximum total onsite energy consumption of 100kWh/m²/yr, averaged over the typical mix of different kinds of commercial buildings, is a challenging but achievable target for new buildings in all climate zones. Modest measures will readily achieve an energy intensity of no more than 200kWh/m²/yr, as seen by Figures 4.80 and 4.81. Table 4.29 provides many examples from around the world of commercial buildings that have achieved energy intensities of 100kWh/m²/yr or less. Given typical energy intensities in various countries of 200–400kWh/m²/yr for new commercial buildings, this represents a factor of two to four reduction compared to current conventional practice. Renovations of existing commercial buildings, especially if involving a curtain wall replacement and significantly upgrading and reconfiguring highly inefficient (but typical) HVAC systems can also typically achieve a factor of two to four reduction in energy use compared to the pre-existing energy use, as shown by many case studies listed in Chapter 4.

The large savings identified above are achievable using generally low-tech, readily available and reliable technologies and techniques. The larger energy savings involve a significant reliance on passive solar energy for one or more of heating, cooling, ventilation and lighting, and also require enlightened occupant behaviour but no sacrifice in comfort or convenience. The discussion in Chapter 4 (subsection 4.11.2) placed great emphasis on the design *process* in achieving dramatic energy savings in commercial buildings.

Because the building stock is slow to be replaced or extensively renovated, the overall decrease in the energy intensity of the building stock will be slow. In regions where large increases in the total building floor area can be expected, absolute increases in energy use by buildings can be expected for some time. It is critical that the energy use of new buildings be reduced rapidly, and that the energy savings routinely achieved through renovations be increased quickly. Otherwise, each new building and renovation is a lost opportunity that will probably lock in high energy use to beyond mid-century. Scenarios that take into account increasing population and GDP per person (both of which lead to an increase in building floor area), the technical potential to reduce energy requirements in new and renovated buildings, and the slow rate of building turnover, are presented later.

10.3.3 Transportation

The potential to improve the fuel economy of gasoline-based, ICE road vehicles was extensively discussed in Chapter 5 and summarized in Tables 5.22 and 5.23. The available measures include improving the thermal and mechanical efficiency of vehicle drivetrains, forming gasoline–electric or diesel–electric hybrids, and reducing the load that the drivetrain must supply. These loads are the aerodynamic resistance, the rolling resistance, the vehicle weight and the energy used by vehicle accessories (such as air conditioning). Advanced gasoline vehicles can reduce the energy requirement by about 40 per cent compared to current gasoline vehicles, while advanced gasoline–electric hybrids can reduce energy requirements by over 50 per cent. A shift to diesel–electric hybrid vehicles provides a further 15 per cent savings and should be acceptable once new stringent pollutant emission standards for diesel vehicles are achieved. To the extent that consumers can be induced to purchase smaller and less powerful vehicles than in the recent past (through financial incentives or higher fuel taxes), further savings will occur. There is room for gradually shifting perhaps 20–30 per cent of urban travel in North America to rail-based transit if the appropriate investments are made and if land use policies that facilitate the use of rail transit are implemented. The potential for shifting urban transportation from automobiles to electric rail transit may be even larger in European cities, in spite of the already high use of public transportation, because the urban form in Europe is more conducive to public transportation.

As shown in Chapter 5 (Table 5.22), the amount of onboard energy required by a hydrogen fuel cell vehicle to drive a given distance is only one quarter that of a comparably sized current gasoline vehicle (or just over half that of an advanced gasoline–electric hybrid vehicle). In PHEVs, the onboard energy (this time in the form of electrochemical energy in a battery) required to drive a given distance is 0.38 times that of an otherwise comparable vehicle using gasoline (19kWh/100km versus 5.6 litres/100km, where 1 litre = 32.4MJ (LHV)). In all cases (gasoline, hydrogen, electricity), the primary energy required to provide a given amount of energy on board needs to be considered.

With regard to urban transit buses, diesel–electric hybrid drivetrains are already being tested and have increased the distance that can be travelled with a given amount of fuel by 8.5 per cent to 75 per cent. This corresponds to a reduction in fuel use to drive a given distance by 8–43 per cent. Other measures should be able to reduce fuel requirements even further. A very important factor in determining energy use per passenger-km is the number of passengers on the bus, so measures that increase the utilization of existing bus services will have a large impact, especially where current passenger loadings are small.

For freight transport by trucks, technologies already under development should lead to a 50 per cent reduction in energy use per tonne-km at constant speed on flat terrain, while a 25–45 per cent saving in urban settings is projected using hybrid diesel–electric technology (see Chapter 5, subsection 5.7.2), so an overall long-term reduction of 50 per cent is reasonable. This would be achieved largely through an increase in the thermal efficiency of diesel engines (from 33 per cent to 55–56 per cent), rather than primarily through load reduction, which plays a significant role in improving passenger-vehicle fuel economy. As a result, the additional efficiency gain from switching to hydrogen-powered fuel cells would be less, since the drivetrain efficiency would be closer to that of a fuel cell vehicle (unlike passenger vehicles, where there is a big jump in switching from a gasoline ICE to a fuel cell).

For transport by rail, the switch from diesel fuel to fuel cells with a DC drive saves about 33 per cent (Chapter 5, subsection 5.7.3), and other measures should in the long term be able to bring the total savings to 40 per cent. Alternatively, freight rail could be converted to run using grid electricity (this is already the case for passenger rail in many parts of the world), with a 60 per cent reduction in secondary energy use. For passenger rail, an average saving of only 20 per cent will be assumed here because much of it is already electrified. For transport by ship (Chapter 5, subsection 5.7.4), measures have already been identified that could save 45 per cent per tonne-km transported over 30 years, assuming no change in ship speed. However, energy use varies with ship speed squared, so a small reduction in ship speed could push the overall savings to 50 per cent.

For air travel (Chapter 5, section 5.6), the next generation of aircraft (already under development) will save about 10–20 per cent compared to older aircraft still in service. A 20–25 per cent reduction in the stock average energy intensity between now and 2030 or so seems to be a reasonable assumption. Hydrogen may become viable as an aircraft fuel, and if it does, then further reductions in energy intensity would be possible through smaller take-off weight (because hydrogen is light) and the ability to induce laminar airflow past the airframe by using the coldness of liquid hydrogen to chill the skin of the aircraft. However, it will probably be necessary to restrain the growth in air travel through taxes on aviation fuel (which, at present, is not taxed), although the likely large increase in the cost of fuel as oil supply peaks will probably significantly slow or even reverse the growth of air travel.

10.3.4 Industrial energy use

The largest physical energy intensity reductions in the industrial sector that have been identified here pertain to the iron and steel and aluminium industries. A combination of efficiency improvements and increasing use of secondary (recycled) materials could lead to reductions in the total onsite energy use (fuels plus electricity) by almost a factor of five for iron and steel and by a factor of seven for aluminium. For cement, a factor of two reduction compared to the present world average should be possible in part through improved manufacturing processes and in part through blending Portland cement with natural cementitious materials. The pulp and paper industry can become energy self-sufficient through much better

use of wood wastes and processing waste (such as black liquor) than at present. For other industrial sectors it is hard to project what the potential efficiency improvement could be, but it is likely to be at least 40 per cent in the long term.

The derivation of an overall physical energy intensity factor for industry is given in Table 10.1. Shown are the energy use in 2005 for the industrial sectors for which separate data are given in IEA (2007a, 2007c), energy intensity factors derived in Chapter 6, the resulting future energy use assuming the same amounts of industrial outputs as today, and the resulting overall intensity factors for fuels and electricity. For non-ferrous metals, I have used the average energy intensity factors from Table 6.37 for aluminium, copper and zinc, weighted by their total global energy use. Cement dominates non-metallic minerals, so I have used the assumed cement energy intensity factor (0.5) for that subsector. The resulting overall energy intensity factors are 0.30 for fuels and 0.46 for electricity.

10.3.5 Agriculture and food systems

Direct energy use on farms is conservatively assumed to decrease by 25 per cent in proportion to food production. Energy is also used in producing food through the production of fertilizers and pesticides. As discussed in Chapter 8, the amount of energy used to produce a given quantity of fertilizer can be reduced by perhaps 20 per

Table 10.1 *Breakdown of global industrial energy use and derivation of future (2050 or 2100) overall industrial energy intensity factors for fuels and electricity*

Sector	Energy use in 2005 (EJ)		Physical intensity factor		Future energy use (scaled to 2005 activity)	
	Fuels	Electricity	Fuels	Electricity	Fuels	Electricity
OECD countries						
Iron and steel	19.2	3.3	0.11	0.71	2.06	2.32
Non-ferrous metals	1.5	2.1	0.28	0.21	0.42	0.44
Non-metallic minerals	9.7	1.3	0.50	0.50	4.83	0.67
Paper	4.8	0.3	0.00	0.00	0.00	0.00
Chemicals	10.7	3.6	0.25	0.25	2.67	0.89
Non-energy	29.4	0.0	0.25	0.25	7.36	0.00
Other	27.0	11.7	0.50	0.50	13.51	5.83
Total	102.3	22.3			30.84	10.15
Overall industrial energy intensity factor relative to 2005					0.30	0.46

Note: The fuel and electricity intensity factors are multiplied by the total industrial fuel and electricity use in 2005, respectively, to get the future fuel and electricity energy use assuming the same amounts and proportions of industrial outputs as in 2005.
Source: Energy use in 2005 is taken from IEA (2007a, 2007c)

cent in the most efficient plants and by perhaps 50 per cent on a worldwide basis (given that much of the world's fertilizer production is relatively inefficient), while the amount of fertilizer used per unit of crop can probably be reduced by 50 per cent compared to current practice in most industrialized countries through more efficient use of fertilizer. In developing countries there is room for increased fertilizer use so as to increase yields, but the current energy intensity of fertilizer production is much greater than in industrialized countries, so potential fractional reductions are larger. Reductions of 50–75 per cent in the fertilizer energy use per unit of food production are therefore plausible. This energy saving would show up in industrial energy use, for which overall savings of 70 per cent for fuels and 54 per cent for electricity have been assumed.

Future diet will have a large impact on agricultural energy use. A low-meat diet, by requiring less phytomass production, will require less energy-intensive fertilizers and pesticides and less of all other energy inputs. As well, a low-meat diet will mean that agricultural yields per unit land area will not need to be pushed to their limits, which in turn will permit lower-energy-input organic farming systems (which generally have lower yields than systems with higher energy input). As well, the meat that is produced can be produced in a less spatially concentrated manner when overall demand is restrained, which will facilitate recycling of nutrients back to the soil, with associated savings in fertilizer energy use (and reduced waste management problems). However, no benefits in terms of reduced energy use from dietary changes are assumed here, although the impact of dietary changes in freeing up land for bioenergy crops will be considered in Volume 2.

10.4 Grand synthesis: Constructing scenarios of future energy demand with aggressive efficiency measures

In this section we develop scenarios of future global demand for electricity and for fuels. The drivers of this demand are growing human population, growing average GDP per person, changes in the mix of economic activities that make up the growing GDP, and changing energy intensities of different economic activities. The purpose here is not to make accurate projections of future energy demand. Rather, it is to show – in an approximately internally consistent manner – the

rough consequences for future demand for electricity and fuels of the various efficiency assumptions adopted here (which are based on the detailed analysis in the preceding chapters) in combination with plausible assumptions concerning future population and GDP per capita. Excel spreadsheets for the complete set of calculations presented in this section are contained in the online supplementary material, permitting the reader to explore the consequences of alternative assumptions of his or her own choosing.

Some of the drivers of energy demand (such as floor space per person, acquisition of consumer goods and annual distances travelled per person) do not increase with income in a linear manner. Instead, they exhibit a saturation behaviour, and furthermore, floor space and annual distance travelled per capita saturate at different values in different regions. This is probably due to differences in physical circumstances (population density in particular) and perhaps also differences in cultural preferences. Thus, we would not obtain the correct variation in global mean energy demand per capita using the variation in global mean per capita income as a driving variable. Rather, the relationships between income and driving factors need to be applied to relatively homogeneous regions with similar incomes. For this reason, the analysis to be presented here adopts the 11 world regions used in the *Special Report on Emission Scenarios* (SRES) of the Intergovernmental Panel on Climate Change (Nakicenovic and Swart, 2000, Appendix 3).[1] The 11 regions (two of which have been combined into one) are shown in Table 10.2 along with population and average GDP per capita in 2005.

As for the future, I use the low and high national population scenarios from the 2008 edition of the United Nations Population Division's *World Population Prospects* (UNPD, 2008), summed over the specific countries that make up each of the SRES regions. The UNDP projections go to 2050 only, so I have extended them to 2100 using the logistic growth function (derived in Box 2.1 of Chapter 2), whereby the population $P(t)$ at some time t after 2050 is given by:

$$P(t) = \frac{P_U}{1 + ((P_U - P_o)/P_o)e^{-a(t-t_o)}} \quad (10.2)$$

where t_o = 2050, P_U is the assumed ultimate (and stable) population in each region and P_o is the population in 2050 according to the chosen UNPD scenario. As time

Table 10.2 *The ten world regions, initial population and GDP, asymptotic population after 2050 for low and high population scenarios, asymptotic GDP/capita for low and high GDP/capita scenarios, and the approach rate tendencies adopted here for population and GDP/P*

Region	Population (millions)			GDP/P ($/person)			Approach rate Tendency	
	2005	Asymptotic		2005	Asymptotic		Pop	GDP/P
		Low	High		Low	High		
PAO (Pacific Asia OECD)	200	135	230	28,789	34,000	50,000	0.02	0.02
NAM (North America)	335	340	575	40,432	34,000	55,000	0.02	0.01
WEU (Western Europe)	475	360	625	27,178	30,000	50,000	0.02	0.02
EEU (Eastern Europe)	119	70	115	12,570	30,000	45,000	0.02	0.04
FSU (former Soviet Union)	280	200	250	8713	27,000	45,000	0.02	0.04
LAM (Latin America)	557	450	1000	8543	25,000	40,000	0.02	0.04
SSA (sub-Saharan Africa)	726	900	2000	1718	22,000	30,000	0.01[a]	0.04
MENA (Middle East and North Africa)	323	400	900	8312	22,000	35,000	0.02	0.04
CPA (centrally planned Asia)	1478	1100	1800	4647	27,000	45,000	0.01	0.06
SAPA (South and Pacific Asia)	1987	1750	3050	2785	22,000	40,000	0.02	0.06

Note: [a]This value is for the low scenario, 0.06 is for the high scenario.

progresses, P asymptotically approaches P_U. Table 10.2 gives the P_U values chosen to extend the UNDP low and high population projections, and Figure 10.5 shows the resulting population scenarios for the ten world regions.

For future GDP per person, I also apply Equation (10.2), this time with P_o representing GDP/capita in 2005, P_U representing the assumed asymptotic GDP/capita value and t_o = 2005. Table 10.2 gives populations and GDP/P in 2005, the asymptotic population and GDP/P values adopted here for low and high growth scenarios, and the parameter a in Equation (10.2), which is assumed to be the same for low and high growth scenarios.[2] Figure 10.6 shows the resulting variation in GDP/capita in each of the ten world regions. For North America, a slight decrease in GDP/capita has been assumed for the low case. The difference in the regional per capita income between the poorest region (sub-Saharan Africa) and the richest region (North America) decreases from a factor of 24 at present to a factor of about 2 by 2100 in both growth scenarios.

Figure 10.7a shows the resulting high and low global population and high and low global mean GPD/capita scenarios. Henceforth, I will consider only the combination of low population with low GDP/capita and high population with high GDP/capita. The resulting variation in the GDP is shown in Figure 10.7b (along with the variation in the annual rate of growth of GDP/capita for the two GDP/capita scenarios). The variation in global population, mean GDP/capita, and

(a)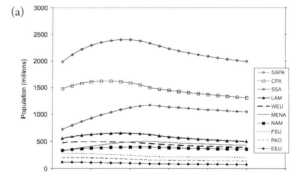

(b)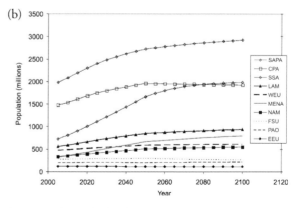

Source: Based on UNPD (2008)

Figure 10.5 *Population scenarios for the ten world regions up to 2050 and extended to 2100 using Equation (10.2) with parameter values from Table 10.2; (a) low case, (b) high case*

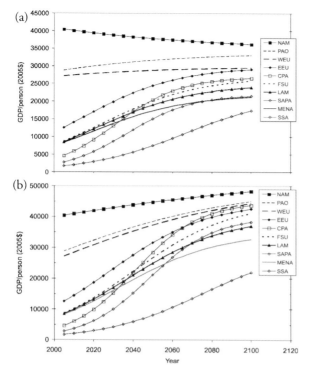

Figure 10.6 *(a) Low and (b) high scenarios of GDP/capita, generated using Equation (10.2) with parameter values from Table 10.2*

GDP roughly matches the corresponding variation for the high and low cases given in Figure 10.1, but now as the result of summing distinct scenarios for different world regions.

10.4.1 Buildings

To explore the net effect of various rates of renovation, replacement of existing buildings, and addition to the total building floor along with various rates of improvement in the energy intensity (energy use per m² of floor area per year) of new and renovated buildings relative to the stock average, a simple accounting procedure, outlined in Box 10.1, is used here. The accounting procedure begins in 2005 but assumes that no change in the energy intensity of new buildings occurs until at least 2010. It is assumed that all buildings undergo a significant renovation between 2005 and 2050 in which maximum advantage is taken of the opportunity to reduce energy use. A transition from the current energy intensity of new and renovated

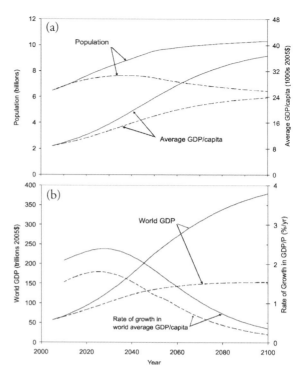

Note: Also shown is the variation in the rate of growth of world average GDP/capita for the low and high GDP/capita scenarios.

Figure 10.7 *(a) Variation in global population and in global average GDP/capita for the high scenario (solid lines) and low scenario (dashed lines); (b) variation in GDP for the combination of high population with high GDP/capita growth (solid line) and low population with low GDP/capita growth (dashed line)*

buildings to a much lower energy intensity is assumed to occur some time between 2010 and 2050. The key parameters in the accounting scheme are:

- the energy intensities of new and renovated buildings at the beginning of the transition period (E_{new-o} and E_{reno-o}) compared to the average energy intensity of all buildings that existed in 2005;
- the energy intensities of new and renovated buildings at the end of the transition period (E_{new-f} and E_{reno-f}) compared to the average energy intensity of all buildings that existed in 2005;
- the years at which the transition from the initial energy intensities to the final energy intensities starts and is completed.

Box 10.1 Procedure for generating scenarios of future average building energy intensity

The following procedure is used to calculate the variation in total building use and in the average energy intensity relative to the values in 2005:

- the entire building stock is divided into 45 equal cohorts of size f_{start} in terms of floor area (f_{start} is given as a fraction of the floor area in 2005, so $f_{start} = 1/45$);
- the initial energy intensity E_{start} of each cohort relative to the average initial energy intensity is assumed to vary across the cohorts from $1+\Delta E$ to $1-\Delta E$, with an equal fraction of the building stock in energy-intensity intervals of equal width;
- the cohorts are replaced or renovated in order of decreasing energy intensity (so in 2006, some portion of the most energy-intensive cohort is renovated and some portion is replaced, while in 2007 the second most energy-intensive cohort is renovated or replaced, and so on);
- the portions of the total building stock that are renovated and replaced by 2050 are F_{reno} and $F_{replaced}$, respectively, so the floor areas renovated or replaced in a given year (as a fraction of the total floor area in 2005) are $f_{reno} = F_{reno}/45$ and $f_{replaced} = F_{replace}/45$, respectively;
- in addition, the building floor area, as a fraction of the initial floor area, is assumed to increase by an annual amount f_{new} that varies over time (based on the regional population and GDP/P scenarios);
- the energy intensity of a new building, or of a renovated building after the renovation, depends on the year in which it is built or renovated and does not change once built or renovated;
- the energy intensity of successive cohorts of new buildings decreases over time from E_{new-0} to E_{new-f}, and the energy intensity of successive cohorts of renovated buildings decreases over time from E_{reno-0} to E_{reno-f} (all these energy intensities are relative to the initial stock average, with the energy intensity of a given cohort fixed until the next renovation);
- the initial cohort energy intensities persist until year I_1, then decrease linearly to their final values by year I_2 and are held constant thereafter.

Thus, the total energy use of the building stock in year n, $E(n)$, relative to energy use in 2005, is given by:

$$E(n) = \sum_{i=1}^{n} \left\{ f_{repl}(i) + f_{new}(i)E_{new}(i) + f_{reno}(i)E_{reno}(i) + f_{remain}(i)E_{start}(i) \right\} + \sum_{i=n+1}^{45} f_{start}(i)E_{start}(i) \qquad (10.3)$$

ΔE is chosen such that the energy intensity of the last cohort, $E_{start}(45)$, is equal to the energy intensity of new buildings built in 2005; that is, $\Delta E = 1.0 - E_{new-o}$.

Electricity use in the residential sector is largely due to consumer goods or furnace fans, which have a rapid (5–20-year) turnover. Thus, in regions where residential electric energy intensity decreases, we always assume a 20-year transition to the reduced energy intensity. This is based on the assumption of a rapid tightening, worldwide, of standards for appliances and consumer electronic goods. In regions where residential electric energy intensity is assumed to increase, the increase is due to increased use of consumer goods partly offset by improved technology (but never reaching the current energy intensity in wealthy countries) and so is always assumed to require 40 years (to 2050). Thus, increasing wealth is accounted for in two ways in poor countries: through the energy use per unit floor area moving toward that of wealthy countries, and through floor area per person increasing with wealth.

After 2050, a second round of renovations occurs, each year involving the buildings that were renovated or added 45 years before. It is arbitrarily assumed that the energy intensity of new and renovated buildings decreases after 2050 at a rate of 0.5 per cent/year.

The application of the accounting scheme is illustrated in Figure 10.8 for a case in which the total floor area doubles between 2005 and 2050, using parameter values given in Table 10.3. Results are shown for three cases: a slow moderate improvement in the energy intensity of new and renovated buildings, a slow strong improvement, and a fast strong improvement. If new buildings already have a 20 per cent lower energy intensity than the existing stock average, if the energy intensity of new buildings decreases at a steady rate to 40 per cent of the starting stock average by 2050 (so that there is a factor of two reduction in the energy intensity of new buildings), and if the energy intensity of buildings after renovations decreases from 90 per cent

Table 10.3 *Parameter values used for the results shown in Figure 10.8, which illustrate the accounting model used to project future energy use by buildings and future average building energy intensity*

Parameter	Case		
	Slow moderate	Slow strong	Fast strong
Start	2010	2010	2010
Finish	2050	2050	2020
E_{new-o}	0.8	0.8	0.8
E_{new-f}	0.4	0.2	0.2
E_{reno-o}	0.9	0.9	0.9
E_{reno-f}	0.5	0.3	0.3

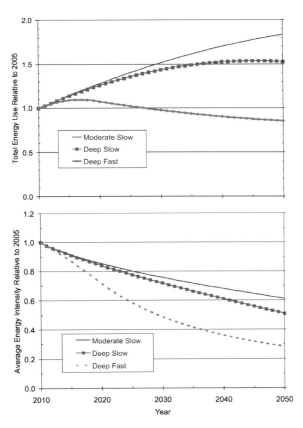

Figure 10.8 *Variation in total energy use and in average building energy intensity, relative to the values in 2005, for a region where total building floor area doubles between 2005 and 2050, as computed using the procedure explained in Box 10.1 and the parameter values given in Table 10.3*

of the starting stock average for buildings renovated in 2010 to 50 per cent of the starting stock average for buildings renovated in 2050, then the total energy use still increases by about 80 per cent. Movement to more stringent performance standards by 2050 (energy intensity of new and renovated buildings equal to 20 per cent and 30 per cent of the starting stock average, respectively) results in a 50 per cent increase in total energy use by 2050. However, if the energy intensity of new buildings is reduced by a factor of four compared to the current practice for new buildings by 2020, and if the quality of retrofits of existing buildings improves to the point where the energy use of any building undergoing a major renovation in 2020 or later is reduced by a factor of three, then by 2050 the total energy use decreases by about 15 per cent in spite of a doubling in building floor area. This example serves to illustrate the magnitude and the rapidity with which the performance of new and renovated buildings must be increased if total energy demand is to be significantly restrained (or slightly reduced). The main factors in limiting this transition are the speed with which the design professions and construction and renovation trades can be retrained in order to achieve these energy intensity reductions and the willingness of governments to require greatly improved building performance, as energy savings of this magnitude can generally be achieved with existing technologies.

In order to generate scenarios of future global energy demand for buildings (separately for fuels and electricity), we first need a scenario for the variation in residential and commercial floor area in each world region. For the commercial sector, floor area can be decomposed as the product of population times the proportion of the population employed in the service sector times the floor area per employee. McNeil et al (2008a) have compiled data on the second and third factors. Floor area per employee increases with mean per capita income in all countries, but at the same income level there is a factor of two to three difference in floor area per employee between countries with the lowest and highest values. Furthermore, floor area per employee tends to saturate rather than increase indefinitely as per capita income increases, and does so at different levels in different countries (the US and Norway, Denmark, UK and Japan all appear to be saturating at different levels). Schipper et al (2001) finds similar behaviour for per capita residential floor area versus per capita income in different countries.

This behaviour can be captured if the residential and commercial building floor area per person increases with increasing income following a logistic function. This is another application of Equation (10.2), except that income rather than time is the independent variable. That is, future floor area per person is given by:

$$A(I) = \frac{A_\infty}{1+((A_\infty - A_{2005})/A_{2005})e^{-a(I-I_{2005})}} \quad (10.4)$$

where I_{2005} and A_{2005} are the regional average GDP and floor area per person in 2005 and A_∞ is the saturation value of the average floor area per person.

Not only does floor area per employee in the commercial sector increase with increasing average income, but the proportion of the population employed in the service sector increases. This is due to a shift in economic output from industrial products to services as an economy develops. This is referred to as a change in the *structure* of consumption but will be considered later in conjunction with industrial sector energy use. Instead, I shall use Equation (10.4) to directly compute floor space per capita in the commercial sector (as well as in the residential sector), using estimates of floor space per capita in 2005 and prescribed asymptotic values. Table 10.4 gives the initial regional per capita floor areas and asymptotic per capita floor areas adopted here. A growth parameter *a* (in Equation (10.4)) of 0.0002 is used. Figure 10.9 shows the resulting low and high scenarios for the variation in global floor area for residential and commercial buildings, obtained by summing the projections made for each of the ten world regions.

The final step in projecting future building energy demand is to specify the initial energy intensities and how they change over time. The starting energy intensities and the assumed future energy intensities to be eventually achieved for new and renovated buildings are given in Table 10.4. Initial estimates of the energy intensities were derived from McNeil et al (2008a) and then modified to take into account biomass energy use (which is very important in Asia and Africa) and further adjusted (using information in Chapter 4) so that the global fuels and electricity uses by residential and commercial buildings closely match the global totals given in Table 2.3.

Figure 10.10 shows the resulting variation in fuel energy and electricity demand for the global residential and commercial building sectors when the energy intensities given in Table 10.4 for new and renovated buildings are assumed to be achieved by 2020 (curves labelled 'fast') or by 2050 (curves labelled 'slow').[3] For the high GDP scenario with slow improvement in building performance, fuel use by the global residential sector peaks in 2035 at only about 40 per cent greater than the 2005 fuel use, in spite of the floor area having almost doubled by then. The relatively small increase in fuel use is due to a disproportionately large increase in residential floor area occurring in warm regions but also due to the substantial reduction in heating energy requirement for new and renovated buildings in cold regions that has already occurred by then. With rapid introduction of tighter performance standards, global

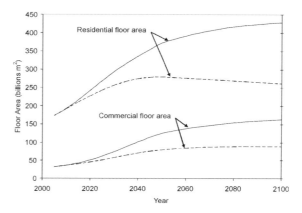

Figure 10.9 *Variation in global residential and commercial building floor area for the low and high scenarios*

Table 10.4 *Initial per capita floor areas, the floor area that would be reached with arbitrarily large per capita income, estimated initial building energy intensities, and energy intensities for renovated or new buildings assumed to be achieved by 2050 at the latest*

Region	Residential buildings								Commercial buildings								
	Floor area per capita (m²)		Energy intensity (kWh/m²/yr)						Floor area per capita (m²)		Energy intensity (kWh/m²/yr)						
			Electricity		Fuels						Electricity			Fuels			
	2005	Asymptotic	Start	Renovation or new	Start	Renovation	New		2005	Asymptotic	Start	Renovation	New	Start	Renovation	New	
PAO	50	55	55	30	70	50	40		26.3	25	150	70	50	150	40	30
NAM	60	60	66	30	100	50	40		24.5	25	150	70	50	150	40	30
WEU	40	45	35	30	140	50	40		13.0	20	100	70	50	130	40	30
EEU	25	45	20	30	180	50	40		2.8	18	100	70	50	200	40	30
FSU	35	45	25	30	230	50	40		2.4	18	80	70	50	240	40	30
LAM	15	40	25	25	50	25	15		3.2	18	100	70	50	40	15	10
SSA	15	40	5	20	220	15	15		1.4	15	40	70	50	5	15	10
MENA	15	40	25	20	10	15	15		2.5	15	100	70	50	10	15	10
CPA	25	40	20	30	80	30	25		3.0	15	120	70	50	160	30	20
SAPA	25	40	10	25	60	25	15		1.8	15	60	70	50	5	15	10

Source: (1) Residential floor area per capita: US, Wilson and Boehland (2005); China, Jin et al (2009); eastern and western Europe, EcoHeatCool (2006); elsewhere, estimates by Maria Sharmina (Central European University) based on scattered data. (2) Commercial floor area: based on total floor area in each region given by McNeil et al (2008a) divided by regional populations from UNPD (2004). (3) Initial residential and commercial energy intensities for electricity and fuels: based on total regional energy use for electricity and fuels divided by regional floor areas. The resulting commercial and residential global totals for electricity and fuels roughly match the corresponding totals given in Table 2.3

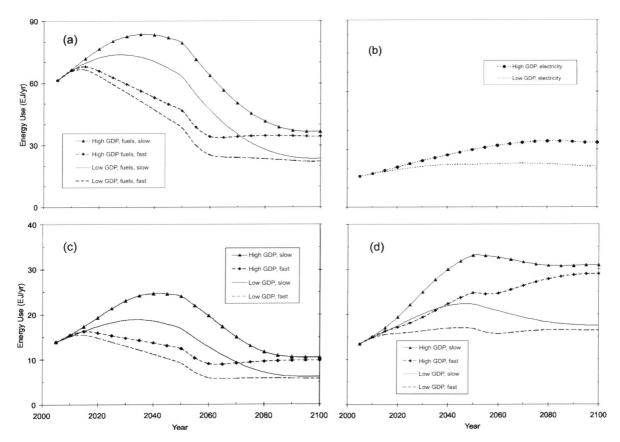

Figure 10.10 *Variation in the following elements of global building energy demand for the low and high GDP scenarios: (a) residential fuels, (b) residential electricity, (c) commercial fuels and (d) commercial electricity*

residential fuel use increases by only about 10 per cent before beginning a decline. Interestingly, global residential fuel use with high GDP and rapid improvement in building standards is less than with low GDP and a slow improvement in building standards until about 2070. After 2070, the high GDP scenario requires greater energy, but this is beyond the time of the peak rate of installation of C-free energy sources required in order to stabilize atmospheric CO_2 at 450ppmv and so could be accommodated by continued high rates of installation of C-free energy sources. Of course, the combination of low GDP and rapid improvement in building standards leads to even lower residential fuel requirements throughout the century.

Similar behaviour is seen for fuel use by commercial buildings. A short-term acceleration in the

rate of decrease in fuel use is seen in both sectors after 2050. This is because buildings that were renovated or built soon after 2005 have a second opportunity to be upgraded, this time to the much better performance standards applicable 50 years later.

Electricity use in the residential sector increases for all scenarios because the efficiency potential assumed here is less than for fuels and is overwhelmed by the projected increase in electricity services. This may be too pessimistic, as McNeil (2008b, Figure 5) projects the beginning of a decrease in global residential electricity consumption in the late 2020s based on an item-by-item analysis and assuming implementation of aggressive standards in line with close to the maximum efficiency potential. In the commercial sector a small dip or pause in the growth of electricity demand is seen after 2050, for the same reason as for the dip in fuel demand but weaker due to the assumed smaller efficiency potential for electricity than for fuels.

10.4.2 Transportation

Transportation energy use is projected separately for movement of people and freight, with transport of people further broken down into transport by light-duty vehicles, by air and by all other means. To do this, I have made use of the extensive data set compiled by the World Business Council for Sustainable Development for its own transportation scenarios (WBCSD, 2004). This data set provides estimates of vehicle stocks, average distance travelled per vehicle per year, average passenger loadings, and average fuel economy in 11 world regions that roughly correspond to the 11 SRES regions utilized here for the building sector scenarios. The vehicles considered for moving people are light-duty vehicles (LDVs), motorized two-wheel and three-wheel vehicles, buses, minibuses, air and rail. Table 10.5 gives the estimates of passenger travel by each mode in each WBCSD region, along with the global mean energy intensity for each mode as calculated from the regional energy intensities for each mode given by WBCSD (2004), weighted by the total travel in each region (the global share of passenger energy use for each mode of travel is shown in Chapter 5, Figure 5.2).

The data presented in Table 10.5 are used to construct scenarios of future transportation energy demand as follows. First, the logistic growth curve (Equation (10.2)) is used to project future average total per capita distances travelled by people in each region as a function of regional average per capita income. Initial fractions of total travel by LDVs and air are also prescribed and optionally allowed to change over time following a logistic curve. Travel not by LDV or air is lumped together as 'Other'. Table 10.6 gives the initial per capita total annual travel, LDV and air fractions, and the initial energy intensity for each mode of travel. The initial total travel and modal fractions are taken from WBCSD (2004) but have been slightly adjusted in some cases so that the total global passenger travel and energy use match the global totals given in WBCSD (2004) when using the SRES world regions. Population times per capita total travel times the fraction of travel in each mode gives future passenger transportation activity (passenger-km travelled) for each mode in each region.

Passenger transportation energy use in any given year is given by the transportation activity for each mode in that year times the energy intensity (MJ/passenger-km) in that year for that mode, summed over all modes. If we assume no change in loading factors, energy intensities per passenger-km for each mode will decrease in proportion to the decrease in vehicle energy intensities. These are

Table 10.5 *Per capita passenger travel in various world regions in 2005*

Region	Per capita passenger-km/yr (1000s)							
	LDVs	Two-wheelers	Three-wheelers	Buses	Minibuses	Rail	Air	Total
OECD North America	16.20	0.08		1.30	0.10	0.12	3.51	21.31
OECD Europe	8.35	0.44		1.74	0.13	0.63	1.98	13.28
OECD Pacific	7.48	0.75		3.41	0.43	1.27	1.84	15.19
FSU	2.77	0.37		1.09	0.55	1.09	0.31	6.18
Eastern Europe	4.31	0.23		1.11	0.56	0.75	0.65	7.62
China	0.32	0.39	0.12	0.43	0.50	0.41	0.15	2.32
Other Asia	0.44	0.70	0.15	1.08	0.82	0.10	0.26	3.55
India	0.18	0.38	0.10	0.58	0.44	0.46	0.06	2.21
Middle East	0.97	0.26		1.33	1.00	0.47	0.63	4.66
Latin America	1.92	0.20		0.86	0.65	0.03	0.60	4.26
Africa	0.38	0.08		0.49	0.57	0.02	0.11	1.65
World average	2.55	0.37	0.06	0.91	0.52	0.35	0.63	5.39
Average energy intensity (MJ/passenger-km)								
World	2.25	0.54	0.66	0.55	0.50	0.30	2.56	1.57

Source: Data from WBCSD (2004)

Table 10.6 *Estimated average total per capita travel in 2005 in various world regions, the value assumed to be approached asymptotically as income becomes arbitrarily large, the initial fractions of total travel in LDVs and by air, and the energy intensities for each mode of travel in each region*

Region	Average per capita travel (km) per year		Initial fractions		Initial MJ/pkm		
	2000	Asymptotic	LDV	Air	LDV	Air	Other
PAO	15,000	15,000	0.50	0.12	2.11	2.56	0.71
NAM	23,000	25,000	0.80	0.15	2.67	2.56	0.98
WEU	13,000	15,000	0.63	0.15	1.67	2.56	0.81
EEU	7600	15,000	0.57	0.09	1.81	2.56	0.54
FSU	6200	15,000	0.45	0.05	2.06	2.56	0.55
LAM	4300	10,000	0.45	0.15	2.23	2.56	0.54
SSA	1500	10,000	0.23	0.05	2.65	2.56	0.43
MEA	3500	10,000	0.21	0.15	2.27	2.56	0.54
CPA	2500	10,000	0.18	0.05	2.15	2.56	0.43
SAPA	3350	10,000	0.18	0.05	2.10	2.56	0.42

Source: Computed from or adjusted from WBCSD (2004)

assumed to gradually approach values that are based on the literature reviewed in Chapter 5 (and summarized in Table 5.22 for LDVs). Future energy intensities of LDVs for fuels and electricity are computed as the present fuel energy intensity times separate energy intensity factors for fuels and electricity. Table 10.7 presents the input data for these calculations along with intermediate and final results. Future LDVs are assumed to be either PHEVs with hydrogen in fuel cells or ethanol in an advanced internal combustion engine (ICE) as a range extender. In the first case, the drivetrain consists of either a fuel cell or battery, followed by an inverter, motor and gears (see Figure 5.27). The fuel cell at 55 per cent average efficiency is replaced with the battery having a 90 per cent charging + discharging efficiency when grid electricity is being used. Thus, the ratio of AC electricity to hydrogen energy needed when the hydrogen PHEV is using grid electricity will be 0.55/0.9 = 0.61. I assume that 75 per cent of urban driving is powered by AC electricity with the balance provided with hydrogen, while highway driving is assumed to be 100 per cent using hydrogen. The resulting energy intensity factors (assuming 70 per cent urban driving, 30 per cent highway driving) are 0.104 for fuels and 0.076 for electricity (Table 10.7), giving a total average onsite energy requirement of 0.180 times that of an average LDV today. If ethanol is used in an ICE as a range extender, the fuel

intensity in a PHEV when powered by ethanol rather than grid electricity is assumed to be 0.9 that given in Table 10.7 for an advanced hybrid vehicle.[4] Assuming the same electricity:fuel proportions for urban and highway driving, and the same urban:highway driving proportions as for the hydrogen PHEV, the resulting fuel energy intensity factor is 0.168 (the electricity energy intensity factor is unchanged).

Also shown in Table 10.7 are the eventual energy intensity factors for air, other modes of ground travel, and freight. The category 'Other' includes two- and three-wheelers, rail, buses and mini-buses, all of which have a similar energy intensity at present (especially compared to the difference between any one of these modes and LDVs or air). It is assumed here that the overall energy intensity in this group decreases by 25 per cent at the same time that 50 per cent of transportation service is shifted to electricity (as rail public transportation).

With regard to the variation in energy intensities over time, a realistic temporal variation would be for the energy intensity to initially decrease slowly (as industry retools to make more efficient vehicles and these vehicles begin to enter the market), to have the most rapid rate of decrease in energy intensity after moving half way from the initial to final energy intensities, and then for energy intensity to decrease ever more slowly as the target final energy intensities are reached. This behaviour can be created if the energy intensity I_e relative to the 2010 energy intensity varies according to:

Table 10.7 *Computation of energy intensity factors for LDVs, and presentation of energy intensity factors adoptedhere for other transportation modes*

	Urban	Highway	Average
LDV energy use (MJ/km)			
Today	2.820	2.060	2.592
Advanced HEV	1.200	0.910	1.113
H_2 in FCV	0.660	0.510	0.615
Electricity in FCV	0.374	0.289	0.349
Proportion of fuel and grid electricity			
Fuel	0.250	1.000	
Grid electricity	0.750	0.000	
Energy intensities (MJ/km) averaged over all driving, H_2 PHEV			
Fuel	0.165	0.510	0.269
Grid electricity	0.281	0.000	0.196
LDV Energy intensity factors, H_2 PHEV			
Fuel			0.104
Electricity			0.076
Results using ethanol in a PHEV			
MJ/km when driving on fuel	1.085	0.823	1.006
Fuel MJ/km averaged over all driving	0.271	0.823	0.437
Fuel energy intensity factor			0.168
Energy intensity factors for other transport modes			
Air	0.65		
Other ground	0.75 overall, with 50% shift to electricity		
Freight	0.470, fuels, 0.024, electricity		

Note: The fuel and electricity intensity factors for LDVs at any given time are multiplied by the initial concurrent fuel energy use. See Figure 10.11b for the temporal variation of the energy intensity factors. An additional 0.5%/yr reduction is assumed after 2050

$$I_e = 1 - \frac{(1 - I_f)}{1 + 100e^{-(t - t_{2010})/\tau}} \qquad (10.5)$$

where τ is a time constant for the transition to more efficient vehicles and is given a value of six years and I_f is the final energy intensity (given in Table 10.7).[5] The fuel intensity factors decrease from their present values to I_f, while the electricity intensity factors start at zero and rise to the final values given in Table 10.7. A further small rate of decrease of 0.5 per cent/year in I_e is assumed after 2050 for all modes of transport, since it is extremely unlikely that technical means for further reductions in energy intensity will not be found.

As seen from Table 10.5, two-wheelers, three-wheelers and buses account for the majority of passenger transportation in much of Asia. If these are replaced with LDVs, even highly efficient LDVs, total energy demand will increase substantially. Alternatively, if these modes of transportation are replaced with rail-based transit, energy intensity could decrease while providing vastly superior service. A critical factor in future transportation energy demand in all regions is the share taken by LDVs. For the base case scenario, it is assumed that the LDV share tends toward 50 per cent in each world region or stays constant at its current value if that value is already greater than 50 per cent. It is also assumed that air travel as a share of total distance travelled tends toward 18 per cent as income and hence total travel increases. Total distance travelled is within 5–20 per cent of the asymptotic limits given in Table 10.6 by 2100 for the low GDP scenario, and within 1–2 per cent for the high GDP scenarios, as are the LDV and air shares.

In contrast to people movement, tonne-km of freight movement and the associated energy use are assumed to increase in direct proportion to total world GDP. This is based on the implicit assumption that global trade increases in proportion to GDP, which is unlikely if (as almost certainly will happen), the price of transportation fuels increases sharply in the long run. This assumption will be relaxed later.

Figure 10.11 shows the resulting scenarios for global passenger and freight movement per year and the relative transport energy intensities for freight and for LDVs. Relative energy intensities are shown for a slow rate of improvement (τ = six years) and for a fast rate of improvement (with τ = four years). Fleet average automobile fuel intensity has dropped by 20 per cent by 2030 for the slow case and by 20 per cent by 2023 for the fast case. Figure 10.12 shows the corresponding global annual fuel and electricity energy use for transportation. For the low GDP scenario and slow reduction in fuel energy intensities, total transportation fuel use peaks in 2030 at a level 38 per cent above the 2005 level, while for the same GDP scenario and fast

reduction in energy intensity, peak demand occurs in 2020 at a level 21 per cent above the 2005 level. For the high GDP scenario, the corresponding peaks are over 83 per cent and over 50 per cent above the 2005 level. Given that global oil supply is likely to peak at a level not substantially above the current rate of supply in the near future (see Chapter 2, subsection 2.5.1), and that ramping up of biofuels this rapidly is unlikely to be feasible (see Volume 2, Chapter 4), neither demand scenario shown in figure 10.12 is likely to be feasible. These scenarios serve only to show the impact of alternative demand and efficiency assumptions.

In the transportation scenarios present here, annual passenger-km travel in LDVs increases by a factor of about nine in Asia for the low GDP scenario and by a factor of 15 for the high GDP scenario. This is almost certainly infeasible due to space constraints (see

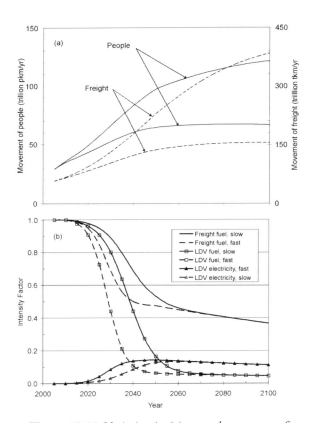

Figure 10.11 *Variation in (a) annual movement of people and freight for the low and high scenarios, and (b) assumed variation in the fuel and electricity energy intensity factors for movement of people and freight*

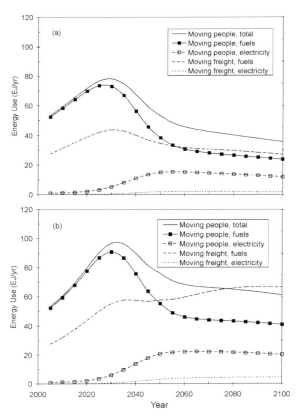

Figure 10.12 *Variation in transportation energy demand for (a) the low GDP scenario and (b) the high GDP scenario, in both cases for the slow efficiency improvement*

Chapter 5, subsection 5.2.3) and, in any case, is not inevitable even with large increases in per capita income (as assumed here). Recall, from Chapter 5 and Table 5.5, that the economic transportation energy intensity (transportation energy use per unit of income) in wealthy Asian cities is about seven times less than that of poor Asian cities (total per capita transportation energy use is only 25 per cent greater, but per capita income is a factor of eight greater). This is because a large fraction of transportation in wealthy Asian cities is through efficient rail-based systems, compared to inefficient personal vehicles in less wealthy Asian cities.

Even for the low GDP scenario with a rapid transition to efficient plug-in hybrid LDVs, transportation fuel demand may exceed global supply in the near term, due to the likely imminent peaking in world oil supply. It is therefore pertinent to consider a scenario of aggressive shifting from LDV to other modes of urban travel beginning in 2010, combined with restraints in air travel.[6] Thus, we consider a scenario called 'fast + green', whereby:

1   annual average distance travelled per person drops from 15,000km to 10,000km in PAO, from 23,000km to 18,000km in NAM, from 13,000km to 10,000km in WEU, is capped at 8000km in EEU and FSU, and is capped at 6000km elsewhere;

2   the share of total travel by LDVs drops from 0.5 to 0.4 in PAO, from 0.8 to 0.7 in NAM, drops to 0.5 in WEU and EEU, drops to 0.4 in FUS and LAM, and is capped at 0.25 elsewhere;

3   the share of travel by air drops to or is capped at 0.1 everywhere; and

4   the ratio of global tonne-km of freight transport to world GDP falls smoothly by a factor of two as world GDP increases by a factor of 2.7 (with the decrease in transport to GDP ratio proportional to the increase in GDP).

The assumption of a LDV share capped at 25 per cent in Africa and Asia assumes that further urban development is strongly oriented toward efficient public transportation systems. The result of these assumptions is that global passenger-km (pkm) of travel peaks at 38 per cent above the 2005 level in 2045 (after which it declines due to falling population), total travel by LDVs is relatively constant at around 16 trillion pkm/yr throughout the century, air travel grows from

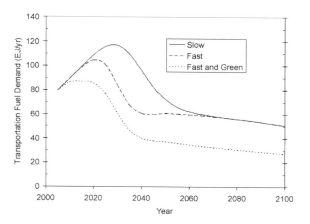

Figure 10.13 *Comparison of the variation in global transportation fuel demand for the slow, fast and fast + green scenarios, all combined with the lower GDP scenario*

3.7 trillion pkm in 2005 to a peak of 4.6 trillion pkm in 2050, and freight movement increases by only 35 per cent from 2005 to 2100.

Figure 10.13 shows the resulting variation in fuel demand for transportation, and compares the fast + green scenario with that of the fast and slow scenarios (all assuming the lower GDP scenario). Slower economic growth combined with fast + green is the only scenario where the peak in transportation fuel demand (7 per cent above the 2005 demand in 2015) is consistent with the geologically based estimate of future oil supply shown in Figure 2.21.

10.4.3 Agriculture and industry

Baseline agricultural energy use is assumed to grow in proportion to global population, which assumes no shift in diet as income increases. The baseline energy use in turn is multiplied by an energy intensity factor that decreases from 1.0 in 2010 toward a final value of 0.75 using Equation (10.5), with an additional decrease of further 0.5 per cent/year after 2050. Baseline industrial energy use is assumed to grow in proportion to GDP, multiplied by an energy intensity parameter that decreases from 1.0 in 2010 toward the values given in Table 10.1, with a further 0.5 per cent/year decrease after 2050. As with transportation, slow and fast cases are considered, with τ = six and four years, respectively.

10.4.4 Aggregate global demand for fuels and electricity, and global primary energy demand

The variation of global demand for fuel energy and for electricity, summed over all the sectors considered above, is shown in Figure 10.14. Results are shown for simultaneous slow or rapid implementation of efficiency measures in all sectors, as well as for another case described later. A faster implementation of efficiency improvements has very little effect on overall electricity demand. This is because electricity demand for passenger transportation increases more quickly (due to the more rapid transition to PHEVs) while electricity demand by other sectors

decreases more quickly. Figure 10.15 shows the sectoral breakdown of fuel and electricity demand in 2100 for the low GDP scenario. Fuel energy used to transport people is relatively small because about half of this energy demand has been shifted to electricity. Industry accounts for about 40 per cent of total fuel use in this scenario.

The fuel and electricity secondary energies can be divided by the corresponding primary-to-secondary conversion efficiencies and added to give the total demand for primary energy. Today, the primary energy equivalent of electricity from sources such as hydropower, wind and solar is determined by dividing by the efficiency of thermal powerplants. This gives the amount of fossil fuel energy that is saved by having non-fossil alternatives. Once there is no fossil fuel generation

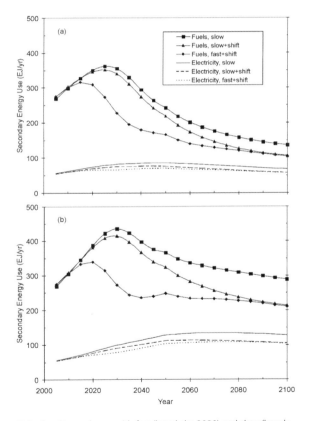

Note: Results are shown with fast (largely by 2020) and slow (largely by 2050) implementation of the efficiency improvements considered here, as well as for fast implementation with account taken of illustrative structural shifts in the economy.

Figure 10.14 *Variation in global demand for fuels and electricity for (a) the low GDP scenario and (b) the high GDP scenario*

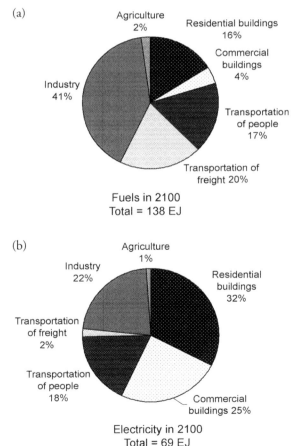

Figure 10.15 *Breakdown of (a) fuel and (b) electricity use in 2100 for the low GDP scenario*

in the mix – which is the ultimate objective – it is perhaps rather academic to compute the primary energy equivalent of C-free electricity in this way. Nevertheless, this procedure places a higher value on electricity than on fuels (which is appropriate), and permits comparison with the preceding fossil fuel era. I assume that the global average efficiency in supplying electricity increases from 28 per cent in 2005 to 60 per cent in 2050 and is held constant thereafter, while the efficiency in producing fuels and district heat from primary energy is assumed to decrease from an average of 88.6 per cent in 2005 to 75 per cent in 2050. The 75 per cent figure is a somewhat arbitrarily chosen conversion efficiency. It could arise as the weighted average of solid fuels for heating and electricity generation supplied from biomass with an efficiency of 95 per cent (i.e. with 5 per cent loss), hydrogen produced by electrolysis of water, compressed and transported with an efficiency of 70–75 per cent (see Volume 2, Chapter 10, subsections 10.3.3 and 10.5.1) and gaseous or liquid fuels produced from biomass at an efficiency of about 65 per cent. In Volume 2, the land areas for biomass and wind or solar capacity required to produce the needed mix of solid, liquid and gaseous fuels

will be determined for various assumptions concerning which kinds of fuels are used for specific end uses.

Figure 10.16 shows the resulting variation in global primary energy demand for the low and high GDP scenarios, and for cases with rapid (largely by 2020) and slow (largely by 2050) implementation of the efficiency measures considered above. Global primary energy demand peaks somewhere between 2020 and 2035 and at levels ranging from about 570EJ/yr to 800EJ/yr, depending on the scenario (in contrast, primary energy demand was 483EJ in 2005).

10.4.5 Rate of reduction in economic primary energy intensity

Energy demand in the Kaya framework (Equation (10.1)) is given by the product of population (P), GDP per person, and economic energy intensity (energy use per unit of income). However, the energy intensities discussed earlier in this book are physical energy activities; that is, energy use per unit of activity (where the activity could be the production of a tonne of steel, the operation

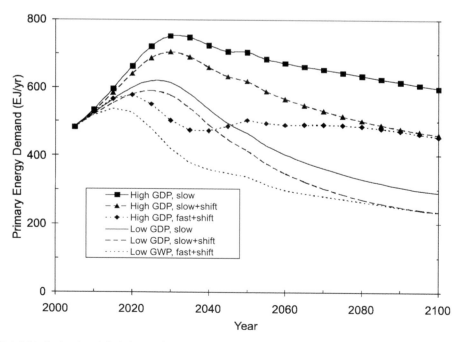

Figure 10.16 *Variation in global demand primary energy demand for the low and high GDP scenarios with fast (largely by 2020) and slow (largely by 2050) implementation of the efficiency improvements considered here*

of a unit area of building floor area, or the travel of a given distance). In terms of physical energy activities, energy use at the national scale can be written as:

(energy use) = population × ($ of GDP/capita) × (activity/$ of GDP) × (MJ/activity) = ($ of GDP) × (activity/$ of GDP) × (MJ/activity)   (10.6)

Thus, if we were to assume that energy use varies in proportion to population and physical energy intensity, we would be implicitly assuming no change in the (activity/$) ratio. That is, we would be assuming that that all energy-using activities increase in proportion to income. This is not the case. Instead, as incomes rise, some activities (such as distances travelled and household floor area) increase in a smaller proportion, while other activities (such as educational and health services) probably increase in greater proportion. There will also be a shift away from industrial output, and a change in the mix of commodities produced within the industrial sector (relatively less energy-intensive materials such as steel are produced, and relatively more high-value manufactured products are produced). This change in the proportions of different kinds of consumption is referred as a change in the *structure* of consumption. Assuming energy use to vary in proportion to population and physical energy intensity also ignores changes in the cost of energy that will be correlated with GDP (both are likely to go up over time, the cost of oil and natural gas probably dramatically) and ignores congestion effects.

Some of these effects are already included in the procedure used here to project future energy demand. In particular, annual distance travelled per person is assumed to increase logistically with income (such that it eventually saturates at some upper limit as income becomes arbitrarily large, rather than increasing in proportion to income). Similarly, floor space per capita in the commercial sector is assumed to follow a logistic relationship. However, we have not accounted for a reduction in the relative importance of industry as the economy grows and in the production of materials that need to be transported, as we have assumed both industrial and freight transportation energy use to increase in proportion to GDP. A partial shift from industry to services as the global economy grows would lead to a reduction in both industrial and freight transportation energy use and an increase in the energy use by commercial buildings (this would be due to an

increasing proportion of the workforce being employed in commercial buildings as average income increases, a factor that was deliberately omitted in the above analysis).

The decrease in industrial energy use due to a shift away from industrial production is equal to the decrease in the dollar value of industrial value-added times the average energy intensity (MJ/$) of the lost industrial production, while the increase in commercial sector energy use is equal to the increase in commercial sector value-added times the energy intensity of new commercial activities. Thus, if the reduction in value-added in the industrial and freight transportation sectors is offset by an equal increase in value-added output from the commercial sector, the net change in energy consumption depends only on the ratio of energy intensities of the sectors involved in the shift and on the magnitude of the shift. Even changes within the industrial sector, from heavy industry to general manufacturing, would entail a significant net reduction in energy use. Schipper et al (2001) indicate the following average industrial energy intensities in the US during the 1990s: ferrous metals, 50MJ/$; non-ferrous metals, 30MJ/$; non-metallic minerals, 27MJ/$; chemicals, 18MJ/$ and general manufacturing, 3MJ/$. The overall average for US industry is 20MJ/$. If the energy intensity of the commercial sector is no more than that of general manufacturing, then a shift from average industrial dollar output to commercial services dollar output entails about 1MJ increase in commercial sector energy use for every 7MJ decrease in industrial sector energy use. To be conservative, I will assume a 1 to 4 ratio so that, if anything, I underestimate the net reduction in energy use due to a partial shift from industrial to commercial activities.

Assume, for the sake of illustration, that industrial and freight activity (in dollar terms) as a fraction of total GDP decrease by 50 per cent by 2100, as world GDP becomes three to five times larger. The corresponding industrial and freight energy uses also decrease by 50 per cent (assuming the reduction in industrial value-added to be distributed across all industries in proportion to their initial energy use), but the compensating increase in commercial sector energy use is only one quarter as large (assuming a 1 to 4 rather than 1 to 7 ratio of energy intensities). The resulting demands for fuels, electricity and primary energy are shown in Figures 10.14 and 10.16. The result of this shift is that global primary energy demand in 2100 for the low GDP scenario drops from 290EJ to 235EJ. The

resulting demand is less than half the global primary
energy demand in 2005 (483EJ) while at the same time
GDP is almost three times larger. As a result, global
mean energy intensity has fallen by almost a factor of
six, from 8.68MJ/$ in 2005 to 1.53MJ/$ in 2100.

It should be stressed that the deep (factor of three to
five) reductions in secondary energy intensity assumed
here for specific end-use sectors depend not only on
vastly improved efficiency of individual energy-using
devices (such as automobiles, appliances, lights, and
heating and cooling equipment), but also on improved
organization of entire energy-using systems (such as the
mechanical and lighting systems in buildings,
transportation systems and alternative manufacturing
processes) and enlightened human behaviour (avoiding
wasteful uses of energy, adapting indoor temperature
settings to changing outdoor conditions, maximal
recycling of material goods).

Figure 10.17 shows the annual rate of decrease in
primary energy intensity averaged over successive decades
for the low GDP scenario (neglecting the additional
structural shifts discussed above), as well as the

cumulative average compounded rate of decrease from
2010 to years up to 2100 with and without the structural
shift given above. The decadal average annual rate of
decrease in energy intensity peaks at about 4.0 per
cent/year in the decade 2030–2040, then gradually
declines toward the residual rate of decrease of 0.5 per
cent/year that is assumed after 2050 for new buildings,
renovations and other sectors. During the 11 energy-
conscious years of 1975–1985, new energy-using
products in the US improved their energy efficiency by an
average of 5 per cent/year (Rosenfeld, 1999). However,
the rates given in Figure 10.17 pertain to the average of
the entire economy, not just new products. The
compounded annual average decrease is 2.4 per cent/year
from 2010 to 2050 and about 1.4 per cent/year from
2010 to 2100, neglecting the additional structural shifts
discussed above. With the additional structural shifts
assumed above, compounded average rates of decrease are
2.7 per cent/year and 1.8 per cent/year to 2050 and 2100,
respectively. The compounded average rates of reduction
in energy intensity for the high GDP scenario are higher
still (3.0 per cent/year to 2050, 2.0 per cent/year to 2100

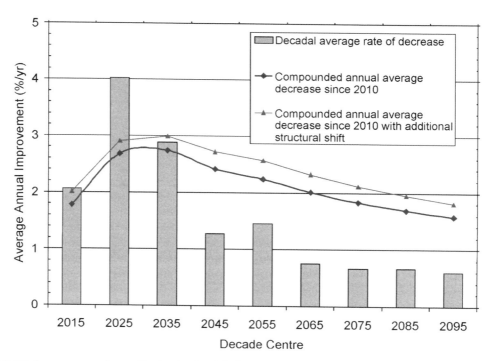

Figure 10.17 *Average annual rate of reduction in global mean primary energy intensity, decade by decade (bars), and the annual average compounded rate of reduction in primary energy intensity from 2010 to 2100 (lines)*

with the same structural shift) due to the greater proportion of building stock built after 2005 and hence with lower energy intensity than in the low GDP scenario. As there will certainly be some shift in the structure of consumption toward less energy-intensive sectors over the coming century, it is concluded here that the achievable average compounded rate of reduction in economic energy intensity between now and 2050 is at least 2.5 per cent/year.

The primary energy requirement in 2050 for the high GDP scenario with rapid and strong improvement in energy efficiency and allowing for structural shifts is 526EJ, the equivalent of 16.7TW of continuous primary power. Given a permitted fossil fuel primary power supply of about 6TW in 2050 (see Figure 1.9b), about 11TW of C-free power would be required in 2050. For the low GDP scenario the primary power requirement is 10.7TW, so the C-free power requirement would be about 4.7TW or just under half that required under the high GDP scenario. These requirements are roughly consistent with the C-free requirements given in Figure 10.4b at 2.7 per cent/year reduction in energy intensity. By comparison, the global primary power associated with C-free energy sources was 3.28TW in 2005, of which nuclear accounted for 0.78TW, hydro 0.90TW, biomass 1.52TW and others (mostly wind, solar and geothermal) 0.08TW.[7] The non-nuclear C-free power supply in 2005 was only 2.50TW, so if nuclear power is phased out by 2050 (or if one advocates a phasing out of nuclear power), we would have to almost double the current non-nuclear C-free power supply by 2050 for the low GDP scenario. This is an extremely challenging task, particularly since most of the increase would have to come from energy sources that constitute a very small fraction of the current C-free power supply. For the high GDP scenario, non-nuclear C-free power supply would need to be quadrupled by 2050. It should be stressed that if the compounded improvement in the global mean energy intensity falls short of the 2.7 per cent/year value assumed here, even greater amounts of C-free power would be required.

Volume 2 systematically examines the ultimate potential and costs of various C-free power sources. C-free energy supply from renewable energy sufficient to meet the separate demands for fuels and for electricity in about 2100 for the low and high demand scenarios is postulated and is assumed to be approached following various logistic curves. The rates of establishment of C-free energy sources and the associated material and energy flows are determined and assessed in terms of feasibility. The supply scenarios are combined with the scenarios for fuel and electricity demand to determine the variation in fossil fuel emissions to the point where they reach zero (around 2100). These emissions are then used as input to a coupled climate–carbon cycle model to determine which combination(s) of demand- and supply-side scenarios, if any, can limit the peak atmospheric CO_2 concentration to no more than 450ppmv.

However, before moving on to Volume 2, it is pertinent to briefly examine what is required, in general terms, to achieve the low population scenario, to bring about a stable transition to slower and eventually zero growth in GDP per person, and to induce the dramatic improvements in energy efficiency assumed here. This is the subject of the final chapter of this book.

Notes

1   The only exception pertains to South Korea, which is placed here in the Pacific-Asia region alongside Japan, Australia and New Zealand.
2   I have termed *a* the *approach rate* tendency, since it governs how quickly GDP/capita or population approaches the asymptotic value, whether this value is less than or greater than the current value.
3   In both cases, residential electricity intensity reaches the target values by 2020 when the target intensities are below 2005 intensities because residential electricity use consists largely of appliances and consumer goods with a rapid replacement rate.
4   According to the analysis of net energy saving with biomass ethanol presented in Volume 2 (Chapter 4, subsection 4.6.1), 1 litre of ethanol displaces about 0.75 litres of gasoline. However, the energy contents of ethanol and gasoline are 23.3 MJ/litre and 32.4 MJ/litre, respectively (see Appendix B), so 1 MJ of gasoline energy requires only 0.9 MJ of ethanol energy to travel the same distance.
5   Equation (10.5) is another form of the logistic equation.
6   If oil supply does indeed begin to fall below demand, then growth of air travel will be curtailed even without government policy intervention, due to rising fuel costs and the greater relative importance of fuel to the cost of travel by air than by other modes.
7   These figures are computed from Table 2.3.

11

Policies to Reduce the Demand for Energy

In this, the concluding chapter of Volume 1, the policies and programmes needed to dramatically reduce the demand for energy relative to business-as-usual scenarios are briefly outlined. The demand for energy depends on the demand for energy services as well as the efficiency with which these services are provided. The demand for energy services in turn depends on the human population and the average GDP per person, with the particular mix of energy services demanded dependent on economic development, energy prices and lifestyle factors. All of these factors are addressed in this chapter. Policies and programmes to promote lower growth in the human population and in average GDP/person are outlined, as well as policies to dramatically reduce the energy intensity of the global economy. Dramatic improvements in energy efficiency can be largely achieved with existing technologies, but important areas for further research and development are identified here.

Energy savings achieved through improved energy efficiency, if unaccompanied by measures to limit overall demand for energy services, tend to be eroded by increased energy use elsewhere in the economy – a phenomenon referred to as the rebound effect. The issue of the rebound effect and the broader question of the links between human happiness, economic growth and sustainable levels of consumption are addressed here.

11.1 Limiting future population

The rate of growth of population depends on the difference between birth rates and death rates. Given the desirability of reducing death rates by increasing the average human lifespan, the key to limiting future increases in human population is to rapidly reduce birth rates or fertility – expressed as lifetime births per woman.[1] The replacement level of fertility is about 2.1 in developed countries and as high as 3.3 in some developing countries, with a global mean of about 2.3. The replacement level is greater than 2.0 due to differences in the male to female sex ratio (more males) at birth and due to mortality of some females before they reach the end of their fertility. Figure 11.1 shows fertility rates across the world. In all OECD countries except the US and in the former Soviet Union, China, Thailand, Vietnam, Iran, Tunisia, Algeria, Cuba, Brazil, Uruguay and Bolivia, fertility rates are now less than 2.0. However, fertility rates are still very high in most of Africa, in parts of Central America and the Caribbean, and in Paraguay. The relatively high fertility rate in the US (2.1) is due to high fertility among Hispanics and other minority groups and to a relatively high fertility among non-Hispanic white people (1.9, compared to 1.3–1.8 in Western Europe). In most countries with sub-replacement levels of fertility, population is still growing because the age structure is such that death rates are still less than birth rates.

Nearly one billion people live in countries in which fertility averages four or more children per women, such that the population would double with current growth rates in less than 35 years (Leahy et al, 2007).[2] It is widely recognized that the keys to limiting fertility rates where they are still high are:

- to reduce the demand for large families by making it socially and economically advantageous to have fewer children;

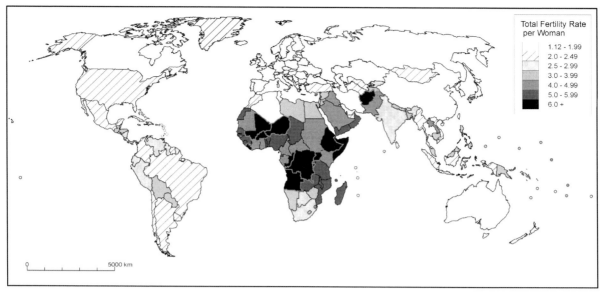

Source: Data from US Census Bureau (www.census.gov/ipc/www)

Figure 11.1 *Fertility rates in the countries of the world*

- to increase the educational level of women so that marriage is delayed and women have alternatives to a life with many children (according to Birdsall (1994) each additional year of female education reduces fertility by 5–10 per cent);
- to provide family planning and contraceptive services so that unwanted pregnancies can be avoided (according to Bongaarts et al, (1997) one in five births in the mid-1990s were unwanted and 25 million abortions were performed annually due to insufficient access to contraceptives).

Two of the keys to reducing the demand for large families are (1) to provide old-age security, and (2) to provide economic incentives of various sorts. The demand for large families decreases on its own with urbanization because extra children become an economic burden rather than an economic asset.

China, Iran, Mexico and Tunisia all provide examples of rapid declines in fertility rate (Leahy et al, 2007). In 1979, China introduced a policy of encouraging (through financial incentives but also sometimes through coercive measures) only one child per family. Chinese fertility rates fell from around 5 children per women in 1970 to an estimated 1.3–2.0 (probably 1.6–1.8) today, with most of the decline

occurring before implementation of the one-child policy (the one-child policy is nevertheless credited with avoiding 300–400 million births up to the present).

In the case of Iran, a national family planning programme was fully implemented in the early 1990s. It involved modern free contraceptives on demand at clinics and mandatory attendance by prospective married couples at a government-sponsored class on family planning prior to receiving a marriage licence. Fertility rate fell from 6.5 children per women in 1980 to 1.71 children per woman by 2007. Iran's population growth rate dropped from an all-time high of 3.2 per cent/year in 1986 to 1.2 per cent/year in 2001 and 0.7 per cent/year in 2007 (due in part to a net out-migration of 4.29/1000/year). The Iranian experience indicates that economic growth is not a prerequisite to large declines in fertility. Mexico also saw a sharp decline in fertility following the introduction of a national family planning programme (in 1972) and later improvements in health services, falling from 6.5 children per women in the early 1970s to 2.15 by 2005. A third example is Tunisia, where fertility fell from more than seven children per women in 1960 to two children per woman by 2005.

Apart from limiting future GHG emissions (and other impacts of humans on the environment) by reducing the future human population, family planning services will improve the health of mothers and those children that are born, and will reduce the dilution of increased services (as the economy grows) among a growing population. A smaller future population will also limit the impact of adverse climatic change in the future and, indeed, will reduce the number of people at risk of hunger even in the absence of climatic change (as discussed by Smil, 1994b, the world agricultural system is likely to be able to feed a world of 9 billion people, but will be hard pressed to feed a substantially larger population, although a sufficient moderation in projected future meat consumption could compensate for an increasing population and adverse climatic change). A decline in fertility rates through an increased supply of condoms will also do much to reduce the spread of HIV/AIDS, which is a particular problem in some of the very countries (in Africa) with the highest fertility rates.

11.2 Limiting economic growth

Stabilization of climate will ultimately require constraints on economic growth (if such constraints do not arise naturally as a result of shortages of materials and cheap energy), that is, in the consumption of goods and services. If per capita consumption of goods and service were to continue to grow exponentially indefinitely, such growth would ultimately swamp the gains that are possible through improved energy efficiency and overwhelm the earth's ability to provide renewable (carbon-free) energy on a sustainable basis.

In this section, three strategies for limiting economic growth in wealthy countries are briefly discussed. These are:

1   to shift attention and policy goals away from maximization of GDP and toward maximization of broader indicators of human happiness;
2   to promote shorter working hours or at least greater choice in the number of hours worked; and
3   to promote greater product longevity.

11.2.1 Limiting economic growth by focusing on determinants of human happiness

GDP per capita has been used as a proxy for human happiness, but growth in GDP has become a goal in and of itself. However, there is growing evidence that, beyond some level of GDP per person, increasing GDP per capita does not increase human happiness and, in many cases, leads to decreased happiness. At any given point in time, individuals with higher income tend to rate themselves as happier than do individuals with lower income, but over time, there has been almost no increase in reported happiness within a given country, and in some cases a decrease, as incomes have increased (Frey and Stutzer, 2002). The fundamental reason why increased consumption does not lead to an increase in happiness is that the happiness associated with consumption largely depends on one's wealth *relative* to other people (Layard, 2003, 2005). Since it is impossible to simultaneously make everyone wealthier relative to everyone else, it is impossible to increase overall happiness by pursing greater GDP per person, yet such pursuit is a prime driver of environmental destruction and increasing GHG emissions. It is thus important to long-term climate stabilization to be able to direct current thinking in society away from the belief that increasing consumption is a source of happiness.

In order to do this, it is necessary to recognize the different kinds of human needs that are satisfied, to varying degrees, by material consumption. As discussed by Maslow (1954), Max-Neef (1992), Michaelis (2003) and Hofstetter et al (2006), among others, these needs include physiological needs (food, clothing, shelter) but also a wide range of non-physiological needs such as social recognition, self-esteem, affection, cultural expression and self-actualization. To achieve sustainable and climate-friendly levels of consumption requires encouraging the production of goods and services that maximize the satisfaction of human needs while minimizing the use of materials and energy. For example, a shift in urban development toward more compact forms that require less time for commuting, and investment in high-quality rail-based rapid-transit infrastructure that ultimately costs less than car-based transportation systems (including the purchase and operating costs of cars) can free up more time to be spent with one's family or for creative leisure activities, both because less time is required for commuting and also because less hours need to be worked due to the reduced expenses if one can forego the purchase of a personal automobile altogether.

Jackson (2009) calls for increasing the means for people to flourish (physically, psychologically and socially) in less materialistic ways so that they are less dependent on their display of consumption as a means of broadcasting their social status. He suggests that a core element of this will include reducing social inequality by reducing huge income disparities – reining in the salaries that excessively reward socially detrimental behaviour while providing better salaries for – and thus better recognition of – those engaged in child care and care for the elderly and disabled. Daly (1996) also argues that a less unequal distribution of income (through maximum as well as minimum income) is a prerequisite for limiting the endless growth in consumption.

To achieve sustainable and climate-friendly levels of consumption also requires educating the public (beginning with the primary education system) into greater reflection on the sources of long-lasting happiness. Kasser (2002), in *The High Price of Materialism*, notes that setting and achieving non-materialistic goals brings a greater sense of well-being than setting and achieving materialistic goals, while Diener and Oishi (2000) find that those valuing love higher than money have a much higher life satisfaction than those who give priority to money. As discussed by Schwartz (2004), excessive choice in what is available for purchase also tends to a decrease happiness and satisfaction with what we do buy (more choices require more time for decision-making, more reasons for regret and more opportunities for people to judge their condition unfavourably by comparisons with others), which in turn fuels more consumption that never makes people as happy as they expected. (At the personal level, Schwartz provides a number of suggestions to avoid becoming a victim of excessive choice.)

Michaelis (2003) recommends the following strategies to reduce consumption:

- promoting diversity and understanding the thinking of those with 'ethical consumption' so that governments can develop effective arguments and stimuli for reduced consumption as part of a public dialogue;
- supporting groups such as Simplicity Circles (discussed by Maniates, 2002), the growing Transition Town initiative (see www.transitiontowns.org) and community EcoTeams, as it is usually difficult for individuals to change consumption patterns unless they belong to a supportive group that affirms their values and beliefs;

- applying the ideas used to promote technological innovation to social innovation, especially supporting new emerging practices;
- promoting and using alternative indicators of 'progress' (rather than just GDP);
- shifting employment law and taxation in order to promote more leisure time.

Marketing, however, constitutes a formidable force encouraging ever more consumption, permeating every element of most people's day-to-day lives. Restrictions on marketing could therefore be quite effective but are difficult to implement. Some countries have tried to limit advertising to children, with mixed success, but in any case they cannot control satellite TV. Increasing public funding of schools and universities to the level needed for them to provide the expected services, so that they are not partly dependent on advertising revenue, could reverse the invasion of marketing into the realm of public education, and government support (or greater government support) for public broadcasting could at least provide the public with television and radio free of advertising.

The following paragraphs focus on ways of promoting shorter working hours and more leisure time.

11.2.2 Limiting economic growth through reduced working hours

A shift toward shorter working hours will reduce the income available to spend on energy-consuming goods and services. As discussed by Schor (2005), annual hours worked in the US and Britain rose to a peak in the mid-19th century, then began a long decline until the 1980s. Figure 11.2 gives the variation in the number of hours worked per year per employee in European countries, Australia, Canada, the US and Japan from 1950 to 2000, while Figure 11.3 gives the percentage decrease by decade and Figure 11.4 gives the overall decrease from 1950 to 2000. In most countries, decadal decreases of about 5 per cent occurred from 1950 to 1980, but during the last two decades there has been a much smaller or no decrease, with the exception of Japan (which still has the highest number of hours worked per employee).

Resistance to shorter working hours comes both from employers and sometimes from employees. As discussed by Schor (2005), firms tend to resist a

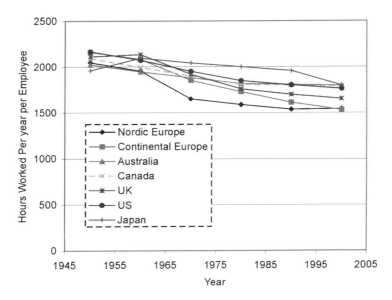

Source: Data from Schor (2005)

Figure 11.2 *Variation in the number of hours worked per year per employee in European countries, Australia, Canada, the US and Japan from 1950 to 2000*

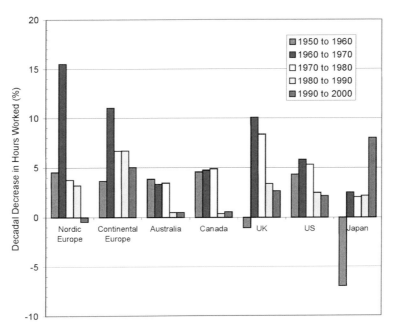

Source: Computed from data in Schor (2005)

Figure 11.3 *Percentage decrease by decade in the number of hours worked per year per employee in European countries, Australia, Canada, the US and Japan from 1950 to 2000*

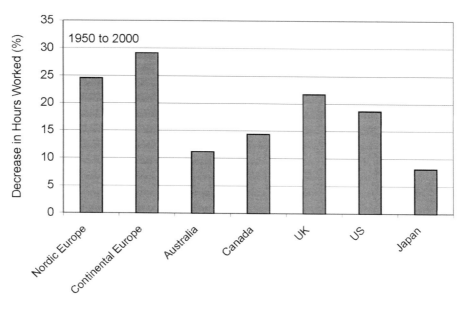

Source: Computed from data in Schor (2005)

Figure 11.4 *Overall decrease from 1950 to 2000 in the number of hours worked per year per employee in European countries, Australia, Canada, the US and Japan*

reduction in the number of working hours per employee compensated by an increase in the number of employees for several reasons. Foremost among these is the fact that employment-related costs (such as employer contributions to health care or disability insurance) tend to be structured on a per-person basis rather than on a per-hour basis (even when these costs vary with hours worked, they tend to be capped at a certain level, introducing a per-person rather than a per-hour component if the cap is sufficiently low that those with reduced working hours still reach it). Thus, greater public financing of costs such as health care and disability insurance will reduce the pressure to maintain the current number of working hours per employee. Schor (2005) argues that there is not a real 'market' in number of hours worked; that is, that current numbers of hours worked do not reflect workers' preferences concerning the balance between income and leisure time. In countries where unions are stronger, the creation of a market in hours worked is easier because unions can bargain for workers as a whole. Some employees will of course prefer to maintain current

numbers of working hours (and pay levels) because of the high cost (for them) of day-to-day living.

A number of strategies have been suggested to make it easier for workers to choose to work less, but not all of these will necessarily be accompanied by a reduction in material consumption, which is the motive here.[3] The strategy most likely to increase the preference of employees for shorter working hours is to gradually restructure society so as to reduce the cost of day-to-day living. This would include measures to encourage the provision of more affordable housing (such as smaller housing with more efficient use of space, as discussed in Chapter 4, subsection 4.2.7) and provision of sufficiently good public transportation alternatives to travel by automobile, so that foregoing the purchase of an automobile becomes a viable option for a larger segment of the population. These strategies will also directly reduce energy consumption.

Finally, Schor (2005) suggests that as the economy shifts to more leisure time, there will be less demand for speed and convenience. As many of the devices that provide speed and convenience are energy intensive, this would be equivalent to a negative rebound on energy use (that is, an initial measure to reduce energy use (reduced working hours) leads to further reductions in energy use, rather than eroding the initial savings in energy use).

A gradual reduction in consumer spending would not lead to a noticeable increase in unemployment if it is brought about by working less. If, over time, there is a societal shift to less emphasis on material consumption as a source of happiness, there would presumably be a desire to work less on average, so the need for work could be redistributed among more people. The result would be lower economic growth (that is the intention), but also, more free time, less stress, less useless and counterproductive material consumption and less waste.

11.2.3 Limiting economic growth through a focus on product longevity

As pointed out by former World Bank economist Herman Daly, GDP is largely a measure of the *throughput* of material goods and energy in the economy, rather than a measure of the physical stock of assets that can be used and enjoyed (Daly, 1996). The faster these assets wear out, the harder we have to work – the greater the GDP – in order to maintain a given quantity of assets. If assets were to last longer, they would not have to be replaced as often and, depending on the effort required (and hence cost) to make longer-lasting goods, we would not have to work as hard. GDP would fall but there would be no reduction in what matters (the stock of assets) while life would be better because we would have more leisure time and less stress. In many instances, goods are discarded not because they have worn out, but because we have grown tired of them or they have gone out of 'fashion'. For example, according to studies cited by Cooper (2005), one third of discarded appliances in the UK are still functional and another third could be repaired, while 77 per cent of discarded upholstered furniture could theoretically be refurbished and reused.

A factor that has led to a decrease in repairing of broken items in countries such as the UK or the US is that repair work is done using high-cost local labour, whereas manufacturing is done on a large scale elsewhere using low-cost labour. The relocation of

manufacturing to low-income countries may also have led to a loss of workers skilled enough to be employed in repair shops. To increase the lifespan of consumer products, there must be a greater emphasis on designing products for longevity, quality and greater ease of repair, and there must be a shift in consumer attitudes. Increased quality of course increases longevity because the product will need repair less frequently and also because purchasing a new high-quality product is more expensive than purchasing a mass-produced replacement, so it is more worthwhile to repair the product. Ax (2001), for example, argues that handcrafted shoes are more likely to be repaired because they are more comfortable and attractive than cheap, mass-produced shoes (which are also inherently difficult to repair). Owners of products will need to develop greater attachment to their possessions, rather than aspiring to update them as soon as new models appear on the market (Cooper, 2005).

11.3 Increasing the price of energy

An important underlying strategy in reducing GHG emissions is to provide a steadily increasing carbon tax that is applied, without exception, to all sources of energy in proportion to their CO_2 emission. The revenues from such a tax can be used to reduce other taxes (such as payroll taxes), to support energy efficiency measures, and to blunt the hardship otherwise imposed on low-income people. Although C-free sources of energy would not be taxed, these energy sources are in general more expensive than fossil fuels. By increasing the cost of fossil fuels to the point where C-free energy sources are competitive, the overall cost of energy will increase and the demand for energy will decrease. Taxing CO_2 emissions is preferable to subsidizing C-free energy sources because subsidization encourages more consumption than if all costs reflected true costs, and so leads to greater overall costs to society (as well as burdening government finances).[4]

The fundamental importance of high energy costs in reducing long-term energy demand is illustrated in Figure 11.5, which shows the relationship between the cost of electricity and the electricity intensity of the economies of wealthy OECD countries. A factor of two greater cost of electricity is roughly associated with a factor of two smaller electricity intensity. This is due in

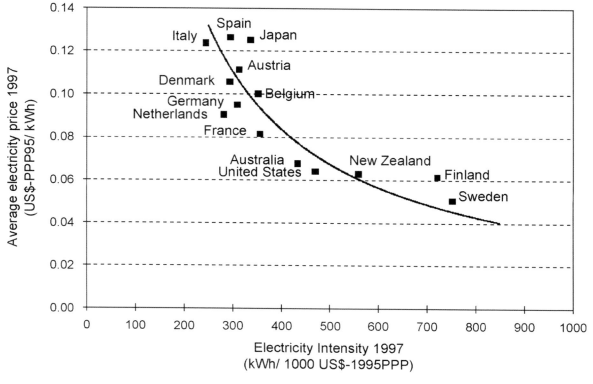

Source: Verbruggen (2006)

Figure 11.5 *Relationship between electricity price and electricity intensity in selected wealthy OECD countries*

part to the decision to locate electricity-intensive industries (such as aluminium smelting) in countries with low electricity costs, with export of electricity-intensive products to countries with higher electricity costs, but this is likely to be a small factor. Rather, as stated by Verbrugeen (2006), one cannot bypass an increase in electricity prices in raising the overall electricity efficiency of a country. High energy costs work to reduce energy demand in two ways: by changing consumer behaviour and purchasing decisions, and by promoting innovations that lead to more energy-efficient equipment.

However, an increase in the cost of energy alone is a crude instrument, imposing unnecessary hardship on some people, sectors and regions and being relatively ineffective if not accompanied by a wide array of supporting measures. Verbruggen (2006) draws an interesting analogy with heat engines: the fundamental limitation in

the efficiency of a heat engine is given by the Carnot Cycle efficiency, which depends on the difference between the input and output temperatures (see Equation (3.16)). Because the minimum output temperature is limited by the ambient temperature, the limit on the efficiency of a heat engine is determined by the temperature of the hot source. The actual efficiency of the heat engine depends on the layout and performance of the machinery in extracting work from the heat flow between the hot source and cold sink. However, even the best-performing machinery cannot provide a high conversion efficiency if the driving temperature difference is small. Similarly, when energy costs are low, there is not enough financial pressure to induce changes, but a well-designed policy apparatus is needed to effectively convert the driving force into real change. In the following sections, we consider elements of well-designed policy packages for promoting energy efficiency.

11.4 Promoting energy efficiency

Across-the-board increases in energy efficiency, both at the scale of individual energy-using devices (for example, boilers and automobiles) and systems (for example, building thermal envelopes and transportation infrastructure) are needed. Extensive discussions of policy options can be found in Geller (2004), Geller et al (2004) and Gillingham et al (2006), as well as in the various chapters of the report of Working Group III of the Fourth Assessment Report of the Intergovernmental Panel on Climate Change (see, in particular, Sims et al (2007) with regard to electricity supply, Ribeiro et al (2007) with regard to transportation, Levine et al (2007) with regard to buildings and Bernstein et al (2007) with regard to industry). Some key elements of effective energy-efficiency policies in different sectors are outlined here.

11.4.1 Buildings and district energy systems

Chapter 4 of this book explained how deep (factors of three to five) reductions in the energy requirements of new buildings can be achieved, compared to current designs, and how factors of two to three (and sometimes much more) reduction in the energy use of existing buildings can be achieved. A more thorough discussion of how to achieve such large energy savings is found in Harvey (2006).

The primary barriers to achieving deep reductions in energy use in new buildings are not technological (because technologies and, more so, system designs, already exist that can provide deep reductions) nor even economic (because fully integrated designs entail very little and often no additional upfront cost compared to current conventional practice, as extensively documented earlier in this book). Rather, the barriers are behavioural in nature: they involve the fragmented nature of the design process and resulting lack of optimization, lack of awareness, time constraints during the design process, and an over-reliance on established ways of doing things. There is a widespread if not universal perception that low-energy buildings must, *of necessity*, entail greater capital costs. This leads to a lack of desire on the part of the client to have a low-energy building, and without a committed client, architects and engineers will usually not undertake the additional design effort required to produce low-energy buildings.

Lack of awareness of energy-savings opportunities among practising architects, engineers, lighting specialists and interior designers is a major impediment to the construction of low-energy buildings. This in part is a reflection of inadequate training at universities and technical schools, where the curricula often reflect the fragmentation seen in the building design profession. There is a significant need, in many countries, to create comprehensive, integrated programmes at universities for training future architects and engineers in the design of low-energy buildings, with parallel programmes at technical schools for training technical specialists. The value of such programmes would be significantly enhanced if they have an outreach component to upgrade the skills and knowledge of practising architects and engineers and to assist in the use of computer simulation tools as part of the integrated design process.

Significant reductions in energy use (factors of two to three overall, up to a factor of 10 in heating requirements) can be achieved in retrofitting existing buildings compared to their pre-existing energy use, but it is generally not possible to achieve as low an absolute energy intensity as can be achieved in new buildings, and such reductions as can be achieved entail greater cost than if buildings are designed from the beginning to require minimal use of energy. Hence, by delaying programmes to dramatically reduce the energy intensity of new buildings, or the energy intensity of old buildings when they require major renovations, significant windows of opportunity will be lost (irreversibly in the case of new buildings).

Some of the key actions required with respect to new buildings are thus:

- upgrading the teaching of building sciences at universities (in architecture and engineering departments), community colleges and vocational schools with, in particular, the creation of comprehensive, integrated programmes that combine all of the elements needed to create truly sustainable buildings;
- developing university-based outreach programmes to improve the design process among practising design firms (architectural and engineering);

- undertaking 'market transformation' programmes so that high-performance buildings, designed using the integrated design process, are increasingly what the market expects. This could entail financial support for high-profile projects that demonstrate large savings at little to no incremental cost (many examples of which already exist and have been documented in this book), incentives to support the additional cost of design using the integrated design process, and training in the integrated design process;
- rapid upgrading of building codes to eventually require a factor of two to three reduction in energy use compared to current practice for new buildings (with incentives to do even better where possible), providing training in meeting upgraded building codes, and providing enhanced inspection ability to increase the degree of compliance with the upgraded building codes;
- providing financial support for continued improvement and reduction in the cost of a wide array of promising advanced technologies that further increase the potential for reducing the energy intensity of buildings, particularly of the large stock of existing buildings.

Policies to support comprehensive retrofits of existing buildings include:

- provision of low- or zero-interest loans to be used for renovations that achieve a given minimum energy performance, or partial rebates based on estimated energy use before and after renovations;
- legislation to permit the involvement of energy service companies (ESCOs) in public sector buildings, including legislation to permit long-term (up to 25-year) contracts and to permit procurement based on lowest lifecycle cost rather than based on lowest investment cost;
- legislation to require upgrading buildings to attain a minimum energy performance at the time of sale.

Barriers to the involvement of ESCOs, ways of overcoming these barriers and various financing mechanisms are discussed in Goldman et al (2005).

For both new and existing buildings, possible policies include:

- partial rebates to reduce the cost of more efficient equipment (such as boilers, cooling systems, motors);
- mandatory rating of buildings in terms of energy use and mandatory supply of this information to prospective buyers of new or existing buildings;
- continuous strengthening of energy efficiency standards pertaining to heating, cooling and ventilation equipment, appliances, consumer electronic goods (such as computers and entertainment systems), office equipment and lighting.

Lebot et al (2004) note that advertising promoting consumption or purchase of larger equipment can (and is) undermining the benefit of energy efficiency programmes. Most efficiency standards are based on different size categories, so that a large piece of equipment with a favourable energy efficiency rating frequently uses more energy than a smaller piece of equipment with a lower rating. They propose that advertisers be required to give the energy consumption of their products in all advertising, and that energy efficiency standards be based on absolute energy use. Energy requirements for housing (in terms of maximum allowed energy use per unit of floor area) also need to be designed so at to be more stringent the larger the house. However, as discussed by Harris et al (2008), the current Home Energy Rating System (HERS) in the US in effect requires smaller houses to have a more efficient furnace than larger houses in order to attain the same score.

Power utilities can be required to provide demand-side management (DSM) programmes, and rate structures (which are subject to government regulatory approval in the case of private utilities) should be designed to reward utilities that achieve deep energy savings for their customers at the lowest possible cost.

At the community scale, government support for district energy systems is required even in circumstances where they make economic and environmental sense.[5] This can entail direct financial involvement, enabling legislation in cases where there are legal barriers and expeditious review and permitting of proposals.

11.4.2 Transportation

The objectives of reducing energy use and GHG emissions from passenger road transportation are likely to be only two of many objectives related to passenger transportation. There are many approaches that can be taken to achieving energy-related and non-energy-related objectives with regard to passenger transportation. The approaches can be grouped into a small number of broad strategies. Box 11.1 outlines 15 possible transportation-related objectives, 4 strategies and 24 specific policy actions. Some strategies may, if carried out in isolation, achieve one objective while hindering other objectives. For example, an increase in fuel economy alone may lead to an increase in driving and an increase in congestion and traffic deaths by

making the cost of driving a given distance less expensive. Imposition of higher fuel economy standards will probably increase the cost of a given vehicle, leading to a delay in the replacement of old (and inefficient) vehicles with newer vehicles. Thus, with regard to the fuel efficiency of the automobile fleet, dramatically higher (factor of two) fuel economy standards should be accompanied by steadily increasing fuel taxes (but perhaps adjustable so as to smooth out the effect of large fluctuations in the price of oil) and possibly some system of rebates on vehicles that are more efficient than the average and surcharges on vehicles that are less efficient. Anecdotal evidence indicates that the sharp spike in the price of oil in 2008 (reaching $150/barrel), which is probably a forewarning of long-term prices, began to affect

---

Box 11.1 Possible policy objectives and approaches related to passenger transportation energy use in urban areas

Transportation policy objectives

1   Reduce air pollution
2   Reduce GHG emissions
3   Reduce traffic congestion
4   Reduce amount of land taken up with parking lots
5   Reduce traffic accidents and deaths
6   Reduce consumer monetary costs for travel (especially as oil prices rise)
7   Reduce commuter stress and commuting time
8   Reduce facility costs (especially for provision of parking)
9   Reduce upstream environmental impacts (oil spills, tar sands in Canada)
10 Reduce potential for war over dwindling oil supplies (more relevant to the US)
11 Conserve non-renewable resources[a]
12 Promote physical fitness and health
13 Improve community liveability
14 Improve the aesthetic appeal of cities
15 Promote local economic development (through more local spending of local income)

Overall strategies related to the objectives outlined above

1   Implement more stringent pollution emission standards
2   Implement more stringent fuel efficiency standards
3   Implement measures to reduce the total number of trips and distance travelled
4   Implement measures to change the modal split (in favour of non-automobile options)

Specific policy options

Related to average fuel economy of the automobile fleet

1   Mandated improvement in average new car fleet fuel economy (CAFE standards)
2   Gas-guzzler purchase tax (to increase market demand for fuel-efficient vehicles)
3   Increase in gasoline tax (to increase market demand for fuel-efficient vehicles and to prevent the effective price of gasoline from falling due to more efficient use)
4   Enforcement of speed limits (improves actual on-road fuel economy)
5   Public education (benefits of engine maintenance, proper tyre pressure, less aggressive driving, no idling)

Related to air pollution emissions

6   Strengthen pollutant emission standards
7   Strengthen fuel quality standards (especially sulphur, cetane and aromatics content)
8   Improve enforcement of emission standards
9   Public education/exhortations

Related to total amount of travel, timing and modal split

10   Urban intensification (smart growth, infill)
11   Major public transportation infrastructure improvements
12   Incremental improvements to public transportation
13   Road pricing
14   Parking management (supply, cost, shared use, unbundling)
15   Pay-as-you-drive insurance
16   Promote car sharing
17   Promote car pooling
18   Promote telecommuting
19   Promote off-peak travel (via variable road pricing, for example)
20   Promote HOV (high-occupancy-vehicle) lanes
21   Improve walking/bicycling environment
22   Promote shift of freight to rail
23   Promote walking school buses
24   Provide student discounts for public transportation

Note: [a] This includes not only oil, but also – in the case of petroleum products made from the Canadian tar sands – the substantial amounts of natural gas that are used in upgrading the tar.

consumer personal vehicle choice, driving, air travel and freight transport. For a further discussion of policy packages that simultaneously achieve multiple objectives, see Litman (2007).

With regard to freight transportation, a combination of mandated fuel efficiency improvements for trucks combined with government support for research and development of new technologies and a steadily increasing carbon tax could achieve significant reductions in energy use per tonne-km of transport. Transport by rail can be encouraged through a steadily increasing carbon tax. At present, aviation fuel is not subject to any tax (and is not subject to any restrictions under the Kyoto Protocol).

Subsidies for the production of ethanol from corn or other starchy crops should cease immediately, as the energy and GHG benefits are small at best, there are other more effective ways to reduce transportation fossil fuel energy use (as outlined in Box 11.1) and the use of biomass for biofuels is the least effective in reducing CO_2 emissions of the various ways in which biomass can be used (see Volume 2, Table 4.56). Biodiesel is potentially attractive but caution is required in stimulating its use due to the potential of negative social and environmental impacts if development is rapid, uncontrolled and not matched to the availability of surplus agricultural land.

A critical factor in reducing long-term emissions from the transportation sector is appropriate urban form (densities and mixes of land use) coupled with the provision of high-quality rapid-transit infrastructure and support for bicycling and walking as alternatives to private motorized transport. For existing urbanized areas, this will require urban intensification, which is subject to a number of non-technical obstacles but is certainly aided by the decreasing economic viability of low-density suburbs in the face of rising transportation fuel costs.

11.4.3 Industry

In the industrial sector, the most effective policies to reduce energy use per unit of output are likely to be negotiated voluntary agreements, government support for long-term research and development, and tax incentives to invest in newer and more efficient equipment. Mandatory standards are likely to be unworkable because of the complexity of industrial energy use and its dependence on highly site-specific conditions (such as the particular mix of products produced at a given facility). Rietbergen et al (2002) deduced that one quarter to one half of the energy savings in Dutch industry since 1992 can be attributed to the long-term agreements that have been the main policy instrument for industrial energy efficiency in Holland since 1992. Social mechanisms, such as the establishment of local social networks with group targets, can be used to motivate management to pay more attention to energy efficiency (Jochem and Gruber, 2007). Information programmes and free energy audits seem to be preferable to long-term agreements for small- and medium-sized industrial firms, as small firms usually do not have the staff or knowledge to design and implement energy efficiency measures (Thollander et al, 2007). Broad support for high rates of recycling throughout society will also contribute to reduced energy use per unit of output, as the energy required to produce metals, plastics and paper from recycled materials is generally much less than making these products from raw materials (see Table 6.4).

11.4.4 Power utilities and cogeneration

In most parts of the world, cogeneration by independent power producers is blocked by electric utilities, or by government regulations (such as the prohibition, in the US, of running private electrical wires across public streets, thereby preventing potential cogenerators of electricity from selling it to potential adjacent customers or even to their own adjacent operations). Opportunities to generate electricity from hot exhaust flows to produce electricity (discussed in Chapter 6, subsection 6.14.1) are also blocked in the same way, and do not qualify in meeting Renewable Portfolio Standards in most US states, even though no incremental fossil fuel energy is required (Casten and Ayres, 2007). In other instances, power utilities can effectively block cogeneration by accepting only highly unfavourable rates for the purchase of cogenerated electricity. Dismantling existing regulatory barriers and the monopoly powers of electric utilities could therefore facilitate the rapid deployment of large amounts of efficient cogeneration capacity.

11.5 Promoting diets low in meat and with less embodied energy

As noted in Chapter 7 (subsection 7.9.3), there is (in the case of Sweden) a factor of four difference in the fossil fuel energy embodied in low-energy and high-energy meals with comparable nutritional value and meat content. Meat itself requires many times more original phytomass than direct consumption of plants for food. The ratio of phytomass energy input to the agricultural system to food energy output is about 1:1 for fresh fruits and vegetables and 1.5:1 for processed foods, but ranges from as little as 6:1 for fatty poultry to over 200:1 for beef (see Table 7.15). Meat-based diets thus require vastly greater amounts of land than a vegetarian diet. In the great majority of cases, this land could instead be used to produce bioenergy crops that could displace fossil fuels. If adjustments are made to account for possible differences in the yield of crops for animal feed versus bioenergy crops, and for differences in the efficiencies with which bioenergy crops and fossil fuels can be used, the phytomass energy input can rightly be added to the direct fossil fuel energy inputs to give the overall potential impact of the consumption of food products on the demand for fossil fuels. When this is done, diet emerges as the single most important factor in determining energy use and GHG emissions that is subject to direct control by consumers.

As discussed in Volume 2 (Chapter 4, subsection 4.9.2), for scenarios with relatively low meat consumption, a significant fraction of future energy needs could be met with bioenergy crops grown on surplus agricultural land. Significant methane and nitrous oxide emissions are associated with livestock (although these can be reduced to some extent through improved feed and better handling of wastes). Promotion of diets with low meat and low embodied energy is thus an important strategy in reducing future GHG emissions. This can be done by:

- eliminating all subsidies for feed grain and livestock producers (equivalent to $2.9 billion/year in the US alone);
- imposing strict environmental and labour standards on livestock operations (thereby increasing market costs while non-market costs decrease);
- perhaps providing financial support for vegetable and fruit production;
- imposing a heavy tax on sugar;
- eliminating all subsidies for fisheries (these currently amount to about $30–40 billion/year worldwide according to Halweil and Nierenberg (2008));
- creating greater public awareness of the health risks of high-meat diets, which have been linked to coronary heart disease, type II diabetes, many forms of cancer and the increase in obesity in the US (Mann, 2002; Key et al, 2004; Chao et al, 2005).

Consumption of food is a social and cultural expression of individuals, but the development of the corporate food economy has increased the separation of people from their sources of food (Levkoe, 2006). In many cities, grassroots organizations have arisen to counter this tendency through the promotion of urban agriculture, markets for local producers and awareness of the environmental and social-justice implications of alternative food choices. The development of youth community gardens has been shown to positively influence dietary behaviour and to enhance environmental awareness and appreciation

(Lautenschlager and Smith, 2007). Government support for such grassroots movements could therefore contribute to a long-term social transition to greater awareness of the environmental (and social) implications of alternative dietary choices, and could contribute to a decline in per capita meat consumption. Such support would parallel the support for Simplicity Circles and similar groups advocated above.

11.6 Research and development in energy efficiency and renewable energy

Most of the technologies needed to achieve dramatic reductions in energy use and a significantly increased use of renewable energy already exist and have been used successfully in at least some jurisdictions. There is, nevertheless, a role for continued technological development in reducing the cost and increasing the market penetration and technical performance of many technologies. However, government support for research and development in the areas of energy efficiency and renewable energy has been stagnant for the past decade, in spite of increasing awareness of the urgency of dealing with the global warming problem. Figure 11.6a shows the variation in total energy research and development spending by selected governments between 1974 and 2006.[6] There was a major spike in research and development during the time of high oil prices in the late 1970s and early 1980s, but (with the exception of Japan) this support fell when the price of oil fell. Total research and development spending (in 2005$) among 26 out of 30 IEA countries[7] peaked at $18.2 billion in 1980, fell to a minimum of $8.3 billion in 1997, then rose modestly to $10.2 billion in 2006. However, during this time period, energy research and development has been overwhelmingly directed toward nuclear energy, as shown in Figure 11.6b. Figure 11.7 shows the relative shares of energy research and development support in 2006 given to nuclear energy, fossil fuels, renewable energy, energy efficiency, hydrogen and fuel cells, electricity (including transmission) and energy storage, and other (which includes energy systems analysis). Nuclear energy (which cannot make a significant contribution to reducing GHG emissions at a global scale according to the analysis presented in Volume 2, Chapter 8, section 8.11) took 39 per cent of total

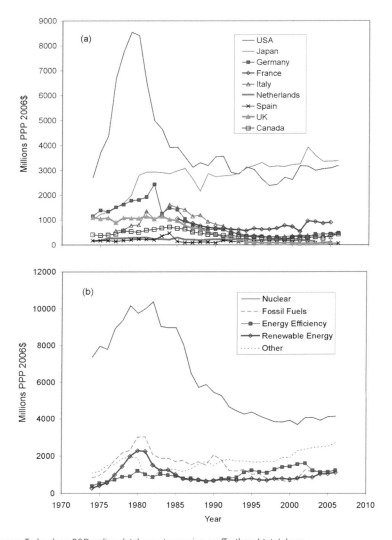

Source: Data from IEA Energy Technology R&D online database at www.iea.org/Textbase/stats/rd.asp

Figure 11.6 *Government support for energy research and development from 1974 to 2006: (a) for selected countries belonging to the IEA, and (b) by energy sector*

energy research and development funding by IEA governments, while energy efficiency and renewable energy (the focus of this book and of Volume 2 because of their overwhelming importance) together took only 24 per cent of total funding.[8]

Given the urgency of the global warming problem and the lack of significant increases in government funding of energy research and development during the past decade (particularly in comparison to the response

to the 1970s' oil price shock), there is a clear need to dramatically increase funding levels, *particularly in the areas of energy efficiency and renewable energy.* Indeed, substantial increases in the funding of these critical areas could be accomplished by simply redirecting a significant fraction of the disproportionately large nuclear support into these areas. Funding increases should be accompanied by measures to increase the rate of uptake and diffusion of innovations in the energy

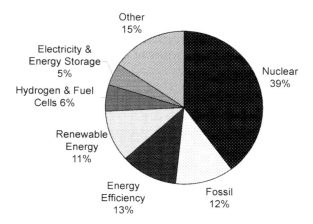

Total = $10.4 billion

Source: Data from IEA Energy Technology R&D online database at www.iea.org/Textbase/stats/rd.asp

Figure 11.7 *Allocation of energy research and development funding by IEA member governments in 2006*

sector, as discussed by Gallagher et al (2006). Kammen and Nemet (2007) argue that increased government research and development support in the energy sector would leverage additional research and development from the energy industry (whose funding has also been falling recently, at least in the US).

For the sake of completeness, an overall discussion of where research priorities should lie both with regard to energy efficiency (the subject of this book) and with regard to renewable energy (the subject of Volume 2) is presented next. The priorities listed below are based on my expectations of where important reductions in the cost and improvements in the technical performance of technologies already under development could occur.

In the buildings sector, priorities are:

• LED lighting;
• vacuum insulation (especially useful in retrofit applications);
• phase-change materials in insulation;
• smart (electrochromatic) windows;
• heat pumps using non-halocarbon working fluids for integrated space and water heating and air conditioning;
• solar air conditioning and combined solar space and hot-water heating;[9]
• solar retrofits of existing buildings.

In the transportation sector, priorities are:

• batteries and ultra-capacitors for use in plug-in hybrid vehicles and as part of a V2G system to provide storage for renewably generated electricity;
• improved performance and emissions of light- and heavy-duty diesel vehicles;
• lightweight and strong materials for road and air transportation;
• fuel cells for various transportation applications;
• development of biofuels from lignocellulosic feedstocks.[10]

In the industrial sector, the priority is fundamental research pertaining to chemical processes and nanotechnology, although there are promising possibilities for further improvement in energy efficiency throughout the industrial sector (such as outlined in Chapter 6, subsections 6.7.7 and 6.9.3).

In the agricultural sector, the priority is research on how to maintain current yields (or close to current yields) in low-energy (such as organic) agricultural systems in all regions of the world.

In the electricity supply sector, priorities are:

• crystalline and thin-film silicon modules;
• concentrating solar thermal electricity production;
• compressed air underground storage for use with wind;
• advanced gasification and cogeneration using biomass fuels;
• enhanced geothermal systems;
• ocean wave, tidal, current and thermal energy systems;
• carbon capture and storage (to accelerate the reduction in CO_2 emissions while fossil fuels are being phased out, and for later application to biomass carbon so as to create negative emissions).

The technology dynamics literature identifies the possibility of two different future technology paths: the first, a carbon-intensive path, in which gaseous and liquid fuels are made from coal and replace conventional gaseous and liquid fuels as they are depleted (this assumes that the available coal resource is large, an assumption that is questioned in Chapter 2, subsection 2.5.3); and the second, a low-carbon path, in which biomass, solar, wind and other renewable energy sources gradually replace fossil fuels as conventional fuels are depleted. Because initial developments along either path will lead to cost reductions through learning-by-doing, it will be very difficult to switch from one path to another once we have started down one path. This phenomenon is referred to as 'carbon lock-in' and already characterizes the present energy system (Unruh, 2000, 2002). Thus, if we do not want a carbon-intensive future, or if we do not want to begin travelling a path that will have to be aborted at great cost, then governments should *not* support research and development of various processes for making gaseous and liquid fuels from coal. If the mineable coal resource is as small as suggested in Chapter 2, this path will have to be aborted in any case, so the investment will largely be wasted, additional CO_2 will have been emitted (even if carbon capture and storage of large point sources can be widely deployed) and the development of a sustainable energy system unnecessarily delayed.

11.7 Rebound effects

It is widely recognized that improvements in energy efficiency, while directly reducing energy use, indirectly lead to an increase in the demand for energy that partly undoes some of the direct energy savings. This indirect increase in energy use is referred to as the *rebound effect*. As discussed by Greening et al (2000), four different rebound effects are recognized:

- a *direct rebound effect*, through the decrease in the cost of energy services due to the need for less energy, leading (usually) to a greater demand for that service;
- an *indirect rebound effect*, whereby the money saved due to less expenditures on energy is spent on other goods and services that in turn require energy;
- a reduction in the *real* price of energy due to less total demand for energy (this is referred to by economists as a *general equilibrium effect*, or economy-wide effect due to changes in the prices and consumption of goods and services throughout the economy);
- *transformational effects*, which are changes in consumer preferences, alterations of social institutions and changes in the organization of production.

Various studies indicate that, in aggregate, these rebound effects are small – reducing the energy savings by at most a few per cent (Laitner, 2000), in the order of 5–15 per cent (Schipper and Grubb, 2000), no more

than 25 per cent (Binswanger, 2001), or around 26 per cent (Barker et al, 2007). This is because, for most activities, energy costs represent only a few per cent of the total costs. However, where energy use is strongly constrained by price (such as heating of poorly insulated houses of poor people), an increase in energy efficiency (such as more insulation and better windows) will be offset to a much greater extent by increased provision of energy services (warmer indoor temperatures in the case of poorly insulated houses).

11.7.1 Reducing or reversing the rebound effect

The analyses of the rebound effect, cited above, assume that the energy savings lead to net cost savings. In reality, equipment or actions to reduce long-term energy use often require a greater upfront investment than less efficient equipment, and payments on this extra investment will offset some or all of the energy cost savings. For example, higher levels of insulation and high-performance windows in residential buildings generally entail greater overall building costs, which can be financed through an increase in the monthly mortgage payments. Similarly, more fuel-efficient automobiles would cost more for the same performance (such as acceleration) and amenity value (such as size) as less fuel-efficient vehicles. If energy-saving measures are pushed to the point where they are just cost effective over the lifetime of the measure (that is, such that annual energy cost savings equal annual financing costs including interest), then the sum of the first two rebound effects will be zero or close to zero. If energy-saving measures are implemented that exceed that which is cost effective in purely monetary terms, then there would be a *negative* net rebound – energy will be saved due to the net reduction in yearly funds available to be spent on other things, in addition to the savings arising from more efficient equipment.

This would occur in the case of high-performance windows, for example, that replace low-performance windows in part for aesthetic reasons. If a consumer replaces windows with high heat loss with high-performance windows because they look better, and spends more than can be economically justified based on energy cost savings, the consumer is spending money to buy an intangible service (better-looking windows). The energy cost of this service is the extra embodied energy beyond that of a window that would just be cost effective. If the ratio of the extra embodied energy to the extra cost (MJ/$) is less than that of the goods and services that the extra money would otherwise be spent on, then there will be a negative rebound effect.

The relationships between energy and financial costs and savings, and the rebound effect, are illustrated in Table 11.1 for the case of the replacement of low-performance windows with high-performance windows. Window costs, embodied energy and annual energy savings roughly correspond to a retrofit recently performed on the author's house. This retrofit was not cost effective from a purely economic point of view at the retail cost of heating energy (natural gas) at the time of the retrofit ($11/GJ), but lead to a marked improvement in the appearance of the house and in convenience (the replacement windows on the second and third floors can be easily cleaned on the outside). A cost recovery factor (see Appendix D) can be computed based on the interest rate and assumed time period over which the purchase of the window is financed; this times the purchase cost of the window gives the annual cost of the window including interest (the interest rate corresponds either to the interest rate on the home

mortgage, or the foregone interest on an alternative risk-free investment). The direct energy savings due to the window is the reduction in heating energy use. The indirect effect of the window on energy use consists of that due to the reduction in annual energy costs (the first rebound effect, which is an increase in energy use because the savings in annual heating costs can be spent in other ways) and that due to the initial purchase cost of the window and the cumulative interest paid until the principle is paid back (money used to pay for the window plus interest cannot be used on plane trips, for example). The reduction in energy use due to the purchase cost plus interest is equal to the annual financing cost of the window times the number of years over which the window is financed (this gives the cumulative cost of the window, including interest) times the energy intensity (MJ/$) of alternative expenditures.[11] For the assumptions given in Box 10.1, and assuming that the money spent on the window would have been spent on other products with the *average* Canadian energy intensity (10MJ/$), the first and second rebound effects together form a negative rebound of about 40 per cent if the window is financed over a 20-year time

Table 11.1 *Analysis of the rebound effect associated with the purchase of energy-efficient windows*

General inputs		
Window cost ($/m^2)	450	
Window embodied energy (GJ/m^2)	2.0	
Reduction in annual heat loss (GJ/m^2/yr)	0.59	
Efficiency of heating system	0.90	
Annual heating energy savings (GJ/m^2/yr)	0.65	
Cost of heating energy ($/GJ)	11	
Annual energy cost savings ($/m^2/yr)	7.18	
Energy intensity of window (MJ/$)	**4.44**	
Canadian GNP in 2005 (trillion $)	1.079	
Canadian primary energy consumption in 2005 (EJ)	10.85	
Average energy intensity of Canadian expenditures (MJ/$)	**10.06**	
Specific inputs and results		
Interest rate	0.04	0.04
Financing period (yr)	20	50
CRF (cost recovery factor)	0.074	0.047
Annual cost of window ($/m^2)	33.11	20.95
Annual energy cost savings ($/m^2)	7.21	7.21
Net annual cash flow ($/m^2)	−25.90	−13.74
Direct energy savings during financing period (GJ/m^2)	13.11	32.78
Indirect energy savings due to purchase of window (GJ/m^2)	6.66	10.53
Indirect energy savings due to reduced annual energy cost (GJ/m^2)	−1.45	−3.63
Indirect energy savings (GJ/m^2):	5.21	6.91
Overall energy savings (GJ/m^2)ª:	16.32	37.68
Rebound effect (%)	−39.73	−21.07

Note: The analysis presented here is based on the total window cost and embodied energy, which is appropriate when it is not strictly necessary to replace the window. Otherwise, the incremental cost and embodied energy when a high-performance window is chosen should be used.

period, and 20 per cent if financed over a 50-year period.[12] That is, the reduction in energy use due to the author having less money available to spend on other things adds 20–40 per cent to the energy savings arising directly from the better windows.

The policy implications of the above examples are that:

- governments should push energy efficiency measures to the point of cost effectiveness (so that there is no net annual cash flow savings and hence no rebound effect);
- governments should promote energy efficiency measures on the basis of their non-economic co-benefits (so that they are done even when there is a net monetary cost, thereby creating a negative rebound), as well as on the basis of their economic benefits; and
- incentives for energy-efficient consumption need to be coupled with incentives to work less (as discussed in subsection 11.2.2) or increases in the cost of energy (through a carbon tax, for example), so that there is a decrease in overall consumption.

Note that if energy efficiency is motivated by an increase in the cost of energy, then energy efficiency measures serve to reduce the impact of this cost increase. If the cost increase is so large that large parts of the population would be forced to do without some energy services, or to cut back in essential consumption (such as food purchases), then the effect of efficiency improvements is to permit maintenance of close to the previous energy service levels while reducing energy use relative to the previous rate of energy use. In this case there would be no rebound relative to previous rates of energy use.

Alfredsson (2004) discusses the rebound on CO_2 emissions associated with a shift to a 'green' diet (i.e. one with less meat and processed foods), which can cost less as well as reduce the energy required to supply and cook the food. He finds, using Swedish data on the marginal propensity to spend (that is, how additional income tends to be spent), that the CO_2 emissions associated with the reallocation of the saved money exceeds the direct reduction in emissions associated with the shift to a green diet. That is, the rebound effect is greater than 100 per cent. This is because the difference in kgC/$ between green and non-green diets is less than the average kgC/$ of the saved money that is spent in other ways. However, this analysis does not take into account

the additional biomass energy that could, in principle, be produced on the surplus land that becomes available with a shift from a high-meat to a low-meat or vegetarian diet (see Volume 2, Chapter 4, subsection 4.9.2).

Furthermore, 'green' diets may cost more (particularly diets consisting of organically certified food) but are also healthier. Thus, promotion of green but more expensive and less energy-intensive diets on the basis of health will provide additional environmental benefits through the reduction in money available to be spent by consumers on other products or services (which unavoidably entail the use of energy, whether direct or indirect).

11.7.2 Non-economic constraints on the use of energy

Hofstetter et al (2006) have pointed out that cost is only one of several factors that constrain energy-using activities. They list five limiting factors that can cause positive or negative rebound effects: cost, time, space or volume, skills and information.

If a low-energy activity requires more time, then the time available for more energy-intensive activities will decrease, leading to a further reduction in energy use. This is a negative rebound effect – subtracting further from energy use (adding to the initial energy savings), rather than adding energy use. Examples include bicycling instead of driving (although sometimes bicycling is faster!) or promotion of leisure activities such as gardening through the provision of rooftop gardens. Rooftop gardens have very little direct impact on the heating and cooling loads of buildings and only a modest impact on transportation energy use if food is grown in place of food imported from elsewhere, but they could induce large indirect impacts if, for example, the owners of the garden tend them on weekends instead of driving to the countryside. Similarly, if parking lots are converted to gardens in order to discourage car use, there could be a sizeable negative rebound. To the extent that greater insulation levels in the walls of houses reduce the available interior space, residents will tend to buy less material goods – or perhaps smaller goods – as there will be less space to store them. Acquisition of a pet such as a dog increases energy use (through the provision of dog food), but this could probably be overwhelmed by the loss of time for energy-intensive activities due to the time required to walk the dog one or two times per day.

11.7.3 Sufficiency as a complement to efficiency

The fundamental problem with a focus solely on efficiency is that, in the face of unlimited human demands, greater efficiency makes it possible to produce or obtain more, rather than reducing environmental impacts and the rate of resource depletion at the pre-existing level of consumption. For this reason, the widespread recognition of the need for *efficiency* needs to be complemented by widespread acceptance of the idea of *sufficiency* – that at some point, we must learn to be satisfied with what we have, an idea that is elaborately developed in Thomas Princen's book *The Logic of Sufficiency* (Princen, 2005). We must learn to strive for what is sufficient rather than the maximum possible that we can, or think we can, attain. Acceptance of sufficiency implies conservation – voluntary consumption of less (in the near term) than one could consume. This in turn implies discipline, an orientation away from material values, and caring for common interests.

For some people, the restrictions on energy use needed to avert future ecological catastrophe might be regarded as a sacrifice. Any required 'sacrifices' will, however, be small and will be compensated by multiple benefits in the present. And certainly, we should not complain about making small 'sacrifices' for the well-being of future generations or for the well-being of the many other species with which we share this planet. In the words of Gandhi in 1925:

> No sacrifice is worth the name unless it is a joy.
> Sacrifice and a long face go ill together.
> He must be a poor specimen of humanity
> who is in need of sympathy for his sacrifice.
> ('Quotes of Gandhi', compiled by Shalu Bhalla, UBS Publishers' Distributors, New Delhi, 2005)

11.8 Reflections on the primal role of economic growth today

The analysis presented in Chapter 10 has underlined the importance of reducing the rate of economic growth in achieving deep reductions in absolute energy use.[13] The need to limit economic growth will be confirmed in Volume 2, in which attempts are made to construct scenarios that meet the global demand for energy and fuels projected under the low and high GDP and the green scenarios of Chapter 10 while reducing CO_2 emissions sufficiently rapidly to limit the peak atmospheric CO_2 concentration to no more than 450ppmv. In order to provide an opportunity for developing countries to improve the material standard of living of their people (which is desperately needed in many parts of the world), economic growth must at least temporarily pause in the already-rich countries. Indeed, a modest downsizing of the economies of the rich countries would be highly beneficial. Earlier in this chapter, a variety of strategies for reducing consumer spending – which is the ultimate driver of economic growth – were outlined. However, the response of the political leaders of the rich countries to the economic crisis that began in late 2008 has been to dramatically and, in some cases, dangerously increase government deficits in an effort to quick-start their economies, hoping to get them back onto a growth trajectory driven by ever more consumer spending.[14] And perhaps without exception, the present world leadership is determined that measures to reduce GHG emissions should not in any way slow down economic growth. Thus, economic growth is primary, climate stabilization – even avoiding future ecological and humanitarian disaster – is secondary.

Why the current fixation on economic growth above all else? Early economists such as John Stuart Mill, and more recently even John Maynard Keynes himself (whose writings are now being invoked to justify massive government stimulus spending) envisaged a time when economic growth would have to stop. Keynes foresaw a time when the 'economic problem' would be solved and 'we prefer to devote our energies to non-economic purposes' (Keynes, 1930).

The reasons for the current universal fixation on economic growth are nicely elucidated in two recent books, *Managing Without Growth: Slower by Design, Not Disaster*, by Peter Victor (2008), and *Prosperity Without Growth: Economics for a Finite Planet*, by Tim Jackson (2009). The fundamental problem is that the current world economy is structurally dependent on growth for stability. If it is not growing, it tends to spiral into depression. Economic growth has been the default mechanism for preventing collapse, and those who question growth have, in the words of Jackson, been regarded as lunatics, idealists or revolutionaries. Because the idea of limits to growth is so foreign to most people, Peter Victor goes to great length – taking up almost half of his book – carefully making the case that the rich countries have little choice but to learn how to manage without growth. His argument is not that we should adopt zero growth as an overarching policy objective, but rather, that we should not bother with growth as a policy objective at all, or at least only as secondary to more specific objectives that have a clearer connection to human well-being. Avoiding catastrophic climatic change should certainly be one of those objectives.

However, for the past few decades, government policies of all sorts have been evaluated in part (some would say largely or entirely) through how they contribute to economic growth. In the words of Tim Jackson (2009, p6):

> the growth imperative has shaped the architecture of the modern economy. It motivated the freedoms granted the financial sector. It stood at least partly responsible for the loosening of regulations and the proliferation of unstable financial derivatives. Continued expansion of credit was deliberately courted as an essential mechanism to stimulate consumption growth.

In words of Peter Victor (2008, pp1–2):

> economic growth is the policy objective against which all other proposals must be judged. Environmental policy must not be allowed to impede growth, and where possible should be advocated because it will boost growth… Education policy must see that students are trained for work in the 'new economy'. Transportation policy should result in a more rapid movement of goods. Immigration policy should attract the most highly educated and the wealthiest to meet the needs of a growing economy. Support for the arts is based on [their] economic contribution. All are judged against their contribution to growth.

What is needed is to move climatic change and reduction of GHG emissions to the place now held by economic growth. All other policies must now be evaluated, at least in part, in terms of how they contribute to the overarching goal of reducing GHG emissions. A price needs to be set on carbon that is initially high enough, and increases at a sufficiently rapid rate, that climate-related targets – such as limiting atmospheric CO_2 concentration to no more than 450ppmv and then drawing down the concentration to 350ppmv or less – are met.[15] How high this price needs to be depends in part on how effective governments are in facilitating the transition to much greater levels of energy efficiency (through the policies outlined in Sections 11.4 and 11.6)

and in facilitating the transition to C-free energy sources (discussed in Volume 2). The economy will then adapt, with the economic impact of a high carbon tax (or revenues from auctioning limited emission permits) strongly dependent on how the revenues are recycled. However, some of the massive carbon revenues will surely be needed to pay down some of the massive government debt that continues to build up.

With economic growth no longer emphasized in government policy, and with both consumers and governments living within their financial means (and moving in the direction of living within the planet's ecological and biophysical means), economic growth will slow. The fundamental problem is that there is very little understanding of how to manage an economy that is not growing, or that grows only slowly. In the words of Tim Jackson (2009, p77)

> virtually no attempt has been made to develop an economic model that doesn't rely on long term growth… we have no model for how common macro-economic 'aggregates' (production, consumption, investment, trade, capital stock, public spending, labour, money supply and so on) behave when capital doesn't accumulate… In short, there is no macro-economics of sustainability and there is an urgent need for one.

Initial, tentative steps in this direction have been carried out by Peter Victor (2008) – one of the few economists to begin to study how a non-growing economy would function and maintain stability. Key elements are for increased productivity to flow into decreased working hours (as argued earlier in this chapter), and for a shift from private investment directed toward more consumption, to public investment directed toward social assets. However, the economics profession as a whole urgently needs to address the question of how to maintain stability, solvent government finances and human prosperity in the broadest sense of the word without depending on never-ending economic growth.

This is part of the New Reality posed by the prospect of large, rapid and catastrophic climatic change.

Notes

1   More precisely, as defined in Wikipedia, the fertility rate is the average number of children that would be born to a woman over her lifetime if (1) she were to experience the exact current age-specific fertility rates through her

lifetime, and (2) she were to survive from birth through to the end of her reproductive life. It is obtained by summing the single-year age-specific rates at a given time.

2   The doubling time in years for any quantity that grows exponentially is equal to roughly 70 divided by the percentage growth rate per year. This can be derived from the expression $P(t)/P(0) = \exp(at) = 2.0$, where a is the fractional growth rate per year ($a = 0.035$ for 3.5 per cent/year growth).

3   Examples include reducing the income gap between rich and poor through a more progressive tax structure, provision of free universal health care where not already available, and greater investment in quality public education (Sanne, 2002). If lower-income people work less because they get to keep more of their income, their total expenditures have not likely decreased (although if the tax reduction for lower-income groups is financed by a tax increase on higher-income groups, they will have less to spend on energy-intensive goods and will have an incentive to work less, so there could be a reduction in overall consumption). Although providing free health care and free quality education are laudable goals (and, indeed, can be regarded as basic human rights), they will not lead to a net reduction in spending on energy-intensive goods unless financed by tax increases (as will usually be required) that are sufficiently large to absorb the private funds formerly spent in these areas.

4   There may, however, be circumstances where narrowly focused and carefully designed short-term subsidies might be justified.

5   District energy systems can be powered by both renewable and conventional energy sources, and so are discussed in Volume 2, Chapter 11, but also affect the demand for energy and so are relevant here.

6   The research and development expenditures shown in Figure 11.6 come from the IEA and have been adjusted by the IEA to be in constant 2005$ and to represent comparable purchasing power in different countries.

7   The IEA countries are also members of the OECD. The countries in the research and development tally are: Australia, Austria, Belgium, Canada, Czech Republic, Denmark, Finland, France, Germany, Greece, Hungary, Ireland, Italy, Japan, Korea, Luxembourg, The Netherlands, New Zealand, Norway, Portugal, Spain, Sweden, Switzerland, Turkey, the UK and the US. Not included are Iceland, Mexico, Poland and the Slovak Republic.

8   When supporting technologies such as electricity storage, hydrogen and fuel cells are included, the total is still only 35% of total funding.

9   Discussed in Volume 2, Chapter 2.

10 Discussed in Volume 2, Chapter 4.

11 If the average energy intensity of alternative financial expenditures is equal to the energy intensity of the window (MJ/$, given by the embodied energy of the window divided by the cost of the window), then the negative rebound due to the purchase cost of the window will exactly offset the embodied energy of the window. If the annual energy cost saving equals the annual financing cost of the window (meaning that the window just pays for itself over its lifetime), then the two components of the indirect effect sum to zero. If the energy intensity of the alternative expenditure is equal to the energy intensity of the money spent on the window, the indirect effect is still non-zero if there is a net cost of the window over its lifetime.

12 Surprisingly, the energy intensity of dollars spent on long-haul airline flights is comparable to the average of the Canadian economy: a round trip between Toronto and Paris is a distance of 11,500km at an energy intensity of about 0.6MJ/passenger-km (according to Table 5.3) times about 1.7 to account for embodied energy (according to Table 5.4), giving a total of about 1.0MJ/passenger-km. An economy class ticket costs about $1000 (including all taxes and fees). This works out to about 11MJ/$. However, the carbon intensity for air travel would be substantially greater than the economy average.

13 Some would go further and say that economic growth eventually must cease. There is no disagreement that material and energy flows must eventually be stabilized. The question as to whether or not money flows (i.e. GDP) must also be stabilized hinges on the question of whether or not ever more or newer services can be provided that require less and less material and energy throughputs. I suspect not.

14 This is all the more surprising given that part of the consumer spending that drove growth in the rich countries was dependent on growing levels of consumer debt (financed in large part by saving in China) and so was clearly not sustainable.

15 At the 2009 G8 summit, the leaders of the G8 countries accepted a target of limiting climatic change to no more than a 2°C warming above pre-industrial levels. The problem with temperature-based targets such as this is that they do not tell us what the concentration limits for GHGs should be, and thus what we should do (within a range of uncertainty governed by uncertain climate–carbon cycle feedbacks) in terms of emissions. This is because there is a factor of three uncertainty in the climate sensitivity, which relates GHG concentrations to long-term temperature change. Given the likely climate sensitivity of 1.5–4.5°C, there is about a 90 per cent probability that the 2°C target will be exceeded even if we succeed in stabilizing atmospheric CO_2 at 450ppmv, and about a 10 per cent probability that we will exceed 4°C. I have previously argued (in Chapter 1, subsection 1.4.1) that a sustained warming of as little as 1°C may be sufficient to provoke an eventual sea level rise of 10–12m.

Appendix A
Units and Conversion Factors

All physical quantities except electric charge can be expressed in the fundamental units consisting of distance, time and mass. In the metric system, two sets of units can be used: the mks system (metres, kilograms, seconds), and the cgs system (centimetres, grams, seconds). The metric system is inherently superior to any other system, and is used as the standard in this book (with occasional reference to the British system for readers who are more familiar with that system).

Any other unit can be expressed as a combination of the fundamental units of distance, mass and time. Some of the key derived units are shown below in Table A.1.

Table A.1 *Fundamental units in the mks system*

Quantity	Definition	Units	
		Metric	Fundamental
Velocity	change in distance/time	m/s	m/s
Acceleration	change in velocity/time	m/s^2	m/s^2
Force	mass × acceleration	newtons	$kg\ m/s^2$
Energy, work done	force × distance	joules	$kg\ m^2/s^2$
Power	energy/time	watts (J/s)	$kg\ m^2/s^3$
Pressure	force/area	pascals	$kg/m/s^2$

Table A.2 gives the prefixes that are used with physical units in the metric system.

Table A.2 *Prefixes used with physical units in the metric system*

Prefix	Factor	Common examples
Nano	10^{-9}	nm (nanometre)
Micro	10^{-6}	μm (micrometre)
Milli	10^{-3}	mm (millimetre)
Kilo	10^3	kW (kilowatt), km (kilometre), kg (kilogram)
Mega	10^6	MW (megawatt), MJ (megajoule)
Giga	10^9	GW (gigawatt), GJ (gigajoule), Gt (gigatonne)
Tera	10^{12}	TW (terawatt), TJ (terajoule)
Peta	10^{15}	PJ (petajoule)
Exa	10^{18}	EJ (exajoule)

The conversion factors between British and metric units are given in Table A.3

Table A.3 *Conversion factors between and among metric and British units*

British to Metric	Metric to British
Energy and power	
1 British Thermal Unit (Btu) = 1055 joules	1000 joules = 0.9486Btu
1 calorie = 4.186 joules	
1 therm = 105.5MJ	
1 quad = 1 quadrillion (10^{15}) Btus	
1 horsepower (Hp) = 746 watts	1000 watts = 1.341Hp
1 ton of cooling = heat removal at a rate of 3.516kW (3.516kW$_c$)[a]	1kW$_c$ = 0.2844 tons
1000Btu/hr (used for heating) = 0.2931kW (0.2931kW$_{th}$)	1kW$_{th}$ = 3412Btu/hr
1 kilowatt-hour (kWh) = 1000 watts × 3600 seconds = 3.6 megajoules = 0.0036 gigajoules	
Pressure (Hastings and Møck, 2000)	
1 atmosphere (atm) = 101.325kPa = 1.01325 bars = 14.696 pounds per square inch (psi)	
1Pa (pascal) = 0.102mm water = 4.02 × 10^{-3} inches water = 7.5 × 10^{-3}mm Hg	
0.1 IWC (inches water column) = 25Pa	1 mb (millibar) = 100Pa = 0.402IWC
Distance, area, volume, mass and weight	
1 foot = 0.30480 metres	1 metre = 3.28083 feet = 39.37 inches
1 mile = 5280 feet = 1.6093km	1 kilometre (km) = 0.6214 miles
1ft^2 = 0.092904m^2	1m^2 = 10.7638ft^2
1 acre (1/640mi^2) = 4046.87m^2 = 0.404687ha	1ha = 100m × 100m = 2.47104 acres
1ft^3 = 28.32 litres = 0.02832m^3	1m^3 = 1000 litres = 35.3145ft^3
1SCF[b] (at 70°F) = 38Nm^3 (at 0°C)[c]	1Nm^3 (at 0°C) = 0.02632SCF (at 70°F)
1 Imperial gallon = 4.546 litres	1 litre = 0.2198 imperial gallons
1 US gallon = 3.785 litres	1 litre = 0.2642 US gallons
1cfm (cubic foot per minute) = 0.4720 litre/s	1 litre/s = 2.119cfm
1 avoirdupois pound (a unit of weight) corresponds to 0.4536 kilograms (kg) on earth. This is what is commonly called 'pound', and is divided into 16 avoirdupois ounces ('ounces'), each corresponding to 28.35gm.	1 kilogram (a unit of mass) corresponds to 2.2046 pounds on earth
1 troy pound corresponds to 0.3732kg on earth, and is divided into 12 troy ounces, each corresponding to 31.10gm.	

Note: [a]The term 'ton of cooling' originates from the time when cooling was measured in terms of the amount of cooling from the melting of a given amount of ice. A ton of cooling is the rate of cooling provided by melting 2000 pounds of ice over a period of 24 hours, which absorbs 12,000Btu of heat per hour. [b] Standard cubic feet, used as a measure of the amount of natural gas. [c] Nm^3 = Normal cubic metres, used as a measure of the amount of natural gas. Pertains to a temperature of 0°C and a pressure of 101.325kPa (1atm). 1Nm^3 ≈ 38MJ.

Appendix B
Heating Values of Fuels and Energy Equivalents

The heating value of a fuel is the heat that is released when the fuel is combusted. The heat released depends on whether the water vapour produced by combustion is condensed (releasing latent heat) or not. The heating value in the absence of condensation is referred to as the lower heating value (LHV) or the net heating value, while the heating value in the presence of condensation is referred to as the higher heating value (HHV) or gross heating value. The difference between the two depends on the hydrogen content of the fuel, as this determines how much water vapour is produced in relation to the energy released by combustion. LHV and HHV values, the per cent of hydrogen on a mass basis and the ratio of LHV to HHV are given in Table B.1 for a variety of fuels.

Table B.1 *LHV, HHV per cent hydrogen by mass and density of various fuels*

	LHV (MJ/kg)	HHV (MJ/kg)	Mass % H	LHV/HHV	Density (kg/L)	Source
			Solid fuels			
Lignite				0.90		
Subbituminous coal		25–30		0.94		
Bituminous coal		30–37	5	0.96		
Anthracite coal		35–37	3–4	0.98		
			Liquid fuels			
Gasoline (C_5H_{12} to $C_{12}H_{26}$) Average: $C_{7.14}H_{14.28}$, 85.5% C	44.0	44.35	15.3–16.7	0.99	0.737	Demirbas (2005)
Diesel ($C_{10}H_{22}$ to $C_{15}H_{32}$) Average: $C_{13.57}H_{27.14}$, 87.0% C	43.0	45.9	15.1–15.5	0.94	0.85	Farrell et al (2006), wikipedia (density)
Kerosene	43.1	46.2		0.92	0.78–0.81	Wikipedia
Methanol (CH_3OH) (37.5% C)	19.9	21.08	12.5	0.94		Demirbas (2005)
Ethanol (C_2H_5OH)	26.7	27.61	13	0.967	0.796	Demirbas (2005)
			Natural gas, natural gas liquids and hydrogen			
Methane (CH_4) (75% C)	50.02	55.50	25	0.900	0.7174	Hiete et al (2001)
Ethane (C_2H_6)	47.48	51.87	20	0.915	1.3553	Hiete et al (2001)
Propane (C_3H_8)	46.35	50.34	18.1	0.921	2.0102	Hiete et al (2001)
n-Butane (C_4H_{10})	45.71	49.50	17.2	0.924	2.7030	Hiete et al (2001)
Iso-butane (C_4H_{10})	45.62	49.40	17.2	0.923	2.6912	Hiete et al (2001)
Hydrogen (H_2)	120.2	141.9	100	0.85	0.08988[a] 70.8[b]	

Note: [a] As a gas at STP. Under compression, density would increase in proportion to the pressure. [b] As a liquid at −253°C.

Energy supply and use statistics are often given in terms of millions of tonnes of coal equivalent (Mtce), or millions or billions of tonnes of oil equivalent (Mtoe or Gtoe). Oil extraction and consumption data are invariably given in terms of barrels of oil. The energy equivalents of these quantities are as follows:

- 1 Mtce ≈ 28 petajoules (PJ);

- 1 Mtoe ≈ 42PJ (41.868PJ according to Hiete et al, 2001);
- 1 billion tonnes of oil equivalent (Gtoe) ≈ 42EJ;
- 1 tonne of crude oil = 7.33 barrels;
- 1 barrel (oil) = 42.0 US gallons = 5.75GJ.

The energy equivalent of a barrel of oil is computed based on the following: 1 barrel = 42.0 US gallons = 159.0 litres, while crude oil has a LHV of 42.8MJ/kg and a density of 0.845kg/litre.

Appendix C
Definitions of Country Groups

The following country groups are referred to at various places in this book:

- OECD (Organisation for Economic Co-operation and Development), consisting of the following 30 countries: Australia, Austria, Belgium, Canada, Czech Republic, Denmark, Finland, France, Germany, Greece, Hungary, Iceland, Ireland, Italy, Japan, Korea, Luxembourg, Mexico, The Netherlands, New Zealand, Norway, Poland, Portugal, Slovak Republic, Spain, Sweden, Switzerland, Turkey, the UK and the US.
- OPEC (Organization of Petroleum Exporting Countries), consisting of the following 12 countries: Algeria, Angola, Ecuador, Iran, Iraq, Kuwait, Libya, Nigeria, Qatar, Saudi Arabia, the United Arab Emirates and Venezuela.
- EU-15 – prior to May 2004, the European Union consisted of the following 15 countries: Austria, Belgium, Denmark, Finland, France, Germany, Greece, Ireland, Italy, Luxembourg, The Netherlands, Portugal, Spain, Sweden and the UK. This group of nations is commonly referred to as the EU-15.
- EU-25 – on 1 May 2004, the EU-15 became the EU-25 with the addition of the following ten countries: Estonia, Malta, Latvia, Lithuania, Poland, Czech Republic, Slovak Republic, Hungary, Slovenia and Cyprus.
- EU-27 – on 1 January 2007, the EU-25 became the EU-27 with the addition of Bulgaria and Romania.
- EU27+2 – this term refers to the EU-27 plus Norway and Switzerland.
- EIT – economies in transition, generally those of eastern Europe and the former Soviet Union

Appendix D
Financial Parameters

In this Appendix a number of parameters relevant to assessing the costs of alternative investments, or the benefits of measures that reduce energy costs, are introduced. For a more detailed discussion with examples, see Chapter 5 of Masters (2004).

Cost recovery factor and annualized investment cost

The *cost recovery factor* (CRF) is the amount of money that must be repaid every year to exactly repay an initial load or investment over a period of N years, including interest on the amount still owing at any given time. It depends on the real interest rate expressed as a fraction, i, and the repayment time N as follows:

$$CRF = i / \left(1 - (1+i)^{-N}\right) \qquad (D.1)$$

CRF times the initial investment I_o gives the fixed annual payment needed to exactly pay back the investment plus interest after N years. This is referred to as the *annualized investment cost*. If one starts with an initial investment, calculates the interest accumulated the first year based on the amount owing at the start of the year, subtracts $CRF \times I_o$ at the end of the year, then recomputes the annual interest at the start of the next year based on the new balance and continues in this way, one will arrive at exactly zero funds owing after N years.

Net present value and net present value factor

The *net present value* (NPV) of some income or cost occurring n years in the future is the amount of money that would have to be invested today, at an interest rate i, in order to have an amount of money equal to that future income or cost after n years. A dollar invested now is worth $1 \times (1+i)$ dollars one year from now, $1 \times (1+i)^2$ dollars two years from now, and so on (multiplying by a factor of $1+i$ for each additional year). X dollars now is worth $X(1+i)^n$ dollars n years in the future, so X dollars n years from now is worth only $X/(1+i)^n$ dollars today.

For a string of N annual costs or savings of equal value spread over N years, the NPV of the string is equal to $N \times F_{NPV}$, where F_{NPV} is the *net present value factor* and is given by

$$F_{NPV} = \frac{(1+i)^N - 1}{(1+i)^N i} \qquad (D.2)$$

where i is the discount (interest) rate. The NPV factor is equal to the reciprocal of the cost recovery factor (given by D.1).

The NPV of N years of savings computed using F_{NPV} is the same as would be obtained if one computed the NPV of each annual savings and added them all up, where the NPV of the savings X in year n is equal to $X/(1+i)^n$.

Simple payback

If an investment in energy efficiency (for example) entails a first cost of C but generates an annual saving in energy costs of S, the simple payback time is C/S. This is the length of time required for the energy cost savings to pay back the original investment, ignoring interest on the initial investment while it is being paid back.

Simple rate of return

The simple rate of return is the reciprocal of the simple payback period; that is, the annual cost savings divided by the initial investment.

Internal rate of return

If an initial investment I_o generates a string of cost savings over a period of N years that has a NPV S, then the NPV of the investment is:

$$NPV = S - I_o \qquad \text{(D.3)}$$

The *internal rate of return* (IRR) is the rate of interest that has to be used in calculating the NPV of all the future savings such that the NPV of the investment (given by Equation D.3) is zero. In this case, the cost savings just pay back the initial investment plus interest (at an interest rate equal to the IRR) over the N years. There is no simple, general formula for calculating the IRR; rather, it has to be determined through trial and error or through an iterative calculation procedure.

For projects with a long lifespan relative to the simple rate of return, the simple return of return is a good approximation to the IRR.

Impact of fuel price escalation different from the general rate of inflation

The interest rate used in the previous formula should always be the real (or nominal) interest rate minus the rate of inflation. This gives the same results as if the real interest were used combined with escalation of future costs and income at the rate of inflation. However, if the future rate of increase in the cost of energy is different from the general rate of inflation (which is used to reduce the effective interest rate), incorrect results will be obtained. Instead, an effective interest or discount rate i' should be used, computed as follows:

$$i' = \frac{i - e}{1 + e} \qquad \text{(D.4)}$$

where e is the rate of escalation in the price of energy.

Appendix E
Alternative Measures of Transportation Fuel Efficiency

From the information in Table A.3, the following conversion factors can be deduced: litres/100km = 235.23/mpg (miles per US gallon) and mpg (miles per US gallon) = 235.23/(litres/100km).

Thus, if you have fuel economy in miles per gallon, divide 235.23 by mpg to get litres/100km. Alternatively, if you have fuel consumption in litres/100km, divide 235.23 by litres/100km to get mpg.

Conversion factors from litres/100km to $gmCO_2$/km can be computed as follows for gasoline:

$$1 \text{ litre} = 32.3\text{MJ (LHV)} = 631.3\text{gmC}$$
$$= 2314\text{gm } CO_2; 1\text{MJ} = 19.6\text{gmC} \quad \text{(E.1)}$$

so:

$$\text{litres/100km} = (gmCO_2/\text{km})/23.14$$
$$= (\text{gmC/km})/6.313$$

$$gm\ CO_2/\text{km} = (\text{litres/100 km}) \times 23.14$$
$$\text{gmC/km} = (\text{litres/100km}) \times 6.313 \quad \text{(E.2)}$$

For diesel fuel:

$$1 \text{ litre} = 35.8\text{MJ (LHV)} = 744.6\text{gmC}$$
$$= 2731\text{gm } CO_2; 1\text{MJ} = 20.8\text{gmC} \quad \text{(E.3)}$$

so:

$$\text{litres/100km} = (gmCO_2/\text{km})/27.31$$
$$= (\text{gmC/km})/7.446$$
$$gmCO_2/\text{km} = (\text{litres/100km}) \times 27.31$$
$$\text{gmC/km} = (\text{litres/100km}) \times 7.446 \quad \text{(E.4)}$$

Appendix F
Websites with More Information

Listed here are websites related to energy that I have found to be particularly useful. In addition, most national governments have general websites related to energy use and their own energy efficiency programmes.

Up-to-date and reliable information on global warming

The IPCC is the definitive source of information on the science of global warming and its impacts. Full assessment reports, special reports, technical reports and summaries for policymakers can be freely downloaded from www.ipcc.ch.

The global warming denial industry has not and probably will not die. This industry continually recycles long-since-discredited arguments against the scientific consensus, and most of its arguments are intellectual fluff, generally relying on misrepresentations, over-simplifications and in some cases outright falsification of the scientific data. One excellent website, continually responding to and correcting the various pseudo-scientific arguments in circulation, is www.realclimate.org.

Energy consumption data

Information for major world regions and countries on extraction and consumption of natural gas, oil and coal over time; refining capacity and output over time; spot prices over time; estimated sizes of regional reserves of natural gas, oil and coal; international trade flows; and hydro, nuclear and total electricity production over time, can be found in the annual *British Petroleum Statistical Review*, which can be obtained (along with data in Excel spreadsheets) from www.bp.com.

Energy-efficient buildings

The IEA coordinates international research in a vast number of areas related to energy supply and demand, including energy-efficient buildings. Relevant websites are: www.ieafuelcell.com (for fuel cell research) www.ecbcs.org (for energy conservation in buildings), www.ieadsm.org (for policies to promote energy efficiency) www.heatpumpcentre.org (for heat pumps), www.iea-eces.org (for energy conservation through energy storage) www.caddet.org (for the Centre for the Analysis and Dissemination of Demonstrated Energy Technologies) Numerous reports and case studies are available on these sites, most of which can be downloaded at no charge.

Some other particularly useful websites are:

*   Advanced Buildings, Technologies and Practices (Canada), www.advancedbuildings.org,
*   Carbon Trust (UK), www.thecarbontrust.co.uk,
*   Energy Design Resources group (California), www.energydesignresources.com,
*   National Association of Home Builders (US), www.toolbase.org, and
*   Solanova ('Solar supported Solar-supported, integrated eco-efficient renovation of large residential buildings and heat-supply-systems', a project of the European Commission in Eastern Europe that started in January 2003), www.solanova.eu.

Websites related to the Passive House Standard

The website of the founding Passive House Institute is www.passiv.de, while further technical information can be found at www.cepheus.de

A database of buildings meeting the Passive House Standard in Austria is at www.HAUSderZukunft.at (many individual case studies have English versions)

Websites of local Passive House institutes are:

*   www.passivehouse.ca in Canada.
*   www.pasivnidomy.cz in the Czech Republic,
*   www.passivehouse.org.uk in the UK, and
*   www.passivehouse.us in the US.

Transportation energy use

For up-to-date developments on fuel-efficient road vehicles, see www.greencarcongress.com

For up-to-date information on policy and trends related to transportation, see the Victoria Transport Policy Institute, www.vtpi.org

References

Accent Refrigeration Systems (2009) *Energy Efficiency Products*, accessed May 2009 from www.accent-refrigeration.com/refrigeration-systems/energy-efficiency-products

Ådahl, A., Harvey, S. and Berntsson, T. (2006) 'Assessing the value of pulp mill biomass savings in a climate change conscious economy', *Energy Policy*, vol 34, pp2330–2343

Adalberth, K. (1997) 'Energy use during the life cycle of single-unit dwellings: Examples', *Building and Environment*, vol 32, pp321–329

ADB (Asian Development Bank) (2001) *Urban Indicators for Managing Cities: Cities Data Book*, ADB, Bangkok, www.adb.org/Documents/Books/Cities_Data_Book/default.asp

Åhman, M. (2001) 'Primary energy efficiency of alternative powertrains in vehicles', *Energy*, vol 26, pp973–989

Akbari, H. and Konopacki, S. (2004) 'Energy effects of heat-island reduction strategies in Toronto, Canada', *Energy*, vol 29, pp191–210

Åkerman, J. and Höjer, M. (2006) 'How much transport can the climate stand? – Sweden on a sustainable path in 2050', *Energy Policy*, vol 34, pp1944–1957

Alfredsson, E. C. (2004) '"Green" consumption – no solution for climate change', *Energy*, vol 29, pp513–524

Alley, R. B., Clark, P. U., Huybrechts, P. and Joughin, I. (2005) 'Ice-sheet and sea-level changes', *Science*, vol 310, pp456–460

Al-Temeemi, A. D. (1995) 'Climatic design techniques for reducing cooling energy consumption in Kuwaiti houses', *Energy and Buildings*, vol 23, pp41–48

Altairnano (2009) *NanoSafe Battery Technology*, product brochure, www.b2i.cc/Document/546/NanoSafeBackgrounder060920.pdf

Alvarado, S., Maldonado, P., Barrios, A. and Jaques, I. (2002) 'Long term energy-related environmental issues of copper production', *Energy*, vol 27, pp183–196

Ampofo, F., Maidment, G. G. and Missenden, J. F. (2006) 'Review of groundwater cooling systems in London', *Applied Thermal Engineering*, vol 26, pp2055–2062

Amstalden, R. W., Kost, M., Nathani, C. and Imboden, D. M. (2007) 'Economic potential of energy-efficient retrofitting in the Swiss residential building sector: The effects of policy instruments and energy price expectations', *Energy Policy*, vol 35, pp1891–1829

Anderson, R., Christensen, C. and Horowitz, S. (2006) 'Analysis of residential system strategies targeting least-cost solutions leading to net zero energy homes', *ASHRAE Transactions*, vol 112, part 2, pp330–341

Ang-Olson, J. and Schroeer, W. (2002) 'Energy efficiency strategies for freight trucking: Potential impact on fuel use and greenhouse gas emissions', *Transportation Research Record*, vol 1815

Annan, J. D. and Hargreaves, J. C. (2006) 'Using multiple observationally-based constraints to estimate climate sensitivity', *Geophysical Research Letters*, vol 33, L06704

Anonymous (1997) 'Tunneling through the cost barrier: Why big savings often cost less than small ones', *Rocky Mountain Institute Newsletter*, vol XIII, no 2, pp1–4

Anonymous (2004) 'Let there be daylight', *ASHRAE Journal*, vol 46, no 1, pp6–8

Anonymous (2006) 'Generation technology', *Power*, vol 150, no 8, pp55–56

Antunes, J. M. G., Mikalsen, R. and Roskilly, A. P. (2008) 'An investigation of hydrogen-fuelled HCCI engine performance and operation', *International Journal of Hydrogen Energy Research*, vol 33, pp5823–5828

Antunes, J. M. G., Mikalsen, R. and Roskilly, A. P. (2009) 'An experimental study of a direct injection compression ignition hydrogen engine', *International Journal of Hydrogen Energy Research*, vol 34, no 15, pp6516–6522

Ardente, F., Beccali, M., Cellura, M. and Mistretta, M. (2008) 'Building energy performance: A LCA case study of kenaf-fibres insulation board', *Energy and Buildings*, vol 40, pp1–10

Argiriou, A. A., Lykoudis, S. P., Balaras, C. A. and Asimakopoulos, D. N. (2004) 'Experimental study of an earth-to-air heat exchanger coupled to a photovoltaic system', *Journal of Solar Energy Engineering*, vol 126, pp620–625

Arkar, C. and Medved, S. (2007) 'Free cooling of a building using PCM heat storage integrated into the ventilation system', *Solar Energy*, vol 81, pp1078–1087

Armstrong, P. R., Leeb, S. B. and Norford, L. K. (2006) 'Control with building mass – Part II: Simulation', *ASHRAE Transactions*, vol 112, part 1, pp462–473

Arnell, N. W., Cannell, M. G. R., Hulme, M., Kovats, R. S., Mitchell, J. F. B., Nicholls, R. J., Parry, M. L., Livermore, M. T. J. and White, A. (2002) 'The consequences of CO_2 stabilization for the impacts of climate change', *Climatic Change*, vol 53, pp413–446

ASHRAE (American Society of Heating, Refrigerating and Air-Conditioning Engineers) (2000) *2000 ASHRAE Handbook, Heating, Ventilating, and Air Conditioning*

Systems and Equipment, SI Edition, American Society of Heating, Refrigeration and Air-Conditioning Engineers, Atlanta

ASHRAE (2001) *2001 ASHRAE Handbook, Fundamentals, SI Edition*, American Society of Heating, Refrigeration and Air-Conditioning Engineers, Atlanta

Atanasiu, B. and Bertoldi, P. (2009) 'Analysis of the electricity end-use in EU-27 households', in P. Bertoldi and R. Werle (eds) *Energy Efficiency in Domestic Appliances and Lighting, Proceedings of the 5th International Conference EEDAL'09*, 16–18 June, Berlin, Vol 1, pp189–210

Atif, M. R. and Galasiu, A. D. (2003) 'Energy performance of daylight-linked automatic lighting control systems in large atrium spaces: Report on two field-monitored case studies', *Energy and Buildings*, vol 35, pp441–461

Atkinson, B., Denver, A., McMahon, J. E., Shown, L. and Clear, R. (1997) 'Energy efficient technologies: Energy efficient lighting technologies and their applications in the commercial and residential sectors', in F. Kreith and R. E. West (eds) *CRC Handbook of Energy Efficiency*, CRC Press, Boca Raton, Florida, pp399–427

Audenaert, A., de Cleyn, S. H. and Vankerckhove, B. (2008) 'Economic analysis of passive houses and low-energy houses compared with standard houses', *Energy Policy*, vol 36, pp47–55

Aunan, K., Fang, J., Vennemo, H., Oye, K. and Seip, H. M. (2004) 'Co-benefits of climate policy – lessons learned from a study in Shanxi, China', *Energy Policy*, vol 32, pp567–581

Ax, C. (2001) 'Slow consumption for sustainable jobs', in M. Charter and U. Tischner (eds) *Sustainable Solutions*, Greenleaf, Sheffield

Axley, J. (1999) 'Passive ventilation for residential air quality control', *ASHRAE Transactions*, vol 105, part 2, pp864–876

Ayres, R. U., Ayres, L. W. and Råde, I. (2003) *The Life Cycle of Copper, Its Co-Products and Byproducts*, Kluwer, Dordrecht

Ayres, R. U., Turton, H. and Casten, T. (2007) 'Energy efficiency, sustainability and economic growth', *Energy*, vol 32, pp634–648

Babikian, R., Lukachko, S. P. and Waitz, I. A. (2002) 'The historical fuel efficiency characteristics of regional aircraft from technological, operational, and cost perspectives', *Journal of Air Transport Management*, vol 8, pp389–400

Baczek, S., Yost, P. and Finegan, S. (2002) 'Using wood efficiently: From optimizing design to minimizing the dumpster', Building Science Corporation, www.buildingscience.com/resources/misc/wood_efficiency.pdf

Badgley, C., Moghtader, J., Quintero, E., Zakem, E., Chappell, M. J., Avilés-Vázquez, K., Samulon, A. and Perfecto, I. (2007) 'Organic agriculture and the global food supply', *Renewable Agriculture and Food Systems*, vol 22, pp86–108

Bähr, A. and Sambale, M. (2009) 'Certification as a European Passive House planner', in *Conference Proceedings, 13th International Passive House Conference 2009, 17–18 April, Frankfurt am Main*, Passive House Institute, Darmstadt, Germany, pp259–263

Baird, G. (2001) *The Architectural Expression of Environmental Control Systems*, Spon Press, London

Baker, N. and Steemers, K. (1999) *Energy and Environment in Architecture: A Technical Design Guide*, E & FN Spoon, London

Baker, N. and Steemers, K. (2002) *Daylight Design of Buildings*, James & James, London

Balaras, C. A. (2001) 'Energy retrofit of a neoclassic office building: Social aspects and lessons learned', *ASHRAE Transactions*, vol 107, part 1, pp191–197

Baldwin, S. F. (1989) 'Energy-efficient electric motor drive systems', in T. B. Johansson, B. Bodlund and R. H. Williams (eds) *Electricity: Efficient End-Use and New Generation Technologies and Their Planning Implications*, Lund University Press, pp21–58

Balmér, P. (1997) 'Energy conscious waste water treatment plant', *CADDET Energy Efficiency Newsletter*, June, pp11–12

Bandivadekar, A., Bodek, K., Cheah, L., Evans, C., Groode, T., Heywood, J., Kasseris, E., Mromer, M. and Weiss, M. (2008) *On the Road in 2035, Reducing Transportation's Petroleum Consumption and GHG Emissions*, MIT LFEE 2008–005 RP, http://web.mit.edu/sloan-auto-lab/research/beforeh2/reports.htm

Banwell, P., Brons, P. J., Freyssinier-Nova, J. P., Rizzo, P. and Figueiro, M. (2004) 'A demonstration of energy-efficient lighting in residential new construction', *Lighting Resources Technology*, vol 36, pp147–164

Bär, S. (2009) 'Actively building Passive Houses in municipal situations – experiences from Hannover', in *Conference Proceedings, 13th International Passive House Conference 2009, 17–18 April, Frankfurt am Main*, Passive House Institute, Darmstadt, Germany, pp75–80

Barker, T., Ekins, P. and Foxon, T. (2007) 'The macro-economic rebound effect and the UK economy', *Energy Policy*, vol 35, pp4935–4946

Barnett, T. P., Adam, J. C. and Lettenmaier, D. P. (2005) 'Potential impacts of a warming climate on water availability in snow-dominated regions', *Nature*, vol 438, pp303–309

Barnitt, R. A. (2008) 'In-use performance comparison of hybrid electric, CNG and diesel buses at New York City transit', presented at the 2008 SAE International Powertrains, Fuels & Lubricants Conference, June 2008, Shanghai, China, www.nrel.gov/vehiclesandfuels/fleettest/research_hybrid.html

Barrett, P. J., Adams, C. J., McIntosh, W. C., Swisher, C. C. and Wilson, G. S. (1992) 'Geochronological evidence supporting Antarctic deglaciation three million years ago', *Nature*, vol 339, pp816–818

Barry, J. A. (2007) *WATERGY: Energy and Water Efficiency in Municipal Water Supply and Wastewater Treatment, Cost-Effective Savings of Water and Energy*, Alliance to Save Energy, Washington, DC, www.ase.org

Bartos, P. J. (2002) 'SX-EW copper and the technology cycle', *Resources Policy*, vol 28, pp85–94

Baruah, D. C. and Dutta, P. K. (2007) 'An investigation into the energy use in relation to yield of rice (*Oryza sativa*) in Assum, India', *Agriculture Ecosystems and Environment*, vol 120, pp185–191

Barwig, F. E., House, J. M., Klaassen, C. J., Ardehali, M. M. and Smith, T. F. (2002) 'The National Building Controls information program', in *Proceedings of the 2002 ACEEE Summer Study on Energy Efficiency in Buildings*, vol 3, American Council for an Energy Efficient Economy, Washington, DC

Bastian, Z. (2009) 'Case study: Modernization in stages, optimizing energy upgrades', in *Conference Proceedings, 13th International Passive House Conference 2009, 17–18 April, Frankfurt am Main*, Passive House Institute, Darmstadt, Germany, pp357–362

Baumgärtner, C. (2009) 'Study of reduced operating costs at an elementary school in Frankfurt/Main', in *Conference Proceedings, 13th International Passive House Conference 2009, 17–18 April, Frankfurt am Main*, Passive House Institute, Darmstadt, Germany, pp307–314

Beeler, A. G. (1998) 'Integrated design team management within the context of environmental systems theory', in *Proceedings of the 1998 ACEEE Summer Study on Energy Efficiency in Buildings*, vol 10, American Council for an Energy Efficient Economy, Washington, DC

Beerkens, R. (2006) 'Energy balances of glass furnaces: Parameters determining energy consumption of glass melt processes', *67th Conference on Glass Problems*, Columbus, Ohio

Beg, N. Morlot, J. C., Davidson, O., Afrane-Okesse, Y., Tyani, L., Denton, F., Sokona, Y., Thomas, J. P., La Rovere, E. L., Parikh, J. K., Parikh, K. and Rahman, A. A. (2002) 'Linkages between climate change and sustainable development', *Climate Policy*, vol 2, pp129–144

Behr, I (2009) 'Utility bills in Passive Houses – doing away with metered billing', in *Conference Proceedings, 13th International Passive House Conference 2009, 17–18 April, Frankfurt am Main*, Passive House Institute, Darmstadt, Germany, pp377–382

Bell, M. and Lowe, R. (2000) 'Energy efficient modernisation of housing: A UK case study', *Energy and Buildings*, vol 32, pp267–280

Bentley, R. W., Mannan, S. A. and Wheeler, S. J. (2007) 'Assessing the date of the global oil peak: The need to use 2P reserves', *Energy Policy*, vol 35, pp6364–6382

Bergerson, J. A. and Lave, L. B. (2005) 'Should we transport coal, gas, or electricity: Cost, efficiency, and environmental implications', *Environmental Science and Technology*, vol 39, pp5905–5910

Bergerson, J. A. and Lave, L. B. (2007) 'Baseload coal investment decisions under uncertain carbon legislation', *Environmental Science and Technology*, vol 41, pp3431–3436

Berghof, R., Schmitt, A., Eyers, C., Haag, K., Middel, J., Hepting, M., Grübler, A. and Hancox, R. (2005) *CONSAVE 2050 Executive Summary*, www.aero-net.org

Bergström, L. and Goulding, K. W. T. (2005) 'Perspectives and challenges in the future use of plant nutrients in tilled and mixed agricultural systems', *Ambio*, vol 34, no 4–5, pp283–287

Bernier, M. and Lemire, N. (1999) 'Non-dimensional pumping power curves for water loop heat pump systems', *ASHRAE Transactions*, vol 105, part 2, pp1226–1232

Bernstein, L. et al (2007) 'Industry', in B. Metz, O. R. Davidson, P. R. Bosch, R. Dave and L. A. Meyer (eds) *Climate Change 2007: Mitigation, Contribution of Working Group III to the Fourth Assessment Report of the Intergovernmental Panel on Climate Change*, Cambridge University Press, Cambridge and New York, pp447–496

Berntsson, T. (2002) 'Heat sources – technology, economy and environment', *International Journal of Refrigeration*, vol 25, pp428–438

Berry, G. D., Martinez-Frias, J., Espinsoa-Loza, F. and Aceves, S. M. (2003) 'Hydrogen storage and transportation', *Encyclopedia of Energy*, pp267–281

Bertoldi, P. and Atanasiu, B. (2007) *Electricity Consumption and Efficiency Trends in the Enlarged European Union*, European Commission and Institute for Environment and Sustainability, EUR 22753 EN

Bertoldi, P., Aebischer, B., Edlington, C., Herschberg, C., Lebot, B., Lin, J., Marker, T., Meier, A., Nakagami, H., Shibata, Y., Siderius, H. P. and Webber, C. (2002) 'Standby power use: How big is the problem? What policies and technical solutions can address it?', in *Proceedings of the 2002 ACEEE Summer Study on Energy Efficiency in Buildings*, vol 7, American Council for an Energy Efficient Economy, Washington, DC

Bevilacqua-Knight Inc. (2000) *Closed-cycle Totally Chlorine Free Bleached Kraft Pulp Production at Louisiana Pacific's Samoa Pulp Mill*, California Energy Commission CEC-400-2000-900, Sacramento, California

Bin, S. and Dowlatabadi, H. (2005) 'Consumer lifestyle approach to US energy use and the related CO_2 emissions', *Energy Policy*, vol 33, pp197–208

Binswanger, M. (2001) 'Technological progress and sustainable development: What about the rebound effect?', *Ecological Economics*, vol 36, pp119–132

Binz, A. and Steinke, G. (2005) 'Applications of vacuum insulation in the building sector', in M. Zimmermann (ed) *7th International Vacuum Insulation Symposium,*

EMPA, Duebendorf, Switzerland, 28–29 September, p43–48, www.empa/ch/VIP-Symposium

Birdsall, N. (1994) 'Another look at population and global warming', in *Population, Environment, and Development*, Proceedings of the United Nations Expert Group Meeting on Population, Environment and Development held at the United Nations Headquarters, 20–24 January, New York, United Nations

Blaustein, A. R. and Dobson, A. (2006) 'A message from the frogs', *Nature*, vol 439, pp143–144

Blaxter, K. (1989) *Energy Metabolism in Animals and Man*, Cambridge University Press, Cambridge

Bleken, M. A. and Bakken, L. R. (1997) 'The nitrogen cost of food production', *Ambio*, vol 26, pp134–142

Blok, K. (2005) 'Improving energy efficiency by five percent and more per year?', *Journal of Industrial Ecology*, vol 8, no 4, pp87–99

Blondeau, P., Sperandio, M. and Allard, F. (1997) 'Night ventilation for building cooling in summer', *Solar Energy*, vol 6, pp327–335

Bodart, M. and De Herde, A. (2002) 'Global energy savings in office buildings by the use of daylighting', *Energy and Buildings*, vol 34, pp421–429

Bodem, M. (2009) 'User-considerate renovation of a multi-family building – complete installation from the outside', in *Conference Proceedings, 13th International Passive House Conference 2009, 17–18 April, Frankfurt am Main*, Passive House Institute, Darmstadt, Germany, pp283–288

Boermans, T. and Petersdorff, C. (2007) *U-values for Better Energy Performance of Buildings*, Report established by Ecofys for EURIMA, www.eurima.org

Bolland, O. and Undrum, H. (1999) 'Removal of CO_2 from gas turbine power plants: Evaluation of pre- and postcombustion methods', in P. Riemer, B. Eliasson and A. Wokaun (eds) *Greenhouse Gas Control Technologies*, Elsevier Science, New York, pp125–130

Bongaarts, J., O'Neill, B. C. and Gaffin, S. R. (1997) 'Global warming policy: Population left out in the cold', *Environment*, vol 39, no 9, pp40–41

Boonstra, C. and Thijssen, I. (1997) *Solar Energy in Building Renovation*, James & James, London

Borchardt, J. K. (2006) 'Recycling, plastics', in *Kirk-Othmer Encyclopedia of Chemical Technology*, John Wiley & Sons, New York, volume 21

Börjesson, P. and Gustavsson, L. (2000) 'Greenhouse gas balances in building construction: Wood versus concrete from life-cycle and forest land-use perspectives', *Energy Policy*, vol 28, pp575–588

Bourassa, N., Haves, P. and Huang, J. (2002) 'A computer simulation appraisal of non-residential low energy cooling systems in California', in *Proceedings of the 2002 ACEEE Summer Study on Energy Efficiency in Buildings*, vol 3, American Council for an Energy Efficient Economy, Washington, DC

Boustead, I. (2008) *Plastics Recycling – An Overview*, Plastics Europe, www.plasticseurope.org

Bower, J. (1995) *Understanding Ventilation: How to Design, Select, and Install Residential Ventilation Systems*, Healthy House Institute, Bloomington, Indiana

BP (British Petroleum) (2007) *BP Statistical Review 2007*, www.bp.com

Brager, G. S. and de Dear, R. (2000) 'A standard for natural ventilation', *ASHRAE Journal*, vol 42, no 10, pp21–28

Brandemuehl, M. J. and Braun, J. E. (1999) 'The impact of demand-controlled and economizer ventilation strategies on energy use in buildings', *ASHRAE Transactions*, vol 105, no 2, pp39–50

Brandon, R. J. and Snoek, C. W. (2000) 'Microturbine cogeneration', *ASHRAE Transactions*, vol 106, part 1, pp669–674

Brandt, A. R. (2008) 'Converting oil shale to liquid fuels: Energy inputs and greenhouse gas emissions of the Shell in situ conversion process', *Environmental Science and Technology*, vol 42, pp7489–7495

Brandt, A. R. (2009) 'Converting oil shale to liquid fuels with the Alberta Taciuk Processor: Energy inputs and greenhouse gas emissions', *Energy and Fuels,* vol 23, no 12, pp6253–6258, DOI:10.1021/ef900678d

Bratt, M. and Persson, A. (2001) 'Future CO_2 savings from on-line shopping jeopardized by bad planning', in *European Council for an Energy Efficient Economy, 2001 Summer Proceedings*, vol 3, pp480–492

Bretzke, A. (2008) 'Planning and construction of passive solar primary school Kalbacher Höhe 15, Frankfurt am Main', www.stadt-frankfurt.de/energiemanagement/english/Passive-house-school-Riedberg.pdf

Bretzke, A. (2009) 'Benefits of the Passive House Standard in schools: Cost-effectiveness and use convenience', in *Conference Proceedings, 13th International Passive House Conference 2009, 17–18 April, Frankfurt am Main*, Passive House Institute, Darmstadt, Germany, pp233–244

Brown, E. and Elliott, R. N. (2005) *Potential Energy Efficiency Savings in the Agriculture Sector*, American Council for an Energy-Efficiency Economy, Washington, DC

Brown, L. F. (2001) 'A comparative study of fuels for on-board hydrogen production for fuel-cell-powered automobiles', *International Journal of Hydrogen Energy*, vol 26, no 4, pp381–397

Brown, L. R. (2001) 'Redesigning cities for people', in W. W. Norton (ed) *Eco-Economy: Building an Economy for the Earth*, Earth Policy Institute, New York, www.earth-policy.org

Brown, M. A., Levine, M. D., Romm, J. P., Rosenfeld, A. H. and Koomey, J. G. (1998) 'Engineering-economic studies of energy technologies to reduce greenhouse emissions:

Opportunities and challenges', *Annual Review of Energy and the Environment*, vol 23, pp287–385

BTM Consult (2001) *International Wind Energy Development, World Market Update 2000*, BTM Consult, www.btm.dk

Buchanan, A. H. and Levine, S. B. (1999) 'Wood-based building materials and atmospheric carbon emissions', *Environmental Science and Policy*, vol 2, pp427–437

Buchmann, I. (2006) 'High Power Lithium-ion, a New Area of Portable Power', www.buchmann.ca/Article27-Page1.asp, accessed September 2008

Burns, L., McCormick, J. and Borroni-Bird, C. (2002) 'Vehicle of change', *Scientific American*, vol 287, no 4, pp64–73

Busch, J. F. (1992) 'A tale of two populations: Thermal comfort in air-conditioned and naturally-ventilated offices in Thailand', *Energy and Buildings*, vol 18, pp235–249

Byström, S. and Lönnstedt, L. (1997) 'Paper recycling: Environmental and economic impact', *Resources, Conservation, and Recycling*, vol 21, pp109–127

Çakmanus, I. (2007) 'Renovation of existing office buildings in regard to energy economy: An example from Ankara, Turkey', *Building and Environment*, vol 42, pp1348–1357

Calise, F., d'Accadia, M. D., Palombo, A. and Vanoli, L. (2006) 'Simulation and exergy analysis of a hybrid solid oxide fuel cell (SOFC) – gas turbine system', *Energy*, vol 31, pp3278–3299

Camilleri, S. (2001) 'Development of a silent, high efficiency ceiling fan', *CADDET Energy Efficiency Newsletter*, June, pp6–8

Campbell, C. J. and Siobhan, H. (2009) *An Atlas of Oil and Gas Depletion*, Jeremy Mills Publishing, UK

Canals, L. M., Cowell, S. J., Sim, S. and Basson, L. (2007) 'Comparing domestic versus imported apples: A focus on energy use', *Environmental Science and Pollution Research*, vol 14, pp338–344

Caneta Research Incorporated (1995) *Commercial/Institutional Ground-Source Heat Pump Engineering Manual*, American Society of Heating, Refrigerating and Air-Conditioning Engineers, Atlanta

Carlson, E. J., Kopf, P., Sinha, J., Sriramulu, S. and Yang, Y. (2005) 'Cost Analysis of PEM Fuel Cell Systems for Transportation', National Renewable Energy Laboratory, Golden, Colorado, NREL SR-56039104

Carlsson, A. (1997) 'Greenhouse gas emissions in the life-cycle of carrots and tomatoes', IMES/EESS Report No. 24, Department of Environment and Energy Systems Studies, Lund University, Sweden

Carlsson-Kanyama, A. (2007) 'Diet, energy, and greenhouse gas emissions', *Encyclopedia of Energy*, vol 1, pp809–816

Carlsson-Kanyama, A., Ekström, M. P. and Shanahan, H. (2003). 'Food and life cycle energy inputs: Consequences

of diet and ways to increase efficiency', *Ecological Economics*, vol 44, pp293–307

Cassman, K. G., Dobermann, A. and Walters, D. T. (2002) 'Agroecosystems, nitrogen-use efficiency, and nitrogen management', *Ambio*, vol 31, no 2, pp132–140

Casten, T. R. and Ayres, R. U. (2007) 'Energy myth eight – Worldwide power systems are economically and environmentally optimal', in B. K. Sovacool and M. A. Brown (eds) *Energy and American Society – Thirteen Myths*, Springer-Verlag, Berlin

CATF (Clean Air Task Force) (2003) *The Last Straw: Water Use by Power Plants in the Arid West*, www.catf.us

CEC (California Energy Commission) (1996) *Source Energy and Environmental Impacts of Thermal Energy Storage*, www.energy.ca.gov

CEC (2000) *Demonstration of Caterpillar C-10 Duel-Fuel Engines in MC1 102DL13 Commuter Buses*, National Renewable Energy Laboratory, Golden, Colorado, NREL/SR-540-26758, www.energy.ca.gov

Cederberg, C., Wivstad, M., Bergkvist, P., Mattsson, B. and Ivarsson, K. (2005) 'Environmental assessment of plant protection strategies using scenarios for pig feed production', *Ambio*, vol 34, no 4–5, pp408–413

CEE (Consortium for Energy Efficiency) (2009) *CEE Super-Efficient Home Appliances Initiative, Clothes Washer Qualifying Product List*, www.cee1.org

Cervero, R. (1998) *The Transit Metropolis: A Global Inquiry*, Island Press, Washington, DC

Çetin, B. and Vardar, A. (2008) 'An economic analysis of energy requirements and input costs for tomato production in Turkey', *Renewable Energy*, vol 33, pp428–433

Chandler, K. and Walkowicz, K. (2006) *King County Metro Transit Hybrid Articulated Buses: Final Evaluation Results*, National Renewable Energy Laboratory, Golden, Colorado, Technical Report NREL/TP-540-40585, www.nrel.gov/vehiclesandfuels/fleettest/research_hybrid.html

Chao, A., Thun, M. J., Connell, C. J., McCullough, M. L., Jacobs, E. J., Flanders, W. D., Rodrigues, C., Sinha, R. and Calle, E. E. (2005) 'Meat consumption and the risk of colorectal cancer', *Journal of the American Medical Association*, vol 293, pp233–234

Charles, R., Jolliet, O., Gaillard, G. and Pellet, D. (2006) 'Environmental analysis of intensity level in wheat crop production using life cycle assessment', *Agriculture Ecosystems and Environment*, vol 113, pp216–225

Chen, X., Zhang, G., Peng, J., Lin, X. and Liu, T. (2006) 'The performance of an open-loop lake water heat pump system in south China', *Applied Thermal Engineering*, vol 26, pp2255–2261

Chester, M. V. and Horvath, A. (2009) 'Environmental assessment of passenger transportation should include infrastructure and supply chains', *Environmental Research Letters*, vol 4, no 2, doi:10.1088/1748-9326/4/2/024008

Chow, T. T., Lin, Z. and Liu, J. P. (2002) 'Effect of condensing unit layout at building re-entrant on split-type air-conditioner performance', *Energy and Buildings*, vol 34, pp237–244

Claridge, D. E., Liu, M., Deng, S., Turner, W. D., Haberl, J. S., Lee, S. U., Abbas, M. and Bruner, H. (2001) 'Cutting heating and cooling use almost in half without capital expenditure in a previously retrofit building', in *European Council for an Energy Efficient Economy, 2001 Summer Proceedings*, vol 4, pp74–85

Cleveland, C. J. (2005) 'Net energy from the extraction of oil and gas in the United States', *Energy*, vol 30, pp769–782

Clodic, D. et al (2005) 'Mobile air conditioning', in B. Metz, L. Kuijpers, S. Solomon, S. O. Andersen, O. Davidson, J. Pons, D. de Jager, T. Kestin, M. Manning and L. Meyer (eds) *Safeguarding the Ozone Layer and the Global Climate System: Issues Related to Hydrofluorocarbons and perfluorocarbons*, Intergovernmental Panel on Climate Change, Cambridge University Press, pp295–314

Cohen, R., Nelson, B. and Wolff, G. (2004) *Energy Down the Drain: The Hidden Costs of California's Water Supply*, Natural Resources Defence Council and Pacific Institute, New York and Oakland, www.nrdc.org

Cohen, R., Bordass, W. and Leaman, A. (2007) 'Evaluations and comparisons of the achieved energy and environmental performance of two library buildings in England and Sweden', *ASHRAE Transactions*, vol 113, part 2, pp14–26

Cole, R. J. and Kernon, P. C. (1996) 'Life-cycle energy use in office buildings', *Building and Environment*, vol 31, pp301–317

Coley, D. A. (2002) 'Emission factors for human activity', *Energy Policy*, vol 30, pp3–5

Coley, D. A. Goodliffe, E. and Macdiarmid, J. (1998) 'The embodied energy of food: The role of diet', *Energy Policy*, vol 26, pp455–459

Consonni, S., Giugliano M. and Grosso, M. (2005) 'Alternative strategies for energy recovery from municipal solid waste, Part A: Mass and energy balances', *Waste Management*, vol 25, pp123–135

Cooper, T. (2005) 'Slower consumption: Reflections on product life spans and the "throwaway society"', *Journal of Industrial Ecology*, vol 9, pp51–67

Corbett, J. J. (2004) 'Marine transportation and energy use', *Encyclopedia of Energy*, vol 3, Elsevier, Amsterdam, pp745–758

Cordell, D. (2006) 'Urine diversion and reuse in Australia: A homeless paradigm or sustainable solution for the future?', Masters Thesis, Department of Water and Environmental Studies, Linköping University, Sweden, http://liu.diva-portal.org/smash/record.jsf?pid=diva2:23137

Cordell, D., Drangert, J. O. and White, S. (2009a) 'The story of phosphorus: Global food security and food for thought', *Global Environmental Change*, vol 19, pp292–305

Cordell, D., Schmid-Neset, T., White, S. and Drangert, J. O. (2009b) 'Preferred future phosphorus scenarios: A framework for meeting long-term phosphorus needs for global food demand', in K. Ashley, D. Mavinic and F. Koch (eds) *International Conference on Nutrient Recovery from Wastewater Streams*, IWA Publishing, London, pp23–43

Costelloe, B. and Finn, D. (2003) 'Indirect evaporative cooling potential in air-water systems in temperate climates', *Energy and Buildings*, vol 35, pp573–591

Counsell, T. A. M. and Allwood, J. M. (2007) 'Reducing climate change gas emissions by cutting out stages in the life cycle of office paper', *Resources Conservation and Recycling*, vol 49, pp340–352

Cownden, R., Hahon, M. and Rosen, M. A. (2001) 'Modelling and analysis of a solid polymer fuel cell system for transportation applications', *International Journal of Hydrogen Energy*, vol 26, pp615–623

Cox, P. M., Betts, R. A., Jones, C. D., Spall, S. A. and Totterdell, I. J. (2000) 'Acceleration of global warming due to carbon-cycle feedbacks in a coupled climate model', *Nature*, vol 408, pp184–187

Cox, P. M., Betts, R. A., Collins, M., Harris, P. P., Huntingford, C. and Jones, C. D. (2004) 'Amazonian forest dieback under climate-carbon cycle projections for the 21st century', *Theoretical Applied Climatology*, vol 78, pp137–156

Croft, G. D. and Patzek, T. W. (2009) 'Potential for coal-to-liquids conversion in the U.S.-resource base', *Natural Resources Research*, vol 18, pp173–180

Crosbie, T. (2008) 'Household energy consumption and consumer electronics: The case of television', *Energy Policy*, vol 36, pp2191–2199

Crowley, T. J. (1996) 'Pliocene climates: The nature of the problem', *Marine Micropaleontology*, vol 27, pp3–12

Cuffey, K. M. and Marshall, S. J. (2000) 'Substantial contribution to sea-level rise during the last glaciation from the Greenland Ice Sheet', *Nature*, vol 404, pp591–594

Curti, V., von Spakovsky, M. R. and Favrat, D. (2000) 'An environomic approach for the modeling and optimization of a district heating network based on centralized and decentralized heat pumps, cogeneration, and/or gas furnace. Part I: Methodology', *International Journal of Thermal Science*, vol 39, pp721–730

da Graça, G. C., Chen, Q., Glicksman, L. R. and Norfold, L. K. (2002) 'Simulation of wind-driven ventilative cooling systems for an apartment building in Beijing and Shanghai', *Energy and Buildings*, vol 34, p1–11

Dalgaard, T., Halberg, N. and Fenger, J. (2002) 'Can organic farming help to reduce national energy consumption and emissions of greenhouse gases in Denmark?', in E. C. van Lerland and A. O. Lansink (eds) *Economics of Sustainable Energy in Agriculture*, Kluwer Academic Publishers, Dordrecht, The Netherlands

Daly, H. E. (1996) *Beyond Growth: The Economics of Sustainable Development*, Beacon Press, Boston

Daly, H. E. and Cobb, J. B. (1994) *For the Common Good: Redirecting the Economy Toward Community, the Environment, and a Sustainable Future*, Beacon Press, Boston

Darley, J. (2004) *High Noon for Natural Gas: The New Energy Crisis*, Chelsea Green, White River, Vermont

Dascalaki, E. and Santamouris, M. (2002) 'On the potential of retrofitting scenarios for offices', *Building and Environment*, vol 37, pp557–567

Davenport, W. G., King, M., Schlesinger, M. and Biswas, A. K. (2002) *Extractive Metallurgy of Copper*, Elsevier Academic Press, Amsterdam

DCE (Digital CEnergy) (2007) 'Discussion paper, television energy rating labels: The case, and proposal', for MEPS and Labelling Televisions, report 2007/10, www.energyrating.gov.au/library/pubs/200710-tv-meps-labelling.pdf

de Almeida, A. T. and Greenberg, S. (1997) 'Energy efficient technologies: Electric motor systems efficiency', in F. Kreith and R. E. West (eds) *CRC Handbook of Energy Efficiency*, CRC Press, Boca Raton, Florida

de Beer, J., Worrell, E. and Blok, K. (1998a) 'Future technologies for energy-efficient iron and steel making', *Annual Review of Energy and the Environment*, vol 23, pp123–205

de Beer, J., Worrell, E. and Blok, K. (1998b) 'Long-term energy-efficiency improvements in the paper and board industry', *Energy*, vol 23, pp21–42

de Carli, M. and Olesen, B. W. (2002) 'Field measurements of operative temperatures in buildings heated or cooled by embedded water-based radiant systems', *ASHRAE Transactions*, vol 108, part 2, pp714–725

de Cicco, J., An, F. and Ross, M. (2001) *Technical Options for Improving the Fuel Economy of US Cars and Light Trucks by 2010–2015*, American Council for an Energy-Efficient Economy, Washington, DC

de Dear, R. J. and Brager, G. S. (1998) 'Developing an adaptive model of thermal comfort and preference', *ASHRAE Transactions*, vol 104, part 1A, pp145–167

de Dear, R. J. and Brager, G. S. (2002) 'Thermal comfort in naturally ventilated buildings: Revisions to ASHRAE Standard 55', *Energy and Buildings*, vol 34, pp549–561

DEG (Davis Energy Group) (2004) *Development of an Improved Two-Stage Evaporative Cooling System*, California Energy Commission Report P500-04-016, www.energy.ca.gov/buildings

Delsante, A. and Vik, T. A. (eds) (2002) *Hybrid Ventilation: State-of-the-Art Review*, IEA Annex 35, International Energy Agency, Paris, www.hybvent.civil.auc.dk/publications/sotar.pdf

Demirbas, A. (2005) 'Biodiesel production from vegetable oils via catalytic and non-catalytic supercritical methanol transesterification methods', *Progress in Energy and Combustion Science*, vol 31, pp466–487

Demirbilek, F. N., Yalçiner, U. G., Inanici, M. N., Ecevit, A. and Demirbilek, O. S. (2000) 'Energy conscious dwelling design for Ankara', *Building and Environment*, vol 35, pp33–40

Den Baars, S. (2008) 'Energy efficient white LEDs for sustainable solid-state lighting', in D. Hafemeister, B. G. Levi, M. D. Levine and P. Schwartz (eds) *Physics of Sustainable Energy, Using Energy Efficiently and Producing it Renewably*, American Institute of Physics, Melville, New York

Denman, K. L. et al (2007) 'Couplings between changes in the climate system and biogeochemistry', in S. Solomon, D. Qin, M. Manning, Z. Chen, M. Marquis, K. B. Averyt, M. Tignor and H. L. Miller, (eds) *Climate Change 2007: The Physical Science Basis. Contribution of Working Group I to the Fourth Assessment Report of the Intergovernmental Panel on Climate Change*, Cambridge University Press, Cambridge and New York, pp499–587

Deru, M., Pless, S. D. and Torcellini, P. A. (2006) 'Bighorn home improvement center energy performance', *AHSRAE Transactions*, vol 112, part 2, pp349–366

Deumling, R. (2004) 'Thinking outside the refrigerator: Shutting down power plants with NAECA?', in *Proceedings of the 2004 ACEEE Summer Study on Energy Efficiency in Buildings*, vol 11, American Council for an Energy Efficient Economy, Washington, DC, pp13–24

de Vries, E. (2007) 'Husum 2007: Wind technology overview', *Renewable Energy World*, vol 10, no 5, pp44–55

Dharmadhikari, S. (1997) 'Consider trigeneration techniques for process plants', *Hydrocarbon Processing*, July, p91–100

Dharmadhikari, S., Pons, D. and Principaud, F. (2000) 'Contribution of stratified thermal storage to cost-effective trigeneration project', *ASHRAE Transactions*, vol 106, part 2, pp912–919

Diaz, R. J. and Rosenberg, R. (2008) 'Spreading dead zones and consequences for marine ecosystems', *Science*, vol 321, pp926–929

Diener, E. and Oishi, S. (2000) 'Money and happiness: Income and subjective well-being across nations', in E. Diener and E. M. Suh (eds) *Cross-cultural Psychology of Subjective Well-being*, MIT Press, Cambridge, MA, pp185–218

Dodbiba, G., Takahashi, K., Sadaki, J. and Fujita, T. (2008) 'The recycling of plastic wastes from discarded TV sets: Comparing energy recovery with mechanical recycling in the context of life cycle assessment', *Journal of Cleaner Production*, vol 16, pp458–470

Donner, S. D. (2008) 'Surf or turf: A shift from feed to food cultivation could reduce nutrient flux to the Gulf of Mexico', *Global Environmental Change*, vol 17, pp105–115

Donner, S. D., Skirving, W. J., Little, C. M., Oppenheimer, M. and Hoegh-Guldberg, O. (2005) 'Global assessment of coral bleaching and required rates of adaptation under climate change', *Global Change Biology*, vol 11, pp2251–2265

Dornburg, V., Faaij, A. P. C. and Meuleman, B. (2006) 'Optimising waste treatment systems, Part A: Methodology and technological data for optimizing energy production and economic performance', *Resources Conservation and Recycling*, vol 49, pp68–88

Dou, Z., Zhang, G. Y., Stout, W. L., Toth, J. D. and Ferguson, J. D. (2003) 'Efficacy of alum and coal combustion by-products in stabilizing manure phosphorus', *Journal of Environmental Quality*, vol 32, pp1490–1497

Douglas, J. (2008) 'Plug-in hybrids on the horizon: Building a business case', *EPRI Journal*, Spring, p6–15

Downey, T. and Proctor, J. (2002) 'What can 13,000 air conditioners tell us?', in *Proceedings of the 2002 ACEEE Summer Study on Energy Efficiency in Buildings*, vol 1, American Council for an Energy Efficient Economy, Washington, DC, p53–67

Dowsett, H. J., Thompson, R., Barron, J., Cronin, T., Fleming, F., Ishman, S., Poore, R., Willard, D. and Holtz, T. (1994) 'Joint investigation of the Middle Pliocene climate I: PRISM paleoenvironmental reconstructions', *Global Planetary Change*, vol 9, pp169–195

Dowsett, H., Barron, J. and Poore, R. (1996) 'Middle Pliocene sea surface temperatures: A global reconstruction', *Marine Micropaleontology*, vol 27, pp13–26

DSS Management Consultants Inc. (2005) Cost Benefit Analysis: Replacing Ontario's Coal-Fired Electricity Generation

Duleep, K. G. (2007) 'Fuel economy of heavy-duty trucks in the USA', presented at the *International Workshop on Fuel Efficiency Policies for Heavy-Duty Vehicles*, 21–22 June 2007, International Energy Agency, Paris, www.iea.org/textbase/work/workshopdetail.asp?WS_ID=306

Durkin, T. H. (2006) 'Boiler system efficiency', *ASHRAE Journal*, vol 48, no 7, pp51–57

Durkin, T. H. and Kinney, L. (2002) 'Two-pipe HVAC makes a comeback: An idea discarded decades ago may be the future of school heating and cooling', in *Proceedings of the 2002 ACEEE Summer Study on Energy Efficiency in Buildings*, vol 3, American Council for an Energy Efficient Economy, Washington, pp93–106

Dutch National Team (1995) 'Energy savings through programme for electronic variable speed drives', *CADDET Energy Efficiency Newsletter*, June, pp8–9

Dutilh, C. E. and Kramer, K. J. (2000) 'Energy consumption in the food chain: Comparing alternative options in food production and consumption', *Ambio*, vol 29, no 2, pp98–101

Dwyer, G. S., Cronin, T. M., Baker, P. A., Raymo, M. E., Buzas, J. S. and Correge, T. (1995) 'North American deepwater temperature change during the late Pliocene and late Quaternary climate cycles', *Science*, vol 270, pp1347–1351

Dyer, J. A. and Desjardins, R. L. (2005) 'Analysis of trends in CO_2 emissions from fossil fuel use for farm fieldwork related to harvesting annual crops and hay, changing tillage practices and reduced summer fallow in Canada', *Journal of Sustainable Agriculture*, vol 25, pp141–155

Easley, W. L., Kapic, A. and Milam, D. M. (2005) 'The path to a 50% thermal efficient engine', DEER Conference, 23 August, www1.eere.energy.gov/vehiclesandfuels/pdfs/deer_2005/session3/2005_deer_easley.pdf

EC (European Commission) (2001a) *Green Paper – Towards a European Strategy for the Security of Energy Supply, Technical Document.* Available from ec.europa.eu.

EC (2001b) *Integrated Pollution Prevention and Control (IPPC), Reference Document on Best Available Techniques in the Glass Manufacturing Industry*, www.eippcb.jrc.es/pages/FActivities.htm

EC (2001c) *Integrated Pollution Prevention and Control (IPPC), Reference Document on Best Available Techniques in the Pulp and Paper Industry*, www.eippcb.jrc.es/pages/FActivities.htm

EC (2001d) *Waste Management Options and Climate Change*, www.ec.europ.eu/environment/waste/studies/climate_change.htm

EC (2005) *European Steel Technology Platform Strategic Research Agenda*, www.eurofer.org

EcoHeatCool (2006) *EcoHeatCool Work Package 2, The European Cold Market, Final Report.* Available from www.ecoheatcool.org

ECF (European Cyclists' Federation) (2009) *Factsheet*, www.velo-city2009.com/assets/files/VC09-ECF-facts-and-figures.pdf

Eicker, U., Huber, M., Seeberger, P. and Vorschulze, C. (2006) 'Limits and potentials of office building climatisation with ambient air', *Energy and Buildings*, vol 38, pp574–581

Emmerich, S. J., McDowell, T. P. and Anis, W. (2007) 'Simulation of the impact of commercial building envelope airtightness on building energy utilization', *ASHRAE Transactions*, vol 113, part 2, pp379–393

Endhardt, M. (2009) 'Savings bank to the Passive House Standard using pre-cast concrete construction', in *Conference Proceedings, 13th International Passive House Conference 2009, 17–18 April, Frankfurt am Main*, Passive House Institute, Darmstadt, Germany, pp333–338

EPRI (Electric Power Research Institute) (2004) *Advanced Batteries for Electric-Drive Vehicles: A Technology and Cost-Effectiveness Assessment for Battery Electric Vehicles, Power Assist Hybrid Electric Vehicles, and Plug-In Hybrid Electric Vehicles. 1009299*, Final Report, May, EPRI, Paolo Alto, California

EPRI (2007) *Environmental Assessment of Plug-In Hybrid Electric Vehicles, Volume 1: Nationwide Greenhouse Gas Emissions. 1015325*, Final Report, July, EPRI, Paolo Alto, California

Eriksson, H. and Harvey, S. (2004) 'Black liquor gasification – consequences for both industry and society', *Energy*, vol 29, pp581–612

Eriksson, O., Finnveden, G., Ekvall, T. and Björklund, A. (2007) 'Life cycle assessment of fuels for district heating: A comparison of waste incineration, biomass- and natural gas combustion', *Energy Policy*, vol 35, pp1346–1362

Ertel, T. (2009) 'Highest energy efficiency in household refrigerating appliances', in P. Bertoldi and R. Werle (eds) *Energy Efficiency in Domestic Appliances and Lighting, Proceedings of the 5th International Conference EEDAL'09,* 16–18 June, Berlin, vol 3, pp1265–1270

Eshel, G. and Martin, P. A. (2006) 'Diet, energy, and global warming', *Earth Interactions*, vol 10, pp1–17

Esper, J., Cook, E. R. and Schweingruber, F. H. (2002) 'Low-frequency signals in long tree-ring chronologies and the reconstruction of past temperature variability', *Science*, vol 295, pp2250–2253

EWG (Energy Watch Group) (2007) *Coal: Resources and Future Production*, www.energywatchgroup.org

Farahani, S., Worrell, E. and Bryntse, G. (2004) 'CO_2-free paper?', *Resources Conservation and Recycling*, vol 42, pp317–336

Farla, J., Blok, K. and Schipper, L. (1997) 'Energy efficiency developments in the pulp and paper industry', *Energy Policy*, vol 25, pp745–758

Farrell, A. E., Plevin, R. J., Turner, B. T., Jones, A. D., O'Hare, M. and Kammen, D. M. (2006) 'Ethanol can contribute to energy and environmental goals', *Science*, vol 311, pp506–508

Farrington, R. and Rugh, J. (2000) *Impact of Vehicle Air-Conditioning on Fuel Economy, Tailpipe Emissions, and Electric Vehicle Range*, National Renewable Energy Laboratory, Golden, Colorado, NREL/CP-540-28960

Fechter, J. V. and Porter, L. G. (1979) *Kitchen Range Energy Consumption*, Office of Energy Conservation, NBSIR 78–1556, US Department of Energy, Washington, DC

Fehrm, M., Reiners, W. and Ungemach, M. (2002) 'Exhaust air heat recovery in buildings', *International Journal of Refrigeration*, vol 25, pp439–449

Feist, W. (1996) 'Life-cycle energy balances compared: Low-energy house, passive house, self-sufficient house', in *Proceedings of the International Symposium of CIB W67*, Vienna, Austria

Feist, W. (2007) 'Tasks – challenges – perspectives', in *Conference Proceedings, 11th International Passive House Conference 2007, Bregenz*, Passive House Institute, Darmstadt, Germany, pp383–392

Feist, W., Schnieders, J., Dorer, V. and Haas, A. (2005) 'Re-inventing air heating: Convenient and comfortable within the frame of the Passive House concept', *Energy and Buildings*, vol 37, pp1186–1203

Feustel, H. E. and Stetiu, C. (1995) 'Hydronic cooling – preliminary assessment', *Energy and Buildings*, vol 22, pp193–205

Fiedler, F. (2004) 'The state of the art of small-scale pellet-based heating systems and relevant regulations in Sweden, Austria and Germany', *Renewable and Sustainable Energy Reviews*, vol 8, pp201–221

Fischer, V., Grüneis, H. and Richter, R. (1997) *Sir Norman Foster and Partners Commerzbank, Frankfurt am Main*, Edition Axel Menges, Stuttgart and London

Fisher, D., Schmid, F. and Spata, A. J. (1999) 'Estimating the energy-saving benefit of reduced-flow and/or multi-speed commercial kitchen ventilation systems', *ASHRAE Transactions*, vol 105, part 1, pp1138–1151

Fixen, P. E. and West, F. B. (2002) 'Nitrogen fertilizers: Meeting contemporary challenges', *Ambio*, vol 31, pp169–176

Florides, G. A., Tassou, S. A., Kalogirou, S. A. and Wrobel, L. C. (2002) 'Review of solar and low energy cooling technologies for buildings', *Renewable and Sustainable Energy Reviews*, vol 6, pp557–572

Ford, B., Patel, N., Zaveri, P. and Hewitt, M. (1998) 'Cooling without air conditioning: The Torrent Research Centre, Ahmedabad, India', *Renewable Energy*, vol 15, pp177–182

Forster, P. et al. (2007) 'Changes in atmospheric constituents and in radiative forcing', in S. Solomon, D. Qin, M. Manning, Z. Chen, M. Marquis, K. B. Averyt, M. Tignor and H. L. Miller (eds) *Climate Change 2007: The Physical Science Basis, Contribution of Working Group 1 to the Fourth Assessment Report of the Intergovernmental Panel on Climate Change*, Cambridge University Press, Cambridge and New York, pp129–234

Francisco, P. W., Palmiter, L. and Davis, B. (1998) 'Modeling the thermal distribution efficiency of ducts: Comparisons to measured results', *Energy and Buildings*, vol 28, pp287–297

Franconi, E. (1998) 'Measuring advances in HVAC distribution system design', in *Proceedings of the 1998 ACEEE Summer Study on Energy Efficiency in Buildings*, vol 3, American Council for an Energy Efficient Economy, Washington, DC

Frankenhaeuser, M., Henricson, A. K., Hakulinen, A. and Mark, F. E. (2008) *Co-combustion of Solid Recovered Fuel and Solid Biofuels in a Combined Heat and Power Plant*, Plastics Europe, www.plasticseurope.org

Frenette, G. and Forthoffer, D. (2009) 'Economic and commercial viability of hydrogen fuel cell vehicles from an automotive manufacturer perspective', *International Journal of Hydrogen Energy*, vol 34, pp3578–3588

Frey, B. S. and Stutzer, A. (2002) 'What can economists learn from happiness research?', *Journal of Economic Literature*, vol 40, pp402–435

Friedman D., Mark, J., Monahan, P., Nach, C. and Ditlow, C. (2001) *Drilling in Detroit: Tapping Automaker Ingenuity to Build Safe and Efficient Automobile*, Union of Concerned Scientists, Cambridge, MA

Fröling, M. and Svanström, M. (2005) 'Life cycle assessment of the district heat distribution system, Part 2: Network construction', *International Journal of Life Cycle Assessment,* vol 10, pp425–435

Fthenakis, V. C., Wang, W. and Kim, H. C. (2009) 'Life cycle inventory analysis of the production of metals used in photovoltaics', *Renewable and Sustainable Energy Reviews*, vol 13, pp493–517

Gallagher, K. S., Holdren, J. P. and Sagar, A. D. (2006) 'Energy-technology innovation', *Annual Review of Environment and Resources*, vol 31, pp193–237

Gallo, C. (1998) 'Chapter 5: The utilization of microclimate elements', *Renewable and Sustainable Energy Reviews*, vol 2, pp89–114

Gamble, D., Dean, B., Meisegeier, D. and Hall, J. (2004) 'Building a path towards zero energy homes with energy efficient upgrades', in *Proceedings of the 2004 ACEEE Summer Study on Energy Efficiency in Buildings*, vol 1, American Council for an Energy Efficient Economy, Washington, DC

Gambogi, J. (2008) 'Titanium and titanium dioxide', *2007 Commodity Summary*, http://minerals.er.usgs.gov/minerals/pubs/commodity

Gan, A. I., Klein, S. A. and Reindl, D. T. (2000) 'Analysis of refrigerator/freezer appliances having dual refrigeration cycles', *ASHRAE Transactions*, vol 106, part 2, pp185–191

García-Rodríguez, L. (2003) 'Renewable energy applications in desalination: State of the art', *Solar Energy*, vol 75, pp381–393

Gardner, G. (1998) 'When cities take bicycles seriously', *World Watch*, September–October, pp17–22, www.worldwatch.org

Gardner, G. (1999) 'Why share?', *World Watch*, July–August 1999, pp10–20, www.worldwatch.org

Garnett, T. (2003) *Wise Moves, Exploring the Relationship between Food, Transport, and CO_2* Transport 2000 Trust, London

Gartner, E. (2004) 'Industrially interesting approaches to "low-CO_2" cements', *Cement and Concrete Research*, vol 34, pp1489–1498

Gartner, E. and Quillin, K. (2007) 'Low-CO_2 cements based on calcium sulfoaluminates', in *Proceedings of the International Conference on Sustainability in the Cement and Concrete Industry*, 16–19 September, Lillehammer, Norway

Gavalos, G. R., Voecks, G. E., Moore, N. R., Ferrall, J. F. and Prokopius, P. R. (1995) *Fuel Cell Locomotive Development and Demonstration Program, Phase One: System Definition, Final Report*, JPL D-12087, Jet Propulsion Laboratory

Geiser, K. (2001) *Materials Matter: Towards a Sustainable Materials Policy*, MIT Press, Cambridge, MA

Geller, H. (2004) *Energy Revolution: Policies for a Sustainable Future*, Island Press, Washington, DC

Geller, H., Schaeffer, R., Szklo, A. and Tlomasquim, M. (2004) 'Policies for advancing energy efficiency and renewable energy use in Brazil', *Energy Policy*, vol 32, pp1437–1450

Genest, F. and Minea, V. (2006) 'High-performance retail store with integrated HVAC systems', *ASHRAE Transactions*, vol 112, part 2, pp342–348

Genon, G., Torchio, M. F., Poggio, A. and Poggio, M. (2009) 'Energy and environmental assessment of small district heating systems: Global and local effects in two case studies', *Energy Conversion and Management*, vol 50, pp522–529

Gensch, C.-O. (2009) 'Environmental impacts and costs of different ways to dry clothes', in presented at *Energy Efficiency in Domestic Appliances and Lighting, Proceedings of the 5th International Conference EEDAL'09*, 16–18 June, Berlin

Gerbens-Leenes, P. W. and Nonhebel, S. (2002) 'Consumption patterns and their effects on land required for food', *Ecological Economics*, vol 42, pp185–199

Gielen, D. (2003) 'CO_2 removal in the iron and steel industry', *Energy Conversion and Management*, vol 44, pp1027–1037

Gielen, D. and Moriguchi, Y. (2002) 'CO_2 in the iron and steel industry: An analysis of Japanese emission reduction potentials', *Energy Policy*, vol 30, pp849–863

Gilbert, R. and Perl, A. (2007) *Transport Revolutions: Moving People and Freight Without Oil*, Earthscan, London

Gillingham, K., Newell, R. and Palmer, K. (2006) 'Energy efficiency policies: A retrospective examination', *Annual Review of Environment and Resources*, vol 31, pp161–192

Gitay, H., Brown, S., Easterling, W. and Jallow, B. (2001) 'Ecosystems and their goods and services', in J. J. McCarthy, O. S. Canziani, N. A. Leary, D. J. Dokken and K. S. White (eds) *Climate Change 2001: Impacts, Adaptation, and Vulnerability*, Cambridge University Press, Cambridge

Giurco, D. (2005) 'Towards sustainable metal cycles: The case of copper', PhD Thesis, Department of Chemical Engineering, University of Sydney, Sydney

Giurco, D. and Petrie, J. G. (2007) 'Strategies for reducing the carbon footprint of copper: New technologies, more recycling or demand management?', *Minerals Engineering*, vol 20, pp842–853

Gleick, P. H. and Cooley, H. S. (2009) 'Energy implications of bottled water', *Environmental Research Letters*, vol 4, no 1, doi:10.1088/1748-9326/4/1/014009

GMC (General Motors Corporation), Argonne National Laboratory, BP, ExxonMobil, Shell. (2001) *Volume 2, Well-to-Wheel Energy Use and Greenhouse Gas Emissions of Advanced Fuel/Vehicle Systems – North American Analysis – North American Analysis*, www.transportation.anl.gov/ttrdc/publications/index.html

Goldberg, J. and Wilkinson, C. (2004) 'Global threats to coral reefs: Coral bleaching, global climate change, disease, predator plagues, and invasive species', in Wilkinson (ed) *Status of Coral Reefs of the World: 2004*, vol 1, Australian Institute of Marine Science, www.aims.gov.au/pages/publications.html

Goldemberg, J. Johansson, T. B., Reddy, A. K. N. and Williams, R. H. (1988) *Energy for a Sustainable World*, Wiley Eastern, New Dehli

Goldman, C. A., Hopper, N. C. and Osborn, J. G. (2005) 'Review of US ESCO industry market trends: An empirical analysis of project data', *Energy Policy*, vol 33, pp387–405

Goldner, F. S. (1999) 'Control strategies for domestic hot water recirculation systems', *ASHRAE Transactions*, vol 105, part 1, 1030–1046

Goldner, F. S. (2000) 'Effects of equipment cycling and sizing on seasonal efficiency', in *Proceedings of the 2000 ACEEE Summer Study on Energy Efficiency in Buildings*, vol 1, American Council for an Energy Efficient Economy, Washington, DC

Goldstein, L., Hedman, B., Knowles, D., Freedman, S. I., Woods, R. and Schweizer, T. (2003) *Gas-Fired Distributed Energy Resource Technology Characterizations, NREL/TP-620–34783*, National Renewable Energy Laboratory, Golden, Colorado, www.nrel.gov/analysis/pdfs/2003/2003_gas-fired_der.pdf

Gómez, F., Guzmán, J. I. and Tilton, J. E. (2007) 'Copper recycling and scrap availability', *Resources Policy*, vol 32, pp183–190

Gordon, P. and Richardson, H. W. (1997) 'Are compact cities a desirable planning goal?', *Journal of the American Planning Association*, vol 63, pp71–82

Gordon, R. B., Graedel, T. E., Bertram, M., Fuse, K., Lifset, R., Rechberger, H. and Spatari, S. (2003) 'The characterization of technological zinc cycles', *Resources Conservation and Recycling*, vol 39, pp107–135

Gordon, R. B., Bertram, M. and Graedel, T. E. (2006) 'Metal stocks and sustainability', *Proceedings of the National Academy of Sciences*, vol 103, pp1209–1214

Gordon, R. B., Bertram, M. and Graedel, T. E. (2007) 'On the sustainability of metal supplies: A response to Tilton and Lagos', *Resources Policy*, vol 32, pp24–28

Gosselin, J. R. and Chen, Q. (2008) 'A computational method for calculating heat transfer and airflow through a dual-airflow window', *Energy and Buildings,* vol 40, pp452–458

Gowdy, J. and Juliá, R. (2007) 'Technology and petroleum exhaustion: Evidence from two mega-oilfields', *Energy*, vol 32, pp1448–1454

GPO (Government Printing Office) (2008) *Code of Federal Regulations*, Title 10, Part 430, www.gpoaccess.gov/CFR

Graboski, M. S. (2002) *Fossil Energy Use in the Manufacture of Corn Ethanol,* Report to the National Corn Growers Association, Golden, Colorado.

Graedel, T. E., van Beers, D., Bertram, M., Fuse, K., Gordon. R. B., Gritsinin, A., Kapur, A., Klee, R. J., Lifset, R. J., Memon, L., Rechberger, H., Spatari, S. and Vexler, D. (2004) 'Multilevel cycle of anthropogenic copper', *Environmental Science and Technology*, vol 38, pp1242–1252

Graedel, T. E., van Beers, D., Bertram, M., Fuse, K., Gordon, R. B., Gritsinin, A., Harper, E. M., Kapur, A., Klee, R. J., Lifset, R. J., Memon, L. and Spatari, S. (2005) 'Multilevel cycle of anthropogenic zinc', *Journal of Industrial Ecology*, vol 9, no 3, 67–90

Graham-Rowe, D. (2008a) 'Four wheels good?', *Nature*, vol 454, p810

Graham-Rowe, D. (2008b) 'Do the locomotion', *Nature*, vol 454, p1036

Grandum, S. and Horntvedt, B. (2000) 'Upgrading industrial waste heat using a hybrid heat pump', *CADDET Energy Efficiency Newsletter*, March, pp406

Gratia, E. and de Herde, A. (2007a) 'Guidelines for improving natural daytime ventilation in an office building with a double-skin façade', *Solar Energy*, vol 81, pp435–448

Gratia, E. and de Herde, A. (2007b) 'Are energy consumptions decreased with the addition of a double skin', *Energy and Buildings*, vol 39, pp605–619

Graus, W. H. J. and Voogt, M. (2005) *Updated Comparison of Power Efficiency on Grid Level*, Ecofys, Project ECS05036, Utrecht, The Netherlands

Graus, W. H. J. and Worrell, E. (2007) 'Effects of SO_2 and NO_x control on energy-efficiency power generation', *Energy Policy*, vol 35, pp3898–3908

Graus, W. H. J., Voogt, M. and Worrell, E. (2007) 'International comparison of energy efficiency of fossil power generation', *Energy Policy*, vol 35, pp3936–3951

Greening, L. A., Greene, D. L. and Difiglio, C. (2000) 'Energy efficiency and consumption – the rebound effect – a survey', *Energy Policy*, vol 28, pp389–401

Griffith, B. and Arasteh, D. (1995) 'Advanced insulations for refrigerator/freezers: The potential for new shell designs incorporating polymer barrier construction', *Energy and Buildings*, vol 22, pp219–231

Grönroos, J., Seppälä, J., Voutilainen, P., Seuri, P. and Koikkalainen, K. (2006) 'Energy use in conventional and organic milk and rye bread production in Finland', *Agriculture, Ecosystems and Environment*, vol 117, pp109–118

Grut, L. (2003) 'Daimler Chrysler Building, Berlin', in B. Edwards (ed) *Green Buildings Pay*, Spon Press, London

Gugliermetti, F. and Bisegna, F. (2003) 'Visual and energy management of electrochromic windows in Mediterranean climate', *Building and Environment*, vol 38, pp479–492

Gulli, F. (2006) 'Small distributed generation versus centralized supply: A social cost-benefit analysis in the residential and service sectors', *Energy Policy*, vol 34, pp804–832

Gündoğmuş, E. (2006) 'Energy use on organic farming: A comparative analysis on organic versus conventional apricot production on small holdings in Turkey', *Energy Conversion and Management*, vol 47, pp3351–3359

Haeseldonckx, D., Peeters, L., Helsen, L. and D'haeseleer, W. (2007) 'The impact of thermal storage on the operational behaviour of residential CHP facilities and the overall

CO_2 emissions', *Renewable and Sustainable Energy Reviews*, vol 11, pp1227–1243

Haller, A., Schweizer, E., Braun, P. O. and Voss, K. (1997) *Transparent Insulation in Building Renovation*, James & James, London

Halozan, H. (1997) 'Residential heat pump systems and controls', *IEA Heat Pump Centre Newsletter*, vol 15, no 2, pp19–21

Halozan, H. and Rieberer, R. (1997) 'Air heating systems for low-energy buildings', *IEA Heat Pump Centre Newsletter*, vol 15, no 4, pp21–22

Halozan, H. and Rieberer, R. (1999) 'Heat pumps in low-heating-energy buildings', in *20th International Congress of Refrigeration, IIR/IIF, Volume V (paper 499)*, Sydney

Halweil, B. (2002) *Home Grown: The Case for Local Food in a Global Market*, Worldwatch Institute, Washington, DC

Halweil, B. (2005) 'Aquaculture pushes fish harvest higher', in *Vital Signs 2005, The Trends that are Shaping Our Future*, Worldwatch Institute, Washington, DC, pp26–27

Halweil, B. and Nierenberg, D. (2008) 'Meat and seafood: The global diet's most costly ingredients', in *2008 State of World, Innovations for a Sustainable Economy*, Worldwatch Institute, Washington, DC, pp61–74

Hamada, Y., Nakamura, M., Ochifuji, K., Yokoyama, S. and Nagano, K. (2003) 'Development of a database of low energy homes around the world and analysis of their trends', *Renewable Energy*, vol 28, pp321–328

Hansen, J. (2005) 'A slippery slope: How much global warming constitutes "dangerous anthropogenic interference"?', *Climatic Change*, vol 68, pp269–279

Hanssen, O. J., Rukke, E. O., Saugen, B., Kolstad, J., Hafrom, P., Von Krogh, L., Raadal, H. L., Rønning, A. and Wigum, K. S. (2007) 'The environmental effectiveness of the beverage sector in Norway in a factor 10 perspective', *International Journal of Life Cycle Analysis*, vol 12, pp257–265

Harrington, L., Brown, J. and Ryan, P. (2006) 'Quantification of standby in Australia and trends in standby for new products', in P. Bertoldi, B. Kiss and B. Atanasiu (eds) *Energy Efficiency in Domestic Appliances and Lighting, Proceedings of the 4th International Conference EEDAL'06*, 21–23 June, London, pp985–997

Harris, J., Diamond, R., Iyer, M., Payne, C., Blumstein, C. and Siderius, H. P. (2008) 'Towards a sustainable energy balance: Progressive efficiency and the return of energy conservation', *Energy Efficiency*, vol 1, no 3, pp175–188

Harvey, L. D. D. (1993) 'A guide to global warming potentials (GWPs)', *Energy Policy*, vol 21, pp24–34

Harvey, L. D. D. (1995) 'Solar-hydrogen electricity generation in the context of global CO_2 emission reduction', *Climatic Change*, vol 29, pp53–89

Harvey, L. D. D. (2000) *Global Warming: The Hard Science*, Prentice Hall, Harlow, UK

Harvey, L. D. D. (2001) 'A quasi-one-dimensional coupled climate-carbon cycle model, Part II: The carbon cycle component', *Journal of Geophysical Research–Oceans*, vol 106, no 10, pp22355–22372

Harvey, L. D. D. (2006) *A Handbook on Low-Energy Buildings and District Energy Systems: Fundamentals, Techniques and Examples*, James & James, London

Harvey, L. D. D. (2007) 'Net climatic impact of solid foam insulation produced with halocarbon and non-halocarbon blowing agent', *Buildings and Environment*, vol 42, pp2860–2879

Harvey, L. D. D. and Kaufmann, R. K. (2002) 'Simultaneously constraining climate sensitivity and aerosol radiative forcing', *Journal of Climate*, vol 15, pp2837–2861

Harvey, L. D. D., Gregory, J., Hoffert, M., Jain, A., Lal, M., Leemans, R., Raper, S., Wigley, T. and de Wolde, J. (1997) *An Introduction to Simple Climate Models used in the IPCC Second Assessment Report*, Technical Paper No. 2, Intergovernmental Panel on Climate Change, Washington, DC

Hastings, R. and Wall, M. (2007a) *Sustainable Solar Housing, Volume 1, Strategies and Solutions*, Earthscan, London

Hastings, R. and Wall, M. (2007b) *Sustainable Solar Housing, Volume 2, Exemplary Buildings and Technologies*, Earthscan, London

Hastings, S. R. (1994) *Passive Solar Commercial and Institutional Buildings: A Sourcebook of Examples and Design Insights*, John Wiley, Chichester

Hastings, S. R. and Mørck, O. (2000) *Solar Air Systems: A Design Handbook*, James & James, London

Hayter, S. J., Torcellini, P. A., Judkoff, R. and Jenior, M. M. (1998) 'Creating low-energy commercial buildings through effective design and evaluation', in *Proceedings of the 1998 ACEEE Summer Study on Energy Efficiency in Buildings*, vol 3, American Council for an Energy Efficient Economy, Washington, DC, pp181–192

He, B. and Setterwall, F. (2002) 'Technical grade paraffin waxes as phase change materials for cool thermal storage and cool storage systems capital cost estimation', *Energy Conversion and Management*, vol 43, pp1709–1723

Hegerl, G. C. et al (2007) 'Understanding and attributing climate change', in S. Solomon, D. Qin, M. Manning, Z. Chen, M. Marquis, K. B. Averyt, M. Tignor and H. L. Miller (eds) *Climate Change 2007: The Physical Science Basis. Contribution of Working Group I to the Fourth Assessment Report of the Intergovernmental Panel on Climate Change*, Cambridge University Press, Cambridge and New York, pp663–745

Hekkert, M. P., van den Reek, J., Worrell, E. and Turkenburg, W. C. (2002) 'The impact of material efficient end-use technologies on paper use and carbon emissions', *Resources Conservation and Recycling*, vol 36, pp241–266

Heller, M. C. and Keoleian, G. A. (2000) *Life Cycle-Based Sustainability Indicators for Assessment of the U.S. Food*

System, Report CSS00-04, Center for Sustainable Systems, University of Michigan

Helser, Z. R. (2006) 'Energy in pesticide production and use', in D. Pimental (ed) *Encyclopedia of Pest Management*, Taylor & Francis, London

Henderson Jr., H. I., Parker, D. and Huang, Y. J. (2000) 'Improving DOE-2's RESYS routine: User defined functions to provide more accurate part load energy use and humidity predictions', in *Proceedings of the 2000 ACEEE Summer Study on Energy Efficiency in Buildings*, vol 1, American Council for an Energy Efficient Economy, Washington, DC, pp113–124

Henderson, S. C. and Wickrama, U. K. (1999) 'Aircraft emissions: Current inventories and future scenarios', in J. E. Penner et al (eds) *Aviation and the Global Atmosphere, A Special Report of IPCC Working Groups I and III*, Cambridge University Press, Cambridge, pp291–331

Hepting, C. and Ehret, D. (2005) *Centre for Interactive Research on Sustainability: Energy Performance Analysis Report*, www.sdri.ubc.ca/cirs

Herrera, A., Islas, J. and Arriola, A. (2003) 'Pinch technology application in a hospital', *Applied Thermal Engineering*, vol 23, pp127–139

Hertwich, E. (2005) 'Life cycle approaches to sustainable consumption: A critical review', *Environmental Science and Technology*, vol 39, pp4673–4684

Herzog, T. (1996) *Solar Energy in Architecture and Urban Planning*, Prestel, Munich

Hesse, M. (2002) 'Shipping news: The implications of electronic commerce for logistics and freight transport', *Resource, Conservation and Recycling*, vol 36, pp211–240

Hestnes, A. G. and Kofoed, N. U. (1997) *OFFICE: Passive Retrofitting of Office Buildings to Improve their Energy Performance and Indoor Environment, Final Report of the Design and Evaluation Subgroup*, European Commission Directorate General for Science Research and Development, JOULE Programme: JOR3-CT96-0034

Hestnes, A. G. and Kofoed, N. U. (2002) 'Effective retrofitting scenarios for energy efficiency and comfort: Results of the design and evaluation activities within the OFFICE project', *Building and Environment*, vol 37, pp569–574

Hestnes, A. G., Hastings, R. and Saxhof, B. (eds) (1997) *Solar Energy Houses: Strategies, Technologies, Examples*, James & James, London

Heywood, J. B., Weiss, M. A., Schafer, A., Bassene, S. A. and Natarajan, V. K. (2003) *The Performance of Future ICE and Fuel Cell Powered Vehicles and their Potential Fleet Impact*, MIT LFEE 2003–004 RP, http://web.mit.edu/sloan-auto-lab/research/beforeh2/reports.htm

Hien, W. N., Liping, W., Chandra, A. D., Pandey, A. R. and Xiaolin, W. (2005) 'Effects of double glazed façade on energy consumption, thermal comfort and condensation for a typical office building in Singapore', *Energy and Buildings*, vol 37, pp563–572

Hiete, M., Berner, U. and Richter, O. (2001) 'Calculation of global carbon dioxide emissions: Review of emission factors and a new approach taking fuel quality into consideration', *Global Biogeochemical Cycles*, vol 15, pp169–181

Hiller, C. C., Miller, J. and Dinse, D. R. (2002) 'Field test comparison of hot water recirculation loop vs. point-of-use water heaters in a high school', *ASHRAE Transactions*, vol 108, part 2, pp771–779

Hirsch, R. L. (2008) 'Mitigation of maximum world oil production: Shortage scenarios', *Energy Policy*, vol 36, pp881–889

Hirschberg, S., Heck, T., Gantner, U., Lu, Y., Spadaro, J. V., Trukenmuler, A. and Zhao, Y. (2004) 'Health and environmental impacts of China's current and future electricity supply, with associated external costs', *International Journal of Global Energy Issues*, vol 22, pp155–179

Hodder, S. G., Loveday, D. L., Parsons, K. C. and Taki, A. H. (1998) 'Thermal comfort in chilled ceiling and displacement ventilation environments: Vertical radiant temperature asymmetry effects', *Energy and Buildings*, vol 27, pp167–173

Hoegh-Guldberg, O. (2005) 'Low coral cover in a high-CO_2 world', *Journal of Geophysical Research–Oceans*, vol 110, no C9, ppC09S06.1–C09S06

Hoffert, M. I. et al (2002) 'Advanced technology paths to global climate stability: Energy for a greenhouse planet', *Science*, vol 298, pp981–987

Hofstetter, P., Madjar, M. and Ozawa, O. (2006) 'Happiness and sustainable consumption: Psychological and physical rebound effects at work in a tool for sustainable design', *International Journal of Life Cycle Analysis*, vol 11, pp105–115

Holton, J. K. and Rittelmann, P. E. (2002) 'Base loads (lighting, appliances, DHW) and the high performance house', *ASHRAE Transactions*, vol 108, part 1, pp232–242

Holtzclaw, J. (2004) 'A vision of energy efficiency', in *Proceedings of the 2004 ACEEE Summer Study on Energy Efficiency in Buildings*, vol 9, American Council for an Energy Efficient Economy, Washington, DC, pp55–62

Höök, M. and Aleklett, K. (2009) 'Historical trends in American coal production and a possible future outlook', *International Journal of Coal Geology*, vol 78, pp201–216

Höök, M., Hirsch, R. and Aleklett, K. (2009) 'Giant oil field decline rates and their influence on world oil production', *Energy Policy*, vol 37, pp2262–2272

Hoppe, K. (2009) 'Energy standards in community building codes: The path to implementation in Freiburg', in *Conference Proceedings, 13th International Passive House Conference 2009, 17–18 April, Frankfurt am Main*, Passive House Institute, Darmstadt, Germany, pp51–56

Horvath, A. and Shehabi, A. (2008) 'Improving the energy performance of data centers', University of California

Energy Institute (UCEI), Berkeley, EDT-104, available from www.ucei.berkeley.edu

Howe, M., Holland, D. and Livchak, A. (2003) 'Displacement ventilation: Smart way to deal with increased heat gains in the telecommunication equipment room', *ASHRAE Transactions*, vol 109, part 1, pp323–327

HPC (Heat Pump Centre) (2000) 'Heat pump working fluids', www.heatpumpcentre.org/about_heat_pumps/hp_working_fluids.asp, accessed January 2010

Huang, S. (2004) 'Merging information from different resource for new insights into climate change in the past and future', *Geophysical Research Letters*, vol 31, L13205, doi:10.1029/2004GL019781

Hughes, T. P., Baird, A. H., Bellwood, D. R., Card, M., Connolly, S. R., Folke, C., Grosberg, R., Hoegh-Guldberg, O., Jackson, J. B. C., Kleypas, J., Lough, J. M., Marshall, P., Nyström, M., Palumbi, S. R., Pandolfi, J. M., Rosen, B. and Roughgarden, J. (2003) 'Climate change, human impacts, and the resilience of coral reefs', *Science*, vol 301, pp929–933

Hülsbergen, K. J., Feil, B., Biermann, S., Rathke, G. W., Kalk, W. D. and Diepenbrock, W. (2001) 'A method of energy balancing in crop production and its application in a long-term fertilizer trial', *Agriculture Ecosystems and Environment*, vol 86, pp303–321

Humphreys, K. and Mahasenan, M. (2002) *Toward a Sustainable Cement Industry, Substudy 8: Climate Change*, World Business Council for Sustainable Development, Cement Sustainability Initiative, www.wbcsdcement.org

Hutson, S. S., Barber, N. L., Kenny, J. F., Linsey, K. S., Lumia, D. S. and Maupin, M. A. (2004) *Estimated Use of Water in the United States in 2000*, United States Geological Survey Circular 1268, available from http://pubs.usgs.gov/circ/2004/circ1268

Huovila, P., Ala-Juusela, M. and Pouffary, S. (2007) *Buildings and Climate Change, Status, Challenges, and Opportunities*, United Nations Environment Programme, Sustainable Buildings and Construction Initiative, Nairobi

Hydromantis (2006) *Energy Consumption Implications for Wastewater Treatment in Canada*, Hydromantis, Hamilton, Ontario, Canada

IAI (International Aluminium Institute) (2004) *Third Bauxite Mine Rehabilitation Survey*, www.world-aluminium.org

IEA (International Energy Agency) (1996) *Solar Collector System for Heating Ventilation air, CADDET Result 228*, International Energy Agency, Paris, www.caddet-ee.org

IEA (1999) *District Cooling, Balancing the Production and Demand in CHP*, Netherlands Agency for Energy and Environment, Sittard

IEA (2000) *Daylighting in Buildings: A Sourcebook on Daylighting Systems and Components*, Lawrence Berkeley National Laboratory, www.eetd.lbl.gov

IEA (2002) *Reclaiming heat from shower water*, CADDET Demo 48, International Energy Agency, Paris, www.caddet-ee.org

IEA (2004a) *Oil Crises and Climate Challenges: 30 Years of Energy Use in IEA Countries*, International Energy Agency, Paris

IEA (2004b) *Coming in from the Cold, Improving District Heating Policy in Transition Economies*, International Energy Agency, Paris

IEA (2005) *Projected Costs of Generating Electricity, 2005 Update*, International Energy Agency, Paris

IEA (2006) *Energy Technology Perspectives 2006, In Support of the G8 Plan of Action, Scenarios and Strategies to 2050*, International Energy Agency, Paris

IEA (2007a) *Energy Balances of OECD Countries 2004–2005*, International Energy Agency, Paris

IEA (2007b) *Energy Statistics of OECD Countries 2004–2005*, International Energy Agency, Paris

IEA (2007c) *Energy Balances of non-OECD Countries 2004–2005*, International Energy Agency, Paris

IEA (2007d) *Energy Statistics of non-OECD Countries 2004–2005*, International Energy Agency, Paris

IEA (2009) *Gadgets and Gigawatts: Policies for Energy Efficiency Electronics*, International Energy Agency, Paris

Igarashi, Y., Daigo, I., Matsuno, Y. and Adachi, Y. (2007) 'Dynamic material flow analysis for stainless steels in Japan – Reductions potential of CO_2 emissions by promoting closed loop recycling of stainless steels', *ISIJ International*, vol 47, pp758–763

IISI (International Iron and Steel Institute) (2007) *World Steel in Figures 2007*. Available from www.worldsteel.org

IMF (International Monetary Fund) (2008) *World Economic Outlook Online Database*, IMF, Washington, DC, www.imf.org

Interface Engineering (2005) *Engineering a Sustainable World: Design Process and Engineering Innovations for the Center for Health and Healing at the Oregon Health and Science University, River Campus*, www.interface-engineering.com

Ishimatsu, A., Hayashi, M., Lee, K. S., Kikkawa, T. and Kita, J. (2005) 'Physiological effects on fishes in a high-CO_2 world', *Journal of Geophysical Research–Oceans*, vol 110, no C09S09, doi:10.1029/2004JC002564

Jaboyedoff, P., Roulet, C. A., Dorer, V., Weber, A. and Pfeiffer, A. (2004) 'Energy in air-handling units: Results of the AIRLESS European project', *Energy and Buildings*, vol 36, pp391–399

Jackson, T. (2009) *Prosperity Without Growth: Economics for a Finite Planet*, Earthscan, London

James, P. W., Sonne, J. K., Vieira, R. K., Parker, D. S. and Anello, M. T. (1996) 'Are energy savings due to ceiling fans just hot air?', in *Proceedings of the 1996 ACEEE Summer Study on Energy Efficiency in Buildings*, vol 8, American Council for an Energy Efficient Economy, Washington, DC, pp89–93

Janson, U. and Wall, M. (2009) 'Renovation to Passive House standard in Brogården, Sweden', in *Conference Proceedings, 13th International Passive House Conference*

2009, 17–18 April, Frankfurt am Main, Passive House Institute, Darmstadt, Germany, pp189–194

Jasinski, S. M. (2009) 'Phosphate Rock', in *2008 Commodity Summary*, http://minerals.er.usgs.gov/minerals/pubs/commodity

Jenkinson, D. S. (2001) 'The impact of humans on the nitrogen cycle, with focus on temperate arable agriculture', *Plant and Soil*, vol 228, pp3–15

Jenks, M., Burton, E. and Williams, K. (eds) (1996) *The Compact City: A Sustainable Urban Form?*, E & FN SPon, London

Jennings, J. D., Rubinstein, F. M., DiBartolomeo, D. and Blanc, S. L. (2000) *Comparison of Control Options in Private Offices in an Advanced Lighting Controls Testbed*, Lawrence Berkeley National Laboratory, Berkeley, California

Jianwei, Y., Guolong, S., Cunjiang, K. and Tianjun, Y. (2003) 'Oxygen blast furnace and combined cycle (OBF-CC) – an efficient iron-making and power generation process', *Energy*, vol 28, pp825–835

Jin, Z., Wu, Y., Li., B. and Gao, Y. (2009) 'Energy efficiency supervision strategy of Chinese large-scale public buildings', *Energy Policy*, vol 37, pp2066–2072

Jochem, E. and Gruber, E. (2007) 'Local learning networks on energy efficiency in industry – successful initiative in Germany', *Applied Energy*, vol 84, pp806–816

Joelsson, J. M. and Gustavsson, L. (2008) 'CO$_2$ emission and oil use reduction through black liquor gasification and energy efficiency in pulp and paper industry', *Resources Conservation and Recycling*, vol 52, pp747–763

Johnson, J., Schewel, L. and Graedel, T. E. (2006) 'The contemporary anthropogenic chromium cycle', *Environmental Science and Technology*, vol 40, pp7060–7069

Johnson, J., Reck, B. K., Wang, T. and Graedel, T. E. (2008) 'The energy benefit of stainless steel recycling', *Energy Policy*, vol 36, pp181–192

Johnson, J. M., Franzluebbers, A. J., Weyers, S. L. and Reicosky, D. C. (2007) 'Agricultural opportunities to mitigate greenhouse gas emissions', *Environmental Pollution*, vol 150, pp107–124

Johnson, S. (2002) *LEDs – An Overview of the State of the Art in Technology and Application*, Lawrence Berkeley National Laboratory, Berkeley, California

Jordaan, S. M., Keith, D. W. and Stelfox, B. (2009) 'Quantifying land use of oil sands production: A life cycle perspective', *Environmental Research Letters,* vol 4, no 2, doi:10.1088/1748-9326/4/2/024004

Jordan, E. M. (2009) 'Passive House kindergarten in record time', in *Conference Proceedings, 13th International Passive House Conference 2009, 17–18 April, Frankfurt am Main*, Passive House Institute, Darmstadt, Germany, pp327–332

Jørgensen, U., Dalgaard, T. and Kristensen, E. S. (2005) 'Biomass energy in organic farming – the potential role of short rotation coppice', *Biomass and Bioenergy*, vol 28, pp237–248

Kaarsberg, T., Elliott, R. N. and Spurr, M. (1999) 'An integrated assessment of the energy savings and emissions-reduction potential of combined heat and power', updated and expanded from the version in the proceedings of the ACEEE 1999 Industrial Summer Study, American Council for an Energy-Efficient Economy, Washington, DC, 20036. Available from www.nemw.org

Kah, O., Schnieders, J. and Feist, W. (2009) 'Conditions and planning aspects of Passive House gymnasiums', in *Conference Proceedings, 13th International Passive House Conference 2009, 17–18 April, Frankfurt am Main*, Passive House Institute, Darmstadt, Germany, pp63–68

Kaiser, J. (2002) 'Breaking up is far too easy', *Science*, vol 297, pp1494–1496

Kalogirou, S. (2005) 'Seawater desalination using renewable energy sources', *Progress in Energy and Combustion Science*, vol 31, pp242–281

Kaltsas, A. M., Mamolos, A. P., Tsatsarelis, C. A., Nanos, G. D. and Kalburtji, K. L. (2007) 'Energy budget in organic and conventional olive groves', *Agriculture Ecosystems and Environment*, vol 122, pp243–251

Kammen, D. M. and Nemet, G. F. (2007) 'Energy myth eleven – Energy R&D investment takes decades to reach market', in B. K. Sovacool and M. A. Brown (eds) *Energy and American Society – Thirteen Myths*, Springer-Verlag, Berlin, p289–309

Karvountzi, G. C., Themelis, N. J. and Modi, V. (2002) 'Maximum distance to which cogenerated heat can be economically distributed in an urban community', *ASHRAE Transactions*, vol 108, part 1, pp334–339

Kasanen, P. (2000) *Efficient Domestic Ovens, Final Report*, Save II Project (4.1031/D/97-047), Tummavuoren Kirjapaino Oy, Helsinki

Kasser, T. (2002) *The High Price of Materialism*, MIT Press, Cambridge, MA

Kato, T., Kubota, M., Kobayashi, N. and Suzuoki, Y. (2005) 'Effective utilization of by-product oxygen from electrolysis hydrogen production', *Energy*, vol 30, pp2580–2595

Kats, G., Alevantis, L., Berman, A., Mills, E. and Perlman, J. (2003) *The Costs and Financial Benefits of Green Buildings: A Report to California's Sustainable Building Task Force*, Sustainable Building Task Force, Sacramento, California

Kaufmann, B., Peper, S., Pfluger, R. and Feist, W. (2009) 'Scientific monitoring of the Tevesstrasse Passive House renovation in Frankfurt a.M.', in *Conference Proceedings, 13th International Passive House Conference 2009, 17–18 April, Frankfurt am Main*, Passive House Institute, Darmstadt, Germany, pp165–170

Kaufmann, R. K. (2006) 'Planning for the peak in world oil production', *World Watch*, vol 19, pp19–21

Kavalov, B. and Peteves, S. D. (2007) *The Future of Coal*, European Commission Joint Research Centre, EUR 22744 EN, www.ie.jrc.cec.eu.int/publications/scientific_publications/2007/EUR22744EN.pdf

Kemp, R. (2007) *T618 – Traction Energy Metrics*, Lancaster University, Lancaster, www.rssb.co.uk

Kemp, R. and Smith, R. (2007) *Technical Issues Raised by the Proposal to Introduce a 500 km/h Magnetically-Levitated Transport System in the UK*, Imperial College London and Lancaster University, London and Lancaster

Keoleian, G. A., Blanchard, S. and Reppe, P. (2001) 'Life-cycle energy, costs, and strategies for improving a single-family house', *Journal of Industrial Ecology*, vol 4, pp135–156

Kesik, T. and Saleff, I. (2009) *Tower Renewal Guidelines for the Comprehensive Retrofit of Multi-Unit Residential Buildings in Cold Climates*, Faculty of Architecture and Landscape, and Design, University of Toronto, Toronto

Key, T. J., Schatzkin, A., Willett, W. C., Allen, N. E., Spencer, E. A. and Travis, R. C. (2004) 'Diet, nutrition and the preventation of cancer', *Public Health Nutrition*, vol 7, pp187–200

Keynes, J. M. (1930) *Economic Possibilities for our Grandchildren, Essays in Persuasion*, W. W. Norton, New York

Khrushch, M., Worrell, E., Price, L., Matrin, N. and Einstein, D. (2001) 'Carbon emissions reduction potential in the US chemicals and pulp and paper industries by applying CHP technologies', *Energy Policy*, vol 29, no 3, pp205–215

Kikegawa, Y., Genchi, Y., Kondo, H. and Hanaki, K. (2006) 'Impacts of city-block-scale countermeasures against urban heat-island phenomena upon a building's energy-consumption for air-conditioning', *Applied Energy*, vol 83, pp649–668

Kilkis, I. B. (1998) 'Rationalization of low-temperature to medium-temperature district heating', *ASHRAE Transactions*, vol 104, no 2, pp565–576

Kim, J., Lee, S. M. and Srinivasan, S. (1995) 'Modeling of proton exchange membrane fuel cell performance with an empirical equation', *Journal of the Electrochemical Society*, vol 142, pp2670–2674

Kim, T. S. (2004) 'Comparative analysis on the part load performance of combined cycle plants considering design performance and power control strategy', *Energy*, vol 29, pp71–85

Kliesch, J. and Langer, T. (2006) *Plug-in Hybrids: An Environmental and Economic Performance Outlook*, American Council for an Energy Efficiency Economy, Washington, DC, www.aceee.org

Klimstra, J. (2007) *On the Values of Local Electricity Generation*, European Local Electricity Production, European Commission, Brussels, Contract EIE/04/175/S07.38664

Klimstra, J. (2008) 'The benefits of flexible local electricity generation in the Middle East', *Power Gen Middle East*, 4–6 February 2008, Bahrain

Klimstra, J. and Hattar, C. (2006) 'Performance of natural-gas fueled engines heading towards their optimum', *Proceedings of ICES06: 2006 Internal Combustion Engine Division Spring Technical Conference*, 7–10 May 2006, Aachen, Germany

Klingenberg, K., Kernagis, M. and James, M. (2008) *Homes for a Changing Climate*, Low Carbon Productions, Larkspur, California

Klugman, S., Karlsson, M. and Moshfegh, B. (2007) 'A Scandinavian chemical wood-pulp mill. Part 2. International and model mills compared', *Applied Energy*, vol 84, pp340–350

Kobayashi, S., Plotkin, S. and Ribeiro, S. K. (2009) 'Energy efficiency technologies for road vehicles', *Energy Efficiency*, vol 2, pp125–137

Koga, N. (2008) 'An energy balance under a conventional crop rotation system in northern Japan: Perspectives on fuel ethanol production from sugar beet', *Agriculture Ecosystems and Environment*, vol 125, pp101–110

Kohli, D. K., Khardekar, R. K., Singh, R. and Gupta, P. K. (2008) 'Glass micro-container based hydrogen storage scheme', *International Journal of Hydrogen Energy*, vol 33, pp417–422

Kolokotroni, M. (2001) 'Night ventilation cooling of office buildings: Parametric analyses of conceptual energy impacts', *ASHRAE Transactions*, vol 107, part 1, pp479–489

Koomey, J. G. (2008) 'Worldwide electricity used in data ceners', *Environmental Research Letters*, vol 3, doi:10.1088/1748-9326/3/3/034008

Konuma, N. (2007) 'Japanese current activities for fuel economy improvement – Top-Runner standard', www.iea.org/Textbase/work/2007/vehicle/Konuma.pdf

Kooijman, J. M. (1993) 'Environmental assessment of packaging: Sense and sensibility', *Environmental Management*, vol 17, pp575–586

Krähling, H. and Krömer, S. (2000) *HFC-365mfc as Blowing and Insulation Agent in polyurethane Rigid Foams for Thermal Insulation, Life Cycle Aassessment Accompanying Application Development and Market Positioning, Final Summary Report*, Solvay Management Group, Hanover, Germany.

Krapmeier, H. and Drössler, E. (eds) (2001) *CEPHEUS: Living Comfort Without Heating*, Springer-Verlag, Vienna

Krishan, A. (1996) 'The habitat of two deserts in India: Hot-dry desert of Jaisalmer (Rajasthan) and the cold-dry high altitude mountainous desert of Leh (Ladakh)', *Energy and Buildings*, vol 23, pp217–229

Kromer, M. and Heywood, J. B. (2007) *Electric Powertrains: Opportunities and Challenges in the U.S. Light-Duty Vehicle Fleet*, Laboratory for Energy and the Environment, Massachusetts Institute of Technology, Cambridge, MA

Laitner, J. A. (2000) 'Energy efficiency: Rebounding to a sound analytical perspective', *Energy Policy*, vol 28, pp471–475

Lam, J. C. (2000) 'Energy analysis of commercial buildings in subtropical climates', *Building and Environment*, vol 35, pp19–26

Lam, J. C. and Li, D. H. W. (1999) 'An analysis of daylighting and solar heat for cooling-dominated office buildings', *Solar Energy*, vol 65, pp251–262

Lammert, M. (2008) *Long Beach Transit: Two-year Evaluation of Gasoline-Electric Hybrid Transit Buses*, National Renewable Energy Laboratory, Golden, Colorado, Technical Report NREL/TP-540-42226, www.nrel.gov/vehiclesandfuels/fleettest/research_hybrid.html

Larsson, N. (2001) 'Canadian green building strategies', in *18th International Conference on Passive and Low Energy Architecture*, Brazil, 7–9 November, pp17–25

Lasher, S., McKenney, K., Sinha, J., Ahluwalia, R., Hua, T. and Peng, J. K. (2009) *Technical Assessment of Compressed Hydrogen Storage Tank Systems for Automotive Applications, Report to the United States Department of Energy, Office of Energy Efficiency and Renewable Energy, Hydrogen, Fuel Cells and Infrastructure Technology Program, Part 1*. TIAX LLC, Cambridge, Massachusetts, Federal Grant Number DE-FC36-04GO14203

Lassaux, S., Renzoni, R. and Germain, A. (2007) 'Life cycle assessment of water from the pumping station to the wastewater treatment plant', *International Journal of Life Cycle Assessment*, vol 12, pp118–126

Laurance, W. and Williamson, G. B. (2001) 'Positive feedbacks among forest fragmentation, drought, and climate change in the Amazon', *Conservation Biology*, vol 15, pp1529–1535

Lautenschlager, L. and Smith, C. (2007) 'Beliefs, knowledge, and values held by inner-city youth about gardening, nutritition, and cooking', *Agriculture and Human Values*, vol 24, pp245–258

Layard, R. (2003) 'Towards a happier society', *New Statesman*, 3 March, pp25–28

Layard, R. (2005) *Happiness: Lessons from a New Science*, Allen Lane, London

Lazzarin, R. and Noro, M. (2006) 'Local or district heating by natural gas: Which is better from energetic, environmental and economic point of views?', *Applied Thermal Engineering*, vol 26, pp244–250

Leahy, E., Engelman, R., Vogel, C. B., Haddock, S. and Preston, T. (2007) *The Shape of Things to Come: Why Age Structure Matters to a Safer, More Equitable World*, Population Action International, Washington, DC, www.populationaction.org

Lebot, B., Bertoldi, P. and Harrington, P. (2004) 'Consumption versus efficiency: Have we designed the right policies and programmes?', in *Proceedings of the 2004 ACEEE Summer Study on Energy Efficiency in Buildings*, vol 7, American Council for an Energy Efficient Economy, Washington, DC, pp206–217

Lee, E. S., DiBartolomeo, D. L. and Selkowitz, S. E. (1998) 'Thermal and daylighting performance of an automated venetian blind and lighting system in a full-scale private office', *Energy and Buildings*, vol 29, pp47–63

Lee, J. J., Lukachko, S. P., Waitz, I. A. and Schafer, A. (2001) 'Historical and future trends in aircraft performance, cost, and emissions', *Annual Review of Energy Environment*, vol 26, pp167–200

Lee, K. H., Han, D. W. and Lim, H. J. (1996) 'Passive design principles and techniques for folk houses in Cheju Island and Ullüng Island of Korea', *Energy and Buildings*, vol 23, pp207–216

Lee, K. H. and Strand, R. K. (2008) 'The cooling and heating potential of an earth tube system in buildings', *Energy and Buildings*, vol 40, pp486–494

Lee, S. and Sherif, S. A. (2001) 'Thermoeconomic analysis of absorption systems for cooling', *ASHRAE Transactions*, vol 107, no 1, pp629–637

Lee, W. L., Yik, F. W. H., Jones, P. and Burnett, J. (2001) 'Energy saving by realistic design data for commercial buildings in Hong Kong', *Applied Energy*, vol 70, pp59–75

Leemans, R. and Eickhout, B. (2004) 'Another reason for concern: Regional and global impacts on ecosystems for different levels of climate change', *Global Environmental Change*, vol 14, pp219–228

Lefèvre, B. (2009) 'Long-term energy consumption of urban transportation: A prospective simulation of "transport-land uses" policies in Bangalore', *Energy Policy*, vol 37, pp940–953

Lekov, A., Lutz, J., Dunham Whitehead, C. and McMahon, J. E. (2000) 'Payback analysis of design options for residential water heaters', in *Proceedings of the 2000 ACEEE Summer Study on Energy Efficiency in Buildings*, vol 1, American Council for an Energy Efficient Economy, Washington, DC, pp163–174

Lemar, P. L. (2001) 'The potential impact of policies to promote combined heat and power in the US industry', *Energy Policy*, vol 29, pp1243–1254

Lemire, N. and Charneux, R. (2005) 'Energy-efficiency laboratory design', *ASHRAE Journal*, vol 47, no 5, pp58–64

Lenarduzzi, F. J. and Yap, S. S. (1998) 'Measuring the performance of a variable-speed drive retrofit on a fixed-speed centrifugal chiller', *ASHRAE Transaction*, vol 104, part 2, pp658–667

Lenzen, M. (1999) 'Total requirements of energy and greenhouse gases for Australian transport', *Transportation Research D*, vol 4, pp265–290

Levermore, G. J. (2000) *Building Energy Management Systems: Applications to Low-Energy HVAC and Natural Ventilation Control*, E & FN Spon, London

Levkoe, C. Z. (2006) 'Learning democracy through food justice movements', *Agriculture and Human Values*, vol 23, pp89–98

Levine, E. and Jamison, K. (2001) 'Oxy-fuel firing for the glass industry: An update on the impact of this successful government-industry cooperative effort', in *Proceedings of the 2001 ACEEE Summer Study on Energy Efficiency in Industry*, vol 1, American Council for an Energy Efficient Economy, Washington, DC, pp375–383

Levine, M., Ürge-Vorsatz, D., Blok, K., Geng, L., Harvey, D., Lang, S., Levermore, G., Mehlwana, A. M., Mirasgedis, S., Novikova, A., Rilling, J. and Yoshino, H. (2007)

'Residential and commercial buildings', in B. Metz, O. R. Davidson, P. R. Bosch, R. Dave and L. A. Meyer (eds) *Climate Change 2007: Mitigation, Contribution of Working Group III to the Fourth Assessment Report of the Intergovernmental Panel on Climate Change*, Cambridge University Press, Cambridge and New York, pp387–446

Lewis, J. S. and Niedzwiecki, R. W. (1999) 'Aircraft technology and its relation to emissions', in J. E. Penner et al. (eds) *Aviation and the Global Atmosphere, A Special Report of IPCC Working Groups I and III,* Cambridge University Press, Cambridge

Lewis, M. (2004) 'Integrated design for sustainable buildings', *Building for the Future, A Supplement to ASHRAE Journals*, vol 46, no 9, pp22–30

Li, D. H. W. and Lam, J. C. (2003) 'An investigation of daylighting performance and energy saving in a daylight corridor', *Energy and Buildings*, vol 35, pp365–373

Li, D. H. W., Lam, J. C. and Wong, S. L. (2005) 'Daylighting and its effects on peak load determination', *Energy*, vol 30, pp1817–1831

Li, H., Nalim, R. and Haldi, P. A. (2006) 'Thermal-economic optimization of a distributed multi-generation energy system – A case study of Beijing', *Applied Thermal Engineering*, vol 26, pp709–719

Lin, F., Yi, J., Weixing, Y. and Xuzhong, Q. (2001) 'Influence of supply and return water temperatures on the energy consumption of a district cooling system', *Applied Thermal Engineering*, vol 21, pp511–521

Lin, J. and Iyer, M. (2007) 'Cold or hot wash: Technological choices, cultural change, and their impact on clothes-washing energy use in China', *Energy Policy*, vol 35, pp3046–3052

Lin, J., Zhou, N., Levine, M. and Fridley, D. (2008) 'Taking out 1 billion tons of CO_2: The magic of China's 11th five-year plan?', *Energy Policy*, vol 36, pp954–970

Lin, Z. and Deng, S. (2004) 'A study on the characteristics of nighttime bedroom cooling load in tropics and subtropics', *Building and Environment*, vol 39, pp1101–1114

Lindenberg, S., Smith, B., O'Dell, K., DeMeo, E. and Ram, B. (2008) *20% Wind Energy by 2030, Increasing Wind Energy's Contribution to U.S. Electricity Supply,* US Department of Energy, Washington

Litman, T. (2007) 'Win-Win transportation solutions: Mobility management strategies that provide economic, social and environmental benefits', Victoria Transport Policy Institute, Victoria, Canada, www.vtpi.org

Little, A. D. (2000) *Cost Analysis of Fuel Cell System for Transportation, Baseline System Cost Estimate, Task 1 and 2 Final Report to Department of Energy*, Cambridge, MA

Liu, H., Ni, W., Li, Z. and Ma, L. (2008) 'Strategic thinking on IGCC development in China', *Energy Policy*, vol 36, pp1–11

Liu, M. and Claridge, D. E. (1999) 'Converting dual-duct constant-volume systems to variable-volume systems without retrofitting the terminal boxes', *ASHRAE Transactions*, vol 105, part 1, pp66–70

Lollini, R., Barozzi, B., Fasano, F., Meroni, I. and Zinzi, M. (2006) 'Optimisation of opaque components of the building envelope; Energy, economic, and environmental issues', *Building and Environment*, vol 41, no 8, pp1001–1013

Lombardi, L., Carnevale, E. and Corti, A. (2006) 'Greenhouse effect reduction and energy recovery from waste landfill', *Energy*, vol 31, pp3208–3219

Long, N., Torcellini, P. A., Pless, S. D. and Judkoff, R. (2006) 'Evaluation of the low-energy design process and energy performance of the Zion National Park Visitor Center', *ASHRAE Transactions*, vol 112, part 1, pp321–340

Loudermilk, K. J. (1999) 'Underfloor air distribution solutions for open office applications', *ASHRAE Transactions*, vol 105, part 1, p605–613

Lowenstein, A. and Novosol, D. (1995) 'The seasonal performance of a liquid-desiccant air conditioner', *ASHRAE Transactions*, vol 101, part 1, pp679–685

Luiten, E. E. M. and Blok, K. (2003) 'Stimulating R&D of industrial energy-efficient technology: The effect of government intervention on the development of strip casting technology', *Energy Policy*, vol 31, pp1339–1356

Lundie, S., Peters, G. M. and Beavis, P. C. (2004) 'Life cycle assessment for sustainable metropolitan water systems planning', *Environmental Science and Technology*, vol 38, pp3465–3473

Lutsey, N., Brodrick, C. J. and Lipman, T. (2007) 'Analysis of potential fuel consumption and emissions reductions from fuel cell auxiliary power units (APUs) in long-haul trucks', *Energy*, vol 32, pp2428–2438

Lutz, J. D., Klein, G., Springer, D. and Howard, B. D. (2002) 'Residential hot water distribution systems: Roundtable session', in *Proceedings of the 2002 ACEEE Summer Study on Energy Efficiency in Buildings*, vol 1, American Council for an Energy Efficient Economy, Washington, DC, pp131–144

Lutz, W., Sanderson, W. and Scherbov, S. (2001) 'The end of world population growth', *Nature*, vol 412, pp543–545

Maaløe, B. and Johansson, M. (1995) 'Variable speed drive for an air compressor', *CADDET Energy Efficiency Newsletter*, June, pp6–7

MacKay, D. J. C. (2007) *Sustainable Energy – Without the Hot Air*, Department of Physics, Cambridge University, Cambridge, www.withouthotair.com

MacRae, M. (1992) *Realizing the Benefits of Community Integrated Energy Systems*, Canadian Energy Research Institute, Calgary

Mäder, P., Fließbach, A., Dubois, D., Gunst, L., Fried, P. and Niggli, U. (2002) 'Soil fertility and biodiversity in organic farming', *Science*, vol 296, pp1694–1697

Maker, T. and Penny, J. (1999) *Heating Communities with Renewable Fuels: The Municipal Guide to Biomass District Energy*, Natural Resources Canada and USA Department of Energy. Available from www.nrcan.gc.ca

Malcolm, J. R., Liu, C., Neilson, R. P., Hansen, L. and Hannah, L. (2006) 'Global warming and extinctions of endemic species from biodiversity hotspots', *Conservation Biology*, vol 20, pp538–548

Malik, A. R. (1997) 'CoChill: A model for evaluating trigeneration options (simultaneous heating, cooling, and electricity production)', Masters Engineering Thesis, University of Toronto, Toronto

Maniates, M. (2002) 'In search of consumptive resistance: The voluntary simplicity movement', in T. Princen, M. Maniates and K. Conca (eds) *Confronting Consumption*, MIT Press, Cambridge, MA

Mann, J. I. (2002) 'Diet and risk of coronary heart disease and type 2 diabetes', *Lancet*, vol 360, pp783–789

Mann, M. E., Zhang, Z., Hughes, M. K., Bradley, R. S., Miller, S. K., Rutherford, S. and Ni, F. (2008) 'Proxy-based reconstructions of hemispheric and global surface temperature variations over the past two millennia', *Proceedings of the National Academy of Sciences*, vol 105, pp13,252–13,257

Manz, H. (2004) 'Total solar energy transmittance of glass double facades with free convection', *Energy and Buildings*, vol 36, pp127–136

Manz, H., Brunner, S. and Wullschleger, L. (2006) 'Triple vacuum glazing: Heat transfer and basic mechanical design constraints', *Solar Energy*, vol 80, pp1632–1642

Marbek Resource Consultants (1992) *City of Toronto Potential for Electricity Conservation, Residential Sector Appendices*, Marbek Resource Consultants, Ottawa, Ontario, Canada

Marland, G., Pielke, R. A., Apps, M., Avissar, R., Betts, R. A., Davis, K. J., Frumhoff, P. C., Jackson, S. T., Joyce, L. A., Kauppi, P., Katzenberger, J., Macdicken, K. G., Neilson, R. P., Niles, J. O., Niyogi, D. S., Norby, R. J., Pena, N., Sampson, N. and Xue, Y. (2003) 'The climatic impacts of land surface change and carbon management, and the implications for climate-change mitigation policy', *Climate Policy*, vol 3, pp149–157

Marland, G., Boden, T. A. and Andres, R. J. (2008) 'Global CO_2 emissions from fossil fuel burning, cement manufacture, and gas flaring, 1751–2005', Carbon Dioxide Information and Analysis Centre, Oak Ridge, www.cdiac.esd.ornl.gov/trends/emis/meth_reg.htm

Martchek, K. J. (2000) 'The importance of recycling to the environmental profile of metal products', in *Proceedings of the Fourth International Symposium on Recycling Metals and Engineered Materials*, TMS (The Mineral, Metals, & Materials Society), October 2000, Pittsburgh, Pensylvania.

Martchek, K. J. (2006) 'Modelling more sustainable aluminium', *International Journal of Life Cycle Assessment*, vol 11, pp34–37

Martinot, E. (2001) 'World Bank energy projects in China: Influences on environmental protection', *Energy Policy*, vol 29, pp581–594

Maruoka, N. and Akiyama, T. (2006) 'Exergy recovery from steelmaking off-gas by latent heat storage for methanol production', *Energy*, vol 31, pp1632–1642

Maslow, A. (1954) *Motivation and Personality*, Harper & Row, New York

Mason, H. E. and Spaner, D. (2006) 'Competitive ability of wheat in conventional and organic management systems: A review of the literature', *Canadian Journal of Plant Science*, vol 86, pp333–343

Mast, B., McCormick, J., Vogt, T., Ignelzi, P., Kolderup, E., Berman, M. and Dimit, M. (2000) 'Carrots or sticks? Policy options for building energy standards', in *Proceedings of the 2000 ACEEE Summer Study on Energy Efficiency in Buildings*, vol 9, American Council for an Energy Efficient Economy, Washington, DC, pp261–274

Masters, G. M. (2004) *Renewable and Efficient Electric Power Systems*, Wiley-Interscience, Hoboken, NJ

Matthews, H. S. and Hendrickson, C. T. (2003) 'The economic and environmental implications of centralized stock keeping', *Journal of Industrial Ecology*, vol 6, pp71–81

Maurer, M., Schwegler, P. and Larsen, T. (2003) 'Nutrients in urine: Energetic aspects of removal and recovery', *Water Science and Technology*, vol 48, pp37–46

Maus, S., Hapke, J., Ranong, C. N., Wüchner, E., Friedlmeier, G. and Wenger, D. (2008) 'Filling procedure for vehicles with compressed hydrogen tanks', *International Journal of Hydrogen Energy*, vol 33, pp4612–4621

Max-Neef, M. (1992) 'Development and human needs', in P. Ekins and M. Max-Neef (eds) *Real-life Economics: Understanding Wealth Creation*, Routledge, London, pp197–213

McCowan, B., Coughlin, T., Bergeron, P. and Epstein, G. (2002) 'High performance lighting options for school facilities', in *Proceedings of the 2002 ACEEE Summer Study on Energy Efficiency in Buildings*, vol 3, American Council for an Energy Efficient Economy, Washington, DC, pp253–268

McCulloch, M. T. and Esat, T. M. (2000) 'The coral record of last interglacial sea levels and sea surface temperatures', *Chemical Geology*, vol 169, pp107–129

McCulloch, M. T., Neabel, D. and Francis, E. (2006) *Greenhouse Gas Emissions Calculations for the Mackenzie Gas Project*, Pembina Institute, Drayton Valley, Alberta

McDonell, G. (2003) 'Displacement ventilation', *The Canadian Architect*, vol 48, no 4, pp32–33

McDougall, T., Nordmeyer, K. and Klaassen, C. J. (2006)
'Low-energy building case study: IAMU office and
training headquarters', *ASHRAE Transactions*, vol 112,
part 1, pp312–320

McGinnis, R. L. and Elimelech, M. (2008) 'Global
challenges in energy and water supply: The promise of
engineered osmosis', *Environmental Science and
Technology*, vol 42, pp8625–8629

McLaughlin, N. B., Hiba, A., Wall, G. J. and King, D. J.
(2000) 'Comparison of energy inputs for inorganic
fertilizer and manure based corn production', *Canadian
Agricultural Engineering*, vol 42, pp9–17

McMahon, J., Toriel, I., Rosenquist, G. J., Lutz, J. D.,
Boghosian, S. H. and Shown, L. (1997) 'Energy-efficient
technologies: Appliances, heat pumps, and air conditioning',
in F. Kreith and R. E. West (eds) *CRC Handbook of Energy
Efficiency*, CRC Press, Boca Raton, Florida, pp429–433

McMullan, J. T., Williams, B. C. and McCahey, S. (2001)
'Strategic considerations for clean coal R&D', *Energy
Policy*, vol 29, pp441–452

McNeil, M. A., Letschert, V. E. and de la Rue du Can, S.
(2008a) *Global Potential of Energy Efficiency Standards and
Labeling Programs*, Lawrence Berkeley National
Laboratory, Berkeley, CA

McNeil, M. A., de la Rue du Can, S. and McMahon, J. E.
(2008b) *Enduse Global Emissions Mitigation Scenarios
(EGEMS): A New Generation of Energy Efficiency Policy
Planning Models*, Lawrence Berkeley National Laboratory,
Berkeley, CA

Mehta, P. K. (1977) 'Properties of blended cements made
from rice husk ash', *Journal of the American Concrete
Institute*, vol 74, pp440–442

Mei, L. and Dai, Y. J. (2008) 'A technical review on use of
liquid-desiccant dehumidification for air-conditioning
application', *Renewable and Sustainable Energy Reviews*,
vol 12, pp662–689

Melet, E. (1999) *Sustainable Architecture: Towards a Diverse
Built Environment*, NAI Publishers, Rotterdam

Mendler, S. and Odell, W. (2000) *The HOK Guidebook to
Sustainable Design*, John Wiley, New York

Meng, Q. Y. and Bentley, R. W. (2008) 'Global oil peaking:
Responding to the case for "abundant supplies of oil"',
Energy, vol 33, pp1179–1184

Mercer, A. (1991) 'Process integration in the UK', *CADDET
Newsletter*, June, p4–7

Metwally, H. M. B. (2001) 'New method for speed control of
single phase induction motor with improved motor
performance', *Energy Conversion and Management*, vol 42,
pp941–950

Meul, M., Nevens, F., Reheul, D. and Hofman, G. (2007)
'Energy use efficiency of specialized dairy, arable, and pig
farms in Flanders', *Agriculture, Ecosystems and
Environment*, vol 119, pp135–144

Michaelis, L. (2003) 'Sustainable consumption and
greenhouse gas mitigation', *Climate Policy*, vol 3, no S1,
ppS135–S146

Minea, V. (2003) 'An exhaust-air heat recovery heat pump
system optimized for use in cold climates', *IEA Heat Pump
Centre Newsletter*, vol 21, no 3, pp14–17

Mitchell, C. and Cleveland, C. J. (1993) 'Resource scarcity,
energy use and environmental impact: A case study of the
New Bedford, Massachusetts, USA, fisheries',
Environmental Management, vol 17, pp305–317

Mitchell-Jackson, J., Koomey, J. G., Blazek, M. and
Nordman, B. (2002) 'National and regional implications
of internet data center growth in the US', *Resources
Conservation and Recycling*, vol 36, pp175–185

Mitchell-Jackson, J., Koomey, J. G., Nordman, B. and Blazek, M.
(2003) 'Data centre power requirements: Measurements
from Silicon Valley', *Energy*, vol 28, pp837–850

Moberg, A., Sonechkin, D. M., Holmgren, K., Datsenko, N. M.
and Karlén (2005) 'Highly variable northern hemisphere
temperatures reconstructed from low- and high-resolution
proxy data', *Nature*, vol 433, pp613–617

Modera, M. P., Brzozowski, O., Carrié, F. R., Dickerhoff, D. J.,
Delp, W. W., Fisk, W. J., Levinson, R. and Wang, D.
(2002) 'Sealing ducts in large commercial buildings with
aerosolized sealant particles', *Energy and Buildings*, vol 34,
pp705–714

Möllersten, K., Yan, J. and Westermark, M. (2003) 'Potential
and cost-effectiveness of CO_2 reductions through energy
measures in Swedish pulp and paper mills', *Energy*, vol 28,
pp691–710

Mollison, B. C. (1990) *Permaculture: A Practical Guide for a
Sustainable Future*, Island Press, Washington, DC

Monni, S. Pipatti, R., Lehtilä, A., Savolainen, I. and Syri, S.
(2006) *Global Climate Change Mitigation Scenarios for
Solid Waste Management*, VTT Publications 603, VTT
Technical Research Centre of Finland, www.vtt.fi/inf/pdf/
publications/2006/P603.pdf

Mosier, A., Kroeze, C., Nevison, C., Oenema, O., Seitzinger,
S. and van Cleemput, O. (1998) 'Closing the global N_2O
budget: Nitrous oxide emissions through the agricultural
nitrogen cycle', *Nutrient Cycling in Agroecosystems*, vol 52,
pp225–248

Mumma, S. A. (2001) 'Ceiling panel cooling system',
ASHRAE Journal, vol 43, no 11, pp28–32

Munasinghe, M. and Swart, R. (2005) *Primer on Climate
Change and Sustainable Development: Facts, Policy Analysis
and Applications*, Cambridge University Press, Cambridge

Muneer, T., Abodahab, N., Weir, G. and Kubie, J. (2000)
*Windows in Buildings: Thermal, Acoustic, Visual and Solar
Performance*, Architectural Press, Oxford

Murakami, S., Levine, M. D., Yoshino, H., Inoue, T., Ikaga,
T., Shimoda, Y., Miura, S., Sera, T., Nishio, M.,
Sakamoto, Y. and Fujisaki, W. (2006) 'Energy

consumption, efficiency, conservation, and greenhouse gas mitigation in Japan's building sector', *Energy and Carbon Emissions Country Studies*, Lawrence Berkeley National Laboratory in collaboration with Japanese institutions, Berkeley, CA

Murphy, P. (ed) (2002) *Solar Energy Activities in IEA Countries*, International Energy Agency, Solar Heating and Cooling Programme, Paris, www.iea-shc.org

Mysen, M., Berntsen, S., Nafstad, P. and Schild, P. G. (2005) 'Occupancy density and benefits of demand-controlled ventilation in Norwegian primary schools', *Energy and Buildings*, vol 37, pp1234–1240

Nakicenovic, N. and Swart, S. (eds) (2000) *Emission Scenarios*, Special Report of the Intergovernment Panel on Climate Change, Cambridge University Press, Cambridge

Nadel, S., deLaski, A., Eldridge, M. and Kleisch, J. (2006) *Leading the Way: Continued Opportunities for New State Appliance and Equipment Efficiency Standards*, American Council for an Energy Efficient Economy, Washington, DC

Neal, C. L. (1998) 'Field adjusted SEER [SEERFA] residential buildings: Technologies, design and performance analysis', in *Proceedings of the 1998 ACEEE Summer Study on Energy Efficiency in Buildings*, vol 1, American Council for an Energy Efficient Economy, Washington, DC, pp197–209

Neelis, M. L., Patel, M., Gielen, D. J. and Blok, K. (2005) 'Modelling CO_2 emissions from non-energy use with the non-energy use emission accounting tables (NEAT) model', *Resources, Conservation and Recycling*, vol 45, pp226–250

Neelis, M., Patel, M., Bach, P. and Blok, K. (2007) 'Analysis of energy use and carbon losses in the chemical industry', *Applied Energy*, vol 84, pp853–862

Nel, W. P. and Cooper, C. J. (2009) 'Implications of fossil fuel constraints on economic growth', *Energy Policy*, vol 37, pp166–180

Newman, P. and Kenworthy, J. (1999) *Sustainability and Cities: Overcoming Automobile Dependence*, Island Press, Washington, DC

Nierenberg, D. (2005) 'Meat production and consumption rise', in *Vital Signs 2005, The Trends that are Shaping Our Future*, Worldwatch Institute, Washington

Nipkow, J. (2009) 'Promotion of energy-efficient heat pump dryers', in P. Bertoldi and R. Werle (eds) *Energy Efficiency in Domestic Appliances and Lighting, Proceedings of the 5th International Conference EEDAL'09*, 16–18 June, Berlin, Vol 3, pp1215–1225

Nishikawa, H., Sasou, H., Kurihara, R., Nakamura, S., Kano, A., Tanaka, K., Aoki, T. and Ogami, Y. (2008) 'High fuel utilization of pure hydrogen fuel cells', *International Journal of Hydrogen Energy*, vol 33, pp6262–6269

Niu, J. L., Zhang, L. Z. and Zuo, H. G. (2002) 'Energy savings potential of chilled-ceiling combined with desiccant cooling in hot and humid climates', *Energy and Buildings*, vol 34, pp487–405

Noland, R. B. (2005) 'Fuel economy and traffic fatalities: Multivariate analysis of international data', *Energy Policy*, vol 33, pp2183–2190

Nordhoff, A. (2009) 'Passive House solutions for nursing homes', in *Conference Proceedings, 13th International Passive House Conference 2009, 17–18 April, Frankfurt am Main*, Passive House Institute, Darmstadt, Germany, pp57–62

Norgate, T. E. and Jahanshahi, S. (2004) 'Routes to stainless steel with improved efficiency', in *Green Processing 2004*, Australian IMM, Cairns, pp97–103

Norgate, T. E. and Rankin, W. J. (2000) 'Life cycle assessment of copper and nickel production', in *Proceedings of MINPREX 2000*, Australian Institute of Mining and Metallurgy, Melbourne, pp133–138

Norgate, T. E. and Rankin, W. J. (2001) 'Greenhouse gas emissions from aluminum production – a life cycle approach', in *Proceedings of the International Symposium on Greenhouse Gases in the Metallurgical Industries: Policies, Abatement, and Treatment*, 26–29 August, Toronto, Metallurgical Society of the Canadian Institute of Mining, Metallurgy, and Petroleum, pp275–290

Norgate, T. E. and Rankin, W. J. (2002) 'An environmental assessment of lead and zinc production processes', in *Green Processing 2002*, Australian IMM, Cairns, pp177–184

Norgate, T. E., Rajakumar, V. and Trang, S. (2004) 'Titanium and other light metals: Technology pathways to sustainable development', in *Green Processing 2004*, Australian IMM, Cairns, pp105–112

Norgate, T. E., Jahanshahi, S. and Rankin, W. J. (2007) 'Assessing the environmental impact of metal production processes', *Journal of Cleaner Production*, vol 15, pp838–848

NRCan (Natural Resourses Canada) (2003) *Technical Fact Sheets on the Impacts of New Energy Efficiency Technologies and Measures in Ice Rinks*, Catalogue No. M93-91/2003E, NRCan, Ottawa

NRCan (2007) *2007 EnerGuide Appliance Directory*, www.oee.nrcan.gc.ca/publications/infosource/pub/applia nces/2007/pdf/acdirectory3.pdf

Nuber, D., Eichberger, H. and Rollinger, B. (2006) 'Circored fine ore direct reduction', *Millennium Steel 2006*, pp37–40

Nussbaumer, T., Wakili K. G. and Tanner, C. (2006) 'Experimental and numerical investigation of the thermal performance of a protected vacuum-insulation system applied to a concrete wall', *Applied Energy*, vol 83, pp841–855

OECD (Organisation for Economic Co-operation and Development) (2007) *OECD Communications Outlook 2007*, OECD, Paris, www.economist.com/research/articlesBySubject/displaystory.cfm?subjectid=7933596&story_id=9527126

Oesterle, E., Lieb, R. D., Lutz, M. and Heusler, W. (2001) *Double-Skin Facades: Integrated Planning: Building Physics, Construction, Aerophysics, Air-conditioning, Economic Viability*, Prestel, Munich

Ogden, J. M. (1999) 'Prospects for building a hydrogen energy infrastructure', *Annual Review of Energy and the Environment*, vol 24, pp227–279

OIDA (Optoelectronics Industry Development Association) (2002) *Light Emitting Diodes (LEDs) for General Illumination, An OIDA Technology Roadmap Update 2002*, www.lighting.sandia.gov/lightingdocs/OIDA_SSL_LED_Roadmap_Full.pdf

Okamura, T., Furukawa, M. and Ishitani, H. (2007) 'Future forecast for life-cycle greenhouse gas emissions of LNG and city gas 13A', *Applied Energy*, vol 84, pp1136–1149

Oktay, D. (2002) 'Design with the climate in housing environments: An analysis in Northern Cyprus', *Building and Environment*, vol 37, pp1003–1012

Onovwiona, H. I. and Ugursal, V. I. (2006) 'Residential cogeneration systems: Review of the current technology', *Renewable and Sustainable Energy Reviews*, vol 10, pp389–431

Orr, J. C., Fabry, V. J., Aumont, O., Bopp, L., Doney, S. C., Feely, R. A., Gnanadesikan, A., Gruber, N., Ishida, A., Joos, F., Key, R. M., Lindsay, K., Maier-Reimer, E., Matear, R., Monfray, P., Mouchet, A., Najjar, R. G., Plattner, G. K., Rodgers, K. B., Sabine, C. L., Sarmiento, J. L., Schlitzer, R., Slater, R. D., Totterdell, I. J., Weirig, M. F., Yamanaka, Y. and Yool, A. (2005) 'Anthropogenic ocean acidification over the twenty-first century and its impact on calcifying organisms', *Nature*, vol 437, pp681–686

Orth, A., Anastasijevic, N. and Eichberger, H. (2007) 'Low CO_2 emission technologies for iron and steelmaking as well as titania slag production', *Minerals Engineering*, vol 20, pp854–861

Osmon, A. and Ries, R. (2007) 'Life cycle assessment of electrical and thermal energy systems for commercial buildings', *International Journal of Life Cycle Analysis*, vol 12, pp308–316

Otto-Bliesner, B. L., Marshall, S. J., Overpeck, J. T., Miller, G. H. and Hu, A. (2006) 'Simulating Arctic climate warmth and icefield retreat in the last interglacial', *Science*, vol 311, pp1751–1753

Ove Arup and Partners Hong Kong (2003) *Implementation Study for a District Cooling Scheme at South East Kowloon Development*, Electrical and Mechanical Services Department, Hong Kong. English summary available from www.emsd.gov.hk/emsd/eng/pee/wacs.shtml.

Overpeck, J. T., Otto-Bliesner, B. L., Miller, G. H., Muhs, D. R., Alley, R. B. and Kiehl, J. T. (2006) 'Paleoclimatic evidence for future ice-sheet instability and rapid sea-level rise', *Science*, vol 311, pp1747–1750

Ozalp, N. and Hyman, B. (2006) 'Calibrated models of on-site power and steam production in US manufacturing industries', *Applied Thermal Engineering*, vol 26, pp530–539

Pany, K. (2009) 'The City of Wels – Passive House Standard for an entire city', in *Conference Proceedings, 13th International Passive House Conference 2009, 17–18 April, Frankfurt am Main*, Passive House Institute, Darmstadt, Germany, pp43–50

Parker, D. S., Sherwin, J. R., Sonne, J. K., Barkaszi, S. F., Floyd, D. B. and Withers, C. R. (1998) 'Measured energy savings of a comprehensive retrofit in an existing Florida residence', in *Proceedings of the 1998 ACEEE Summer Study on Energy Efficiency in Buildings*, vol 1, American Council for an Energy Efficient Economy, Washington, DC, pp235–251

Parker, D. S., Callahan, M. P., Sonne, J. K. and Su, G. H. (1999) 'Development of a high efficiency ceiling fan', www.fsec.ucf.edu/bldg/pubs, accessed October 2003

Parker, D. S., Dunlop, J. P., Barkaszi, S. F., Sherwin, J. R., Anello, M. T. and Sonne, J. K. (2000) 'Towards zero energy demand: Evaluation of super efficient building technology with photovoltaic power for residential housing', in *Proceedings of the 2000 ACEEE Summer Study on Energy Efficiency in Buildings*, vol 1, American Council for an Energy Efficient Economy, Washington, DC, pp207–224

Parker, D. S., Sonne, J. K. and Sherwin, J. R. (2002) 'Comparative evaluation of the impact of roofing systems on residential cooling energy demand in Florida', in *Proceedings of the 2002 ACEEE Summer Study on Energy Efficiency in Buildings*, vol 1, American Council for an Energy Efficient Economy, Washington, DC, pp219–234

Parry, M., Arnell, N., McMichael, T., Nicholls, R., Martens, P., Kovats, S., Livermore, M., Rosenzweig, C., Iglesias, A. and Fischer, G. (2001) 'Millions at risk: Defining critical climate change threats and targets', *Global Environmental Change*, vol 11, pp181–183

Parry, M. L., Rosenzweig, C., Iglesias, A., Livermore, M. and Fischer, G. (2004) 'Effects of climate change on global food production and socio-economic scenarios', *Global Environmenal Change*, vol 14, pp53–67

Parsons Brinkerhoff Asia (2003) *Territory-Wide Implementation Study for Water-cooled Air Conditioning Systems in Hong Kong*, Electrical and Mechanical Services Department, Hong Kong. English summary available from www.emsd.gov.hk/emsd/eng/pee/wacs.shtml

Pasquay, T. (2004) 'Natural ventilation in high-rise buildings with double facades, saving or waste of energy', *Energy and Buildings*, vol 36, pp381–389

Patel, M. (2003) 'Cumulative energy demand (CED) and cumulative CO_2 emissions for products of the organic chemical industry', *Energy*, vol 28, pp721–740

Patel, M. and Mutha, N. (2004) 'Plastics production and energy', *Encyclopedia of Energy*, vol 5, pp81–91

Patz, J. A., Campbell-Lendrum, D., Holloway, T. and Foley, J. A. (2005) 'Impact of regional climate change on human health', *Nature*, vol 438, pp310–317

Pedersen, S. and Peuhkuri, R. (2009) 'A real Passive House in Finland', in *Conference Proceedings, 13th International Passive House Conference 2009, 17–18 April, Frankfurt am Main*, Passive House Institute, Darmstadt, Germany, pp177–182

Peeters, P. M., Middel, J. and Hoolhorst, A. (2005) *Fuel Efficiency of Commercial Aircraft: An Overview of Historical Future Trends*, National Aerospace Laboratory, The Netherlands, www.transportenvironment.org/docs/Publications/2005pubs/2005-12_nlr_aviation_fuel_efficiency.pdf

Peoples, M. B., Freney, J. R. and Mosier, A. R. (1999) 'Minimizing gaseous losses of nitrogen', in W.-Y. Huang and N. D. Uri (eds) *The Economic and Environmental Consequences of Nutrient Management in Agriculture*, Nova Science Publishers, Commack, New York

Peper, S. (2009) 'Low energy versus passive: Better air quality', in *Conference Proceedings, 13th International Passive House Conference 2009, 17–18 April, Frankfurt am Main*, Passive House Institute, Darmstadt, Germany, pp295–300

Peper, S. and Grove-Smith, J. (2009) 'Building renovation: From 290 to 16 – in 17 months', in *Conference Proceedings, 13th International Passive House Conference 2009, 17–18 April, Frankfurt am Main*, Passive House Institute, Darmstadt, Germany, pp171–176

Perzon, M., Johansson, K. and Fröling, M. (2007) 'Life cycle assessment of district heat distribution in suburban areas using PEX pipes insulated with expanded polystyrene', *International Journal of Life Cycle Assessment*, vol 12, pp317–327

Peters, C. J., Wilkins, J. L. and Fick, G. W. (2007) 'Testing a complete-diet model for estimating the land resource requirements of food consumption and agricultural carrying capacity: The New York State example', *Renewable Agriculture and Food Systems*, vol 22, pp145–153

Peters, C. J., Bills, N. L., Lembo, A. J., Wilkins, J. L. and Fick, G. W. (2009) 'Mapping potential foodsheds in New York State: A spatial model for evaluating the capacity to localize food production', *Renewable Agriculture and Food Systems*, vol 24, pp72–84

Petersdorff, C., Boermans, T., Harnisch, J., Joosen, S. and Wouters, F. (2002) *The Contribution of Mineral Wool and other Thermal Insulation Materials to Energy Saving and Climate Protection in Europe*, ECOFYS, Cologne. Available from www.ecofys.com.

Petersdorff, C., Boermans, T., Joosen, S., Kalacz, I., Jakubowska, B., Scharte, M., Stobbe, O. and Harnisch, J. (2005a) *Cost-effective Climate Protection in the EU Building Stock*, Report established by Ecofys for EURIMA, www.eurima.org

Petersdorff, C., Boermans, T., Harnisch, J., Stobbe, O., Ullrich, S. and Wortmann, S. (2005b) *Cost-effective Climate Protection in the Building Stock of the New EU Members: Beyond the EU Energy Performance of Buildings Directive*, Report established by Ecofys for EURIMA, www.eurima.org

Petersen, A. K. and Solberg, B. (2002) 'Greenhouse gas emissions, life-cycle inventory and cost-efficiency of using laminated wood instead of steel construction. Case: beams at Gardermoen airport', *Environmental Science and Policy*, vol 5, pp169–182

Petit, J. R. Jouzel, J., Raynaud, D., Barkov, N. I., Barnola, J. M., Basile, I., Bender, M., Chappellaz, J., Davis, J. Delaygue, G., Delmotte, M. Kotlyakov, V. M., Legrand, M., Lipenkov, V. M., Lorius, C., Pépin, L., Ritz, C., Saltzman, E. and Stievenard, M. (1999) 'Climate and atmospheric history of the past 420,000 years from the Vostok ice core, Antarctica', *Nature*, vol 399, p429–436

Pfeffer, W. T., Harper, J. T. and O'Neel, S. (2008) 'Kinematic constraints on glacier contributions to 21st-century sea level rise', *Science*, vol 321, pp1340–1343

Phair, J. W. (2006) 'Green chemistry for sustainable cement production and use', *Green Chemistry*, vol 8, pp763–780

Phylipson, D., Blok, K., Worrell, E. and de Beer, J. (2002) 'Benchmarking the energy efficiency of Dutch industry: An assessment of the expected effect on energy consumption and CO_2 emissions', *Energy Policy*, vol 30, pp663–679

Piette, M. A., Kinney, S. K. and Haves, P. (2001) 'Analysis of an information monitoring and diagnostic system to improve building operations', *Energy and Buildings*, vol 33, pp783–791

Pimentel, D. (2004) 'Livestock production and energy use', in *Encyclopedia of Energy*, vol 3, Elsevier

Pimentel, D. and Pimentel, M. H. (1996) 'Food processing, packaging, and preparation', in D. Pimentel and M. Pimentel (eds) *Food, Energy, and Society*, 2nd Edition, University Press of Colorado, pp186–201

Pimentel, D. and Pimentel, M. H. (2008) *Food, Energy, and Society*, 3rd Edition, CRC Press, Boca Raton

Pimentel, D., Houser, J., Preiss, E., White, O., Fang, H., Mesnick, L., Barsky, T., Tariche, S., Schreck, J. and Alpert, S. (1997) 'Water resources: Agriculture, the environment, and society', *BioScience*, vol 47, pp97–106

Piringer, G. and Steinberg, L. J. (2006) 'Reevaluation of energy use in wheat production in the United States', *Journal of Industrial Ecology*, vol 10, pp149–167

Placet, M. and Fowler, K. (2002) *Toward a Sustainable Cement Industry, Substudy 7: How Innovation Can Help the Cement Industry Move Toward More Sustainable Practices,*

World Business Council for Sustainable Development, Cement Sustainability Initiative, www.wbcsdcement.org

Plastics Europe (2008) *The Compelling Facts about Plastics 2007: An Analysis of Plastics Production, Demand and Recovery for 2007 in Europe*, www.plasticseurope.org

Pless, S. D., Torcellini, P. A. and Petersen, J. E. (2006) 'Energy performance evaluation of a low-energy building', *ASHRAE Transactions*, vol 112, part 1, pp295–311

Ploss, M. (2008) 'Passive house retrofit: Taking it easy', *Renewable Energy World*, vol 11, no 2, pp56–62

Poirazis, H., Blomsterberg, Å. and Wall, M. (2008) 'Energy simulations for glazed office buildings in Sweden', *Energy and Buildings*, vol 40, pp1161–1170

Porta-Gándara, M. A., Rubio, E. and Fernández, J. L. (2002) 'Economic feasibility of passive ambient comfort in Baja California dwellings', *Building and Environment*, vol 37, pp993–1001

Pörtner, H. O., Langenbuch, M. and Michaelidis, B. (2005) 'Synergistic effects of temperature extremes, hypoxia, and increases in CO_2 on marine animals: From Earth history to global change', *Journal of Geophysical Research–Oceans*, vol 110, no C09S10, doi:10.1029/2004JC002561

Poullikkas, A. (2005) 'An overview of current and future sustainable gas turbine technologies', *Renewable and Sustainable Energy Reviews*, vol 9, pp409–433

Pounds, J. A., Bustamante, M. R., Coloma, L. A., Consuegra, J. A., Fogden, M. P. L., Foster, P. N., La Marca, E., Masters, K. L., Merino-Viteri, A., Puschendorf, R., Ron, S. R., Sánchez-Azofeifa, G. A., Still, C. J. and Young, B. E. (2006) 'Widespread amphibian extinctions from epidemic disease driven by global warming', *Nature*, vol 439, pp161–167

Prasad, D. and Snow, M. (2005) *Designing with Solar Power: A Source Book for Building Integrated Photovoltaics (BiPV)*, James & James, London

Prentice, I. C., Farquhar, G. D., Fasham, M. J. R., Goulden, M. L., Heimann, M., Jaramillo, V. J., Kheshgi, H. S., Le Quéré, C., Scholes, R. J. and Wallace, D. W. R. (2001) 'The carbon cycle and atmospheric carbon dioxide', in J. T. Houghton, Y. Ding, D. J. Griggs, M. Noguer, P. J. van der Linden, X. Dai, K. Maskell and C. A. Johnson (eds) *Climate Change 2001: The Scientific Basis*, Cambridge University Press, Cambridge

Presutto, M. (2009) 'Spinning speed of washing machines: An analysis of the tradeoff with the penetration and use of tumble dryers', in P. Bertoldi and R. Werle (eds) *Energy Efficiency in Domestic Appliances and Lighting, Proceedings of the 5th International Conference EEDAL'09*, 16–18 June, Berlin, Vol 3, pp1194–1205

Pretty, J. (2005) 'Sustainability in agriculture: Recent progress and emergent challenges', in R. E. Hester and R. E. Harrison (eds) *Issues in Environmental Science and Technology, No 21, Sustainability in Agriculture*, Royal Society of Chemisty, London

Price, W. and Hart, R. (2002) 'Bulls-eye commissioning: Using interval data as a diagnostic tool', in *Proceedings of the 2002 ACEEE Summer Study on Energy Efficiency in Buildings*, vol 3, American Council for an Energy Efficient Economy, Washington, DC, pp295–307

Price, H., Mehos, M., Kutscher, C. and Blair, N. (2007) 'Current and future economics of parabolic trough technology', in *Proceedings of Energy Sustainability 2007*, 27–30 June, Long Beach, CA

Princen, T. (2005) *The Logic of Sufficiency*, MIT Press, Cambridge, MA

Proc, K., Chaney, L. and Sailor, E. (2008) 'Thermal load reduction of truck tractor sleeper cabins', presented at the *2008 SAE Commercial Vehicle Engineering Congress & Exhibition, October 2008, Chicago, Illinois*, National Renewable Energy Laboratory, Golden, Colorado, NREL/CP-540-43402, www.nrel.gov/vehiclesandfuels/fleettest/research_hybrid.html

Pucher, J. (1997) 'Bicycling boom in Germany: A revival engineered by public policy', *Transportation Quarterly*, vol 51, no 4, pp31–46

Rabl, A. and Spadaro, J. V. (2000) 'Public health impact of air pollution and implications for the energy system', *Annual Review of Energy and the Environment*, vol 25, pp601–627

Racoviceanu, A. I., Karney, B. W., Kennedy, C. A. and Colombo, A. F. (2007) 'Life-cycle energy use and greenhouse gas emissions inventory for water treatment systems', *Journal of Infrastructure Systems*, vol 13, pp261–270

Råde, I. and Andersson, B. A. (2001) 'Platinum group metal resource constraints for fuel-cell electric vehicles', Report 2001:01, Chalmers University of Technology, Department of Physical Resource Theory, Gothenberg, Sweden

Rafiqul, I., Weber, C., Lehmann, B. and Voss, A. (2005) 'Energy efficiency improvements in ammonia production – perspectives and uncertainties', *Energy*, vol 30, pp2487–2504

Rahmstorf, S. (2007) 'A semi-empirical approach to projecting future sea level rise', *Science*, vol 315, pp368–370

Ramírez, C. A. and Worrell, E. (2006) 'Feeding fossil fuels to the soil: An analysis of energy embedded and technological learning in the fertilizer industry', *Resources Conservation and Recycling*, vol 46, pp75–93

Randall, D. A., Wood, R. A., Bony, S., Colman, R., Fichefet, T., Fyfe, J., Kattsov, V., Pitman, A., Shukla, J., Srinivasan, J., Stouffer, R. J., Sumi, A. and Taylor, K. E. (2007) 'Climate models and their evaluation', in S. Solomon, D. Qin, M. Manning, Z. Chen, M. Marquis, K. B. Averyt, M. Tignor and H. L. Miller (eds) *Climate Change 2007: The Physical Science Basis. Contribution of Working Group 1 to the Fourth Assessment Report of the Intergovernmental Panel on Climate Change*, Cambridge University Press, Cambridge and New York, pp589–662

Rappa, R. F. and Lawas, A. (2005) 'Industrial process improvements and energy efficiency', in *Proceedings of the 2005 ACEEE Summer Study on Energy Efficiency in Industry*, vol 6, American Council for an Energy Efficient Economy, Washington, DC, pp151–162

Rawitscher, M. and Mayer, J. (1977) 'Nutritional outputs and energy inputs in seafoods', *Science*, vol 198, pp261–264

Ray-Jones, A. (2000) *Sustainable Architecture in Japan: The Green Buildings of Nikken Sekkei*, Wiley-Academic, Chichester, UK

Reaka-Kudla, M. L. (1996) 'The global biodiversity of coral reefs: A comparison with rain forests', in M. L. Reaka-Kudla, D. E. Wilson and E. O. Wilson (eds) *Biodiversity II: Understanding and Protecting Our Biological Resources*, Joseph Henry Press, Washington, DC, pp83–108

Reay, D. A. and MacMichael, D. B. A. (1988) *Heat Pumps*, 2nd Edition, Pergamon Press, Oxford

Reck, B. K., Müller, D. B., Rostkowski, K. and Graedel, T. E. (2008) 'Anthropogenic nickel cycle: Insights into use, trade, and recycling', *Environmental Science and Technology*, vol 42, pp3394–3400

Reijnders, L. (2007) 'The cement industry as a scavenger in industrial ecology and the management of hazardous substances', *Journal of Industrial Ecology*, vol 11, pp15–25

Reinberg, G.W. (2009) 'Protecting historic buildings: Passive House renovations in Purkersdorf, Vienna', in *Conference Proceedings, 13th International Passive House Conference 2009, 17–18 April, Frankfurt am Main*, Passive House Institute, Darmstadt, Germany, pp153–158

Reinhart, C. F. (2002) 'Effects of interior design on the daylight availability in open plan offices', in *Proceedings of the 2002 ACEEE Summer Study on Energy Efficiency in Buildings*, vol 3, American Council for an Energy Efficient Economy, Washington, DC, pp309–322

Reinhart, C. F., Voss, K., Wagner, A. and Löhnert, G. (2000) 'Lean buildings: Energy efficient commercial buildings in Germany', in *Proceedings of the 2000 ACEEE Summer Study on Energy Efficiency in Buildings*, vol 3, American Council for an Energy Efficient Economy, Washington, DC, pp257–298

Ren, K. B. and Kagi, D. A. (1995) 'Upgrading the durability of mud bricks by impregnation', *Building and Environment*, vol 30, pp433–440

Ren, T., Patel, M. and Blom, K. (2006) 'Olefins from conventional and heavy feedstocks: Energy use in steam cracking and alternative processes', *Energy*, vol 31, pp425–451

Reynolds, C. and Kandlikar, M. (2007) 'How hybrid-electric vehicles are different from conventional vehicles: The effect of weight and power on fuel consumption', *Environmental Research Letters*, vol 2, doi:10.1088/1748-9326/2/1/014003

Ribeiro, S. K., Kobayashi, S., Beuthe, M., Gasca, J., Greene, D., Lee, D. S., Muromachi, Y., Newton, P. J., Plotkin, S., Sperling, D., Wit, R. and Zhou, P. J. (2007) 'Transport and its infrastructure', in B. Metz, O. R. Davidson, P. R. Bosch, R. Dave and L. A. Meyer (eds) *Climate Change 2007: Mitigation, Contribution of Working Group III to the Fourth Assessment Report of the Intergovernmental Panel on Climate Change*, Cambridge University Press, Cambridge and New York, pp323–385

Richard Rogers Architects (1996) 'Three demonstration buildings, Potsdamer Platz, Berlin', in Fitzgerald, E. and Lewis, J. O. (eds) *European Solar Architecture: Proceedings of a Solar House Contractors' Meeting, Barcelona 1995*, Energy Research Group, University College Dublin, pp198–207

Rietbergen, M. G., Farla, J. C. M. and Blok, K. (2002) 'Do agreements enhance energy efficiency improvement? Analysing the actual outcome of long-term agreements on industrial energy efficiency improvement in The Netherlands', *Journal of Cleaner Production*, vol 10, pp153–163

Rishel, J. B. and Kincaid, B. L. (2007) 'Reducing energy costs with condensing boilers and heat recovery chillers', *ASHRAE Journal*, vol 49, no 3, pp46–55

Ritchie, A. and Howard, W. (2006) 'Recent developments and likely advances in lithium-ion batteries', *Journal of Power Sources*, vol 162, pp809–812

Rogner, H. H. (1997) 'An assessment of world hydrocarbon resources', *Annual Review of Energy and the Environment*, vol 22, pp217–262

Rosenfeld, A. H. (1999) 'The art of energy efficiency: Protecting the environment with better technology', *Annual Review of Energy and the Environment*, vol 24, pp33–82

Rosenfeld, A. H., Akbari, H., Romm, J. J. and Pomerantz, M. (1998) 'Cool communities: Strategies for heat island mitigation and smog reduction', *Energy and Buildings*, vol 28, pp51–62

Rosenquist, G., McNeil, M., Iyer, M., Meyers, S. and McMahon, J. (2006) 'Energy efficiency standards for equipment: Additional opportunities in the residential and commercial sectors', *Energy Policy*, vol 34, pp3257–3267

Ross, C. P. (2005) 'Recycling, glass', in Kirk-Othmer (ed) *Encyclopedia of Chemical Technology*, John Wiley & Sons, New York

Ross, M. (1994) 'Automobile fuel consumption and emissions: Effects of vehicle and driving characteristics', *Annual Review of Energy and the Environment*, vol 19, pp75–112

Ross, M. and.Wenzel, T. (2002) *An Analysis of Traffic Deaths by Vehicle Type and Model*, American Council for an Energy-Efficient Economy, Washington, DC

Ross, M., Patel, D. and Wenzel, T. (2006) 'Vehicle design and the physics of traffic safety', *Physics Today*, January, pp49–54

Roth, K. W., Goldstein, F. and Kleinman, J. (2002) *Energy Consumption by Office and Telecommunications Equipment in Commercial Buildings. Volume 1: Energy Consumption Baseline*, Arthur D. Little Inc., Cambridge, MA, www.eren.doe.gov/buildings/documents/pdfs/office_telecom_vol1_final.pdf

Roy, A. N., Mahmood, A. R., Baslev-Olesen, O., Lojuntin, S., Tang, C. K. and Kannan, K. S. (2005) 'Low energy office building in Putrajaya, Malaysia. Case studies and innovations', in *Proceedings of Conference on Sustainable Building South Asia, 11–13 April, Malaysia*, pp223–230

Roy, R. N., Misra, R. V. and Montanez, A. (2002) 'Decreasing reliance on mineral nitrogen – yet more food', *Ambio*, vol 31, pp177–183

Royer, D. L., Berner, R. A. and Park, J. (2007) 'Climate sensitivity constrained by CO_2 concentrations over the past 420 million years', *Nature*, vol 446, pp530–532

Rubin, E. S., Berkenpas, M. B., Farrell, A., Gibbon, G. A. and Smith, D. N. (2001) 'Multi-pollutant emission control of electric power plants', in *The EPA/DOE/EPRI Mega Symposium, 20–23 August 2001*, www.iecm-online.com/publications.html.

Rubinstein, F. and Johnson, S. (1998) *Advanced Lighting Program Development (BG9702800) Final Report*, Lawrence Berkeley National Laboratory, Berkeley, CA

Rubinstein, F., Jennings, H. J. and Avery, D. (1998) *Preliminary Results from an Advanced Lighting Controls Testbed*, Lawrence Berkeley National Laboratory, Berkeley, CA

Rudd, A., Kerrigan Jr., P. and Ueno, K. (2004) 'What will it take to reduce total residential source energy use by up to 60%?', in *Proceedings of the 2004 ACEEE Summer Study on Energy Efficiency in Buildings*, vol 1, American Council for an Energy Efficient Economy, Washington, DC, pp293–305

Rugh, J. O., Hovland, V. and Andersen, S. O. (2004) 'Significant fuel savings and emissions reductions by improving vehicle air conditioning', Mobile Air Conditioning Summit, Washington, DC, 14–15 April 2004.

Ruhrberg, M. (2006) 'Assessing the recycling efficiency of copper from end-of-life products in Western Europe', *Resources Conservation and Recycling*, vol 48, pp141–165

Rumsey, P. and Weale, J. (2007) 'Chilled beams in labs: Eliminating reheat and saving energy on budget', *ASHRAE Journal*, vol 49, no 1, pp18–25

Ruth, M. and Dell'Anno, P. (1997) 'An industrial ecology of the US glass industry', *Resources Policy*, vol 23, pp109–124

Rutherford, S., Mann, M. E., Osborn, T.J., Bradley, R.S., Briffa, K. R., Hughes, M. K. and Jones, P. D. (2005) 'Proxy-based northern hemisphere surface temperature reconstructions: Sensitivity to method, predictor network, target season, and target domain', *Journal of Climate*, vol 18, pp2308–2329

Rutledge, D. (2007) *Hubbert's Peak, the Coal Question, and Climate Change*, powerpoint presentation and data files available at www.rutledge.caltech.edu

(S&T)² Consultants Inc. (2005) *Documentation for Natural Resources Canada's GHGenius Model 3.0*, www.ghgenius.ca

Sachs, H. M., Nadel, S., Amann, J. T., Tuazon, M., Mendelsohn, E., Rainer, L., Todecso, G., Shipley, D. and Adelaar, M. (2004) *Emerging Energy-Saving Technologies and Practices for the Buildings Sector as of 2004*, American Council for an Energy Efficient Economy, Washington, DC

Saelens, D., Roels, S. and Hens, H. (2008) 'Strategies to improve the energy performance of multiple-skin facades', *Building and Environment*, vol 43, pp638–650

Sakintuna, B., Lamari-Darkrim, F. and Hirscher, M. (2007) 'Metal hydride materials for solid hydrogen storage: A review', *International Journal of Hydrogen Energy*, vol 32, pp1121–1140

Sanne, C. (2002) 'Willing consumers – or locked in? Policies for a sustainable consumption', *Ecological Economics*, vol 42, pp273–287

Sarkar, A. and Banerjee, R. (2005) 'Net energy analysis of hydrogen storage options', *International Journal of Hydrogen Energy*, vol 30, pp867–877

Sathaye, J. and Phadke, A. (2006) 'Cost of electric power sector carbon mitigation in India: International implications', *Energy Policy*, vol 34, pp1619–1629

Sathre, R. and Gustavsson, L. (2006) 'Energy and carbon balances of wood cascade chains', *Resources Conservation and Recycling*, vol 47, pp332–355

Saunders, C. and Hayes, P. (2007) *Air Freight Transport of Fresh Fruit and Vegetables, Research Report No 299*, Lincoln University, Lincoln, New Zealand

Saunders, C., Barber, A. and Taylor, G. (2006) *Food Miles – Comparative Energy/Emissions Performance of New Zealand's Agriculture Industry*, Research Report 285, AERU, Lincoln University, Lincoln, New Zealand

Sausen, R., Isaksen, I., Grewe, V., Hauglustaine, D., Lee, D. S., Myhre, G., Köhler, M. O., Pitari, G., Schumann, U., Stordal, F. and Zerefos, C. (2005) 'Aviation radiative forcing in 2000: An update on IPCC (1999)', *Meteorologische Zeitschrift*, vol 14, pp555–561

Sawhney, R. L., Buddhi, D. and Thanu, N. M. (1999) 'An experimental study of summer performance of a recirculation type underground airpipe air conditioning system', *Building and Environment*, vol 34, pp189–196

Scambos, T. A., Bohlander, J. A., Shuman, C. A. and Skvarca, P. (2004) 'Glacier acceleration and thinning after ice shelf collapse in the Larsen B embayment, Antarctica', *Geophysical Research Letters*, vol 31, no 18, L18402

Scartezzini, J. L. and Courret, G. (2002) 'Anidolic daylighting systems', *Solar Energy*, vol 73, pp123–135

Schell, M. B., Turner, S. C. and Shim, R. O. (1998) 'Application of CO_2-based demand-controlled ventilation using ASHRAE standard 62: Optimizing energy use and ventilation', *ASHRAE Transactions*, vol 104, part 2, pp1213–1225

Schild, P. and Blom, P. (2002) *Pilot Study Report: Jaer School, Nesodden Municipality, Norway*, International Energy Agency, Energy Conservation in Buildings and Community Systems, Annex 35. Available from http://hybvent.civil.auc.dk

Schimel, D. (2006) 'Climate change and crop yields: Beyond Cassandra', *Science*, vol 312, pp1889–1890

Schipper, L. and Grubb, M. (2000) 'On the rebound? Feedback between energy intensities and energy use in IEA countries', *Energy Policy*, vol 28, pp367–388

Schipper, L., Unander, F., Murtishaw, S. and Ting, M. (2001) 'Indicators of energy use and carbon emissions: Explaining the energy economy link', *Annual Review of Energy and the Environment*, vol 26, pp49–81

Schipper, L., Marie-Lilliu, C. and Fulton, L. (2002) 'Diesels in Europe: Analysis of characteristics, usage patterns, energy savings and CO_2 emission implications', *Journal of Transport Economics and Policy*, vol 36, pp305–340

Schleisner, L. (2000) 'Life cycle assessment of a wind farm and related externalities', *Renewable Energy*, vol 20, pp279–288

Schlesinger, M. E. (2007) *Aluminum Recycling*, CRC Press, Boca Raton

Schmidt, D. (2002) 'The Centre for Sustainable Building (ZUB), A Case Study', presented at *Sustainable Buildings 2002*, Oslo, Norway, International Initiative for a Sustainable Built Environment, www.iisbe.org

Schneider, S. H., Semenov, S., Patwardhan, A., Burton, I., Magadza, C. H. D., Oppenheimer, M., Pittock, A. B., Rahman, A., Smith, J. B., Suarez, A. and Yamin, F. (2007) 'Assessing key vulnerabilities and the risk from climate change', in M. L. Parry, O. F. Canziani, J. P. Palutikof, P. J. van der Linden and C. E. Hansen (eds) *Climate Change 2007: Impacts, Adaptation and Vulnerability. Contribution of Working Group II to the Fourth Assessment Report of the Intergovernmental Panel on Climate Change*, Cambridge University Press, Cambridge and New York, pp799–810

Schneider, U. H. (2009) 'Energy base: Renewable energy and use comfort in a Passive House office complex', in *Conference Proceedings, 13th International Passive House Conference 2009, 17–18 April, Frankfurt am Main*, Passive House Institute, Darmstadt, Germany, pp301–306

Schnepf, R. (2004) *Energy Use in Agriculture: Background and Issues*, CRS Report for Congress RL32677, www.nationalaglawcenter.org/assets/crs/RL32677.pdf

Schnieders, J. and Hermelink, A. (2006) 'CEPHEUS results: Measurements and occupants' satisfaction provide evidence for Passive Houses being an option for sustainable building', *Energy Policy*, vol 34, pp151–171

Schnieders, J., Wagner, A. and Heinrich, H. (2009) 'Passive Houses in south-west Europe – highlights from a theoretical analysis', in *Conference Proceedings, 13th International Passive House Conference 2009, 17–18 April, Frankfurt am Main*, Passive House Institute, Darmstadt, Germany, pp81–86

Schor, J. B. (2005) 'Sustainable consumption and worktime reduction', *Journal of Industrial Ecology*, vol 9, pp37–50

Schossig, P., Henning, H.-M., Gschwander, S. and Haussmann, T. (2005) 'Micro-encapsulated phase-change materials integrated into construction materials', *Solar Energy Materials and Solar Cells*, vol 89, pp297–306

Schulz, T. (2009) 'The Passive House Standard for indoor swimming pools', in *Conference Proceedings, 13th International Passive House Conference 2009, 17–18 April, Frankfurt am Main*, Passive House Institute, Darmstadt, Germany, pp123–128

Schwartz, B. (2004) *The Paradox of Choice: Why More is Less*, Ecco, New York

Schwarz, H. G., Briem, S. and Zapp, P. (2001) 'Future carbon dioxide emissions in the global material flow of primary aluminum', *Energy*, vol 26, pp775–795

Scott, D. S., Rogner, H. H. and Scott, M. B. (1993) 'Fuel cell locomotives in Canada', *International Journal of Hydrogen Energy*, vol 18, pp253–263

Sekhar, S. C. and Phua, K. J. (2003) 'Integrated retrofitting strategy for enhanced energy efficiency in a tropical building', *ASHRAE Transactions*, vol 109, part 1, pp202–214

Sellers, D. and Williams, J. (2000) 'A comparison of the ventilation rates established by three common building codes in relationship to actual occupancy levels and the impact of these rates on building energy consumption', in *Proceedings of the 2000 ACEEE Summer Study on Energy Efficiency in Buildings*, vol 3, American Council for an Energy Efficient Economy, Washington, DC, pp299–313

Service, R. F. (2006) 'Desalination freshens up', *Science*, vol 313, pp1088–1090

Sheppard, C. R. C. (2003) 'Predicted recurrences of mass coral mortality in the Indian Ocean', *Nature*, vol 425, pp294–297

Sherman, M. H. and Jump, D. A. (1997) 'Thermal energy conservation in buildings', in F. Kreith and R. E. West (eds) *CRC Handbook of Energy Efficiency*, CRC Press, Boca Raton

Shiau, C. S. N., Samaras, C., Hauffe, R. and Michalek, J. J. (2009) 'Impact of battery weight and charging patterns on the economic and environmental benefits of plug-in hybrid vehicles', *Energy Policy*, vol 37, pp2653–2663

Shirayama, Y. and Thornton, H. (2005) 'Effects of increased atmospheric CO_2 on shallow water marine benthos', *Journal of Geophysical Resources–Oceans*, vol 110, no C09, doi:10.1029/2004JC002618

Short, C. A. and Lomas, K. J. (2007) 'Exploiting a hybrid environmental design strategy in a US continental climate', *Building Research and Information*, vol 35, pp119–143

Short, W. and Denholm, P. (2006) *A Preliminary Assessment of Plug-In Hybrid Electric Vehicles on Wind Energy Markets*, National Renewable Energy Laboratory, Golden, Colorado, NREL/TP-620-39729

Sibley, S. F. and Butterman, W. C. (1995) 'Metals recycling in the United States', *Resources Conservation and Recycling*, vol 15, pp256–267

Siikavirta, H., Punakivi, M., Kärkkäinen, M. and Linnanen, L. (2003) 'Effects of e-commerce on greenhouse gas emissions', *Journal of Industrial Ecology*, vol 6, pp83–97

Simmler, H. and Brunner, S. (2005) 'Vacuum insulation panels for building application: Basic properties, aging mechanisms and service life', *Energy and Buildings*, vol 37, pp1122–1131

Simpson, A. (2006) *Cost-benefit Analysis of Plug-in Hybrid Electric Vehicle Technology*, National Renewable Energy Laboratory, Golden, CO

Sims, R. E. H., Schock, R. N., Adegbululgbe, A., Fenhann, J., Konstantinaviciute, I., Moomaw, W., Nimir, H. B., Schlamadinger, B., Torres-Martínez, J., Turner, C., Uchiyama, Y., Vuori, Seppo, Wamukonya, N. and Zhang, X. (2007) 'Energy Supply', in B. Metz, O. R. Davidson, P. R. Bosch, R. Dave and L. A. Meyer (eds) *Climate Change 2007: Mitigation, Contribution of Working Group III to the Fourth Assessment Report of the Intergovernmental Panel on Climate Change*, Cambridge University Press, Cambridge and New York, pp251–322

Singh, H., Singh, A. K., Kushwaha, H. L. and Singh, A. (2007) 'Energy consumption pattern of wheat production in India', *Energy*, vol 32, pp1848–1854

Sinton, C. W. (2004) 'Glass and energy', in *Encyclopedia of Energy*, vol 3, Elsevier, New York, pp1–10

Skjølsvik, K. O., Andersen, A. B., Corbett, J. J. and Skjelvik, J. M. (eds) (2000a) *Study of Greenhouse Gas Emissions from Ships, Final Report to the International Maritime Organization*, International Maritime Organization, London, www.unfccc.int/files/methods_and_science/ emissions_ from_intl_ transport/application/pdf/imoghgmain.pdf

Skjølsvik, K. O., Andersen, A. B., Corbett, J. J. and Skjelvik, J. M. (eds) (2000b) *Study of Greenhouse Gas Emissions from Ships, Appendices*, International Maritime Organization, London, www.unfccc.int/files/methods_and_science/emissions_ from_intl_transport/application/pdf/imoghgapp.pdf

Small, K. and van Dender, K. (2005) *The Effect of Improved Fuel Economy on Vehicle Miles Traveled: Estimating the Rebound Effect Using U.S. State Data, 1966–2001*, University of California Energy Institute, Working Paper 014, www.ucei.org

Smil, V. (1994a) *Energy in World History*, Westview Press, Boulder, Colorado

Smil, V. (1994b) 'How many people can the Earth feed?', *Population and Development Review*, vol 20, pp255–292

Smil, V. (2000) 'Phosphorus in the environment: Natural flows and human interferences', *Annual Review of Energy and the Environment*, vol 25, pp53–88

Smith, C. B. (1997) 'Electrical energy management in buildings', in F. Kreith and R. E. West (eds) *CRC Handbook of Energy Efficiency*, CRC Press, Boca Raton, pp305–336

Smith, P. et al (2007) 'Agriculture', in B. Metz, O. R. Davidson, P. R. Bosch, R. Dave and L. A. Meyer (eds) *Climate Change 2007: Mitigation, Contribution of Working Group III to the Fourth Assessment Report of the Intergovernmental Panel on Climate Change*, Cambridge University Press, Cambridge and New York, pp497–540

Söderbergh, B., Robelius, F. and Aleklett, K. (2007) 'A crash programme scenario for the Canadian oil sands industry', *Energy Policy*, vol 35, pp1931–1947

Solaini, G., Dall'o', G. and Scansani, S. (1998) 'Simultaneous application of different natural cooling technologies to an experimental building', *Renewable Energy*, vol 15, pp277–282

Sovacool, B. and Hirsh, R. F. (2009) 'Beyond batteries: An examination of the benefits and barriers to plug-in hybrid electric vehicles (PHEV) and a vehicle-to-grid (V2G) transition', *Energy Policy*, vol 37, pp1095–1103

Sperling, D. and Cannon, J. S. (2004) *The Hydrogen Energy Transition: Moving Toward the Post Petroleum Age in Transportation*, Elsevier Academic Press, Amsterdam

Springer, D., Loisos, G. and Rainer, L. (2000) 'Non-compressor cooling alternatives for reducing residential peak load', in *Proceedings of the 2000 ACEEE Summer Study on Energy Efficiency in Buildings*, vol 1, American Council for an Energy Efficient Economy, Washington, DC, pp319–330

Srinivasan, S., Mosdale, R., Stevens, P. and Yang, C. (1999) 'Fuel cells: Reaching the era of clean and efficient power generation in the twenty-first century', *Annual Review of Energy and the Environment*, vol 24, pp281–328

Stec, W. J. and van Paassen, A. H. C. (2005) 'Symbiosis of the double skin façade with the HVAC system', *Energy and Buildings*, vol 37, pp461–469

Steinbock, J., Eijadi, D. and McDougall, T. (2007) 'Net zero energy building case study: Science house', *ASHRAE Transactions*, vol 113, part 1, pp26–35

Steinfeld, H. and Wassenaar, T. (2007) 'The role of livestock production in carbon and nitrogen cycles', *Annual Review of Environment and Resources*, vol 32, pp271–294

Stern, D. I. (2006) 'Reversal of the trend in global anthropogenic sulfur emissions', *Global Environmental Change*, vol 16, pp207–220

Stetiu, C. and Feustel, H. E. (1999) 'Energy and peak power savings potential of radiant cooling systems in US commercial buildings', *Energy and Buildings*, vol 30, pp127–138

Stirling, C. H., Esat, T. M., Lambeck, K. and McCulloch, M. T. (1998) 'Timing and duration of the last interglacial: Evidence for a restricted interval of widespread coral reef growth', *Earth and Planetary Science Letters*, vol 160, pp745–762

Strachan, N. and Farrell, A. (2006) 'Emissions from distributed vs centralized generation: The importance of system performance', *Energy Policy*, vol 34, pp2677–2689

Strapatsa, A. V., Nanos, G. D. and Tsatsarelis, C. A. (2006) 'Energy flow for integrated apple production in Greece', *Agriculture Ecosystems and Environment,* vol 116, pp176–180

Strasser, P., Fan, Q., Devenney, M., Weinberg, W. H., Liu, P. and Nørskov, J. K. (2003) 'High throughput experimental and theoretical predictive screening of materials – a comparative study of search strategies for new fuel cell anode catalysts', *Journal of Physical Chemistry B*, vol 107, pp11013–11021

Strickler, B. (1997) 'Car sharing in Switzerland – 21st century mobility takes shape', *CADDET Energy Efficiency*, September, pp25–27

Sun, H. S. and Lee, S. E. (2006) 'Case study of data centers' energy performance', *Energy and Buildings*, vol 38, pp522–533

Sundqvist, J. O. (2006) 'Assessment of organic waste treatment', in J. Dewulf and H. van Langenhove (eds) *Renewables-Based Technology*, John Wiley & Sons, Chichester, pp247–263

Suzuki, M. and Oka, T. (1998) 'Estimation of life cycle consumption and CO_2 emission of office buildings in Japan', *Energy and Buildings*, vol 28, pp33–41

Swanson, M. D., Miller R. A., Aardsma, J. J. and Chimack, M. J. (2005) 'E2 and P2 improvement opportunities in secondary aluminum processing: A case study', in *Proceedings of the 2005 ACEEE Summer Study on Energy Efficiency in Industry*, vol 1, American Council for an Energy Efficient Economy, Washington, DC, pp179–191

Swart, R., Robinson, J. and Cohen, S. (2003) 'Climate change and sustainable development: Expanding the options', *Climate Policy*, vol 3, no S1, ppS19–S40

Takagi, R. (2006) 'Development of low-energy consumption trains in Japan – A literature survey', University of Birmingham, Birmingham

Tantasavasdi, C., Srebric, J. and Chan, Q. (2001) 'Natural ventilation for houses in Thailand', *Energy and Buildings*, vol 33, pp815–824

Tao, Z. and Li, M. (2007) 'What is the limit of Chinese coal supplies? A STELLA model of Hubbert peak', *Energy Policy*, vol 35, pp3145–3154

Tenorio, R. (2007) 'Enabling the hybrid use of air conditioning: A prototype on sustainable housing in tropical regions', *Building and Environment*, vol 42, pp605–613

Thanu, N. M., Sawhney, R. L., Khare, R. N. and Buddhi, D. (2001) 'An experimental study of the thermal performance of an earth-air-pipe system in single pass mode', *Solar Energy*, vol 71, pp353–364

Thekdi, A. (2003) 'Review of opportunities and activities to improve energy efficiency in the aluminum industry', in S. K. Das (ed) *Aluminum 2003*, The Minerals, Metals & Materials Society, pp225–237

Thollander, P., Danestig, M. and Rohdin, P. (2007) 'Energy policies for increased industrial energy efficiency: Evaluation of a local energy programme for manufacturing SMEs', *Energy Policy*, vol 35, pp5774–5783

Thomas, C. D., Cameron, A., Green, R. E., Bakkenes, M., Beaumont, L. J., Collingham, Y. C., Erasmus, B. F. N., de Siqueira, M. F., Grainger, A., Hannah, L., Hughes, L., Huntley, B., van Jaarsveld, A. S., Midgley, G. F., Miles, L., Ortega-Huerta, M. A., Peterson, A. T., Phillips, O. L. and Williams, S. E. (2004) 'Extinction risk from climate change', *Nature*, vol 427, pp145–147

Thomas, C. E., James, B. D., Lomax Jr., F. D. and Kuhn Jr., I. F. (2000) 'Fuel options for the fuel cell vehicle: Hydrogen, methanol, or gasoline?', *International Journal of Hydrogen Energy*, vol 25, pp551–567

Thomas, R. (1999) *Environmental Design: An Introduction for Architects and Engineers*, E & FN Spon, London

Thormark, C. (2002) 'A low energy building in a life cycle – its embodied energy, energy need for operation and recycling potential', *Building and Environment*, vol 37, pp429–435

Tidåker, P., Kärrman, E., Baky, A. and Jönsson, H. (2006) 'Wastewater management integrated with farming – an environmental systems analysis of a Swedish country town', *Resources Conservation and Recycling*, vol 47, pp295–315

Tilten, J. E. and Lagos, G. (2007) 'Assessing the long-run availability of copper', *Resources Policy*, vol 32, pp19–23

Tippenhauer, H. (2009) 'How compatible are consumer electronics and sustainability', in P. Bertoldi and R. Werle (eds) *Energy Efficiency in Domestic Appliances and Lighting, Proceedings of the 5th International Conference EEDAL'09*, 16–18 June, Berlin, Vol 3, pp1123–1131

Tiwari, P. (2001) 'Energy efficiency and building construction in India', *Building and Environment*, vol 36, pp1127–1135

Tiwari, P., Parikh, J. and Sharma, V. (1996) 'Performance evaluation of cost effective buildings – a cost, emissions and employment point of view', *Building and Environment*, vol 31, pp75–90

Tokimatsu, K. (2007) 'Latest HDV fuel efficiency technologies in Japan', presented at the *International Workshop on Fuel Efficiency Policies for Heavy-Duty Vehicles*, 21–22 June, International Energy Agency, Paris, www.iea.org/textbase/work/workshopdetail.asp?WS_ID=306

Tomić, J. and van Amburg, B. (2007) 'Medium- and heavy-duty hybrids – The HTUF process from hybrids to ultra efficient trucks', presented at the *International Workshop on Fuel Efficiency Policies for Heavy-Duty Vehicles*, 21–22 June, International Energy Agency, Paris, www.iea.org/textbase/work/workshopdetail.asp?WS_ID=306

Tommerup, H. and Svendsen, S. (2006) 'Energy savings in Danish residential building stock', *Energy and Buildings*, vol 38, pp618–626

Tonner, J. B. and Tonner, J. (2004) 'Desalination and energy use', *Encyclopedia of Energy*, vol 1, pp791–799

Torcellini, P. A. and Crawley, D. B. (2006) 'Understanding zero-energy buildings', *ASHRAE Journal*, vol 48, pp62–69

Torcellini, P. A., Judkoff, R. and Hayter, S. J. (2002) 'Zion National Park Visitor Center: Significant energy savings achieved through a whole-building design process', in *Proceedings of the 2002 ACEEE Summer Study on Energy Efficiency in Buildings*, vol 3, American Council for an Energy Efficient Economy, Washington, DC, pp361–372

Torcellini, P. A., Deru, M., Griffith, B., Long, N., Pless, S., Judkoff, R. and Crawley, D. B. (2004a) 'Lessons learned from field evaluation of six high-performance buildings', in *Proceedings of the 2004 ACEEE Summer Study on Energy Efficiency in Buildings*, vol 3, American Council for an Energy Efficient Economy, Washington, DC, pp325–337

Torcellini, P. A., Judkoff, R. and Crawley, D. B. (2004b) 'High-performance buildings: Lessons learned. Buildings for the future (supplement)', *ASHRAE Journal*, vol 46, no 9, ppS4–S11

TRB (Transportation Research Board, National Research Council) (2002) *Effectiveness and Impact of Corporate Average Fuel Economy (CAFE) Standards*, National Academies Press, Washington, DC

Trivedi, K. K. (1997) 'Process energy efficiency: Pinch technology', in F. Kreith and R. E. West (eds) *CRC Handbook of Energy Efficiency*, CRC Press, Boca Raton

Tschudi, W. and Fok, S. (2007) 'Best practices for energy-efficient data centers identified through case studies and demonstration projects', *ASHRAE Transactions*, vol 113, part 1, pp450–456

Turner, C. H. and Tovey, N. K. (2006) 'Case study on the energy performance of the Zuckerman Institute for Connective Environmental Research (ZICER) building', *ASHRAE Transactions*, vol 113, part 2, pp320–329

Ullah, M. B. and Lefebvre, G. (2000) 'Estimation of annual energy-saving contribution of an automated blind system', *ASHRAE Transactions*, vol 106, part 2, pp408–418

UN (United Nations) (1992) *United Nations Framework Convention on Climate Change*, United Nations, Treaty Series, vol 1771, p107

UN (2008) *2005 Energy Statistics Yearbook*, United Nations, Department of Economic and Social Affairs, New York

UNPD (United Nations Population Division) (2008) *World Population Prospects: The 2008 Revision Population Database*, UNPD, www.esa.un.org/unpp

Unruh, G. C. (2000) 'Understanding carbon lock-in', *Energy Policy*, vol 28, pp817–830

Unruh, G. C. (2002) 'Escaping carbon lock-in', *Energy Policy*, vol 30, pp317–325

US Census Bureau (2008) 'International Database, International Programs', www.census.gov/ipc/www/idb/worldpopinfo.php

US DOE (United States Department of Energy) (2002) *Efficient Lighting Strategies, Technology Fact Sheet*, www.toolbase.org

US DOE (2004) *2004 Buildings Energy Databook*, http://buildings databook.eere.energy.gov

US DOE (2009) *Fuel Economy Guide 2009*, www.fuel.economy.gov

US EPA (United States Environmental Protection Agency) (1995) *Compilation of Air Pollutant Emission Factors AP-42, Fifth Edition, Volume 1: Stationary Point and Area Sources*. Available from www.epa.gov/oms/ap42.html

US EPA (2003) *Municipal Solid Waste in the United States: 2001 Facts and Figures*, www.epa.gov/waste/nonhaz/municipal/pubs/msw2001.pdf

US EPA (2006) *Light-Duty Automotive Technology and Fuel Economy Trends: 1975 through 2006*, www.epa.gov/otaq/fetrends.htm

USGS (United States Geological Survey) (2000) *US Geological Survey World Petroleum Assessment 2000 – Description and Results*, http://energy.usgs.gov/search.html

USGS (United States Geological Survey) (2009) 'Lithium', *2009 Commodity Summary*, http://minerals.er.usgs.gov/minerals/pubs/commodity

US NAS (United States National Academy of Science) (2007) *Coal Research and Development to Support National Energy Policy*, National Academies Press, Washington, DC

van Hauwermeiren, A., Coene, H., Engelen, G. and Mathijs, E. (2007) 'Energy lifecycle inputs in food systems: A comparison of local versus mainstream cases', *Journal of Environmental Policy and Planning*, vol 9, pp31–51

van Oss, H. G. (2005) *Background Facts and Issues Concerning Cement and Cement Data*, Open File Report 2005–1152, US Geological Survey, Reston, Virginia

van Oss, H. G. and Padovani, A. C. (2002) 'Cement manufacture and the environment. Part I: Chemistry and Technology', *Journal of Industrial Ecology*, vol 6, pp89–105 van Vuuren, D. P., Cofala, J., Eerens, H. E., Oostenrijk, R., Heyes, C., Klimont, Z., den Elzen, M. G. J. and Amann, M. (2006) 'Exploring the ancillary benefits of the Kyoto Protocol for air pollution in Europe', *Energy Policy*, vol 34, pp444–460

Vasile, C. (1997) 'Residential waste water heat-recovery system: GFX', *CADDET Energy Efficiency Newsletter*, December, pp15–17

Verbeeck, G. and Hens, H. (2005) 'Energy savings in retrofitted dwellings: Economically viable?', *Energy and Buildings*, vol 37, pp747–754

Verbruggen, A. (2006) 'Electricity intensity backstop level to meet sustainable backstop supply technologies', *Energy Policy*, vol 34, pp1310–1317

Victor, P. A. (2008) *Managing Without Growth: Slower by Design, Not Disaster*, Edward Elgar, Cheltenham

Villatico, F. and Zuccari, F. (2008) 'Efficiency comparison between FC and ICE in real urban driving cycles', *International Journal of Hydrogen Energy*, vol 33, pp3235–3242

Vine, E., Lee, E., Clear, R., DiBartolomeo, D. and Selkowitz, S. (1998) 'Office worker response to an automated venetian blind and electric lighting system: A pilot study', *Energy and Buildings*, vol 29, pp205–218

Viridén, K., Ammann, T., Hartmann, P. and Huber, H. (2003) *P+D – Projekt Passivhaus im Umbau* (in German). Available from www.viriden-partner.ch

Vitins, J. (2009) 'Reducing energy costs with electric, diesel and dual-powered locomotives', in *Proceedings of the 2009 IEE/ASME Joint Rail Conference*, 4–5 March, Pueblo, CO

von Medeazza, G. and Moreau, V. (2007) 'Modelling of water-energy systems. The case of desalination', *Energy*, vol 32, pp1024–1031

Voorspools, K. R. and D'haeseleer, W. D. (2006) 'Reinventing hot water? Toward optimal sizing and management of cogeneration: A case study for Belgium', *Applied Thermal Engineering*, vol 26, pp1972–1981

Voorspools, K. R., Brouwers, E. A. and D'haeseleer, W. D. (2000) 'Energy content and indirect greenhouse gas emissions embedded in "emission-free" power plants: Results for the Low Countries', *Applied Energy*, vol 67, pp307–330

Voss, K. (2000) 'Solar energy in building renovation – results and experience of international demonstration buildings', *Energy and Buildings*, vol 32, pp291–302

Voss, K., Herkel, S., Pfafferott, J., Löhnert, G. and Wagner, A. (2007) 'Energy efficient office buildings with passive cooling – Results and experiences from a research and demonstration programme', *Solar Energy*, vol 81, pp424–434

Vringer, K. and Blok, K. (2000) 'The energy requirement of cut flowers and the consumer options to reduce it', *Resources, Conservation and Recycling*, vol 28, pp3–28

Vuorinen, A. (2007) *Planning of Optimal Power Systems*, Vammalan Kirjapaino Oy, Vammala, Finland

Vyas, A., Saricks, C. and Stodolsky, F. (2002) *The Potential Effect of Future Energy-Efficiency and Emissions-Improving Technologies on Fuel Consumption of Heavy Trucks*, Center for Transportation Research, Argonne National Laboratory, www.transportation.anl.gov/pdfs/TA/102.pdf

Wagner, A., Herkel, S., Löhnert, G. and Voss, K. (2004) 'Energy efficiency in commercial buildings: Experiences and results from the German funding program SolarBau', presented at EuroSolar 2004, Freiburg, www.solarbau.de

Waide, P., Guertler, P. and Smith, W. (2006) *High-Rise Refurbishment: The Energy-efficient Upgrade of Multi-story Residences in the European Union*, International Energy Agency, Paris

Wakili, K. G., Bundi, R. and Binder, B. (2004) 'Effective thermal conductivity of vacuum insulation panels', *Building Research and Information*, vol 32, no 4, pp293–299

Walker, C. E., Glicksman, L. R. and Norford, L. K. (2007) 'Tale of two low-energy designs: Comparison of mechanically and naturally ventilated office buildings in temperate climates', *ASHRAE Transactions*, vol 113, part 1, pp36–50

Walker, G. (2007) 'A world melting from the top down', *Nature*, vol 446, pp718–721

Walker, I. S. and Mingee, D. (2003) 'Reducing air handler electricity use: More than just a better motor', *Home Energy*, November/December, pp8–9

Walker, S. (1996) 'Cost and resource estimating', in G. Boyle (ed) *Renewable Energy, Power for a Sustainable Future*, Oxford University Press, Oxford, pp435–458

Wallgren, C. (2006) 'Local or global food markets: A comparison of energy use for transport', *Local Environment*, vol 11, no 2, pp233–251

Wang, T., Müller, D. B. and Graedel, T. E. (2007a) 'Forging the anthropogenic iron cycle', *Environmental Science and Technology*, vol 41, pp5120–5129

Wang, M., Wu, M. and Huo, H. (2007b) 'Life-cycle energy and greenhouse gas emission impacts of different corn ethanol plant types', *Environmental Research Letters*, vol 2, April–June, p024001

WBCSD (World Business Council for Sustainable Development) (2004) *Mobility 2030: Meeting the Challenges to Sustainability*, World Business Council for Sustainable Development, Geneva, Switzerland, www.wbcsd.org/web/publications/mobility/mobility-full.pdf and www.wbcsd.org/web/publications/mobility/smp-model-spreadsheet.xls

WBCSD (2008) *Energy Efficiency on Buildings, Business Realities and Opportunity*, World Business Council for Sustainable Development, Geneva, Switzerland.

Weiss, M. A., Heywood, J. B., Drake, E. M., Schafer, A. and AuYeung, F. F. (2000) *On the Road in 2020: A Life-cycle*

Analysis of New Automobile Technologies, Massachusetts Institute of Technology, Cambridge, MA, www.web.mit.edu/energylab

Weitzmann, P., Kragh, J., Roots, P. and Svendsen, S. (2005) 'Modelling floor heating systems using a validated two-dimensional ground-coupled numerical model', *Building and Environment*, vol 40, pp153–163

West, T. O. and Marland, G. (2002) 'A synthesis of carbon sequestration, carbon emissions, and net carbon flux in agriculture: Comparing tillage practices in the United States', *Agriculture, Ecosystems and Environment*, vol 91, pp217–232

White, A., Melvin, G. R. C. and Friend, A. D. (1999) 'Climate change impacts on ecosystems and the terrestrial carbon sink: A new assessment', *Global Environmental Change*, vol 9, ppS21–S30

White, C. M., Steeper, R. R. and Lutz, A. E. (2006) 'The hydrogen-fueled internal combustion engine: A technical review', *International Journal of Hydrogen Energy*, vol 31, pp1292–1305

Wigginton, M. and Harris, J. (2002) *Intelligent Skins*, Butterworth-Heinemann, Oxford

Williams, E. and Tagami, T. (2003) 'Energy use in sales and distribution via e-commerce and conventional retail', *Journal of Industrial Ecology*, vol 6, pp99–114

Williams, R. H. and Larson, E. D. (1989) 'Expanding roles for gas turbines in power generation', in T. B. Johansson, B. Bodlund and R. H. Williams (eds) *Electricity: Efficient End-Use and New Generation Technologies and Their Planning Implications*, Lund University Press, Lund

Williams, R. H., Bunn, M., Consonni, S., Gunter, W., Holloway, S., Moore, R. and Simbeck, D. (2000) 'Advanced energy supply technologies', in *World Energy Assessment: Energy and the Challenge of Sustainability*, United Nations Development Programme, New York

Wilson, A. and Boehland, J. (2005) 'Small is beautiful: US house size, resource use, and the environment', *Journal of Industrial Ecology*, vol 9, pp277–287

Winther, B. N. and Hestnes, A. G. (1999) 'Solar versus green: The analysis of a Norwegian row house', *Solar Energy*, vol 66, pp387–393

Wirsenius, S. (2000) 'Human use of land and organic materials', Ph D Thesis, Chalmers University of Technology, Göteborg, Sweden

Wirsenius, S. (2003) 'The biomass metabolism of the food system: A model-based survey of the global and regional turnover of food biomass', *Journal of Industrial Ecology*, vol 7, pp47–80

Withers, C. R. and Cummings, J. B. (1998) 'Ventilation, humidity and energy impacts of uncontrolled airflow in a light commercial building', *ASHRAE Transactions*, vol 104, part 2, pp733–742

Wohlmeyer, H. (1998) 'Agro-eco-restructuring: Potential for sustainability', in R. U. Ayres and P. M. Weaver (eds) *Ecorestructuring: Implications for Sustainable Development*, United Nations University Press, Tokyo, pp176–310

Worm, B. et al (2006) 'Impact of biodiversity loss on ocean ecosystem services', *Science*, vol 314, pp787–790

Worrell, E. (2004) 'Cement and energy', in *Encyclopedia of Energy*, vol 1, Elsevier, London, pp307–315

Worrell, E. (2005) 'Energy efficiency improvement in the petroleum refining industry', in *Proceedings of the 2005 ACEEE Summer Study on Energy Efficiency in Industry*, vol 4, American Council for an Energy Efficient Economy, Washington, DC, pp158–169

Worrell, E., Meuleman, B. and Blok, K. (1995) 'Energy savings by efficient application of fertilizer', *Resources, Conservation and Recycling*, vol 13, pp233–250

Worrell, E., Martin, N. and Price, L. (2000) 'Potentials for energy efficiency improvement in the US cement industry', *Energy*, vol 25, pp1189–1214

Worrell, E., Price, L., Martin, N., Hendriks, C. and Media, L. O. (2001) 'Carbon dioxide emissions from the global cement industry', *Annual Review of Energy and the Environment*, vol 26, pp303–329

WRAP (Waste and Resources Action Programme) (2006) *Environmental Benefits of Recycling: An International Review of Life Cycle Comparisons for Key Materials in the UK Recycling Sector*, www.wrap.org.uk

Wright, A. (1999) 'Natural ventilation or mixed mode? An investigation using simulation', *Sixth International IBPSA Conference*, Kyoto, Japan, 13–15 September, pp449–455

Wulfinghoff, D. R. (1999) *Energy Efficiency Manual*, Energy Institute Press, Wheaton, Maryland

WWI (Worldwatch Institute) (2008) *Vital Signs 2007–2008*, WWI, Washington, DC, www.worldwatch.org

Xiong, W., Zhang, Y. and Yin, C. (2009) 'Optimal energy management for a series-hybrid electric bus', *Energy Conversion and Management*, vol 50, pp1730–1738

Xu, P., Huang, J., Jin, R. and Yang, G. (2007) 'Measured energy performance of a US-China demonstration energy-efficient office building', *ASHRAE Transactions*, vol 113, part 1, pp56–64

Xu, W. C., Takahashi, K., Matsuo, Y., Hattori, Y., Kumagai, M., Ishiyama, S., Kaneko, K. and Iijima, S. (2007) 'Investigation of hydrogen storage capacity of various carbon materials', *International Journal of Hydrogen Energy*, vol 32, pp2504–2512

Yang, C. J. (2009) 'An impending platinum crisis and its implications for the future of the automobile', *Energy Policy*, vol 37, pp1805–1808

Yik, F. W. H., Burnett, J. and Prescott, I. (2001) 'A study on the energy performance of three schemes for widening application of water-cooled air-conditioning systems in Hong Kong', *Energy and Buildings*, vol 33, pp167–182

Yonehara, T. (1998) 'Ventilation windows and automatic blinds help to control heat and lighting', *CADDET Energy Efficiency Newsletter*, December, pp9–11

Yoshikawa, S. (1997) 'Japanese DHC system uses untreated sewage as a heat source', *CADDET Energy Efficiency Newsletter*, June, pp8–10

Yürüm, Y., Taralp, A. and Veziroglu, N. (2009) 'Storage of hydrogen in nanostructured carbon materials', *International Journal of Hydrogen Energy*, vol 34, pp3784–3798

Yushi, M., Hong, S. and Fuqiang, Y. (2008) *The True Cost of Coal*, Greenpeace, Energy Foundation (San Francisco) and World Wildlife Fund

Zachariadis, T. (2006) 'On the baseline evolution of automobile fuel economy in Europe', *Energy Policy*, vol 34, pp1773–1785

Zentner, R. P., McConkey, B. G., Stumborg, M. A., Campbell, C. A. and Selles, F. (1998) 'Energy performance of conservation tillage management for spring wheat production in the Brown soil zone', *Canadian Journal of Plant Science*, vol 78, pp553–563

Zhang, Q., Weili, T., Yumei, W. and Yingxu, C. (2007) 'External costs from electricity generation of China up to 2030 in energy and abatement scenarios', *Energy Policy*, vol 35, pp4295–4304

Zhang, X. and Muneer, T. (2002) 'A design guide for performance assessment of solar light-pipes', *Lighting Research and Technology*, vol 34, pp149–169

Zhao, L. and Gallagher, K. S. (2007) 'Research, development, demonstration, and early deployment policies for advanced-coal technology in China', *Energy Policy*, vol 35, pp6467–6477

Zhao, P. C., Zhao, L., Ding, G. L. and Zhang, C. L. (2003) 'Temperature matching method of selecting working fluids for geothermal heat pumps', *Applied Thermal Engineering*, vol 23, pp179–195

Zhen, B., Shanhou, L. and Weifeng, Z. (2005) 'Energy efficient techniques and simulation of energy consumption for the Shanghai ecological building', in *Proceedings 2005 World Sustainable Building Conference, Tokyo, 27–29 September*, pp1073–1078

Zhivov, A. M. and Rymkevich, A. A. (1998) 'Comparison of heating and cooling energy consumption by HVAC system with mixing and displacement air distribution for a restaurant dining area in different climates', *ASHRAE Transactions*, vol 104, part 2, pp473–484

Zimmermann, M. (2004) 'ECBCS building retrofit initiative', *ECBCS News*, October 2004, pp11–14. Available from www.ecbcs.org

Zogg, R., Sriramulu, S., Carlson, E., Roth, K. and Brodick, J. (2006) 'Using solid-oxide fuel cells for distributed generation', *ASHRAE Journal*, vol 48, no 12, pp116–118

Zogg, R., Carlson, E., Roth, K. and Brodick, J. (2007) 'Using molten-carbonate fuel cells for distributed generation', *ASHRAE Journal*, vol 49, no 2, pp62–63

Index